REGULATION OF OVARIAN AND TESTICULAR FUNCTION

ADVANCES IN EXPERIMENTAL MEDICINE AND BIOLOGY

Recent Volumes in this Series

REGULATION OF OVARIAN AND TESTICULAR FUNCTION

Edited by

Virendra B. Mahesh
Medical College of Georgia
Augusta, Georgia

Dharam S. Dhindsa
National Institutes of Health
Bethesda, Maryland

Everett Anderson
Harvard Medical School
Boston, Massachusetts

and

Satya P. Kalra
University of Florida College of Medicine
Gainesville, Florida

PLENUM PRESS • NEW YORK AND LONDON

Library of Congress Cataloging in Publication Data

Regulation of ovarian and testicular function.

(Advances in experimental medicine and biology; v. 219)
Based on a workshop held in Augusta, Ga., Feb. 7–9, 1987.
Includes bibliographies and index.
1. Human reproduction—Endocrine aspects—Congresses. 2. Gonadotropin—Secre-
tion—Regulation—Congresses. 3. Ovulation—Regulation—Congresses. 4. Corpus
luteum—Congresses. 5. Spermatogenesis—Congresses. I. Mahesh, Virendra B. II.
Series. [DNLM: 1. Ovary—physiology—congresses. 2. Reproduction—congresses. 3.
Testis—physiology—congresses. WP 520 R344 1987]
QP252.R44   1987   612'.6   87-22073
ISBN 0-306-42676-5

Proceedings of a workshop on Regulation of Ovarian and Testicular Function,
sponsored by the National Institutes of Health,
held February 7–9, 1987, in Augusta, Georgia

© 1987 Plenum Press, New York
A Division of Plenum Publishing Corporation
233 Spring Street, New York, N.Y. 10013

Printed in the United States of America

PREFACE

Rapid advances have taken place in various aspects of reproductive biology during the last decade. These advances have centered around several organ systems that comprise the reproductive system and encompass molecular events and structure-function relationships. It becomes important to review these advances in knowledge, at periodic intervals, with respect to feedback systems and regulatory loops that control reproductive processes in vivo. Towards this end, a workshop entitled "Functional Correlates of Hormone Receptors in Reproduction" sponsored by the National Institute of Child Health and Human Development and the Reproductive Biology Study Section of the Division of Research Grants, National Institutes of Health was held in October 1980. The proceedings of the workshop were published by Elsevier Biomedical/New York. This workshop was followed by two workshops sponsored by the Reproductive Biology Study Section of the Division of Research Grants, National Institutes of Health entitled "Role of Peptides and Proteins in Control of Reproduction" in February 1982 and published by Elsevier Biomedical and "Molecular and Cellular Aspects of Reproduction" in October 1985 and published by Plenum Press. It was, therefore, timely to review the current state of knowledge regarding the regulation of ovarian and testicular function by bringing together scientists working in separate and discrete aspects of reproduction to review the functional implications of their research on the regulation of function within the same tissue and also in relationship to feedback systems and regulatory loops with other tissues.

This book is the proceedings of the workshop on the "Regulation of Ovarian and Testicular Function" held in Augusta, Georgia, February 7-9, 1987. The workshop was arranged by an organizing committee composed of Drs. Everett Anderson, Dharam S. Dhindsa, Satya P. Kalra and Virendra B. Mahesh and supported by the National Institute of Child Health and Human Development and Reproductive Biology Study Section of the Division of Research Grants, National Institutes of Health. The program consisted of invited papers from leaders in the field in four major areas and a number of short papers presented as posters. The major areas covered were 1) neuroendocrine regulation of gonadotropin secretion, 2) regulation of folliculogenesis and ovulation, 3) regulation of corpus luteum function in pregnancy, and 4) regulation of testicular cell function and spermatogenesis. The four major areas covered are presented as four separate sections in this book, followed by a fifth section containing the short papers.

The Organizing Committee gratefully acknowledges the encouragement and support provided by Dr. William A. Sadler, National Institute of Child Health and Human Development and the Reproductive Biology Study Section, during the conceptualization of the workshop. The success of the workshop was critically dependent upon the contributions of leading scientists invited for major presentation and review of their fields. Their participation was made possible by partial financial support provided by Dr. William A. Sadler, Chief, Reproductive Sciences Branch and Dr. Florence P. Hazeltine, Director, Center for Population Research, National Institute for Child Health and Human Development, and the Division of Research Grants. The Organizing Committee also thanks the members of the Reproductive Biology, Reproductive Endocrinology, and Human Embryology and Development Study Sections for attending the workshop and contributing to the scientific discussion of various papers. The Organizing Committee gratefully acknowledges the contributions of Mrs. Jean Bennett and Ms. Andrea Magoulas in helping to organize the workshop and their valuable assistance in editing the proceedings.

Virendra B. Mahesh

Dharam S. Dhindsa

Everett Anderson

Satya P. Kalra

CONTENTS

PART I

NEUROENDOCRINE REGULATION OF GONADOTROPIN SECRETION

PEPTIDE-STEROID INTERACTIONS IN THE MODULATION OF

THE HYPOTHALAMIC-PITUITARY AXIS

L. Martini, D. Dondi, P. Limonta, R. Maggi,
M. Motta and F. Piva

Department of Endocrinology
University of Milano
21, Via A. del Sarto
20129 - Milano, Italy

INTRODUCTION

It is now clear that the hypothalamic-pituitary axis is controlled not only by the classical neurotransmitters (e.g., norepinephrine, dopamine, serotonin, etc.), but also by a large series of newly discovered peptides. Among the peptides involved in the control of the hypothalamic-pituitary axis, a special role is certainly played by the opioids.

It is now generally accepted that the endogenous opioids exert a stimulatory effect on growth hormone (GH) and prolactin secretion (reviewed by Clement-Jones and Rees, 1982). Evidence has also accumulated which suggests that brain opioids may participate in the control of gonadotropin secretion, acting either at hypothalamic or at anterior pituitary level (reviewed by Kalra, 1986). However, the data available are not as clear-cut as those regarding the regulation of the release of GH and prolactin. The majority of the results obtained with the administration of different types of opioids or opioid antagonists (e.g., naloxone, naltrexone, etc.) suggests that brain opioids inhibit LH and/or FSH secretion (reviewed by Meites et al., 1979). However, some recent reports underline the possibility that these principles, in particular conditions, might stimulate rather than inhibit gonadotropin release (reviewed by Piva et al., 1984, 1986).

Recent evidence suggests that the endocrine effects of the opioids and of their antagonists may depend on the endocrine status existing at the time of the experiment, and especially on the circulating levels of sex steroids (reviewed by Piva et al., 1984, 1985, 1986). It is also known that the effects of the naturally occurring opioids (met-enkephalin, leu-enkephalin, beta-endorphin, dynorphin, etc.) are exerted through the interaction with specific binding sites; different classes of opioid receptors (named respectively, mu, kappa, delta, etc.) have been described (Paterson et al., 1983). Among these, the mu receptors appear to be particularly relevant for the control of gonadotropin secretion (Pfeiffer et al., 1983), while the mu and kappa receptors seem to be especially involved in the control of prolactin secretion (Spiegel et al., 1982; Krulich et al., 1986a,b).

The present report will summarize the data obtained in the authors' laboratory on the interactions between brain opioids and sex steroids in the control of gonadotropin and prolactin secretion. In particular the paper will describe:

(a) The effects of intraventricular (i.v.t.) injections of an opioid agonist on LH, FSH and prolactin secretion in normal and castrated male rats (Sections 1 and 2).

(b) The effects exerted, in the adult female rat, by an opioid antagonist on LH, FSH and prolactin release during the different phases of the estrous cycle, which are characterized by changing levels of endogenous sex steroids (Section 3).

(c) The effects exerted by estrous cyclicity on the number of brain mu opioid receptors in the female rat (Section 4).

(d) The effects exerted by aging on the number of brain mu and kappa receptors in the male rat (Section 5).

RESULTS AND DISCUSSION

1. Effects of Intraventricular Injections of Morphine on LH and FSH Release in Normal and Castrated Male Rats

Fig. 1 (upper panel) shows that morphine when injected i.v.t. into normal adult male rats in the dose of 200 μg/rat, induces a conspicuous and significant elevation of serum LH levels at 10 and 20 min. Fig. 1 (lower panel) also shows the results obtained following the i.v.t. administration of the same dose of morphine to adult castrated male rats. In castrated animals, the i.v.t. injection of morphine is not followed by any significant modification of serum LH levels 10 and 20 min after treatment. However, morphine administration brings about a significant decrease of serum LH levels at 40 and 60 min. This inhibitory effect of morphine on LH secretion lasts up to 180 min after injection (data not shown). FSH secretion was not affected by i.v.t. injections of morphine either in normal or in castrated males at any time considered (data not shown).

The finding that, in normal male rats, morphine may activate LH release, even if it appears in contrast with several literature data (reviewed by Meites et al., 1979), finds support in a series of recent studies, which have been recently reviewed by Piva et al. (1984, 1986). The present data suggest the presence in the brain of some opioid-modulated system(s) which exert(s) stimulatory influences on LH secretion. This activatory mechanism is probably complementary to the inhibiting one described in the literature (reviewed by Meites et al., 1979; Kalra, 1986) and confirmed in previous studies of this laboratory (Piva et al., 1985) (also below, Section 3). In this connection, it is interesting to underline that a dual opiatergic control (stimulatory and inhibitory) has emerged from some recent studies dealing with the regulation of GH (Spiegel et al., 1982; Krulich et al., 1986a,b) and antidiuretic hormone secretion (Iovino et al., 1983). Several hypotheses may be put forward for explaining the stimulatory effect exerted by morphine and other opioids on LH secretion. It may be postulated that, in the brain, there are, in addition to opioid pathways inhibiting LH secretion, also opioid pathways which stimulate the release of this hormone. It is possible that morphine, when given i.v.t., might more easily reach the stimulatory than the inhibitory pathways. This hypothesis is supported by the finding that the administration of morphine into the third ventricle or in the median eminence decreases the stimulatory effect of cold on TSH release, while the administration of morphine into the posterior hypothalamus potentiates the effect of cold on TSH secretion (Mannisto et al., 1984). Another possibility is that the inhibiting and the stimulatory effects of the opioids on LH secretion are mediated by different classes of opioid

4

receptors. This hypothesis is indirectly supported by a few recent findings which suggest that the opioids may stimulate prolactin and GH release acting at different receptorial sites (Spiegel et al., 1982; Delitala et al., 1983; Krulich et al., 1986a,b). Only further studies

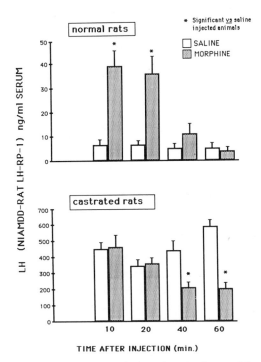

Fig. 1. Effects of intraventricular injections of morphine (200 µg/rat) on serum LH levels of adult normal and castrated (4 weeks) male rats.

will permit to establish whether the hypothesis that morphine exerts respectively its inhibitory and stimulatory effects on LH release acting on different opioid receptors is correct or not. Studies performed with agonists which bind specifically to one single class of receptors will be particularly useful; these types of studies are presently underway in this laboratory.

The data here presented have clearly indicated that the stimulatory effect of i.v.t. injections of morphine on LH release disappears following castration. Actually, the data have shown that morphine, when given i.v.t. to long-term castrated male rats, significantly decreases serum LH levels from 40 to 180 min following injection. The finding that morphine may decrease LH release in castrated male rats is in agreement with a series of previous studies, in which morphine or the opioid peptides have been given systemically or i.v.t. to castrated male rats (reviewed by Meites et al., 1979; Piva et al., 1984, 1986).

The opposite results obtained in normal and in castrated male rats show very clearly that the effect of the opioids on gonadotropin release may be modulated by the endocrine "milieu" existing in the animal at time of the experiment, and especially by the levels of circulating androgens.

2. Effects of Intraventricular Injections of Morphine on Prolactin Release in Normal and Castrated Male Rats

Fig. 2 (upper and lower panels) shows the effects on serum levels of prolactin of morphine (200 μg/rat) administered i.v.t. to normal and to long-term castrated adult male rats. The dose of morphine selected for these experiments (200 μg/rat) significantly enhanced prolactin release in normal animals at all times considered. On the contrary, the same dose of morphine was totally unable to increase prolactin release in castrated male rats; serum prolactin concentrations in the orchidectomized animals treated i.v.t. with morphine were not significantly different from those found in the corresponding control animals injected with saline.

The results of the present study show that when morphine given i.v.t. to normal adult male rats is able to increase serum prolactin levels, are in agreement with data obtained by previous authors following systemic (reviewed by Meites et al., 1979; Clement-Jones and Rees, 1982), or i.v.t. injections of morphine, the endogenous opioid peptides and their synthetic analogues (reviewed by Giudici et al., 1984; Limonta et al., 1986).

The experiments performed in long-term castrated rats clearly show that even the i.v.t. injection of a rather elevated dose of morphine is totally unable to significantly increase serum prolactin concentrations in the total absence of androgens. The present observation is reminiscent of the data shown in the previous section of this paper (Section 1) and of reports in the literature, showing that morphine, the opioids and their antagonists may lose their ability to influence LH release after orchidectomy, if steroid replacement therapy is not provided (reviewed by Piva et al., 1984, 1986).

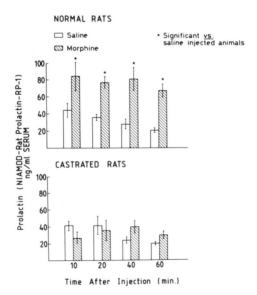

Fig. 2. Effects of intraventricular injections of morphine (200 μg/rat) on serum prolactin levels of adult normal and castrated (4 weeks) male rats.

There are only a few data in the literature supporting the view that the absence of testicular steroids might alter the prolactin response to the administration of morphine or other opioids. Goldberg et al. (1982)

have shown that the i.v.t. administration of 500 µg of met-enkephalin to normal adult male rats causes a significant rise in serum prolactin levels, and that the same treatment results in a much lower release of prolactin in orchidectomized rats. Kato et al. (1982) have found that beta-endorphin induces an increase of plasma prolactin in castrated male rats, which is almost 10 times lower than that observed in normal animals (Kato et al., 1978). Moreover, Forman et al. (1981) have reported that the stimulatory effect of morphine on prolactin release is decreased in aged male rats, which are known to have serum testosterone levels lower than those of younger animals.

As previously mentioned, it is presently believed that the opioids modify prolactin release through the activation of specific brain receptors of the mu and kappa families (Spiegel et al., 1982; Krulich et al., 1986a,b). Consequently, one might hypothesize that the loss of effectiveness of morphine on prolactin secretion induced by orchidectomy might be due to an alteration, induced by castration, of the number or of the binding characteristics of brain mu and kappa receptors. Unfortunately, the literature on the effects exerted by castration on brain opioid receptors is scantly and controversial. Hahn and Fishman (1985) have found in the rat that the binding of naltrexone, naloxone and diprenorphine to opioid receptors in the whole brain is increased by orchidectomy; on the contrary, Wilkinson et al. (1981), Diez and Roberts (1982), Cicero et al. (1983) and Olasmaa et al. (1987) have found that neither castration nor the subsequent androgen replacement therapy modify the binding capacity of the whole brain and of the hypothalamus for a number of opioid ligands. An alternative possibility is that castration might have modified the bioavailability of other brain neurotransmitters involved in the control of prolactin secretion. There is indeed plenty of evidence indicating that the opioids may modulate the activity of catecholaminergic, serotoninergic and cholinergic systems in the brain (reviewed by Kalra, 1986).

3. Effects Exerted by Naloxone on LH, FSH and Prolactin Release During the Different Phases of the Estrous Cycle of the Adult Female Rat

Relatively little information is available on the effects exerted by opioid antagonists on LH, FSH and prolactin secretion when administered during the different phases of the ovulatory cycle of the female rat. Moreover the results reported so far have been rather conflicting (Blank et al., 1979; Gabriel et al., 1983). The present investigation has been performed in order to solve this controversy. To this purpose, the effects exerted by naloxone on LH, FSH and prolactin secretion were analyzed during the different phases of the estrous cycle in normal adult female rats.

It is evident from Fig. 3 that, in our colony of rats and under the illumination schedule adopted (14 h light, 10 h dark; lights on at 0630 h), the control animals (normally cycling adult female rats treated subcutaneously with saline) have low levels of serum LH during the days of diestrus 1 and 2 during the day of estrus. In the experimental conditions adopted, the expected LH surge of the day of proestrus begins after 1600 h, reaches a maximum at 1800 h and starts declining at 2000 h. This is confirmatory of previous evidence (reviewed by Limonta et al., 1981; Piva et al., 1982; Kalra and Kalra, 1983). It is also evident from Fig. 3 that the s.c. administration of naloxone (2.5 mg/kg) is followed by an increase of serum LH levels in every day of the estrous cycle. However, the magnitude of the effects exerted by naloxone on LH release is different in the various days of the estrous cycle and, on the days of proestrus and estrus, at different hours of the day. This appears clearly from Fig. 4,

in which the data have been plotted as the percent increase of the LH
values obtained in the naloxone-treated animals vs those obtained in the
saline-treated controls. The administration of naloxone brings about a

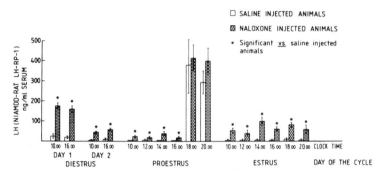

Fig. 3. Effects of subcutaneous injections of naloxone (2.5 mg/kg) on
serum LH levels of adult cycling female rats treated at different
hours of their estrous cycle.

moderate increase of serum LH during the days of diestrus 1 and 2; during
these two days, the effects exerted by naloxone on LH secretion are
quantitatively similar at 1000 and at 1600 h. During the day of proestrus,
naloxone exerts a stimulatory effect on LH release quantitatively similar
to that observed during the days of diestrus 1 and 2 at 1000 and at 1400
h; at 1200 h the effect of naloxone appears to be lower. On the contrary,
a major increase of the stimulatory effect of naloxone on LH release is
observed at 1600 h; the increase in serum LH levels induced by naloxone at
1600 h is significantly higher than those found at 1000, 1200 and 1400 h.
After 1600 h on the day of proestrus the effect of naloxone on LH
secretion seems to disappear; the values of serum LH in the naloxone-
treated animals recorded at 1800 and 2000 h are not significantly
different from those found in the corresponding saline-treated controls.

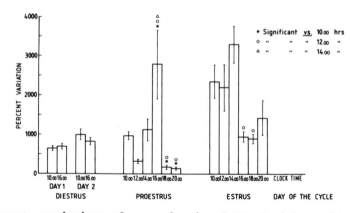

Fig. 4. Percent variations of serum levels of LH of adult cycling female
rats treated with naloxone at different hours of their estrous
cycle.

During the day of estrus, a very significant LH response to naloxone occurred at 1000, 1200 and 1400 h; at these time intervals, the increases in serum LH levels induced by the treatment with the drug are similar to that observed at 1600 h of the day of proestrus. At later times of observation, the effects of naloxone on LH secretion appear to diminish; the increases observed at 1600, 1800 and 2000 h are similar to those found during the day of diestrus 2. The patterns of FSH and prolactin secretion in saline-treated animals were similar to those previously reported by this and other laboratories (reviewed by Limonta et al., 1981; Piva et al., 1982; Kalra and Kalra, 1983). Throughout the present experiments, there was no effect of the administration of naloxone on either FSH or prolactin release. In no instance, the values of serum FSH and prolactin recorded after naloxone administration were significantly different from those of the corresponding controls (data not shown).

The present data then show that naloxone is able to induce significant increases of serum LH levels during every day of the estrous cycle. As previously mentioned, the results obtained by previous authors have been conflicting on this point (Blank et al., 1979; Gabriel et al., 1983). This finding suggests that, in normally cycling adult females, the endogenous opioids exert an inhibitory effect on LH release.

The data of the present study clearly indicate that the stimulatory effect naloxone exerted on LH release exhibits conspicuous and significant quantitative variations throughout the 4 days of the estrous cycle. The response of serum LH levels to naloxone appears to be moderate during the days of diestrus 1 and 2, at 1000, 1200 and 1400 h on the day of proestrus, and at 1600, 1800 and 2000 h on the day of estrus. On the contrary, major LH responses after the administration of naloxone appear to occur at 1600 h on the day of proestrus and during the earlier hours (1000, 1200 and 1400 h) on the day of estrus.

It is interesting to underline that, during the 2 days of diestrus, no significant differences have been recorded between the effects naloxone exerts on LH secretion in the morning and in the afternoon. A diurnal periodicity in the responses to naloxone has been found in pharmacological and behavioral studies (Kavaliers and Hirst, 1983). Blank and Mann (1981) have also reported a diurnal variation of the LH responses to naloxone in immature female rats.

In the present experiments, it has been confirmed (Gabriel et al., 1983) that the LH responses to naloxone are quantitatively similar in the second day of diestrus and on the day of proestus (at 1000, 1200 and 1400 h), prior to the preovulatory LH surge. The conclusion we derive from these findings is similar to that proposed by Gabriel et al. (1983). There is now convincing evidence indicating that naloxone facilitates LH release mainly via a hypothalamic mechanism, which results in a stimulation of LHRH secretion. If one keeps in mind that the responsiveness of the anterior pituitary to LHRH progressively increases from the second day of diestrus until the afternoon of the day of proestrus, as a consequence of the increased secretion of estradiol occurring during this period (Kalra and Kalra, 1983), the present data certainly suggest that the effects of naloxone on LHRH release gradually decrease between the day of diestrus 2 and the day of proestrus. It remains to be established whether this decrease of the effectiveness of naloxone is a primary phenomenon due to a spontaneous change of the opiatergic tone controlling LHRH secretion, or whether it is induced by the reported changes of estrogen levels.

As previously mentioned, the present data have clearly shown that naloxone becomes much more effective in facilitating the release of LH at 1600 h on the day of proestrus, i.e. just before the initiation of the spontaneous proestrous LH surge. The increased sensitivity to naloxone of the mechanisms controlling LH secretion during the day of proestrus appears to be of short duration, since the drug totally loses its ability to stimulate LH secretion at 1800 and 2000 h on the same day. We believe that the increased sensitivity to naloxone does not underline a change in the opiatergic tone controlling LHRH secretion, but only reflects the increased responsiveness of the anterior pituitary to endogenous LHRH induced by the elevated estrogenic secretion present at this time (Kalra and Kalra, 1983). At present, we do not have a definite explanation for the disappearance of the naloxone effect during the evening hours of the day of proestrus. However, it might be suggested that this is due to the effect of progesterone and of other progestogens (20alpha-OH-progesterone, etc.), whose secretion increases very significantly between 1700 h and 2400 h on the day of proestrus (Barraclough et al., 1971). In this connection, it is interesting to recall that Gabriel et al. (1983) reported that, in ovariectomized rats pretreated with estradiol, the administration of progesterone is followed by a total blockade of the naloxone effect on LH secretion from 400 to 800 h after the administration of progesterone.

In the present study, the LH sensitivity to naloxone reappeared during the day of estrus; at 1000, 1200 and 1400 h on that day the increases of LH secretion induced by naloxone were not only major, but actually the highest seen in the present study. We believe that the increased sensitivity to naloxone appearing on the day of estrus may result from the summation of two factors: a good sensitivity of the anterior pituitary to LHRH, and a change occurring in the opiatergic tone controlling LHRH secretion. This possible change of the opiatergic tone controlling LHRH release is probably more important than the pituitary factor in explaining the return of the response to naloxone. All the data available show indeed that the responses of the anterior pituitary to LHRH on the day of estrus are greatly reduced when compared to those found during the day of proestrus (Kalra and Kalra, 1983), even if they are higher than those on the day of diestrus. Consequently, lower effects of naloxone would have been obtained during the day of estrus than during the day of proestrus, in case the pituitary factor was of primary importance. Obviously this was not the case. The change of the opiatergic tone controlling LH secretion on the morning of estrus might be a primary phenomenon aimed at inhibiting LH secretion at this time, or might be secondary to the priming effects of ovarian steroids secreted during the previous days. It is indeed possible that the endogenous secretion of estrogens and/or progestogens modify the synthesis, the storage and the release of the opiates, or the receptorial mechanisms through which the opiates exert their effects. Unfortunately, the information available on the modifications of these parameters during the estrous cycle, and after castration or sex steroid administration is scanty and rather contro-versial.

In the present experiments, naloxone was found unable to modify serum prolactin levels either when these were basal (diestrous days 1 and 2 and day of estrus) or when they were elevated (prolactin proestrus surge). The observation that naloxone does not influence basal prolactin levels con-firms the results of previous studies of this (Giudici et al., 1984; Limonta et al., 1986) and other laboratories (Muraki et al., 1979). However, in some previous studies the administration of opioid antagonists has been reported to result in a depression of baseline serum prolactin levels (Meites et al., 1979). The reasons for these divergent literature findings are unclear at present, but may involve differences in the doses and in the route of administration of the drug used, as well as in the time of sampling. The failure of naloxone to produce any effect on basal prolactin secretion observed in the present study argues against the existence of a permanent stimulating activity of endogenous opioid-like peptides on the release of this hormone. It is also obvious that the present data clearly show that the opiatergic mechanisms controlling LH and prolactin secretion are substantially different.

Throughout the experiments reported in Sections 1 and 3 of this paper it has been observed that treatments with either morphine or naloxone do not result in any modification of FSH secretion. The literature on the effects of the opioids and of their antagonists on FSH secretion is still controversial. Some authors have found that morphine, the opioid peptides and their antagonists may simultaneously alter LH and FSH secretion (Meites et al., 1979; Muraki et al., 1979; Bhanot and Wilkinson, 1983), while others have reported that these treatments modify preferentially LH without affecting FSH secretion (Marton et al., 1981; Motta and Martini, 1982; Piva et al., 1984). The present data are in line with those reports indicating that brain opioids apparently do not participate in the control of FSH secretion. The divergent effects exerted by morphine and naloxone on LH and FSH secretion appear difficult to reconcile with the hypothesis that one single hypothalamic releasing factor might simultaneously control the secretion of the two gonadotropins. Several hypotheses may be put forward for explaining the dichotomy of the effects exerted by the two drugs on LH and FSH secretion. First of all, it may be postulated that brain opioids, while modulating LHRH secretion, do not intervene in the control of the release of FSH-RF; even if the existence of this factor in the brain remains to be fully demonstrated, several lines of evidence point to its existence as a separate entity (Lumpkin and McCann, 1984). Secondly, it might be proposed that the treatment with morphine or naloxone has changed the rhythmicity of the secretion of LHRH in a way which specifically facilitates LH release. Wildt et al. (1981) have shown that the pulsatile administration of LHRH to Rhesus monkeys bearing extended hypothalamic lesions results in an elevation of LH secretion when the frequency of the pulses is rather elevated and, on the contrary, in a specific release of FSH when the frequency of the pulses is decreased. There is ample evidence, especially derived from clinical studies, which indicates that the administration of opioid antagonists may influence the pulsatility of gonadotropin secretion (reviewed by Clement-Jones and Rees, 1982).

4. Effects Exerted by Estrous Cyclicity on the Number of Brain Mu Opioid Receptors

Several data in the literature indicate that, in female animals and in women, the effects of the opioids and of their antagonists on gonadotropin secretion largely depend on the levels of estrogens and progesterone present at time of administration (Blank et al., 1979; Bhanot and Wilkinson, 1983; Gabriel et al., 1983; Piva et al., 1985; Allen and Kalra, 1986; Kalra, 1986; Petraglia et al., 1986).

To the authors' knowledge, no information is available on possible changes of the number and binding characteristics of opioid receptors of the mu type in the brain of female rats during the different phases of the estrous cycle, which are characterized by significant alterations of circulating levels of both estrogens and progesterone (Barraclough et al., 1971; Kalra and Kalra, 1983). The present experiments have been conducted in order to analyze whether the number and the affinity of brain receptors of the mu type change during the various phases of the rat estrous cycle.

In all studies, dihydromorphine, a typical mu receptor ligand, has been used in a receptor binding assay described in detail elsewhere (Piva et al., 1987).

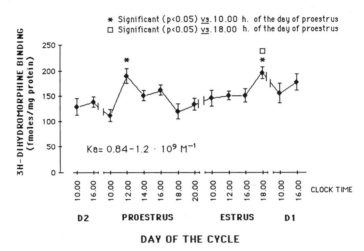

Fig. 5. Number of brain mu opioid receptors during the various phases of the estrous cycle of adult cycling female rats.

It is clear from Fig. 5 that the number of opioid receptors of the mu type in the whole brain shows significant variations during the different phases of the estrous cycle. Their number is rather low at 1000 h and 1600 h of the second day of diestrus and at 1000 h of the day of proestrus. A significant increase of the number of brain mu receptors is observed at 1200 h on this day. This increase is significant vs the values found at 1000 h on the day of proestrus. After this time there is a progressive decline of the number of mu receptors which return to the levels found at 1000 h on the day of proestrous at 1800 h. The number of brain mu receptors then remains low and rather constant up to 1600 h on the day of estrus; a significant increase occurs again at 1800 h on this day. The values recorded at this time interval are significantly different from those detected at 1000 h and 1800 h on the day of proestrus. This second increase is then followed by a decrease, which, during the first day of diestrus, gradually brings the number of receptors to the low levels observed during the second day of diestrus. During all these studies, the changes in the number of mu receptors were not accompanied by any change in the affinity constant.

The present data clearly show that the number of the opioid receptors binding dihydromorphine shows conspicuous changes in the whole brain of the female rat during the different phases of the estrous cycle. In particular, an increase in the number of these receptors is observed at 1200 h on the day of proestrus and at 1800 h on the day of estrus.

It is not known, at present, whether these estrus-linked modifications of the number of brain mu opioid receptors are to be considered at primary, or whether they are induced by the changes of the steroid "milieu" occurring during the different phases of the estrous cycle. If one accepts this second hypothesis, it might be speculated that the number of brain receptors binding dihydromorphine increases during the day of proestrus under the influence of estrogens secreted during this phase of the estrous cycle (Kalra and Kalra, 1983), and that their decline during the late phases on the day of proestrus and during estrus are the consequence of the concomitant effects of estrogens and progesterone secreted during these phases (Barraclough et al., 1971; Kalra and Kalra, 1983). However, the increase of the number of brain opioid receptors occurring on the evening of the day of estrus appears difficult to explain on the basis of changes of sex steroid secretion since, at this time, both estrogens and progesterone remain constant.

In agreement with the hypothesis that circulating estrogens might modify the number of brain mu opioid receptors, Wilkinson et al. (1985) have demonstrated a significant increase in naloxone binding in the anterior hypothalamus of ovariectomized female rats following the chronic implantation (12 weeks) of Silastic capsules filled with estradiol or the subcutaneous administration of estradiol valerate. In this study, estradiol treatment did not modify naloxone binding in the cerebral cortex and in the amygdala. However, the study of Wilkinson et al. (1985) suffers from the fact that these authors were unable to define whether the increase in naloxone binding was due to a change of the Bmax or of the affinity of the ligand for the receptors. Apparently there are no data in the literature regarding the effects of progesterone on the number of mu receptors in the brain.

It is obviously difficult to strictly compare the results here obtained with those previously reported (Section 3 of this paper) on the effects of naloxone on LH release during the different phases of the estrous cycle. As previously mentioned, naloxone induces a moderate LH response at 1200 h on the day of proestrus and at 1800 h on the day of estrus, i.e. at two times at which the number of brain mu receptors is elevated. The apparent discrepancy between the two groups of results may be due to several reasons. First of all, in the present study the concentrations of opioid receptors have been studied only in the whole brain; obviously, this has prevented a detailed analysis of changes of opioid mu receptors possibly occurring in the hypothalamus and in other CNS structures involved in the control of gonadotropin secretion. Second, as already discussed, the changes of the effects naloxone exerts on serum LH levels are probably due not only to the liberation of LHRH induced by the suppression of the opiatergic tone, but also to the estrogen—induced modifications of the sensitivity of the anterior pituitary to the liberated LHRH. Third, even if the majority of the data available suggests that the mu opioid receptors are particularly relevant for the control of gonadotropin secretion (Pfeiffer et al., 1983) one cannot exclude the possibility that also other types (kappa and delta) of opioid receptors might participate in such a control. Finally, it must be underlined that the changes of the number of brain opioid receptors are not the only factor which may modulate the effects of the opioids in regulating gonadotropin secretion. Also changes of the synthesis, storage and degradation of the various brain opioids may be of significance. The information available on these processes is scanty and controversial.

5. Effects Exerted by Aging on the Number of Brain Mu and Kappa Opioid Receptors in the Male Rat

Evidence is accumulating which indicates that age exerts profound influences on the hypothalamic-pituitary-gonadal axis of the male rat. In particular, it has been observed that old male rats have decreased serum levels of testosterone, LH and FSH (Meites, 1982). Moreover, the magnitude and the frequency of the LH pulses are altered in old male animals (Karpas et al., 1983). These alterations have been attributed either to an age-dependent damage of the neurons synthesizing LHRH (Hoffman and Sladek, 1980), or to changes in the synthesis, metabolism and release of the brain neurotransmitters (e.g., serotonin, catecholamines, opioids, etc.) which control LHRH secretion (Simpkins et al., 1977). It has also been reported that the pituitary of old male rats is less responsive to the administration of exogenous LHRH than that of younger animals (Miller and Riegle, 1982).

Very little information is available on possible age-related changes of the number and of the binding characteristics of brain opioid receptors (Messing et al., 1980, 1981). The experiments here to be described have been designed in order to analyze whether the number and the affinity of brain mu and kappa receptors is modified by age in the male rat. In the first experiment, the concentration of brain mu receptors has been measured in the whole brain of male rats of 2, 15 and 22 months of age. In the second experiment, the concentration of mu receptors has been studied in the hypothalami of male rats of 2 and 22 months of age; in this experiment, it was also investigated whether the administration of exogenous testosterone via Silastic capsules might modify the number of mu binding sites in the hypothalamus of old animals. As in the studies reported in Section 4, dihydromorphine has been used as the in vitro ligand. In the third experiment the concentration of kappa receptors was investigated in the following brain structures: hypothalamus, amygdala, mesencephalon, corpus striatum, hippocampus, thalamus, frontal poles, anterior and posterior cortex, collected from male rats of 2 and 19 months of age. To evaluate the binding characteristics of kappa receptors, ^{3}H-bremazocine was used as the ligand, after protection of mu and delta receptors respectively with dihydromorphine and d-ala-d-leu-enkephalin.

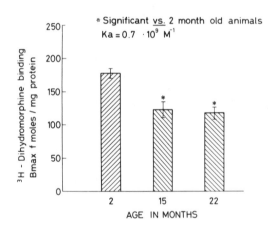

Fig. 6. Effect of aging on the number of mu opioid receptors in the whole brain of male rats.

Experiment 1   Fig. 6 shows the results of the first experiment, in which opioid mu receptors have been measured in the whole brain of male rats of 2, 15 and 22 months of age. It is apparent that the concentration of opioid receptors binding dihydromorphine significantly decreases between 2 and 15 months of age. The number of dihydromorphine binding sites does not seem to further decrease in animals of 22 months of age. The decrease in the number of mu receptors was not accompanied by any change in the affinity constant.

Experiment 2   Fig. 7 shows that, also at hypothalamic level, a conspicuous decrease of the number of mu receptors is observed at 22 months of age. The data also indicate that the administration of exogenous testosterone does not bring back to normal the number of opioid receptors in the hypothalami of 22 month old animals. Also in these experiments, the affinity constant of the mu receptors did not show any change in the various groups of animals.

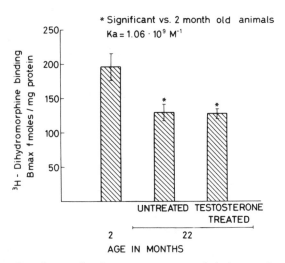

Fig. 7. Effect of aging and of testosterone administration on the number of mu opioid receptors in the hypothalamus of male rats.

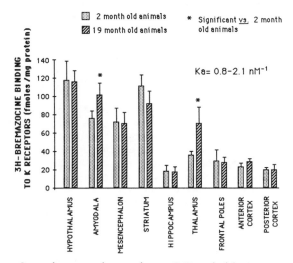

Fig. 8. Effect of aging on the number of K opioid receptors in discrete brain regions of male rats.

Experiment 3. Fig. 8 shows, first of all, that the concentration of kappa receptors in the different areas investigated is extremely variable both in young and in old animals. In young animals, particularly elevated concentrations of kappa receptors have been found (in descending order) in the hypothalamus, in the striatum, in the mesencephalon and in the amygdala; the number of kappa receptors found in the mesencephalon and in the amygdala is significantly lower than that detected in the hypothalamus and in the striatum, in which the concentrations of kappa receptors appear to be very similar. The concentrations of kappa receptors in the thalamus, frontal poles, hippocampus, anterior and posterior cortex are also similar among them, but are significantly lower than those found in the hypothalamus, the striatum, the amygdala and the mesencephalon.

Age does not seem to influence the number of kappa receptors in the hypothalamus, corpus striatum, mesencephalon, frontal poles, hippocampus, anterior and posterior cortex. Only in the amygdala and in the thalamus, the number of kappa receptors was significantly increased in aged animals. The increase in the number of kappa receptors was not accompanied by any change in the affinity constant.

The present results show, first of all, that in the male rat, the number of mu opioid receptors decreases with age, both in the whole brain and in the hypothalamus. The data show in addition that the administration of testosterone, in a dose able to bring back to normal serum testosterone levels, is ineffective in re-establishing the number of mu opioid receptors in the hypothalamus of old animals. The present results agree, in general, with those reported in the only other study in which dihydro-morphine binding has been studied in the brain of old male rats (Messing et al., 1981). It is interesting to note that, in the present study, a decrease of the number of mu receptors was shown to occur in the whole brain already at 15 months. Consequently, the present data indicate that the fall of opioid receptors in the brain of male rats is a rather precocious phenomenon. Unfortunately, it was not possible to ascertain whether the same phenomenon occurs also at hypothalamic level.

One might hypothesize that the age-linked decrease of the number of brain opioid receptors here reported might be linked to the decline of serum testosterone levels occurring in old male rats (Meites, 1982). It is still controversial whether serum testosterone may influence the number of brain opioid receptors in young animals (Wilkinson et al., 1981; Diez and Roberts, 1982; Cicero et al., 1983; Hahn and Fishman, 1985; Olasmaa et al., 1987) (see Section 2). Whatever the effects of testosterone on the number of brain opioid receptors in young animals, the present findings clearly show that the administration of exogenous testosterone does not re-establish to normal the number of opioid receptors at hypo-thalamic level in old animals. This appears particularly surprising, since it has been repeatedly reported that testosterone is more effective in old than in young animals, at least as far as the inhibition of gonado-tropin secretion is concerned (Shaar et al., 1975; Pirke et al., 1978; Gray et al., 1980). The present data then suggest that the decline of the number of opioid receptors observed with increasing age is not due to the decreased availability of testosterone. On the basis of the present findings, the age-linked decrease of the number of the brain receptors binding dihydromorphine in the whole brain and in the hypothalamus appears to be an autonomous phenomenon.

The present data have also shown that aging exerts little influence on the number of kappa receptors in the majority of the brain structures considered. The only exceptions were represented by the amygdala and by the thalamus, in which the number of kappa receptors was increased in aged animals. The effect of age on brain kappa receptors appears then to be totally different from that reported for the mu receptors, both in terms of the direction of the effect (increase vs decrease), and of the brain structures involved.

It is possible that the age-linked changes in the number of brain opioid receptors reported in this paper to occur in brain structures which, like the hypothalamus and the amygdala (Borrell et al., 1979; Piva et al., 1980) are involved in the neuroendocrine control, play a role in inducing the deterioration of the function of the hypothalamic-pituitary-gonadal axis observed in old male rats.

ACKNOWLEDGEMENTS

The experiments reported in this paper have been supported by the Consiglio Nazionale delle Ricerche (Target Projects "Preventive and Rehabilitative Medicine" and "Endocrinology", contracts no. 86.01902.56, 86.01871.56 and 86.00539.04) and by the Ministero della Pubblica Istruzione.

REFERENCES

Allen LG, Kalra SP, 1986. Evidence that a decrease in opioid tone may evoke preovulatory luteinizing hormone release in the rat. Endocrinology 118:2375-2381

Barraclough CA, Collu R, Massa R, Martini L, 1971. Temporal inter-relationships between plasma LH, ovarian secretion rates and peripheral plasma progestin concentrations in the rat: effects of nembutal and exogenous gonadotropins. Endocrinology 88:1437-1447

Bhanot R, Wilkinson M, 1983. Opiatergic control of LH secretion is eliminated by gonadectomy. Endocrinology 112:399-401

Blank MS, Mann DR, 1981. Diurnal influences on serum luteinizing hormone responses to opiate receptor blockade with naloxone or to luteinizing hormone-releasing hormone in the immature female rat. Proc Soc Exp Biol Med 168:338-343

Blank MS, Panerai AE, Friesen HG, 1979. Opioid peptides modulate luteinizing hormone secretion during sexual maturation. Science 203:1129-1131

Borrell J, Piva F, Martini L, 1979. Adrenergic inputs to the amygdala and the control of gonadotropin release. Acta Endocr 90:385-393

Cicero TJ, Newmann KS, Meyer ER, 1983. Testosterone does not influence opiate binding sites in the male rat brain. Life Sci 33:1231-1239

Clement-Jones V, Rees LH, 1982. Neuroendocrine correlates of the endorphins and enkephalins. In: Besser GM, Martini L (eds.), Clinical Neuroendocrinology, Vol. 2. New York:Academic Press, pp. 139-203

Delitala G, Grossman A, Besser GM, 1983. Differential effects of opiate peptides and alkaloids on anterior pituitary hormone secretion. Neuroendocrinology 37:275-279

Diez JA, Roberts DL, 1982. Evidence contradicting the notion that gonadal hormones regulate brain opioid receptors. Biochem Biophys Res Commun 108:1313-1319

Forman LJ, Sonntag WE, Miki N, Ramos T, Meites J, 1981. Comparison of the effects of central acting drugs on prolactin release in young and old male rats. Proc Soc Exp Biol Med 167:354-358

Gabriel SM, Simpkins JW, Kalra SP, 1983. Modulation of endogenous opioid influence on luteinizing hormone secretion by progesterone and estrogen. Endocrinology 113:1806-1811

Giudici D, D'Urso R, Falaschi P, Negri L, Melchiorri P, Motta M, 1984. Dermorphin stimulates prolactin secretion in the rat. Neuroendocrinology 39:236-244

Goldberg R, Conforti N, Spitz IN, 1982. The effect of methionine enkephalin on prolactin and luteinizing hormone levels in intact and castrated rats. Hor Metab Res 14:89-92

Gray GD, Smith ER, Davidson JM, 1980. Gonadotropin regulation in middle-aged male rats. Endocrinology 107:2021-2026

Hahn EF, Fishman J, 1985. Castration affects male rat brain opiate receptors content. Neuroendocrinology 41:60–63

Hoffman GE, Sladek JR, Jr, 1980. Age–related changes in dopamine, LHRH and somatostatin in the rat hypothalamus. Neurobiol Aging 1:27–37

Iovino M, Rouhani S, Polnaru S, 1983. Dual effect of met–enkephalin on plasma ADH concentration in the rat. Neuroendocr Lett 5:239–243

Kalra SP, 1986. Neural circuitry involved in the control of LHRH secretion: a model for preovulatory LH release. In: Ganong WF, Martini L (eds.), Frontiers in Neuroendocrinology, Vol. 9. New York: Raven Press, pp. 31–75

Kalra SP, Kalra PS, 1983. Neural regulation of luteinizing hormone secretion in the rat. Endocr Rev 4:311–351

Karpas AE, Bremner WJ, Clifton DK, Steiner RA, Dorsa DM, 1983. Diminished luteinizing hormone pulse frequency and amplitude with aging in the male rat. Endocrinology 112:788–792

Kato Y, Iwasaki Y, Abe H, Ohgo S, Imura H, 1978. Effect of endorphin on prolactin and growth hormone secretion in rats. Proc Soc Exp Biol Med 158:431–436

Kato Y, Hiroto S, Katakami H, Matsushita N, Shimatsu A, Imura H, 1982. Effects of a synthetic Met[5]–enkephalin analog on plasma luteinizing hormone and prolactin levels in conscious orchidectomized rats. Proc Soc Exp Biol Med 169:95–100

Kavaliers M, Hirst M, 1983. Daily rhythms of analgesia in mice: effects of age on photoperiod. Brain Res 279:387–393

Krulich L, Koenig JI, Conway S, McCann SM, Mayfield MA, 1986a. Opioid K receptors and the secretion of prolactin (Prl) and growth hormone (GH) in the rat. I. Effect of opioid K receptor agonists Bremazocine and U–50,488 on secretion of Prl and GH: comparison with morphine. Neuroendocrinology 42:75–81

Krulich L, Koenig JI, Conway S, McCann SM, Mayfield MA, 1986b. Opioid receptors and the secretion of prolactin (Prl) and growth hormone (GH) in the rat. II. GH and Prl release–inhibiting effects of the opioid K receptor agonists bremazocine and U–50,488. Neuroendocrinology 42:82–87

Limonta P, Maggi R, Giudici D, Martini L, Piva F, 1981. Role of the subfornical organ (SFO) in the control of gonadotropin secretion. Brain Res 229:75–84

Limonta P, Piva F, Maggi R, Dondi D, Motta M, Martini L, 1986. Morphine stimulates prolactin release in normal but not in castrated male rats. J Reprod Fertil 76:745–750

Lumpkin MD, McCann SM, 1984. Effect of destruction of the dorsal anterior hypothalamus on follicle–stimulating hormone secretion in the rat. Endocrinology 115:2473–2480

Mannisto PT, Rauhala P, Tuominen R, Mattila J, 1984. Dual action of morphine on cold–stimulated thyrotropin secretion in male rats. Life Sci 35:1101–1107

Marton J, Molnar J, Koves K, Halasz B, 1981. Naloxone reversal of the pentobarbital induced blockade of ovulation in the rat. Life Sci 28:737–743

Meites J, 1982. Changes in neuroendocrine control of anterior pituitary function during aging. Neuroendocrinology 34:151–156

Meites J, Bruni IF, Van Vugt DA, Smith AF, 1979. Relations of endogenous opioid peptides and morphine to neuroendocrine functions. Life Sci 24:1325–1336

Messing RB, Vasquez BJ, Spiehler VR, Martinez JL, Jr, Jensen RA, Rigter H, McGaugh JL, 1980. ^{3}H–Dihydromorphine binding in brain regions of young and aged rats. Life Sci 26:921–927

Messing RB, Vasquez BJ, Samaniego B, Jensen RA, Martinez JL, Jr, McGaugh JL, 1981. Alterations in dihydromorphine binding in cerebral hemispheres of aged male rats. J Neurochem 36:784–787

Miller AE, Riegle GD, 1982. Temporal patterns of serum luteinizing hormone and testosterone and endocrine response to luteinizing hormone releasing hormone in aging male rats. J Gerontol 37:522-528

Motta M, Martini L, 1982. Effect of opioid peptides on gonadotropin secretion. Acta Endocr 99:321-325

Muraki T, Nakadate H, Tokunaga Y, Kato R, Makino T, 1979. Effect of narcotic analgesics and naloxone on proestrous surges of LH, FSH and prolactin in rats. Neuroendocrinology 28:241-247

Olasmaa M, Limonta P, Maggi R, Dondi D, Martini L, Piva F, 1987. Further evidence that gonadal steroids do not modulate brain opiate receptors in male rats. Endocr Jap, in press

Paterson SJ, Robson LE, Kosterlitz HW, 1983. Classification of opioid receptors. Br Med Bull 39:31-36

Pfeiffer DG, Pfeiffer A, Shimohigashi Y, Merriam GR, Loriaux DL, 1983. Predominant involvement of mu- rather than delta- or K-opiate receptors in LH secretion. Peptides 4:647-649

Petraglia F, Locatelli V, Facchinetti F, Bergamaschi M, Genazzani AR, Cocchi D, 1986. Oestrous cycle-related LH responsiveness to naloxone: effect of high oestrogen levels on the activity of opioid receptors. J Endocrinol 108:89-94

Pirke KM, Geiss M, Sintermann R, 1978. A quantitative study on feedback control of LH by testosterone in young adult and old male rats. Acta Endocr 89:789-795

Piva F, Borrell J, Limonta P, Gavazzi G, Martini L, 1980. Cholinergic inputs to the amygdala and the control of gonadotropin release. Acta Endocr 93:1-6

Piva F, Limonta P, Martini L, 1982. Role of the organum vasculosum laminae terminalis in the control of gonadotrophin secretion in rats. J Endocrinol 93:355-364

Piva F, Limonta P, Maggi R, Martini L, 1984. Dual effects of opioids in the control of gonadotropin secretion. In: Delitala G, Motta M, Serio M (eds.), Opioid Modulation of the Endocrine Function. New York: Raven Press, pp. 155-169

Piva F, Maggi R, Limonta P, Motta M, Martini L, 1985. Effect of naloxone on luteinizing hormone, follicle-stimulating hormone, and prolactin secretion in the different phases of the estrous cycle. Endocrinology 117:766-772

Piva F, Limonta P, Maggi R, Martini L, 1986. Stimulatory and inhibitory effects of the opioids on gonadotropin secretion. Neuroendocrinology 42:504-512

Piva F, Maggi R, Limonta P, Dondi D, Martini L, 1987. Decrease of mu opioid receptors in the brain and in the hypothalamus of the aged male rat. Life Sci 40:391-398

Shaar CJ, Euker JS, Riegle GD, Meites J, 1975. Effects of castration and gonadal steroids on serum luteinizing hormone and prolactin in old and young rats. J Endocrinol 66:45-51

Simpkins JW, Mueller GP, Huang HH, Meites J, 1977. Evidence for depressed catecholamine and enhanced serotonin metabolism in aging male rats: possible relation to gonadotropin secretion. Endocrinology 100:1672-1678

Spiegel K, Kourides IA, Pasternak GW, 1982. Different receptors mediate morphine-induced prolactin and growth hormone release. Life Sci 31:2177-2180

Wildt L, Hausler A, Marshall G, Hutchinson JS, Plant TM, Belchetz PE, Knobil E, 1981. Frequency and amplitude of gonadotropin-releasing hormone stimulation and gonadotropin secretion in the Rhesus monkey. Endocrinology 109:376-385

Wilkinson M, Herdon H, Wilson CA, 1981. Gonadal steroid modification of adrenergic and opiate receptor binding in the central nervous system. In: Fuxe K, Gustafsson JA, Wetterburg L (eds.), Steroid Hormone Regulation of the Brain. Oxford: Pergamon Press, pp. 253–263

Wilkinson M, Brawer Jr, Wilkinson DA, 1985. Gonadal steroid-induced modification of opiate binding sites in anterior hypothalamus of female rats. Biol Reprod 32:501–506

LOCALIZATION OF NEUROACTIVE SUBSTANCES IN THE HYPOTHALAMUS WITH

SPECIAL REFERENCE TO COEXISTENCE OF MESSENGER MOLECULES

T. Hokfelt, Y. Tsuruo, B. Meister, T. Melander
M. Schalling and B. Everitt

Department of Histology
Karolinska Institutet
Stockholm, Sweden

INTRODUCTION

Harris (1955) formed the concept that the hormone secretion from the anterior pituitary is controlled by the brain via chemical messengers released from the hypothalamus into the hypophysial portal vessels. At the anatomical level Szentagotai and collaborators (1962) demonstrated that the medio-basal hypothalamus was an important brain area in this function, and that tubero-infundibular neurons with cell bodies in the arcuate nucleus and fibers in the median eminence could represent a morphological correlate for this control. Dopamine (DA) was the first compound identified with histochemical techniques in the tubero-infundibular region and was found in fibers in the external layer of the median eminence (Fuxe, 1964), arising from cell bodies in the arcuate nucleus (Fuxe and Hökfelt, 1966). More recently an increasing number of compounds have been discovered in this brain region including amines, amino acids and peptides (Everitt et al., 1986).

Here we would like to briefly outline the localization of some transmitters and peptides in the arcuate nucleus in relation to this dopaminergic system (A12 cell group, Dahlstrom and Fuxe, 1964) and discuss studies attempting to trace the localization of cell bodies projecting to the median eminence. Furthermore, coexistence of several peptides in a paraventricular-median eminence system and its possible functional implications will be discussed. The possibility that LHRH neurons contain a further messenger molecule, possibly related to leukotrienes, will be raised. In view of the limitations on the length of the articles for this volume, only a limited number of references can be included.

ASPECTS ON METHODOLOGY

Initial histochemical studies were based on the formaldehyde induced technique, based on the finding of Eränkö (1955) that formalin reacts with catecholamines in the adrenal medulla to form fluorescent products. Falck, Hillarp and their collaborators developed this method into a reliable technique with the various reaction products thoroughly characterized (Jonsson, 1971; Björklund, 1985). Semiquantitative subjective evaluations could be performed comparing, for example, control and hormone treated groups (Fuxe and Hökfelt, 1969; Hökfelt and Fuxe, 1972). Today this technique has been further advanced using microspectrofluorometry combined with internal standards which allow monitoring of actual catecholamine levels (Löfström et al., 1976a,b; Andersson et al., 1985).

For further characterization of neurons, immunohistochemical methodology has gained an increasing popularity. Based on the direct and indirect immunofluorescence methods developed by Coons and his collaborators (Coons, 1958), several technical advancements have been introduced, for example, including the use of peroxidase as marker (Nakane and Pierce, 1966; Avrameas, 1969; Sternberger et al., 1970). This technique has general applicability and allows demonstration of virtually every compound against which antisera can be raised, thus providing a powerful tool.

Immunohistochemistry (Coons, 1958; Nairn, 1969; Sternberger, 1979; Cuello, 1983; Polak and van Noorden, 1983) has turned out to be a good choice for analyzing multiple antigens in one and the same neuron, and several approaches are now available for this task. They include the 'adjacent section' method, where consecutive sections are incubated with different primary antisera. No cross-reaction between antisera can occur, and consequently there are no problems of specificity due to interference between antibodies. Only large objects such as cell bodies can be studied, but with sufficiently thin sections a cell body can often be identified in two or even more consecutive sections. When epoxy resin embedded material is used, sections can be cut at 1 μm or thinner, and then numerous sections through a single cell body can be analyzed (Berod et al., 1984). 'Elution-restaining methods' (Nakane, 1968; Tramu et al., 1978) have been extensively utilized. After photography of the first staining pattern, the antibodies are eluted with acid solutions, the sections are then reincubated with another antiserum, and the new staining patterns are compared with the previously taken photographs. This method has limitations, since the elution procedure seems to damage some antigens. The third approach is 'direct double-staining', which is based on availability of antisera raised in different species. Secondary antibodies labelled with different chromogens (e.g., green fluorescent fluorescein isothiocyanate, FITC, and red fluorescent tetramethyl rhodamine isothiocyanate, TRITC) and directed against IgG from the two respective species then allow visualization of the two antigens in the same section by switching between proper filter combinations (Nairn, 1969; Wessendorf and Elde, 1985). It has recently been shown that three antigens can be visualized in a single section using a third, blue fluorescent dye conjugated to an appropriate secondary antibody (Staines et al., 1987). By combining this triple staining technique with elution-restaining, it should be possible to visualize four or even more antigens in a section. The final analysis of coexistence will, however, include electron microscopic studies. It has, for example, been shown that 5-hydroxytryptamine (5-HT) and substance P (SP) are stored in the same vesicles in some nerve endings in the spinal cord (Pelletier et al., 1981), and also at the ultrastructural level there are now methods to demonstrate three antigens in one section (Doerr-Schott, 1986).

Biochemistry has only been used to a limited extent to analyze coexistence, probably because the nervous system is an extremely hetero-geneous tissue. This makes it difficult to study coexistence with the present status of sensitivity of biochemical techniques. An exception may be some invertebrate neurons, which are so large that they can be isolated individually and that their content of neuroactive compounds possibly can be determined biochemically (Osborne, 1979). Biochemistry can be used however to demonstrate coexistence indirectly. For example, there are 'specific' neurotoxins such as 6-hydroxydopamine (Thoenen and Tranzer, 1968) and 5,6-dihydroxytryptamine (Baumgarten et al., 1971) which destroy catecholamine and 5-HT neurons, respectively. With the latter compound, a concomitant depletion of 5-HT, SP and thyrotropin releasing hormone (TRH) has been shown and was interpreted to indicate coexistence of these compounds in single neurons (Gilbert et al., 1982).

Immunohistochemical methods should, in spite of their power and usefulness, be considered with considerable caution both with regard to specificity and their sensitivity. Thus, it can not be excluded that the antisera cross-reacts with compounds which are structurally similar to the immunogens. Recently, evidence has been presented that one single amidated amino acid in the C-terminal position may be sufficient to cause cross-reactivity (Berkenbosch et al., 1986; Ju et al., 1986). Therefore, expressions such as 'substance P-like immunoreactivity' (LI), 'substance P-immunoreactive' - etc., should be used.

The sensitivity problem should also be emphasized and it is important to point out that negative results should be interpreted with great caution. It has been demonstrated repeatedly that improvement of the fixation technique and/or production of antibodies with a higher affinity and/or higher avidity reveal a certain antigen in places where it had not be demonstrated earlier. In Fig. 1 such an example is shown. Thus, with improved methodology it is now possible to demonstrate more TRH-positive structures in the rat brain than so far achieved, for example, in the hypothalamus (Tsuruo, Hökfelt and Visser, in preparation). Also, peptide levels in cell bodies are often too low to be visualized in central neurons, but can be increased by pretreatment of experimental animals with a mitosis inhibitor, colchicine (Dahlstrom, 1971) and in this way visualized (Fig. 1).

COEXISTENCE IN THE ARCUATE NUCLEUS - MEDIAN EMINENCE COMPLEX

The arcuate nucleus - median eminence complex has long been considered as a final common pathway for output of information from brain to the periphery, particularly to the anterior pituitary gland (Szentágothai et al., 1962). In addition to dopamine neurons (Fig. 2a), more recent work has demonstrated a variety of transmitter- and peptide-containing cell bodies in the arcuate nucleus. This issue has recently been analyzed in detail as well as reviewed by Everitt et al. (1986), Chronwall (1985), Hokfelt et al. (1986). Briefly, there is strong evidence for the existence of numerous γ-aminobutyric acid (GABA) neurons at all levels of the arcuate nucleus (Vincent et al., 1982b; Tappaz et al., 1983; Mugnaini and Oertel, 1985), and in several cases arcuate neurons contain both the catecholamine-synthesis enzyme tyrosine hydroxylase (TH) (Fig. 2a) and the GABA synthesizing enzyme glutamic acid decarboxylase (GAD), suggesting that some arcuate neurons may produce two classical transmitters, GABA and dopamine (Everitt et al., 1984). The arcuate nucleus may contain at least one more classical transmitter, acetylcholine (ACh) (Carson et al., 1977).

The detailed analysis of TH-positive cells in the arcuate nucleus has revealed that it may be possible to distinguish two subgroups, one dorsal, periventricular group and a second group located in the ventrolateral parts of the nucleus extending laterally along the basal part of the brain. Under certain experimental conditions (low antibody concentrations, no colchicine treatment), the ventrolateral group seems to contain only low levels of TH-LI. These neurons can be shown with several TH antisera (Chan-Palay et al., 1984; Hökfelt et al., 1984b; Van den Pool et al., 1985), and the immunoreactivity disappears after preabsorption of the antiserum with TH enzyme. However, it has so far not been possible to demonstrate this ventrolateral group with antisera raised to bovine aromatic acid decarboxylase (AADC), the enzyme converting L-DOPA to dopamine. The identity of the ventrolateral TH-positive arcuate cell group is therefore not fully elucidated, and the possibility exists that it does not produce

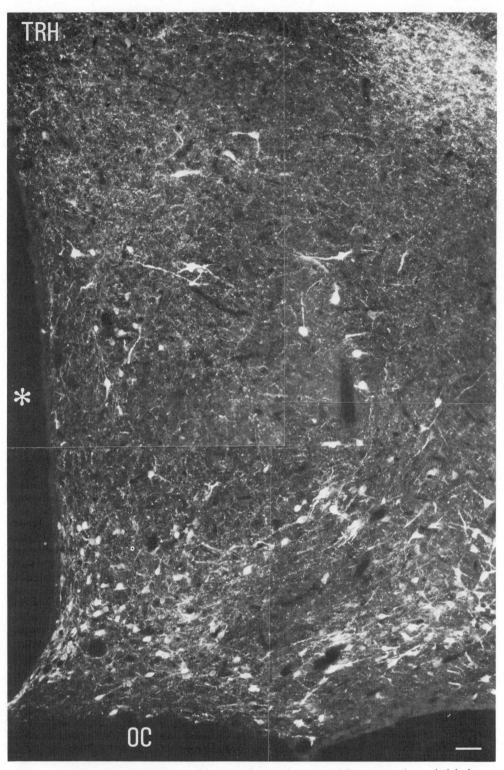

Fig. 1. Immunofluorescence micrograph of the preoptic area of colchicine treated rat after incubation with TRH antiserum. Note many cell bodies and dense fiber networks. Asterisk denotes third ventricle and OC = optic chiasm. Bar indicates 50 μm.

dopamine. The lack of AADC-LI may, however, be due to the fact that this antibody was not raised against the rat enzyme, and that it was produced more than a decade ago, i.e., this may simply represent a sensitivity problem.

Arcuate neurons also contain a wealth of peptides which include neurotensin (Uhl et al., 1979; Kahn et al., 1981; Jennes et al., 1982; Ibata et al., 1983b, 1984a,b; Hökfelt et al., 1984a), some times together with TH-LI (Ibata et al., 1983a; Hökfelt et al., 1984a), and galanin (Skofitsch and Jacobowitz, 1985; Melander et al., 1986a,b), and many of these cells are TH-immunoreactive (Everitt et al., 1986; Melander et al., 1986b). Numerous growth hormone releasing factor (GRF)-positive cell bodies are present extending into adjacent areas (Bloch et al., 1983a,b, 1984; Bugnon et al., 1983; Fellmann et al., 1983; Jacobowitz et al., 1983; Lechan et al., 1984; Merchenthaler et al., 1984; Smith et al., 1984), and most of these cell bodies are TH-positive (Meister et al., 1985, 1986; Okamura et al., 1985). In fact, the TH- and GRF-positive cell bodies are often also galanin- (cf. Fig. 2a and 2b; Fig. 3) and probably neurotensin-containing. Opioid peptides are found in the arcuate nucleus, with products from all three precursors, proenkephalin A, proenkephalin B and proopiomelanocortin (PMOC) expressed (McGinty and Bloom, 1983; Everitt et al., 1986). Thus, both met-enkephalin (Hokfelt et al., 1977; Sar et al., 1978; Wamsley et al., 1980; Khachaturian et al., 1983; Williams and Dockray, 1983; Krukoff et al., 1984; Kachaturian et al., 1985a,b; Merchenthaler et al., 1986), the enkephalin heptapeptide met-enkephalin-Arg-Phe (Williams and Dockray, 1983; Everitt et al., 1986), the enkephalin octapeptide met-enkephalin-Arg-Glu-Leu (Everitt et al., 1986), as well as the proenkephalin B product dynorphin (Vincent et al., 1982a; Watson et al., 1982, 1983; Weber et al., 1982) are present in arcuate neurons. Some of these cells, including some enkephalin octapeptide- and dynorphin-positive ones are TH-positive, since it has long been known that products of the POMC precursor occur in a substantial number of arcuate neurons at all levels extending laterally into adjacent areas (Watson et al., 1977; Bloch et al., 1978, 1979; Bugnon et al., 1979; Sofroniew, 1979; Finely et al., 1981; Ibata et al., 1980; Khachaturian et al., 1985a,b). These neurons to a large extent overlap with the ventrolateral group of TH-positive cells, but so far no coexistence between ACTH/β-endorphin- and TH-positive neurons have been observed (Bugnon et al., 1979; Ibata et al., 1980; Everitt et al., 1986).

In the ventromedial aspects of the arcuate nucleus two major peptides containing cell populations can be found containing, respectively, somatostatin, (Shiosaka et al., 1982; Ibata et al., 1983b; Ohtsuka et al., 1983; Chronwall et al., 1984a; Johansson et al., 1984; Chronwall, 1985; Vincent et al., 1985; Everitt et al., 1986) and neuropeptide tyrosine (NPY)-LI (Chronwall et al., 1984a, 1985; Chronwall, 1985; De Quidt and Emson, 1986; Everitt et al., 1986) (Fig. 2c), with the latter located ventral to the somatostatin cells. These neurons are numerous and small in size and have so far not been shown to contain any other peptide, but a very small proportion contains both somatostatin- and NPY-LI (Chronwall et al., 1984a).

The arcuate nucleus contains some immunoreactivities which may represent cross-reactivities. Thus, FMRFamide-positive neurons have been described in the arcuate nucleus (Chronwall et al., 1984b), and this may represent cross-reactivity with NPY (Lundberg et al., 1984a; Everitt et al., 1986). Furthermore, metorphamide-LI may also represent cross-reactivity with NPY (Everitt et al., 1986). Finally, earlier described avian and bovine pancreatic polypeptide (APP, BPP)-immumoreactive neurons (Lorén et al., 1979a; Olschowka et al., 1981; Card et al., 1983) are identical to the NPY neurons mentioned above, and it is likely that in these neurons the APP and BPP antisera have cross-reacted with NPY (Allen et al., 1983; Lundberg et al., 1984b; DiMaggio et al., 1985).

In conclusion, the arcuate nucleus exhibits a large degree of heterogeneity with regard to content of neurotransmitters and peptides (Fig. 3). Several examples of coexistence of TH- plus GRF- plus GAL-LI and TH- plus NT- plus GAL-LI have been observed in the ventrolateral area of the arcuate nucleus, and TH- plus NT-LI and TH-LI plus GAD-LI in the dorsal aspects, but presumably also in the entire nucleus. It is likely that some neurons in fact contain at least five compounds, dopamine, GABA, galanin, GRF and neurotensin.

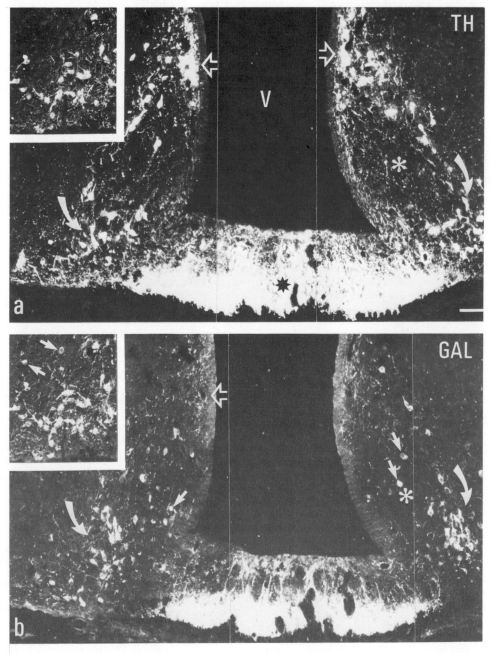

Fig. 2a,b   (see next page for legend)

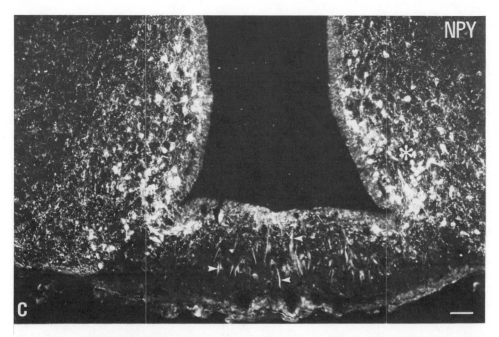

Fig. 2a–c. Immunofluorescence micrographs of three consecutive sections
the median eminence–arcuate nucleus after incubation with antiserum to tyrosine hydroxylase (TH) (a), galanin (GAL) (b) and
neuropeptide Y (NPY) (c). Insets show the same incubated with GAL
antiserum (inset in b) followed by elution and restaining with TH
antiserum (inset in a). TH-positive cell bodies are mainly present
in the dorsal periventricular (open arrows) and in the ventrolateral (curved arrows) parts of the arcuate nucleus (a), whereas
GAL-positive cells are mainly found in the ventro-lateral region
(b). As seen in the insets, most TH-positive cells are also GAL-
immunoreactive. Small arrows point to some GAL cells lacking TH-
like immunoreactivity. NPY-positive cells are mainly found in the
medial, periventricular zone (asterisks), which mostly lack TH-
positive cells (c). In the median eminence (star) dense networks
of TH- and GAL-positive fibers (a,b), but only a few NPY-immuno-
reactive fibers are seen (c). Arrow heads in c point to unspecific
stain in tanycytes in the median eminence. V = third ventricle.
Bar indicates 50 μm. All micrographs have the same magnification.

PARAVENTRICULAR NEURONS CONTAINING CRF-, VIP/PHI- AND ENK-LI

It has been reported that a population of neurons in the parvocellular part of the paraventricular nucleus contains a peptide histidine
isoleucine (PHI)-like peptide (Hökfelt et al., 1982a; Hökfelt et al.,
1983a) (Fig. 4a), which is a recently isolated 27 amino acid peptide
(Tatemoto and Mutt, 1981). It has a considerable similarity to vasoactive
intestinal polypeptide (VIP), and these two peptides have been shown to be
derived from the same precursor peptide (Itoh et al., 1983). In spite of
this, we were at that time not able to demonstrate VIP-LI in the paraventricular neurons which contained PHI-LI, although PHI-VIP coexistence
could be demonstrated in other types of neurons (Christofides et al.,
1982; Lundberg et al., 1984a). In fact, VIP had at that time not been
demonstrated immunohistochemically in the external layer of the median
eminence in any of the major mapping papers (Fuxe et al., 1977; Lorén et
al., 1979b; Sims et al., 1980; Hökfelt et al., 1982b). Subsequently it was

27

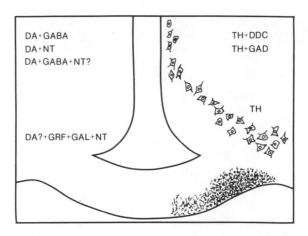

Fig. 3. Schematic illustration of coexistence situations in various parts
of the arcuate nucleus. In the dorsal aspects the cell bodies contain tyrosine hydroxylase (TH) and aromatic amino acid decarboxylase (dopadecarboxylase; DDC) and in many cases glutamic acid
decarboxylase (GAD).   These neurons thus presumably produce
dopamine and at least some of them, and may in addition have the
peptide neurotensin (NT). Whether all three compounds are
present, the same neurons have not been definitely shown. In the
ventrolateral group only TH—like immunoreactivity (LI) has been
demonstrated, and these neurons have therefore not yet been shown
to be truely dopaminergic.   They contain at least three peptides,
growth hormone—releasing factor (GRF), galanin (GAL) and
neurotensin (NT).

recognized that these neurons were also corticotropin releasing factor
(CRF)—immunoreactive (Fig. 4b), and that many of them in addition
contained an enkephalin—like peptide (Fig. 4c) (Hökfelt et al., 1983a).
These three compounds were all located in parvocellular neurons, clearly
distinguished from the magnocellular, neurophysin—positive cells (Fig.
4d).   It was proposed that these neurons could represent a morphological
substrate for the well known concomitant stress—induced release of adrenocorticotropic hormone (ACTH) and prolactin (Ganong, 1963; Meites et al.,
1963; Brown and Martin, 1974; Harms et al., 1975), since it could be
demonstrated that not only VIP (Kato et al., 1978; Ruberg et al., 1978;
Shaar et al., 1979; Vijayan et al., 1979; Abe et al., 1985; Ohta et al.,
1985a,b) but also PHI has prolactin releasing activity (Samson et al.,
1983; Werner et al., 1983; Kaji et al., 1984; Ohta et al., 1985a,b).
Moreover, since enkephalin has been shown to inhibit dopamine release
(Ferland et al., 1977; Meites et al., 1979), and since dopamine is a well
known inhibitor of prolactin release (McLeod and Lehmeyer, 1974; Neill,
1980), enkephalin could at the median eminence level reduce dopamine
release and thus indirectly enhance prolactin release from the lactotrops.

    The fact that CRF and PHI exhibited a complete overlap suggested the
possibility of cross—reactivity, and this was tested for by incubating PHI
antiserum with ovine CRF peptide (Hökfelt et al., 1983a), the only CRF
available at that time.   Our conclusion was that PHI antiserum does not
cross—react with CRF (Hökfelt et al., 1983a). Recently Berkenbosch et al.
(1986) have, however, demonstrated that several PHI antisera cross—react

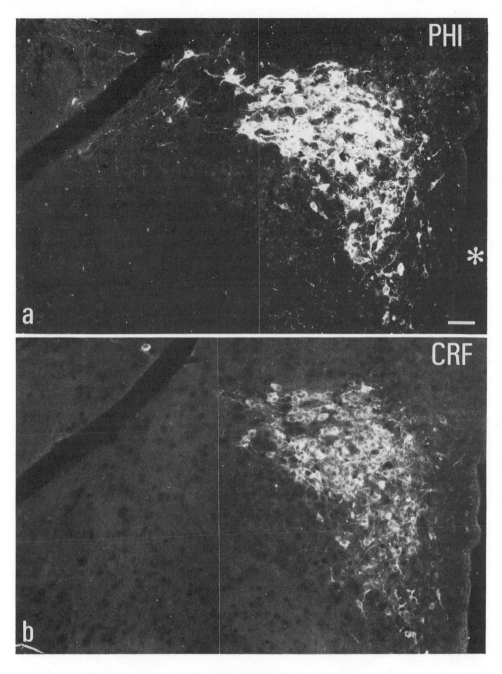

Fig. 4a–d. Immunofluorescence micrographs of four semiconsecutive sections
of the rat paraventricular nucleus after incubation with antiserum
to PHI (a), CRF (b), enkephalin (ENK) (c) and neurophysin (NF)
(d). Note a similar distribution of PHI-, CRF- and ENK-like
immunoreactivities. However, whereas the PHI and CRF have a
perfect overlap, the distribution of ENK-positive cells is somewhat different. NF-positive cells are mainly located in the
magnocellular part of the nucleus. Asterisk shows third
ventricle. Bar indicates 50 μm. All micrographs have the same
magnification. (Fig. 4c and d on next page).

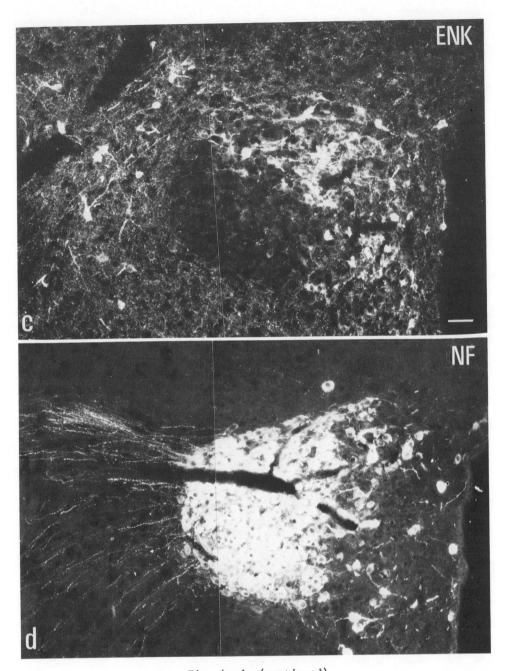

Fig. 4c,d (continued)

with <u>rat</u> CRF when absorbed with high peptide concentrations (10^{-5}M and higher). They have pointed out that the last amino acid in rat CRF is isoleucine (in ovine CRF the last amino acid is arginine), and isoleucine is the last amino acid also in PHI. In both cases isoleucine is amidated, and Berkenbosch et al. (1986) have suggested that due to this one-amino acid similarity, there is a risk for cross-reactivity, pointing out that CRF in the median eminence is present in very high concentrations, probably much higher than the 10^{-5}M necessary to show cross-reactivity <u>in vitro</u>.

Our PHI antisera used both in the radioimmunoassay and immunohisto-chemical model have therefore been reanalyzed, and they appear to be N-terminally and C-terminally directed, respectively (Hökfelt et al., 1987). The antiserum demonstrating numerous cells in the paraventricular nucleus (Fig. 4a) is PHI Ab 3668-5, which is a primarily N-terminal antibody (in the following called PHI-N antiserum). Fragments PHI (1-15), -(3-15) and -(7-15) strongly react with the antiserum, whereas the C-terminal fragments PHI (14-27), -(20-27) and -(23-27) do not show any appreciable cross reactivity, nor does VIP, but an interaction with CRF can be seen. Thus this antiserum is primarily N-terminally directed, but may have a population of antibodies exhibiting a C-terminally directed activity. This could account for the immunoreaction seen in the CRF cells. This is supported by absorption experiments, where a complete disappearance of PHI-N-immunoreactive fibers could be seen in the external layer of the median eminence after absorption with $\underline{rat}$ CRF at 10^{-5} M and a marked decrease with 5×10^{-5} M, whereas no certain effect could be recorded at 10^{-6} M. As in previous experiments (Hökfelt et al., 1983a), $\underline{ovine}$ CRF 10^{-4} M had no effect. Similar results were obtained in the analysis of PHI immunoreactive cell bodies in the parvocellular paraventricular nucleus. These findings confirm the results of Berkenbosch et al. (1986). Also our PHI-N antiserum thus cross-reacts with CRF and our earlier demonstration of extensive coexistence between CRF and PHI may at least in part be artefactual. The second PHI antibody, 3692-3, is clearly C-terminally directed but does not react with the C-terminal isoleucine and not with CRF.

The question is then whether or not there are any PHI/VIP-containing neurons in the paraventricular nucleus. Their existence is indirectly supported by the fact that both PHI and VIP can be measured in portal blood (Said and Porter, 1979; Shimatsu et al., 1981, 1983; Brar et al., 1986). In fact, it has recently been shown that it is possible to visualize VIP-positive neurons in parvocellular neurons in the para-ventricular nucleus under certain conditions, such as adrenalectomy combined with colchicine treatment (Mezey and Kiss, 1985; Mezey, 1986). The number of positive cells is, however, low as compared to the CRF positive neurons. Moreover, according to Mezey and Kiss (1985) 'double staining with CRF and VIP antibodies suggests that although the two peptides are present in the same subdivision of the paraventricular nucleus, they do not coexist in the same cells'. This question was reinvestigated using immobilization stress combined with colchicine treatment (Hökfelt et al., 1987), and absorptions experiments have also been done, i.e., the N-terminally directed PHI antiserum has been preabsorbed with varying concentrations of CRF. We could confirm the findings of Mezey and Kiss (1985) and Mezey (1986) on the presence of VIP-positive neurons in this nucleus. In addition, an approximate similar number of PHI-positive cells were observed in the same area after immobilization stress, in both cases combined with colchicine treatment. Moreover, using the elution-restaining technique (Tramu et al., 1978), it has been possible to demonstrate that some VIP- and PHI-positive neurons also contain CRF-LI and some also ENK-LI. Finally, it could also be shown that preabsorption of PHI-N antiserum with rat CRF at 10^{-5} M abolished staining of cells in the parvocellular part of the paraventricular nucleus, but that some immunoreactive cells remained and some of these were CRF immunoreactive. It may therefore be justified to state, as done in the previous study (Hökfelt et al., 1983a), that there are neurons in the paraventricular nucleus, presumably sending axons to the external layer of the median eminence which contain both PHI/VIP-, CRF- and ENK-LI, and that the con comitant release of these two peptides may induce parallel release of ACTH and prolactin (Hökfelt et al., 1983a). The number of such cells is, however, much smaller than suggested by the original experiments. The present situation with a restricted number of CRF/PHI-positive neurons may indicate that only a subpopulation of CRF neurons are involved in the events controlling stressed-induced, concomitant release of ACTH and CRF. The CRF neurons represent a heterogenous population and consists of chemically coded neurons (Swanson et al., 1986).

ORIGIN OF NERVE FIBERS IN THE EXTERNAL LAYER OF THE MEDIAN EMINENCE

The median eminence is built up of only few neuronal cell bodies but mainly dense networks of nerve endings and tanycytes (Kobayashi and Matusi, 1969). Therefore the vast number of nerve endings in this area probably originate outside the median eminence. To elucidate the origin of these fibers various methods can be used. Classically, lesions of various types, for example, electrolytic lesions or knife cuts have been carried out. More recently novel techniques for tracing neuronal projections have been introduced. We have combined retrograde tracing of fluorescent compounds, such as Fast Blue (FB) (Kuypers and Huisman, 1983) with transmitter histochemistry (Hokfelt et al., 1983b). Routinely this technique employs injection of a fluorescent tracer into specific brain regions. However, injection of tracers into the median eminence represents a particularly difficult problem (Lechan et al., 1982), and we have therefore taken a different approach (Fig. 5). Since the median eminence is located outside the blood brain barrier, it may be anticipated that intravenously injected tracer, in this case FB, is taken up only by nerve endings in the median eminence and the pituitary gland (and, of course, also in other brain areas located outside the blood brain barrier, e.g., area postrema). Preliminary experiments have revealed retrogradely labelled cells in several hypothalamic nuclei. Thus, for example, most magnocellular and parvocellular neurons in the paraventricular nucleus and many neurons in the anterior periventricular hypothalamus and in the arcuate nucleus were strongly FB-labelled. Immunohistochemical processing of such sections revealed that the majority of FB labelled neurons in the arcuate nucleus were TH-positive, i.e., represent DA neurons belonging to the A12 group. In the magnocellular paraventricular nucleus virtually all cell bodies were labelled. In the parvocellular paraventricular nucleus, for example, PHI- and CRF-IR FB labelled neurons were identified. Many FB-positive neurons in the anterior periventricular nucleus were somatostatin-IR, but so far no FB-positive DA cells of the A14 group were seen. No or only single FB-positive cells were observed in the medial parts of the arcuate nucleus suggesting that neither NPY- nor somatostatin-IR cell bodies project to the external layer of the median eminence. Also ACTH-IR positive neurons have so far not been found to accumulate retrogradly transported FB. This work is now in progress to further outline and define the projection neurons to the median eminence.

These findings confirm the present general view on neuronal projections to the median eminence (Fig. 6). Taken together the 'classical' releasing and inhibitory neurons (containing TRH, luteinizing hormone releasing hormone (LHRH), somatostatin or CRF) have their cell bodies outside the medial basal hypothalamus with the exception of the GRF neurons, which in part are identical to TH-positive, presumably DA neurons. In the arcuate nucleus some neurons containing peptides (NPY, somatostatin, ACTH) seem to project centrally to other parts of the brain, whereas other neurons contain one (ACh, GABA or DA) or even two (DA + GABA) classical transmitters, sometimes together with one or more peptides (DA + GAL, DA + GRH, DA + NT). Thus, the latter classical transmitter neurons in the arcuate nucleus occupy a strategic position, through which they at the median eminence level may control the secretion of releasing and inhibitory factors into the portal vessels. In this way the tubero-infundibular neurons may resemble interneurons in the dorsal horn, which

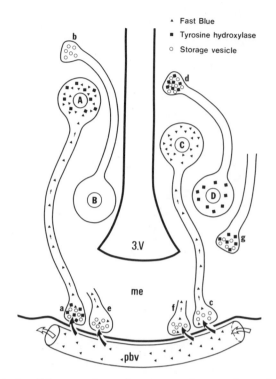

Fig. 5. Schematic illustration of retrograde tracing of neurons projecting to the external layer of the median eminence by intravenous injection of fluorescent dye (Fast Blue). The dye (triangles) are taken up into nerve endings in the median eminence, since they are located outside the blood brain barrier. Fast Blue is transported retrogradely to the cell bodies. Subsequently the section is processed for immunohistochemistry using antiserum to tyrosine hydroxylase (TH) (squares). Neuron A is double-labelled, whereas neuron B contains neither Fast Blue nor TH, neuron C only the retrograde tracer and neuron D only TH. Thus, only one of these neurons is a dopaminergic neuron projecting to the external layer of the median eminence. 3.V = third ventricle; pbv = portal blood vessel. From Hökfelt et al. (1986).

according to the gate theory of pain transmission (Melzack and Wall, 1965) control sensory <u>influx</u> to the central nervous system. The short tuberoinfundibular neurons may, instead, partly act as gating neurons to control the <u>outflow</u> of neuroendocrine information to the pituitary gland.

EVIDENCE FOR A SECOND LH RELEASING FACTOR

The release of luteinizing hormone (LH) from the anterior pituitary is controlled by a decapeptide, LHRH (Blackwell and Guillemin, 1973; Schally et al., 1973). Neurons producing LHRH have been defined with immunohistochemical techniques (Barry et al., 1986). Novel compounds belonging to the arachidonic acid metabolites have been discovered, the leukotrienes (Samuelsson, 1983). Using a variety of biochemical techniques, including radioimmunoassay, reversed high performance liquid chromatography and spectral analysis, evidence has been obtained that leukotrine C_4 (LTC$_4$) can be produced in slices from the rat brain under

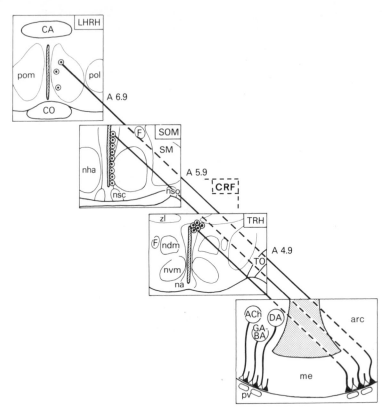

Fig. 6. Schematic illustration of the origin of LHRH, somatostatin (SOM), CRF and TRH neurons. These hypothalamic hormone-containing neurons have their cell bodies in rostral areas with the CRF neurons in the paraventricular nucleus, where also TRH-positive neurons are located. All these systems project to the external layer of the median eminence (me). In the arcuate nucleus (arc) several systems containing classical transmitters, acetycholine (ACh), GABA and dopamine (DA) are present, also projecting to the external layer of the median eminence. The dopamine and GABA cells in the ventrolateral part of the arcuate nucleus are to a large extent identical to the GRF neurons. From Hökfelt et al. (1986).

stimulation with ionophor and in the presence of arachidonic acid, with the highest concentrations in the hypothalamus (Lindgren et al., 1984). Using an antiserum raised to a LTC_4-bovine serum albumin (BSA)-conjugate, an attempt has been made to localize LTC_4 in the nervous system. LTC_4-IR nerve fibers were observed in the external layer of the median eminence (Lindgren et al., 1984). This immunoreactivity could be adsorbed out with LTC_4-BSA conjugate and also in about 10x higher concentrations with glutathione, but not with LTC_4 alone nor with high concentrations (10^{-4}M) of LHRH. The identity of the LTC_4-like immunoreactivity is thus uncertain, especially against the background of considerable specificity problems inherent to the immunohistochemical technique (see above).

The distribution of LTC_4-immunoreative fibers is similar to the one described for LH-positive fibers in the median eminence (Barry et al., 1986). In fact, elution-restaining experiments have revealed that these fibers to a large part are identical (Hulting et al., 1985). This suggests that an LTC4-like compound is present in LHRH neurons and that the two compounds may be released together. Against this background we have tested the effect of LTC4 on release of hormones from dispersed anterior pituitary cells in culture. It was observed that LTC4 in very low concentrations (down to 10^{-15}M) can release LH (Hulting et al., 1984, 1985). The release caused by LTC4 had a more rapid onset than LHRH-induced release and was of a short duration (Hulting et al., 1985). No effect of LTC4 could be observed on growth hormone release. The LTC4-precursor LTB4 did not exert any effect on LH release. The effects of leukotrienes on LHRH have also been analyzed by Gerozissis et al. (1985, 1986). They observed a stimulation of LHRH release at very low concentrations suggesting that leukotrienes may also exert effects on the gonadotropic axis at the hypothalamic level. A messenger role for leukotrienes in the CNS is also suggested by electrophysiological studies (Palmer et al., 1980, 1981) and the demonstration of LTC4 binding sites (Schalling et al., 1985).

The release of LH may thus be under control of more than one compound. The time course, when comparing LHRH and LTC4, suggests that the latter may be responsible for a more rapid phase of LH release, whereas LHRH evokes a sustained release with a slower onset. It will be important to clarify the identify of the compound visualized in the LHRH neurons with immunohistochemistry.

CONCLUSION

The present findings indicate that hypothalamic neurons are highly differentiated with regard to the content of messenger molecules and that neurons often contain more than one such compound. The physiological significance of the histochemical demonstration of coexistence of multiple messengers seen in the hypothalamus and in many other parts of the nervous system, is at present difficult to evaluate. From studies in the peripheral nervous system, evidence has been obtained that classical transmitters and peptides are co-released and interact in a cooperative way on effector cells, but other types of interaction may also occur. For example, peptides have been shown to inhibit the release of the co-existing classical transmitter. In the central nervous system the situation is even less clear but similar mechanisms may operate. Indirect evidence suggests that peptides may in some cases strengthen transmission at synaptic (or non-synaptic) sites, and in other cases inhibit release of the co-existing classical transmitter. Thus, multiple messengers may provide a mechanism for relaying differential responses and for increasing the amount of information transmitted at synapses.

Co-existing messengers could also exert other types of actions, for example, have trophic effects or induce other types of long-term events in neurons and effectors cells. It has been shown that SP exerts growth stimulatory effects on smooth muscle cells (Nilsson et al., 1985), and another peptide, calcitonin gene-related peptide (CGRP) may be involved in regulation of expression of transmitter receptors (Fontaine et al., 1986; New and Mudge, 1986). It may be argued that the co-existence phenomenon as such, that neurons in addition to classical transmitter(s) contain other compounds, suggests that peptides are involved in other functions, since the neurons already have a classical transmitter at their disposal for accomplishing the task of fast cell-to-cell communication.

An important feature of the central nervous system is redundancy, and one may ask the question, why should it not be sufficient with one trans-mitter at each synapse when there are so many nerve cells? An answer could be that redundancy is also present at the level of the individual synapse. Thus, in addition to the aspect that a highly differentiated transmission process may be necessary to achieve the enormous operational capacity of our brain, several types of messengers may work in parallel and thus provide a 'safety device' should one of them fail.

The coexistence phenomenon should also be looked upon with a critical view, since physiological implications so far are very little elucidated. It cannot be excluded that coexistence of multiple messengers is a para-phenomenon representing a consequence of evolution. Peptides may have been important messengers in lower species, but may have been replaced by more 'efficient', small molecule transmitters, especially in phylogene-tically young areas of the brain such as cortex. Therefore peptides may at least in some places be carried along more or less as 'silent passengers'. It will be an important task to establish in the future whether or not peptides and classical transmitters are in fact released from the same nerve endings and under which conditions this may occur. Furthermore, it will be important to determine the models of action and interaction of the different messenger molecules.

ACKNOWLEDGEMENTS

This study was supported by the Swedish MRC (04X-2887) and Alice och Knut Wallenbergs Stiftelse. We thank Ms. W. Hiort, Ms. S. Nilsson, Ms. S. Soltesz-Mattisson and Ms. K. Åman for excellent technical assistance and Ms. E Björklund for expert secretarial help.

We thank Drs. J. Fahrenkrug, Copenhagen, Denmark, M. Goldstein, New York, NY, USA; Lars Terenius, Uppsala, Sweden; T. Visser, Rotterdam, The Netherlands; R.P. Elde, Minneapolis, MN, USA and W. Vale, La Jolla, CA, USA for supplying antisera used in the illustrations of this paper. For details, we refer to the original papers cited in the text and the reference list.

REFERENCES

Abe H, Engler D, Molitch ME, Bollinger-Gruber J, Reinchlin S, 1985. Vaso-active intestinal peptide is a physiological mediator of prolactin release in the rat. Endocrinology 116:1383-1390

Allen YS, Adrian TE, Allen JM, Tatemoto K, Crow TJ, Bloom SR, Polak JM, 1983. Neuropeptide Y distribution in the rat brain. Science 221:877-879

Andersson K, Fuxe K, Agnati LF, 1985. Determinations of catecholamine half-lives and turnover rates in discrete catecholamine nerve terminal systems of the hypothalamus, the preoptic region and the forebrain by quantitative histofluorimetry. Acta Physiol Scand 123:411-426

Avrameas S, 1969. Coupling of enzymes to proteins with glutaraldehyde. Use of the conjugates for the detection of antigens and antibodies. Immunochemistry 6:43-47

Barry J, Hoffman GE, Wray S, 1985. LHRH-containing systems. In: Bjorklund A and Hokfelt T (eds.), Handbook of Chemical Neuroanatomy, Vol. 4: GABA and Neuropeptides in the CNS. Amsterdam: Elsevier, pp. 166-215

Baumgarten HG, Björklund A, Lachenmayer L, Nobin A, Stenevi U, 1971. Long-lasting selective depletion of brain serotonin by 5,6-dihydroxy-tryptamine. Acta Physiol Scand Suppl 373:1-16

Berkenbosch F, Linton EA, Tilders FJH, 1986. Colocalization of PHI- and CRF-immunoreactivity in neurons of the rat hypothalamus: a surprising artefact. Neuroendocrinology 44:338-346

Berod A, Chat M, Paut L, Tappaz M, 1984. Catecholaminergic and GABAergic anatomical relationship in the rat substantia nigra, locus coeruleus, and hypothalamic median eminence: Immunocytochemical visualization of biosynthetic enzymes on serial semithin plactic-embedded sections. J Histochem Cytoche 32:1331-1338

Björklund A, 1985. Fluorescence histochemistry of biogenic monoamines. In: Björklund A and Hökfelt T. (eds.), Handbook of Chemical Neuroanatomy. Vol. 1: Methods in Chemical Neuroanatomy, Amsterdam, Elsevier, pp 50-121

Blackwell RE, Guillemin R, 1973. Hypothalamic control of adenohypophysial secretions. Ann Rev Physiol 35:357-390

Bloch B, Bugnon C, Fellmann D, Lenys D, 1978. Immunocytochemical evidence that the same neurons in the human infundibular nucleus are stained with anti-endorphins and antisera of other related peptides. Neurosci Lett 10:147-152

Bloch B, Bugnon C, Fellman D, Lenys D, Gouget A, 1979. Neurons of the rat hypothalamus reactive with antisera against endorphins, ACTH, MSH and B-LPH. Cell Tissue Res 204:1-15

Bloch B, Brazeau P, Bloom F, Ling N, 1983a. Topographical study of the neurons containing hpGRF immunoreactivity in monkey hypothalamus. Neurosci Lett 37:23-28

Bloch B, Brazeau P, Ling N, Bohlen P, Esch F, Wehrenberg WB, Benoit R, Bloom F, Guillemin R, 1983b. Immunohistochemical detection of growth hormone-releasing factor in brain. Nature (Lond.) 301:607-609

Bloch B, Ling N, Benoit R, Wehrenberg WB, Guillemin R, 1984. Specific depletion of immunoreactive growth hormone-releasing factor by monosodium glutamate in rat median eminence. Nature (Lond.) 307:272-274

Brar AK, Fink G, Maletti M, Rostene W, 1986. Vasoactive intestinal peptide in rat hypophysial portal blood: effects of electrical stimulation of various brain areas, the oestrous cycle and anaesthetics. J Endocrinol 106:275-280

Brown GM, Martin JB, 1974. Corticosterone, prolactin, and growth hormone responses to handling and new environment in the rat. Psychosom Med 36:241-247

Bugnon C, Bloch B, Lenys D, Gouget A, Fellmann D, 1979. Comparative study of the neuronal populations containing β-endorphin, corticotrophin and dopamine in the arcuate nucleus of the rat hypothalamus. Neurosci Lett 14:43-48

Bugnon C, Gouget A, Fellmann D, Clavequin MC, 1983. Immunocytochemical demonstration of a novel peptidergic neurone system in the cat brain with an antigrowth hormone-releasing factor serum. Neurosci Lett 38:131-137

Card JP, Brecha N, Moore R, 1983. Immunohistochemical localization of avian pancreatic polypeptide-like immunoreactivity in the rat hypothalamus. J Comp Neurol 217:123-136

Carson KA, Nemeroff CB, Rone MS, Youngblood WW, Prange AJ, Hanker JS, Kizer JS, 1977. Biochemical and histochemical evidence for the existence of a tuberoinfundibular cholinergic pathway in the rat. Brain Res 129:169-173

Chan-Palay V, Zaborszky L, Kohler C, Goldstein M, Palay S, 1984. Distribution of tyrosine hydroxylase-immunoreactive neurons in the hypothalamus of rats. J Comp Neurol 227:467-496

Christofides ND, Yangori Y, Blank MA, Tatemoto K, Polak JM, Bloom SR, 1982. Are peptide histidine isoleucine and vasoactive intestinal peptide co-synthesized in the same prohormone? Lancet ii:1398

Chronwall BM, 1985. Anatomy and physiology of the neuroendocrine arcuate nucleus. Peptides 6:1-11

Chronwall BM, Chase TN, O'Donohue TL, 1984. Coexistence of neuropeptide Y and somatostatin in rat and human cortical and rat hypothalamic neurons. Neurosci Lett 52:213-217

Chronwall BM, Olschowka JA, O'Donohue TL, 1984. Histochemical localization of FMRFamide-like immunoreactivity in the rat brain. Peptides 5:569–584

Chronwall BM, DiMaggio DA, Massari VJ, Pickel VM, Ruggiero DA, O'Donohue TL, 1985. The anatomy of neuropeptide-Y containing neurons in rat brain. Neuroscience 15:1159–1181

Coons AH, 1958. Fluorescent antibody methods. In: Danielli JF (ed.), General Cytochemical Methods. New York: Academic Press, pp. 399–422

Cuello AC (ed.), 1983. Immunohistochemistry, IBRO Handbook Series: Methods in the Neurosciences, Vol 3, Chichester: John Wiley & Sons.

Dahlström A, 1971. Effects of vinblastine and colchicine on monoamine containing neurons of the rat with special regard to the axoplasmic transport of amine granules. Acta Neuropathol, Suppl 5:226–237

Dahlström A, Fuxe K, 1964. Evidence for the existence of monoamine containing neurons in the central nervous system. I. Demonstration of monoamines in the cell bodies of brain stem neurons. Acta Physiol Scand 62, Suppl 232:1–55

De Quidt ME, Emson PC, 1986. Distribution of neuropeptide Y-like immunoreactivity in the rat central nervous system II. Immunohistochemical analysis. Neuroscience 18:545–618

DiMaggio DA, Chronwall BM, Buchanan K, O'Donohue TL, 1985. Pancreatic polypeptide immunoreactivity in rat brain is actually neuropeptide Y. Neuroscience 15:1149–1157

Doerr-Schott J, 1986. Multiple immunocytochemical labelling methods for the simultaneous ultrastructural localization of various hypophysial hormones. In: Yoshimura F and Gorbman A (eds.), Pars Distalis of the Pituitary Gland – Structure, Function and Regulation. Excerpta Medica Int Congr Ser 673, Amsterdam: Elsevier Science Publ, pp 95–106

Eränkö O, 1955. Histochemistry of noradrenaline in the adrenal medulla of rats and mice. Endocrinology 57:363–368

Everitt BJ, Hökfelt T, Wu J-Y, Goldstein M, 1984. Coexistence of tyrosine hydroxylase-like and gamma-aminobutyric acid-like immunoreactivities in neurons of the arcuate nucleus. Neuroendocrinology 39:189–191

Everitt BJ, Meister B, Hökfelt T, Melander T, Terenius L Rökaeus Å, Theodorsson-Norheim E, Dockray G, Edwardson J, Cuello C, Elde R, Goldstein M, Hemmings H, Ouimet C, Walaas I, Greengard P, Vale W, Weber E, Wu J-Y, Chang K-J, 1986. The hypothalamic arcuate nucleus-median eminence complex: immunohistochemistry of transmitters, peptides and DARPP-32 with special reference to coexistence in dopamine neurons. Brain Res Rev 11:97–155

Fellmann D, Gouget A, Bugnon C, 1983. Miss en évidence d'un nouveau système neuronal peptidergique immunoreactif à un immun-serum anti-hpGRF 44 dans le cerveau du lerot (Eliomys quercinus) CR Acad Sci 296:487–492

Ferland L, Fuxe K, Eneroth P, Gustafsson J-Å, Skett P, 1977. Effects of methionine-enkephalin on prolactin release and catecholamine levels and turnover in the median eminence. Eur J Pharmacol 43:89–90

Finley JCW, Lindström P, Petrusz P, 1981. Immunocytochemical localization of β-endorphin-containing neurons in the rat brain. Neuroendocrinology 33:28–42

Fontaine B, Klarsfeld A, Hökfelt T, Changeux J-P, 1986. Calcitonin gene-related peptide, a peptide present in spinal cord motoneurons, increases the number of acetylcholine receptors in primary cultures of chick embryo myotubes. Neurosci Lett 71:59–65

Fuxe K, 1964. Cellular localization of monoamines in the median eminence and infundibular stem of some mammals. Z Zellforsch 61:710–724

Fuxe K, Hökfelt T, 1966. Further evidence for the existence of tubero-infundibular dopamine neurons. Acta Physiol Scand 66:245–246

Fuxe K, Hökfelt T, 1969. Catecholamines in the hypothalamus and the pituitary gland. In: Ganong WF and Martini L (eds.), Frontiers in Neuroendocrinology. New York: Oxford University Press, pp. 47–96

Fuxe K, Hökfelt T, Said SI, Mutt V, 1977. Vasoactive intestinal poly-peptide and the nervous system: immunohistochemical evidence for localization in central and peripheral neurons, particularly intra-cortical neurons of the cerebral cortex. Neurosci Lett 5:241-246

Ganong WF, 1963. The central nervous system and the synthesis and release of adrenocorticotropic hormone. In: Nalbandov AV (ed.), Advances in Neuroendocrinology, Urbana: Univ of Illinois Press, pp. 92-149

Gerozissis K, Rougeot C, Dray F, 1986. Leukotriene C_4 is a potent stimulator of LHRH secretion. European J Pharmacol 121:159-160

Gerozissis K, Vulliez-le-Normand B, Saavedra JM, Murphy R, Dray F, 1985. Lipoxygenase products of arachidonic acid stimulate LHRH release from rat median eminence. Neuroendocrinology 40:272-276

Gilbert RFT, Emson PC, Hunt SP, Bennett GW, Marsden CA, Sandberg BEB, Steinbusch H, Verhofstad AAJ, 1982. The effects of monoamine neuro-toxins on peptides in the rat spinal cord. Neuroscience 7:69-88

Harms PG, Langlier P, McCann SM, 1975. Modification of stress-induced prolactin release by dexamethasone or adrenalectomy. Endocrinology 96:475-478

Harris GW, 1955. Neural Control of the Pituitary Gland. E. Arnold, London

Hökfelt T, Fuxe K, 1972. On the morphology and the neuroendocrine role of the hypothalamic catecholamine neurons. Brain-Endocrine Interaction. Int Symp Munich, 1971. Median Eminence: Structure and Function. Basel: Karger, pp. 181-223

Hökfelt T, Elde R, Johansson O, Terenius L, Stein L, 1977. The distri-bution of enkephalin immunoreactivity in the central nervous system. Neurosci Lett 5:25-31

Hökfelt T, Fahrenkrug J, Tatemoto K, Mutt V, Werner S, 1982a. PHI, a VIP-like peptide, is present in the rat median eminence. Acta Physiol Scand 116:469-471

Hökfelt T, Schultzberg M, Lundberg JM, Fuxe K, Mutt V, Fahrenkrug J, Said SI, 1982b. Distribution of vasoactive intestinal polypeptide in the central and peripheral nervous systems as revealed by immunocyto-chemistry. In: Said SI (ed.), Vasoactive Intestinal Peptide. New York: Raven Press, pp. 65-90

Hökfelt T, Fahrenkrug J, Tatemoto K, Mutt V, Werner S, Hulting A-L, Terenius L, Chang KJ, 1983a. The PHI (PHI-27)/corticotropin-releasing factor/enkephalin-immunoreactive hypothalamic neurons: Possible morphological basis for integrated control of prolactin, cortico-tropin, and growth hormone secretion. Proc Natl Acad Sci USA 80:895-898

Hökfelt T, Skagerberg G, Skirboll L, Björklund A, 1983b. Combination of retrograde tracing and neurotransmitter histochemistry. In: Björklund A and Hökfelt T (eds.), Handbook of Chemical Neuroanatomy, Vol. 1, Amsterdam: Elsevier, pp. 228-285

Hökfelt T, Everitt BJ, Theodorsson-Norheim E, Goldstein M, 1984a. Occurrence of neurotensin-like immunoreactivity in subpopulations of hypothalamic, mesencephalic and medullary catecholamine neurons. J Comp Neurol 222:543-549

Hökfelt T, Mårtensson R, Björklund A, Kleinau S, Goldstein M, 1984b. Distributional maps of tyrosine hydroxylase-immunoreactive neurons in the rat brain. In: Björklund A and Hökfelt T (eds.), Handbook of Chemical Neuroanatomy, Vol. 2: Classical Transmitters in the CNS, Part 1. Amsterdam: Elsevier, pp. 277-379

Hökfelt T, Everitt B, Meister B, Melander T, Schalling M, Johansson O, Lundberg JM, Hulting A-L, Werner S, Cuello C, Hemmings H, Ouimet C, Walaas I, Greengard P, Goldstein M, 1986. Neurons with multiple messengers with special reference to neuroendocrine systems. Recent Progr Horm Res 42:1-70

Hökfelt T, Fahrenkrug J, Ju G, Ceccatelli S, Tsuruo Y, Meister B, Mutt V, Rundgren M, Brodin E, Terenius L, Hulting A-L, Werner S, Björklund H, Vale W, 1987. Analysis of PHI/VIP-immunoreactive neurons in the central nervous system with special reference to their relation to CRF- and enkephalin-like immunoreactivities in the paraventricular hypothalamic nucleus. Neuroscience, in press

Hulting A-L, Lindgren J-Å, Hokfelt T, Heidvall K, Eneroth P, Werner S, Patrono C, Samuelsson B, 1984. Leukotriene C_4 stimulates LH secretion from rat pituitary cells in vitro. Europ J Pharmacol 106:459-460

Hulting A-L, Lindgren J-Å, Hökfelt T, Eneroth P, Werner S, Samuelsson B, 1985. Leukortine C_4 as a mediator of LH release from rat anterior pituitary cells. Proc Natl Acad Sci USA 52:3834-3838

Ibata Y, Watanabe K, Kinoshita H, Kubo S, Sano N, Yanaihara C, Yanaihara N, 1980. Dopamine and β-endorphin are contained in different neurons of the arcuate nucleus of the hypothalamus as revealed by combined fluorescence histochemistry and immunohistochemistry. Neurosci Lett 17:185-189

Ibata Y, Fukui K, Okamura H, Kawakami T, Tanaka M, Obata HL, Tsuto T, Terubayashi H, Yanaihara C, Yanaihara N, 1983a. Coexistence of dopamine and neurotensin in hypothalamic arcuate and periventricular neurons. Brain Res 269:177-179

Ibata Y, Obata HL, Kubo S, Fukui K, Okamura H, Ishigami T, Imagawa K, Sin S, 1983b. Some cellular characteristics of somatostatin neurons and terminals in the periventricular nucleus of the rat hypothalamus and median eminence. Electron microscopic immunohistochemistry. Brain Res 258:291-295

Ibata Y, Kawakami F, Fukui K, Obata-Tsuto HL, Tanaka M, Kubo T, Okamura H, Morimoto N, Yanaihara C, Yanaihara N, 1984a. Light and electron microscopic immunocytochemistry of neurotensin-like immunoreactive neurons in the rat hypothalamus. Brain Res 302:221-230

Ibata Y, Kawakami F, Fukui K, Okamura H, Obata-Tsuto HL, Tsuto T, Terubayashi H, 1984b. Morphological survey of neurotensin-like immunoreactive neurons in the hypothalamus. Peptides 5, Suppl 1:109-120

Itoh N, Obata K, Yanaihara N, Okamoto H, 1983. Human prepro-vasoactive intestinal polypeptide contains a novel PHI-27-like peptide, PHM-27. Nature 304:547-549

Jacobowitz DM, Schulte H, Chrousos GP, Loriaux DL, 1983. Localization of GRF-like immunoreactive neurons in the rat brain. Peptides 4:521-524

Jennes L, Stumpf WE, Kalivas PW, 1982. Neurotensin: topographical distribution in rat brain by immunohistochemistry. J Comp Neurol 210:211-224

Johansson O, Hökfelt T, Elde R, 1984. Immunohistochemical distribution of somatostatin-like immunoreactivity in the central nervous system of the adult rat. Neuroscience 13:265-339

Jonsson G, 1971. Quantitation of fluorescence of biogenic monoamines. Prog Histochem Cytochem 2:299-334

Ju G, Hökfelt T, Fischer JA, Frey P, Rehfeld JF, Dockray GJ, 1986. Does cholecystokinin-like immunoreactivity in rat primary sensory neurons represent calcitonin gene related peptide? Neurosci Lett 68:305-310

Kahn D, Abrams G, Zimmerman EA, Carraway R, Leeman SE, 1981. Neurotensin neurons in the rat hypothalamus: an immunocytochemical study. Endocrinology 107:47-53

Kaji H, Chihara K, Abe H, Minamitani N, Kodama H, Kita T, Fujita T, Tatemoto K, 1984. Stimulatory effect of peptide histidine isoleucine amide 1-27 on prolactin release in the rat. Life Sci 35:641-647

Kato Y, Iwasaki Y, Iwasaki J, Abe H, Yanaihara N, Imura H, 1978. Prolactin release by vasoactive intestinal polypeptide in rats. Endocrinology 103:554-558

Khachaturian H, Lewis ME, Watson SJ, 1983. Enkephalin systems in diencephalon and brainstem of the rat. J Comp Neurol 220:310-320

Khachaturian H, Lewis ME, Tsou K, Watson SJ, 1985a. β-Endorphin, α-MSH, ACTH and related peptides. In: Björklund A and Hökfelt T (eds.), Handbook of Chemical Neuroanatomy, Vol. 4: GABA and Neuropeptides in the CNS. Amsterdam: Elsevier, pp. 216-272

Kachaturian H, Lewis ME, Schäfer MK-H, Watson S, 1985b. Anatomy of the CNS opioid systems. TINS 8:111-119

Kobayashi H, Matsui T, 1969. Fine structure of the median eminence and its functional significance. In: Ganong WF and Martini L (eds.), Frontiers in Neuroendocrinology, New York: Oxford University Press, pp. 3-46

Krukoff TL, Calaresu FR, 1984. A group of neurons highly reactive for enkephalins in the rat hypothalamus. Peptides 5:931-936

Kuypers HEJM, Huisman AM, 1984. Fluorescent tracers. In: Fedoroff S (ed.), Advances in Cellular Neurobiology, Vol. 5, California: Academic Press, pp. 307-340

Lechan RM, Nestler JL, Jacobson S, 1982. The tuberoinfundibular system of the rat as demonstrated by immunohistochemical localization of retrogradely transported wheat germ agglutinin (WGA) from the median eminence. Brain Res 245:1-15

Lechan RM, Lin HD, Ling N, Jackson IM, Jacobsen IMD, Jacobson S, Reichlin S, 1984. Distribution of immunoreactive growth hormone-releasing factor (1-44)NH2 in the tuberoinfundibular system of the rhesus monkey. Brain Res 309:55-61

Lindgren J-Å, Hökfelt T, Dahlén S-E, Patrono C, Samuelsson B, 1984. Leukotrienes in the rat central nervous system. Proc Natl Acad Sci USA 81:6212-6216

Löfström A, Jonsson G, Fuxe K, 1976a. Microfluorimetric quantitation of catecholamine fluorescence in rat median emience. I. Aspects on the distribution of dopamine and noradrenaline nerve terminals. J Histochem Cytoehcm 24:415-429

Löfström A, Jonsson G, Wiesel FA, Fuxe K, 1976b. Microfluorimetric quantitation of catecholamine fluorescence in rat median eminence. II. Turnover changes in hormonal states. J Histochem Cytochem 24:430-442

Lorén IJ, Alumets R, Håkanson R, Sundler F, 1979. Immunoreactive pancreatic polypeptides (PP) occurs in the central and peripheral nervous system: preliminary immunocytochemical observations. Cell Tiss Res 200:179-186

Lorén I, Emson PC, Fahrenkrug J, Björklund A, Alumets J, Håkanson R, Sundler F, 1979. Distribution of vasoactive intestinal polypeptide in the rat and mouse brain. Neuroscience 4:1953-1976

Lundberg JM, Fahrenkrug J, Hökfelt T, Martling C-R, Larsson O, Tatemoto K, Änggård A, 1984a. Co-existence of peptide HI (PHI) and VIP in nerves regulating blood flow and bronchial smooth muscle tone in various mammals including man. Peptides 5:593-605

Lundberg Jm, Terenius L, Hökfelt, T, Tatemoto K, 1984b. Comparative immunocytochemical and biochemical analysis of pancreatic polypeptide-like peptides with special reference to presence of neuropeptide Y in central and peripheral neurons. J Neurosci 4:2376-2386

MacLeod RM, Lehmeyer JE, 1974. Studies on the mechanism of dopamine-mediated inhibition of prolactin secretion. Endocrinology 94:1077-1085

McGinty JF, Bloom F, 1983. Double immunostaining reveals distinctions among opioid peptidergic neurons in the medial basal hypothalamus. Brain Res 278:145-153

Meister B, Hökfelt T, Vale WW, Goldstein M, 1985. Growth hormone-releasing factor (GRF) and dopamine coexist in hypothalamic arcuate neurons. Acta Physiol Scand 124:133-136

Meister B, Hökfelt T, Vale WW, Sawchenoko PE, Swanson LW, Goldstein M, 1986. Coexistence of tyrosine hydroxylase and growth hormone-releasing factor in a subpopulation of tuberoinfundibular neurons of the rat. Neuroendocrinology 42:237–247

Meites J, Nicoll CS, Talwalker PK, 1963. The central nervous system and the secretion and release of prolactin. In: Nalbandov AV (ed.), Advances in Neuroendocrinology. Urbana: Univ of Illinois Press, pp. 238–277

Meites J, Bruni JF, Van Vugt DA, Smith AF, 1979. Relation of endogenous opioid peptides and morphine to neuroendocrine functions. Life Sci 24:1324–1336

Melander T, Hökfelt T, Rökaeus Å, 1986a. Distribution of galanin-like immunoreactivity in the rat central nervous system. J Comp Neurol 248:475–517

Melander T, Hökfelt T, Rökaeus Å, Cuello AC, Oertel WH, Verhofstad A, Goldstein M, 1986b. Coexistence of galanin-like immunoreactivity with catecholamines, 5-hydroxytryptamine, GABA and Neuropeptides in the Rat CNS. J Neurosci 6:3640–3654

Melzack R, Wall PD, 1965. Pain mechsnisms: a new theory. Science (Wash.) 150:971–979

Merchenthaler I, Vigh S, Schally AV, Petrusz P, 1984. Immunocytochemical localization of growth hormone-releasing factor in the rat hypothalamus. Endocrinology 114:1082–1085

Merchentaler I, Maderdrut JL, Dockray GJ, Altschuler RA, Petrusz P, 1986. Immunocytochemical localization of proenkephalin-derived peptides in the central nervous system of the rat. Neuroscience 17:325–348

Mezey E, 1986. Vasoactive intestinal polypeptide immunopositive neurons in the paraventricular nucleus of homozygous Brattleboro rats. Neuroendocrinology 42:88–90

Mezey E, Kiss JZ, 1985. Vasoactive intestinal peptide-containing neurons in the paraventricular nucleus may participate in regulating prolactin secretion. Proc Natl Acad Sci USA 82:245–247

Mugnaini E, Oertel WH, 1985. An atlas of the distribution of GABAergic neurons and terminals in the rat CNS as revealed by GAD immunohistochemistry. In: Björkllund A and Hökfelt T (eds.), Handbook of Chemical Neuroanatomy, Vol. 4: GABA and Neuropeptides in the CNS. Part 1. Amsterdam: Elsevier, pp. 436–608

Nairn RC, 1969. Immunological tracing: general considerations. In: Nairn RC (ed.), Fluorescent Protein Tracing, 3rd edn. Edinburgh and London: Livingstone, pp. 111–151

Nakane PK, 1968. Simultaneous localization of multiple tissue antigens using the peroxidase-labeled antibody method: a study in pituitary glands of the rat. J Histochem Cytochem 16:557–560

Nakane PK, Pierce GB, 1966. Enzyme-labeled antibodies: preparation and application for the localization of antigens. J Histochem Cytochem 14:929–931

Neill JD, 1980. Neuroendocrine regulation of prolactin secretion. In: Martini L and Ganong WF (eds.), Frontiers in Neuroendocrinology, Vol. 6. New York: Raven Press, pp. 129–155

New HV, Mudge AW, 1986. Calcitonin gene-related peptide regulates muscle acetylcholine receptor synthesis. Nature 323:809–811

Nilsson J, von Euler AM, Dalsgaard C-J, 1985. Stimulation of connective tissue cell growth by substance P and substance K. Nature 325:61–63

Ohta H, Kato Y, Shimatsu A, Tojo K, Kabayama Y, Inoue T, Yanaihara N, Imura H, 1985a. Inhibition by antiserum to vasoactive intestinal polypeptide (VIP) of prolactin secretion induced by serotonin in the rat. Eur J Pharmacol 109:409–412

Ohta H, Kato Y, Tojo K, Shimatsu A, Inoue T, Kabayama Y, Imura H, 1985b. Further evidence that peptide histidine isoleucine (PHI) may function as a prolactin releasing factor in rats. Peptides 6:709–712

Ohtsuka M, Hisano S, Daikoku S, 1983. Electronmicroscopic study of somatostatin-containing neurons in rat arcuate nucleus with special reference to neuronal regulation. Brain Res 263:191-199

Okamura H, Murakami S, Chihara K, Nagatsu I, Ibata Y, 1985. Coexistence of growth hormone releasing factor-like and tyrosine hydroxylase-like immunoreactivities in neurons of the rat arcuate nucleus. Neuroendocrinology 41:177-179

Olschowka JA, O'Donohue TL, Jacobowitz DM, 1981. The distribution of bovine pancreatic polypeptide-like immunoreactive neurons in rat brain. Peptides 2:309-331

Osborne NN, 1979. Is Dale's principle valid? TINS 2:73-75

Palmer MR, Mathews R, Murphy RC, Hoffer BJ, 1980. Leukotriene C elecits a prolonged excitation of cerebellar Purkinje neurons. Neurosci Lett 18:173-180

Palmer MR, Mathews WR, Hoffer BJ, Murphy RC, 1981. Electrophysiological response of cerebellar Purkinje neurons to leukotriene C_4 and B_4. J Pharmacol Exp Ther 219:91-96

Pelletier G, Steinbusch HW, Verhofstad A, 1981. Immunoreactive substance P and serotonin present in the same dense core vesicles. Nature 293:71-72

Polak JM, Van Noorden S (eds.), 1983. Immunocytochemistry. Practical Applications in Pathology and Biology. Bristol: Wright - PSG

Ruberg M, Rotsztejn W, Arancibia S, Besson J, Enalbert A, 1978. Stimulation of prolactin release by vasoactive intestinal peptide. Europ J Pharmacol 51:319-320

Said SI, Porter JC, 1979. Vasoactive intestinal polypeptide: release into hypophyseal portal blood. Life Sci 24:227-230

Samuelsson B, 1983. Leukotrienes: mediators of immediate hypersensitivity reactions and inflammation. Science 220:227-230

Samson WK, Lumpkin MD, McDonald JK, McCann SM, 1983. Prolactin-releasing activity of porcine intestinal peptide (PHI-27). Peptides 4:817-819

Sar M, Stumpf WE, Miller RJ, Chang K-J, Cuatrecasas P, 1978. Immunohistochemical localization of enkephalin in rat brain and spinal cord. J Comp Neurol 182:17-38

Schalling M, Neil A, Terenius L, Hökfelt T, Lindgren J-Å, Samuelsson B, 1985. Leukotriene C_4 binding sites in the rat central nervous system. Europ J Pharmacol 122:251-257

Schally A, Arimura A, Kastin AJ, 1973. Hypothalamic regulatory hormones. Science 179:341-350

Shaar CJ, Clemens JA, Dininger NB, 1979. Effect of vasoactive intestinal polypeptide on prolactin release in vitro. Life Sci 25:2071-2074

Shimatsu A, Kato Y, Matsushita N, Katakami H, Yanaihara N, Imura H, 1981. Immunoreactive vasoactive intestinal polypeptide in rat hypophysial portal blood. Endocrinology 108:395-398

Shimatsu A, Kato Y, Inoue T, Christofides ND, Bloom SR, Imura H, 1983. Peptide histidine isoleucine- and vasoactive intestinal polypeptide-like immunoreactivity coexist in rat hypophysial portal blood. Neurosci Lett 43:259-262

Sims KB, Hoffman DL, Said SI, Zimmerman EA, 1980. Vasoactive intestinal polypeptide (VIP) in the mouse and rat brain: an immunocytochemical study. Brain Res 186:165-183

Shiosaka S, Takatsuki K, Sakanaka M, Inagaki S, Takagi H, Senba E, Kawai Y, Iida H, Minagawa H, Matsuzaki T, Tohyama M, 1982. Ontogeny of somatostatin-containing neuron system of rat: immunohistochemical analysis. II. Forebrain and diencephalon. J Comp Neurol 204:211-224

Skofitsch G, Jacobowitz DM, 1985. Immunohistochemical mapping of galanin-like neurons in the rat central nervous system. Peptides 6:509-546

Smith RM, Howe PRC, Oliver JR, Willoughby JE, 1984. Growth hormone releasing factor immunoreactivity in rat hypothalamus. Neuropeptides 4:109-115

Sofroniew MV, 1979. Immunoreactive β-endorphin and ACTH in the same neurons of the hypothalamic arcuate nucleus in the rat. Am J Anat 154:283-289

Staines WA, Meister B, Melander T, Nagy JI, Hökfelt T. Three-colour immunofluorescence allowing triple labelling within a single section. J Histochem Cytochem, in press

Sternberger LA (ed.), 1979. Immunocytochemistry, 2nd Ed. New York: John Wiley

Sternberger LA, Hardy PH, Cuculis JJ, Meyer HG, 1970. The unlabelled antibody-enzyme method of immunohistochemistry. Preparation and properties of soluble antigen-antibody complex (horserdsish peroxidase-antihorseradish peroxidase) and its use in identification of spirochetes. J Histochem Cytochem 18:315-333

Swanson LW, Sawchenko PE, Lind RW, 1986. Regulation of multiple peptides in CRF parvocellular neurosecretory neurons: implications for the stress response. In: Hökfelt T, Fuxe K and Pernow B (eds.), Progress in Brain Res, Vol 68. Amsterdam: Elsevier, pp. 169-190

Szentágothai J, Flerkó B, Mess B, Halász B, 1962. Hypothalamic control of the anterior pituitary. Akadémiai Kiadó

Tappaz ML, Wassef M, Oertel WH, Paut L, Pujol JF, 1983. Light- and electron-microscopic immunocytochemistry of glutamic acid decarboxylase (GAD) in the basal hypothalamus: morphological evidence for neuroendocrine gamma aminobutyrate (GABA). Neuroscience 9:271-287

Tatemoto K, Mutt V, 1981. Isolation and characterization of the intestinal peptide porcine PHI (PHI-27), a new member of the glucagon-secretion family. Proc Natl Acad Sci USA 78:6603-6607

Thoenen H, Tranzer JP, 1968. Chemical sympathectomy by selective destruction of adrenergic nerve endings with 6-hydroxydopamine. Arch Pharmak Exp Path 261:271-288

Tramu G, Pillez A, Leonardelli J, 1978. An efficient method of antibody elution for the successive or simultaneous location of two antigens by immunocytochemistry. J Histochem Cytochem 26:322-324

Uhl GR, Goodman RR, Snyder SH, 1979. Neurotensin-containing cell bodies, fibers and nerve terminals in the brainstem of the rat: immunohistochemical mapping. Brain Res 167:77-91

Van den Pool AN, Herbst RS, Powell JF, 1985. Tyrosine hydroxylase-immunoreactive neurons of the hypothalamus: a light and electron microscopic study. Neuroscience 13:1117-1156

Vijayan E, Samson WK, Said SI, McCann SM, 1979. Vasoactive intestinal peptide: evidence for a hypothalamic site of action to release growth hormone, luteinizing hormone and prolactin in conscious ovariectomized rats. Endocrinology 104:53-57

Vincent SR, Hökfelt T, Christensson I, Terenius L, 1982a. Dynorphin-immunoreactive neurons in the central nervous system of the rat. Neurosci Lett 33:185-190

Vincent SR, Hökfelt T, Wu J-Y, 1982b. GABA neuron systems in hypothalamus and the pituitary gland. Neuroendocrinology 34:117-125

Vincent SR, McIntosh CHS, Buchan AMJ, Brown JC, 1985. Central somatostatin systems revealed with monoclonal antibodies. J Comp Neurol 238:169-186

Wamsley JK III, Young WS, Kuhar MJ, 1980. Immunohistochemical localization of enkephalin in rat forebrain. Brain Res 190:153-174

Watson SJ, Barchas JD, Li CH, 1977. β-lipoprotein. Localization of cells and axons in rat brain by immunohistochemistry. Proc Natl Acad Sci USA 74:5155-5158

Watson SJ, Akil H, Fischli W, Goldstein A, Zimmerman E, Nilaver G, van Wimersma Greidanus TJB, 1982. Dynorphin and vasopressin: common localization in magnocellular neurons. Science 216:85-87

Watson SJ, Khachaturian H, Taylor L, Fischli W. Goldstein A, Akil H, 1983. Pro-dynorphin peptides are found in the same neurons throughout the rat brain: immunocytochemical study. Proc Natl Acad Sci USA 80:891-894

Weber E, Roth KA, Barchas JD, 1982. Immunohistochemical distribution of α-neoendorphin/dynorhin neuronal systems in rat brain: evidence for colocalization. Proc Natl Acad Sci USA 79:3062-3066

Werner S, Hulting AL, Hökfelt T, Eneroth P, Tatemoto K, Mutt V, Maroder L, Wünsch E, 1983. Effect of the peptide PHI-27 on prolactin release in vitro. Neuroendocrinology 37:476-478

Wessendorf MW, Elde RP, 1985. Characterization of an immunofluorscence technique for the demonstration of coexisting neurotransmitters within nerve fibers and terminals. J Histochem Cytochem 33:984-994

Williams RG, Dockray GJ, 1983. Distribution of enkephlin-related peptides in rat brain: immunohistochemical studies using antisera to Met-enkephalin and Met-enkephalin Arg Phe. Neuroscience 9:563-586

RECEPTOR-WEIGHTED MECHANISTIC APPROACH TO ANALYSIS OF THE ACTIONS

OF ESTROGEN AND PROGESTERONE ON GONADOTROPIN SECRETION

T.G. Muldoon and V.B. Mahesh

Department of Physiology and Endocrinology
Medical College of Georgia
Augusta, GA 30912

INTRODUCTION

Synthesis and secretion of gonadotropins from the anterior pituitary, and the end-product regulation of these processes, represents the corner-stone of reproductive physiology; to the extent that we are lacking in a thorough basic understanding of these phenomena, the superstructure of endocrine-based manipulation of reproductive function collapses. Real-ization of this fact accounts for the vast number of experimental approaches which have been taken to delve ever deeper into the intricacies of the complex interrelationships involved. In the early 1970's, we joined the ranks of a small number of groups who were actively investigating the occurrence and nature of steroid hormone receptors in the anterior pituitary (Eisenfeld, 1970; Kato, 1970; Notides, 1970; Korach and Muldoon, 1973). These analyses rapidly developed into collaborative ventures between our two laboratories (Mahesh et al., 1972), investigating correlative dependencies between estrogen receptor dynamics and gonado-tropin fluctuations under similarly manipulated experimental conditions. These studies were furthered by subsequent description of the estrogen retention system in the hypothalamus (Korach and Muldoon, 1974). Over the intervening years, we have continued to examine a number of features of interplay between manifest gonadotropin fluctuations and neural steroid hormone receptor turnover (Muldoon, 1980a; Muldoon et al., 1984; Mahesh et al., 1980; Mahesh, 1984; Mahesh and Muldoon, 1987). It is the purpose of this report to summarize the current status of these studies and to consider them in the perspective of the nature and extent of information to be expected from this approach to basic questions in endocrine physiology.

Initial indications that estrogen receptor changes in the anterior pituitary might be functionally related to gonadotropin secretion were obtained from an analysis of temporal fluctuations during the rat estrous cycle. As shown in Fig. 1, cytosolic estrogen receptors (presumably those not strongly bound to chromatin effector sites) rise gradually from a nadir at noon of proestrus to a peak at noon of diestrus-2. Nuclear receptor levels remain very low through diestrus-1 and only begin to rise after noon on diestrus-2, reaching a maximal level at noon on proestrus; the level is decreasing prior to the late afternoon surge of gonado-tropins. The synchrony of nuclear receptor content with responsiveness to LHRH stimulation of LH release is striking. Interestingly, the trans-location of receptor from the cytosol to the nucleus lags somewhat behind

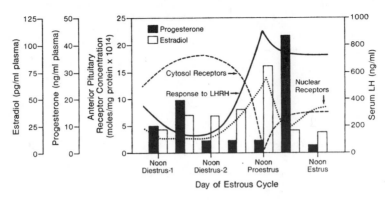

Fig. 1. Correlations among estrogen receptor levels, circulating estradiol
and progesterone levels, and pituitary responsiveness to LHRH
during the rat estrous cycle. The proestrous surge of gonadotropins occurs on the early evening of proestrus.

the rising pattern of serum estradiol, and the reason for this has not yet
been ascertained. Estrogen receptor patterns in the hypothalamus show
similar characteristics to those in the anterior pituitary (Muldoon et
al., 1982).

Estradiol manifests a biphasic action on gonadotropin secretion,
being acutely inhibitory and chronically stimulatory. These sequential
opposing responses have been clearly described in the intact animal (Yen
and Tasi, 1971), with a separation of several hours between the opposing
actions. While it appears clear that both hypothalamic and anterior
pituitary sites of estrogen action are involved in this feedback mechanism
(Negro-Vilar et al., 1973; Gross, 1980; Kalra and Kalra, 1982), the
precise nature of the individual tissue responses has not been resolved.
The secondary positive effect at the hypothalamic level may actually
reflect a release of the initial negative action on the synthesis and
secretion of LHRH. The observation that implants of estradiol can blunt
the ovulatory surge of gonadotropins (Turgeon and Barraclough, 1977;
Turgeon, 1979) provides indirect evidence supporting this idea. A
conceivable mechanism for such a self-directed regulation is estrogen
stimulation of estrogen-2/4-hydroxylase activity in neural tissue, thus
forming the metabolite, 2-hydroxyestrone, which is antiestrogenic at this
level; as recently noted (Mondschein et al., 1986), such putative local
modulation presupposes this as-yet undemonstrated metabolic conversion by
brain tissue in vivo. It is clear that the catecholaminergic neurons
which regulate LHRH secretion are themselves negatively regulated by
neurons of the central opiatergic system, and evidence is accumulating
that gonadal steroids have a tropic effect on endogenous opioid peptides
(Kalra and Kalra, 1983; Negro-Vilar et al., 1986).

Little is known about the primary negative effect of estrogens at the
anterior pituitary level, although a study of the system in vitro strongly
indicates desensitization of the tissue to LHRH (Moll and Rosenfield,
1984). Here again, the subsequent positive action at the pituitary level
might represent, at least partially, release of this desensitization;
however, there is good evidence to indicate a true positive enhancement of
sensitivity by induction of an LHRH self-priming effect (Aiyer et al.,
1974). The mechanism of this action, insofar as it involves changes in
LHRH receptor activity, is not clearly understood, and the action may well
occur at a mechanistic site distal to the primary receptor interaction
(Giguere et al., 1981).

It has been known for many years that castration of the adult female rat releases the negative component of estrogen feedback, resulting in chronic elevation of serum LH and FSH. Moreover, persistent exposure of such animals to the estrogen maintains the gonadotropin suppression; thus, there is no biphasic action under these conditions. Using this animal model, we have examined the kinetics of serum LH fluctuations in unison with anterior pituitary and hypothalamic estrogen receptor turnover in response to a single injection of estradiol (Eldridge et al., 1986). As shown in Fig. 2, estrogen receptors are depleted from the cytosol maximally within the first hour, following which they are gradually replenished over a 15-h period, involving a complex combination of receptor synthesis, recycling and regeneration (Muldoon, 1980b). During this 15-h interval, serum LH is depressed at a slower rate than the receptors, and gradually rises again to untreated levels in temporal concert with the recovery of cytosol receptors. This pattern of cytosolic receptor turnover reflects the inverse pattern of strong nuclear interactions (Table 1), which are generally thought to be those directly implicated in the hormonal activity at the genomic level.

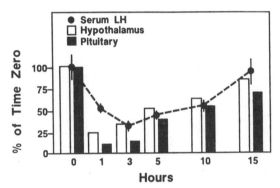

Fig. 2. Pattern of estradiol suppression of post-castration serum LH levels, aligned with concomitant changes in cytosolic estrogen receptors of the anterior pituitary and hypothalamus. Adult female rats were ovariectomized two weeks prior to the administration of 10 g of estradiol at time zero. Receptor levels are single results of tissues pooled from 4-6 animals at each interval, analyzed at 4 dilutions in duplicate. LH values are the mean $\pm$ SEM from 4-6 individual animal specimens.

Table 1. Representative Distribution of Anterior Pituitary Estrogen Receptors Between Cytosol and Nuclear Fractions as a Function of Time After Administration of 10 μg of Estradiol to Adult Ovariectomized Rats

Hours after estradiol	Femtomoles/Pituitary	
	Cytosol Receptors	Nuclear Receptors
0	66.2	2.3
1	5.0	39.7
3	8.6	21.0
5	15.1	14.2
10	28.2	15.8
15	44.4	8.8

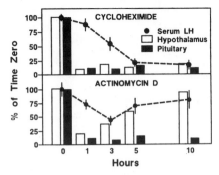

Fig. 3. Response of pituitary and hypothalamic estrogen receptors and
serum LH titer to a single injection of estradiol in the presence
of cycloheximide (1 mg/animal) or actinomycin D (600 μg).
Conditions as described for Fig. 2.

The protein synthesis inhibitor, cycloheximide, permits depletion of
cytosol receptors by estradiol, but totally blocks their subsequent
replenishment in this castrate animal model (Fig. 3). In the presence of
this drug, estradiol is capable of suppressing LH, but the release of this
suppression does not occur over 15 h. Thus, prolongation of the nuclear
receptor occupancy concomitantly prolongs the suppression of LH.
Actinomycin D is also effective in preventing receptor replenishment in
the anterior pituitary following estrogen-stimulated depletion, but this
inhibitor of transcription does not gain physical access to the hypothalamus (Cidlowski and Muldoon, 1974) and therefore does not alter the
receptor response in this tissue. In the presence of this metabolic
inhibitor, LH undergoes the expected suppression within 3 h, and then does
recover as the hypothalamic cytosol receptors replenish (Fig. 3). At
surface value, these results with the two inhibitors seem to suggest that
the suppressibility phenomenon of LH correlates only with the turnover of
hypothalamic receptors. It is crucial to keep in mind, however, that the
depletion phenomenon occurs in both tissues in the face of the actinomycin
block.

We performed experiments in which we investigated these correlations
using the antiestrogen, CI-628, which we have shown to be similar to
actinomycin D in its inability to traverse the blood-brain barrier
(Cidlowski and Muldoon, 1976). CI-628, as opposed to estradiol, caused
depletion of cytosolic receptors in the pituitary with no subsequent
replenishment, and caused no changes in hypothalamic receptor levels
(Fig. 4). Serum LH levels showed a somewhat blunted response, although
the suppression and its subsequent release were statistically
demonstrable events. This experiment seemed to indicate that pituitary
receptor turnover alone is responsible for this action, and this is in
exact contradiction to the results with actinomycin D, which implicated
the hypothalamus as the sole correlative responsive tissue.

The model presented as Fig. 5 encompasses these observations and
serves as the basis for a possible interpretation of the negative feedback
control of LH secretion. At the level of the anterior pituitary, uptake
of the steroid-receptor complex into the nucleus is sufficient to trigger

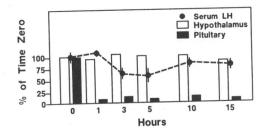

Fig. 4. The effect of the antiestrogen CI-628 (50 µg/animal) on the estradiol-induced patterns of serum LH and pituitary/hypothalamic receptors in the ovariectomized rat. Conditions as described for Fig. 2.

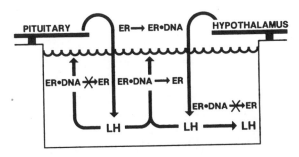

Fig. 5. Schematic representation of the dependence of serum LH fluctuations on pituitary/hypothalamic estrogen receptor dynamics in the castrate female rat. Nuclear accumulation of receptor in either tissue is followed by a sharp decline in serum LH levels. Subsequently, the release of receptors from their tight nuclear binding in either tissue allows recovery of serum LH levels. If the nuclear retention is prolonged in the hypothalamus, the LH levels remain suppressed. On the other hand, considering the pituitary, the suppression is released independently of whether these tissue receptors are retained in the nucleus or are released from their nuclear binding sites.

gonadotropin suppression. However, the suppression is released, irrespective of whether the complexes remain in the nucleus or are recycled from their nuclear sites of action. In the hypothalamus, receptor depletion also occasions LH release. In this instance, however, the serum LH levels remain depressed as long as the steroid-receptor complex remains in the nucleus. Thus, we distinguish between long-term hypothalamic negative feedback and acute anterior pituitary negative feedback by estrogen.

In terms of the positive effect of estradiol on LH secretion, we decided to investigate, at the receptor locus, the mechanism whereby estrogen enhances pituitary sensitivity to LHRH. Reasoning that an increased level of nuclear estrogen receptor would increase the sensitivity of a tissue to estrogen, and being aware of the self-priming activity of LHRH at the pituitary level, we explored the possibility that LHRH could enhance estrogen receptor activity. Using dispersed or cultured anterior pituitary cells from ovariectomized, estrogen-replaced rats, we showed that incubation in the presence of LHRH resulted in a rapid increase in cellular estrogen receptor binding capacity, and that the stimulation was limited to the receptors of the nuclear compartment (Fig. 6). Cytosol receptor levels fell, but the decrease was not dependent on the level of LHRH in the medium (as was the nuclear effect), so induced receptor translocation was not a viable explanation. At longer intervals of incubation, the nuclear receptors underwent redistribution to an equilibrium situation with the cytosol; these phenomena were entirely reproducible upon _in vivo_ administration of LHRH (Singh and Muldoon, 1982). The influence of the endogenous estrogen level on the subsequent ability of the pituitary cells to respond to LHRH was examined by comparing intact, castrate and castrate-replaced groups of animals (Fig. 7) and by comparing high-dose and low-dose regimens of estrogen replacement (Fig. 8). In all instances, the relative degree to which nuclear receptor activity was elevated was consistent with an increasing level of circulating estrogen.

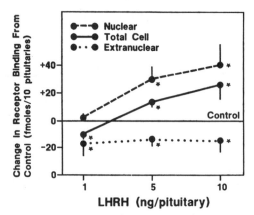

Fig. 6. Estrogen receptor levels following incubation of intact cells with LHRH for 30 min. Whole anterior pituitary cells were incubated with 1, 5 or 10 ng of LHRH per pituitary for 30 min at 37 °C, and estrogen receptor binding was estimated in the nuclear and extranuclear compartments of the cells. Control receptor levels were arbitrarily designated as zero and the positive or negative change in concentration was plotted at each dose level. Values are the mean ± SEM of 4–6 separate experiments, with triplicate determinations at each point. Asterisks designate statistical significance of differences from control values at P < 0.05.
(From Singh and Muldoon, 1982)

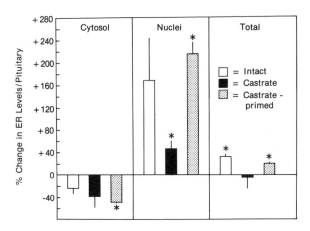

Fig. 7. Effect of endogenous estrogen level on the response of pituitary
cell estrogen receptors to LHRH. Pituitary cells from either
intact, castrate (nonprimed) or castrate estradiol-primed (1.0 μg/
day) rats were incubated with LHRH (5 pmoles/pituitary) for 30 min
at 37 °C, and estrogen receptors were assayed in the cytosol and
nuclear fractions of the cells. The % change in cytosol (A),
nuclear (B) and total (C) receptor per pituitary equivalent is
shown in LHRH-treated vs control cells. Each value is the mean +
SEM of 4 observations within 2 experiments. Asterisk: P < 0.05 vs
control. (From Singh and Muldoon, 1983)

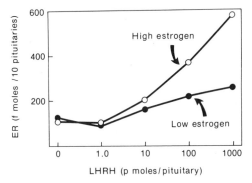

Fig. 8. Nuclear estrogen receptor sensitivity to LHRH as a function of
priming dose of estrogen. Castrate adult (60-days) animals were
primed following castration with 0.1 (low dose) or 1.0 (high dose)
μg of estradiol per day for 3 days and killed 18 h after the final
injection. Pituitary cells were prepared and incubated with
increasing levels of LHRH for 30 min at 37 °C. The change in
nuclear estrogen receptor level in cells from the two groups of
rats is shown. Each point is the mean of 3 observations and the
standard deviation at each point is less than 5%. (From Singh
and Muldoon, 1983)

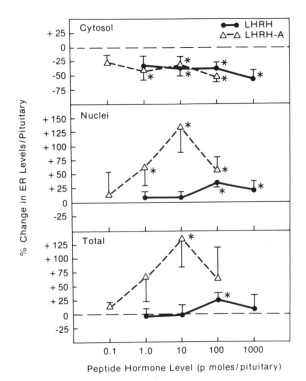

Fig. 9. Response of estrogen receptors to LHRH and the agonist, (D-Ala[6], des-Gly[10]) LHRH-N-ethylamide (LHRH-A) in intact anterior pituitary cells. Following incubation for 30 min at 37°C in the presence of varying amounts of either compound, the change in cytosol (A), nuclear (B) and total (C) receptor content was determined. Each value is the mean ± SEM of 10–24 observations within 6–18 separate experiments. Asterisks: significant differences at P < 0.05 level from control (designated zero on the ordinate).
(From Singh and Muldoon, 1983)

The effects we observed in the pituitary cells were specific, being hyperstimulated by an LHRH agonist of enhanced activity (Fig. 9) and inhibited by incubation in the presence of an LHRH antagonist. Thyrotropin releasing hormone, over a wide range of concentrations, did not stimulate estrogen receptor activity (Singh and Muldoon, 1983). A surprising observation was that LHRH incubation with isolated pituitary cell nuclei was as effective in stimulating the nuclear receptor activity as was incubation with the whole cells, suggesting that plasma membrane LHRH receptor interactions were not necessary to elicitation of the nuclear effect. A series of experiments, exemplified by the findings shown in Table 2, clearly showed that cAMP was not a mediator of the LHRH action on the nuclear estrogen receptors, although it might be involved in the nonspecific decrease seen in cytosol receptor activity (Singh et al., 1985). Very recently, we have demonstrated that the subpopulation of nuclear estrogen receptors which most avidly associates with nuclear components is the fraction of cellular receptors stimulated by LHRH (Singh and Muldoon, 1986); this is shown in Fig. 10. Detergent or nuclease treatment did not effect the extraction of these stimulated nuclear estrophiles, strongly suggesting their occurrence as nuclear matrix components, and implicating the action of LHRH as a true modulator of estrogenic responsiveness of these cells (Barrack and Coffey, 1982). It is pertinent

Table 2. Effects of LHRH or Dibutyryl cAMP on Cytosol and Nuclear Estrogen Receptor Levels in Whole Pituitary Cells in Suspension

Agent	% of Control Level	
	Cytosol Receptors	Nuclear Receptors
LHRH, pmol/pit		
1.0	76 + 3	102 + 1
10.0	62 + 2	166 + 5
100.0	56 + 3	192 + 6
dbcAMP, nM		
0.1	66 + 3	95 + 1
10.0	34 + 2	98 + 4
100.0	20 + 3	116 + 2
1,000.0	30 + 5	97 + 6
10,000.0	26 + 3	112 + 2

Freshly-prepared pituitary cells were incubated for 30 min at 37°C in the presence or absence (controls) of the indicated amounts of LHRH or dbcAMP. Values are the mean ± SEM of triplicate determinations, using groups of 20 animals per experiment.

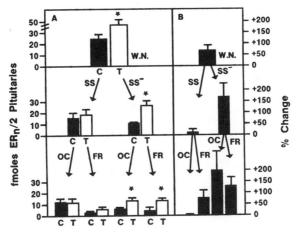

Fig. 10. Effect of LHRH on nuclear estrogen receptor levels and distribution in isolated anterior pituitary cell nuclei. Animals were castrate, estrogen-primed female rats. Pituitary nuclear fractions were prepared and incubated in the presence (treated; T) or absence (control; C) of LHRH for 30 min at 37°C. At the end of the incubation, nuclear estrogen receptor levels were measured in whole nuclei (W.N.) and in salt-soluble (SS) and salt-insoluble (SS⁻) fractions as ligand-occupied (OC) or ligand-free (FR) sites. Data are presented as femtomoles of nuclear estrogen receptor per two pituitaries (panel A) and as the percent change in receptor levels in the treated vs control groups (the latter arbitrarily assigned a value of zero) in panel B. Each data point is the mean ± SEM of 8 determinations from 4 experiments. Asterisk, P < 0.05, control vs treated groups. (From Singh and Muldoon, 1986)

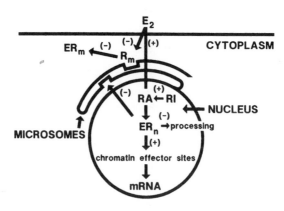

Fig. 11. Intracellular loci at which LHRH might act to elevate nuclear estrogen receptor activity in isolated anterior pituitary cells. Sites of stimulation include: (1) enhanced uptake of receptor onto functional nuclear effector sites or (2) conversion of receptor from an inactive to an active, steroid-binding form. Inhibitory actions resulting in augmented receptor activity include: (1) depression of steroid uptake onto microsomal receptors leading to decreased mobilization of these receptors and decreased extraction of nuclear receptor complexes onto the microsomal acceptor sites (2) inhibition of receptor processing, leading to a longer viable half-life of the nuclear complex.

that significant levels of matrix-associated receptor were measurable in the anterior pituitary, challenging the proposed notion that this receptor population is limited to cells which show a growth response to the steroid hormone.

Our interpretation of these findings is summarized diagrammatically in Fig. 11. Current information suggests that estradiol interacts initially with receptors in the nucleus which have little affinity for DNA. These receptors may be free or loosely associated with chromatin, and represent the unoccupied population of sites normally found in cytosol. The formed steroid-receptor complexes may not all bind to functional effector sites on chromatin, and some may therefore also be leached out into the cytosol. We have recently described a unique set of microsomal estrogen receptors (Watson and Muldoon, 1985) which may actually represent the initial acceptors for estrogen entering the cell; the concentration of these receptors at any time appears to regulate the nuclear content of receptor. The effects of LHRH on this system must be viewed in terms of our observations that it causes an increase in nuclear receptor activity without an accompanying decrease in the cytosolic binding capacity. This could be occasioned by increasing accessibility of a sequestered pool of receptor or enhancing the steroid-induced conversion of an inactive receptor to an active form. Alternatively, LHRH may decrease the mobility of microsomal receptors, having the effect of diminishing exchange with the nuclear form and increasing nuclear receptor content. This action could be mediated through inhibition of the interaction between estradiol and the microsomal receptors. A third possible mechanism would involve attenuation of nuclear protease activity,

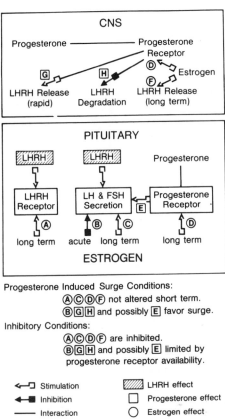

Fig. 12. A model to explain stimulatory and inhibitory effects of
progesterone on gonadotropin secretion. At this point, the model
does not encompass dose-dependency of progesterone action or its
effect on pulsatile release of LHRH.

prolonging the nuclear half-life of the receptors. Finally, LHRH could
alter chromatinic distribution of receptors to favor a nuclear matrix
localization. All these mechanisms have in common the result of
increasing the observed nuclear binding capacity for estrogen. Since
estrogen sensitizes the pituitary to LHRH-stimulation of LH release, an
increase in estrogen responsiveness induced by LHRH through enhancement of
functional nuclear binding would be reflected in an increase in sensitivity to LHRH. This is precisely the situation which defines the process
of LHRH self-priming, and our results therefore present a reasonable
molecular explanation for this phenomenon.

Consideration of the effects of progesterone on gonadotropin
secretion is complicated by its interplay with the actions of estradiol,
apparently at both the hypothalamic and pituitary sites of action. The
interactions between these two hormones are depicted at their current
state of understanding in Fig. 12. The effects of progesterone appear to
depend heavily upon the availability of sufficient progesterone receptor
to elicit a given response. This translates to a dependency on the
estrogen priming level of the endogenous milieu, since progesterone
receptor activity is critically regulated by estrogen (O'Malley et al.,
1971; Vu Hai et al., 1977). During the rat estrous cycle, progesterone
levels begin to rise just as estradiol levels have peaked following a long
gradual rise. This represents a situation of strong estrogen priming and

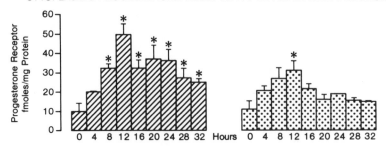

CYTOPLASMIC PROGESTERONE RECEPTOR AFTER ESTRADIOL INJECTION

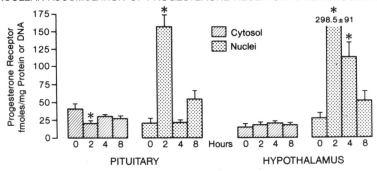

NUCLEAR ACCUMULATION OF PROGESTERONE RECEPTOR AFTER PROGESTERONE

Fig. 13. Appearance and responsiveness of progesterone receptors in response to estrogen administration. Upper panel: Immature rats were ovariectomized and given a single injection of 2 µg of estradiol on the following day. Cytosol progesterone receptor content of the anterior pituitary and hypothalamus was then monitored at 4-h intervals as shown. Lower panel: Ovariectomized animals were primed with estradiol to induce progesterone receptor activity. At 12 h after estradiol, a single injection of progesterone (0.8 mg/kg body weight) was given. Groups of animals were killed at 2-h intervals and the level of progesterone receptor in the cytosol and nuclei of the anterior pituitary and hypothalamus were measured. Asterisk: P < 0.05 difference from control values.

elevated progesterone receptor levels. Under these conditions (Fig. 12), progesterone can act immediately and has its most pronounced effects on the acute responses of the tissues to estrogen. Since estrogen is acutely inhibitory to gonadotropin secretion, progesterone would tend to be facilitory. In addition, progesterone independently appears to directly stimulate LHRH release, inhibit its degradation, and perhaps alter pituitary sensitivity to LHRH, all of which would intensify a positive feedback effect of this hormone. On the other side of the coin, appearance of progesterone under conditions of relative estrogen deficiency would result in limited rapid actions because of the low progesterone receptor activity. In this setting, progesterone is antagonistic to the long-term positive feedback effects of estrogen, and is thereby inhibitory to gonadotropin secretion.

Examination of the kinetics of progesterone receptor activity in response to a single injection of estradiol into an immature female rat yielded the patterns shown in Fig. 13 for the anterior pituitary and

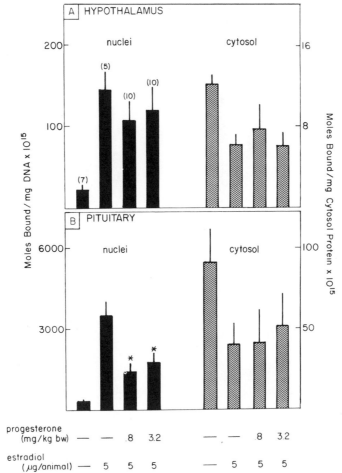

Fig. 14. Receptor distribution in hypothalami and pituitary glands from groups of ovariectomized estradiol–primed immature rats. Nuclear binding is expressed as femtomoles/mg DNA; cytosol binding, as femtomoles/mg protein. Where appropriate, the designated dose of progesterone was administered 2 h prior to killing; the estradiol, 1 h before killing. Values are the mean ± SEM of the number of determinations shown in parentheses. Asterisk: P < 0.01, progesterone–estradiol vs estradiol treatment. (From Smanik et al., 1983)

hypothalamus. In either tissue, receptor rose to a maximal level at about 12 h, fell somewhat and remained at an elevated plateau level for as long as 32 h. Using the 12–h interval of maximal activity, we showed (Fig. 13) that these receptors were capable of responding to progesterone, using nuclear translocation as the criterion.

In the ovariectomized immature rat primed daily with estradiol, progesterone administration selectively reduces nuclear estrogen receptor activity in the anterior pituitary, without affecting cytosol receptor levels (Fig. 14); this correlates with the antiestrogenicity of pro-gesterone in this model system. The estrogen receptor inhibition is not seen in the hypothalamus, in accord with progesterone's action being

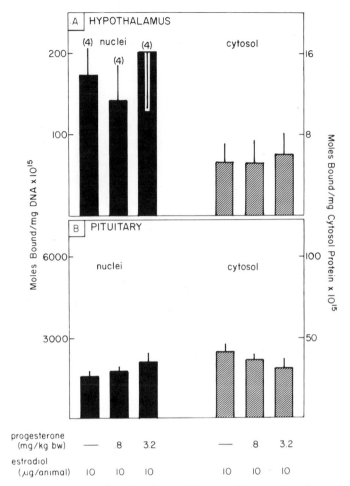

Fig. 15. Estrogen receptor levels in nuclear and cytoplasmic samples from tissues of ovariectomized non-primed immature rats. Hypothalamic and anterior pituitary binding capacities were measured 7 days after ovariectomy. Where appropriate, progesterone was given 2 h before death, estradiol 1 h before death. Each bar represents the mean ± SEM of 4 binding capacity measurements derived from Scatchard plot analysis. (From Smanik et al., 1983)

direct at this level. If the regimen shown in Fig. 14 is altered by doubling the priming dose of estradiol, progesterone is inhibitory at the 3.2 mg level, but no longer at the 0.8 mg dose; thus, the effect can be at least partially blocked by magnifying the estrogenic background (Smanik et al., 1983). As may be seen in Fig. 15, the ability of progesterone to depress pituitary nuclear estrogen receptors is not seen if the animals are not primed in vivo with estradiol, indicating that the action is progesterone receptor-mediated, as suspected. If the inhibition of nuclear estrogen receptor by progesterone were of functional significance, it should have resulted in diminished capability for response to a second estradiol stimulus. To test this, we used the protocol shown in Fig. 16. Animals were treated with estradiol followed by progesterone at 12 h to depress nuclear estrogen receptor. At 1 or 4 h after progesterone, a second injection of estradiol was administered, and all animals were killed 12 h later. Progesterone receptor activity was measured as an

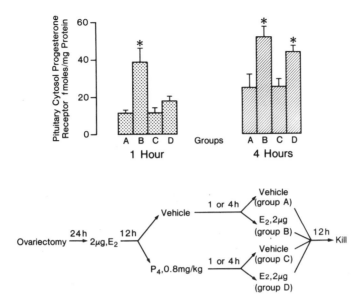

Fig. 16. Induction of cytosol progesterone receptors by a second injection of estradiol in ovariectomized estrogen-treated rats with (Group D) or without (Group B) progesterone pretreatment 1 or 4 h before the second estradiol injection. Asterisk: P < 0.05 difference from other values.

indicator of estrogen functionality. At 1 h after progesterone, the system does not respond to estrogen; by 4 h, the progesterone receptor is largely restored.

SUMMARY

This presentation has touched upon three experimental designs which we have used to examine the single premise that analysis of steroid hormone receptor activity is a viable approach to an increased understanding of steroid-gonadotropin feedback mechanisms and reproductive function.

In the study of LH suppressibility in the castrate animal, we observed strikingly similar patterns of receptor turnover and LH suppression-recovery. By discriminate choice of stimulating and blocking agents which had differential access to hypothalamic sites of action, we accumulated data suggesting that the inhibitory effects of estradiol on LH secretion are transient at the pituitary level and of longer duration in the hypothalamus. This was based on the different responses to alterations in nuclear receptor retention time in the two tissues. This initial analysis of the negative segment of the feedback will need to be extended to an animal model in which positive feedback comes into play before a full picture of the receptor correlations can be drawn.

The ability of LHRH to selectively stimulate estrogen receptor activity in the nucleus using isolated pituitary cells provides a mechanistic explanation for the self-priming action of LHRH. By enhancing receptor activity, LHRH increases the responsiveness of the pituitary to estrogen. This in turn intensifies the sensitivity to LHRH, and a secondary LHRH stimulus will elicit a magnified response. The ability to demonstrate selective augmentation of the functional matrix-associated receptor population, and our recent results showing that gonadotropes are indeed the responsive cells (Singh P, Muldoon TG, unpublished observations) speak to the specificity and relevance of these findings. The equal ability of LHRH to reduce nuclear estrogen receptor binding in whole cells and in isolated nuclei indicates that internalized LHRH may be capable of recognizing secondary intracellular receptors, and the system is amenable to exploration of this phenomenon.

Using a combination of estrogen receptor and progesterone receptor measurements, we have begun to unravel the interrelationships between these two hormones at this level apropos of their interplay in gonadotropin regulation. The kinetics of progesterone receptor induction by estrogen are consistent with an inability of the system to respond acutely to progesterone in the absence of strong estrogen backing, but with the capability of inhibiting long-term responses to estrogen. The acute capability of progesterone, acting through its receptors, to depress nuclear estrogen receptor activity, correlates with its acute antiestrogenic stimulation of gonadotropin secretion. The lack of such a receptor effect in the hypothalamus is consonant with a lack of acute modulation of estrogen action at this feedback site. The transiency of the progesterone block of estrogen action gives a clue to the dynamics of these interactions.

Some of our findings are supportive of the observations of others in the field using different approaches. A number of our results arise uniquely from the manner in which the experimental conditions can be manipulated, and will await confirmation of such findings from other avenues of investigation. It is clear that our studies represent only a beginning of the many possible designs which can be constructed around the receptor analyses of the hypothalamic-pituitary-ovarian system and further extrapolation to other physiological processes involving similar mechanisms of action.

ACKNOWLEDGEMENT

These studies were supported by NIH grants DK32046 and HD16688.

REFERENCES

Aiyer MS, Chiappa SA, Fink G, 1974. A priming effect of luteinizing hormone releasing factor on the anterior pituitary gland in the female rat. J Endocrinol 62:573-588

Barrack ER, Coffey DS, 1982. Biological properties of the nuclear matrix: steroid hormone binding. Recent Prog Horm Res 38:133-195

Cidlowski JA, Muldoon TG, 1974. Estrogenic modulation of cytoplasmic receptor populations in estrogen-responsive tissues of the rat. Endocrinology 95:1621-1629

Cidlowski JA, Muldoon TG, 1976. Dissimilar effects of antiestrogens upon estrogen receptors in responsive tissues of male and female rats. Biol Reprod 15:381-389

Eisenfeld AJ, 1970. ^{3}H-estradiol: in vitro binding to macromolecules from the rat hypothalamus, anterior pituitary and uterus. Endocrinology 86:1313-1318

Eldridge JC, Cidlowski JA, Muldoon TG, 1986. Correlation between LH and estrogen receptor turnover in pituitary and hypothalamus of castrate rats following estrogen agonists and antagonists. J Steroid Biochem 24:623-628

Giguere V, Lefebvre F, Labrie F, 1981. Androgens decrease LHRH binding sites in rat pituitary cells in culture. Endocrinology 108:350-352

Gross D, 1980. Effect of castration and steroid replacement on immuno-reactive gonadotropin-releasing hormone in the hypothalamus and preoptic area. Endocrinology 106:1442-1450

Kalra SP, Kalra PS, 1982. Stimulatory role of gonadal steroids on luteinizing hormone releasing hormone secretion. In: Motta M, Zanisi M, Piva F (eds.), Pituitary Hormones and Related Peptides. New York: Academic Press, pp. 157-170

Kalra SP, Kalra PS, 1983. Neural regulation of luteinizing hormone secretion in the rat. Endocr Rev 4:311-351

Kato J, 1970. In vitro uptake of tritiated oestradiol by the anterior hypothalamus and hypophysis of the rat. Acta Endocrinol 64:687-695

Korach KS, Muldoon TG, 1973. Comparison of specific 17β-estradiol-receptor interactions in the anterior pituitary of male and female rats. Endocrinology 92:322-326

Korach KS, Muldoon TG, 1974. Studies on the nature of the hypothalamic estradiol-concentrating mechanism in the male and female rat. Endocrinology 94:785-793

Mahesh VB, 1984. Nature and mechanism of steroid modulation of gonado-tropin secretion. In: Saxena BB, Catt KJ, Birnbaumer L, Martini L (eds.), Hormone Receptors in Growth and Reproduction. New York: Raven Press, pp. 201-225

Mahesh VB, Muldoon TG, 1987. Integration of the effects of estradiol and progesterone in the modulation of gonadotropin secretion. J Steroid Biochem, in press

Mahesh VB, Muldoon TG, Eldridge JC, Korach KS, 1972. Studies on the regulation of FSH and LH secretion by gonadal steroids. In: Saxena BB, Beling CG, Gandy HM (eds.), Gonadotropins. New York: Wiley and Sons, pp. 730-748

Mahesh VB, O'Conner JL, Allen MB, 1980. In vivo methods for the study of pituitary-gonadal relationships. In: Mahesh VB, Muldoon TG, Saxena BB, Sadler WA (eds.), Functional Correlates of Hormone Receptors in Reproduction. North Holland, New York: Elsevier, pp. 45-61

Moll GW, Rosenfield RL, 1984. Direct inhibitory effects of estradiol on pituitary luteinizing hormone responsiveness to luteinizing hormone releasing hormone is specific and of rapid onset. Biol Reprod 30:59-66

Mondschein JS, Hersey RM, Weisz J, 1986. Purification and characterization of estrogen-2/4-hydroxylase activity from rabbit hypothalami: peroxidase-mediated catechol estrogen formation. Endocrinology 119:1105-1112

Muldoon TG, 1980a. Role of receptors in the mechanism of steroid hormone action in the brain. In: Motta M (ed.), The Endocrine Functions of the Brain. New York: Raven Press, pp. 51-93

Muldoon TG, 1980b. Regulation of steroid hormone receptor activity. Endocrine Rev 1:339-364

Muldoon TG, Singh P, Watson GH, 1982. Steroid hormone receptors and gonadotropin secretion. In: Muldoon TG, Mahesh VB, Perez-Ballester B (eds.), Recent Advances in Fertility Research, Part A. New York: Alan Liss, pp. 95-109

Muldoon TG, Singh P, Watson GH, Harper SA, 1984. Molecular determinants of pituitary responsiveness to estrogen and luteinizing hormone-releasing hormone. In: Saxena BB, Catt KJ, Birnbaumer L, Martini L (eds.), Hormone Receptors in Growth and Reproduction. New York: Raven Press, pp. 237-248

Negro-Vilar A, Culler MD, Masotto C, 1986. Peptide-steroid interactions in brain regulation of pulsatile gonadotropin secretion. J Ster Biochem 25:741-747

Negro-Vilar A, Orias R, McCann SM, 1973. Evidence for a pituitary site of action for the acute inhibition of LH release by estrogen in the rat. Endocrinology 92:1680-1684

Notides AC, 1970. The binding affinity and specificity of the estrogen receptor of the rat uterus and anterior pituitary. Endocrinology 87:987–992

O'Malley BW, Sherman MR, Toft DO, Spelsberg TC, Schrader WT, Steggles AW, 1971. A specific oviduct target tissue receptor for progesterone. Adv Biosci 7:213–231

Singh P, Bhalla VK, Muldoon TG, 1985. Apparent lack of involvement of cAMP as a mediator of LHRH stimulation of nuclear estrogen receptor activity in the rat anterior pituitary. Neuroendocrinology 40:430–437

Singh P, Muldoon TG, 1982. A direct effect of LHRH on anterior pituitary estrogen receptors in the female rat. J Steroid Biochem 16:31–37

Singh P, Muldoon TG, 1983. Specific, estrogen-sensitive alterations in anterior pituitary cytoplasmic and nuclear estrogen receptors actuated by LHRH. Neuroendocrinology 37:98–105

Smanik EJ, Young HK, Muldoon TG, Mahesh VB, 1983. Analysis of the effect of progesterone in vivo on estrogen receptor distribution in the rat anterior pituitary and hypothalamus. Endocrinology 113:15–22

Turgeon JL, 1979. Estradiol–luteinizing hormone relationship during the proestrous gonadotropin surge. Endocrinology 105:731–736

Turgeon JL, Barraclough CA, 1977. Regulatory role of estradiol in pituitary responsiveness to LHRH on proestrus in the rat. Endocrinology 101:548–554

Vu Hai MT, Logeat F, Warembourg M, Milgrom E, 1977. Hormonal control of progesterone receptors. Ann NY Acad Sci 286:199–209

Watson GH, Muldoon TG, 1985. Specific binding of estrogen and estrogen-receptor complex by microsomes from estrogen-responsive tissues of the rat. Endocrinology 117:1341–1349

Yen SSC, Tasi CC, 1971. The biphasic pattern in the feed-back actions of ethinyl estradiol and the release of pituitary FSH and LH. J Clin Endocrinol Metab 33:882–887

THE STEROID-NEUROPEPTIDE CONNECTION IN THE CONTROL OF LHRH SECRETION

S.P. Kalra, P.S. Karla, A. Sahu, L.G. Allen and W.R. Crowley*

Department of Obstetrics and Gynecology
University of Florida College of Medicine
Gainesville, FL 32610
*Department of Pharmacology
University of Tennesse Memphis, TN 38163

INTRODUCTION

The concept that hormonal secretions of the gonads are the primary signals that drive and sustain the intricate balance within the closed feedback loop of the hypothalamo-pituitary-gonadal axis is well established. In 1932, Moore and Price proposed that a tight reciprocal relationship between the gonad and the pituitary was responsible for maintaining gonadal function in the adult rat (Fig. 1). Intensive investigations over the intervening years have affirmed the broad outline of this gonad-pituitary "push-pull" operation. The secretion of luteinizing hormone (LH) and follicle stimulating hormone (FSH) is generally restrained to a low basal range in intact rats, and in the absence of the gonadal feedback signals, as exemplified by gonadectomy, gonadotropins are secreted at augmented rates. It is apparent that gonadal steroids do not inhibit LH release by directly restraining the release of LH from the pituitary gonadotrophs but they monitor the response of gonadotrophs to luteinizing hormone releasing hormone (LHRH), the hypothalamic humoral signal secreted in episodic fashion. Further, the evidence suggests that androgens in male rats and a combined milieu of estradiol-17β (E_2) and progesterone (P) in female rats diminish, while E_2 in the absence of P, facilitate the LHRH-induced LH release. On the other hand, also in 1932, Höhlweg and Junkman proposed that the site of gonadal steroid action in the regulation of reproductive function may be the central nervous system (Fig. 1). The conceptual substance of this view was strongly supported by the ingenious studies of Harris and his colleagues (1955) who convincingly demonstrated that neurohumoral substances of hypothalamic origin were discharged into the hypophysial portal circulation for passage to the pituitary wherein they stimulated gonadotropin secretion. The chemical composition of the hypothalamic neurohumoral substance was elucidated (Matsuo et al., 1971), and LHRH is recognized as the primary hypothalamic neural signal responsible for controlling LH release from the pituitary gonadotrophs. Interestingly, the basic premises of the indirect gonadal steroid feedback concept has been extended over the years to include the notion that sex-steroids inhibit LHRH secretion and this, in turn, maintained gonadotropin secretion at a low basal range. Unfortunately, this assumption has proved to be difficult to substantiate because strict proof of this idea requires that changes in the levels of circulating gonadal steroids should evoke appropriate changes, on a moment-to-moment basis, in the rate of LHRH

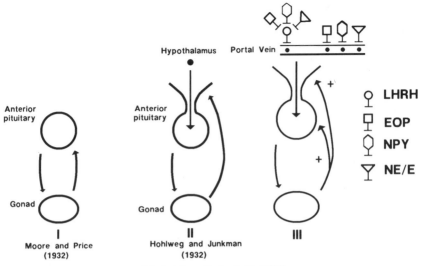

Adapted from Harris, G. W. (1955)

Figure 1   Diagramatic representation of the reciprocal
            relationship between the gonads and anterior
            pituitary as advanced by Moore and Price
            (1932); between the gonads, hypothalamus and
            pituitary as advanced by Höhlweg and Junkman
            (1932) and supported by Harris (1955); and the
            recent view of the facilitatory (+) nature of
            gonadal steroids on a variety of hypothalamic
            neuropeptidergic and classical aminergic neuro-
            transmitters in the control of gonadotropin
            secretion. For details see text. LHRH =
            luteinizing hormone releasing hormone; EOP =
            endogenous opioid peptides; NPY = neuropeptide
            Y; NE = norepinephrine; E = epinephrine.

secretion. A variety of attempts has been made in diverse species to
discern a relationship between gonadal steroids and LHRH secretion. However, a perusal of the available information has failed to reveal a tight
reciprocal relationship between gonadal steroids and LHRH secretion. In
the monkey, estrogen treatment failed to acutely decrease LHRH concentrations in the hypophysial portal plasma (Carmel et al., 1976). Similarly,
Clarke and Cummins (1985) found that while estradiol benzoate (EB) suppressed LH release for 18 h in ovariectomized sheep, pulsatile LHRH secretion into the hypophysial portal plasma continued unabated. In the rat
(Table I), attempts to lower acutely the portal plasma LHRH levels in response to estrogen have yielded contradictory results (Sarkar and Fink,
1980; Sherwood and Fink, 1980). Conversely, hypersecretion of LHRH, a
predicted consequence of opening the negative feedback loop by gonadectomy, has not been observed both in vivo and in vitro experiments (Sarkar
and Fink, 1980, Brar et al., 1985; Kalra, 1986). In fact gonadectomy reduced LHRH output in the basal hypothalamus as measured by push-pull cannula technique in the rat (Levine and Ramirez, 1980,1982; Dluzen and
Ramirez, 1983) and sheep (Schillo et al., 1985). Also, in vitro LHRH secretion from the hypothalami of castrated rats was drastically reduced as
compared to that from hypothalami of intact rats (Table I; Kalra et al.,
1984; Kalra et al., 1987). In the wake of these findings, it has not been
possible to satisfactorily reconcile the differential LHRH secretion, if

any, with the changing gonadal steroidal milieu and the varied patterns of LH secretion either on each day of the estrous cycle in the female rat or the train of robust episodes, resembling the castrate pattern, interspersed with relatively quiescent phases of LH secretion in the intact male rat. Essentially, the primary issue of how gonadal steroids act in the brain to modulate pituitary gonadotropin secretion has remained unresolved.

Recently, precise knowledge of the distribution patterns of steroid concentrating neurons and LHRH producing neurons and their fiber systems coupled with the revelations that a variety of neuropeptidergic circuits in the hypothalamus can influence LH release, has opened new avenues to examine the long standing issue of indirect steroid feedback from a different perspective. The availability of specific radioimmunoassays to accurately measure neuropeptide levels in microdissected loci innervated by the neuropeptidergic circuits and their secretion both in vivo and in vitro have facilitated a greater understanding of the functional relationships between LHRH neurons and other peptidergic systems in the vicinity, and how gonadal steroids may modulate their interaction. In this communication, our findings suggesting that gonadal steroids may exert a trophic influence on the hypothalamic neural circuitry that controls LHRH secretion are briefly presented (Fig. 1). Further, they underscore the idea that LHRH synthesis and release may be controlled by distinct hypothalamic mechanisms and that neuropeptides may act in multiple ways to influence LH release. Finally, an attempt is made to develop a working model to explain the puzzling question of how gonadal steroids both facilitate and inhibit LH release.

LUTEINIZING HORMONE RELEASING HORMONE

Effects of Gonadal Steroids in Male Rats

With the development of radioimmunoassays for the measurement of LH-RH, it became clear that orchidectomy decreased hypothalamic LHRH levels (Rev. Kalra and Kalra, 1983; Kalra, 1986). Since this depletion was invariably accompanied by hypersecretion of LH it was reasoned that castration increased LHRH release and the augmented rate of secretion unmatched by a corresponding replenishment process was responsible for the gradual depletion in hypothalamic LHRH contents. Implicit in this line of thinking is the assumption that gonadal steroids restrain LHRH release. Interestingly, as briefly presented in the preceding section, the assumption that androgens inhibit LHRH release has not yet been experimentally substantiated (Table I; Dluzen and Ramirez, 1983, Brar et al., 1985). On the other hand, evidence has accumulated to indicate that androgens may facilitate the neurosecretory function of LHRH neurons. Various lines of evidence obtained from in vitro and in vivo studies in support of this notion have been reviewed extensively (Kalra, 1985; Kalra and Kalra, 1986). Only a few salient features of these findings will be presented here. Our studies show that the castration-induced depletion in the medial basal hypothalamus (MBH) LHRH stores can be reinstated by androgens at levels that may or may not inhibit LH release, and the time course of the androgen-induced MBH LHRH and LH responses were markedly different suggesting that the increase in the MBH LHRH levels may be due to de novo synthesis rather than a mere consequence of diminution in LHRH release. Recent evidence that gonadotropin-releasing hormone Associated Peptide, a cleavage peptide produced during the biosynthesis of LHRH, decreased in castrated rats (Negro-Vilar et al., 1986) and mRNA encoded in LHRH neurons was higher in intact than in castrated rats (Rothfeld et al., 1986), is in accord with our observations that androgens may increase the rate of LHRH synthesis (Kalra, 1985; Kalra and Kalra, 1986).

TABLE I HYPOTHALAMIC LHRH SECRETION IN THE RAT: EFFECTS OF GONADECTOMY AND STEROID REPLACEMENT

TREATMENT	PROCEDURE	RESPONSE	INVESTIGATORS
ORCHI.	IN VITRO	↓	KALRA AND KALRA, 1983 KALRA, 1986 KALRA ET AL., 1987
ORCHI. + T	IN VITRO	↑	KALRA ET AL., 1985 KALRA ET AL., 1987 KALRA, 1986
OVX	IN VITRO	↓	FIG. 2
OVX + E_2	IN VITRO	↑	FIG. 2 DROUVA ET AL., 1983,1984
OVX + E_2 + P	IN VITRO	↑	KIM AND RAMIREZ, 1985 LEADEM AND KALRA, 1984
ORCHI.	PPC PORTAL PLASMA	↓ NC	DLUZEN ET AL., 1983 BRAR ET AL., 1985
OVX	PPC PORTAL PLASMA	↓ NC, ↑	LEVINE ET AL., 1982 SARKAR AND FINK, 1980
OVX + E_2	PPC PORTAL PLASMA PORTAL PLASMA	NC ↓ NC	LEVINE ET AL., 1980 SARKAR AND FINK, 1980 SHERWOOD AND FINK, 1980

OVX = OVARIECTOMIZED: ORCHI. = ORCHIDECTOMIZED; T = TESTOSTERONE;
PPC = PUSH-PULL CANNULA PROCEDURE;
NC = NO CHANGE; ↑ = INCREASE; ↓ = DECREASE

With respect to the effects of androgens on LHRH release, there is general consensus that as compared to that from intact rats, the basal and evoked LHRH release in vitro from the hypothalami of castrated rats was decreased (Table I; Rev. Kalra and Kalra, 1986). The in vivo evidence is not as clear-cut perhaps due to the technical difficulties encountered in measuring LHRH. Nevertheless, Dluzen et al. (1983) reported enhanced LHRH release from the basal hypothalamus of intact as compared to castrated rats. On the other hand, Brar et al. (1985) failed to observe any change in portal plasma concentrations of LHRH after orchidectomy. Collectively then, although there is little evidence to date to suggest that androgens exert a negative feedback effect on LHRH neurosecretion, the data are in agreement with the view that androgens may exert a trophic effect on LHRH neurosecretion (Kalra, 1985; Kalra and Kalra, 1983).

Effects of Gonadal Steroids in Female Rats

Because estrogen treatment induces daily LH surges in ovariectomized rats and there are dynamic changes in hypothalamic LHRH levels tightly coupled with LH surges, it has not been possible to carefully examine the effect of ovarian steroids on LHRH secretion (Kalra and Kalra, 1983;

Kalra, 1985). The only evidence to support the view that ovarian steroids may facilitate LHRH accumulation in the hypothalamus, possibly due to increased de novo synthesis, is the demonstrated increase in LHRH levels preceding the onset of the preovulatory LH surge (Kalra et al., 1973; Kalra and Kalra, 1977); a similar response can be readily induced by P in estrogen-primed ovariectomized rats (Kalra and Kalra, 1979; Kalra et al., 1978; Simpkins et al., 1980; Crowley et al., 1982).

However, similar to the action of androgens in males, either estrogen alone (Drouva et al., 1983, 1984) or P in estrogen-primed rats augmented basal and evoked LHRH release from the hypothalamus in vitro (Leadem and Kalra, 1984; Kim and Ramirez, 1985; Table I). In addition, our recent in vitro studies show a correlation between basal LHRH release and E_2 titers during the estrous cycle. As evident in Fig. 2, basal LHRH release was low on estrus when circulating E_2 levels ranged between 5-10 pg/ml. As the serum E_2 concentrations rose on diestrus I, basal LHRH release was also significantly increased, a response reflected in increased LH secretion (Kalra and Kalra, 1974; Gallo, 1981). Presently, we cannot explain a fall in the basal LHRH secretion on diestrus II despite the sustained rise in E_2 titers. However, on the morning of proestrus when E_2 secretion peaked, the rate of LHRH release increased correspondingly. Additionally, the basal rate of LHRH output from the hypothalami of ovariectomized rats was significantly less than that from the hypothalami of diestrous I and proestrous rats. Clearly then, these observations are in line with the possibility that ovarian steroids may augment LHRH secretion during the estrous cycle over the rate normally present in ovariectomized rats.

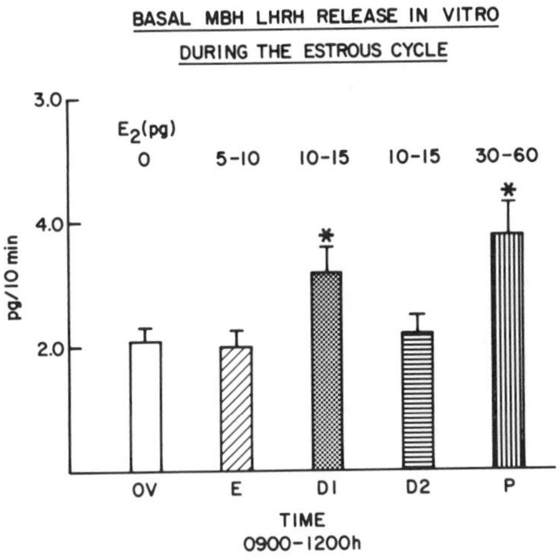

BASAL MBH LHRH RELEASE IN VITRO
DURING THE ESTROUS CYCLE

Figure 2   Correlation between the circulating concentrations of estradiol 17β (E_2) and in vitro basal LHRH output from the medial basal hypothalamus (MBH) of rats during each day (0900 - 1200 h) of the estrous cycle. OV = ovariectomized; E = estrus; D_1 = diestrus I; D_2 = diestrus II; P = proestrus. * $p < 0.05$ vs OV, E or D_2 groups.

A number of investigators have assessed the effects of estrogen on _in vivo_ LHRH secretion in the ovariectomized rat, sheep and monkey. As expected, estrogen treatment rapidly decreased LH release but a corresponding decrease in LHRH release was not detected. The failure to observe a tight reciprocal relationship between estrogen and LHRH release was apparent when LHRH release was assessed either by the push-pull cannula technique in the rat (Levine and Ramirez, 1980), monkey (Levine et al., 1985) and sheep (Schillo et al., 1985), or in the hypophysial portal plasma in monkey and sheep (Carmel et al., 1976; Clark and Cummins, 1985). Despite these consistent results, we cannot preclude the possibility that further refinement of the technology to measure accurately the total LHRH signal arriving at the pituitary may demonstrate a reciprocal relationship between gonadal steroids and LHRH secretion. However, it is noteworthy that a decrease in LHRH secretion is not essential to suppress LH release after estrogen treatment in monkeys, because LH secretion decreased to a low basal range despite the presence of uninterrupted LHRH signals (Knobil, 1980). Thus the _in vitro_ evidence and failure to document a reciprocal relationship between ovarian steroids and LHRH release _in vivo_ favor the thesis that estrogen may not adversely affect LHRH release but that the ensuing decrease in LH release may be due to a decrease in the pituitary responsiveness to episodic LHRH signals and/or due to the release of hypothalamic factor(s) that inhibit LH release (Kalra, 1976; Leadem and Kalra, 1984; Kalra, S., 1986, Kalra, P., 1986).

NEUROPEPTIDE Y

In 1982, Tatemoto reported the amino acid sequence of neuropeptide Y (NPY) isolated from the porcine brain and soon it became apparent that the hypothalamus was extremely rich in immunoassayable NPY (Allen et al., 1983). Due to our long-standing interest in adrenergic involvement in the control of LH release (Kalra et al., 1972; Kalra and McCann, 1973, Crowley et al., 1978, Adler et al., 1983), the report that NPY coexisted with adrenergic transmitters in some neurons in the brain stem (Hökfelt et al., 1984) suggested the possibility that NPY and adrenergic transmitters may be coreleased and coact to excite LH release. The source of NPY in the hypothalamus appeared to be the NPY immunopositive perikarya in the arcuate (ARC) and periventricular nucleus and cells in the brain stem (Hökfelt et al., 1984; Chronwall et al., 1985). These NPY containing cells densely innervate the preoptic-tuberal pathway which was previously implicated in the control of LH release (Everett, 1976; Kalra et al., 1971) and is also the projection site of the LHRH (Kelly et al., 1982; King et al., 1982) and steroid concentrating neurons (Pfaff, 1980; Sar and Stumpf, 1973). In early 1984, we reported that human pancreatic peptide (hPP), a peptide structurally related to NPY, inhibited LH release in ovariectomized rats but stimulated LH release if the ovariectomized rats were primed with ovarian steroids (Kalra and Crowley, 1984a). These effects of hPP paralleled those of norepinephrine (NE) on LH release (Gallo and Drouva, 1979). These results provided the first experimental evidence showing that NE and a coexisting neuropeptide can elicit qualitatively similar LH responses and raised the possibility that a corelease of adrenergic transmitters and NPY within the preoptic-tuberal pathway may be of physiological significance. As soon as NPY became available to us for testing in 1983, we reported that NPY evoked similar differential LH responses in the rat (Kalra and Crowley, 1984b). Since it stimulated LH release only in rats that were primed either exogenously (Kalra and Crowley, 1984b) or endogenously (Allen et al., 1985b) with gonadal steroids, it was of interest to investigate the mechanism whereby gonadal steroid action facilitated the excitatory expression of NPY.

There are several ways by which gonadal steroids may accomplish this reversal of NPY-induced LH response. First, since the NPY-containing and steroid-concentrating neurons are coextensive in the hypothalamus, it is possible that gonadal steroids may modulate the neurosecretion of NPY by influencing one or more loci in the biosynthesis and storage of NPY in the nerve terminals. Second, gonadal steroids may facilitate postsynaptic neural processes in a manner that allows stimulation of LHRH release by either NPY alone or in concert with the coexisting adrenergic transmitters. Third, since NPY-containing neurons innervate the median eminence (ME) extensively, it is possible that like many other neuropeptides and biogenic amines, NPY may be transported to the pituitary via the hypophysial portal system, where gonadal steroids may facilitate LH release in response to either a direct action of NPY or in interaction with LHRH. What follows is a summary of our findings to highlight the multifaceted influence of gonadal steroids on hypothalamic NPY function, a neuropeptide shown recently to modulate a multitude of brain functions related to reproduction (Kalra et al., 1986).

The Effects of Gonadal Steroids on NPY Neurosecretion

We have utilized two experimental models previously used to delineate the facilitatory nature of gonadal steroids on LHRH levels and release (Kalra and Kalra, 1979; Kalra et al., 1978; Crowley et al., 1982). Crowley et al. (1985) studied the effects of sequential EB and P treatment of ovariectomized rats on NPY levels in several microdissected sites in the preoptic-tuberal pathway and correlated them on a time sequence basis with changes in LHRH levels in these sites and with the rise in LH secretion. As expected (Kalra and Kalra, 1979; Simpkins et al., 1981; Crowley et al., 1982), administration of P to EB-primed rats induced a sequential rise and fall in the ME LHRH levels which preceded the LH surge. Quite unexpectedly, P treatment also induced a rise and a fall in the ME NPY concentrations with a time course similar to that of LHRH. Since these antecedent dynamic shifts in the ME LHRH levels appear to be responsible for stimulation of LH surge and since NPY can stimulate LHRH release from the ME nerve terminals (Crowley and Kalra, 1987), these observations imply that ovarian steroids can independently facilitate NPY neurosecretion which, in turn, may be involved in the hypersecretion of LHRH. Also, parallel alterations of NPY and LHRH concentrations in the ME are suggestive of synchronizing effects of this steroid regimen on the neurosecretory function of two disparate but functionally linked neuropeptidergic systems.

Additional evidence suggests that testicular steroids may promote some aspect of NPY neurosecretion in the male rat. Sahu et al. (1987) observed that the removal of circulating gonadal steroids following orchidectomy, resulted in significant decreases in NPY concentrations in the ME, ARC, and ventromedial nucleus (VMN). The NPY response in the ME and ARC is of interest since these sites in the MBH are intimately involved in LHRH release and a similar decrease in LHRH concentrations occurred after orchidectomy (Kalra and Kalra, 1983; Kalra, 1985). Our preliminary findings also show that testosterone replacement therapy largely reinstated NPY levels in the ARC and VMN (unpublished). Further, not only did the levels of NPY in the ME and ARC decrease, but the in vitro hypothalamic NPY release evoked by K$^+$ was also diminished after castration (Sahu et al., 1987). Seemingly, gonadal steroids (androgens, estrogen and progesterone) modulate key aspects of NPY neurosecretion. Based on the regional specificity of the castration-induced NPY and LHRH responses, it is tempting to propose that a subset of strategically located gonadal steroid concentrating neurons in the ARC and VMN may locally facilitate the availability for release of the two neurohormones in the ME nerve terminals (Sahu et al., 1987).

Facilitation of the Postsynaptic NPY Expression by Gonadal Steroids

Since NPY stimulated LH release only in steroid-primed rats, we investigated the site and mode of gonadal steroid action in facilitating the excitatory expression of NPY. Crowley and Kalra (1987) observed that NPY (10^{-6}M) failed to stimulate in vitro LHRH release from the MBH of ovariectomized, hormonally untreated rats. On the other hand, under identical in vitro conditions, NPY readily stimulated LHRH release from the MBH of ovariectomized rats pretreated with estrogen. Also, NPY (10^{-9} — 10^{-6} M) evoked a dose-related stimulation of LHRH from the MBH of EBP-primed, ovariectomized rats. In fact, maximal stimulation of LHRH release in response to 10^{-6} M NPY was seen from the MBH of rats receiving the combined ovarian steroid treatment. Thus, stimulation of LH release in steroid-primed rats evoked by intracerebroventricular administration of NPY (Kalra and Crowley, 1984b) appeared, in part, to be due to LH-RH release. It seems that the excitatory NPY effects are exerted via an axo-axonic link between NPY and LHRH neurons in the preoptic-tuberal pathway and that the excitatory expression of the functional link is gonadal steroid-dependent.

Allen et al. (1985a; 1987) showed that under the influence of gonadal steroids NPY may interact with adrenergic transmitters. The NPY-induced increase in plasma LH levels was not affected by prior blockade of either α_1 or β-adrenoreceptors, or dopamine receptors. However, the α_2-adrenoreceptor antagonist, yohimbine, attenuated the NPY-induced LH response, suggesting an α_2-adrenoreceptor mediation of NPY action.

In addition, Allen et al. (1987) observed that NPY and NE acted in concert in ovarian steroid-primed, ovariectomized rats. The results showed that NPY and NE in doses, which separately were only minimally effective in stimulating LH release, when administered together, induced a LH response greater than the sum of the individual responses. On the other hand, when NE and NPY were administered together in doses that alone were either ineffective or maximally stimulatory, a similar additive or synergistic effect was not observed.

Collectively, these findings document that gonadal steroids facilitate the release of LHRH by adrenergic transmitters and NPY and that each may act via α_2-adrenoreceptors. Thus it appears that gonadal steroids coordinate the function of several aminergic and peptidergic systems in the regulation of gonadotropin secretion in the rat.

Action of NPY on LHRH-Induced LH Release

Immunocytochemical mapping studies show dense NPY-immunopositive nerve terminals in the proximity of hypophysial portal veins in the ME (Hökfelt et al., 1984; Chronwall et al., 1985). It is also known that peptidergic and aminergic systems that project into the ME are likely to release their products into the hypophysial portal veins, and in some instances, this has been shown to be of physiological significance in the regulation of pituitary function (Fig. 1). Also, intracerebroventricularly administered peptides invariably reach the pituitary by this route (Ben-Jonathan et al., 1974; Tannebaum and Patel, 1986) in concentrations large enough to influence pituitary hormone secretion. Our observations of parallel changes in NPY and LHRH levels in the ME in association with the LH surge raised the possibility of a simultaneous discharge of LHRH and NPY into the portal circulation for an action at the level of the pituitary. McDonald et al. (1985) reported increased LH release when NPY was delivered to anterior pituitary cells in an in vitro perifusion system. However, Kerkerian et al. (1985) did not observe any effect of NPY on LH release in the pituitary cell incubation system.

Crowley et al. (1987) observed that addition of NPY _in vitro_ to hemipituitaries from ovariectomized rats only marginally stimulated the release of LH. However, when added with LHRH (10^{-9} M), NPY (10^{-6} M) significantly potentiated the LHRH-induced LH release. Additionally, NPY (10^{-9}, 10^{-7} M) on its own failed to activate LH release from dispersed pituitary cells in culture, but it markedly potentiated LH release in response to LHRH (10^{-9}, 10^{-8} M).

Consequently, when these observations are considered together with the earlier findings that intracerebroventricular NPY stimulated LH release (Kalra and Crowley, 1984b), and NPY stimulated LHRH release from the hypothalamus _in vitro_ (Crowley et al., 1987), it appears that NPY may have dual sites of action, one within the hypothalamus and the other at the level of pituitary gonadotrophs. Under the physiological conditions characterized by LH release, NPY may be released in the hypothalamus to induce LHRH secretion, and into the hypophysial portal veins, to potentiate the LHRH-induced LH response. Thus, a two-pronged action of NPY, stimulation of LHRH release in concert with adrenergic transmitters in the hypothalamus and stimulation of LH release in concert with LHRH at the level of the pituitary, may be important in evoking the preovulatory and ovarian steroid-induced LH surges lasting for several hours.

ENDOGENOUS OPIOID PEPTIDES

Endogenous opioid peptides (EOP) are the best example of hypothalamic neuropeptides shown to exercise an inhibitory influence on LH release in gonadectomized and intact rats (Kalra, 1983; Kalra and Leadem, 1984). Of the three classes of EOP, whose distribution in the hypothalamus has been described in detail (Finley, 1981; Watson et al., 1982), β-endorphin neurons are anatomically best suited to interact with LHRH neurons. Indeed, β-endorphin readily suppressed LH release in gonadectomized rats (Kinoshita et al., 1980; Leadem and Kalra, 1985b) and blocked preovulatory LH release and ovulation (Leadem and Kalra, 1985a). Dynorphin, another EOP abundant in the hypothalamus but with a distribution pattern distinct from that of β-endorphin (Watson et al., 1982), suppressed LH release to a lesser degree (Leadem and Kalra, 1985b). The enkephalins were ineffective (Leadem and Kalra, 1985b) but the methionine-enkephalin analogs that presumably defy rapid inactivation after intracerebroventricular injections, produced a sustained decrease in plasma LH levels in ovariectomized rats and also blocked ovulation (Leadem and Kalra, 1985b; Köves et al., 1981). Because of the possibility that each of these EOP may employ specific opiate receptor types, one suspects that either μ or ε receptor activation by β-endorphin, κ receptor activation by dynorphin and α receptor activation by methionine-enkephalin analogs result in suppression of LH release (Leadem and Kalra, 1985b). Evidence that μ receptors may be preferentially involved in the control of LH release has been presented (Pfeiffer et al., 1983). However, since specific receptor antagonists for each of these receptor subtypes are not currently available, identification of the opiate receptor subtypes involved in controlling the two modalities of LH secretion in rat can at best be considered tentative.

The possibility that EOP or opiate-induced decrease in LH release may primarily be a result of diminution in the LHRH output has not been experimentally substantiated. In fact, morphine infusion _in vitro_ failed to decrease hypothalamic LHRH release (Kalra et al., 1987). However, the observations that morphine blocked the progesterone-induced increase in the MBH LHRH levels (Kalra and Simpkins, 1981), the opiate receptor antagonist, naloxone, stimulated _in vitro_ hypothalamic LHRH release (Leadem et al., 1985; Kalra et al., 1987; Kalra et al., 1984) and when microinfused in the proximity of LHRH neurons in the hypothalamus, naloxone increased

LH release (Kalra, 1981), suggest that EOP release in the hypothalamus may suppress LHRH release. On the other hand, contrary to previous reports (Cicero et al., 1977), a number of laboratories have recently shown that morphine and β-endorphin not only decreased basal LH release from cultured pituitary cells but also attenuated the LHRH-induced LH release response (Matteri et al., 1985; Cacicedo et al., 1986; Blank et al., 1986). It seems that EOP may have a dual site of action; one in the hypothalamus to inhibit LHRH release and the other at the level of the pituitary to attenuate the LHRH-induced LH release.

Despite the voluminous experimental evidence suggesting that the EOP neurons may be a hypothalamic inhibitory neural circuit, the precise physiological significance of EOP in the regulation of gonadotropin secretion is imprecise. EOP may participate in the dynamics of the hypothalamic neural circuitry which controls episodic basal and preovulatory LH release in two possible ways. 1) EOP may mediate the inhibitory feedback effects of gonadal steroids on LH release (Cicero et al., 1979, 1980; Kalra, 1983). This view implies that gonadal steroids may monitor the inhibitory opioid tone on LH secretion by either regulating EOP output in the vicinity of LHRH neurons or by modulating EOP receptor function. 2) EOP may exert an inhibitory tone on LH release but the stimulatory feedback action of gonadal steroids may curtail the inhibitory influence to allow the preovulatory LH surge to occur on proestrus. The evidence in support of these two possible modes of EOP involvement in the regulation of LH secretion are critically examined in the following section.

Inhibitory Steroid Feedback Mediation by EOP

Since the blockade of opiate receptors by naloxone transiently increased LH release in intact rats, it was advocated that EOP neurons may exercise a tonic inhibition on gonadotropin secretion and that gonadal steroids may monitor the opioid restraint. This hypothesis also implied that the EOP restraint may be absent or drastically reduced in steroid-deficient rats. Indeed, it was noted that a bolus injection of naloxone within a week after gonadectomy failed to stimulate LH release probably due to elimination of the opioid restraint on LH release (Bhanot and Wilkinson, 1983, 1984). In contrast, recent studies showed that a similar bolus injection or a slow intravenous infusion of naloxone readily stimulated LH release in rats castrated for up to 4 weeks (Fournet et al., 1985 and unpublished). Also, naloxone was found to stimulate LHRH release from the MBH of long-term castrated rats and due to diminished LHRH stores, the amount released was considerably smaller than that from the MBH of intact rats (Kalra et al., 1987). Consequently, the naloxone-induced LH response may not be a reliable index of central opioid tone in intact and castrated rats because the availability of LHRH stores for release and the effects of gonadal steroids on LHRH-induced LH release are important factors that may profoundly affect the LH response. The findings that despite marked changes in the gonadal steroid milieu, naloxone evoked a similar LH response on each day of the estrous cycle are in accord with this assumption (Gabriel et al., 1983). Also, whether castration induces changes in opiate receptor number and affinity is not settled (Hahn and Fishman, 1979; Wilkinson et al., 1982). Therefore, the thesis that the negative feedback action of gonadal steroids on LH release is mediated by EOP and that gonadal steroids regulate the inhibitory EOP tone is not supported by experimental findings to date.

Opiates Monitor the Negative Feedback Action of Gonadal Steroids

Cicero et al. (1980) proposed that morphine may act like testosterone in evoking the central LHRH response. However, unlike testosterone, morphine could not raise the MBH LHRH levels in 2-week castrated rats (unpublished). In the course of these investigations, we unexpectedly discovered a novel interaction between testosterone and opiate receptor activation with morphine (Gabriel et al., 1985,1986). We found that long-term activation of opiate receptors with morphine profoundly affected the response of the hypothalamo-pituitary axis to gonadal steroids. In accord with the findings of Cicero et al. (1980), continuous opiate receptor stimulation by subcutaneous morphine pellets failed to alter LH secretion in castrated rats. However, when these rats were primed with either testosterone, 5α-dihydrotestosterone or E_2, at levels which on their own were ineffective (Kalra, 1985), there was a drastic suppression in plasma LH levels (Gabriel et al., 1985; 1986). In fact, plasma LH levels were suppressed to the low basal range seen characteristically in intact male rats. Apparently, continuous stimulation of opiate receptors rendered rats extremely sensitive to the negative feedback action of gonadal steroids on LH release and further revealed that within the hypothalamo-pituitary axis, there may be an opioid-sensitive neural component which has the potential to monitor the negative feedback action of gonadal steroids on LH secretion. Additional studies showed that combined morphine and testosterone treatment, a regimen that drastically reduced LH release in vivo, had no effect on in vitro LHRH release from the hypothalami of these rats (Kalra, P., 1986), but it diminished the stimulation of LH in response to LHRH in physiological doses (Kalra and Sahu, 1987). Presumably, combined action of morphine and testosterone at the level of the pituitary may be responsible for the marked decrease in LH release. One cannot discount the alternate possibility that the combined treatment stimulated the release of hypothalamic hormone(s) which inhibited LH release (Kalra, 1976; Leadem and Kalra, 1984; Kalra, 1983; Kalra 1986). Nevertheless, the revelation that opioid receptor activation on a continuous basis can profoundly alter the feedback effects of gonadal steroids on LH release has numerous clinical and basic implications. It provides a provocative underlying avenue to explore the etiology of increased sensitivity to gonadal steroids seen in deafferented (Kalra et al., 1970), hyperprolactinemic (McNeilly et al., 1983) and senescent rats (Gray et al., 1980), and that observed at the onset of puberty and following drastic shifts in photoperiod in seasonal breeders (Sisk and Turek, 1983). It is also of interest to note that narcotic addicts display increased sensitivity towards gonadal steroids in a fashion somewhat similar to what we have observed experimentally.

Effects of Gonadal Steroids on EOP Secretion

A number of investigators have studied the effects of gonadal steroids on the hypothalamic levels and release of EOP. Gonadectomy decreased the levels of β-endorphin in the hypophysial portal plasma of monkey (Ferin et al., 1983) and rat (Sarkar and Yen, 1985). Conversely, combined treatment of estrogen and progesterone reinstated β-endorphin levels in the portal plasma of ovariectomized monkeys to the range normally seen during the luteal phase of the menstrual cycle (Ferin et al., 1983). On the other hand, long-term steroid replacement therapy with either testosterone in orchidectomized rats (Wardlaw et al., 1986) or estrogen in ovariectomized rats (Wardlaw et al., 1982) decreased hypothalamic β-endorphin levels. Intriguingly, 3 days of estrogen treatment of rats ovariectomized for 2 weeks, decreased hypothalamic proopiomelanocortin mRNA, an effect also induced by long-term ovariectomy alone (Wilcox and Roberts, 1985). At present, it is difficult to correlate the observed shifts in the hypothalamic levels of β-endorphin and proopiomelanocortin mRNA, or β-endorphin changes in the hypophysial portal blood with the patterns of LH secretion in these rats. However, these studies strongly imply that gonadal steroids have the capability to affect some aspect of β-endorphin neurosecretion, presumably by acting within the β-endorphin producing perikarya in the arcuate nucleus (Morrell et al., 1985).

EOP Mediation of Stimulatory Feedback Effects of Gonadal Steroids

EOP may participate in the regulation of steroid-dependent induction of preovulatory LH release on proestrus. A number of studies showed that morphine and β-endorphin inhibited ovulation, primarily by blocking the preovulatory LH and LHRH hypersecretion (Ching, 1985; Kalra, 1986). Our studies (Gabriel et al., 1983) and those of Piva et al. (1985) showed that naloxone failed to stimulate LH release during the LH surge occurring either spontaneously on proestrus or that induced by sequential estrogen, progesterone therapy. These observations together with the demonstration that intraventricular β-endorphin administration decreased the high amplitude, high frequency LH pulses in ovariectomized rats (Leadem and Kalra, 1985b), led us to propose that sometime prior to and during the critical period, a decrease in opioid tone in the preoptic-tuberal pathway may trigger the preovulatory LH surge (Kalra, 1983). This hypothesis was tested by infusing naloxone between 1000 - 1200 h on proestrus, to see whether the ensuing decrease in opioid tone would advance the LH surge (Allen and Kalra, 1986; Allen et al., 1986). Indeed, our findings showed that a sustained decrease in opioid tone induced LH hypersecretion, and in the majority of rats the pattern and magnitude of LH surge resembled that normally seen on the afternoon of proestrus. We also noted that the naloxone-induced LH surge occurred in incremental episodes, and in several instances, episodic LH discharge was accelerated from a normal proestrous morning frequency of 1 pulse/h to 2-3 pulses/h (Allen et al., 1986). Further, similar naloxone infusion on diestrus II, the day preceding proestrus, failed to evoke LH surge (unpublished). Since it is well known that a neural clock triggers the preovulatory LH and LHRH hypersecretion on the afternoon of proestrus, and the ovarian steroid milieu in the interval preceding the critical period on proestrus facilitates the timing of the neural clock, we proposed that prior to the critical period on proestrus the neural clock either directly or via other neuronal systems may transiently curtail the tonic inhibitory EOP influence locally in the preoptic-tuberal pathway. The progressive decrease in the inhibitory EOP tone may gradually accelerate the frequency of episodic LHRH discharge from the circhoral to that approximating the inherent frequency of 2-3 pulses/h as seen in ovariectomized rats. This high frequency LHRH discharge may eventually activate and sustain a prolonged LH surge from the pituitary in the afternoon of proestrus.

CONCLUDING REMARKS: A MODEL

In this discussion we have identified several features of the steroid-neuropeptide connection in the control of LH secretion. It is clear that a variety of neuropeptidergic circuits either connected to or in close apposition either with the LHRH neurons or with the hypophysial portal vessels are potentially important in the central control of LH secretion (Fig. 1). Also, the evidence reviewed lends credence to our view that gonadal steroids generally exert a facilitatory effect on the neuropeptidergic systems in a complex fashion. Recent findings illustrate that gonadal steroids may modify neurosecretion (biosynthesis, storage and release) and adjust the postsynaptic responses of the peptidergic signals in the hypothalamus. Within this framework is an intriguing revelation that some neuropeptidergic circuits also may modulate the central and pituitary action of steroids. Presumably, the gonadal steroid concentrating neurons are the anatomical substrate for steroid action. An understanding of precisely how gonadal steroid signals are coded and transduced therein to generate a new set of neurochemical signals which, in turn, excite or inhibit LH release remains to be delineated.

MODE OF ACTION OF GONADAL STEROIDS IN THE REGULATION OF LH RELEASE:

A HYPOTHESIS

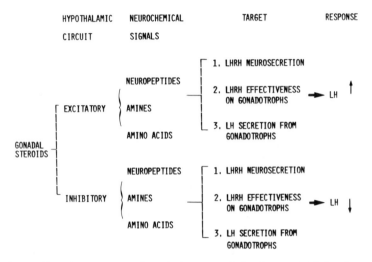

Figure 3   A conceptual model to explain how gonadal steroids may employ excitatory and inhibitory hypothalamic neural circuits to stimulate or inhibit
LH release, respectively. For details see text.

However, the idea of a trophic interaction between steroids and
neuropeptidergic signals and the suggestion that gonadal steroids modify,
and perhaps, control the delivery of the messages has provided us a basis
for formulating a working model (Fig. 3). This model attempts to unify
the intricate functional interrelationships between gonadal steroid concentrating neurons, the central sensors of the steroidal milieu, and the
peptidergic circuits. It incorporates the view that gonadal steroids may
interact with two distinct hypothalamic circuits, an excitatory and an inhibitory circuit, each composed of one or more of the three major classes
of identified neurochemical signals. The three classes of signals are the
classical aminergic neurotransmitters, the neuropeptides and the amino
acids. These circuits may be connected with one another by synapses or
may be in close apposition to affect the neighboring systems. The model
assumes that to stimulate LH release, the excitatory circuits either
directly activate the release of LHRH, or they may potentiate the action
of LHRH at the pituitary level. They may also independently stimulate LH
release from the pituitary gland. Our findings on the multifaceted actions of NPY fit into this scheme of interactions within the gonadal
steroid-excitatory neuropeptide circuit. Similarly, the model entails
that to inhibit LH release, gonadal steroids may mobilize the inhibitory
neural circuits that either inhibit LHRH release, attenuate the effectiveness of LHRH at the level of the pituitary or directly inhibit LH release.
So far the known actions of the inhibitory EOP system appear to fulfill

some key requirements of this scheme. The versatility of this skeletal model allows the accomodation of other neurochemical signals that undoubtedly will be identified in the near future. Already the newly discovered brain peptides, such as galanin and neuropeptide K, are likely candidates for excitatory and inhibitory roles, respectively (Sahu and Kalra, unpublished). The ultimate challenge is to accurately define the spatial arrangements among the components of the circuits and how they actually collaborate under the influence of gonadal steroids to produce the two modalities of gonadotropin secretion.

ACKNOWLEDGEMENTS

The research described in the review was supported by grants from the National Institutes of Health (HD 08634; HD 11362 and HD 13703). Thanks are due to Ms. Sally McDonell for untiring secretarial assistance.

REFERENCES

Adler BA, Johnson MD, Lynch CO, Crowley WR, 1983. Evidence that norepinephrine and epinephrine systems mediate the stimulatory effects of ovarian hormones on luteinizing hormone and luteinizing hormone releasing hormone. Endocrinology 113:1431-38

Allen LG, Kalra SP, 1986. Evidence that a decrease in opioid influence evoke preovulatory LH release. Endocrinology 118:1275-81

Allen LG, Crowley WR, Kalra SP, 1985a. Role of catecholamines and opiates in stimulation of LH release by neuropeptide Y (NPY), (Abst), 15th Ann Mtg Soc Neurosci, Dallas, Texas, October 20-25

Allen LG, Kalra PS, Crowley WR, Kalra SP, 1985b. Comparison of the effect of Neuropeptide Y and adrenergic transmitters on LH release and food intake in male rats. Life Sci 37:617-23

Allen LG, Caton D, Kalra SP, 1986. Preovulatory LH surge in the rat: pulse analysis and advancement of opioid receptor blockade. 1st Internat Congr Neuroendocrinology (Abst), San Francisco, CA, July 9-11

Allen LG, Crowley WR, Kalra SP, 1987. Interactions between Neuropeptide Y and adrenergic systems in stimulation of LH release in steroid-primed ovariectomized rats. Endocrinology (under review).

Allen YS, Adrian TE, Allen JM, Tatemoto K, Crow TJ, Bloom SR, Polak JM, 1983. Neuropeptide Y distribution in the rat brain. Science 221: 877-79

Ben-Jonathan N, Mical RS, Porter JC, 1974. Transport of LRF from CSF to hypophysial portal and systemic blood and release of LH. Endocrinology 95:18-25

Bhanot R, Wilkinson M, 1983. Opiatergic control of LH secretion is eliminated by gonadectomy. Endocrinology 112:399-01

Bhanot R, Wilkinson M, 1984. The inhibitory effect of opiates on gonadotropin secretion is dependent upon gonadal steroids. J Endocrinol 102:133-41

Blank MS, Fabbri A, Catt K, Dufau M, 1986. Inhibition of luteinizing hormone release by morphine and endogenous opiates in cultured pituitary cells. Endocrinology 118:2097-02

Brar AK, McNeilly AS, Fink G, 1985. Effects of hyperprolactinemia and testosterone on the release of LH-releasing hormone and gonadotrophins in intact and castrated rats. J Endocrinol 104:35-43

Cacicedo F, Sanchez-Franco F, 1986. Direct action of opioid peptides and naloxone on gonadotropin secretion by cultured rat anterior-pituitary cells. Life Sci 38:617-23

Carmel PW, Araki S, Ferin M, 1976. Pituitary stalk portal blood collection in rhesus monkeys: evidence for pulsatile release of gonadotropin-releasing hormone (GnRH). Endocrinology 99:243-48

Ching JCL, Christofides ND, Anand P, Gibson SJ, Allen YS, Su HC, Tatemoto K, Morrison JFB, Polak JM, Bloom SR, 1985. Distribution of galanin immunoreactivity in the central nervous system and the responses of galanin-containing neuronal pathways to injury. Neuroscience 16: 343-54

Chronwall BM, DiMaggio DA, Massari VJ, Vickel VM, Ruggiero DA, O'Donohue TL, 1985. The anatomy of neuropoeptide Y-containing neurons in the rat brain. Neuroscience 15:1159-81

Cicero, TJ, Badger TM, Wilcox CE, Bell RD, Meyer ER, 1977. Morphine decreases luteinizing hormone by an action on the hypothalamic pituitary axis. J Pharmac Exp Ther 203:548-55

Cicero TJ, Shainker BA, Meyer ER, 1979. Endogenous opioids participate in the regulation of the hypothalamic-pituitary-luteinizing hormone axis and testosterone's negative feedback control of luteinizing hormone. Endocrinology 104:1286-91

Cicero TJ, Meyer ER, Gabriel SM, Ball RD, Wilcox CE, 1980. Morphine exerts testosterone-like effects in the hypothalamus of the castrated male rat. Brain Res 202:151-64

Clarke IJ, Cummins JT, 1985. Increased GnRH pulse frequency associated with estrogen-induced LH surge in ovariectomized ewe. Endocrinology 116:2376-83

Crowley WR, Kalra S, 1987. Neuropeptide Y stimulates the release of luteinizing hormone-releasing hormone from medial basal hypothalamus in vitro: Modulation by ovarian hormones. Neuroendocrinology (in press)

Crowley WR, O'Donohue TL, Wachslicht H, Jacobowitz DM, 1978. Effects of estrogen and progesterone on plasma gonadotropins and on catecholamine levels and turnover in discrete brain regions of ovariectomized rats. Brain Res 154:345-57

Crowley WR, Terry LC, Johnson MD, 1982. Evidence for the involvement of central epinephrine systems in the regulation of luteinizing hormone, prolactin, and growth hormone release in female rats. Endocrinology 110:1102-07

Crowley WR, Tessel RE, O'Donohue TL, Adler BA, Kalra SP, 1985. Effects of ovarian hormones on the concentrations of immunoreactive neuropeptide Y in discrete brain regions of the female rat: correlation with serum LH and median eminence LHRH. Endocrinology 117:1151-55

Crowley WR, Hassid A, Kalra SP, 1987. Neuropeptide Y enhances the release of luteinizing hormone induced by luteinizing hormone-releasing hormone. Endocrinology 120:941-45

Dluzen DE, Ramirez VD, 1983. In vivo LHRH release from the median eminence of conscious, unrestrained, intact, acute castrate and longterm castrate male rats as determined with push-pull perfusion (PPP). (Abst), 65th Ann Mtg Endocrine Society, San Antonio TX, June 8-10, p. 156

Drouva S, Laplante E, Kordon C, 1983. Effects of ovarian steroids on in vitro release of LHRH from mediobasal hypothalamus. Neuroendocrinology 37:336-41

Drouva S, Laplante E, Gautron J, Kordon C, 1984, Effects of 17β-estradiol on LH-RH release from rat mediobasal hypothalamus slices. Neuroendocrinology 38:152-57

Everett JW, 1977. The timing of ovulation. J Endocrinol 75:1P-13P

Ferin M, Van Vugt D, Wardlaw S, 1983. The hypothalamic control of the menstrual cycle and the role of endogenous opioid peptides. Rec Progr Horm Res 40:441-80

Finley J, Lindstrom P, Petrusz P, 1981. Immunocytochemical localization of β-endorphin-containing neurons in the hypothalamus. Neuroendocrinology 33:28-42

Fournet N, Allen LG, Kalra PS, Kalra SP, 1985. Effects of a sustained decrease in opioid tone on LH secretion in intact and long-term castrated rats. (Abst) 15th Ann Mtg Soc Neurosci, October 20-25, Dallas, TX

Gabriel SM, Simpkins JW, Kalra SP, 1983. Modulation of endogenous opioid influence on luteinizing hormone secretion by progesterone and estradiol. Endocrinology 113:1806-11

Gabriel SM, Simpkins JW, Kalra SP, Kalra PS, 1985. Chronic morphine treatment induces hypersensitivity to testosterone negative feedback in castrated male rats. Neuroendocrinology 40:39-44

Gabriel SM, Berglund LA, Kalra SP, Kalra PS, Simpkins JW, 1986. The influence of chronic morphine treatment on the negative feedback regulation of gonadotropin secretion by gonadal steroids. Endocrinology 119:2762-67

Gallo RV, 1981. Pulsatile LH release during periods of low level LH secretion in the rat estrous cycle. Biol Reprod 24:771-77

Gallo RV, Drouva SV, 1979. Effect of intraventricular infusion of catecholamines on luteinizing hormone release in ovariectomized and ovariectomized steroid-primed rats. Neuroendocrinology 29:149-62

Gray GD, Smith ER, Davidson JM, 1980. Gonadotrophin regulation in middle-aged male rats. Endocrinology 107:2021-26

Hahn EF, Fishman J, 1979. Changes in rat brain opiate receptor content upon castration and testosterone replacement. Biochem Biophys Res Commun 90:819-23

Harris GW, 1955. Neural control of the pituitary gland, London: Arnold Ltd

Höhlweg W, Junkman K, 1932. Die hormonal nervose regulierung der funktion des hypophysenborderlappens. Klin Wsch 11:321-23

Hökfelt T, Johansson O, Goldstein, M, 1984. Chemical anatomy of the brain. Science 225:1326-34

Kalra PS, 1985. Stimulation of hypothalamic LHRH levels and release by gonadal steroids. J Steroid Biochem 23:725-31

Kalra PS, 1986. Interaction between steroids and opiates on LH and LHRH release. Ann Mtg Soc Neurosci, Washington DC, Nov 9-14, p. 1384

Kalra PS, Kalra SP, 1977. Temporal changes in the hypothalamic and serum leuteinizing hormone-releasing hormone (LH-RH) levels and the circulating ovarian steroids during the rat oestrous cycle. Acta Endocrinol 85:449-55

Kalra, PS, Kalra SP, 1986. Steroidal modulation of the regulatory neuropeptides: lutenizing hormone releasing hormone, neuropeptide Y and endogenous opioid peptides. J Steroid Biochem 25:733-40

Kalra PS, Sahu A, 1987. Mode of opiate-androgen interaction in the control of LH secretion. (Abst), Soc Gyn Invest, Atlanta, GA, March 18-21

Kalra PS, Kalra SP, Krulich L, Fawcett CP, McCann SM, 1972. Involvement of norepinephrine in transmission of the stimulatory influence of progesterone on gonadotropin release. Endocrinology 90:1168-76

Kalra PS, Leadem CA, Kalra SP, 1984. Effects of Testosterone (T) and naloxone on LHRH secretion in vitro - relationship with hypothalamic LHRH concentration. (Abst) VII Int Congr Endocrinol, Quebec, July 1-8

Kalra PS, Crowley WR, Kalra SP, 1987. Differential in vitro stimulation by naloxone of LHRH and catecholamine release from the hypothalami of intact and castrated rats. Endocrinology 120:178-85

Kalra SP, 1976. Ovarian steroids differentially augment pituitary FSH release in deafferented rats. Brain Res 144:541-44

Kalra SP, 1981. Neural loci involved in naloxone-induced luteinizing hormone release: Effects of a norepinephrine synthesis inhibitor. Endocrinology 109:1805-10

Kalra SP, 1983. Opioid peptides - Inhibitory neuronal systems in regulation of gonadotropin secretion. In: McCann SM and Dhindsa DS (eds.) Role of Peptides and Proteins in Control of Reproduction. New York: Elsevier Biomedical, pp. 63-87

Kalra SP, 1986. Neural circuitry involved in control of LHRH secretion: a model for the preovulatory LH release. In: Ganong WF and Martini L (eds.), Frontiers in Neuroendocrinology, Vol. 9. New York: Raven Press, pp. 31-75

Kalra SP, Crowley WR, 1984a. Differential effects of pancreatic polypeptide on luteinizing hormone release in female rats. Neuroendocrinology 38:511-13

Kalra SP, Crowley WR, 1984b. Norepinephrine-like effects of neuropeptide Y on LH release in the rat. Life Sci 35:1173-76

Kalra SP, Kalra PS, 1974. Temporal interrelationships among circulating levels of estradiol, progesterone and LH during the rat estrous cycle: effects of exogenous progesterone. Endocrinology 95:1711-18

Kalra SP, Kalra PS, 1979. Dynamic changes in hypothalamic LH-RH levels associated with the ovarian steroid-induced gonadotropin surge. Acta Endocrinol 92:1-7

Kalra SP, Kalra PS, 1983. Neural regulation of luteinizing hormone secretion in the rat. Endocr Rev, 4:311-51

Kalra SP, Leadem CA, 1984. Control of luteinizing hormone secretion by endogenous opioid peptides. In: Delitala G, Motta M, Serio M (eds.), Opioid Modulation of Endocrine Function. New York: Raven Press, pp. 171-84

Kalra SP, McCann SM, 1973. Effects of drugs modifying catecholamine synthesis on LH release induced by preoptic stimulation in the rat. Endocrinology 93:356-62

Kalra SP, Simpkins JW, 1981. Evidence for noradrenergic mediation of opioid effects on luteinizing hormone secretion. Endocrinology 109:776-82

Kalra SP, Krulich L, McCann SM, 1973. Changes in gonadotropin-releasing factor content in the rat hypothalamus following electrochemical stimulation of anterior hypothalamic area and during the estrous cycle. Neuroendocrinology 12:321-33

Kalra SP, Velasco ME, Sawyer, 1970. Influences of hypothalamic deafferentation on pituitary FSH release and estrogen feedback in immature female parabiotic rats. Neuroendocrinology 6:228-35

Kalra SP, Ajika K, Krulich L, Fawcett CP, Quijada M, McCann SM, 1971. Effects of hypothalamic and preoptic electrochemical stimulation on gonadotropins and prolactin release in proestrous rats. Endocrinology 8:1150-58

Kalra SP, Kalra PS, Chen CL, Clemens JA, 1978. Effect of norepinephrine synthesis inhibitors and a dopamine agonist on hypothalamic LHRH, serum gonadotropins and prolactin levels in gonadal steroid treated rats. Acta Endocrinol 89:1-9

Kalra SP, Allen LG, Clark JT, Crowley WR, Kalra PS, 1986. Neuropeptide Y - an integrator of reproductive and appetitive functions. In: Moody TW (ed.), Neural and Endocrine Peptides and Receptors. New York: Plenum Press, pp. 353-66

Kelly MJ, Ronnekleiv O, Eskay RL, 1982. Immunocytochemical localization of luteinizing hormone-releasing hormone in neurons in the medial basal hypothalamus of the female rat. Exp Brain Res 48:97-06

Kerkerian L, Guy J, Lefevre G, Pelletier G, 1985. Effects of neuropeptide Y (NPY) on the release of anterior pituitary hormones in the rat. Peptides 6:1201-04

Kim K, Ramirez VD, 1985. In vitro luteinizing hormone-releasing hormone release from superfused rat hypothalami: Site of action of progesterone and effect of estrogen priming. Endocrinology 116:252-58

King JC, Tobet SA, Snavely FL, Arimura AA, 1982. LHRH immunopositive cells and their projections to the median eminence and organum vasculosum of the Lamina Terminalis. J Comp Neurol 209:287-00

Kinoshita F, Nakai Y, Katakami H, Kato Y, Yajima H, Imura H, 1980. Effect of -endorphin on pulsatile luteinizing hormone release in conscious castrated rats. Life Sci 27:843-46

Knobil E, 1980. The neuroendocrine control of the menstrual cycle. Rec Progr Horm Res 36:53-88

Köves K, Marton J, Molnar J, Halasz B, 1981. (D-Met^2,Pro^5)-enkephalin-amide-induced blockade of ovulation and its reversal by naloxone in the rat. Neuroendocrinology 32:82-86

Leadem CA, Kalra SP, 1984. Stimulation with estrogen and progesterone of LHRH release from perifused adult female rat hypothalami: correlation with the LH surge. Endocrinology 114:51-56

Leadem CA, Kalra SP, 1985a. Reversal of β-endorphin-induced blockade of ovulation and LH surge with prostaglandin E$_2$. Endocrinology 117: 684-89

Leadem CA, Kalra SP, 1985b. Effects of endogenous opioid peptides and opiates on luteinizing hormone and prolactin secretion in ovariec-tomized rats. Neuroendocrinology 41:342-52

Leadem CA, Crowley WR, Simpkins JW, Kalra SP, 1985. Effects of naloxone on catecholamine and LHRH release from the perifused hypothalamus of the steroid-primed rat. Neuroendocrinology 40:497-00

Levine JE, Ramirez VD, 1980. In vivo release of luteinizing hormone - re-leasing hormone estimated with push-pull cannulae from the mediobasal hypothalami of ovariectomized, steroid-primed rats. Endocrinology 107:1782-90

Levine JE, Ramirez VD, 1982. Luteinizing hormone-releasing hormone re-lease during the rat estrous cycle and after ovariectomy, as estimated with push-pull cannulae. Endocrinology 111:1439-48

Levine J, Norman RL, Gleissman PM, Oijama T, Bangsberg DR, Spies H, 1985. In vivo gonadotropin-releasing hormone release and serum luteinizing hormone measurements in ovariectomized, estrogen-treated rhesus mon-keys. Endocrinology 117:711-21

Matsuo H, Baba Y, Nair RM, Arimura A, Schally AV, 1971. Structure of the porcine LH and FSH releasing hormone. Biochem Biophys Res Comm 43: 134-39

Matteri RL, Moberg GP, 1985. The effect of opioid peptides on ovine pi-tuitary gonadotropin secretion in vitro. Peptides 6:957-63

McDonald J, Lumpkin MD, Samson W, McCann SM, 1985. Neuropeptide Y affects secretion of luteinizing hormone and growth hormone in ovariectomized rats. Proc Natl Acad Sci 82:561-64

McNeilly AS, Sharpe RM, Fraser HH, 1983. Increased sensitivity to the negative feedback effects of testosterone induced by hyperprolactine-mia in the adult male rat. Endocrinology 112:22-28

Moore CR, Price D, 1932. Gonadal hormone functions and the reciprocal in-fluence between gonads and hypophysis. Am J Anat 50:13-71

Morrell J, McGinty F, Pfaff DW, 1985. A subset of β-endorphin- or dynor-phin-containing neurons in the medial basal hypothalamus accumulates estradiol. Neuroendocrinology 41:417-26

Negro-Vilar A, Culler MD, Johnston CA, Nikolics K, Seeburg P, Mastotta C, Valenca MM, 1986. Orchidectomy and hyperprolactinemia induce marked changes in hypothalamic and preoptic LHRH precursor levels. (Abst), 68th Ann Mtg Endocr Soc, Anaheim, CA, p. 150

Pfaff DW, 1980. Estrogens and Brain Function. New York: Springer-Verlag.

Pfeiffer DG, Pfeiffer A, Shimohigashi Y, Merriam GR, Loriaux DL, 1983. Predominant involvement of Mu- rather than delta- or kappa-opiate re-ceptors in LH secretion. Peptides 4:647-49

Piva F, Maggi R, Limonta P. Motta M, Martini L, 1985. Effects of naloxone on luteinizing hormone, follicle-stimulating hormone and prolactin secretion in the different phases of the estrous cycle. Endocrinology 117:766-72

Rothfeld JM, Shivers BD, Hejtmancik JF, Conn PM, Pfaff DW, 1986. Quanti-tation of LHRH mRNA in neurons in the intact and castrate male rat forebrain. (Abst) 16th Ann Mtg Soc Neurosci, Washington, DC, Nov 9-14, p. 3

Sahu A, Kalra SP, Crowley WR, O'Donohue TL, Kalra PS, 1987. Neuropeptide Y levels in microdissected regions of the hypothalamus and in vitro release in response to KCl and prostaglandin E₂: effects of castration. Endocrinology (in press)

Sar M, Stumpf WE 1975. Distribution of androgen-concentrating neurons in rat brain. In: Stumpf WE, Grant LD (eds.), Anatomical Neuroendocrinology. New York: S Karger, pp. 120-33

Sarkar DK, Fink G, 1980. Luteinizing hormone releasing factor in pituitary stalk plasma from long-term ovariectomized rats: effects of steroids. J Endocrinol 86:511-24

Sarkar DK, Yen SSC, 1985. Changes in β-endorphin-like immunoreactivity in pituitary portal blood during the estrous cycle and after ovariectomy in rats. Endocrinology 116:2075-79

Schillo KK, Leshin LS, Kuehl D, Jackson GL, 1985. Simultaneous measurement of luteinizing hormone-releasing hormone and luteinizing hormone during estradiol-induced luteinizing hormone surges in the ovariectomized ewe. Biol Reprod 33:644-52

Sherwood NM, Fink G, 1980. Effect of ovariectomy and adrenalectomy on luteinizing hormone-releasing hormone in pituitary stalk blood from female rats. Endocrinology 106:363-67

Simpkins JW, Kalra PS, Kalra SP, 1980. Temporal alterations in luteinizing hormone-releasing hormone concentrations in several discrete brain regions: effects of estrogen-progesterone and norepinephrine synthesis inhibition. Endocrinology 107:573-77

Sisk CL, Turek FW, 1983. Developmental time course of pubertal and photoperiodic changes in testosterone negative feedback on gonadotropin secretion in the golden hamster. Endocrinology 112:1208-16

Tannebaum G, Patel Y, 1986. On the role of centrally administered somatostatin in the rat. Massive hypersomatostatinemia resulting from leakage into the peripheral circulation has effects on growth hormone secretion and glucoregulation. Endocrinology 118:2137-42

Tatemoto K, 1982. Neuropeptide Y: complete amino acid sequence of the brain peptide. Proc Natl Acad Sci USA 79:5485-89

Wardlaw S, 1986. Regulation of β-endorphin, corticotropin-like intermediate lobe peptide and α-melanotropin-stimulating hormone in the hypothalamus by testosterone. Endocrinology 119:19-24

Wardlaw SC, Thoron L, Frantz AG, 1982. Effects of sex steroids on brain β-endorphin. Brain Res 245:327-31

Watson SJ, Khachaturian H, Akil H, Coy DH, Goldstein A, 1982. Comparison of the distribution of dynorphin system and enkephalin system in brain. Science 218:1134-36

Wilcox JN, Roberts JL, 1985. Estrogen decreases rat hypothalamic proopiomelanocortin messenger ribonucleic acid levels. Endocrinology 117: 2392-96

Wilkinson M, Herden M, Wilson CA, 1982. Gonadal steroid modification of adrenergic and opiate receptor binding in the central nervous system. In: Füxe K (ed.) Steroid Hormone Regulation of Brain. Oxford: Pergamon Press, pp. 253-63

TRANSMEMBRANE SIGNALS AND INTRACELLULAR MESSENGERS MEDIATING

LHRH AND LH SECRETION

A. Negro-Vilar, M.M. Valenca and M.D. Culler

Reproductive Neuroendocrinology Section, LRDT
National Institute of Environmental Health Sciences
National Institutes of Health
Research Triangle Park, NC 27709

INTRODUCTION

The gametogenic and endocrine functions of the gonads are both under the integrative control of the central nervous system. The primary regulatory component in this system is the hypothalamic peptide luteinizing hormone-releasing hormone (LHRH), a decapeptide produced by neurons located in the hypothalamic-preoptic region. After synthesis, LHRH is transported to nerve terminals located in the median eminence, where it is released in close proximity to portal capillaries. The decapeptide is then transported to the anterior pituitary gland through the hypophyseal portal circulation, where, acting on specific membrane receptors, it stimulates LH and FSH secretion. These gonadotropins, in turn, profoundly affect gonadal function.

It is clear from evidence accumulated during the last 15 years that LHRH plays a major modulatory role in the regulation of gonadal function by acting at different levels within the hypothalamic-pituitary-gonadal axis. A better knowledge of the intrinsic regulatory mechanisms that modulate the release and action of LHRH is therefore essential to our understanding of reproductive physiology.

Many studies indicate that feedback interactions play a major role in the regulation of the LHRH-gonadotropin system (Negro-Vilar, 1986; Advis et al., 1980; Kalra and Kalra, 1983). Steroid hormones are known to be important modulators of LHRH and gonadotropin secretion, although the relative contribution of brain and pituitary sites in mediating the effects of the steroid hormones on gonadotropin secretion are still not clearly defined (Kalra and Kalra, 1983; Negro-Vilar and Ojeda, 1981; McEwen et al., 1982; Knobil, 1980). It is clear that gonadal steroids interact with specific cellular receptors at select brain and pituitary sites and, thereby, modify the function of the LHRH neuronal system as well as the responsiveness of the pituitary to the decapeptide. Priming with gonadal steroids results in biphasic alterations in pituitary responsiveness to LHRH; after an initial, inhibitory phase is observed, a period of enhanced responsiveness ensues which will lead, if conditions are appropriate, to a surge of gonadotropin secretion (Negro-Vilar and Ojeda, 1981; Knobil, 1980) and ovulation. Gonadectomy or diminished steroidal output from the testis or ovaries results in an increased release of gonadotropins from the pituitary. The increase in gonadotropin release is believed to be the result of a modification in the LHRH input signal and/or changes in responsiveness of the gonadotrophs. The exact nature of the change in the LHRH input signal, i.e., how the pattern of LHRH secretion into the portal vasculature varies after alterations in the feedback signal are introduced, remains largely unknown (see chapter by Valenca et al., this volume). This deficit in our information is mainly due to the difficulty in collecting sequential blood samples from the hypophyseal portal circulation at a frequency that would allow full characterization of the LHRH secretory pattern under different physiological conditions.

The stimulatory and inhibitory effects of gonadal steroids upon the LHRH neuronal system are thought to be mediated by interactions of the steroids with a variety of central neurotransmitter systems. The best studied among these systems are the central monoaminergic neuronal systems that innervate the hypothalamic-preoptic region; the main areas where LHRH neuronal cell bodies, fibers and terminals are located (Negro-Vilar and Ojeda, 1981; Kalra and Kalra, 1984). Several studies indicate that steroids are concentrated by the monoamine-producing cells, as well as by cells that are contiguous to the LHRH neurons, and that these steroids can profoundly affect monoamine metabolism and neuronal activity (Sar and Stumpf, 1981; McEwen et al., 1982; Morrel et al., 1984). The LHRH neurons, however, do not seem to directly take up estradiol or other steroids.

It is now widely accepted that central noradrenergic (and perhaps also adrenergic) neurons play a major role in regulating the LHRH neuronal system, a concept that is supported by studies from many laboratories involving both animals and human subjects. Interference with normal noradrenergic neuronal activity in both rodents and subhuman primates results in an inhibition of LH release and, under some conditions, in abnormal gonadal function (Negro-Vilar and Ojeda, 1981; Negro-Vilar et al., 1982; Negro-Vilar, 1986; Kalra and Kalra, 1983; Gallo, 1982). Other monoaminergic pathways, including the epinephrine, dopamine, and serotonin neuronal systems, as well as the cholinergic system, have also been implicated in the regulation of LHRH and LH release and have been demonstrated to be modulated by gonadal steroid hormones (Negro-Vilar and Ojeda, 1981; Negro-Vilar et al., 1982; Kalra and Kalra, 1983; Crowley et al., 1978; Morrel et al., 1984).

In recent years it has become clear that endogenous opiate peptides are important modulators of the various hypothalamic peptidergic systems that regulate pituitary function (Meites et al., 1983). In the reproductive area, endogenous opiates have been found to provide a tonic inhibitory tone over gonadotropin secretion (Meites et al., 1983; Kalra and Kalra, 1984). Blockade of opiate receptors results in a rapid rise in LH secretion in both animals and humans. The available evidence suggests that this effect is centrally mediated and that catecholamines may be involved in mediating this response since inhibition of norepinephrine and epinephrine synthesis nullifies the effect of the opiate receptor blocker, naloxone (Kalra and Kalra, 1984).

The examples discussed above represent only a portion of the large number of neurotransmitter and blood-borne chemical signals reaching the LHRH neurons and interacting with specific membrane or intracellular receptors to modify the activity of these neurons and, ultimately, LHRH secretion. It is obvious that an understanding of the intracellular events that mediate the actions of the various neurotransmitters and hormones upon the LHRH neuron is required to unravel the complex interactions between amines, peptides and steroids that are so essential to the integrated function of the hypothalamic-pituitary-gonadal axis.

SIGNAL TRANSDUCTION AND REGULATION OF LHRH SECRETION

After interacting with specific membrane receptors, many hormones, neurotransmitters, growth factors and other biologically active substances generate cellular responses by eliciting signals that are carried from the cell membrane into the cell interior by a variety of different routes. Production of cyclic nucleotides, calcium mobilization, protein kinase C activation and inositol phospholipid breakdown are among the major intracellular events through which signals are transduced and converted into specific cellular responses. Some additional systems also play an important role in mediating hormone secretion, most notably arachidonic acid and the cascade of metabolites generated after its liberation from phospholipids.

Over the last decade our laboratory has been involved in a systematic exploration of a) the specific receptors mediating the actions of amines, peptides and other neurally active substances on LHRH release, and b) the role of specific intracellular messenger systems involved in the secretion of LHRH (Negro-Vilar et al., 1980).

For that purpose we have developed, characterized and made extensive use of an in vitro system which consists of incubating the isolated median eminence (ME) in appropriate medium containing different established or putative neurotransmitters, neuromodulators, their analogs and/or antagonists. Using microdissection procedures described elsewhere (Chiocchio et al., 1976; Negro-Vilar et al., 1979), we were able to obtain a tissue fragment which contained the entire ME and which was free of surrounding structures that contain nerve cell bodies, such as the arcuate and ventromedial nuclei. As previously mentioned, this structure contains mainly axonal fibers and nerve terminals, and therefore the anatomical relationships at that level are preserved in this preparation. Any effects observed under these conditions would, of necessity, have to take place at a presynaptic level and would not reflect effects that may occur at the perikarya of the peptidergic neuron or of an interneuron. Presynaptic regulation of neurotransmitter release has been described for the adrenergic system both in peripheral and central neurons (Langer, 1978) and appears to be an important site of regulatory control. From ultrastructural and neurochemical studies, it seems logical to infer that similar neurochemical regulatory mechanisms may be operative at the presynaptic peptidergic nerve terminals in the ME. A schematic model of this system is presented in Fig. 1.

Work from our laboratory presented the first evidence that norepinephrine (NE) could directly stimulate LHRH release from rat ME nerve terminals incubated in vitro (Negro-Vilar et al., 1979). Dopamine (DA) was also found to have a facilitatory effect on LHRH release under similar conditions (Negro-Vilar et al., 1979). The effect of NE is mediated by specific adrenergic receptors since phentolamine, an α-adrenergic blocker, nullified the action of NE (Negro-Vilar et al., 1979; Ojeda et al., 1982). A recent study of the distribution of α_2-agonist binding sites in brain using ^{3}H-para-aminoclonidine revealed the topographical distribution of these receptors within the hypothalamus. Moderate to high levels of specific binding were associated with the arcuate nucleus and the ME, as well as with a number of hypothalamic and preoptic structures known to be involved in the regulation of LHRH release (Unnerstall et al., 1984). Recent studies in our laboratory using the α_2-receptor agonist, para-aminoclonidine, suggest that activation of α_2-receptors can increase secretion of LHRH in vitro (Valenca and Negro-Vilar, unpublished). Others have reported that activation of α_1-receptors also enhances the release of LHRH from ME's in vitro (Heaulme and Dray, 1984). Alpha$_1$-adrenergic receptors are usually negatively coupled to adenylate cyclase (Yamura et

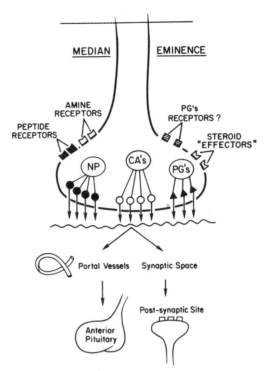

Fig. 1. Schematic diagram representing an axon terminal from a neuron projecting to the median eminence. The secretory product of this neuron may be, for example, a neuropeptide (NP), one of the catecholamines (CA's) and/or one of the prostaglandins (PG's). The secretory products may be released and transported via the portal vessels to the anterior pituitary (as in the case of many neural peptides and of dopamine) or released to act at a post-synaptic site on another neuronal terminal. Several peptide and amine receptors depicted on the left have been described and characterized. PG's receptors in the ME have not yet been described, although PG's actions at this level are well known (see text). Steroid "effectors" represent possible sites at the membrane level (not necessarily classical receptors) at which steroids may act to elicit rapid and local responses not involving genomic activation. From Negro-Vilar, A., 1982.

al., 1977; Michel and Lefkowitz, 1982) and positively coupled to guanylate cyclase (Jaiswal and Sharma, 1986). Alpha$_1$-receptors, on the other hand, activate Ca^{2+} gating through voltage-sensitive Ca^{2+} channels (Exton, 1982) and also enhance inositol phospholipid turnover (Michell, 1975; Berridge, 1981). Interaction of a single secretagogue (NE) with these two receptors alone will potentially generate signals involving all the major intracellular messenger systems described above. We will discuss below studies from our own laboratory as well as those of others that clearly indicate the involvement of these intracellular messenger systems in the release of LHRH.

CALCIUM AS A SECOND MESSENGER IN THE SECRETION OF LHRH

A number of reports have appeared indicating that depolarization of hypothalamic cells by high concentrations of K^+ results in an enhanced output of LHRH in vitro (Bigdeli and Snyder, 1978; Drouva et al., 1981; Gallardo and Ramirez, 1977; Kao and Weisz, 1977; Rotsztejn et al., 1976). The effects of K^+ can be prevented by omitting Ca^{2+} from the incubation medium and/or by the addition of chelating agents such as EGTA (Bigdeli and Snyder, 1978; Rotsztejn et al., 1976). The above observations using large hypothalamic fragments were corroborated and expanded by studies in which we analyzed in detail the role of Ca^{2+} in LHRH secretion from ME nerve terminals incubated in vitro (Ojeda and Negro-Vilar, 1984; Ojeda and Negro-Vilar, 1985). Incubation of ME tissues in the presence of a Ca^{2+} ionophore, A23187, resulted in a concentration-related increase in LHRH release. The effect of A23187 was dependent on the concentration of Ca^{2+} in the incubation medium. Similarly, varying the concentration of Ca^{2+} in the medium affected the ability of high K^+ to release LHRH (Ojeda and Negro-Vilar, 1984). The presence of chelating agents (EGTA, EDTA) in the medium completely prevented the release of LHRH induced by a maximal concentration of K^+. Verapamil, a calcium-entry blocker, decreased the effect of the A23187 ionophore in a concentration-dependent manner. In subsequent studies we examined the efflux of Ca^{2+} by preloading the nerve terminals in the ME with $^{45}Ca^{2+}$ and placing the tissues in perifusion chambers. The use of this perifusion system allowed us to examine the dynamics of $^{45}Ca^{2+}$ efflux. Application of a depolarizing pulse of K^+ (56mM) produced a rapid, very marked increase in $^{45}Ca^{2+}$ efflux with a pulse-like configuration. Release of LHRH was also rapidly (within 1 min) and greatly (4-fold) enhanced by K^+. The dynamics and profile of the two events indicates a close relationship between Ca^{2+} release and peptide secretion in this system. In additional experiments, depletion of intracellular Ca^{2+} stores was attempted using a procedure described by Blaustein et al. (1980) which consisted of incubating the tissues (ME's) in Ca^{2+}-free, EGTA containing media and stimulating repeatedly with the ionophore A23187. Under these conditions, the release of LHRH from the nerve terminals, elicited by a specific secretagogue like prostaglandin E_2 (see below) was completely obliterated, indicating that normal intracellular Ca^{2+} stores are essential for a secretory response (Ojeda and Negro-Vilar, 1985).

From the above studies it can be concluded that Ca^{2+} is an important intracellular messenger involved in the secretion of LHRH. Linkage of a specific receptor to the activation of intracellular Ca^{2+} release, however, remains to be established in this system. Because the populations of nerve terminals present in the ME are heterogeneous, it has not been possible, to date, to obtain direct confirmation of the presence of a specific receptor type in identified LHRH nerve terminals.

ROLE OF ARACHIDONIC ACID AND ITS METABOLITES ON THE RELEASE OF LHRH

Arachidonic acid, an important constituent of biological membranes, is almost invariably esterified in the 2-position of several cellular phospholipids. The liberation of arachidonic acid from these sites is usually mediated by the action of phospholipases, particularly phospholipase A_2 (Flower and Blackwell, 1976), in a Ca^{2+}-dependent reaction. The release of arachidonic acid may also occur as a consequence of the sequential action of phospholipase C and diacylglycerol lipase (Bell et al., 1979). In many systems, activation of phospholipase C (a phosphodiesterase that cleaves the inositol phosphates from the glycerol backbone) initiates the breakdown of polyphosphoinositides into different inositols and diacylglycerol.

Upon release, arachidonic acid is oxidized either through lipoxygenase to hydroperoxides (HPETE'S) (Borgeat, 1981; Hammarström, 1982) or through cyclooxygenase to an endoperoxide (Samuelsson et al., 1978). Both kinds of intermediates are precursors to a variety of compounds with potent physiological and pharmacological properties (Samuelsson 1979-80; Borgeat, 1981; Hammarström, 1982). The products derived through the cyclooxygenase pathway are prostaglandins (PG), prostacyclins and thrombo-xanes (Samuelsson, 1979–80). The series of products originating from the lipoxygenase pathway includes hydroxyeicosatetraenoic acid (HETE's) and leukotrienes (Samuelsson et al., 1978; Samuelsson, 1980). In addition to these two major metabolic routes, recent work has described the existence of a new pathway in the kidney and liver which oxidizes arachidonic acid through the catalytic action of microsomal cytochrome P–450 (Capdevila et al., 1981a; Capdevila et al., 1981b). Cytochrome P–450 actively catalyzes the NADPH–dependent oxidation of arachidonic acid to form epoxyeico-satrienoic acid derivatives (EET's). We have recently reported that this pathway also operates in the hypothalamus (Capdevila et al., 1983) and in the neural lobe of the pituitary gland (Negro–Vilar et al., 1985). Products from this novel pathway (EET's) have been found to be active secretagogues for several brain peptides including somatostatin (Capdevila et al., 1983), vasopressin and oxytocin (Negro–Vilar et al., 1985). The release of LHRH, however, is only marginally stimulated by the different EET's tested.

Addition of arachidonic acid to our in vitro ME incubation system elicits a significant increase in the release of LHRH. This effect can be completely prevented by the addition of 5,8,11,14–eicosatetraynoic acid (ETYA), an inhibitor of all pathways of arachidonic acid metabolism, or, more selectively, by the cyclooxygenase inhibitor, indomethacin. The latter suggests that the main stimulatory effect of arachidonic acid is mediated by its metabolism through the cyclooxygenase pathway. Indeed, we have presented clear evidence to suggest that PGE_2 is a specific secretagogue for LHRH (Ojeda et al., 1979; Negro–Vilar et al., 1980). Addition of increasing concentrations of PGE_2 stimulated LHRH release in a dose–dependent manner. Of all the prostaglandins tested, PGE_2 was the most effective secretagogue for LHRH. The effect of PGE_2 was found to be only partially dependent on the availability of extracellular Ca^{2+} (Ojeda and Negro–Vilar, 1984) and was also, at least, partly dependent on the ability of PGE_2 to mobilize intracellular Ca^{2+} (Ojeda and Negro–Vilar, 1985). The evidence from our in vitro studies indicated that stepwise decreases in the concentration of Ca^{2+} in the medium progressively reduced the LHRH secretory response to PGE_2 stimulation. Neither Ca^{2+}–free medium nor the addition of the chelating agent EGTA, however, completely suppressed PGE_2 –induced LHRH release. Depletion of intracellular Ca^{2+} stores by repeated stimulation with the ionophore A23187 in Ca^{2+} –free, EGTA containing medium, as described above, completely suppressed the response to PGE_2, suggesting that normal concentrations of intracellular Ca^{2+} in the nerve terminal are required for PGE_2 to be an effective secretagogue for LHRH. PGE_2 was also shown to stimulate Ca^{2+} efflux in temporal correlation with the release of LHRH. Taken together, these data suggest that Ca^{2+} is required for the stimulatory effect of PGE_2 to occur, and that translocation of Ca^{2+} from intracellular stores is an important event associated with the action of PGE_2 (Ojeda and Negro–Vilar, 1985).

The release of LHRH and PGE seems to be intrinsically linked in many pharmacological and physiological paradigms that we have explored over the years. Norepinephrone stimulates both LHRH and PGE_2 release from the ME, the latter being a putative intracellular messenger mediating the action of NE on LHRH secretion. The evidence to support this concept includes: a) a concomitant and parallel increase in LHRH and PGE_2 release upon stimulation with NE; b) the dependency of these effects on the activation of α–adrenergic receptors (Negro–Vilar et al., 1979; Ojeda et al., 1979); and c) the complete inhibition of the effect of NE on LHRH secretion if PG's synthesis is inhibited either in vivo or in vitro (Ojeda et al., 1979).

Further support for the role of PGE_2 as an important intracellular messenger for the action of NE on the LHRH neuron is afforded by studies exploring the $NE-PGE_2-LHRH$ pathway in different physiological situations. In particular, we evaluated the sensitivity of ME nerve terminals to NE and PGE_2 stimulation at different times of the day during each phase of the estrous cycle of the rat, and observed clear changes in the responsiveness of the nerve terminals to both NE and PGE_2 on the day of proestrus, preceding the predicted time of the preovulatory LHRH surge (DePaolo et al., 1982). Moreover, our studies show that in vivo treatment with pentobarbital (nembutal), which has been shown to block the LHRH/LH surge, results in a complete inhibition of the release of PGE_2 and LHRH in response to NE stimulation in vitro, providing a neurochemical and molecular substrate to explain the action of the drug (DePaolo et al., 1982).

Very recent studies in our laboratory link the inhibitory effects of the opiates to the $NE-PGE_2-LHRH$ pathway. In vitro treatment with the opiate receptor blocker, naloxone, results in an enhanced release of LHRH concomitantly with an increased output of PGE_2. The effect of naloxone can be blocked by the addition of the α-adrenergic receptor blocker, phentolamine (Valenca, Masotto and Negro-Vilar, submitted), suggesting the activation of α-receptors by endogenously released NE (Leadem et al., 1985). Moreover, the effect of naloxone can also be blocked by inhibiting prostaglandin synthesis with indomethacin, indicating that the action of the opiate receptor blocker is mediated by a $NE-PGE_2-LHRH$ interaction. Tissues obtained from orchidectomized animals failed to respond to naloxone treatment in vitro, suggesting that the steroids play a modulatory role in the opiate-LHRH connection. A schematic diagram depicting these interactions is presented in Fig. 2.

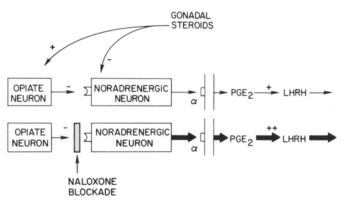

Fig. 2. Schematic diagram depicting opiate-norepinephrine-PGE_2 -LHRH interaction. Blockade of opiate receptors by naloxone removes the tonic inhibitory tone exerted by opiate neurons and increases neuronal activity in noradrenergic neurons. Increased NE release, acting through α-adrenergic receptors, enhances prostaglandin E_2 formation and, in turn, increases LHRH secretion. Gonadal steroids play a modulatory role in this interaction.

The fact that the full secretory LHRH response elicited by arachidonic acid can be prevented by the cyclooxygenase inhibitor, indomethacin, indicates that this is the major pathway from which the stimulatory arachidonate metabolites (PG's) are derived. Some reports have indicated, however, that products of the lipoxygenase pathway, particularly leukotriene C_4 (Gerozissis et al., 1986) and 5- and 12-HETE (Gerozissis et al., 1985) can stimulate LHRH release. We have obtained stimulation of LHRH release using high concentrations (10^{-6} M) of 12-HETE (Negro-Vilar et al., 1986); however, lower concentrations of 12-HETE as well as 5- and 15-HETE, at any concentration, failed to produce consistent or significant increments in LHRH release. Very recently we have obtained evidence suggesting that a metabolite of arachidonic acid derived from the lipoxygenase pathway may be an inhibitor of LHRH release upon stimulation of the nerve terminals with a synthetic diacylglycerol (Valenca et al., 1985; Negro-Vilar et al., 1986). The implications of these findings will be discussed in more detail in the section dealing with diacylglycerol/protein kinase C pathways (see below).

The overall conclusion from these studies is that different arachidonic acid metabolites play a selective role in the events leading to LHRH secretion after agonist-induced receptor activation. A link between α-adrenergic receptor activation, PGE_2 generation and LHRH release seems to be firmly established.

ROLE OF SIGNAL TRANSDUCTION BY DIACYLGLYCEROL-PROTEIN KINASE C IN LHRH SECRETION

As indicated above, it is now apparent that in numerous hormone-mediated control systems the transfer of signals from the cell membrane to the cell interior is effected by mechanisms that lead to Ca^{2+} mobilization and activation of a specific kinase, protein kinase C (Nishizuka, 1984). Receptors in this system are coupled to a specific enzyme, phospholipase C, which catalyzes the hydrolysis of membrane inositol phospholipids thereby generating two intracellular messengers, inositol triphosphate, involved in intracellular Ca^{2+} mobilization, and diacylglycerol, responsible for the activation of protein kinase C. Coupling of the receptor to phospholipase C is thought to involve one or more G proteins (Williamson, 1986). Protein kinase C is a calcium-activated, phospholipid-dependent protein kinase, which can either be associated or not associated with membrane phospholipids. In the non-associated form, the enzyme has very low activity and is relatively insensitive to Ca^{2+}. Upon activation by diacylglycerol and in the presence of phospholipids (phosphatidyl-serine, phosphatidylcholine, etc.) the enzyme associates with the membrane, becomes exquisitively sensitive to Ca^{2+} and its activity increases many fold (Nishizuka, 1984).

In a variety of tissues, stimulation of α-adrenergic receptors leads to enhanced inositol phospholipid turnover, which in turn sets in motion the intracellular cascade of events leading to Ca^{2+} mobilization, protein kinase C activation and hormone secretion (Nishizuka, 1984). Since activation of α-adrenergic receptors in the ME leads to secretion of LHRH, as discussed above, it was of importance to determine whether, in this system, the diacylglycerol-protein kinase system is an active intracellular messenger pathway involved in LHRH secretion.

Using the in vitro ME nerve terminal preparation, we evaluated the effects of protein kinase C activation on LHRH release with three different probes, diacylglycerol, a phorbol ester and phospholipase C. A synthetic diacylglycerol, 1,2 didecanoylglycerol (DiC_{10}), which has a shortened fatty acid chain that enables it to penetrate the cell membrane, was employed to mimic the effects of endogenous diacylglycerol. This compound has been found to be effective in releasing LH from dispersed pituitary cells (Negro-Vilar and Lapetina, 1985). A phorbol ester, 12,13-dibutyrate (PDBu), was used to directly activate protein kinase C (Castagna et al., 1982). Phospholipase C (from Clostridium perfringens) has been shown to be a potent secretagogue in many other systems, presumably acting in a similar fashion to the endogenous, membrane-bound enzyme.

Since activation of this system also leads to the liberation of arachidonic acid and to the subsequent formation of arachidonate products via the three main pathways described above, we monitored, in conjunction with the secretion of LHRH, the changes in PGE_2 release that might occur the different treatments.

In an initial study (Valenca et al., 1985), we presented the first evidence that activation of protein kinase C by either a phorbol ester (PDBu) or a synthetic diacylglycerol (DiC_{10}) resulted in stimulation of LHRH secretion. The stimulatory effect of DiC_{10} on LHRH secretion required the presence of the lipoxygenase inhibitor, nordihydroguaiaretic acid (NDGA) in the medium, suggesting that a compound derived from the lipoxygenase pathway may be inhibitory to the secretion of LHRH (Valenca et al., 1985). In a recent report, the release of LHRH from the hypothalamus of immature female rats was also found to be stimulated by activation of protein kinase C with phorbol ester or diacylglycerol (Ojeda et al., 1986).

In a more recent and comprehensive study we expanded these observations (Negro-Vilar et al., 1986) and explored in greater detail the potential role of arachidonate products in the secretory response elicited by protein kinase C activation. Our observations indicate that synthetic diacylglycerol (DiC_{10}) significantly enhanced PGE_2 release in a concentration-dependent fashion. Blockade of phospholipase A_2 activity with quinacrine completely prevented the DiC_{10}-induced release of PGE_2. DiC_{10} did not affect LHRH release unless the lipoxygenase inhibitor, NDGA, was added to the medium. Under those conditions, the DiC_{10} produced a concentration-related increase in LHRH release of several fold over baseline (Fig. 3, left panels). Phospholipase C enhanced the secretion of both LHRH and PGE (Fig. 4). Here again, blockade of the lipoxygenase pathway with NDGA enhanced the release of LHRH by phospholipase C without affecting the stimulated secretion of PGE_2. The phorbol ester PDBu also stimulated LHRH secretion in a concentration-dependent manner with only a marginal effect on PGE release (Fig. 3, right panels). The presence of NDGA did not modify the action of PDBu on either LHRH or PGE secretion.

The above data suggest that the phorbol ester-induced response may reflect primarily protein kinase C activation, whereas the effects of diacylglycerol and phospholipase C may, in addition to the protein kinase C activation, reflect other actions of these compounds, particularly involving arachidonate metabolites. In order to clarify this issue further, we performed experiments to evaluate the LHRH releasing activity of all three compounds, phorbol ester, diacylglycerol and phospholipase C, in a situation where metabolism of arachidonic acid to different products would be impeded, i.e., in the presence of indomethacin and/or NDGA. DiC_{10} significantly enhanced LHRH release in the presence of both indomethacin and NDGA (Fig. 5). PDBu-induced secretion of LHRH was unaffected by blockade of arachidonic acid metabolism (Fig. 6). Stimulation of LHRH by phospholipase C was unaffected by blockade of the cyclooxygenase pathway and was enhanced in the presence of NDGA, as described earlier (Fig. 4).

Taken together, the results of these experiments suggest that activation of protein kinase C can stimulate LHRH secretion from nerve terminals in vitro and, further, that diacylglycerol may represent an important intracellular messenger participating in the events leading to LHRH secretion. In addition, stimulation with diacylglycerol or phospholipase C resulted in activation of arachidonic acid metabolism.

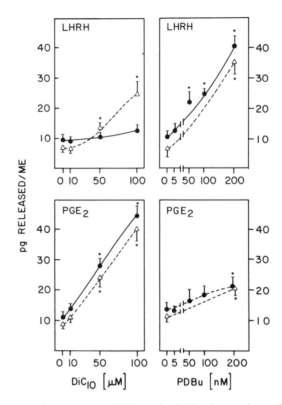

Fig. 3. <u>In vitro</u> release of LHRH and PGE from hypothalamic median eminence (ME) nerve terminals, as stimulated by synthetic 1,2-diacylglycerol (DiC$_{10}$) or phorbol 12,13-dibutyrate (PDBu), during a 30 min incubation in Krebs Ringer bicarbonate glucose buffer, either alone (● —— ●) or containing 10 μM NDGA (Δ----Δ). Both LHRH and PGE$_2$ were measured by specific RIAs.

The observations reported in that study (Negro-Vilar et al., 1986) uncovered a component of the arachidonic acid cascade which appears to be inhibitory to LHRH secretion and which seems to be derived from the lipo-xygenase pathway. Coupled with the well known stimulatory pathway (PGE$_2$), the evidence suggests a dual stimulatory and inhibitory role for arachidonate metabolites on LHRH secretion. This dual system may be part of a complex feedback system that is involved in the modulation of the secretory response. Other examples have been reported in the literature to support this concept. In platelets and neutrophils, the signal-induced inositol phospholipid turnover elicits a series of cellular responses

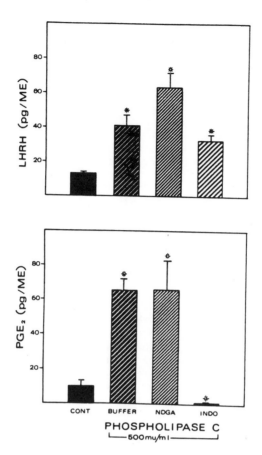

Fig. 4. Effects of specific blockers of arachidonic acid metabolism on phospholipase C-induced LHRH and PGE$_2$ release. Indomethacin (INDO), a cyclooxygenase inhibitor, and NDGA, a lipoxygenase inhibitor, and NDGA, a lipoxygenase inhibitor, were both used at a concentration of 10 μM. *, $P < 0.05$ vs. control (CONT). From Negro-Vilar et al., 1986.

which can be blocked by prostaglandins and cyclic AMP (Takai et al., 1981; Kaibuchi et al., 1982). These feedback effects of arachidonic acid metabolites can also be stimulatory, as has been shown in the case of thromboxanes and platelet activation (MacIntyre et al., 1983). Moreover, the PG—cAMP system is not always inhibitory to events stimulated by inositol phospholipid breakdown (Kaibuchi et al., 1982). Therefore, it seems obvious that in each particular cell a series of positive and negative intracellular feedback systems may be operative, utilizing different arachidonic acid metabolites and/or other second messenger systems. In the case of the LHRH neuron, it is clear that PGE$_2$ is an important positive signal in the secretory process, and that another, probably lipoxygenase-derived arachidonoyl metabolite may play an important negative feedback role to modulate the secretion of this neural peptide.

ROLE OF CYCLIC NUCLEOTIDES IN LHRH RELEASE

Despite early observations that cAMP and adenylate cyclase are present in measurable concentrations in hypothalamic tissue (Kant et al., 1981; Naor et al., 1979) and that the levels as well as the activity of these factors can be modified by sex steroids, gonadectomy, phase of the estrous cycle and other physiological or pharmacological manipulations, the exact role of cyclic nucleotides on LHRH secretion has not been the subject of many studies. Kim and Ramirez (1985) provided the first comprehensive analysis of the role of cAMP on LHRH secretion from the rat median eminence. Using tissues obtained from immature rats, these authors explored the effects of dibutyryl cyclic AMP (dbcAMP) on the release of LHRH. LHRH secretion was stimulated by dbcAMP and the responsiveness of the tissues were modified, in the case of female rats, by estradiol priming. Intermittent pulses of dbcAMP stimulated LHRH release from superfused ME's of ovariectomized-estrogen primed rats. The episodes of LHRH secretion in response to discrete pulses of dbcAMP are consistent with the characteristic rapid on-and-off type of response that cAMP displays when it mediates a sustained secretory response. Additional studies employing forskolin support the idea that generation of cAMP leads to the release of LHRH from ME's of immature female rats (Ojeda et al., 1985). The stimulation of LHRH secretion by cAMP was not accompanied by increased release of PGE_2.

No secretagogue has yet been shown to specifically stimulate cAMP production in LHRH neurons, nor is there direct evidence for what type of receptors may be linked to this second messenger system in these cells.

INTRACELLULAR MESSENGERS MEDIATING LH SECRETION

As discussed earlier in this chapter, LHRH is the primary regulator of LH secretion. Numerous studies have been published describing the intracellular events that occur after LHRH binds to its specific receptor on the gonadotroph cell membrane. In an excellent recent review, Conn (1986) describes in detail the molecular basis for LHRH action on the gonadotroph. Ionic calcium has clearly been established as a second messenger for the action of LHRH. In the absence of Ca^{2+} or in the presence of EGTA, LHRH-induced LH release is completely suppressed even though binding of LHRH to its receptors still occurs. Calmodulin has been shown to be present in membrane patches containing the LHRH receptor (Jennes et al., 1985), and it appears to be an intracellular receptor for Ca^{2+}, mobilized by LHRH stimulation (Conn et al., 1981). Inositol phospholipid breakdown has also been shown to be part of the early action of LHRH (Snyder and Bleasdale, 1982; Raymond et al., 1984). Within seconds after LHRH stimulation, there is an immediate incorporation of $^{32}PO_4$ into phosphatidyl inositol (PI) and phosphatidic acid (PA), with subsequent, rapid changes in the pools of labeled PI, PI-4-phosphate and PI-4,5-bisphosphate (Andrews and Conn, 1986). Additional observations suggest that the turnover of inositol phospholipids in response to LHRH stimulation may be uncoupled from the Ca^{2+}-calmodulin-mediated pathway, stimulated by LHRH, to enhance LH release (Huckle and Conn, 1985).

Activation of protein kinase C may also be involved in the LH secretory process. Both diacylglycerol (Negro-Vilar and Lapetina, 1985; Conn et al., 1985) and phorbol ester (Negro-Vilar and Lapetina, 1985; Smith and Vale, 1980) have been shown to stimulate LH secretion. The stimulatory activity of diacylglycerols and phorbol ester is time- and concentration-dependent and has been shown to include all pituitary hormones (Negro-Vilar and Lapetina, 1985), suggesting that this is a general mechanism for secretagogue-activated pituitary hormone secretion.

As in many other hormonally-controlled tissues, arachidonic acid and its metabolites play a role in the regulation of anterior pituitary hormone secretion. Early studies indicated that LHRH released arachidonic acid from gonadotroph phospholipids (Naor and Catt, 1981). Moreover, inhibition of arachidonic acid release, by blocking phospholipase A_2 activity, prevents LHRH-stimulated LH secretion (Naor and Catt, 1981).

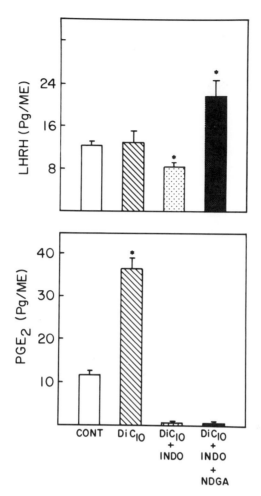

Fig. 5. Effects of specific blockers of arachidonic acid metabolism on diacylglycerol-induced LHRH and PGE_2 release. Indomethacin (INDO), a cyclooxygenase inhibitor, and NDGA, a lipoxygenase inhibitor, were both added at a concentration of 10 µM. DiC_{10} was added at a concentration of 100 µM. *, $P < 0.05$ vs. control (CONT). From Negro-Vilar et al., 1986.

Arachidonic acid promotes LH release when added to pituitary cells in vitro (Naor and Catt, 1981; Berridge, 1981; Snyder et al., 1983). Early reports clearly established that prostaglandins are not involved in stimulating the release of LH (Chobsieng et al., 1975; Ojeda et al., 1979). On the other hand, the lipoxygenase pathway has been reported to modulate, at least in part, the action of LHRH on pituitary gonadotrophs (Naor et al., 1983; Snyder et al., 1983). Naor et al. (1983) have shown that 5-HETE is particularly active in promoting LH secretion. Blockade of lipoxygenase activity with NDGA or other blockers greatly reduces LHRH-stimulated LH secretion (Naor et al., 1983; Snyder et al., 1983). In an interesting study, Snyder et al. (1983) reported that 5,6-epoxyicosatri-enoic acid, an arachidonate product derived from the NADPH-supported cytochrome P-450-dependent epoxygenase pathway (Capdevila et al., 1981a; Capdevila et al., 1983), stimulated LH secretion with a potency at least equal to that of LHRH. These series of studies suggest that arachidonic acid and its lipoxygenase- and epoxygenase-derived metabolites might be a component of the cascade of events initiated by LHRH to promote LH secretion.

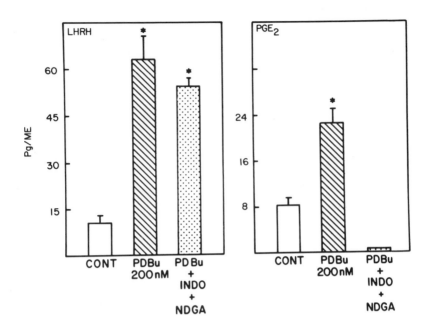

Fig. 6. Effects of specific arachidonic acid metabolism inhibitors on phorbol ester-stimulated LHRH and PGE_2 release. Indomethacin (INDO), a cyclooxygenase inhibitor, and NDGA, a lipoxygenase inhibitor, were both added at a concentration of 10 μM. PDBu was added at a concentration of 200 nM. CONT, control. From Negro-Vilar et al., 1986.

Several studies have eliminated a role for the major cyclic nucleotides (cAMP, cGMP) in the action of LHRH on gonadotropin release (Naor et al., 1978; Conn et al., 1979). This is consistent with the above observations suggesting that the major, if not all, actions of LHRH are involved with Ca^+ mobilization and PI turnover and are independent of cAMP generation.

STUDIES OF THE INTRACELLULAR MESSENGER REGULATION OF PULSATILE LH SECRETION

During the last decade, it has become firmly established that all of the anterior pituitary hormones, as well as many extrapituitary hormones, are secreted in a rhythmic, pulsatile manner. These pulses are believed to be caused by the pulsatile secretion of the various pituitary-regulating peptides from the hypothalamus. In the instance of pulsatile LH secretion, the responsible stimulatory, hypothalamic secretagogue has been clearly demonstrated to be LHRH (Ellis et al., 1983; Culler and Negro-Vilar, 1986).

The significance of a pulsatile pattern of stimulation has been elucidated by studies demonstrating that a continuous stimulation, or an incorrect pulsatile pattern of stimulation, will result in the target cells becoming refractory to further stimulation, a phenomenon known as desensitization (Knobil, 1980; Smith and Vale, 1981; Negro-Vilar and Culler, 1986). Conversely, administration of a proper pulsatile pattern of stimulation will maintain or even enhance target cell responsiveness (Knobil, 1980; Negro-Vilar and Culler, 1986). Despite the recognized importance of pulsatile stimulation, very little is known about how the characteristics of a pulsatile signal affect the final target cell response, and almost nothing is known about the intracellular pathways that respond to and/or require a pulsatile stimulation. The major reason for this lack of information has been the limitation of conventional perifusion systems in producing a pulsatile signal. For this reason, as well as for convenience and methodological logistics, most studies of intracellular regulators of LH secretion have relied on static incubation or cultured cell models in which the cells receive a constant exposure to the stimulatory substance. Recently, however, a computer-assisted perifusion apparatus was developed, which can control and/or manipulate all parameters of a pulsatile hormone signal. Using this system, we have designed an LHRH pulse which rises from zero to peak concentration in 30 secs and then decays exponentially, according to the perifusion flow rate (Culler et al., 1986; Negro-Vilar and Culler, 1986). When perifused hemipituitaries, taken from intact male rats, are exposed to this LHRH signal design, the result is an LH secretory response which closely mimics the shape of LH pulses that are observed in vivo (Culler and Negro-Vilar, 1985; Negro-Vilar and Culler, 1986). Through experimentation to determine the proper pattern (frequency and amplitude) of pulse administration, we have found that LHRH pulses of 10 ng total mass, delivered once each 40 min, can maintain the responsiveness of the perifused hemipituitaries (Negro-Vilar and Culler, 1986). Under these conditions, the first two pulses of LHRH delivered consistently stimulate less LH secretion than the subsequent pulses. This may represent a "priming" action of LHRH, in which the pituitary initially must be sensitized to the peptide before establishing a stable pattern of release (Culler et al., 1986; Negro-Vilar and Culler, 1986).

In order to utilize this system to study the intracellular messengers mediating LH secretion in response to a dynamic pulsatile LHRH stimulation, we designed an experimental model (illustrated diagramatically in Fig. 7) in which perifused hemipituitaries from intact male rats are exposed to three 10 ng total mass "priming" pulses of LHRH followed by a 50 ng total mass LHRH pulse of similar design. After 100 min, to allow time for the resultant rise in LH secretion to return to baseline levels, a second 50 ng total mass pulse is delivered, either alone or concomitantly with a constant infusion of a given drug or inhibitor known to affect a specific intracellular messenger system. The drug/inhibitor infusion is initiated 30 mins prior to the generation of the second pulse. The dynamics of LH secretion, in response to the 50 ng LHRH pulses, are examined in detail by collecting 30 sec fractions of perifusate. A typical LH secretion response to this regimen, in the absence of any intracellular messenger affecting drug, is illustrated in Fig. 8. Both the LHRH signal and the LH response were measured in the perifusate fractions by radioimmunoassay. The "priming" effect of the three 10 ng LHRH pulses, which serve to sensitize and stabilize the LH secretory response in preparation for the later large pulses, can clearly be observed. As revealed by the frequent sampling procedure, a sharp rise in LH secretion, which would reach 300% baseline within 10 min, was evident within the first fraction in which the LHRH signal was detected (Fig. 9). Although the LHRH pulse was completely washed out of the perifusion chamber within 9 min of initiation, the LH secretory response was greatly protracted, requiring 70-80 min to fully return to baseline levels (Fig. 9). During the gradual decay of the LH response to baseline levels, LH appeared to be secreted in episodic waves (Culler et al., 1986) (Fig. 8, 9).

The LH secretory response to the second 50 ng total mass pulse (approximately 1.72 µg rLH) was not significantly different in magnitude or shape as compared with the LH response to the first 50 ng LHRH pulse (approximated 1.65 µg rLH). To further demonstrate the validity of the test model, trials were conducted in which the hemipituitaries were perifused during the same time frame and with the same fraction collection intervals but were exposed either to only the three 10 ng LHRH "priming" pulses (no large LHRH pulses) or to medium alone (no LHRH). In both

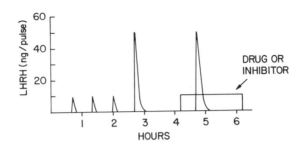

Fig. 7. Experimental design: Perifused hemipituitaries were first sensitized with 3 "priming" LHRH pulses of 10 ng total mass followed by two "test" pulses of 50 ng LHRH total mass. For certain experiments EGTA or NDGA were added to the perifusion medium 30 min prior to the second "test" pulse. Both LH and LHRH were measured in 0.5 min perifusate fractions.

instances, LH secretion, during the time intervals in which the 50 ng LHRH pulses would have normally been delivered, was characterized by a stable baseline pattern of secretion (data not shown) (Valenca et al., 1986).

As discussed earlier for LHRH secretion, the receptor-activated enzyme, phospholipase C, liberates two putative intracellular messengers, inositol triphosphate, which initiates mobilization of intracellular Ca^+, and diacylglycerol, which activates protein kinase C. As a requirement for a substance to be considered as an intracellular mediator of an extra-cellular primary messenger, in this instance LHRH, the putative intra-cellular messenger should be able to induce the same cellular response as the primary messenger. Computer-assisted perifusion, in combination with the described model for observing secretion dynamics, is an ideal system to examine this question because of the ability to visualize both the magnitude and shape of the responses.

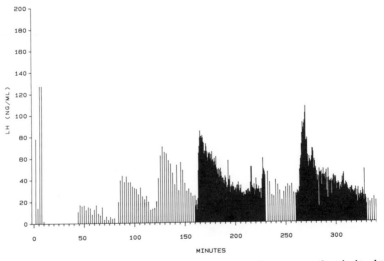

Fig. 8. The _in vitro_ LH secretory pattern from a hemipituitary perifused using the experimental design shown above in Fig. 7. LH was measured by RIA in 0.5 min perifusate fractions.

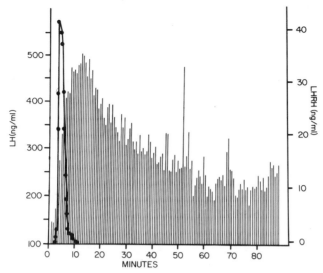

Fig. 9. Effect of a single, computer-designed, LHRH pulse of 50 ng total mass on LH secretion from a hemipituitary _in_ _vitro_ using computer-assisted perifusion. The 50 ng LHRH "test" pulse was delivered after three 10 ng total mass "priming" pulses of LHRH. LH and LHRH were measured by RIA in 0.5 min perifusate fractions.

In order to compare the LH secretory response induced by phospholipase C with that induced by the 50 ng LHRH pulses, a similarly designed 415 mU total content pulse of phospholipase C was delivered to perifused hemipituitaries after the three 10 ng total mass "priming" pulses of LHRH. As shown in Fig. 10, the pulse of phospholipase C stimulated a rapid rise in LH secretion similar to that induced by LHRH in both time sequence and shape. Like the LHRH-induced LH secretory response, the gradual decay of the LH response to phospholipase C was composed of small episodic waves of LH secretion, which continued long past the calculated washout time of the phospholipase C pulse (Valenca et al., 1986) (Fig. 10). The

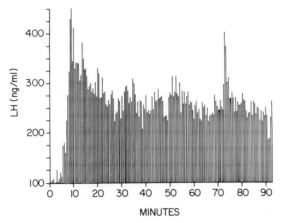

Fig. 10. Effect of a single, computer-designed, 415 mU phospholipase C (PL-C) pulse on LH secretion from a hemipituitary _in_ _vitro_ using computer-assisted perifusion. The phospholipase C was delivered in a pulse of similar design to the LHRH "test" pulse shown in Fig. 9. LH was measured in 0.5 min perifusate fractions.

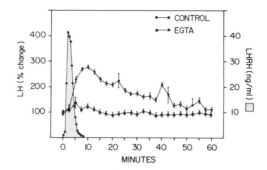

Fig. 11. Effect of EGTA (5 mM) on the LHRH induced LH secretory response (shaded area) from hemipituitaries in vitro using computer-assisted perifusion. ●---●Represents the LH response to LHRH in the absence of EGTA (first 50 ng LHRH "test" pulse). ▲--▲Represents the LH response to LHRH in the presence of EGTA (second 50 ng LHRH "test" pulse). See text and Fig. 7 for model.

similarities between the LHRH- and phospholipase C-induced LH secretory pattern suggest that LHRH may act through a phospholipase C-mediated intracellular system.

In order to test the ionic Ca^{2+} dependency of LH secretion in response to a rapid pulse of LHRH or phospholipase C, a constant infusion of 5 mM EGTA was initiated 30 mins prior to the second 50 ng total mass LHRH pulse or 415 mU phospholipase C pulse. In the presence of 5 mM EGTA, neither the LHRH nor the phospholipase C pulse produced a normal LH secretory response (Fig. 11, 12), but rather produced only a small, initial increase in LH secretion during the first few min (Valenca et al., 1986). These results confirm earlier studies indicating the Ca^{2+} - dependency of both LHRH- and phospholipase C-induced secretory mechanisms (Conn, 1986; and references therein) and expand this Ca^{2+}-dependency to include the pulsatile LH secretory response.

An additional consequence of the inositol phospholipid metabolism initiated by phospholipase C is the liberation of arachidonic acid from the cellular membrane. Free arachidonic acid may then be further metabolized, via the cyclooxygenase pathway or via the lipoxygenase pathway, into a variety of possible metabolites as previously discussed. In order to initiate studies of the role of arachidonate metabolism in pulsatile LH secretion, we tested the ability of the second 50 ng-total mass LHRH pulse to stimulate LH secretion 30 mins after the initiation of a constant infusion of 20 µM NDGA, a lipoxygenase pathway blocker. While the presence of NDGA in the media did not significantly alter the shape of the LH secretory response, the magnitude of the response was decreased by approximately 50% as compared with the LH secretory response induced by the first 50 ng LHRH pulse (Valenca et al., 1986) (Fig. 13). These results suggest that certain arachidonic acid metabolites, derived from the lipoxygenase pathway, may act to amplify the stimulatory action of LHRH. An alternative explanation is that the blockade of the lipoxygenase pathway, by increasing the available substrate, may increase arachidonic acid metabolism through a different metabolic pathway that may be inhibitory to

LH secretion.   Additional studies are underway to further elucidate these possibilities.

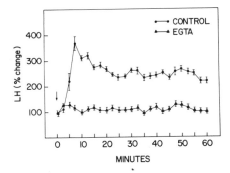

Fig. 12.   Effect of EGTA (5 mM) on the LH response to a single, 415 mU pulse of phospholipase C (arrow). Hemipituitaries were stimulated _in vitro_ using computer-assisted perifusion. ●--●Represents the LH response to phospholipase C in the absence of EGTA (first pulse). ▲--▲Represents the LH response to phospholipase C in the presence of EGTA (second pulse). See text and Fig. 7 for model.

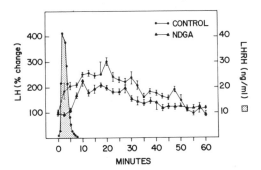

Fig. 13.   Effect of nordihydroguiaretic acid (NDGA, 20 μM), a lipo-xygenase inhibitor, on the LH secretory response to a single, computer-designed LHRH pulse (shaded area). Hemipituitaries were stimulated _in vitro_ using computer-assisted perifusion. ●--● Represents the LH secretory response to the first LHRH "test" pulse, in the absence of NDGA. ▲--▲Represents the LH secretory response to the second LHRH "test" pulse, in the presence of NDGA. See text and Fig. 7 for model.

CONCLUSIONS

The data summarized in this review clearly indicate the complexity of the intracellular pathways mediating LHRH and LH secretion. In the case of LHRH, it is now clear that neurotransmitters such as norepinephrine may stimulate LHRH secretion by activating different and complimentary intracellular systems which cause Ca^{2+} mobilization, arachidonic acid release and metabolism to active products, formation of diacylglycerol and activation of protein kinase C. These events, in turn, lead to increased secretion of the neuropeptide. The results from the studies employing computerized perifusion for the study of pulsatile LH secretion <u>in vitro</u> provide valuable information on both the importance of signal configuration to the secretory output and on the intracellular mechanisms that may be activated by the pulsatile LHRH input.

REFERENCES

Advis JP, McCann SM, Negro-Vilar A, 1980. Evidence that catecholaminergic and peptidergic (LHRH) neurons in suprachiasmatic-medial preoptic, medial basal hypothalamus and median emience are involved in estrogen negative feedback. Endocrinology 107:892-901

Andrews WV, Conn PM, 1986. GnRH stimulates mass changes in phosphoinositides and diacylglycerol accumulation in purified gonadotrope cell culture. Endocrinology 118:1148-58

Bell RL, Kennerly DA, Stanford N, Majerus PW, 1979. Diglyceride lipase: a pathway for arachidonate release from human platelets. Proc Natl Acad Sci USA 76:3238-41

Berridge MJ, 1981. Phosphatidylinositol hydrolysis: a multifunctional transducing mechanism. Mol Cell Endocrinol 24:115-40

Bigdeli H, Snyder PJ, 1978. Gonadotropin-releasing hormone release from the rat hypothalamus: dependence on membrane depolarization and calcium influx. Endocrinology 103:281-86

Blaustein M, McGraw FC, Somlyo AV, Schweitzer ES, 1980. How is the cytoplasmic calcium concentration controlled in nerve terminals? J Physiol (Paris) 76:459-70

Borgeat P, 1981. Leukotrienes: a major step in the understanding of immediate hypersensitivity reactions. J Med Chem 24:121-26

Capdevila J, Chacos N, Falck JR, Manna S, Negro-Vilar A, Ojeda SR, 1983. Novel hypothalamic arachidonate products stimulate somatostatin release from the median eminence. Endocrinology 113:421-23

Capdevila J, Chacos N, Werringloer J, Prough RA, Estabrook RW, 1981a. Liver microsomal cytochrome P-450 and the oxidative metabolism of arachidonic acid. Proc Natl Acad Sci 78:5362-66

Capdevila J, Parkhill L, Chacos N, Okita R, Masters BSS, Estabrook RW, 1981b. The oxidative metabolism of arachidonic acid by purified cytochrome P-450. Biochem Biophy Res Comm 101:1357-63

Castagna M, Takai Y, Kaibuchi K, Sano K, Kikkawa U, Nishizuka Y, 1982. Direct activation of calcium-activated, phospholipid-dependent protein kinase by tumor-promoting phorbol esters. J Biol Chem 257:7847-51

Chiocchio SR, Negro-Vilar A, Tramezzani JH, 1976. Acute changes in norepinephrine content in the median eminence induced by orchidectomy or testosterone replacement. Endocrinology 99:589-95

Chobsieng P, Naor Z, Koch Y, Zor U, Linder HR, 1975. Stimulatory effect of prostaglandin E on LH release in the rat: evidence for hypothalamic site of action. Neuroendocrinology 17:12-17

Conn PM, 1986. The molecular basis of gonadotropin releasing hormone action. Endoc Rev 7:3-10

Conn PM, Chafouleas J, Rogers D, Means AR, 1981. Gonadotropin releasing hormone stimulates calmodulin redistribution in the rat pituitary. Nature 292:264-67

Conn PM, Dufau ML, Catt KJ, 1979. GnRH-stimulated release of LH from rat pituicytes does not require production of cyclic AMP. Endocrinology 104:448-53

Conn PM, Ganong BR, Ebeling J, Staley D, Niedel J, Bell RM, 1985. Diacylclycerols release LH: structure-activity relations reveal a role for protein kinase C. Biochem Biophys Res Commun 126:532-39

Crowley WR, O'Donohue TL, Wachslicht H, Jacobowitz DM, 1978. Effects of estrogen and progesterone on plasma gonadotropins and on catecholamine levels and turnover in discrete brain regions of ovariectomized rats. Brain Res 154:633-35

Culler MD, Negro-Vilar, 1986. Evidence that pulsatile follicle-stimulating hormone secretion is independent of endogenous luteinizing hormone-releasing hormone. Endocrinology 118:609-12

Culler MD, Valenca MM, Romanelli F, Negro-Vilar A. Computer-controlled perifusion: characterization for studies of pulsatile gonadotropin secretion. Proc. of 16th Annual Meeting of Society for Neuroscience, (Abstract 282.7), pp. 1025

DePaolo LV, Ojeda SR, Negro-Vilar A, McCann SM, 1982. Alterations in the responsiveness of median eminence luteinizing hormone-releasing hormone nerve terminals to norepinephrine and prostaglandin E_2 in vitro during the rat estrous cycle. Endocrinology 110:1999-2005

Drouva SV, Epelbaum J, Hery M, Tapia-Arancibia L, Laplante E, Kordon C, 1981. Ionic channels involved in the LHRH and SRIF release from rat mediobasal hypothalamus. Neuroendocrinology 32:155-62

Ellis GB, Desjardins C, Fraser HM, 1983. Control of pulsatile LH release in male rats. Neuroendocrinology 37:177-83

Exton JH, 1982. Molecular mechanisms involved in α-adrenergic response. Trends Pharmacol Sci 3:111-15

Flower RJ, Blackwell GJ, 1976. The importance of phospholipase-A_2 in prostaglandin biosynthesis. Biochem Pharmacol 25:285-91

Gallardo E, Rameriz VD, 1977. A method for the superfusion of rat hypothalami: secretion of luteinizing hormone-releasing hormone (LH-RH). Proc Soc Exp Biol Med 195:79-84

Gallo RV, 1982. Luteinizing hormone secretion during continuous or pulsatile infusion of norepinephrine: central nervous system desensitization to constant norepinephrine input. Neuroendocrinology 35:380-87

Gerozissis K, Rougert C, Dray F, 1986. Leukotriene C_4 is a potent stimulator of LHRH secretion. Eur J Pharmacol 121:159-60

Gerozissis K, Vulliez B, Saavedra JM, Murthy RC, Dray F, 1985. Lipoxygenase products of arachidonic acid stimulate LHRH release from rat median eminence. Neuroendocrinology 40:272-76

Hammarstrom S, 1982. Biosynthesis and biological action of prostaglandins and thromboxanes. Arch Biochem Biophy 214:431-45

Heaulme M, Dray F, 1984. Noradrenaline and prostaglandin E_2 stimulate LH-RH release from rat median eminence through distinct 1-alpha-adrenergic and PGE_2 receptors. Neuroendocrinology 39:403-07

Huckle WR, Conn PM, 1985. PI turnover in response to GnRH: independence of Ca^{2+}-calmodulin and LH release. J Cell Biol (Abstract 14) 101:4a

Jaiswal N, Sharma RK, 1986. Dual regulation of adenylate cyclase and guanylate cyclase: α_2-adrenergic signal transduction in adrenocortical carcinoma cells. Arch Bioch Bioph 249:616-19

Jennes L, Brouson D, Stumpf WE, Conn PM, 1985. Evidence for an association between calmodulin and membrane patches containing GnRH receptor in cultured pituitary gonadotropes. Cell Tiss Res 239:311-15

Kaibuchi K, Takai Y, Ogawa Y, Kimura S, Nishizuka Y, Nakamura T, Tomomura A, Ichihara A, 1982. Inhibitory action of adenosine 3'5'monophosphate on phosphatidylinositol turnover: difference in tissue response. Biochem Biophys Res Commun 104:105-12

Kalra SP, Kalra PS, 1983. Neural regulation of LH secretion in the rat. Endocr Rev 4:311-51

Kalra SP, Kalrs PS, 1984. Opioid-adrenergic-steroid connection in regulation of luteinizing hormone secretion in the rat. Neuroendocrinology 38:418-26

Kant GJ, Sessions GR, Lenox RA, Meyerhoff JL, 1981. The effects of hormonal and circadian cycles, stress, and activity on levels of cyclic AMP and cyclic GMP in pituitary, hypothalamus, pineal gland and cerebellum of female rats. Life Sci 29:2491-99

Kao LWL, Weisz J, 1977. Release of gonadotropin-releasing hormone (GnRH) from isolated perifused medial-basal hypothalamus by melatonin. Endocrinology 100:1723-26

Kim K, Ramirez VD, 1985. Dibutyryl cyclic adenosine monophosphate stimulates in vitro luteinizing hormone-releasing hormone release only from median eminence derived from ovariectomized, estradiol-primed rat. Brain Res 342:154-57

Knobil E, 1980. The neuroendocrine control of the menstrual cycle. Rec Progr Horm Res 36:53-58

Langer SZ, 1978. Presynaptic adrenoceptors and regulation of release. In: Paton DM (ed.), The Release of Catecholamines from Adrenergic Neurons. England: Pergamon Press, pp. 59-85

Leadem CA, Crowley WR, Simpkins JW, Kalra SP, 1985. Effects of naloxone on catecholamine and LHRH release from the perifused hypothalamus of the steroid-primed rat. Neuroendocrinology 40:497-500

MacIntyre DE, Shaw AM, Pollock WK, Marks G, Westwisk J, 1983. Role of endogenous arachidonate metabolites in phospholipid-induced human platelet activation. In: Samuelsson B, Paoletti R, Ramwell P (eds.), Advances in Prostaglandin, Thromboxane, and Leukotriene Research. New York: Raven Press, 11:423-28

McEwen B, Biegon A, David P, Krey LC, Luine VN, McGinnis MY, Paden CM, Parson SB, Rainbow TC, 1982. Steroid hormones: humoral signals which alter brain cell properties and functions. Recent Prog Horm Res 38:41-92

Meites J, Van Vugt DA, Forman LJ, Sylvester PW, Ierie T, Sonntag W, 1983. Evidence that endogenous opiates are involved in control of gonadotropin secretion. In: Bhatnager AS (ed.), The Anterior Pituitary Gland. New York: Raven Press, pp. 327-40

Michel T, Lefkowitz RJ, 1982. Hormonal inhibition of adenylate cyclase. J Biol Chem 257:13557-63

Michell RH, 1975. Inositol phospholipids and cell surface receptor function. Biochim Biophy Acta 415:81-147

Morrell JI, Schwanzel-Fukuda M, Fahrbach SE, Pfaff DW, 1984. Axonal projections and peptide content of steroid hormone concentrating neurons. Peptides 5:227-39

Naor Z, Catt KJ, 1981. Mechanism of action of gonadotropin releasing hormone, involvement of phospholipid turnover in luteinizing hormone release. J Biol Chem 256:2226-29

Naor Z, Ojeda SR, Negro-Vilar A, McCann SM, 1979. Cyclic GMP and cyclic AMP levels in median eminence, hypothalamus and pituitary gland of the rat after decapitation or microwave irradiation. Neurosc Lett 13:189-94

Naor Z, Vanderhoek JY, Linder HR, Catt KJ, 1983. Arachidonic and products as possible mediators of the action of gonadotropin-releasing hormone. In: Samuelsson B, Paoletti R, Ramwell P (eds.), Advances in Prostaglandin, Thromboxane, and Leukotriene Research, New York: Raven Press, 12:259-63

Naor Z, Zor V, Meidan R, Koch Y, 1978. Sex differences in pituitary cyclic AMP response to gonadotropin releasing hormone. Am J Physiol 235:E37-E41

Negro-Vilar A, 1986. LHRH: Physiology, pharmacology and its role in fertility regulation. In: Paulson JD, Negro-Vilar A, Lucena E, Martini L (eds.), Andrology - Male Fertility and Sterility. Orlando: Academic Press, pp. 3-14

Negro-Vilar A, Conte M, Valenca MM, 1986. Transmembrane signals mediating neural peptide secretion: role of protein kinase C activators and arachidonic acid metabolites in luteinizing hormone-releasing hormone secretion. Endocrinology 119:2796-802

Negro-Vilar A, Culler M, 1986. Computer-controlled perifusion system for neuroendocrine tissues: Development and applications. In: Conn PM (ed.), Methods in Enzymology: Neuroendocrine peptides. Academic Press, pp. 67-9

Negro-Vilar A, DePaolo L, Tesone M, Johnston CA, 1982. Role of brain peptides in the regulation of gonadotropin secretion. In: Menchini-Fabris F, Pasini W, Martini L (eds.), Therapy in Andrology: Pharmacological, Surgical, and Physiological Aspects. Amsterdam: Excerpta Medica, pp. 115-23

Negro-Vilar A, Lapetina E, 1985. 1,2-Didecanoylglycerol and phorbol 12,13-dibutyrate enhance anterior pituitary hormone secretion in vitro. Endocrinology 117:1559-64

Negro-Vilar A, Ojeda SR, 1981. Hypophysiotrophic hormones of the hypothalamus. In: McCann SM (ed.), Endocrine Physiology III, International Review of Physiology. Baltimore: University Park Press, pp. 97-156

Negro-Vilar A, Ojeda SR, McCann SM, 1979. Catecholaminergic modulation of luteinizing hormone-releasing hormone release by median eminence terminals in vitro. Endocrinology 104:1749-57

Negro-Vilar A, Ojeda SR, McCann SM, 1980. Hypothalamic control of LHRH and somatostatin: role of central neurotransmitters and intracellular messengers. In: Litwack G (ed.), Biochemical Actions of Hormones. New York: Academic Press, pp. 245-84

Negro-Vilar A, Snyder GD, Falck JR, Manna S, Chacos N, Capdevila J, 1985. Involvement of eicosanoids in release of oxytocin and vasopressin from the neural lobe of the rat pituitary. Endocrinology 116:2663-68

Nishizuka Y, 1984. Turnover of inositol phospholipids and signal transduction. Science 225:1365-70

Ojeda SR, Naor Z, Negro-Vilar A, 1979. The role of prostaglandins in the control of gonadotropin and prolactin secretion. Prostaglandins Med 2:249-75

Ojeda SR, Negro-Vilar A, 1984. Release of prostaglandin E_2 from the hypothalamus depends on extracellular Ca^{2+} availability: relation to LHRH release. Neuroendocrinology 39:442-47

Ojeda SR, Negro-Vilar A, 1985. Prostaglandin E_2 -induced luteinizing hormone releasing hormone release involves mobilization of intracellular Ca^{2+}. Endocrinology 116:1763-70

Ojeda SR, Negro-Vilar A, McCann SM, 1982. Evidence for involvement of α-adrenergic receptors in norepinephrine-induced prostaglandin E_2 and luteinizing hormone-releasing hormone release from the median eminence. Endocrinology 110:409-12

Ojeda SR, Urbanski HF, Katz KH, Costa ME, 1985. Stimulation of cyclic adenosine 3',5'-monophosphate production enhances hypothalamic luteinizing hormone-releasing hormone releasing without increasing prostaglandin E_2 synthesis: studies in prepubertal female rats. Endocrinology 117:1175-78

Ojeda SR, Urbanski HF, Katz KH, Costa ME, Conn PM, 1986. Activation of two different but complementary biochemical pathways stimulates release of hypothalamic luteinizing hormone-releasing hormone. Proc Natl Acad Sci USA 83:4932-36

Raymond V, Leung PCK, Veilleux R, Lefevre G, Labrie F, 1984. LHRH rapidly stimulates phosphatidylinositol metabolism in enriched gonadotrophs. Mol Cell Endocrinol 36:157-64

Rotsztejn NH, Charli JL, Pattou E, Epelbaum J, Kordon C, 1976. In vitro release of luteinizing hormone releasing hormone (LHRH) from rat mediobasal hypothalamus: effects of potassium, calcium, and dopamine. Endocrinology 99:1663-66

Samuelsson B, 1979-80. Prostaglandins, thromboxanes, and leukotrienes: formation and biological roles. Harvey Lect 75:1-40

Samuelsson B, 1980. The leukotrienes: a new group of biologically active compounds including SRS-A. Trends Pharmacol Sci 1:227-30

Samuelsson B, Goldyne M, Granstrom E, Hamberg M, Hammarstrom S, Malmsten C, 1978. Prostaglandins and thromboxanes. Ann Rev Biochem 47:997-1029

Sar M, Stumpf WE, 1981. Central noradrenergic neurones concentrate ^{3}H-oestradiol. Nature 289:500-02

Smith MA, Vale WW, 1981. Desensitization to gonadotropin-releasing hormone observed in superfused pituitary cells on cytodex beads. Endocrinology 108:752-59

Snyder GD, Bleasdale JE, 1982. Effect of LHRH on incorporation of ^{32}P-orthophosphate into phosphatidyl inositol by dispersed anterior pituitary cells. Mol Cell Endocrinol 28:55-63

Snyder GD, Capdevila J, Chacos N, Manna S, Falck JR, 1983. Action of luteinizing hormone releasing hormone: involvement of novel arachidonic acid metabolites. Proc Natl Acad Sci USA 80:3504-07

Takai Y, Kishimoto A, Kawahara Y, Minakuchi R, Sano K, Kikkawa U, Mori T, Yu B, Kaibuchi K, Nishizuka Y, 1981. Calcium and phosphatidylinositol turnover as signaling for transmembrane control of protein phosphory-lation. Adv Cyclic Nucleotide Res 14:301-13

Unnerstall JR, Kopajtic TA, Kuhar MJ, 1984. Distribution of α_2 agonist binding sites in the rat and human central nervous system: analysis of some functional, anatomic correlates of the pharmacologic effects of clonidine and related adrenergic agents. Brain Res Rev 7:69-101

Valenca MM, Conte D, Negro-Vilar A, 1985. Diacylglycerol and phorbol esters enhance LHRH and prostaglandin E secretion from median eminence nerve terminals in vitro. Brain Res Bull 15:657-59

Valenca MM, Culler MD, Romanelli F, Negro-Vilar A, 1986. Evaluation of the intracellular events leading to pulsatile LH secretion using computer-designed input signal and controlled perifusion. Proc. of 16th Annual Meeting of Society for Neuroscience, (Abstract 282.8), pp. 1025

Williamson JR, 1986. Role of inositol lipid breakdown in the generation of intracellular signals. State of the art lecture. Hypertension 8:II 140-56

Yamura H, Lad PM, Rodbell M, 1977. GTP stimulates and inhibits adenylate cyclase in fat cell membranes through distinct regulatory processes. J Biol Chem 252:7964-66

REGULATION OF PULSATILE LUTEINIZING HORMONE RELEASE

DURING THE ESTROUS CYCLE AND PREGNANCY IN THE RAT

R.V. Gallo, G.N. Babu, A. Bona-Gallo, E. Devorshak-Harvey,
R.E. Leipheimer and J. Marco

Department of Physiology and Neurobiology
The University of Connecticut
Storrs, CT 06268

Luteinizing hormone (LH) secretion during the rat estrous cycle (Gallo, 1981; Higuchi and Kawakami, 1982; Fox and Smith, 1985) and pregnancy (Gallo et al., 1985) occurs in a pulsatile manner. Moreover, changes in the amplitude and frequency characteristics of this secretion take place with different stages of the 4-day estrous cycle and during different stages of gestation in this species. Plasma levels of estradiol (E_2) and progesterone (P) also vary from stage to stage in the estrous cycle (Butcher et al., 1974; Kalra and Kalra, 1974b; Smith et al., 1975; Nequin et al., 1979) as well as during the different stages of pregnancy in the rat (Morishige et al., 1973; Taya and Greenwald, 1981; Bridges, 1984). The emphasis in our work is to study the neuroendocrine regulation of pulsatile LH secretion in a physiological context. Therefore, the aim of our experiments has been to first describe the normal changes that occur in LH pulse amplitude and frequency throughout the estrous cycle and pregnancy, and then to define the regulatory negative feedback roles of E_2 and/or P during each 24-h interval of the rat estrous cycle, and during specific stages of gestation. Secondly, several lines of evidence support the hypothesis that endogenous opioid peptides (EOPs) acting centrally suppress LH release, and that they may be involved in mediating negative feedback actions of gonadal steroids on LH secretion (reviewed by Meites et al., 1979, 1983; Kalra, 1983; Kalra and Kalra, 1984; Ferin et al., 1984). Experiments currently ongoing in our laboratory are therefore aimed at defining the role of EOPs in regulating pulsatile LH secretion during various stages of the estrous cycle as well as pregnancy. This chapter will focus first on the regulation of basal pulsatile LH secretion during each stage of the estrous cycle, and secondly on the regulation of this process during pregnancy.

ESTROUS CYCLE

During the rat estrous cycle blood LH levels are the lowest on estrus, owing in large part to the longest LH interpulse interval of the cycle (Gallo, 1981); in fact, two other laboratories have reported an absence of LH pulses on estrus (Higuchi and Kawakami, 1982; Fox and Smith, 1985). Between estrus and diestrus 1 (D1), when plasma levels of both E_2 and P are low, there is a marked increase in LH pulse amplitude and frequency (Gallo, 1981; Gallo and Bona-Gallo, 1985; Babu et al., 1987.) The amplitudes of these LH pulses are the largest of any period of basal pulsatile LH release in the rat cycle (Gallo, 1981). Reduction in the amplitude of these pulses, produced by LH-releasing hormone (LHRH)antagonist administration, without any associated change in frequency, is nevertheless sufficient to increase the rate of follicular atresia on D1 (Devorshak-Harvey et al., 1985). This indicates that ovarian follicles on D1 are sensitive to alterations in the amplitude of the LH pulsatile signal, and demonstrates the physiological importance of these large amplitude pulses in maintaining normal folliculogenesis. Between D1 and diestrus 2 (D2), when the secretion of both ovarian E_2 and P is increasing, LH pulse frequency remains stable while pulse amplitude decreases (Gallo, 1981; Fox and Smith, 1985). In the next 24-h interval between the mornings of D2 and proestrus, a second increase in E_2 secretion occurs, and the corpus luteum regresses, resulting in a decline in plasma P levels. Three studies in our laboratory (Gallo, 1981; Leipheimer et al., 1985; Babu et al., 1986) and another (Higuchi and Kawakami, 1982) have demonstrated that during this period LH pulse amplitude and frequency remain stable. Finally, in the remaining 24-h interval between the mornings of proestrus and estrus, E_2 secretion is maintained for the initial 10 h, P secretion increases for about 10 h, and LH pulse frequency declines while pulse amplitude remains stable (Gallo, 1981; Leipheimer et al., 1986a).

Prior to our analyses of ovarian steroid regulation of these changes in amplitude and frequency characteristics of pulsatile LH secretion, we found that ovariectomy (OVX) on the morning of D1 resulted within 24 h in marked increases in the amplitude and frequency of pulsatile LH release (Leipheimer and Gallo, 1983). Thus, during the estrous cycle one could see the effects of removal of ovarian steroids on pulsatile LH secretion within 24 h. This allowed us to compare the normal change in pulsatile release that occurred over a specific 24-h interval in the rat cycle, with that which occurred over the same time span in ovariectomized animals, as well as in ovariectomized rats in which physiological levels of E_2 and/or P appropriate to that 24-h interval were restored through the use of steroid-containing silastic capsules. These studies have clearly shown that the changing plasma levels of E_2 and P that occur in the rat estrous cycle are critically important in regulating the variations in basal pulsatile LH release that occur at each cycle stage.

ESTRUS-DIESTRUS 1: From estrus through the morning of D1 plasma levels of E_2 and P are the lowest of the entire estrous cycle (Butcher et al., 1974; Kalra and Kalra, 1974b; Smith et al., 1975; Nequin et al., 1979), and there is a marked increase in the amplitude and frequency of pulsatile LH secretion (Gallo, 1981; Gallo and Bona-Gallo, 1985; Babu et al., 1987). In contrast to the presence of prominent ovarian steroid negative feedback systems operative in each of the three succeeding 24-h intervals of the rat estrous cycle, our studies demonstrated that the plasma levels of E_2 and P achieved in this 24-h period are too low to exert any prominent negative feedback effect on pulsatile LH release. Thus, while OVX on estrus decreased plasma levels of both steroids within 24 h when compared to D1 control values, it failed to augment the increases in pulse amplitude and frequency that normally occur between estrus and D1 (Gallo and Bona-Gallo, 1985; Babu et al., 1987; Fig. 1). This demonstrated that the increase in pulsatile LH secretion occurring in the estrous-D1 interval takes place in the absence of ovarian steroid negative feedback. Similarly, the rise in blood LH levels on D1 in the 5-day rat estrous cycle was also suggested to represent an escape from ovarian steroid negative feedback (Goodman, 1978). Utilizing push-pull perfusion of the rostrolateral median eminence (ME), Levine and Ramirez (1982) demonstrated an increase in LHRH secretion from estrus to D1. Our data would suggest that this increase is not under negative feedback control by the low plasma levels of E_2 and P found in this interval of the cycle.

EOPs and morphine suppress LH release (Bruni et al., 1977; Cicero et al., 1980; Kinoshita et al., 1980; Wiesner et al., 1984; Leadem and Kalra, 1985),

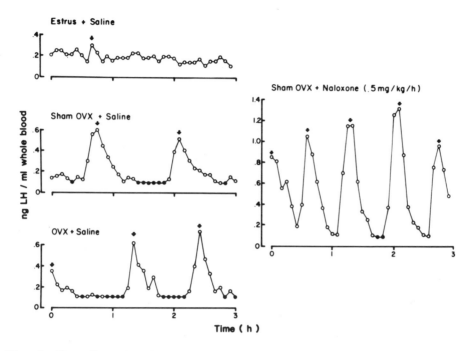

Fig. 1. Examples of pulsatile LH release in rats during a 4-h continuous
infusion of 0.9% saline (0.5 ml/h) on the morning of estrus, or in rats
sham ovariectomized or ovariectomized on estrus and infused with
saline or naloxone (0.5 mg/kg/h) while being bled 24 h later.
Continuous bleeding (40 or 50 ul/5 min) began 1 h after the onset of
infusion. Arrows indicate defined LH pulses, and filled circles
indicate undetectable LH levels. LH values are expressed in terms of
the NIADDK rat LH RP-2. (From Babu et al., 1987).

and naloxone and naltrexone, opiate receptor antagonists which displace EOPs
from receptor sites, increase LH secretion (Blank et al., 1979; Ieiri et al.,
1979; Quigley and Yen, 1980; Kalra and Simpkins, 1981; Ropert et al., 1981;
Sylvester et al., 1982; Gabriel et al., 1983; Van Vugt et al., 1983; Adler and
Crowley, 1984; Veldhuis et al., 1984; Fox and Smith, 1985). Previous studies
have shown that the ability of naloxone to increase LH secretion is lost with
time following OVX (Bhanot and Wilkinson, 1983, 1984; Barkan et al., 1984;
Petraglia et al., 1984) or orchidectomy (Cicero et al., 1980) in rats, or
following menopause in humans (Melis et al., 1984; Casper and Alapin-
Rubillovitz, 1985) unless gonadal steroids are initially given and negative
feedback restored. Similarly, the LH releasing effect of naloxone was seen in
sheep only when P levels were elevated (Malven et al., 1984). One
interpretation of these reports is that EOP involvement in LH secretion is lost
coincident with the disappearance of gonadal steroid negative feedback, i.e.,
that there is no steroid-independent, EOP suppression of LH release. In
contrast, naloxone has been shown to increase LHRH release in vitro from the
medial basal hypothalamus-medial preoptic area (MPOA) of long-term
gonadectomized male rats (Kalra et al., 1987). Moreover, increases in plasma
LH levels (Panerai et al., 1983) and LH pulse amplitude (Sylvester et al.,
1982) have been found in long-term, ovariectomized rats in response to
naloxone. These reports suggest that there is a steroid-independent, EOP
suppression of LH secretion, but do not indicate under which physiological
conditions it may be expressed. During the rat estrous cycle EOP receptor
blockade, resulting from infusion of naloxone, stimulated pulsatile LH
secretion on estrus (Fox and Smith, 1985). Bolus injections of naloxone

acutely increased LH release during the estrous-D1 interval (Gabriel et al., 1983; Petraglia et al., 1986; Piva et al., 1986), a physiological state in which ovarian secretion of E_2 and P occurs at a low level, and as we have shown (Gallo and Bona-Gallo, 1985), plasma steroid levels are ineffective in exerting negative feedback on pulsatile LH release. Therefore, we postulated that there may be an EOP suppression of pulsatile LH release between estrus and D1 which acts independent of the E_2 and P that is secreted in this 24-h interval of the rat estrous cycle, and recent studies in our laboratory support this hypothesis (Babu et al., 1987). While OVX did not augment the increase in pulsatile LH release between estrus and D1, infusions of naloxone did, by increasing both LH pulse amplitude and frequency on the morning of D1 (Fig. 1).

This stimulatory effect of naloxone on pulsatile LH secretion was blocked by simultaneous infusion of morphine, demonstrating that the effect was mediated by EOP receptors. Moreover, in confirmation of previous reports (Cicero et al., 1979, 1980; Chao et al., 1986), naloxone had no effect on in vivo pituitary responsiveness to LHRH, demonstrating that the EOP receptors blocked by naloxone were not located at the pituitary level. Since LH pulses are preceded by LHRH pulses (Clarke and Cummins, 1982; Levine et al., 1982), our data suggested that naloxone increased pulsatile LHRH release. Although naloxone or opiates have been reported to influence basal pituitary LH secretion in vitro (Cacicedo and Franco, 1986; Chao et al., 1986; Blank et al., 1987), other studies do not support this finding (Cicero et al., 1979; Wiesner et al., 1984). Moreover, LHRH release is important for the stimulatory action of naloxone, since antagonists of LHRH block naloxone-induced LH secretion (Blank and Roberts, 1982; Cicero et al., 1985). Naloxone has also been shown to increase LHRH release from superfused rat (Wilkes and Yen, 1981; Leadem et al., 1985; Kalra et al., 1987) and human (Rasmussen et al., 1983) hypothalami, and to increase LH release when implanted into the MPOA or arcuate nucleus (ARH)-ME region, but not closely adjacent areas of the rat brain (Kalra, 1981).

While the stimulatory effect of naloxone on pulsatile LH release on the morning of D1 may represent blockade of a steroid-independent EOP mechanism inhibitory to pulsatile LH release between estrus and D1, an alternative explanation is that it represents blockade of an EOP suppressive system that was activated by ovarian steroids secreted prior to estrus. Nevertheless, increases in the amplitude and frequency of pulsatile LH release occur in the estrous-D1 interval in the absence of ovarian steroid negative feedback, but these increases would be even greater were it not for the existence of a centrally-occurring, EOP suppression of both parameters of pulsatile LH secretion during this 24-h period of the rat estrous cycle.

DIESTRUS 1-DIESTRUS 2: Between D1 and D2, as E_2 and P secretion increase, LH pulse amplitude decreases and frequency remains stable (Gallo, 1981; Fox and Smith, 1985; unpublished data). However, both parameters of pulsatile secretion increase within 24 h following OVX on D1, indicating removal of ovarian steroid negative feedback (Leipheimer and Gallo, 1983; Leipheimer et al., 1984; unpublished data). E_2 alone was found to have no regulatory role in pulsatile LH release at this stage of the cycle, as replacement of D1-D2 levels of this steroid failed to alter this OVX-induced increase in pulsatile LH secretion. The reduction in LH pulse amplitude in this interval is due either to increasing plasma levels of P alone (Leipheimer et al., 1984), or to the combined effect of both E_2 and P (unpublished data). Neither steroid alone altered LH pulse frequency; rather, replacement of both E_2 and P was required to reduce the increase in pulse frequency caused by OVX. This demonstrated that both steroids combined act to maintain a stable LH pulse frequency between D1 and D2 by exerting a restraining, negative feedback action on this parameter of pulsatile LH secretion. The increased effectiveness of E_2 and P combined agrees with previous studies demonstrating a synergistic interaction between E_2 and P in decreasing tonic LH release in ovariectomized rats (McCann, 1962; Blake et al., 1972; McPherson et al., 1975;

Weick, 1977; Goodman, 1978; Weick and Noh, 1984; Goodman and Daniel, 1985).

EOPs have been suggested to be involved in mediating the negative feedback actions of gonadal steroids on LH secretion. Thus, naloxone has been shown to reverse the inhibitory effects of gonadal steroids on LH release (Cicero et al., 1979; Sylvester et al., 1982; Van Vugt et al., 1982; Bhanot and Wilkinson, 1984; Petraglia et al., 1984; Melis et al., 1984; Malven et al., 1984; Veldhuis et al., 1984; Casper and Alapin-Rubillovitz, 1985). In addition, gonadal steroids modify EOP levels in the hypothalamus and MPOA (Dupont et al., 1980; Wardlaw et al., 1982a, 1985; Forman et al., 1985; Bridges and Ronsheim, 1987) and in portal blood (Wardlaw et al., 1982b; Wehrenberg et al., 1982; Sarkar and Yen, 1985), and proopiomelanocortin levels in the hypothalamus (Wilcox and Roberts, 1985). However, the reversal of steroid-induced suppression of LH release by naloxone does not prove that EOPs mediate steroid-induced negative feedback. There may well be both steroid-independent and steroid-dependent EOP suppressions of LH secretion. In agreement with earlier reports from this laboratory (Gallo, 1981) and another (Fox and Smith, 1985), our recent studies (unpublished data) showed that mean blood LH levels decreased between D1 and D2 due to a decrease in LH pulse amplitude as pulse frequency remained stable, but both parameters increased within 24 h following OVX on D1. These latter increases, which occurred in response to removal of the negative feedback provided by the E_2 and P that is secreted in this interval of the cycle, were further augmented by infusion of naloxone 24 h following OVX on D1. Moreover, restoration of LH pulse amplitude and frequency to the levels normally found on D2 was achieved by reinstituting the negative feedback action of E_2 and P combined, and this suppressive action was not associated with any change in in vivo pituitary responsiveness to LHRH. Following restoration of this ovarian steroid negative feedback, naloxone infusion increased both LH pulse amplitude and frequency. Importantly, however, the magnitudes of the increments in pulse amplitude and frequency produced by naloxone in steroid- versus nonsteroid-treated rats were similar, and the mean values for both parameters of pulsatile LH release in response to naloxone were significantly less in steroid- than in nonsteroid-treated animals. Thus, the LH response to naloxone was similar in the presence or absence of the suppressive action of D1-D2 plasma levels of E_2 and P, indicating that at least in this 24-h interval of the rat estrous cycle the negative feedback action of E_2 and P on pulsatile LH release is not mediated by EOPs whose actions are blocked by naloxone.

Since E_2 + P do not act via EOPs to suppress pulsatile LH release between D1 and D2, the stimulatory effect of naloxone on LH pulse amplitude and frequency in ovariectomized, steroid-treated rats suggests that the EOPs act independently to suppress both parameters of pulsatile LH secretion in this 24-h interval. Studies demonstrating a naloxone-induced stimulation of LH or LHRH release following the prolonged absence of gonadal steroids support this conclusion (Sylvester et al., 1982; Panerai et al., 1983; Kalra et al., 1987). Alternatively, this stimulatory action of naloxone may represent blockade of an EOP suppressive mechanism that was activated by ovarian steroids secreted prior to D1. If the increased LH secretion seen in response to naloxone infusion 24 h following OVX represents blockade of an EOP mechanism activated prior to OVX, one might expect the magnitude of this action to gradually diminish with time following loss of ovarian steroid negative feedback. In contrast, Petraglia et al. (1984) found that the stimulatory effect of naloxone on LH secretion progressively increased within the first 48 h following OVX, before a lack of response developed at one week. Moreover, in our own studies when naloxone was administered on D1, following a 24-h period in which plasma steroid levels were too low to exert negative feedback (Gallo and Bona-Gallo, 1985; Babu et al., 1987), the magnitudes of the amplitude and frequency responses to EOP receptor blockade were similar to those seen when naloxone infusion was done 24 h following OVX on D1, i.e., following an additional 24-h period in which ovarian steroid negative feedback was not exerted (unpublished data). In addition, other ongoing studies in our laboratory suggest that during pregnancy in the rat an EOP system is activated by P to suppress LH pulse frequency, and within 24 h following the decline in plasma P levels between days 21 and 22 of gestation the stimulatory effect of naloxone on LH pulse frequency is markedly reduced or eliminated (Devorshak-Harvey et al., 1987a, b; unpublished data). This suggests that steroid-dependent EOP suppression of LH release does not necessarily have to persist for several days following loss of ovarian steroids. Thus, one cannot exclude the possibility of a steroid-independent effect of the EOPs in suppressing pulsatile LH secretion between D1 and D2.

DIESTRUS 2-PROESTRUS: The fact that a parameter of pulsatile LH secretion does not change from one stage of the estrous cycle to the next does not mean that this parameter is not under negative feedback control. This point was clearly seen with respect to LH pulse frequency between D1 and D2 and is further demonstrated between D2 and the morning of proestrus (Leipheimer et al., 1985; Babu et al., 1986). During this period no change occurs in mean blood LH levels, pulse amplitude or frequency, yet all parameters increase within 24 h following OVX on D2, indicating removal of a restraining action on pulsatile LH secretion. Examples are given in Fig. 2. The ovarian factors exerting these negative feedback actions were found to be both steroidal and nonsteroidal. P alone plays no role in regulating pulsatile secretion between D2 and proestrus, since the corpus luteum regresses and P secretion declines early on D2, and there was no significant difference in plasma P levels in the afternoon of D2 in rats ovariectomized or sham ovariectomized earlier that day (Leipheimer et al., 1985). However, restoration of proestrous levels of E_2 reduced the OVX-induced increase in LH pulse frequency to values seen on the morning of proestrus, and markedly but not completely reduced LH pulse amplitude. The restraining action of E_2 alone on LH pulse amplitude between D2 and proestrus is very likely centrally-mediated. E_2 increased pituitary responsiveness to LHRH in preparation for the proestrous LH surge (Leipheimer et al., 1985), in agreement with other reports (Cooper et al., 1973; Negro-Vilar, 1973; Kalra and Kalra, 1974a), but at the same time decreased LH pulse amplitude, suggesting that the E_2-induced suppression of LH pulse amplitude was probably mediated by a decrease in LHRH release. E_2 has been shown to acutely decrease LHRF activity in portal blood (Sarkar and Fink, 1980), and to suppress pulsatile LH release through an action on the medial basal hypothalamus (Blake et al., 1974; Akema et al., 1983). Moreover, as determined by push-pull perfusion, while OVX increased pulsatile LHRH release in the ARH-ME region in rabbits, replacement of physiological levels of E_2 prevented this increase (Pau et al., 1986). LH pulse frequency was also decreased by E_2 implants in the MPOA in rats (Akema et al., 1983, 1984). The possibility that the decrease in hypothalamic levels of LHRH following castration represents increased release of the decapeptide has been questioned, however, by studies demonstrating increases in LHRH synthesis and release in response to ovarian or testicular hormone replacement (Kalra, 1986). Moreover, other reports have not detected increased LHRH release in portal blood following castration (Eskay et al., 1977), nor was an increase in LHRH release detected by push-pull perfusion in the ARH-ME of rats following castration (Levine and Ramirez, 1982). Interestingly, Dluzen and Ramirez (1986) have recently shown increased LHRH levels in perfusate samples obtained from push-pull cannulae located in the anterior pituitary gland of castrate compared to intact rats, and suggested that this may be a better site for determination of LHRH released in response to various stimuli when utilizing this technique.

The full maintenance of a stable LH pulse amplitude in the D2- proestrous interval was found to involve an interaction between E_2 and a nonsteroidal factor of ovarian origin. Thus, while administration of a charcoal-extracted, essentially steroid free preparation of porcine follicular fluid (PFF) on D2 to rats ovariectomized earlier on D2 had no effect on pulsatile LH

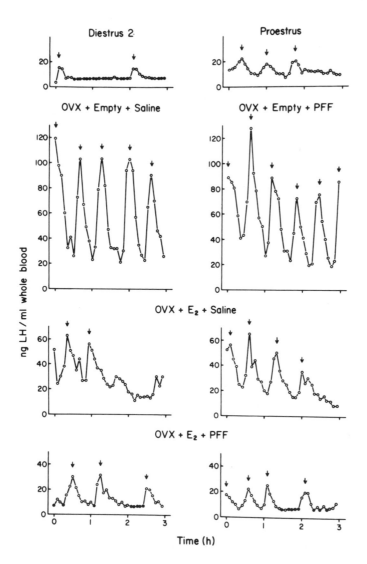

Fig. 2. Examples of pulsatile LH release in rats continuously bled (50 ul/5 min) between 07.30 and 10.30 h on diestrus 2 (D2); proestrus (i.e., sham ovariectomized between 08.30 and 09.30 h on D2, immediately implanted with an empty silastic capsule, given 1 ml saline ip at 12.00 and 24.00 h, and bled on proestrous AM); or in animals ovariectomized on D2, implanted with an empty or E_2-containing capsule, given 1 ml saline or porcine follicular fluid (PFF) at 12.00 and 24.00 h, and bled 24 h following OVX. Plasma E_2 levels in steroid-implanted rats were similar to proestrous AM values. Arrows indicate defined LH pulses, and filled circles indicate undetectable LH levels. LH values are expressed in terms of the NIADDK rat LH RP-1. (From Babu et al., 1986).

release the following morning, PFF given in combination with E_2 was able to further reduce LH pulse amplitude to proestrous morning values (Babu et al., 1986; Fig. 2). Furthermore, this additional restraint on LH pulse amplitude elicited by PFF in E_2-treated animals is exerted, at least in part, at the pituitary level. PFF reduced the extent of the LH response to LHRH in E_2-

treated rats, as had previously been shown in long-term ovariectomized rats (Lumpkin et al., 1984). The administration of PFF was also associated with a reduction in plasma FSH levels, suggesting as have others (Lumpkin et al., 1984) that inhibin, or a factor with inhibin-like activity, is the active compound. Therefore, E_2 alone does not return LH pulse amplitude to control proestrous morning values because in the absence of the ovaries, and this nonsteroidal factor, pituitary responsiveness to LHRH is greater in E_2-treated rats than normally would be found between D2 and proestrus. Inhibin activity has been detected in rat ovarian venous effluent between D2 and proestrus (DePaolo et al., 1979), in rat follicular fluid (Fujii et al., 1983), and in medium derived from rat follicular granulosa cell cultures (Erickson and Hsueh, 1978; Sander et al., 1984). With recent advances in our understanding of the structure of inhibin (Rivier et al., 1985), future studies will be possible to determine if this is the active factor present in PFF.

An additional suppressive influence on pulsatile LH release between D2 and proestrus is exerted by EOPs (unpublished data). OVX on D2 again resulted in increases in both the amplitude and frequency of LH pulses within 24 h when compared to sham ovariectomized controls. In such animals lacking ovarian negative feedback, naloxone infusion further augmented both these increases. Furthermore, in rats in which the OVX-induced increases in amplitude and frequency were suppressed by E_2, naloxone infusion increased both parameters of pulsatile LH secretion. As in the period between D1 and D2, the naloxone-induced increment in LH pulse amplitude in E_2-treated rats was similar to that observed in response to naloxone infusion in nonsteroid-treated animals. Pituitary responsiveness to LHRH has previously been shown to be increased in these E_2-treated rats (Leipheimer et al., 1985). Thus, these data indicated that the E_2-induced restraint on LH pulse amplitude between D2 and proestrus is not mediated by EOPs whose actions are blocked by naloxone. In contrast, however, the naloxone-induced increment in LH pulse frequency in E_2-treated rats was of a sufficiently large magnitude to suggest that at least a part of the E_2-induced restraint on pulse frequency in this interval may be mediated by EOPs. Moreover, since the negative feedback effect of E_2 on LH pulse amplitude is not mediated by EOPs, and that exerted on frequency may be only partially mediated by EOPs, the stimulatory effect of naloxone on both parameters of pulsatile LH secretion in E_2-treated rats may represent blockade of an EOP mechanism suppressing pulsatile LH release, and acting independent of the negative feedback action of E_2 secreted in this cycle interval. Therefore, between D2 and the morning of proestrus LH pulse amplitude is under the suppressive influence of at least three factors: a centrally-mediated action of E_2 alone, an additional interaction between E_2 and a nonsteroidal ovarian factor exerted at the pituitary level, and a centrally-occurring, E_2-independent EOP mechanism. LH pulse frequency in this interval is also regulated by E_2 and EOPs. The EOPs may act both independently and as mediators of E_2-induced negative feedback to restrain changes in LH pulse frequency.

PROESTRUS-ESTRUS: Mean blood LH levels decrease between the mornings of proestrus and estrus due to a reduction in LH pulse frequency, while pulse amplitude remains stable (Gallo, 1981; Leipheimer et al., 1986a). However, OVX on the morning of proestrus resulted within 24 h in increases in both parameters of pulsatile LH secretion, when compared with estrous controls (Leipheimer et al., 1986a). High blood levels of E_2, which increased between D2 and proestrus, and acted centrally to restrain changes in LH pulse amplitude and frequency, and at the pituitary to increase responsiveness to LHRH, remain elevated for the first 9-10 h of this next 24-h interval. This latter period of E_2 secretion on proestrus is critical for the suppression of LH pulse frequency on the morning of estrus. By itself, however, E_2 has no influence on LH pulse amplitude on estrus. E_2 is also the ovarian steroid responsible for stimulating the ovulatory LH surge on the afternoon of proestrus (Ferin et al., 1969; Legan et al., 1975; Legan and Karsch, 1975; Kalra, 1975; Krey and

Parsons, 1982). Therefore, the net effect of increasing blood levels of E_2 characteristic of D2 and proestrus is not only to trigger the ovulatory LH surge, but also to exert restraining or suppressive effects on pulsatile LH release before and after this stimulatory action. Restoration of the temporal pattern and magnitude of the proestrous changes in plasma P levels had no effect on LH pulse amplitude, but slightly reduced the OVX-induced increase in LH pulse frequency, although not to estrous values. However, this small effect of P on pulse frequency is likely of only limited physiological importance between proestrus and estrus, since E_2 restoration alone suppressed LH pulse frequency to estrous values, and P did not. While neither steroid alone had any significant effect on LH pulse amplitude, restoration of the proestrous changes in plasma levels of both steroids reduced LH pulse amplitude to estrous levels. Therefore although P secretion is only elevated for a 9-10 h period on proestrus, it acts to amplify E_2-induced LH surges (Camp and Barraclough, 1985; Camp et al., 1985), to limit these surges to the afternoon of proestrus (Freeman et al., 1976), and, combined with E_2, to restrain any increase in basal LH pulse amplitude on the morning of estrus.

While increases in LH pulse frequency occurred in response to naloxone infusion on the morning of estrus (Fox and Smith, 1985), as described above the low level pulsatile LH secretion found on the morning of estrus is the result of a prominent ovarian steroid negative feedback system operative between proestrus and estrus (Leipheimer et al., 1986a). Additional study will therefore be necessary to determine whether this naloxone-induced increase in pulsatile LH release is due to blockade of steroid-independent and/or steroid-dependent EOP suppression of pulsatile LH secretion.

All of the above studies in our laboratory were done during the physiological context of the estrous cycle, and demonstrated the changing role of ovarian steroids in regulating LH pulse amplitude and frequency in successive intervals of the cycle. For example, in two studies D1-D2 levels of E_2 alone had no effect on pulsatile LH secretion (Leipheimer et al., 1984; unpublished data), and we have shown (Leipheimer et al., 1985) and confirmed (Babu et al., 1986; unpublished data) that between D2 and proestrus E_2 alone exerts suppressive effects on LH pulse amplitude and frequency. However, the literature is not in agreement as to whether E_2 decreases only LH pulse amplitude (Turgeon and Barraclough, 1974; Higuchi and Kawakami, 1982: Goodman and Daniel, 1985), pulse frequency (Lumpkin et al., 1984), or both parameters of pulsatile LH secretion (Akema et al., 1983; Weick and Noh, 1984), but these studies were done in long-term ovariectomized rats. In a recent study (Leipheimer et al., 1986b) we found that if one week is allowed to elapse following OVX before E_2 is replaced for a 24-h period, D2-proestrous levels of this steroid retain their suppressive action on LH pulse amplitude, and D1-D2 levels remain ineffective in altering pulse frequency. However, in contrast, D2-proestrous E_2 levels lose their ability to restrain LH pulse frequency, and D1-D2 levels of E_2 now suppress LH pulse amplitude. The literature has demonstrated that other E_2-influenced processes also change with time following OVX. Injections of estradiol benzoate (EB) alone will stimulate LH surges in rats ovariectomized for two weeks (Legan et al., 1975), but supplemental injections of EB or P are required for this action in rats ovariectomized four weeks or longer (Caligaris et al., 1971; Brown-Grant, 1974; Crowley, 1982). E_2-containing silastic capsules are also less effective in inducing an LH surge as time elapses following OVX (McGinnis et al., 1981). A reduced negative feedback on LH secretion resulted from MPOA implants of EB in long-term versus short-term ovariectomized rats (Docke et al., 1984). Thus, both positive and negative feedback actions of E_2 are reduced in long-term ovariectomized rats. Moreover, the number of cell nuclear E_2 receptors decreased (McGinnis et al., 1981) and catecholamine turnover rates changed (Advis et al., 1980) in the hypothalamus-POA with time after OVX. An important implication here is that one must be cautious with respect to drawing definitive conclusions about the effects of ovarian steroids on LH pulse amplitude or frequency at specific estrous cycle stages, based solely on studies done on long-term ovariectomized rats, especially considering that the actions of these steroids on different parameters of pulsatile release change from stage to stage in the cycle.

In summary, our studies have demonstrated that the daily changes in secretory patterns of E_2 and P in the rat estrous cycle are decisively important in regulating the differences in basal pulsatile LH release that occur at each cycle stage. Moreover, there may well be both steroid-independent and steroid-dependent EOP suppressions of pulsatile LH secretion during the estrous cycle. These relationships are summarized in Table 1. The increases in LH pulse amplitude and frequency that occur between estrus and D1 are not under negative feedback control by the low level secretion of E_2 and P that occurs in this 24-h interval. However, these increases would be even greater were it not for the existence of a centrally-occurring, EOP suppression of both parameters of pulsatile LH release, acting independent of the E_2 and P secreted in this cycle interval. Between D1 and D2 negative feedback becomes operative. The stable LH pulse frequency at this time is due to the restraining effect of E_2 and P combined, while pulse amplitude declines either due to the effect of E_2 and P combined, or to the action of rising blood levels of P alone. E_2 by itself has no regulatory influence on pulsatile LH release between D1 and D2. In this cycle interval the negative feedback actions of E_2 and P are not mediated by EOPs whose actions are antagonized by naloxone, but EOPs may very likely act independent of the ovarian steroids

Table 1: A summary of the effects of estradiol (E_2), progesterone (P), and endogenous opioid peptides (EOPs) on changes in LH pulse amplitude and frequency that occur at each interval of the rat estrous cycle. EOPs referred to are those whose actions are blocked by naloxone.

Estrus–diestrus 1:
  Plasma E_2 and P levels are low
  Amplitude and frequency increase; no steroid negative feedback
  EOP suppression of amplitude and frequency acts independent of E_2 and P
    secreted in this interval

Diestrus 1–diestrus 2:
  Plasma E_2 and P levels increase
  E_2 alone has no regulatory influence
  Frequency stable due to an effect of E_2 + P
  Amplitude decreases due to an effect of E_2 + P, or P alone
  No steroid-dependent EOP suppression of pulsatile LH release
  EOP suppression of amplitude and frequency acts independent of E_2 and P
    secreted in this interval

Diestrus 2–proestrus:
  Plasma E_2 levels increase further; P levels decline on diestrus 2
  Frequency stable due to an effect of E_2
  Amplitude stable due to an effect of E_2 alone, and an interaction between E_2
    and a nonsteroidal ovarian factor
  No E_2-dependent EOP suppression of amplitude
  E_2-dependent EOP suppression of frequency is possible
  EOP suppression of amplitude and frequency acts independent of E_2 secreted
    in this interval

Proestrus–estrus:
  Plasma E_2 levels remain elevated for 10 h; P levels increase for 10 h
  Frequency decreases due to an effect of E_2
  Amplitude stable due to an effect of E_2 + P

secreted in this 24-h period to suppress both LH pulse amplitude and frequency. Between D2 and the morning of proestrus, E_2 becomes the main ovarian steroid which by itself can exert a restraining action on both pulse amplitude and frequency, while also increasing pituitary responsiveness to LHRH preparatory to the ovulatory LH surge. Complete regulation of LH pulse amplitude in this interval also involves an interaction between E_2 and a nonsteroidal ovarian factor, thus far shown to be present in PFF, which is exerted at least in part at the pituitary level. While there is no E_2-dependent, EOP suppression of pulse amplitude at this time of the cycle, E_2 suppression of pulse frequency may be mediated, at least in part, by EOPs. Moreover, EOPs may act independent of the E_2 secreted between D2 and proestrus to suppress both parameters of pulsatile LH secretion in this interval. Finally, in the interval between the mornings of proestrus and estrus, E_2 continues to suppress LH pulse frequency, while E_2 and P act together to restrain any change in pulse amplitude.

PREGNANCY

In order to increase our understanding of the regulation of pulsatile LH release in different physiological conditions, we examined whether LH release was pulsatile at days 6-8, 14-16, or 22 of pregnancy. These times correspond to three stages of gestation in the rat during which changes occur in plasma levels of E_2 and P (Morishige et al., 1973; Taya and Greenwald, 1981; Bridges, 1984). Our studies confirmed that plasma P levels, representing the functional activity of the corpus luteum, increase and then plateau between days 4-10, when a second rise occurs due to placental P secretion. P levels then plateau until about days 20- 21 when values markedly decline. In contrast, plasma E_2 levels increase between days 6-8 and 14-16, but remain high through day 22 (Gallo et al., 1985; Devorshak-Harvey et al., 1987a, b).

The amplitudes and frequencies of pulsatile LH release during gestation in the rat are similar to those which occur during periods of basal LH release in the estrous cycle (Gallo, 1981; Higuchi and Kawakami, 1982; Fox and Smith, 1985), and are clearly reduced compared with values in both acutely and long-term ovariectomized rats (Leipheimer and Gallo, 1983). As in the estrous cycle, the characteristics of pulsatile LH secretion change as pregnancy progresses. LH secretion is clearly more important for the maintenance of pregnancy on days 6-8 than on days 14-16, as abortion will result following hypophysectomy or in vivo neutralization of plasma LH before but not after day 12 of pregnancy (Pencharz and Long, 1933; Greenwald and Johnson, 1968; Loewit et al., 1969; Moudgal, 1969; Raj and Moudgal, 1970; Akaka et al., 1977; Raj et al., 1979; Terranova and Greenwald, 1981). Our initial studies indicated (Gallo et al., 1985), and subsequent studies confirmed (Devorshak-Harvey et al., 1987b), that pulsatile LH secretion is present on days 6-8, and that there is a clearcut trend toward an absence of pulsatile LH release on days 14-16. This correlates well with the diminished role for LH in pregnancy maintenance after day 12. Another prominent change in pulsatile LH secretion occurs on day 22 of pregnancy. Mean plasma LH levels increase at the end of gestation just prior to parturition (Linkie and Niswender, 1972; Morishige et al., 1973; Taya and Greenwald, 1981). Our studies have extended that finding by demonstrating that this increase is due to an increase in the frequency of pulsatile LH secretion (Gallo et al., 1985; Devorshak-Harvey et al., 1987a, b). This increase in LH release is physiologically important since it enhances thecal cell function and E_2 production in preovulatory follicles in preparation for the first postpartum estrous cycle (Bogovich et al., 1981; Carson et al., 1981).

Further study demonstrated that the increase in LH pulse frequency on day 22 of gestation occurs due to the decline in plasma levels of P, and the resultant loss of the negative feedback action of P on pulsatile LH release (Devorshak-Harvey et al., 1987a). Plasma P levels were maintained at or above

119

day 20 values in the last few days of gestation by means of implantation of P-containing silastic capsules on day 20. As a result, in these rats mean blood LH levels and LH pulse frequency on day 22 were equal to day 20 values, and the increases in both parameters observed on day 22 in empty capsule-implanted rats were blocked. Examples are shown in Fig. 3. Therefore, preventing a decline in plasma P values to the low levels found on day 22 prevented the increase in LH pulse frequency and mean blood LH levels normally seen at this time. Other reports in the literature have also suggested that P can decrease LH secretion during gestation in the rat. Administration of antisera to P increased plasma LH levels during mid-pregnancy (Cheesman and Chatterton, 1982), and medroxyprogesterone, a synthetic progestational agent, decreased plasma LH levels during early pregnancy (Brown-Grant et al., 1972).

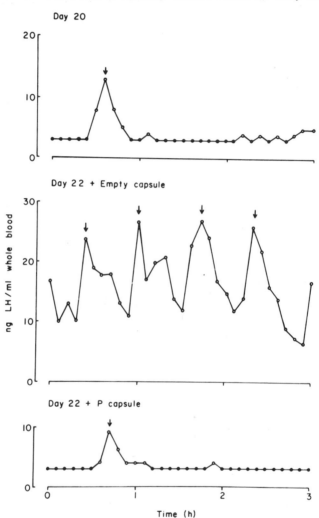

Fig. 3. Examples of pulsatile LH release on day 20 of pregnancy (plasma P levels are elevated), on day 22 in rats implanted with empty silastic capsules on day 20 (plasma P levels have greatly declined from day 20 values), and on day 22 in rats implanted with P-containing capsules on day 20. In this latter group plasma P levels were maintained for 2 days at or above day 20 levels. Arrows indicate defined LH pulses, and filled circles indicate undetectable LH levels. LH values are expressed in terms of the NIADDK rat LH RP-1. (From Devorshak-Harvey et al., 1987a).

The length of the luteal phase during D1 and early D2 in the rat estrous cycle, and therefore the period during which the hypothalamic- pituitary axis is exposed to increased plasma P levels, is very short compared to the period of luteal P secretion in the reproductive cycles of other species, such as sheep and humans. Our previous studies showed that corpus luteum secretion of P alone has no effect on LH pulse frequency in the rat estrous cycle (Leipheimer et al., 1984). In addition, while the 10-h increase in P secretion on proestrus slightly reduced LH pulse frequency on estrus (Leipheimer et al., 1986a), E_2 secretion alone during this interval was sufficient to fully account for the low frequency pulsatile LH release normally seen on estrus. Thus this P-induced suppression of pulse frequency in the proestrous-estrous interval is limited in its physiological importance. In contrast, the hypothalamic-pituitary axis is exposed to elevated plasma P levels for much longer periods of time during the extended luteal phases in sheep and humans, and P-negative feedback on LH pulse frequency is an important regulatory mechanism in these species (Goodman and Karsch, 1980; Soules et al., 1984). Plasma P levels are also elevated for a prolonged period during pregnancy in the rat, increasing by days 4-6 and remaining elevated for nearly three weeks (Morishige et al., 1973; Taya and Greenwald, 1981; Bridges, 1984). Therefore, the suppressive action of P on LH pulse frequency is expressed in rats, just as it is in sheep and humans, so long as elevated plasma P levels are maintained for a prolonged period of time.

Recent studies in our laboratory have also shown that EOPs exert a suppressive influence on pulsatile LH release during pregnancy (Devorshak-Harvey et al., 1987b). Blockade of opiate receptors with naloxone at days 6-8, 14-16, and 22 of pregnancy increased mean blood LH levels at each stage, indicating removal of EOP suppression of pulsatile LH secretion. EOP receptor blockade increased both the frequency and amplitude of pulsatile LH secretion on days 6-8, and stimulated frequency on days 14-16. The effect of naloxone on pulse amplitude on days 14-16 could not be determined, since too few rats had LH pulses prior to naloxone infusion. LH pulse amplitude and frequency increased during EOP receptor blockade on day 22, but these increases were significantly less than those seen on days 6-8 and 14-16, respectively. As we had observed during the estrous cycle, these naloxone-induced increases in pulsatile LH secretion were exerted by way of centrally-located EOP receptors. Simultaneous infusion with morphine prevented the stimulatory effect of naloxone. In addition, naloxone had no effect on pituitary responsiveness to LHRH at each stage of pregnancy, demonstrating that the EOP receptors blocked were not located at the pituitary level.

This study also showed that the pituitary was much less responsive to LHRH during pregnancy than in ovariectomized rats, and that responsiveness progressively declined during the course of gestation, reaching its lowest level on day 22, just prior to parturition (Devorshak-Harvey et al., 1987b). Similarly, Mano (1983) reported that pituitary responsiveness to LHRH decreased between days 10 and 21 of gestation in the rat. Prolonged exposure of the pituitary to elevated plasma E_2 and P levels is a likely factor in the reduced responsiveness of this gland to LHRH. Prolactin may also contribute, in part, since prolactin can directly diminish pituitary responsiveness to LHRH (Cheung, 1983; Miyake et al., 1985), and prolactin secretion is increased during the first week of pregnancy (Smith and Neill, 1976; Voogt et al., 1982) and again on day 22 (Linkie and Niswender, 1972; Bridges and Goldman, 1975; Ben-Jonathan et al., 1980). The lesser effect of naloxone on LH pulse amplitude on day 22 is therefore due, at least in part, to the diminished pituitary responsiveness to LHRH. Further study will be necessary to determine whether the amount of LHRH released in response to naloxone is also reduced on day 22, and therefore whether this is a contributory factor to the diminished naloxone-induced increase in pulse amplitude at this time of gestation. Nevertheless, since naloxone did increase pulse amplitude on day 22, an EOP mechanism suppressing this parameter of pulsatile LH release is still active at this stage of pregnancy.

Naloxone also increased LH pulse frequency on day 22, but this effect was significantly less than that seen on days 14-16, suggesting the possibility of a decreased EOP suppression of LH pulse frequency on day 22. Plasma levels of P and LH are changing at this time. Plasma E_2 levels remain elevated on day 22, but plasma P values, which are high on days 20-21, have greatly declined by day 22. As a result of this decline in plasma P levels, LH pulse frequency and mean blood LH levels increase on day 22. Therefore, our working hypothesis is that during pregnancy P suppresses LH pulse frequency through an EOP-mediated mechanism, and that as plasma P levels decline on day 22, EOP suppression of pulse frequency is diminished. This leads to an increase in LH pulse frequency just prior to parturition, and a reduced effect of naloxone on this parameter of pulsatile LH secretion. The possibility also exists that the decreased pituitary responsiveness to LHRH on day 22 may prevent some naloxone-induced LHRH pulses from stimulating LH pulses, and that this may contribute, in part, to the reduced effectiveness of naloxone in increasing pulse frequency at this time.

Others have also shown that the LH releasing action of naloxone is influenced by the ovarian steroid milieu at the time of naloxone administration. Increases in LH release in response to naloxone only occurred in sheep (Malven et al., 1984) or in the postmenopausal state in women (Casper and Alapin-Rubillovitz, 1985) when progestin-induced negative feedback was reinstituted prior to blockade of EOP receptors. During the menstrual cycle, EOP receptor blockade increased pulsatile LH secretion in the late follicular and luteal phases, but not in the early follicular stage (Quigley and Yen, 1980; Ropert et al., 1981; Blankstein et al., 1981; Snowden et al., 1984). Similarly, LH pulse frequency was increased by naloxone in the luteal but not the follicular stage of the menstrual cycle in monkeys (Van Vugt et al., 1983, 1984). In addition, immunoreactive β-endorphin- and dynorphin-containing neurons in the rat brain contain P or E_2 receptors (Morrell et al., 1985; Fox et al., 1986). While elevated levels of β-endorphin and metenkephalin are found in the medial basal hypothalamus, MPOA and midbrain during pregnancy in the rat (Wardlaw and Frantz, 1983; Petraglia et al., 1985; Bridges and Ronsheim, 1987), the levels of β-endorphin in the MPOA decrease on day 22 compared to earlier stages (Bridges and Ronsheim, 1987). Additional studies are necessary to determine the relationship between this decline in β-endorphin levels, the changing plasma steroid environment on day 22, and the diminished effectiveness of naloxone in increasing LH pulse frequency at this stage of pregnancy.

As summarized in Table 2, basal LH secretion varies during the course of pregnancy in the rat. Plasma E_2 and P levels increase early in gestation. Pulsatile LH release is evident on days 6-8, and is under the influence of a centrally-active EOP suppressive system. Pituitary responsiveness to LHRH is at its highest point during this period of gestation, and LH secreted at this time is vital for the maintenance of the corpus luteum. By days 14-16, plasma E_2 and P levels have increased further, pituitary responsiveness to LHRH has declined, and EOPs continue to exert suppressive effects, at least on pulse frequency. The incidence of pulsatile LH secretion is greatly decreased, correlating well with the diminished role for this hormone in pregnancy maintenance after day 12. Plasma E_2 levels remain elevated through day 22. However, plasma P levels plateau until days 20-21, and then markedly decline by day 22, just prior to parturition. Pituitary responsiveness to LHRH reaches its lowest point during gestation on day 22, and EOPs continue to suppress LH pulse amplitude. Nevertheless, as plasma P levels decline on day 22, EOP suppression of LH pulse frequency is reduced, and mean blood LH levels rise due to an increased frequency of pulsatile LH release. This increase in LH secretion is important for the ovarian changes that will occur in preparation for the first postpartum estrous cycle.

Table 2: A summary of the changes in plasma levels of estradiol (E_2) and progesterone (P), LH pulse amplitude and frequency, pituitary responsiveness to LHRH, and the influence of endogenous opioid peptides (EOPs) on pulsatile LH secretion, during pregnancy in the rat.

Days 6-8:
    Plasma E_2 and P levels are elevated
    Pulsatile LH release is present
    EOP suppression of amplitude and frequency
    Highest pituitary responsiveness to LHRH of 3 stages examined, but
        reduced compared to ovariectomized rats

Days 14-16:
    Plasma E_2 and P levels increase further
    Reduced incidence of pulsatile LH release
    EOP suppression of frequency
    Pituitary responsiveness to LHRH is less than on days 6-8

Day 22:
    Plasma E_2 levels remain elevated; P levels plateau through days
        20-21, then markedly decline
    Increased frequency due to absence of P negative feedback
    EOP suppression of amplitude
    Diminished EOP suppression of frequency
    Lowest pituitary responsiveness to LHRH during pregnancy

ACKNOWLEDGMENTS

RG would like to acknowledge the National Institutes of Health for research grant HD17728 which provided support for the studies done in his laboratory, and Janice Bittner for her excellent assistance in typing this manuscript.

REFERENCES

Adler BA, Crowley WR, 1984. Modulation of luteinizing hormone release and catecholamine activity by opiates in the female rat. Neuroendocrinology 38:248-53

Advis JP, McCann SM, Negro-Vilar A, 1980. Evidence that catecholaminergic and peptidergic (luteinizing hormone-releasing hormone) neurons in suprachiasmatic-medial preoptic, medial basal hypothalamus and median eminence are involved in estrogen negative feedback. Endocrinology 107:892-901

Akaka J, O'Laughlin-Phillips E, Antczak E, Rothchild I, 1977. The relationship between the age of the corpus luteum (CL) and the luteolytic effect of an LH antiserum: comparison of hysterectomized pseudopregnant rats with intact pregnant rats for their response to LHAS treatment at four stages of CL activity. Endocrinology 100:1334-40

Akema T, Tadokoro Y, Kawakami M, 1983. Changes in the characteristics of pulsatile LH secretion after estradiol implantation into the preoptic area and the basal hypothalamus in ovariectomized rats. Endocrinol Jpn 30:281-87

Akema T, Tadokoro Y, Kimura F, 1984. Regional specificity in the effect of estrogen implantation within the forebrain on the frequency of pulsatile luteinizing hormone secretion in the ovariectomized rat. Neuroendocrinology 39:517–23

Babu GN, Bona–Gallo A, Gallo RV, 1986. Interaction between estradiol and a nonsteroidal factor in porcine follicular fluid in regulating LH pulse amplitude between the mornings of diestrus 2 and proestrus in the rat. Neuroendocrinology 44:8–14

Babu GN, Marco J, Bona–Gallo A, Gallo RV, 1987. Steroid–independent endogenous opioid peptide suppression of pulsatile LH release between estrus and diestrus 1 in the rat estrous cycle. Brain Res (in press)

Barkan A, Regiani S, Duncan J, Papavasiliou S, Marshall JC, 1984. Opioids modulate pituitary receptors for gonadotropin–releasing hormone. Endocrinology 112:387–89

Ben–Jonathan N, Neill MA, Arbogast LA, Peters LL, Hoefer MT, 1980. Dopamine in hypophysial portal blood: relationship to circulating prolactin in pregnant and lactating rats. Endocrinology 106:690–96

Bhanot R, Wilkinson M, 1983. Opiatergic control of LH secretion is eliminated by gonadectomy. Endocrinology 112:399–401

Bhanot R, Wilkinson M, 1984. The inhibitory effect of opiates on gonadotrophin secretion is dependent upon gonadal steroids. J Endocrinol 102:133–41

Blake CA, Norman RL, Sawyer CH, 1972. Effects of estrogen and/or progesterone on serum and pituitary gonadotropin levels in ovariectomized rats. Proc Soc Exp Biol Med 141:1100–03

Blake CA, Norman RL, Sawyer CH, 1974. Localization of the inhibitory actions of estrogen and nicotine on release of luteinizing hormone in rats. Neuroendocrinology 16:22–35

Blank MS, Fabbri A, Catt KJ, Dufau ML, 1987. Inhibition of luteinizing hormone release by morphine and endogenous opiates in cultured pituitary cells. Endocrinology 118:2097–2101

Blank MS, Panerai AE, Friesen HG, 1979. Opioid peptides modulate luteinizing hormone secretion during sexual maturation. Science 203:1129–31

Blank MS, Roberts DL, 1982. Antagonist of gonadotropin–releasing hormone blocks naloxone–induced elevations in serum LH. Neuroendocrinology 35:309–12

Blankstein J, Reyes FI, Winter JSD, Faiman C, 1981. Endorphins and the regulation of the menstrual cycle. Clin Endocrinol (Oxf) 14:287–91.

Bogovich K, Richards JS, Reichert Jr LE, 1981. Obligatory role of luteinizing hormone (LH) in the initiation of preovulatory follicular growth in the pregnant rat: specific effects of human chorionic gonadotropin and follicle–stimulating hormone on LH receptors and steroidogenesis in theca, granulosa and luteal cells. Endocrinology 109:860–67

Bridges RS, 1984. A quantitative analysis of the roles of dosage, sequence, and duration of estradiol and progesterone exposure in the regulation of maternal behavior in the rat. Endocrinology 114:930–40

Bridges RS, Goldman BD, 1975. Ovarian control of prolactin secretion during late pregnancy in the rat. Endocrinology 97:496–98

Bridges RS, Ronsheim PM, 1987. Immunoreactive beta endorphin concentrations in brain and plasma during pregnancy in rats: Possible modulation by progesterone and estradiol. Neuroendocrinology (in press)

Brown–Grant K, 1974. Steroid hormone administration and gonadotrophin secretion in the gonadectomized rat. J Endocrinol 62:319–32

Brown–Grant K, Corker CS, Naftolin F, 1972. Plasma and pituitary luteinizing hormone concentrations and peripheral plasma estradiol concentration during early pregnancy and after the administration of progestational steroids in the rat. J Endocrinol 53:31–35

Bruni JF, Van Vugt D, Marshall S, Meites J, 1977. Effects of naloxone, morphine and methionine enkephalin on serum prolactin, luteinizing hormone, follicle stimulating hormone, thyroid stimulating hormone and growth hormone. Life Sci 21:461–66

Butcher RL, Collins WE, Fugo NW, 1974. Plasma concentration of LH, FSH, prolactin, progesterone, and estradiol-17β throughout the 4-day estrous cycle of the rat. Endocrinology 94:1704-08

Cacicedo L, Franco FS, 1986. Direct action of opioid peptides and naloxone on gonadotropin secretion by cultured rat anterior pituitary cells. Life Sci 38:617-25

Caligaris L, Astrada JJ, Taleisnik S, 1971. Release of luteinizing hormone induced by estrogen injection into ovariectomized rats. Endocrinology 88:810-15

Camp P, Akabori A, Barraclough CA, 1985. Correlation of luteinizing hormone surges with estrogen nuclear and progestin cytosol receptors in the hypothalamus and pituitary gland II. Temporal estradiol effects. Neuroendocrinology 40:54-62

Camp P, Barraclough CA, 1985. Correlation of luteinizing hormone surges with estrogen nuclear and progestin cytosol receptors in the hypothalamus and pituitary gland I. Estradiol dose-response effects. Neuroendocrinology 40:45-53

Carson RS, Richards JS, Kahn LE, 1981. Functional and morphological differentiation of theca and granulosa cells during pregnancy in the rat: dependence on increased basal luteinizing hormone activity. Endocrinology 109:1433-41

Casper RF, Alapin-Rubillovitz S, 1985. Progestins increase endogenous opioid peptide activity in postmenopausal women. J Clin Endocrinol Metab 60:34-36

Chao CC, Moss GE, Malven PV, 1986. Direct opioid regulation of pituitary release of bovine luteinizing hormone. Life Sci 39:527-34

Cheesman KL, Chatterton Jr RT, 1982. Effects of antiprogesterone antiserum on serum and ovarian progesterone, gonadotropin secretion, and pregnancy in the rat. Endocrinology 111:564-71

Cheung CY, 1983. Prolactin suppresses luteinizing hormone secretion and pituitary responsiveness to luteinizing hormone-releasing hormone by a direct action at the pituitary. Endocrinology 113:632-38

Cicero TJ, Schainker BA, Meyer ER, 1979. Endogenous opioids participate in the regulation of the hypothalamic-pituitary-luteinizing hormone axis and testosterone's negative feedback control of luteinizing hormone. Endocrinology 104:1286-91

Cicero TJ, Schmoeker PF, Meyer ER, Miller BT, 1985. Luteinizing hormone releasing hormone mediates naloxone's effects on serum luteinizing hormone levels in normal and morphine-sensitized male rats. Life Sci 37:467-74

Cicero TJ, Wilcox CE, Bell RD, Meyer ER, 1980. Naloxone-induced increases in serum luteinizing hormone in male: mechanisms of action. J Pharm Exp Ther 212:573-78

Clarke IJ, Cummins JT, 1982. The temporal relationship between gonadotropin releasing hormone (GnRH) and luteinizing hormone (LH) secretion in ovariectomized ewes. Endocrinology 111:1737-39

Cooper KJ, Fawcett CP, McCann SM, 1973. Variations in pituitary responsiveness to luteinizing hormone-releasing factor during the rat estrous cycle. J Endocrinol 57:187-88

Crowley WR, 1982. Effects of ovarian hormones on norepinephrine and dopamine turnover in individual hypothalamic and extrahypothalamic nuclei. Neuroendocrinology 34:381-86

DePaolo LV, Shander D, Wise PM, Barraclough CA, Channing CP, 1979. Identification of inhibin-like activity in ovarian venous plasma of rats during the estrous cycle. Endocrinology 105:647-54

Devorshak-Harvey E, Bona-Gallo A, Gallo RV, 1987a. The relationship between declining plasma progesterone levels and increasing LH pulse frequency in late gestation in the rat. Endocrinology 120:1597-1601

Devorshak-Harvey E, Bona-Gallo A, Gallo RV, 1987b. Endogenous opioid peptide regulation of pulsatile luteinizing hormone secretion during pregnancy in the rat. Neuroendocrinology (in press)

Devorshak-Harvey E, Peluso JJ, Bona-Gallo A, Gallo RV, 1985. Effect of alterations in pulsatile luteinizing hormone release on ovarian follicular atresia and steroid secretion on diestrus 1 in the rat estrous cycle. Biol Reprod 33:103-11

Dluzen DE, Ramirez VD, 1986. In vivo measurement of LHRH and neurotransmitters from the anterior pituitary gland of castrated rats with push-pull cannulae (PPC). 1st International Congress of Neuroendocrinology, San Francisco, CA, p. 26 (Abst 13)

Docke F, Rohde W, Gerber P, Chaovi R, Dorner G, 1984. Varying sensitivity to the negative oestrogen feedback during the ovarian cycle of female rats: evidence for the involvement of oestrogen and the medial preoptic area. J Endocrinol 102:287-94

Dupont A, Barden N, Cusan L, Merand Y, Labrie F, Vaudry H, 1980. β-endorphin and met-enkephalin: their distribution, modulation by estrogens and haloperidol, and role in neuroendocrine control. Fed Proc 39:2544-50

Erickson, GF, Hsueh AJW, 1978. Secretion of 'inhibin' in rat granulosa cells in vitro. Endocrinology 103:1960-63

Eskay RL, Mical RS, Porter JC, 1977. Relationship between luteinizing hormone releasing hormone concentration in hypophysial portal blood and luteinizing hormone release in intact, castrated, and electrochemically stimulated rats. Endocrinology 100:263-70

Ferin MA, Tempone A, Zimmering PA, Vande Wiele RL, 1969. Effect of antibodies to 17β-estradiol and progesterone on the estrous cycle of the rat. Endocrinology 85:1070-78

Ferin M, Van Vugt DA, Wardlaw S, 1984. The hypothalamic control of the menstrual cycle and the role of endogenous opioid peptides. In: Greep RO (ed.), Recent Progress in Hormone Research, vol 40. New York: Academic Press, pp. 441-85

Forman LJ, Marquis DE, Stevens R, 1985. The effect of chronic estrogen treatment on immunoreactive β-endorphin levels in intact female rats. Proc Soc Exp Biol Med 179:365-72

Fox SR, Shivers BD, Harlan RE, Pfaff DW, 1986. Gonadotrophs and β-endorphin-immunoreactive neurons contain progesterone receptors, but luteinizing hormone releasing hormone-immunoreactive neurons do not. Biol Reprod 34:62 (Abst 25)

Fox SR, Smith MS, 1985. Changes in the pulsatile pattern of luteinizing hormone secretion during the rat estrous cycle. Endocrinology 116:1485-92

Freeman MC, Dupke KC, Croteau CM, 1976. Extinction of the estrogen-induced daily signal for LH release in the rat: a role for the proestrous surge of progesterone. Endocrinology 99:223-29

Fujii T, Hoover DJ, Channing CP, 1983. Changes in inhibin activity, and progesterone, oestrogen, and androstenedione concentrations in rat follicular fluid throughout the oestrous cycle. J Reprod Fertil 69:307-14

Gabriel SM, Simpkins JW, Kalra SP, 1983. Modulation of endogenous opioid influence on luteinizing hormone secretion by progesterone and estrogen. Endocrinology 113:1806-11

Gallo RV, 1981. Pulsatile LH release during periods of low level LH secretion in the rat estrous cycle. Biol Reprod 24:771-77

Gallo RV, Bona-Gallo A, 1985. Lack of ovarian steroid negative feedback on pulsatile LH release between estrus and diestrus 1 in the rat estrous cycle. Endocrinology 116:1525-28

Gallo RV, Devorshak-Harvey E, Bona-Gallo A, 1985. Pulsatile luteinizing hormone release during pregnancy in the rat. Endocrinology 116:2637-42

Goodman RL, 1978. A quantitative analysis of the physiological role of estradiol and progesterone in the control of tonic and surge secretion of luteinizing hormone in the rat. Endocrinology 102:142-50

Goodman RL, Daniel K, 1985. Modulation of pulsatile luteinizing hormone secretion by ovarian steroids in the rat. Biol Reprod 32:217-25

Goodman RL, Karsch FJ, 1980. Pulsatile secretion of luteinizing hormone: differential suppression by ovarian steroids. Endocrinology 107:1286-90

Greenwald GS, Johnson DC, 1968. Gonadotropic requirements for the maintenance of pregnancy in the hypophysectomized rat. Endocrinology 83:1052–64

Higuchi T, Kawakami M, 1982. Changes in the characteristics of pulsatile luteinizing hormone secretion during the oestrous cycle and after ovariectomy and estrogen treatment in female rats. J Endocrinol 94:177–82

Ieiri T, Chen HT, Meites J, 1979. Effects of morphine and naloxone on serum levels of luteinizing hormone and prolactin in prepubertal male and female rats. Neuroendocrinology 29:288–92

Kalra PS, Crowley WR, Kalra SP, 1987. Differential in vitro stimulation by naloxone and K$^+$ of luteinizing hormone–releasing hormone and catecholamine release from the hypothalami of intact and castrated rats. Endocrinology 120:178–85

Kalra SP, 1975. Observation on the facilitation of the preovulatory rise of LH by estrogen. Endocrinology 96:23–28

Kalra SP, 1981. Neural loci involved in naloxone-induced luteinizing hormone release: effects of a norepinephrine synthesis inhibitor. Endocrinology 109:1805–10

Kalra SP, 1983. Opioid peptides-inhibitory neuronal systems in regulation of gonadotropin secretion. In: McCann SM, Dhindsa DS (eds.), Role of Peptides and Proteins in Control of Reproduction. New York: Elsevier Biomedical, pp. 63–88

Kalra SP, 1986. Neural circuitry involved in the control of LHRH secretion: a model for preovulatory LH release. In: Ganong WF, Martini L (eds.), Frontiers in Neuroendocrinology, vol. 9, New York: Raven Press, pp. 31–75

Kalra SP, Kalra PS, 1974a. Effects of circulating estradiol during the rat estrous cycle on LH release following electrochemical stimulation of preoptic brain or administration of synthetic LRF. Endocrinology 94:845–51

Kalra SP, Kalra PS, 1974b. Temporal interrelationships among circulating levels of estradiol, progesterone and LH during the rat estrous cycle: effects of exogenous progesterone. Endocrinology 95:1711–18

Kalra SP, Kalra PS, 1984. Opioid-adrenergic-steroid connections in regulation of luteinizing hormone secretion in the rat. Neuroendocrinology 38:418–26

Kalra SP, Simpkins JW, 1981. Evidence for noradrenergic mediation of opioid effects on luteinizing hormone secretion. Endocrinology 109:776–82

Kinoshita F, Nakai Y, Katakami H, Kato Y, Yajima H, Imura H, 1980. Effect of β-endorphin on pulsatile luteinizing hormone release in conscious castrated rats. Life Sci 27:843–46

Krey LC, Parsons B, 1982. Characterization of estrogen stimuli sufficient to initiate cyclic luteinizing hormone release in acutely ovariectomized rats. Neuroendocrinology 34:315–22

Leadem CA, Crowley WR, Simpkins JW, Kalra SP, 1985. Effects of naloxone on catecholamine and LHRH release from the perifused hypothalamus of the steroid-primed rat. Neuroendocrinology 40:497–500

Leadem CA, Kalra SP, 1985. Effects of endogenous opioid peptides and opiates on luteinizing hormone and prolactin secretion in ovariectomized rats. Neuroendocrinology 41:342–52

Legan S, Coon GA, Karsch FJ, 1975. Role of estrogen as initiator of daily LH surges in the ovariectomized rat. Endocrinology 96:50–56

Legan S, Karsch F, 1975. A daily signal for the LH surge in the ovariectomized rat. Endocrinology 96:57–62

Leipheimer RE, Bona-Gallo A, Gallo RV, 1984. The influence of progesterone and estradiol on the acute changes in pulsatile luteinizing hormone release induced by ovariectomy in diestrous day 1 in the rat. Endocrinology 114:1605–12

Leipheimer RE, Bona-Gallo A, Gallo RV, 1985. Ovarian steroid regulation of pulsatile LH release during the interval between the mornings of diestrus 2 and proestrus in the rat. Neuroendocrinology 41:252–57

Leipheimer RE, Bona-Gallo A, Gallo RV, 1986a. Ovarian steroid regulation of basal pulsatile LH release between the mornings of proestrus and estrus in the rat. Endocrinology 118:2083-90

Leipheimer RE, Bona-Gallo A, Gallo RV, 1986b. Influence of estradiol and progesterone on pulsatile LH secretion in 8-day ovariectomized rats. Neuroendocrinology 43:300-07

Leipheimer RE, Gallo RV, 1983. Acute and long-term changes in central and pituitary mechanisms regulating pulsatile luteinizing hormone secretion after ovariectomy in the rat. Neuroendocrinology 37:421-26

Levine JE, Pau K-YF, Ramirez VD, Jackson GL, 1982. Simultaneous measurement of luteinizing hormone-releasing hormone and luteinizing hormone release in unanesthetized, ovariectomized sheep. Endocrinology 111:1449-55

Levine JE, Ramirez VD, 1982. Luteinizing hormone-releasing hormone release during the rat estrous cycle and after ovariectomy, as estimated with push-pull cannulae. Endocrinology 111:1439-48

Linkie DM, Niswender GD, 1972. Serum levels of prolactin, luteinizing hormone and follicle stimulating hormone during pregnancy in the rat. Endocrinology 90:632-37

Loewit K, Badawy S, Laurence K, 1969. Alteration of corpus luteum function in the pregnant rat by antiluteinizing serum. Endocrinology 84:244-51

Lumpkin MD, DePaolo LV, Negro-Vilar A, 1984. Pulsatile release of follicle-stimulating hormone in ovariectomized rats is inhibited by porcine follicular fluid (inhibin). Endocrinology 114:201-06

Malven PV, Bossut DFB, Diekman MA, 1984. Effects of naloxone and electroacupuncture treatment on plasma concentrations of LH in sheep. J Endocrinol 101:75-80

Mano A, 1983. Pituitary responsiveness to luteinizing hormone releasing hormone during pregnancy in the rat. Acta Obstet Gynec Jpn 35:1649- 52

McCann SM, 1962. Effect of progesterone on plasma luteinizing hormone activity. Am J Physiol 202:601-04

McGinnis MY, Krey LC, MacLusky NJ, McEwen BS, 1981. Steroid receptor levels in intact and ovariectomized estrogen-treated rats: an examination of quantitative, temporal and endocrine factors influencing the efficacy of an estradiol stimulus. Neuroendocrinology 33:158-65

McPherson III JC, Costoff A, Mahesh VB, 1975. Influence of estrogen-progesterone combinations on gonadotropin secretion in castrate female rats. Endocrinology 97:771-79

Meites J, Bruni JF, Van Vugt DA, Smith AF, 1979. Relation of endogenous opioid peptides and morphine to neuroendocrine functions. Life Sci 24:1325-36

Meites J, Van Vugt DA, Forman LJ, Sylvester Jr PW, Ieiri T, Sonntag W, 1983. Evidence that endogenous opiates are involved in control of gonadotropin secretion. In: Bhatnager AS (ed.), Anterior Pituitary Gland. New York: Raven Press, pp 327-40

Melis GB, Paoletti AM, Gambacciani M, Mais V, Fioretti P, 1984. Evidence that estrogens inhibit LH secretion through opioids in postmenopausal women using naloxone. Neuroendocrinology 39:60-63

Miyake A, Terakawa N, Tasaka K, Shimizu I, Ohtsuka S, Lee J-W, Aono T, 1985. Prolactin inhibits oestradiol-induced luteinizing hormone release at the pituitary level. Acta Endocrinol 109:204-07

Morishige WK, Pepe GJ, Rothchild I, 1973. Serum luteinizing hormone, prolactin and progesterone levels during pregnancy in the rat. Endocrinology 92:1527-30

Morrell JI, McGinty JF, Pfaff DW, 1985. A subset of β-endorphin- or dynorphin-containing neurons in the medial basal hypothalamus accumulates estradiol. Neuroendocrinology 41:417-26

Moudgal NR, 1969. Effect of ICSH on early pregnancy in hypophysectomized pregnant rats. Nature 222:286-87

Negro-Vilar A, 1973. Interaction between gonadal steroids and LH-releasing hormone to control gonadotropin secretion at the pituitary level. Acta Physiol Latinoam 23:494-96

Nequin LG, Alvarez J, Schwartz NB, 1979. Measurements of serum steroid and gonadotropin levels and uterine and ovarian variables throughout 4-day and 5-day estrous cycles in the rat. Biol Reprod 20:659-70

Panerai AE, Martini A, Casanueva F, Petraglia F, DiGiulio AM, Mantegazza P, 1983. Opiates and their antagonists modulate luteinizing hormone acting outside the blood brain barrier. Life Sci 32:1751-56

Pau K-YF, Orstead KM, Hess DL, Spies HG, 1986. Feedback effects of ovarian steroids on the hypothalamic-hypophyseal axis in the rabbit. Biol Reprod 35:1009-23

Pencharz RI, Long JA, 1933. Hypophysectomy in the pregnant rat. Am J Anat 53:117-39

Petraglia F, Baraldi M, Giarre G, Facchinetti F, Santi M, Volpe A, Genazzani AR, 1985. Opioid peptides of the pituitary and hypothalamus: changes in pregnant and lactating rats. J Endocrinol 105:239-45

Petraglia F, Locatelli V, Facchinetti F, Bergamaschi M, Genazzani AR, Cocchi D, 1986. Oestrous cycle-related LH responsiveness to naloxone: effect of high oestrogen levels on the activity of opioid receptors. J Endocrinol 108:89-94

Petraglia F, Locatelli V, Penalva A, Cocchi D, Genazzani AR, Muller EE, 1984. Gonadal steroid modulation of naloxone-induced LH secretion in the rat. J Endocrinol 101:33-39

Piva F, Maggi R, Limonta P, Martini L, 1986. Effect of naloxone on luteinizing hormone, follicle stimulating hormone, and prolactin secretion in the different phases of the estrous cycle. Endocrinology 117:766-72

Quigley ME, Yen SSC, 1980. The role of endogenous opiates on LH secretion during the menstrual cycle. J Clin Endocrinol Metab 51:179-81

Raj HGM, Moudgal NR, 1970. Hormonal control of gestation in the rat. Endocrinology 86:874-89

Raj HGM, Talbert LM, Easterling WE, Dym RC, 1979. Role of pituitary LH and placenta in luteal progesterone production and maintenance of pregnancy in the rat. Adv Exp Med Biol 112:535-40

Rasmussen DD, Liu JH, Wolf PL, Yen SSC, 1983. Endogenous opioid regulation of gonadotropin-releasing hormone release from the human fetal hypothalamus in vitro. J Clin Endocrinol Metab 57:881-84

Rivier J, Spiess J, McClintock R, Vaughan J, Vale W, 1985. Purification and partial characterization of inhibin from porcine follicular fluid. Biochem Biophys Res Commun 133:120-27

Ropert JF, Quigley ME, Yen SSC, 1981. Endogenous opiates modulate pulsatile luteinizing hormone release in humans. J Clin Endocrinol Metab 52:583-85

Sander HJ, Van Leeuwen ECM, Jong FH de, 1984. Inhibin-like activity in media from cultured rat granulosa cells collected throughout the estrous cycle. J Endocrinol 103:77-84

Sarkar DK, Fink G, 1980. Luteinizing hormone-releasing factor in pituitary stalk plasma from long-term ovariectomized rats: effects of steroids. J Endocrinol 86:511-24

Sarkar DK, Yen SSC, 1985. Changes in β-endorphin-like immunoreactivity in pituitary portal blood during the estrous cycle and after ovariectomy in rats. Endocrinology 116:2075-79

Smith MS, Freeman ME, Neill JD, 1975. The control of progesterone secretion during the estrous cycle and early pseudopregnancy in the rat: prolactin, gonadotropin, and steroid levels associated with rescue of the corpus luteum of pseudopregnancy. Endocrinology 96:219-26

Smith MS, Neill JD, 1976. Termination at mid-pregnancy of the two daily surges of plasma prolactin initiated by mating in the rat. Endocrinology 98:696-701

Snowden EU, Khan-Dawood FS, Dawood MY, 1984. The effect of naloxone on endogenous opioid regulation of pituitary gonadotropins and prolactin during the menstrual cycle. J Clin Endocrinol Metab 59:298-302

Soules MR, Steiner RA, Clifton DK, Cohen NL, Askel S, Bremner WJ, 1984. Progesterone modulation of pulsatile luteinizing hormone secretion in normal women. J Clin Endocrinol Metab 58:378-83

Sylvester PW, Van Vugt DA, Aylsworth CA, Hanson EA, Meites J, 1982. Effects of morphine and naloxone on inhibition by ovarian hormones of pulsatile release of LH in ovariectomized rats. Neuroendocrinology 34:269-73

Taya K, Greenwald G, 1981. In vivo and in vitro ovarian steroidogenesis in the pregnant rat. Biol Reprod 25:683-91

Terranova PF, Greenwald GS, 1981. Acute effects of an antiserum to luteinizing hormone on ovarian steroidogenesis in the pregnant rat. J Endocrinol 90:19-30

Turgeon JL, Barraclough CA, 1974. Pulsatile plasma LH rhythms in normal and androgen-sterilized ovariectomized rats: effects of estrogen treatment. Proc Soc Exp Biol Med 145:821-25

Van Vugt DA, Bakst G, Dyrenfurth I, Ferin M, 1983. Naloxone stimulation of luteinizing hormone secretion in the female monkey: influence of endocrine and experimental conditions. Endocrinology 113:1858-64

Van Vugt DA, Lam NY, Ferin M, 1984. Reduced frequency of pulsatile luteinizing hormone secretion in the luteal phase of the Rhesus monkey: involvement of endogenous opiates. Endocrinology 115:1095-01

Van Vugt DA, Sylvester PW, Aylsworth CF, Meites J, 1982. Counteraction of gonadal steroid inhibition of luteinizing hormone release by naloxone. Neuroendocrinology 34:274-78

Veldhuis JD, Rogol AD, Samojlik E, Ertel NH, 1984. Role of endogenous opiates in the expression of negative feedback actions of androgen and estrogen on pulsatile properties of luteinizing hormone secretion in man. J Clin Invest 74:47-55

Voogt J, Robertson M, Friesen H, 1982. Inverse relationship of prolactin and placental lactogen during pregnancy. Biol Reprod 26:800-05

Wardlaw SL, Frantz AG, 1983. Brain β-endorphin during pregnancy, parturition and the postpartum period. Endocrinology 113:1664-68

Wardlaw SL, Thoron L, Frantz AG, 1982a. Effects of sex steroids on brain β-endorphin. Brain Res 245:327-31

Wardlaw SL, Wang PJ, Frantz AG, 1985. Regulation of β-endorphin and ACTH in brain by estradiol. Life Sci 37:1941-47

Wardlaw SL, Wehrenberg WB, Ferin M, Antunes JL, Frantz AG, 1982b. Effect of sex steroids on β-endorphin in hypophyseal portal blood. J Clin Endocrinol Metab 55:877-81

Wehrenberg WB, Wardlaw SL, Frantz AG, Ferin M, 1982. β-endorphin in hypophyseal portal blood: variations throughout the menstrual cycle. Endocrinology 111:879-81

Weick RF, 1977. Effects of estrogen and progesterone on pulsatile discharges of luteinizing hormone in the ovariectomized rat. Can J Physiol Pharmacol 55:226-33

Weick RF, Noh KA, 1984. Inhibitory effects of estrogen and progesterone on several parameters of pulsatile LH release in the ovariectomized rat. Neuroendocrinology 38:351-56

Wiesner JB, Koenig JI, Krulich L, Moss RL, 1984. Site of action for β-endorphin-induced changes in plasma luteinizing hormone and prolactin in the ovariectomized rat. Life Sci 34:1463-73

Wilcox JN, Roberts JL, 1985. Estrogen decreases rat hypothalamic proopiomelanocortin messenger ribonucleic acid levels. Endocrinology 117:2392-96

Wilkes MM, Yen SSC, 1981. Augmentation by naloxone of efflux of LRF from superfused medial basal hypothalamus. Life Sci 28:2355-59

SELECTIVE MODULATION OF FSH AND LH SECRETION BY STEROIDS

V.B. Mahesh, L.L. Murphy and J.L. O'Conner

Department of Physiology and Endocrinology
Medical College of Georgia
Augusta, Georgia 30912

INTRODUCTION

Follicle-stimulating hormone (FSH) and luteinizing hormone (LH) secreted by the anterior pituitary are important regulators of male and female reproductive processes. The hypothalamic hormone luteinizing hormone releasing hormone (LHRH) has been shown to regulate the release of LH and FSH from the anterior pituitary (Schally et al., 1971). Extensive work has been done on the interaction between gonadal steroids, hypothalamic LHRH and the anterior pituitary in the synthesis and release of LH and this topic has been reviewed by Kalra and Kalra (1983). The regulation of the secretion of FSH, however, cannot be explained adequately with LHRH as the sole regulatory hypothalamic hormone; this is due to a significant divergence between FSH and LH secretion during the ovulatory cycle (Bast and Greenwald, 1974; Smith et al., 1975), after ovariectomy in the rat (Mahesh et al., 1972; Tapper et al., 1972; Zanisi and Martini, 1975a) and during puberty (Dohler and Wuttke, 1975; Payne et al., 1977). This chapter will review the current concepts concerning the role of (a) specific FSH releasing and inhibiting peptides, (b) neuroendocrine control of FSH secretion and (c) steroid hormones in the divergent secretion of FSH and LH.

SPECIFIC FSH RELEASING AND INHIBITING PEPTIDES

The most direct way to explain divergent secretion of FSH and LH under selected physiological conditions would be to postulate the presence of a distinct FSH releasing factor (FSH-RF). Such a postulate was made repeatedly (Bowers et al., 1973; Igarashi et al., 1973; Igarashi and McCann, 1974) but has not been supported adequately by experimental evidence (Shahmanesh and Jeffcoate, 1976; Schally et al., 1976). More recently, the concept of a FSH-RF is being revived by using a combination of bioassays and radioimmunoassays (Lundanes et al., 1980; Mizunuma et al., 1983). Nevertheless the presence of a specific FSH-RF is still not widely accepted.

The concept of a gonadal peptide responsible for the modulation of FSH secretion was first proposed by McCullagh in 1932. Extensive work on the presence of such a peptide of ovarian origin has been carried out by several investigators (Schwartz and Channing, 1977; Rush et al., 1981) and cyclic fluctuations of inhibin-like material has been reported in ovarian tissue and ovarian venous blood (Chappel, 1979; DePaolo et al., 1979). The ovarian peptides have now been isolated and characterized (Ling et al., 1985; Mason et al., 1985; Miyamoto et al., 1985; Rivier et al., 1985;

Robertson et al., 1985, 1986). Porcine inhibin occurs as two heterodimers with a common α -subunit and a different but related β -subunit. An infusion of antibodies to inhibin on the day of proestrus and estrus in the rat enhanced the magnitude of FSH secretion without altering LH (Rivier et al., 1986). However it did not alter the pattern and duration of FSH elevation indicating the existence of other controlling mechanisms for FSH release in addition to inhibin. The β-subunit of inhibin has been found to have FSH releasing activity (Ling et al., 1986; Vale et al., 1986) and requires a period of 4 to 24 h to manifest its biological activity. Further evaluation is needed to determine if this peptide plays a physiological role in the regulation of FSH secretion.

NEUROENDOCRINE CONTROL OF FSH SECRETION

It is well recognized that electrical stimulation of the preoptic suprachiasmatic nuclear region of the hypothalamus on the day of proestrus brings about a proestrus type of LH and FSH surge. Electrical stimulation of the dorsal anterior hypothalamic area (DAHA) elicits preferential FSH release (Kalra et al., 1971; Chappel and Barraclough, 1976). Bilateral lesions of the DAHA resulted in a reduction of serum FSH levels during proestrus (Lumpkin and McCann, 1982). Selective release of FSH was also shown to occur after prostaglandin E_2 was implanted stereotaxically in the DAHA region (Ojeda et al., 1972). These observations indicate the presence of separate neural pathways involved in the control of FSH and LH and anatomical evidence for these pathways was provided by Kimura and Kawakami (1978). Further work by several investigators has shown that deafferentation of neurosecretory cells in the arcuate-median eminence region from those in the preoptic-anterior hypothalamic area prior to 1800 h of proestrus reduced or abolished the FSH release on estrus but deafferentation after 1800 h did not have any effect (Chappel et al., 1979; Rush et al., 1980, 1982; Blake et al., 1982). The absence of the requirement for LHRH in the secretion of FSH on estrus was further confirmed by demonstrating that antibodies to LHRH administered on the day of estrus did not have any effect on FSH secretion (Hasegawa et al., 1981; Rush et al., 1982). Blockade of the proestrus surge of FSH and LH by the administration of phenobarbital also resulted in the abolition of the elevated secretion of FSH at estrus. Experiments with deafferentation before 1800 h and phenobarbital blockade of the surge suggest that FSH and LH secretion at proestrus may be essential for the secretion of FSH on estrus. Several investigators have studied the effect of LH and FSH administration on subsequent secretion and release of FSH (Ojeda and Ramirez, 1969, 1970; Schwartz and Talley, 1978; Ashiru and Blake, 1979, 1980; Ashiru et al., 1981; Coutifaris and Chappel, 1982). Both hormones were able to stimulate FSH release by the anterior pituitary. Binding sites for FSH in the hypothalamus have also been reported (Davies et al., 1975).

It is now well established that the secretion of LHRH by the hypothalamus is pulsatile in nature and this pulsatibility is essential for maintaining the sensitivity of the anterior pituitary to LHRH (Belchetz et al., 1978). The mode of administration of LHRH (Pickering and Fink, 1977; Wise et al., 1979; Turgeon and Waring, 1982) or the pulse intervals (Lincoln, 1979; Wildt et al., 1981; Pohl et al., 1983) has been shown to induce a divergence in FSH and LH secretory patterns.

REGULATION OF FSH AND LH SECRETION BY ESTROGENS: EVIDENCE OF DIFFERENTIAL REGUALTION IN THE ABSENCE OF GONADAL PEPTIDES

Estrogens exert both a positive and a negative feedback effect on gonadotropin secretion. The presence of the negative feedback system was established by classical experiments demonstrating that ovariectomy resulted in secretion of elevated levels of FSH and LH which could be suppressed by estrogen replacement. The manifestation of the positive feedback effect requires 12 to 24 h and is mediated through hypothalamic actions through the release of LHRH and direct pituitary actions by increasing pituitary responsiveness to LHRH. This subject has been reviewed recently by Mahesh and Muldoon (1987).

After gonadectomy in male and female rats, the rise in serum FSH levels was generally found to be more rapid than LH (Mahesh et al., 1972; Tapper et al., 1972; Eldridge et al., 1974; Zanisi and Martini, 1975a). The more rapid rise in serum FSH levels as compared to LH levels may have been due to the removal of gonadal peptides that suppress the secretion of FSH. The administration of gonadal steroids to male and female rats after gonadectomy brought about a suppression of elevated levels of serum FSH and LH. The suppression of serum LH levels, however, was more rapid than the suppression of FSH levels (Eldridge and Mahesh, 1974; McPherson et al., 1974b). The more rapid suppression of serum LH as compared to serum FSH could also be explained due to the absence of a gonadal peptide that suppressed FSH secretion. However, of interest was a remarkable difference in the ability of different natural and synthetic estrogens to suppress serum FSH levels as compared to serum LH levels (Table 1). The synthetic estrogen 11β-acetoxy estradiol was 4.4 fold more potent than estradiol in the restoration of uterine weight and suppression of serum LH in the ovariectomized rat. At the 0.05 µg/kg body weight dose level this steroid suppressed serum LH levels to 2.6% of that found in the ovariectomized rat while it stimulated serum FSH secretion (Uberoi et al., 1985). The above mentioned results indicate differential regulation of FSH and LH secretion by estrogens independent of the presence or absence of gonadal regulatory peptides.

Table 1*. Relative Ability of Various Natural and Synthetic Estrogens in Suppressing Serum FSH and LH Levels in 26-Day-Old Ovariectomized Immature Rats at Dose Levels Required to Give Uterine Weights of 100 mg/100 g BW

Hormone	Serum FSH	Serum LH
	(percentage of ovariectomized control)	
Estradiol	17.2	8.0
Estradiol Benzoate	29.0	7.3
Estrone	19.5	11.4
Equilenin	50.0	9.4
Ethynyl Estradiol	50.0	12.0
Mestranol	32.5	10.8
11β-acetoxy extradiol	71.4	2.6

*Data compiled from McPherson et al., 1974b and Uberoi et al., 1985.

MODULATION OF FSH AND LH SECRETION BY PROGESTERONE

The experiments of Everett in 1948 clearly show that progesterone could have a stimulatory effect or an inhibitory effect on gonadotropin secretion relative to the time of administration of progesterone in the ovulatory cycle. When progesterone was given within a few hours of the preovulatory gonadotropin surge in the cycling animal or the estrogen induced LH surge in the ovariectomized animal, its effects were stimulatory. Progesterone given at earlier time periods in the cycle or in estrogen treated ovariectomized rats was inhibitory to the gonadotropic surge. The facilitative (Everett, 1948; Nallar et al., 1966; Caligaris et al., 1968; Krey et al., 1973; Martin et al., 1974; Greeley et al., 1975;

McPherson et al., 1975; McPherson and Mahesh, 1979) and inhibitory (Everett, 1948; Martin et al., 1974; McPherson et al., 1975; Freeman et al., 1976; Banks and Freeman, 1978; McPherson and Mahesh, 1979; Banks et al., 1980) effects of progesterone on gonadotropin secretion are now well recognized in laboratory animals. Estrogen priming of the ovariectomized animal is essential for progesterone to exert its action on gonadotropin secretion (McCann, 1962; Nallar et al., 1966; McPherson et al., 1974a, 1975). The manifestation of the full gonadotropin surge in ovariectomized animals treated with estrogen is dependent upon progesterone (Mann and Barraclough, 1973; Krey et al., 1973; Aiyer and Fink, 1974). A physiological role of progesterone in bringing about a rapid release of LH and FSH and manifestation of the full gonadotropin surge during the process of ovulation has been indicated by studies in the rat, monkey and the human (Leyendecker et al., 1972; Shaw et al., 1975; Chang and Jaffe, 1978; March et al., 1979; Helmond et al., 1980; Liu and Yen, 1983; Rao and Mahesh, 1986). The facilitative and inhibitory effects of progesterone of gonadotropin secretion have been reviewed recently by Mahesh and Muldoon (1987) and a model has been proposed to explain these effects. When progesterone is administered either in the presence of endogenous estrogens or in conjugation with estrogen priming under controlled conditions, a greater amount of FSH is secreted as compared with FSH secretion in animals not treated with progesterone. This effect of progesterone has been demonstrated in the rat (McPherson et al., 1975; McPherson and Mahesh, 1979), the hamster (Shander and Goldman, 1978), the human (Leyendecker et al., 1972; Chang and Jaffe, 1978; March et al., 1979; Lutjen et al., 1986) and the monkey (Helmond et al., 1981). Differential effects of progesterone metabolites on FSH and LH secretion have also been reported in estrogen primed ovariectomized rats (Zanisi and Martini, 1975b, 1979; Nuti and Karavolas, 1977; Gilles and Karavolas, 1981).

AN EXPERIMENTAL MODEL FOR DETAILED STUDIES OF PROGESTERONE AND ITS METABOLITES ON FSH AND LH SECRETION

Detailed studies of the effects of steroid hormones on gonadotropin secretion independent of the modulatory effects of gonadal peptides can only be carried out in vivo by using ovariectomized rats. In the absence of estrogens due to ovariectomy, progesterone is unable to exert significant effects on gonadotropin secretion (McCann, 1962; Nallar et al., 1966; McPherson et al., 1974a, 1975; McPherson and Mahesh, 1979). This is due to the fact that estrogens are required for the synthesis of anterior pituitary and hypothalamic progesterone receptors. Estrogen priming of the ovariectomized rat, either in the form of single injections of 5–50µg of estradiol or estradiol benzoate or by using Silastic implants of estradiol for varying periods of time, result in daily surges of LH for several days (Caligaris et al., 1971; Norman et al., 1973; Ramirez and Sawyer, 1974; Legan et al., 1975; Legan and Karsch, 1975; Freeman et al., 1976; Goodman, 1978). Therefore by using the above regimens of estrogen administration, the effect of progesterone on gonadotropin secretion can only be studied in terms of the advancement, delay or modification of the estrogen induced gonadotropin surge.

An animal model which does not have gonadal peptide modulation of gonadotropin secretion and which is sensitive to progesterone in the absence of estrogen induced gonadotropin surges would be highly desirable to study the effects of progesterone on gonadotropin secretion. The immature rat appeared to be ideal for such studies because the inhibitory effects of progesterone on the estrogen induced daily LH surges have been shown to persist for 8–10 days in the adult rat (Banks et al., 1980). The approach developed in our laboratory for such a model was to use the 26-day-old ovariectomized rat; the animals were primed with a low dose of estradiol (0.1 μg/kg BW) for 4 days given as 2 divided doses starting on the day of ovariectomy (McPherson et al., 1975; McPherson and Mahesh, 1979; Mahesh et al., 1980; Peduto and Mahesh, 1985). In this model, serum LH levels were reduced to approximately 25% of that found in the vehicle treated ovariectomized animal and an estrogen induced surge of LH did not take place in the afternoon. The progesterone receptors induced by the 0.1 μg/kg BW dose of estradiol in the pituitary and the hypothalamus were comparable to those induced by the larger dose of estradiol (0.2 μg/kg BW) that resulted in an estrogen induced LH surge (Mahesh et al., 1980; Smanik et al., 1983; Peduto and Mahesh, 1985). Thus this animal model appeared to be ideal for the study of the effects of progesterone on the modulation of gonadotropin secretion.

Extensive studies have been carried out to examine the role of progesterone in modulating gonadotropin secretion using the low dose estradiol primed ovariectomized immature rat. When progesterone was administered in combination with estradiol, its effects were facilitative or inhibitory to gonadotropin release depending upon the dose of progesterone used (McPherson et al., 1975). The dose related effect of progesterone was confirmed in the ovariectomized immature rat primed for 4 days with 0.1 μg/kg BW of estradiol given a single injection of progesterone (McPherson and Mahesh, 1979). Studies to determine the mechanism of action of progesterone in gonadotropin release in the low dose estrogen primed immature rat have shown that progesterone administration resulted in (a) reduction in occupied estradiol receptors of the anterior pituitary but not the hypothalamus (Smanik et al., 1983; Mahesh and Muldoon, 1987), (b) stimulation of the 17β −hydroxysteroid dehydrogenase activity in the anterior pituitary but not in the hypothalamus (El Ayat and Mahesh, 1984), and (c) induction of a dose related rapid release of LHRH by the hypothalamus (Peduto and Mahesh, 1985). These results and their implications have been discussed recently by Mahesh and Muldoon (1987).

SELECTIVE EFFECT OF PROGESTERONE AND ITS METABOLITES ON FSH AND LH SECRETION

Progesterone   The administration of 0.8 mg/kg BW of progesterone to the 26−day−old ovariectomized rat primed with 0.1 μg/kg BW of estradiol at 9:30 AM brought about an LH and FSH surge in the early afternoon while the 3.2 mg/kg BW dose of progesterone was either suppressive or had no effect (McPherson and Mahesh, 1979). The serum LH levels returned to basal levels by 6 PM while serum FSH levels remained elevated when the 0.8 mg/kg BW dose of progesterone was administered and were suppressed by the 3.2 mg/kg BW dose of progesterone (Fig. 1). The stimulatory dose of progesterone resulted in elevated levels of both pituitary LH and pituitary FSH. The selective secretion of FSH over LH during the late evening and night is of considerable interest because a single injection of progesterone to the ovariectomized estrogen primed immature rat reproduced the FSH and LH secretory patterns found in the cycling rat or hamster on the day of proestrus (Bast and Greenwald, 1974; Smith et al., 1975). Because ovariectomized rats were used, this selective effect of progesterone on FSH secretion occurred in the absence of gonadal peptides. Evidence for a greater release of FSH as a result of progesterone administration in the rat, the hamster, the human, and the monkey has been cited earlier in this paper.

Progesterone Metabolites   The role of progesterone in manifesting the full gonadotropin surge leading to ovulation is now being recognized. The ovary secretes a number of progesterone metabolites during the rat estrus cycle and metabolites such as 5α−pregnane−3,20−dione (5α−dihydroprogesterone; 5α−DHP) and 3α−hydroxy−5α−pregnan−20−one (3α,5α−tetrahydroprogesterone;

3α,5α–THP) have been isolated from peripheral blood and ovarian venous blood in the rat on the day of proestrus (Ichikawa et al., 1971, 1974; Holzbauer, 1975). The presence of 3α–5α– and 20α–reductase activities in the hypothalamus and the pituitary have also been reported by several investigators (Cheng and Karavolas, 1973; Tabei et al., 1974; Nowak et al., 1976; Stupnicka et al., 1977; Krause et al., 1981). Among other times in the ovulatory cycle, an increase in 3α–and 5α–reduction occurs in the hypothalamus and the pituitary on the day of proestrus (Cheng and Karavolas, 1973; Nowak et al., 1976; Stupnicka et al., 1977; Krause et al., 1981). 5α–dihydroprogesterone is also concentrated in the brain and the pituitary in vivo (Karavolas et al., 1979). Therefore, a detailed study of the effects of 5α–dihydroprogesterone and 3α,5α –tetrahydro- progesterone on gonadotropin secretion was undertaken by Murphy and Mahesh (1984a,b).

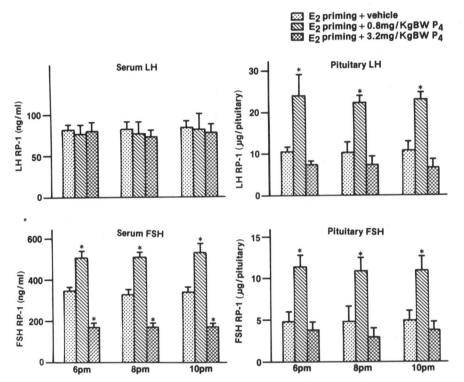

Fig. 1.   26–day–old ovariectomized female rats primed with 0.1 µg/kg BW of estradiol for 4 days and given a single injection of either 0.8 or 3.2 mg/kg BW of progesterone at 9:30 a.m. on day 5, were sacrificed at 6 p.m., 8 p.m. and 10 p.m. for the measurement of serum and pituitary FSH and LH levels. Note elevation of serum FSH and pituitary FSH and LH in animals treated with 0.8 mg/kg BW of progesterone (* P < 0.01) and suppression of serum FSH by the 3.2 mg/kg BW dose of progesterone (* P < 0.01) (Data from McPherson and Mahesh, 1979).

Various doses of 5α–dihydroprogesterone (5α–DHP) were administered in combination with 0.1 µg/kg BW of estradiol to 26–day–old ovariectomized rats for 4 days (Murphy and Mahesh, 1984a). Serum FSH levels were elevated by all doses of 5α–DHP while serum LH levels were unaltered. The 0.4 mg/kg BW of 5α–DHP gave the greatest elevation in serum FSH and was used for further studies. Female rats were ovariectomized at 26 days of age and administered estradiol (0.1 µg/kg BW) for 4 days starting with the day of ovariectomy. At 9:30 a.m. on day 5, a single dose of 5α–DHP (0.4 mg/kg BW) was administered. A significant elevation of serum FSH levels occurred at 12 n and 6 p.m. while serum LH levels remained unchanged by the treatment (Fig. 2). Of considerable interest was the fact that both pituitary FSH and LH levels were elevated at 4 p.m. and 6 p.m. but only FSH was secreted in increased quantities at 6 p.m. (Murphy and Mahesh, 1984a).

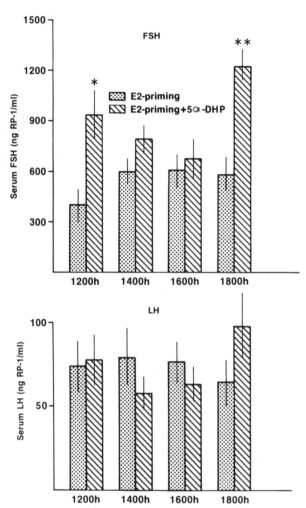

Fig. 2. Serum FSH and LH levels in female rats ovariectomized on 26 days of age and primed with 0.1 µg/kg BW of estradiol for 4 days. On the morning of day 5 the rats received their last injection of estradiol at 8 a.m. and at 9:30 a.m. were administered a single injection of corn oil or 0.4 mg/kg BW of 5α–DHP. Rats were sacrificed at 12 n, 2 p.m., 4 p.m. and 6 p.m. Solid verticle lines and bars indicate SEM of means of not less than 6 rats. * = P < 0.05 and ** = P < 0.01 vs estradiol primed controls (From Murphy and Mahesh, 1984a).

The mechanism involved in the selective release of FSH by 5α–DHP in unclear at the moment. It is possible that the enhanced secretion of FSH at 6 p.m. was related to its initial release at 12 n. Both FSH and LH have been shown to be able to stimulate the release of FSH (Ojeda and Ramirez, 1969, 1970; Schwartz and Talley, 1978; Ashiru and Blake, 1979, 1980;

Ashiru et al., 1981; Coutifaris and Chappel, 1982). In order to test the dependency of the 6 p.m. release of FSH on the 12 n release vehicle or phenobarbital was administered at 12 n, 2 p.m. and 4 p.m. to estrogen primed ovariectomized rats treated with 5α-DHP and the animals were killed at 6 p.m. for the measurement of serum FSH levels. The results in Fig. 3 show that phenobarbital administered at 12 n resulted in the attenuation of increased FSH secretion at 6 p.m. However if the 12 n surge was not interferred with by the administration of phenobarbital at either 2 p.m. or 4 p.m., the increased FSH secretion was comparable to the vehicle treated group which was not administered phenobarbital. Thus the 12 n release of FSH appeared to be necessary for the 6 p.m. release. There is, however, no clear explanation for the initial FSH release at 12 n and it is possible that 5α-DHP induced a pattern of LHRH release conducive to selective secretion of FSH (Lincoln, 1978; Wise et al., 1979; Wildt et al., 1981; Pohl et al., 1983).

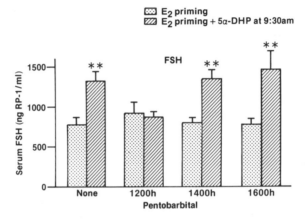

Fig. 3. Effect of pentobarbital administered at 12 n, 2 p.m. or 4 p.m. on serum FSH levels at 6 p.m. in ovariectomized estrogen–primed rats given corn oil or 0.4 mg/kg BW of 5α-DHP (See Fig. 2 for details). Each bar represents mean $\pm$ SEM of not less than 6 rats. ** = P < 0.01 vs estrogen–primed controls (From Murphy and Mahesh, 1984a).

Since progesterone stimulates the release of both FSH and LH and 5α – DHP appears to bring about the selective release of FSH, the modulatory effect of another progesterone metabolite 3α,5α-tetrahydroprogesterone (3α,5α –THP) which is also found in significant quantities during the rat estrus cycle was studied by Murphy and Mahesh (1984b). Initial experiments using 3α,5α-THP in combination with the 0.1 μg/kg BW dose of estradiol in immature ovariectomized rats showed that the 0.2 and the 0.4 mg/kg BW dose of the steroid was effective in inducing an elevation in serum LH without altering serum FSH levels. Of the two doses the 0.4 mg/kg BW dose was found to be more effective. When this dose of 3 ,5α-THP was administered as a single injection at 9:30 a.m. to estrogen primed immature ovariectomized rats, it brought about a selective release of serum LH at 12 n and 3 p.m. without altering serum FSH levels (Murphy and Mahesh, 1984b). When two injections of 3α,5α-THP were given, one at 9:30 a.m. and the other at 12:30 p.m., the LH increase was extended to 6 p.m. without any alterations in serum FSH levels (Fig. 4). The selective release of FSH induced by 5α-DHP and the selective release of LH by 3α,5α-THP raises the possibility that progesterone modulates the secretion of FSH and LH by conversion to its 5α- and 3α,5α-reduced metabolites.

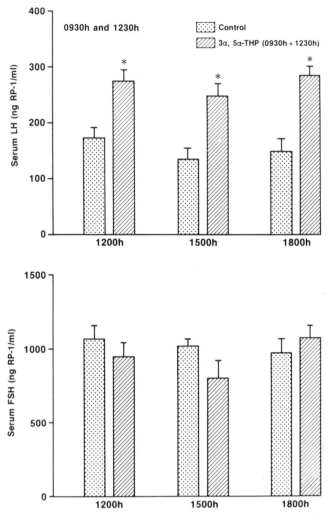

Fig. 4.   Serum LH and FSH levels in female rats ovariectomized at 26 days
of age and primed with 0.1 μg/kg estradiol for 4 days. The rats
received injections of 0.4 mg/kg BW of 3α,5α-THP at 9:30 a.m. and
12:30 p.m. on day 5 and sacrificed at 12 n, 3 p.m. and 6 p.m.
* = P < 0.05 vs estrogen-primed controls (From Murphy and Mahesh,
1984b).

In view of the novel nature of our findings concerning the selective
release of FSH and LH by progesterone metabolites, it was felt necessary
to verify the results in another experimental model. The immature
female rat maintained in constant light from day 22 of age and given 8 IU
of pregnant mare's serum gonadotropin (PMSG) on day 28 of age was used in
this study (Murphy and Mahesh, 1985).   In the PMSG-treated rat maintained
in constant light, the gonadotropin surge that normally occurs on day 30
of age (day 2) after the administration of PMSG on day 28 of age (day 0),
is delayed (McCormack and Bennin, 1970). As expected, the FSH and LH surge
did not occur on day 2 after PMSG due to exposure to constant light and a
delayed and attenuated surge occurred on day 3 (Fig. 5a).   Administration
of 5α-DHP at 9:30 a.m. on day 2 of PMSG resulted in a significant increase

in serum FSH levels on day 3 (day 31 of age) with no significant change in serum LH levels (Fig. 5b). Administration of 3α,5α-THP on the other hand brought about an increase in serum LH levels but not serum FSH levels (Fig. 5c). Thus the selective effects of progesterone metabolites 5α-DHP and 3α,5α-THP on FSH and LH release were confirmed.

CONSTANT LIGHT CONTROL

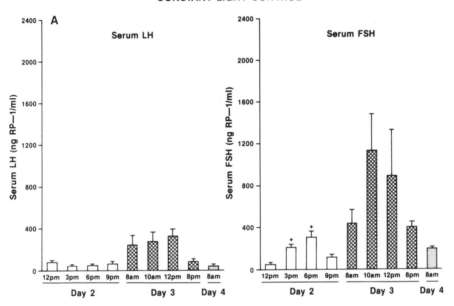

5α—DIHYDROPROGESTERONE ADMINISTRATION

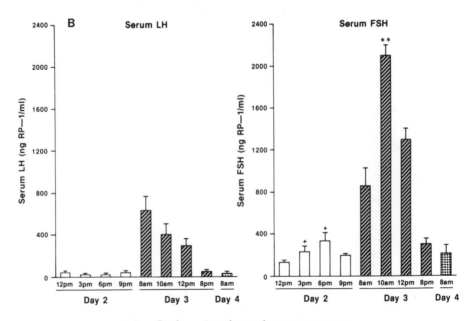

Fig. 5a,b. See legend on next page.

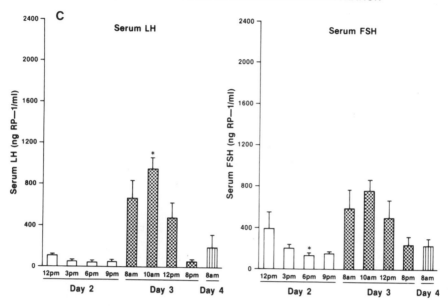

Fig. 5. Serum LH and FSH levels in female rats exposed to constant light starting on day 22 of age and treated with 8 IU of PMSG on day 28 of age (day 0). The rats received either vehicle (constant light control; Fig. 5a) or 0.4 mg/kg BW of 5α–DHP (Fig. 5b) or 3α,5α– THP (Fig. 5c) at 9:30 a.m. on day 30 of age (day 2). The secretion of FSH was enhanced on day 3 by 5α–DHP (Fig. 5b) as compared to levels in the constant light control (** = P < 0.01) while serum LH levels were enhanced by 3α,5α–THP (Fig. 5c) (* = P < 0.05) when compared to levels in the constant light control (Data from Murphy and Mahesh, 1985).

Final confirmation of the effects of 5α–DHP on the selective release of FSH was obtained in the cycling adult rat by Mahesh et al. (1984). When the adult female rat ovariectomized at 9 a.m. on the day of proestrus was administered 0.4 mg/kg BW of 5α–DHP at 9:30 a.m. on the day of proestrus, the endogenous LH surge remained unchanged when compared to the surge levels in vehicle treated controls (Fig. 6). However the serum FSH levels were significantly elevated at 5, 7 and 9 p.m. as a result of 5α–DHP treatment. These results taken together support the concept of different control mechanisms for FSH and LH release and their modulation by progesterone metabolites.

PRIMING OF THE PITUITARY FOR SELECTIVE FSH RELEASE

The regulation of the selective release of FSH is still poorly understood. The presence of an FSH–releasing factor, gonadal FSH inhibiting and releasing peptides, autoregulation by FSH and changes in the pulsatile patterns of LHRH secretion are among the postulates used to explain this phenomenon. A direct effect of steroids such as 5α – dihydrotestosterone and progesterone on the pituitary responsiveness to LHRH in the release of greater amounts of FSH as compared to non–steroid treated controls has been demonstrated in in vitro studies using pituitary tissue or pituitary cell cultures (Schally et al., 1973; Kao and Weisz, 1975; Tang and Spies, 1975; Hsueh et al., 1979; Lagace et al., 1980;

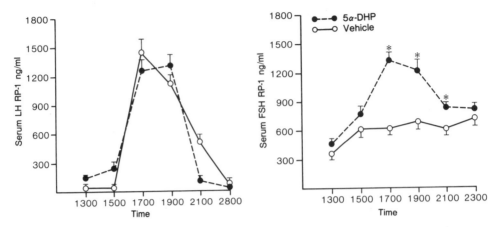

Fig. 6. Serum LH and FSH levels in adult 4-day cycling rats ovariectomized at 9 a.m. of proestrus and injected with vehicle or 0.4
mg/kg BW of 5α-DHP at 9:30 a.m. Note increase in serum FSH
levels in 5α-DHP treated rats with no change in serum LH levels
(* = P < 0.05).

Drouin and Labrie, 1981). Such an effect has been shown to occur in vivo
in the immature ovariectomized estrogen primed rat treated with a single
injection of 5α-DHP (Murphy and Mahesh, 1984a). When 0.4 mg/kg BW of 5α-
DHP was administered to the low dose estrogen primed animal model used in
our study, serum FSH levels were elevated as compared to vehicle treated
controls while LH levels showed no change (Fig. 7). An injection of 20 ng
of LHRH brought about a greater release of FSH in the 5α-DHP treated group
at the 10 min time interval, whereas the rise in LH was comparable in the
steroid-treated and the vehicle-treated control.

The selectively greater release of FSH as compared to LH in response
to LHRH as a result of 5α-DHP treatment was confirmed by Murphy, Kalra and
Mahesh (unpublished data) by electrochemical stimulation studies in the
region of the preoptic-dorsal anterior hypothalamic area in the
ovariectomized immature rat treated with low dose estrogens. Sham
stimulation did not bring about any changes in serum FSH and LH levels
(Fig. 8). The baseline serum FSH levels but not the serum LH levels were
elevated in 5α-DHP-treated animals. Electrochemical stimulation in the
preoptic-dorsal anterior hypothalamic area brought about a greater release
of FSH in the 5α-DHP-treated animals as compared to vehicle treated
controls. The serum LH response to electrochemical stimulation was
similar in the 5α-DHP-treated and control group. These observations
indicate that 5α-DHP either by a direct effect on the pituitary or through
changes in the pulsatile pattern of endogenous LHRH secretion or through
both these mechanisms was able to prime the pituitary gonadotropes in such
a manner that greater amounts of FSH were released in response to LHRH as
compared to non-5α-DHP-treated controls while no differences occurred in
the LH secretory patterns. Furthermore, these effects of 5α-DHP occurred
in the absence of gonadal FSH releasing and inhibitory peptides as the
animals used were ovariectomized.

Examples of the priming of the pituitary gonadotrope for a greater
release of FSH can also be found in the rat during the estrus cycle. In a
study by O'Conner et al. (unpublished data) anterior pituitaries were
collected from 4-day-cycling rats at 8 a.m. on estrus, diestrus 1,
diestrus 2 and proestrus and at 3 and 7 p.m. on proestrus. The anterior

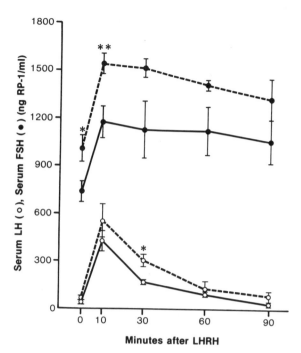

Fig. 7. Serum FSH and LH levels in female rats ovariectomized on day 26 of age and treated with 0.1 µg/kg BW of estradiol for 4 days. On the fifth day all rats received a final injection of estradiol at 8 a.m. and either vehicle (solid line) or 5α-DHP (broken line) at 9:30 a.m. Groups of rats were sacrificed at 6 p.m., before (0 min) and at various intervals after a 20 ng injection of LHRH. Note greater effect of LHRH on FSH release (* = P < 0.05; ** = P < 0.01) than LH release when the results were compared to estrogen-primed controls (From Murphy and Mahesh, 1984).

pituitary cells were dispersed by trypsin digestion and allowed to attach to cytodex beads in Dulbecco's modified Eagle medium (DMEM) plus 10% NuSerum for 48 h. The microcarrier attached pituitary cells were super-fused at a flow rate of 0.125 ml/min and 1 ml fractions of the perfusate were collected for the measurement of FSH and LH levels. After establishing baseline secretion, a pulse of 4 ng LHRH in 100 µl was applied and further fractions collected at 1 min intervals. The LH levels are shown in Fig. 9 and the FSH levels are shown in Fig. 10. There appeared to be no remarkable change in the baseline LH release patterns at the six time points studied. However, FSH baseline release appeared to be higher at 3 p.m. proestrus and 8 a.m. estrus as compared to other times. This secretory pattern of the pituitary gonadotropes is of considerable interest because the pituitary cells were maintained for 48 h in vitro in the absence of endogenous regulatory factors. Also of interest are the changes in responsiveness to LHRH in the release of LH in pituitary cells harvested at different times in the estrus cycle. The pituitary cells released more LH at 7 p.m. proestrus, and estrus and diestrus 1, morning samples. Generally a greater amount of LHRH is required for FSH release as compared to LH release. Since the dose of LHRH was standardized for opti-mal LH release, in all probability it was insufficient to show differences in sensitivity to FSH release in cells harvested at different times in the estrus cycle. Nevertheless, the presence of low serum level LH and high

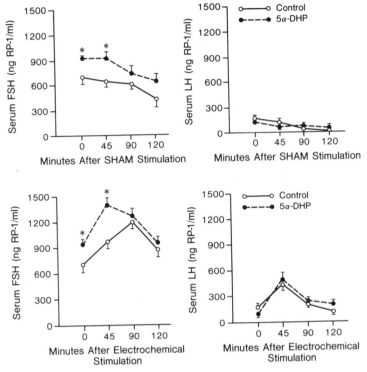

Fig. 8.   26–day–old female rats were ovariectomized and treated with 0.1 μg/kg BW of estradiol for 4 days.   On the morning of day 5, they received a final injection of estradiol and either vehicle or 0.4 mg/kg BW of 5α–DHP at 9:30 a.m.   The animals were subjected to either sham stimulation or electrochemical stimulation at 6 p.m.   * = P < 0.05. Note greater release of serum FSH in 5α–DHP treated rats as compared to sham controls. There was no difference in 5α–DHP treated rats and controls with respect to serum LH.

serum FSH level in the cycling rat on the morning of estrus coupled with high basal release of FSH by pituitary cells and high responsiveness to LHRH in the release of LH, clearly indicates that priming of the pituitary gonadotrope for FSH release rather than LHRH secretion is responsible for elevated FSH secretion at estrus.

———————————————▶

Fig. 9.   4–day–cycling adult rats were sacrificed at various times shown on proestrus and the morning of estrus, diestrus 1 and diestrus 2.   After dispersal of the pituitary cells by trypsin digestion, the cells were allowed to attach to cytodex beads in Dulbecco's modified Eagle medium containing 10% NuSerum for 48 h. The microcarrier attached pituitary cells were superfused at a flow rate of 0.125 ml/min and 1 ml fractions of the perfusate were collected.   After establishing baseline secretion in 51 ml fractions, a pulse of 4 ng LHRH in 100 μl was applied and further 1 ml fractions collected. This figure shows the LH concentrations in the medium.   See Fig. on top of next page.

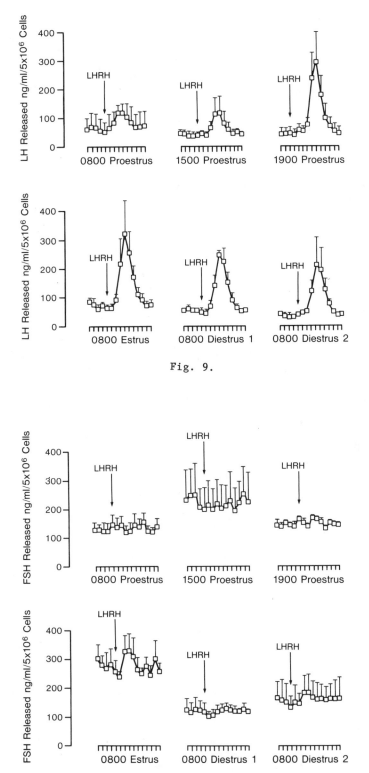

Fig. 9.

Fig. 10. Perfusate FSH concentrations in preparation described in Fig. 9.

SUMMARY

Significant divergence between the pattern of FSH and LH secretion has been observed in the ovulatory cycle, after ovariectomy and during puberty. The presence of an FSH-releasing factor, gonadal FSH inhibiting and releasing peptides and changes in the pulsatile pattern of LHRH secretion are among the postulates used to explain the divergent secretion of FSH and LH. Experiments in our laboratory have shown considerable evidence of differential regulation of FSH and LH secretion by steroids in the absence of gonadal regulatory peptides. Natural and synthetic estrogens show significant differences in the suppression of FSH and LH in the ovariectomized rat using a standard uterine response to the estrogen as the end point. In the immature ovariectomized rat treated with a low dose of estradiol that is sufficient for the synthesis of progesterone receptors to ensure progesterone sensitivity, but not large enough to induce estrogen triggered LH surges, progesterone administration resulted in a pattern of LH and FSH secretion similar to that observed on the day of proestrus in the cycling rat. Selective secretion of FSH was induced in the estrogen primed immature rat model by the administration of progesterone metabolite 5α-dihydroporgesterone (5α-DHP) while selective LH secretion was induced by 3α,5α-tetrahydroprogesterone (3α,5α-THP). The selective secretion of FSH and LH induced by progesterone metabolites was confirmed in the immature female rat primed with PMSG and maintained in constant light. 5α-DHP was also able to induce a greater release of FSH when administered to the adult cycling rat on proestrus. The priming of the pituitary gonadotrope in secreting a high baseline level of FSH or responding to LHRH in releasing a greater amount of FSH appeared to be an important factor in selective FSH release and such priming can be brought about by 5α-DHP in the absence of gonadal regulatory peptides.

ACKNOWLEDGEMENT

This investigation was supported by research grant HD 16688 from the National Institute of Child Health and Human Development, National Institutes of Health, Bethesda, MD 20205.

REFERENCES

Aiyer MS, Fink G, 1974. The role of sex steroid hormones in modulating the responsiveness of the anterior pituitary gland to LHRF in the female rat. J Endocrinol 62:553-572

Ashiru OA, Blake CA, 1979. Stimulation of endogeneous FSH release during estrus by exogenous FSH or LH at proestrus in the phenobarbital-blocked rat. Endocrinology 105:1162-1167

Ashiru OA, Blake CA, 1980. Variations in the effectiveness with which rat follicle-stimulating hormone can stimulate its own secretion during the rat estrous cycle. Endocrinology 106:476-480

Ashiru OA, Rush ME, Blake CA, 1981. Further studies on the effectiveness with which exogenous LH and FSH stimulate the release of endogenous FSH during the rat estrous cycle. J Endocrinol 88:103-113

Banks JA, Freeman ME, 1978. The temporal requirement of progesterone on proestrus for the extinction of the estrogen induced daily signal controlling luteinizing hormone release in the rat. Endocrinology 102:426-432

Banks JA, Mick C, Freeman ME, 1980. A possible cause for the differing responses of the LH surge mechanisms of ovariectomized rats to short-term exposure to estradiol. Endocrinology 106:1677-1681

Bast JD, Greenwald GS, 1974. Serum profiles of follicle stimulating hormone, luteinizing hormone and prolactin during the estrous cycle of the hamster. Endocrinology 94:1295-1299

Belchetz PE, Plant TM, Nakai Y, Keogh EJ, Knobil E, 1978. Hypophysial responses to continuous and intermittent delivery of hypothalamic gonadotropin releasing hormone. Science 202:631-633

Blake CA, Metcalf JP, Hendricks SE, 1982. Anterior pituitary gland secretion after forebrain ablation: periovulatory gonadotropin release. Endocrinology 111:789-793

Bowers CY, Currie BL, Johansson KNG, Folkers K, 1973. Biological evidence that separate hypothalamic hormones release the follicle stimulating and luteinizing hormones. Biochem Biophys Res Commun 50:20-26

Caligaris L, Astrada JJ, Taleisnik S, 1968. Stimulating and inhibiting effects of progesterone on the release of luteinizing hormone. Acta Endocrinol 59:177-185

Caligaris L, Astrada JJ, Taleisnik S, 1971. Release of LH induced by estrogen injection in ovariectomized rats. Endocrinology 88:810-815

Chang RJ, Jaffe RB, 1978. Progesterone effects on gonadotropin release in women pretreated with estradiol. J Clin Endocrinol Metab 47:119-125

Chappel SC, Barraclough CA, 1976. Hypothalamic regulation of pituitary FSH secretion. Endocrinology 98:927-935

Chappel SC, 1979. Cyclic fluctuations in ovarian FSH-inhibiting material in golden hamsters. Biol Reprod 21:447-453

Chappel SC, Norman RL, Spies HG, 1979. Evidence for a specific neural event that controls the estrous release of FSH in golden hamsters. Endocrinology 104:169-174

Cheng YJ, Karavolas HJ, 1973. Conversion of progesterone to 5α-pregnane-3,20-dione and 3α-hydroxy-5α-pregnan-20-one by rat medial basal hypothalami and the effects of estradiol and stage of estrous cycle on the conversion. Endocrinology 93:1157-1162

Coutifaris C, Chappel SC, 1982. Intraventricular injection of follicle stimulating hormone (FSH) during proestrus stimulates the rise in serum FSH on estrus in phenobarbital-treated hamsters through a central nervous system dependent mechanism. Endocrinology 110:105-113

Davies AG, Duncan IF, Lynch SS, 1975. Autoradiographic localization of ^{125}I-labelled FSH in the rat hypothalamus. J Endocrinol 66:301-302

DePaolo LV, Shander D, Wise PM, Barraclough CA, Channing CP, 1979. Identification of inhibin-like activity in ovarian venous plasma of rats during the estrous cycle. Endocrinology 105:647-654

Dohler KD, Wuttke W, 1975. Changes with age in levels of serum gonadotropins, prolactin and gonadal steroids in prepubertal male and female rats. Endocrinology 97:898-907

Drouin J, Labrie F, 1981. Interactions between 17β-estradiol and progesterone in the control of luteinizing hormone and follicle-stimulating hormone release in rat anterior pituitary cells in culture. Endocrinology 108:52-57

El Ayat AAB, Mahesh VB, 1984. Stimulation of 17β-hydroxy steroid dehydrogenase in the rat anterior pituitary gland by progesterone. J Steroid Biochem 20:1141-1145

Eldridge JC, Dmowski WP, Mahesh VB, 1974. Effects of castration of immature rats on serum FSH and LH, and of various steroid treatments after castration. Biol Reprod 10:438-446

Eldridge JC, Mahesh VB, 1974. Pituitary-gonadal axis before puberty: evaluation of testicular steroids in the male rat. Biol Reprod 11:385-397

Everett JW, 1948. Progesterone and estrogens in the experimental control of ovulation time and other features of the estrous cycle in the rat. Endocrinology 43:389-405

Freeman MC, Dupke KC, Croteau CM, 1976. Extinction of the estrogen induced daily signal for LH release in the rat: a role for the proestrous surge of progesterone. Endocrinology 99:223-229

Gilles PA, Karavolas HJ, 1981. Effect of 20α-dihydroprogesterone, progesterone, and their 5α-reduced metabolites on serum gonadotropin levels and hypothalamic LHRH content. Biol Reprod 24:1088-1097

Goodman RL, 1978. A quantitative analysis of the physiological role of estradiol and progesterone in the control of tonic and surge secretion of luteinizing hormone in the rat. Endocrinology 102:142–150

Greeley GH, Allen MB, Mahesh VB, 1975. Potentiation of LH release by estradiol at the level of the pituitary. Neuroendocrinology 18:233–241

Hasegawa Y, Miyamoto K, Yazaki C, Igarashi M, 1981. Regulation of the second surge of FSH; effects of anti–LHRH serum and pentobarbital. Endocrinology 109:130–135

Helmond FA, Simons PA, Heim PR, 1980. The effect of progesterone on estrogen induced LH and FSH release in the female rhesus monkey. Endocrinology 107:478–485

Helmond FA, Simons PA, Heim PR, 1981. Strength and duration characteristics of the facilitory and inhibitory effects of progesterone on the estrogen induced gonadotropin surge in the female rhesus monkey. Endocrinology 108:1837–1842

Holzbauer M, 1975. Physiological variations in the ovarian production of 5α–pregnane derivatives with sedative properties of the rat. J Steroid Biochem 6:1307–1310

Hseuh AJW, Erickson GF, Yen SSC, 1979. The sensitizing effect of estrogens and catechol estrogen on cultured pituitary cells to luteinizing hormone–releasing hormone: its antagonism by progestins. Endocrinology 104:807–813

Ichikawa S, Sawada T, Nakamura Y, Morioka H, 1974. Ovarian secretion of pregnane compounds during the estrous cycle and pregnancy in rats. Endocrinology 94:1615–1620

Ichikawa S, Morioka H, Sawada T, 1971. Identification of the neutral steroids in the ovarian venous plasma of LH–stimulated rats. Endocrinology 88:372–383

Igarashi M, Nallar R, McCann SM, 1973. Existence of FSH–RF distinct from LHRH. In: Hatotani, H. (ed.), Psychoneuroendocrinology. Basel:Karger pp. 178–186

Igarashi M, McCann SM, 1974. A hypothalamic follicle stimulating hormone releasing factor. Endocrinology 74:446–452

Kalra SP, Ajika K, Krulich L, Fawcett CP, Quijada M, McCann SM, 1971. Effects of hypothalamic and preoptic electrochemical stimulation on gonadotropin and prolactin release in proestrous rats. Endocrinology 88:1150–1158

Kalra SP, Kalra PS, 1983. Neural regulation of luteinizing hormone secretion in the rat. Endocr Rev 4:311–351

Karavolas HJ, Hodges DR, O'Brien DJ, MacKenzie KM, 1979. In vivo uptake of [3H] progesterone and [3]–5α–dihydroprogesterone by rat brain and pituitary and effects of estradiol and time: tissue concentration of progesterone itself or specific metabolites? Endocrinology 104:1418–1425

Kao LWL, Weisz J, 1975. Direct effect of testosterone and its 5α–reduced metabolites on pituitary LH and FSH release in vitro: change in pituitary responsiveness to hypothalamic extract. Endocrinology 96:253–260

Kimura F, Kawakami M, 1978. Reanalysis of the preoptic afferent involved in the surge of LH, FSH and prolactin release in the proestrous rat. Neuroendocrinology 27:74–85

Krause JE, Bertics PJ, Karavolas HJ, 1981. Ovarian regulation of hypothalamic and pituitary progestin–metabolizing enzyme activities. Endocrinology 108:1–7

Krey LC, Tyrey L, Everett JW, 1973. The estrogen–induced advance in the cyclic LH surge in the rat: dependency on ovarian progesterone secretion. Endocrinology 93:385–390

Lagace L, Massicotte J, Labrie F, 1980. Acute stimulatory effects of progesterone on luteinizing hormone and follicle-stimulating hormone release in rat anterior pituitary cells in culture. Endocrinology 106:684-689

Legan SJ, Coon GA, Karsch FJ, 1975. Role of estrogen as initiator of daily LH surges in the ovariectomized rat. Endocrinology 96:50-56

Legan SJ, Karsch FJ, 1975. A daily signal for the LH surge in the rat. Endocrinology 96:57-62

Leyendecker G, Wardlaw S, Nocke W, 1972. Experimental studies on the endocrine regulations during the periovulatory phase of the human menstrual cycle. Acta Endocrinol 71:160-178

Lincoln GA, 1979. Differential control of luteinizing hormone and follicle stimulating hormone by luteinizing hormone releasing hormone in the ram. J Endocrinol 80:133-140

Ling N, Ying S-Y, Veno XI, Esch F, Denoroy L, Guillemin R, 1985. Isolation ad partial characterization of a Mr 32,000 protein with inhibin activity from follicular fluid. Proc Natl Acad Sci 82:7217-7221

Ling N, Ying S-Y, Veno N, Shimasaki S, Esch F, Hotta M, Guillemin R, 1986. Pituitary FSH is released by a heterodimer of the α-subunits from the two forms of inhibin. Nature 321:779-782

Liu JH, Yen SSC, 1983. Induction of midcycle gonadotropin surge by ovarian steroids in women: a critical evaluation. J Clin Endocrinol Metab 57:797-802

Lumpkin MD, McCann SM, 1982. Effect of hypothalamic lesions and steroids on the selective control of FSH secretion. Fed Proc 41:984

Lundanes E, Tsuji K, Rampold G, Ohta M, Folkers K, 1980. Studies on the follicle stimulating hormone releasing hormone by radioimmunoassay and bioassays. Biochem Biophys Res Commun 94:827-836

Lutjen PJ, Findlay JK, Trownson AO, Leeton JF, Chan LK, 1986. Effect of plasma gonadotropins on cyclic steroid replacement in women with premature ovarian failure. J Clin Endocrinol Metab 62:419-423

Mahesh VB, Muldoon TG, Eldridge JC, Korach KS, 1972. Studies on the regulation of FSH and LH secretion by gonadal steroids. In: Saxena BB, Beling CG, Gandy HM, (eds.), Gonadotropins. New York: Wiley-Interscience, pp. 730-742

Mahesh VB, O'Conner JL, Allen MB, 1980. In vivo models for the study of pituitary-gonadal relationships. In: Mahesh VB, Muldoon TG, Saxena BB, Sadler WA, (eds.), Functional Correlates of Hormone Receptors in Reproduction. New York:Elsevier, pp. 45-61

Mahesh VB, Murphy LL, Peduto JC, 1984. Modulation of FSH and LH secretion by progesterone and its metabolites. In: Martini L, Gordon GS, Sciarra F, (eds.), Sterols, Steroids and Bone Metabolism, Steroid Modulation of Neuroendocrine Function Research on Steroids. New York: Elsevier, pp. 71-88

Mahesh VB, Muldoon TG, 1987. Integration of the effects of estradiol and progesterone in the modulation of gonadotropin secretion. J Steroid Biochem, In Press

Mann DR, Barraclough CA, 1973. Role of estrogen and progesterone in facilitating LH release in 4-day cyclic rats. Endocrinology 93:694-699

March CM, Goebelsmann U, Nakamura RM, Mishell DR, 1979. Roles of estradiol and progesterone in eliciting the midcycle LH and FSH surges. J Clin Endocrinol Metab 49:507-513

Martin JE, Tyrey L, Everett JW, 1974. Estrogen and progesterone modulation of the pituitary response to LRF in the cyclic rat. Endocrinology 95:1664-1673

Mason AJ, Haylfick JS, Ling N, Esch F, Veno N, Yung SY, Guillemin R, Niall H, Seeburg PH, 1985. Complementary DNA sequences of ovarian follicular inhibin show precursor structure and homology with transforming growth factor-β. Nature 318:659-663

McCann SM. 1962. Effect of progesterone on plasma luteinizing hormone activity. Am J Physiol 202:601–604

McCormack CE. Bennin B. 1970. Delay of ovulation caused by exposure to continuous light in immature rats treated with pregnant mare's serum gonadotropin. Endocrinology 86:611–619

McCullagh DR. 1932. Dual endocrine activity of the testis. Science 76:19–20

McPherson JC. Costoff A. Eldridge JC. Mahesh VB. 1974a. Effects of various progestational preparations on gonadotropin secretion in ovariectomized immature female rats. Fertil Steril 25:1063–1070

McPherson JC. Eldridge JC. Costoff A. Mahesh VB. 1974b. The pituitary-gonadal axis before puberty: effects of various estrogenic steroids in the ovariectomized rat. Steroids 24:41–56

McPherson JC. Costoff A. Mahesh VB. 1975. Influence of estrogen-progesterone combinations on gonadotropin secretion in castrate female rats. Endocrinology 91:771–779

McPherson JC. Mahesh VB. 1979. Dose-related effect of a single injection of progesterone on gonadotropin secretion and pituitary sensitivity to LHRH in estrogen-primed castrated female rats. Biol Reprod 20:763–772

Miyamoto K. Hasegawa Y. Fukuda M. Nomura M. Igarashi M. Kangawa K. Matuso H. 1985. Isolation of porcine follicular fluid inhibin of 32K daltons. Biochem Biophys Res Commun 129:396–403

Mizunuma H. Samson WK. Lumpkin MD. Moltz JH. Kawcett CP. McCann SM. 1983. Purification of a bioactive FSH-releasing factor (FSH-RF). Brain Res Bull 10:623–629

Murphy LL. Mahesh VB. 1984a. Selective release of follicle-stimulating hormone by 5α-dihydroprogesterone in immature ovariectomized estrogen-primed rats. Biol Reprod 30:594–602

Murphy LL. Mahesh VB. 1984b. Selective release of luteinizing hormone by 3α-hydroxy-5α-pregnan-20-one in immature ovariectomized estrogen-primed rats. Biol Reprod 30:795–803

Murphy LL. Mahesh VB. 1985. Selective release of follicle-stimulating hormone and luteinizing hormone by 5α-dihydroprogesterone and 3α,5α-tetrahydro-progesterone in pregnant mare's serum gonadotropin primed immature rats exposed to constant light. Biol Reprod 32:795–803

Nallar R. Antunes-Rodrigues J. McCann S. 1966. Effects of progesterone on the level of plasma luteinizing hormone (LH) in normal female rats. Endocrinology 79:907–911

Norman RL. Blake CA. Sawyer CH. 1973. Estrogen-dependent twenty-four hour periodicity in pituitary LH release in the female hamster. Endocrinology 93:965–970

Nowak FV. Nuti KM. Karavolas HJ. 1976. Quantative changes in the metabolism of 20α-hydroxy-4-pregnen-3-one by rat hypothalamus and pituitary during proestrus. Steroids 28:509–520

Nuti KM. Karavolas HJ. 1977. Effect of progesterone and its 5α-reduced metabolites on gonadotropin levels in estrogen primed ovariectomized rats. Endocrinology 100:777–81

Ojeda SR. Ramirez VD. 1969. Automatic control of LH and FSH secretion by short feedback circuits in immature rats. Endocrinology 84:786–797

Ojeda SR. Ramirez VD. 1970. Failure of estrogen to block compensatory ovarian hypertrophy in prepubertal rats bearing medial basal hypothalamic FSH implants. Endocrinology 86:50–56

Ojeda SR. Jameson HE. McCann SM. 1972. Hypothalamic areas involved in prostaglandin (PG)-induced gonadotropin release. II. Effect of PGE$_2$ and PGE$_{2a}$ implants on follicle stimulating hormone release. Endocrinology 100:1595–1603

Payne AH. Kelch RP. Murono EP. Kerlan JT. 1977. Hypothalamic, pituitary and testicular function during sexual maturation of the male rat. J Endocrinol 72:17–26

Peduto JC, Mahesh VB. 1985. Effects of progesterone on hypothalamic and plasma LHRH. Neuroendocrinology 40:238–245

Pickering AJMC, Fink G. 1977. A priming effect of luteinizing hormone releasing factor with respect to release of follicle–stimulating hormone in vitro and in vivo. Endocrinol 75:155–159

Pohl CR, Richardson DW, Hutchinson JS, Germak JA, Knobil E. 1983. Hypophysiotropic signal frequency and the functioning of the pituitary–ovarian system in the rhesus monkey. Endocrinology 112:2076–2080

Ramirez VD, Sawyer CH. 1974. Differential dynamic responses of plasma LH and FSH to ovariectomy and to a single injection of estrogen in the rat. Endocrinology 94:987–993

Rivier J, Spiess J, McClintock R, Vaughan J, Vale W. 1985. Purification and partial characterization of inhibin from follicular fluid. Biochem Biophys Res Commun 133:120–127

Rivier C, Rivier J, Vale W. 1986. Inhibin–mediated feedback control of follicle stimulating hormone secretion in the female rat. Science 234:205–208

Robertson DM, Foulds LM, Leversha L, Morgan FJ, Hearn MTW, Burger HG, Wettenhall REH, deKrester DM. 1985. Isolation of inhibin from bovine follicular fluid. Biochem Biophys Res Commun 126:220–226

Robertson DM, deVos FL, Foulds LM, Mchachlan RI, Burger HG, Morgan FJ, Hearn MTW, deKrester DM. 1986. Isolation of a 31 kDa form of inhibin from bovine follicular fluid. Molec Cell Endocr 44:271–277

Rush ME, Ashiru OA, Blake CA. 1980. Hypothalamic–pituitary interactions during the periovulatory secretion of FSH in the rat. Endocrinology 107:649–655

Rush M, Ashiru OA, Lipner H, Williams AT, McRae C, Blake CA. 1981. The actions of porcine follicular fluid and estradiol on periovulatory secretion of gonadotropin hormones in rats. Endocrinology 108:2316–2323

Rush ME, Ashiru OA, Blake CA. 1982. Effects of complete hypothalamic deafferentation on the estrous phase of FSH release the cyclic rat. Biol Reprod 26:399–403

Schally AV, Arimura A, Kastin AJ, Matsuo H, Baba Y, Redding TW, Nair RMG, Debeljuk L, White WF. 1971. Gonadotropin releasing hormone: one polypeptide regulates secretion of luteinizing and follicle stimulating hormone. Science 173:1036–1038

Schally AV, Redding JW, Arimura A. 1973. Effect of sex steroids on pituitary responses to LH– and FSH–releasing hormone in vitro. Endocrinology 93:893–902

Schally AV, Arimura A, Redding TW, Debeljuk L, Carter W, DuPont A, Vilchez–Martinez JA. 1976. Reexamination of porcine and bovine hypothalamic fractions for additional luteinizing hormone and follicle stimulating hormone releasing activities. Endocrinology 98:380–391

Schwartz NB, Channing CP. 1977. Evidence of ovarian "inhibin": suppression of the secondary rise in serum follicle stimulating hormone levels in proestrous rats by injection of porcine follicular fluid. Proc Natl Acad Sci USA 74:5721–5724

Schwartz NB, Talley WL. 1978. Effects of exogenous LH or FSH on endogenous FSH, progesterone and estradiol secretion. Biol Reprod 17:820–828

Shahmanesh M, Jeffcoate SL. 1976. Absence of an immunologically distinct FSH–RF in rat stalk median eminence extracts. J Endocrinol 68:89–94

Shander D, Goldman B. 1978. Ovarian steroid modulation of gonadotropin secretion and pituitary responsiveness to luteinizing hormone releasing hormone in female hamsters. Endocrinology 103:1383–1393

Shaw RW, Butt WD, London DR. 1975. The effect of progesterone on LH and FSH response to LHRH in normal women. Clin Endocrinol 4:543–550

Smanik EJ, Young HK, Muldoon TG, Mahesh VB, 1983. Analysis of the effect of progesterone *in vivo* on estrogen receptor distribution in the rat anterior pituitary and hypothalamus. Endocrinology 113:15–22

Smith MS, Freeman ME, Neill JD, 1975. The control of progesterone secretion during the estrous cycle and early pseudopregnancy in the rat–prolactin, gonadotropin and steroid levels associated with the rescue of the corpus luteum of pseudopregnancy. Endocrinology 96:219–226

Stupnicka E, Massa R, Zanisi M, Martini L, 1977. Role of anterior pituitary and hypothalamic metabolism of progesterone in the control of gonadotropin secretion. In: Hubinont PO, L'Hermite M, Robyne CS, (eds.), Clinical Reproductive Neuroendocrinology. Basel: Karger AG, pp. 88–95

Tabei T, Haga H, Heinrichs WL, Herman WL, 1974. Metabolism of progesterone by rat brain, pituitary gland and other tissues. Steroids 23:651–666

Tang LKL, Spies HG, 1975. Effect of gonadal steroids on the basal and LRF–induced gonadotropin secretion by cultures of rat pituitary. Endocrinology 96:349–356

Tapper CM, Naftolin F, Brown–Grant K, 1972. Influence of the reproductive state at the time of operation on the early response to ovariectomy in the rat. J Endocrinol 53:47–57

Turgeon JL, Waring DW, 1982. Differential changes in the rate and pattern of follicle–stimulating hormone secretion from pituitaries of cyclic rats superfused *in vitro*. Endocrinology 111:66–73

Uberoi NK, Hendry LB, Muldoon TG, Myers RB, Segaloff A, Bransome ED, Mahesh VB, 1985. Structure-activity relationships of some unique estrogens related to estradiol are predicted by fit into DNA. Steroids 45:325–340

Vale W, Rivier J, Vaughan J, McClintock R, Corrigan A, Woo W, Karr D, Spiess J, 1986. Purification and characterization of an FSH–releasing protein from porcine ovarian follicular fluid. Nature 321:776–779

Wildt L, Hansler A, Marshall G, Hutchinson JS, Plant TM, Belchetz PE, Knobil E, 1981. Frequency and amplitude of GnRH stimulation and gonadotropin secretion in the rhesus monkey. Endocrinology 109:376–385

Wise PM, Rance N, Barr GD, Barraclough CA, 1979. Further evidence that LHRH also is FSH-RH. Endocrinology 104:940–947

Zanisi M, Martini L, 1975a. Differential effects of castration on LH and FSH secretion in male and female rats. Acta Endocrinol 78:683–688

Zanisi M, Martini L, 1975b. Effect of progesterone metabolites on gonadotropin secretion. J Steroid Biochem 6:1021–1023

Zanisi M, Martini L, 1979. Interaction of oestrogen and of physiological progesterone metabolites in the control of gonadotropin secretion. J Steroid Biochem 11:855–862

ROLE OF PROLACTIN IN THE REGULATION OF SENSITIVITY OF THE HYPOTHALAMIC-PITUITARY SYSTEM TO STEROID FEEDBACK

A. Bartke, K.S. Matt[1], R.W. Steger, R.N. Clayton[2],
V. Chandrashekar and M.S. Smith[3]

Department of Physiology, Southern Illinois University
School of Medicine, Carbondale, IL 62901; [1]Department
of Physiology, Northeastern Ohio Universities, College
of Medicine, Rootstown, OH 44272; [2]Clinical Research
Center, Watford Road, Harrow, Middlesex, England;
[3]Department of Physiology, University of Pittsburgh,
School of Medicine, Pittsburgh, PA 15261.

INTRODUCTION

Biosynthesis of gonadal steroids is regulated primarily by stimulatory action of pituitary gonadotropins, while gonadotropin release is subject to powerful inhibitory control by gonadal steroids and their metabolites. Implicit in the existence of this mechanism for maintaining homeostasis is the requirement for some means of control of the sensitivity to negative steroid feedback. During sexual maturation, increased secretion of gonadotropins stimulates the gonads to produce progressively increasing amounts of steroids without inhibiting further secretion of LH and FSH. As a result of this decrease in the sensitivity of the hypothalamic pituitary system to negative steroid feedback, the concentrations of both pituitary gonadotropins and gonadal steroids rise in concert, increasing gradually from low (prepubertal) to higher (adult) levels. Numerous studies showed that much lower doses of steroids are required to prevent post-castrational rise in peripheral LH and FSH levels in prepubertal as compared to adult individuals thus providing direct evidence for resetting of the sensitivity to steroid negative feedback during sexual maturation. Indeed, this change in the response of the hypothalamic-pituitary system to gonadal steroids is believed to represent one of the mechanisms of sexual maturation in men and animals (the "gonadostat" theory) (McCann et al., 1974; Odell and Swerdloff, 1976; Styne and Grumbach, 1978).

In the overwhelming majority of animal species, reproductive functions are restricted to a well defined and often very brief period of the year. Annual transitions between the periods of gonadal activity and quiescence in seasonal breeders require corresponding changes in the sensitivity to negative feedback of gonadal steroids and there is a great deal of evidence that such changes indeed take place. In one such seasonal breeder, the male golden hamster, reproductive activity is confined to spring and summer and reduction in the length of photoperiod to less than 12.5 hrs of light per day triggers suppression of plasma levels of LH, FSH and testosterone, cessation of spermatogenesis, involution of the gonads, inhibition of sexual behavior, and a prolonged period of infertility (Gaston and Menaker, 1967; Desjardins et al., 1971; Turek et al., 1975a; Morin et al., 1977; Bex et al., 1978; reviews in Reiter, 1980; Bartke, 1985). These changes are accompanied by an impressive increase in the effectiveness of testosterone to inhibit LH and FSH release. Thus, a dose of testosterone which is too low to inhibit plasma LH in castrated adult male hamsters maintained in a long photoperiod is sufficient to suppress plasma LH levels to non-detectable values in castrated adult male hamsters exposed to short photoperiods (Tamarkin et al., 1976; Turek, 1977; Ellis and Turek, 1979). In ovariectomized ewes in which plasma estradiol concentration is artificially maintained at a constant level by means of subcutaneous implants of estradiol-containing capsules, LH levels rise in the fall when breeding season starts in intact females and decline several months later at the time when breeding season normally ends (Legan et al., 1977).

The occurrence of important shifts in the sensitivity of the hypo-thalamic-pituitary system to inhibitory effects of gonadal steroids is not limited to sexual maturation and seasonal transitions between reproductive activity and quiescence. Suppression of gonadal activity observed during aging (Sharr et al., 1975; Gray et al., 1980), malnutrition (Gauthier and Coulaud, 1986; Piacsek et al., 1986) and in certain disease states (details in further section of this review) are associated with increased sensitivity to sex steroid feedback.

INFLUENCE OF PROLACTIN ON THE SENSITIVITY TO STEROID FEEDBACK

We hypothesize that peripubertal and seasonal changes in plasma prolactin (PRL) levels represent one of the mechanisms responsible for concomitant alterations in the sensitivity to steroid feedback. Plasma PRL levels typically increase during sexual maturation with a striking steep increase being characteristic of several laboratory species (Dohler and Wuttke, 1974; Vomachka and Greenwald, 1979; Barkley, 1979). This increase in plasma PRL levels parallels the increase in gonadal steroid output and has been viewed both as a consequence of and as one of the stimuli for gonadal activitation. Thus, estradiol is among the most effective stimuli for PRL release, (Shull and Gorski, 1984), and testosterone increases peripheral PRL levels in the male rat (Nolin et al., 1977), while PRL enhances gonadal responsiveness to gonadotropin stimulation by increasing the number of LH receptors (Bartke and Dalterio, 1976; Aragona et al., 1977; Bex et al., 1978; Purvis et al., 1979), and treatment with PRL was reported to advance puberty via stimulation of FSH release (Clemens et al., 1969; Voogt et al., 1969).

In adult individuals, peripheral PRL levels undergo significant seasonal fluctuations with long photoperiod and high ambient temperature promoting, and short photoperiod and low temperatures inhibiting PRL release. Thus, PRL levels increase in the spring and decline in the fall (Buttle, 1974; Ravault, 1976; Goldman et al., 1981). It is interesting to note that while effects of photoperiod on reproductive function vary quite drastically across species, the pattern of seasonal changes in plasma PRL levels is almost identical in all mammalian species examined to date (Buttle, 1974; Schams and Reinhardt, 1974; Tucker et al., 1974; Ravault, 1976; Beck and Wuttke, 1979; Muduuli, 1979; Goldman et al, 1981; Schulte et al., 1981; Carr and Land, 1982; Ravault et al., 1982; Schams and Barth, 1982; Murphy et al., 1983; Duncan and Goldman, 1984; Maurel et al., 1984; Yellon and Goldman, 1984). However, with very few exceptions, there is a clear temporal relationship between the changes in the sensitivity to gonadal steroid feedback and the changes in plasma PRL levels.

Apart from these temporal associations, there is considerable evidence to implicate PRL as one of the factors responsible for the changes in feedback sensitivity. Hyperprolactinemia in women is associated with increased susceptibility to negative estrogen feedback (Baird et al., 1977; L'Hermite et al., 1978) while positive feedback action of estradiol which usually causes ovulatory LH release is either reduced or absent (Aono et al., 1976; Baird et al., 1977; L'Hermite et al., 1978; Aono et al., 1979). In hyperprolactinemic men, serum levels of both LH and testosterone are typically below or near the lower limit of the normal range thus implying existence of a primary defect in the mechanisms controlling LH release, possibly including increased sensitivity to testosterone feedback (Carter et al., 1978; Perryman and Thorner, 1981; Winters and Troen, 1984). Medical treatment of hyperprolactinemia by oral dosing with bromocriptine causes concomitant increase in peripheral LH and testosterone levels (Ambrosi et al., 1977; Thorner and Besser, 1978; Thorner et al., 1980). Thus, the endocrine response to bromocriptine therapy is compatible with the assumption that hyperprolactinemia increases the sensitivity to testosterone negative feedback and provides indirect support for this possibility. More direct evidence on this point was obtained in studies of male rats with experimentally-induced hyperprolactinemia. McNeilly et al. (1983) examined effects of various sizes of Silastic testosterone implants on the post-castration rise in serum LH and FSH levels in hyperprolactinemic and control rats. Small implants which had no effect on this response in control rats, abolished or greatly delayed it in animals rendered hyperprolactinemic by two ectopic pituitary transplants.

It is of particular interest that the ability of hyperprolactinemia to increase the responsiveness to testosterone negative feedback in the rat may represent an extension of the physiological action of PRL on the pituitary-gonadal axis in this species rather than a purely pathological phenomenon. Thus, suppression of PRL release in normal animals often leads to a significant increase in peripheral LH levels without altering the levels of testosterone (Purvis et al., 1979; Kovacevic et al., 1982; Rao et al., 1984). In a recent experiment, we have demonstrated that hypoprolactinemia induced in immature male rats by treatment with bromocriptine significantly attenuated the ability of injected testosterone propionate to suppress the post-castration rise of plasma LH and FSH levels (Chandrashekar et al., 1987). It can be concluded from these findings that the amounts of PRL normally secreted by the male pituitary act to increase the sensitivity of the hypothalamic-pituitary system to testosterone feedback and that this action is exaggerated in individuals with pathological or experimentally-induced hyperprolactinemia.

Old and middle-aged male rats, who often exhibit mild hyperprolactinemia and decreased basal gonadotropin levels are more sensitive to the negative feedback effects of physiological levels of testosterone than young rats (Shaar et al., 1975; Gray et al., 1980). Aging female rats on the other hand show a slight decrease or no change in the negative feedback response to estradiol despite increases in basal PRL and decreases in basal LH levels (Shaar et al., 1975; Steger and Peluso, 1979; Huang et al., 1980; Gray and Wexlar, 1980; Steger et al., 1980). The positive feedback response to ovarian steroids is also lost or severely attenuated with age (Lu et al., 1977; Peluso et al., 1977; Steger et al., 1980).

Feedback relationships have not been carefully studied in the aging mouse but would be of interest since female C57BL/6 mice show a decrease in PRL and an increase in LH levels with age in direct contrast to what is seen in aging rats (Parkening et al., 1980). Aging C57BL/6 male mice show no changes in basal hormone levels (Finch et al., 1977).

We also have evidence that the seasonal, photoperiod-related alterations in PRL release in male golden hamsters contribute to the concomitant changes in the sensitivity to testosterone feedback. Surprisingly,
the effects of PRL in the hamster are opposite to those described in the
rat and in the human. Thus, mild hyperprolactinemia induced by transplantation of one pituitary gland under the renal capsule was associated
with significant increases in plasma LH and FSH levels in adult,
castrated male hamsters injected with various doses of testosterone (Fig.
1.; Bartke et al., 1984b). Conversely, photostimulation of gonadotropin
release in castrated testosterone-treated hamsters was significantly
attentuated by treatment with bromocriptine, an inhibitor of PRL release
(Fig. 2; Matt et al., 1984). Treatment of adult male hamsters with
multiple ectopic pituitary transplants leads to significant increases in
both gonadotropin and testosterone levels (Bartke et al., 1982), thus
further confirming the ability of PRL to reduce the sensitivity to
negative androgen feedback in this species.

MECHANISMS OF PROLACTIN ACTION ON STEROID FEEDBACK

From evidence reviewed above it can be concluded that in male golden
hamsters there are major seasonal changes in the susceptibility of LH and
FSH release to androgen feedback and that PRL can be suspected of mediating these changes. Consequently, we felt that it would be interesting
to attempt to identify the mechanisms of PRL action on the regulation of
gonadotropin secretion in this species. Because of the known involvement
of catecholamines in the regulation of LHRH and, consequently, LH and FSH
release, we examined the interaction of testosterone and PRL on the
activity of dopaminergic and noradrenergic neurons in the hypothalamus.
The animals were injected with tyrosine hydroxylase blocker, α-methyl-
paratyrosine and turnover rates of dopamine (DA) and norepinephrine (NE)
were calculated from the rate of disappearance of these neurotransmitters
from the median eminence and MBH. Treatment with testosterone enhanced
DA turnover and reduced NE turnover and these effects were significantly
greater in animals that had been exposed to short photoperiods than in
long photoperiod controls (Fig. 3). It is well known that short periods
suppress plasma PRL levels in male hamsters (Bex et al., 1978; Goldman et
al., 1981). In short photoperiod-exposed animals in which plasma PRL
levels were artificially maintained by means of ectopic pituitary transplants, the effects of testosterone on DA and NE turnover were significantly attenuated (Fig. 3; Matt et al., 1985; Steger et al., 1986).
These results provide a very plausible explanation for the observed
effects of PRL on testosterone feedback. From results obtained in
several species, it can be assumed that noradrenergic neurons provide
major stimulatory input to LHRH secreting cells. Therefore in the
hamster, the ability of PRL to interfere with inhibitory effects of
testosterone on NE turnover in these regions would favor increased LH and
FSH release, as was observed in vivo (Bartke et al., 1981; Bartke et al.,
1982; Bartke et al., 1984b; Matt et al., 1984). The significance of
testosterone and PRL-induced alterations in hypothalamic DA turnover is
more difficult to assess because the role of DA in the regulation of the
LHRH neurons is poorly understood and indeed controversial (Kamberi et
al., 1970; Rotsztejn et al., 1977; Huseman et al., 1980; Nicoletti et
al., 1984). However, it is generally believed that DA can suppress LHRH
release, at least under certain circumstances, and thus the ability of
PRL to reduce the stimulatory effect of testosterone on DA turnover could
result in stimulation of LHRH release.

Extrapolation of the results obtained in other species suggests that
the effects of testosterone and PRL on hypothalamic dopaminergic and
noradrenergic neurons could be both direct and indirect. Neurons secre-

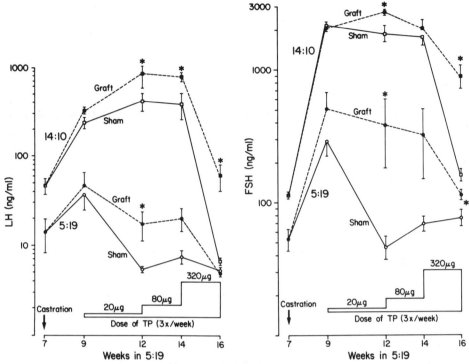

Fig. 1.   Effects of PRL-secreting ectopic pituitary transplants
on the ability of exogenous testosterone to suppress
plasma LH and FSH levels in adult gonadectomized male
hamsters.   The animals were housed in a long photoperiod (14 hrs of light: 10 hrs of darkness) or in a
short photoperiod (5 hrs of light: 19 hrs of darkness).   Short photoperiod is known to reduce peripheral PRL levels and to increase the sensitivity of
mechanisms regulating gonadotropin release to negative
androgen feedback.   Twenty-four animals housed in
14:10 and 22 animals that had been exposed to 5:19
were castrated and either implanted with one pituitary
from an adult female or sham implanted.   Starting two
weeks after gonadectomy, all animals were injected
three times a week with testosterone propionate (TP).
The dose of TP was increased 4-fold after three and
again after two additional weeks of treatment.   Blood
samples for RIA determinations of LH and FSH were
collected before castration, before start of TP
treatment and after treatment with each of the three
doses of TP.   Values are mean ± SEM; asterisks signify
values significantly different from the corresponding
sham-operated controls (P <0.05).   Please note that
both LH and FSH levels are plotted on a logarithmic
scale.   The results indicate that in adult male
hamsters PRL significantly reduces the sensitivity of
LH and FSH release to negative testosterone feedback.
(From Bartke et al., 1984b)

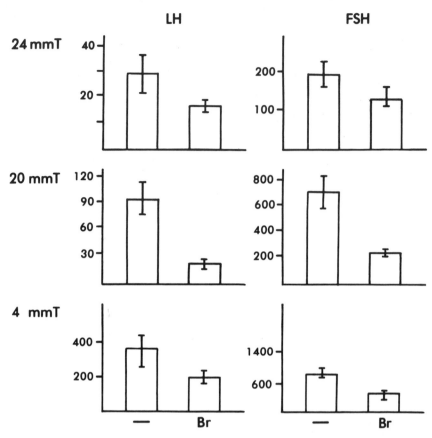

Fig. 2. Effects of hypoprolactinemia induced by bromocriptine (Br) on the ability of different doses of testosterone to suppress plasma LH and FSH levels in castrated adult male hamsters during photostimulation. The animals were exposed to short photoperiods (5 hrs of light: 19 hrs of darkness) for 12 weeks, castrated and implanted with Silastic capsules containing testosterone. Each animal received either one (20 mm) or two (20 mm and 4 mm) capsules. Ten days later, the animals were transferred to long photoperiods (14:10) to stimulate gonadotropin release. Starting on the day of transfer from 5:19 to 14:10, half of the animals in each group were injected daily with 600 μg of bromocriptine in order to prevent plasma PRL levels from increasing in response to long photoperiods. The bars represent plasma levels of LH and FSH (ng/ml; mean ± SEM) after four weeks of photostimulation in animals which had been implanted with two testosterone capsules (24 mm) or one capsule (20 mm) and in animals which were originally implanted with two capsules but had the larger capsule removed two days before collection of blood samples (4 mm). Statistical analysis was based on comparison of gonadotropin levels shown above with values measured earlier during photostimulation and revealed that testosterone was significantly more effective in inhibiting plasma LH and FSH levels in bromocriptine-treated than in control animals. (From Matt et al., 1984)

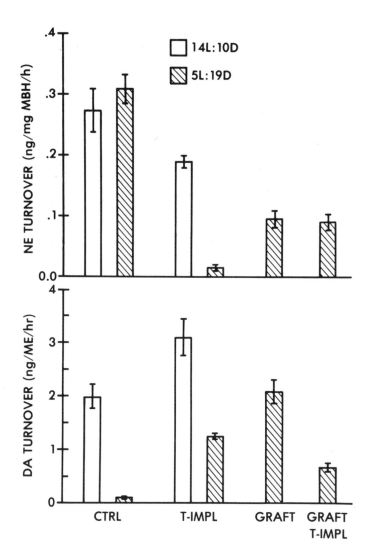

Fig. 3.   The influence of prolactin (PRL) on the ability of
testosterone (T) to modify hypothalamic norepinephrine
(NE) and dopamine (DA) activity.   Adult male hamsters
were maintained in a long photoperiod (14 L : 10 D) or
transferred to a short photoperiod (5 L : 19 D) for 12
weeks to suppress endogenous PRL release.   Subsequently both long photoperiod and short photoperiod
exposed animals were castrated and implanted with a 6
mm empty or T-filled Silastic capsule.   Two days
later, half of the short photoperiod exposed hamsters
received ectopic pituitary grafts as a source of PRL.
Twelve days later (i.e., after 14 weeks in 5 L : 19 D)
the animals were injected with α-methyl-para-tyrosine
or saline and sacrificed for determinations of NE and
DA activities (turnover rates).   (Rates are presented
as mean ± SEM; please see Steger et al., 1985; 1986;
for details of the methodology.)   The results indicate
that the ability of T to suppress NE activity and to
stimulate DA activity is enhanced when PRL levels are
low (5 L : 19 D) and reduced when PRL levels are
elevated (grafts).

ting opioid peptides are among the most likely mediators of steroid hormone action on catecholaminergic neuronal systems. Endogenous opioids are believed to suppress both DA and NE release (Kalra and Kalra, 1984) and stimulation of opioid activity is suspected of mediating the negative feedback action of gonadal steroids on gonadotropin secretion (Bhanot and Wilkinson, 1984; Kalra and Kalra, 1984).

It has been reported that experimentally-induced hyperprolactinemia is associated with increased activity of opiatergic neurons in female rats (Panerai et al., 1980) and that anovulatory sterility of hyperprolactinemic women is believed to be due to opioid-mediated inhibition of LH release (Quigley et al., 1980; Grossman et al., 1982). As an initial step in probing the possibility that the effects of PRL on the regulation of gonadotropin release in male hamsters are mediated via opiatergic neurons, we have examined the effects of pharmacologic blockade of opiate receptors on plasma LH levels in mildly hyperprolactinemic and in normal control hamsters. The animals received one homologous pituitary transplant or were sham operated, were given a single dose of an opiate receptor blocker, naloxone or naltrexone, and bled 10 min (naloxone) or 30 min (naltrexone) later for determination of plasma LH levels. In control animals, treatment with an opiate antagonist produced a sharp rise in plasma LH levels and a small reduction in plasma PRL levels (Table 1). These results are consistent with the previously documented role of endogenous opioids in providing a tonic inhibitory influence on LH release and a mild stimulatory influence on PRL release. Similar to the results obtained in control animals, treatment of pituitary-grafted hyperprolactinemic males with naloxone or naltrexone produced a significant increase in plasma LH levels (Table 1). Plasma PRL levels were not affected, most likely due to the fact that most of the PRL present in the circulation of pituitary grafted animals is derived from the transplanted rather than the in situ pituitary and thus is not subject to the control by hypothalamic neurons, including the opiatergic system. Similarity of LH responses to opioid receptor blockers in the control and pituitary grafted animals indicates that either PRL did not alter the activity of opiatergic neurons, or its effects were masked by effects of PRL on the pituitary or on other neuronal systems involved in the control of LH release. While further work is necessary to arrive at firm conclusions on this point, these results argue against the possibility that alterations in the function of opiatergic neurons mediate the effects of PRL on the sensitivity to steroid feedback in the hamster. These results would also seem to imply that the previously reported absence of LH response to naloxone in short photoperiod-exposed gonadally regressed hamsters (Roberts et al., 1985) may have been due to reduced levels of gonadal steroids rather than to suppression of PRL release in these animals. It is of interest that in contrast to these results obtained in the hamster, hyperprolactinemia abolishes the LH response to naloxone in the male rat (Sweeney et al., 1985).

We have also addressed the possibility that PRL may modulate the responsiveness of gonadotropin release to steroid feedback at the pituitary level. We were particularly interested in the effects of PRL on adenohypophyseal LHRH and androgen receptors. There is a considerable amount of evidence that experimentally-induced hyperprolactinemia in the rat is associated with significant reduction in the number of LHRH receptors (Marchetti and Labrie, 1982; Duncan et al., 1986; Fox et al., 1986), presumably due to suppression of LHRH release (Clayton, 1982). The reduction in LHRH receptors is reflected in a decreased responsiveness of the pituitary to LHRH stimulation. There is also evidence that PRL can increase uptake and binding of testosterone in androgen target tissues (Lloyd et al., 1973; Baker et al., 1977; Prins, 1986).

Table 1.   Effects of an Opiate Antagonist, Naltrexone, on
Plasma LH and PRL Levels (ng/ml) in Control and
Hyperprolactinemic Male Hamsters. The Animals were
given Transplants of three Pituitaries under the
Renal Capsule or were Sham-operated. Blood Samples
for RIA Measurements of LH and PRL Levels were
Collected 30 min after Injection of Naltrexone (5
mg/kg body weight) or Saline. Mean ± SEM.

	Sham-operated		Pituitary-grafted	
	Saline	Naltrexone	Saline	Naltrexone
LH	1.6±0.3	6.3±0.7[1]	1.0±0.3	3.8±0.8[1]
PRL[2]	10.9±1.4	7.7±1.0	51.1±6.0	47.1±6.6

[1]Significantly different from values measured in corresponding
saline-injected animals. (P<0.05).
[2]Measured in samples collected from the same animals on another
occasion.

The possibility that changes in pituitary LHRH binding may mediate
the effects of PRL on regulation of gonadotropin release appeared to be
consistent with the report of significant reduction in the number of LHRH
receptors in the pituitaries of short photoperiod-exposed hamsters
(Pieper, 1984). The reader will recall that in these animals plasma PRL
levels are greatly reduced (Bex et al., 1978; Goldman et al., 1981) and
sensitivity of LH and FSH release to negative testosterone feedback is
increased (Bartke et al., 1984b; Matt et al., 1984). Although short
photoperiod exposure was effective in reducing pituitary content of LHRH
receptors also in our animals (Fig. 4) treatment of adult male hamsters
with bromocriptine or bromocriptine plus PRL failed to modify LHRH
binding in their pituitaries (Table 2). These results argue against a
role of PRL in the control of pituitary LHRH binding in this species.
Effects of short photoperiod exposure on the number of LHRH receptors can
thus be assumed to represent consequences of suppressed LHRH release
and/or altered secretion of gonadal steroids (Clayton, 1982).

Table 2.   Effects of Treatment with Bromocriptine (600 µg/day)
and oPRL (250 µg/day) on Pituitary LHRH Receptors in
Adult Male Syrian Hamsters Maintained in Long
Photoperiod (14 h of light : 10 h of darkness). Mean
± SEM.

	LHRH Receptors (fm/pituitary)
Controls	59.1 ± 5.5
Bromocriptine	66.2 ± 5.5
Bromocriptine + PRL	67.8 ± 8.1

In order to test the possibility that PRL can modify androgen
receptors in the pituitary, we have examined the effects of suppression
of plasma PRL _in vivo_ on the binding of dihydrotestosterone in the
pituitary. In adult castrated hamsters given testosterone implants,
suppression of plasma PRL levels with bromocriptine was associated with a
significant increase in adenohypophyseal cytoplasmic androgen receptors

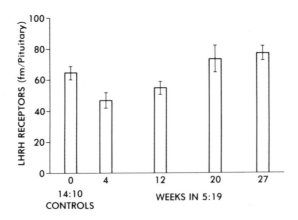

Fig. 4.   Effects of exposure to short photoperiod (5 h : 19 h
           D) on the content of LHRH receptors in the pitui-
           taries of adult male Syrian (golden) hamsters.
           Control animals were maintained in a long (14 L : 10
           D) photoperiod.   The LHRH receptors were quantitated
           in homogenates of individual pituitaries using
           iodinated agonistic LHRH analog [D-Ser(tBu)[6]]des
           Gly[10]LHRH N-ethylamide (Hoechst, A. G. Frankfurt,
           Federal Republic of Germany) as described previously
           (Clayton, 1982; Young et al., 1983).   Suppression of
           plasma PRL and gonadotropin levels after 4 and 12
           weeks in 5 L : 19 D (data not shown) was accompanied
           by a decline in LHRH receptors.   Spontaneous recovery
           of adenohypophyseal and testicular function after 20
           and 27 weeks in 5 L : 19 D was associated with a
           significant (P<0.05) increase in pituitary content of
           LHRH receptors.   (Mean ± SEM).

(16±5 vs 38±3 fm/mg; P<0.005).   In order to further explore the apparent
ability of PRL to modify pituitary androgen binding in the golden
hamsters, we decided to examine the effects of treatment with
bromocriptine or ectopic pituitary transplants on cytoplasmic androgen
receptors in the pituitaries of immature male hamsters.   The onset of
these treatments was timed to advance or to prevent the increase in
plasma PRL levels which normally takes place during sexual maturation
(Vomachka and Greenwald, 1979).   Thus, male hamsters were given ectopic
pituitary transplants at the age of 6 days, i.e. approximately 12 days
before plasma PRL levels would normally begin to rise, or were treated
with injections of bromocriptine starting at the age of 14 days, shortly
before the pubertal increase in serum PRL levels (Stallings et al.,
1985).   Treatment with pituitary transplants produced a significant
advancement of sexual maturation as assessed from measurements of
testicular weight and concentration of LH(hCG) receptors at the age of 20

days (Stallings et al., 1985). These changes were associated with an apparent reduction in pituitary DHT binding (Fig. 5). Conversely, treatment with bromocriptine produced a delay of sexual development as indicated by significantly reduced weights of the testes and the seminal vesicles and reduced levels of testicular LH(hCG) receptors at the age of 31 days (Stallings et al., 1985). These changes were associated with significant ($P<0.05$) increase in pituitary androgen binding (Fig. 5). Plasma gonadotropin levels appeared slightly reduced but this apparent effect of bromocriptine was not statistically significant. We interpret these findings as tentative evidence that the sexual development in male golden hamsters is critically dependent on the increase in peripheral PRL levels and that this increase may be specifically required for the reduction in adenohypophyseal androgen binding and the associated decrease in the ability of testicular androgens to suppress LH and FSH release. While changes in sensitivity to steroid feedback and the suspected role of PRL during this period were not specifically examined

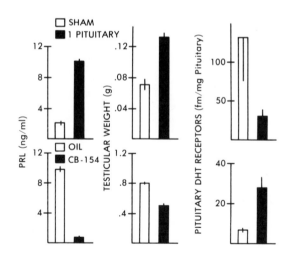

Figure 5. Effects of ectopic pituitary transplants and daily injections of bromocriptine (CB-154) on plasma prolactin (PRL) levels, testicular weight, and pituitary cytosolic androgen (DHT) receptors in juvenile Syrian hamsters. The animals were given pituitary transplants at the age of 6 days and sacrificed at the age of 20 days (top panel) or were treated with CB-154 starting at the age 14 days and sacrificed at the age of 31 days (bottom panel) (Stallings et al., 1985). Mean ± SEM; details in the text.

in the golden hamster, data obtained by Vomachka and Greenwald (1979) indicate striking temporal correlations between the attainment of high adult serum PRL levels and the onset of gradual increases in serum concentrations of LH, FSH and androgen in the developing male hamster.

We realize that the relationship of the number of cystolic androgen receptors to the number of nuclear androgen receptors and to androgen action within gonadotrophs is unknown. Much further work will be necessary to establish (or refute) the suspected causal relationship between cytoplasmic DHT binding measured in these experiments and pituitary responsiveness to gonadal androgens in vivo. However, we would like to emphasize the observed correlation between decreases in plasma PRL levels, increases in feedback sensitivity and increases in concentration of cytosolic androgen receptors in the pituitary.

Recent observations of Negri-Cesi and Martini (1986) raise an intriguing possibility that seasonal changes in the responsiveness of the pituitary to steroid feedback are due not only to differences in androgen binding but also to alterations in its metabolism. These investigations reported that exposure of adult male hamsters to short photoperiod increased the ability of their pituitaries to convert testosterone to androstenedione in vitro.

It is unclear which of the findings obtained in male golden hamsters may prove applicable to females or to other species. Thus, Baum and Schretlen (1979) found no seasonal differences in pituitary or hypothalamic cytoplasmic estradiol binding in the female ferret, while rams were reported to have fewer cytosolic estrogen receptors in the anterior pituitary during the non-breeding, as compared to the breeding season (Pelletier and Caraty, 1981; Glass et al., 1984). The effects of season on hypothalamic and pituitary estrogen binding in the ewe are a matter of controversy (Clarke et al., 1981; Wise et al., 1975; Glass et al., 1984).

DISCUSSION

From the results discussed in this article we would like to conclude that, (1) alterations in peripheral PRL levels constitute one of the control mechanisms responsible for pubertal and seasonal changes in the sensitivity of hypothalamic-pituitary system to androgen feedback, (2) that PRL may modify feedback sensitivity at both pituitary and hypothalamic levels and, (3) that mechanisms of PRL action on feedback sensitivity in the male hamster include modulation of the activity of hypothalamic noradrenergic neurons and the number of cytoplasmic androgen receptors in the pituitary (Fig. 6).

In order to put these conclusions in the proper perspective, we need to address two questions. First of these concerns is the role of other, possibly PRL-unrelated, mechanisms in the control of sensitivity to gonadal steroid feedback and the second relates to the physiological importance of changes in feedback sensitivity in the processes of sexual maturation and in seasonal breeding.

It is clear that in addition to mechanisms discussed above, alterations in feedback sensitivity are temporarily associated with a host of other physiological changes including shifts in the profile of steroids produced by the testes (Steinberger and Ficher, 1969; Tahka et al., 1982), steroid binding in plasma (Whitehead and West, 1977; Cheng et al., 1983), metabolism of steroids in target organs and peripheral tissues (Horst, 1979; Callard et al., 1986), frequency and amplitude of LHRH, LH and testosterone pulses (Lincoln et al., 1977; Winters and Troen, 1984),

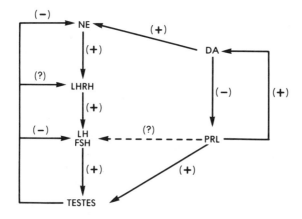

Fig. 6.   Schematic representation of the effects of prolactin
(PRL) on the hypothalamic-pituitary-testicular axis
in male golden hamsters. Prolactin acts directly on
the testes to increase the number of LH receptors and
promote testosterone synthesis. Prolactin also acts
on the pituitary to decrease androgen receptors and
reduce the LH and FSH response to negative feedback
of testosterone.   In addition, PRL increases
norepinephrine (NE) activity in the hypothalamus.
This latter effect is probably mediated through the
ability of PRL to stimulate dopamine (DA) activity.
We have recently reported that dopaminergic agonists
stimulate hypothalamic NE activity in the male golden
hamster (Steger and Matt, 1986). Evidence for the
known and suspected relationships shown in this
figure and additional references are discussed in the
text.

and responsiveness of various central neuronal systems to steroid action
(Ellis and Turek, 1979; Morin et al., 1977).

In the male golden hamster, short photoperiod-induced suppression of
the pituitary-testicular axis is associated with reduced activity of 5α
reductase (Frehn et al., 1974) and aromatase (Callard et al., 1986) in
the hypothalamus. Concomitant with these changes and apparently related
to reduced aromatase activity is a reduction in nuclear estrogen receptors in the same area of the brain (Callard et al., 1986). These changes
would be expected to favor reduced responsiveness of the hypothalamus to
testosterone and thus clearly are not related to the observed increased
sensitivity of LH and FSH release to testosterone inhibition under these
circumstances. However, they may offer a tentative explanation of
reduced behavioral responses to testosterone (Morin et al., 1977;
Campbell et al., 1978) and perhaps also for the reduced number of nuclear

estradiol receptors in the limbic area (Callard et al., 1986) in short photoperiod-exposed hamsters. We are not aware of any evidence directly linking these changes to photoperiod-induced alterations of PRL release. However, results obtained in other species demonstrate the ability of PRL to modify steroid metabolism in the central nervous system (Martini et al., 1978).

Regardless of the mechanisms involved in the apparent ability of PRL to modulate responsiveness to testosterone feedback, it is quite evident that PRL-independent phenomena are also involved. Thus, experimental maintenance of high serum PRL levels in short photoperiod-exposed castrated, testosterone-treated hamsters by means of ectopic pituitary transplants significantly increased plasma LH and FSH levels but failed to restore them to the values measured in long photoperiod controls (Fig. 1; Bartke et al., 1984b).

In discussing the physiological significance of changes in the sensitivity to steroid feedback, it is important to mention other mechanisms involved in the control of sexual maturation and seasonal breeding. Thus, both pubertal and seasonal stimulation of gonadotropin synthesis and release reflects not only reduced sensitivity to inhibitory influences of gonadal steroids but also increases in the intrinsic hypothalamic stimulatory input, the so called steroid-independent drive. For example, plasma gonadotropin levels increase at the normal time of puberty in agonadal children (Grumbach and Conte, 1985) and seasonal fluctuations of peripheral LH and FSH levels have been noted in castrated seasonal breeders not receiving steroid replacement therapy (Davis and Meyer, 1973; Garcia and Ginther, 1976; Tate-Ostroff and Stetson, 1981; Day et al., 1986). Our demonstration that ectopic pituitary transplants can increase plasma FSH levels in adult castrated male hamsters not given concomitant testosterone treatment (Bartke et al., 1981) provides an indication that PRL could be involved also in these mechanisms, most likely due to its ability to increase the activity of hypothalamic noradrenergic neurons (Steger et al., 1985).

Future studies in this area will have to consider recent evidence that the steroid-dependent and steroid-independent mechanisms of the seasonal changes in gonadotropin release are mediated by separate neuronal systems. Meyer and Goodman (1986) reported that steroid-dependent suppression of LH pulse frequency in anestrous ewes appears to be mediated by dopaminergic and noradrenergic neurons, while steroid-independent seasonal suppression of the frequency of LH pulses is due to serotonergic mechanisms.

It has also been argued that the resetting of the gonadostat may occur as a response to puberty rather than as a cause of puberty (Ojeda et al., 1980). In the female rat, it has been shown that a major part of the prepubertal change in the sensitivity to steroid negative feedback occurs shortly after the first preovulatory LH surge and may be due to the activation of ovarian progesterone secretion. Adult progesterone secretion is markedly higher than that seen prepubertally (1 ng/ml in 25 day old rats vs 20 ng/ml in adult diestrous rats). Furthermore, if ovariectomized rats are implanted with progesterone containing Silastic capsules, a prepubertal-like response to estrogen negative feedback is induced (Andrews et al., 1979). Although PRL levels were not reported, it is unlikely that progesterone's actions were mediated by an increase in circulating PRL levels since progesterone alone has little effect on PRL secretion and when given with estrogen, it antagonizes the estrogen mediated PRL release (Chen and Meites, 1970).

Pubertal and seasonal shifts in the activity of the hypothalamic-pituitary-gonadal axis involve yet another important mechanism, namely fluctuations in the responsiveness of the gonads to LH and FSH. In the male golden hamster, PRL is suspected of having a major role in the control of these changes due to its ability to increase the number of LH and PRL receptors in the testis (Bex and Bartke, 1977; Bex et al., 1978; Klemcke et al., 1984; Klemcke et al., 1986). It is interesting that in the adult male rat the suppressive effects of PRL on gonadotropin release and the ability of PRL to augment gonadal responsiveness exert opposite effects on testicular testosterone secretion and thus appear to represent one of the mechanisms of maintaining homeostasis (Dalterio and Bartke, 1976; Bartke et al., 1977b; Bartke et al., 1984a). In contrast, in male hamsters and mice, the effects of PRL on the hypothalamus, pituitary and gonads appear to sum up in terms of the resulting increase in testicular activity (Bartke et al., 1979; Bartke et al., 1977a). Thus, effects of PRL on the regulation of gonadotropin release by androgen feedback in some species may serve as one of the means for facilitating rapid activitation of the gonads during sexual maturation or annual recrudescence (Bex et al., 1978; Stallings et al., 1985) while in others they may provide a tonic inhibitory influence in adult males (Kovacevic et al., 1982; Rao et al., 1984).

SUMMARY

During sexual maturation, pituitary gonadotropins stimulate the gonads to produce increasing amounts of biologically active steroids and yet gonadotropin release does not become suppressed until concentrations of sex hormones, LH and FSH, in peripheral circulation stabilizes at a higher adult level. There is a substantial amount of evidence that in many mammals, this transition from prepubertal to adult level of activity of the pituitary-gonadal axis is associated with a reduction in the sensitivity of the hypothalamic-adenohypophyseal system to negative feedback of gonadal steroids. In the female, these changes are accompanied by the appearance of positive estrogen feedback on gonadotropin release. In seasonal breeders, annual transitions between the periods of gonadal activity and quiescence are associated with corresponding shifts in the sensitivity to steroid feedback.

Peripheral levels of pituitary prolactin (PRL) typically increase during sexual maturation and exhibit large seasonal fluctuations in response to changes in photoperiod and ambient temperature. We propose that PRL is one of the factors which regulate the sensitivity of gonadotropin release to gonadal steroid feedback. In hyperprolactinemic women, responsiveness to negative estrogen feedback increases, while LH response to positive estrogen feedback is reduced or absent. In hyperprolactinemic men, both LH and testosterone levels are reduced, implying increased sensitivity of LH release to negative testosterone feedback. In the male rat, both physiological amounts of PRL and experimentally-induced hyperprolactinemia increase the ability of exogenous testosterone to suppress LH and FSH release. Different regulatory mechanisms appear to operate in the seasonally breeding male golden hamster, in which short photoperiod causes concomitant suppression of PRL, LH, FSH and testosterone release. In this species, pharmacologic suppression of PRL release leads to increased responsiveness of plasma gonadotropin levels to negative feedback effects of testosterone, while PRL-secreting ectopic pituitary transplants exert an opposite effect.

We have examined some of the suspected mechanisms of PRL modulation of testosterone feedback in male golden hamsters. In immature animals, the amount of cytoplasmic androgen receptors in the anterior pituitary was decreased by mild hyperprolactinemia and increased by treatment with bromocriptine, an inhibitor of PRL release. Bromocriptine increased pituitary androgen binding also in adult hamsters. These findings would imply that PRL modulates the responsiveness to negative steroid feedback at the pituitary level. However, PRL appears to act also at the hypothalamic level. Suppression of endogenous PRL release by exposure to short photoperiod increased the ability of testosterone to suppress noradrenaline turnover and to stimulate dopamine turnover in the hypothalamus of adult hamsters. Concomitant treatment with PRL secreting pituitary transplants prevented these changes in the responsiveness of noradrenergic and dopaminergic neurons to testosterone. In contrast to the situation in hyperprolactinemic male rats, effects of PRL on the sensitivity of male golden hamsters to testosterone feedback do not appear to involve alterations in pituitary LHRH binding. Total content of LHRH receptors in hamster adenohypophysis declines after exposure to short photoperiod but was not affected by treatment with PRL or bromocriptine. We also have evidence that PRL does not modify the LH response to blockade of opioid receptors in male hamsters. Thus, the suspected function of endogenous opioids in mediating the effects of gonadal steroids on LHRH release does not appear to be modified by PRL.

We conclude that PRL is involved in altering the sensitivity of the hypothalamic-pituitary unit to steroid feedback during sexual maturation and during seasonal transitions between the active and inactive state of the gonads. Modulation of steroid feedback by PRL involves action at both pituitary and hypothalamic levels. In the male hamster, these effects of PRL include reduction in the number of cytoplasmic androgen receptors in the pituitary and increase in the activity of noradrenergic neurons in the hypothalamus.

ACKNOWLEDGMENTS

These studies were supported by NICHHD through grants HD20033, P30HD10202, and HD06380. We thank Dr. Gary C. Chamness for his invaluable help in measurements of DHT receptors, Sandoz Pharmaceuticals for generous supply of bromocriptine, National Hormone and Pituitary Program for hormones and RIA reagents, and Ms. Marion V. Cathey for typing the manuscript. We also wish to apologize to those whose work relevant to this topic may have been inadvertently omitted.

REFERENCES

Ambrosi B, Bara R, Travaglini P, Weber G, Beck Peccoz P, Rondena M, Elli R, Faglia P, 1977. Study of the effects of bromocriptine on sexual impotence. Clin Endocrinol 7:417-21
Andrews WW, Advis JP, Ojeda SR, 1979. Progesterone restores prepubertal effectiveness of estrogen negative feedback on gonadotropin release in adult female rats. Fed Proc 38:985
Aono T, Miyake A, Shioji T Kinugasa T Onishi T, Kurachi K, 1976. Impaired LH release following exogenous estrogen administration in patients with amenorrhea-galactorrhea syndrome. J Clin Endocrinol Metab 42:696-702
Aono T, Miyake A, Shioji T, Yasuda M, Koike K, Kurachi K, 1979. Restoration of oestrogen positive feedback effect on LH release by bromocriptine in hyperprolactinemic patients with galactorrhoea-amenorrhoea. Acta Endocrinol 91:591-600
Aragona C, Bohnet HG, Friesen HG, 1977. Localization of prolactin binding in prostate and testis. The role of serum prolactin concentration on the testicular LH receptor. Acta Endocrinol 84:402-09
Baird DT, McNeilly AS, Sawyers RW, Sharpe RM, 1977. Failure of positive

feedback response to oestrogen during post-partum amenorrhaea in women. Soc Study Fert Annual Conference, July 77, Dublin, Ireland. Abst 43

Baker HWG, Worgul TJ, Santen RJ, Jefferson LS, Bardin CW, 1977. Effect of prolactin on nuclear androgens in perfused male accessory sex organs. In: Troen P, Nankin HR (eds), The Testis in Normal and Infertile Men. New York: Raven Press, pp. 379-85

Barkey RJ, Shani J, Barzilai D, 1979. Regulation of prolactin binding sites in the seminal vesicle, prostate gland, testis and liver of intact and castrated adult rats: effect of administration of testosterone, 2-Bromo-α-ergocryptine and fluphenazine. J Endocrinol 81:11-18

Bartke A, Dalterio S, 1976. Effects of prolactin on the sensitivity of the testis to LH. Biol Reprod 15:90-93

Bartke A, Goldman BD, Bex F, Dalterio S, 1977a. Effects of prolactin (PRL) on pituitary and testicular function in mice with hereditary PRL deficiency. Endocrinology 101:1760-66

Bartke A, Smith MS, Michael SD, Peron FG, Dalterio S, 1977b. Effects of experimentally-induced chronic hyperprolactinemia on testosterone and gonadotropin levels in male rats and mice. Endocrinology 100:182-86

Bartke A, Smith MS, Dalterio S, 1979. Reversal of short photoperiod-induced sterility in male hamsters by ectopic pituitary homografts. Int J Androl 2:257-62

Bartke A, Siler-Khodr TM, Hogan MP, Roychoudhury P, 1981. Ectopic pituitary transplants stimulate synthesis and release of FSH in hamsters. Endocrinology 108:133-39

Bartke A, Steger RW, Klemcke HG, Siler-Khodr TM, Goldman BD, 1982. Effects of experimentally induced hyperprolactinemia on the hypothalamus, pituitary and testes in the golden hamster. J Androl 3:172-77

Bartke A, Doherty PC, Steger RW, Morgan WW, Amador AG, Herbert DC, Siler-Kohdr TM, Smith MS, Klemcke HG, Hymer WC, 1984a. Effects of estrogen-induced hyperprolactinemia on endocrine and sexual functions in adult male rats. Neuroendocrinology 39:126-35

Bartke A, Matt KS, Siler-Khodr TM, Soares MJ, Talamantes F, Goldman BD, Hogan MP, Hebert A, 1984b. Does prolactin modify testosterone feedback in the hamster? Pituitary grafts alter the ability of testosterone to suppress LH and FSH release in castrated male hamsters. Endocrinology 115:1506-10

Bartke A, 1985. Male hamster reproductive endocrinology. In: Siegel HI (ed), The Hamster. Plenum Publishers, pp. 73-98

Baum MJ, Schretlen PJM, 1979. Cytoplasmic binding of oestradiol-17β in several brain regions, pituitary and uterus of ferrets ovariectomized while in or out of oestrus. J Reprod Fertil 55:317-21

Beck W, Wuttke W, 1979. Annual rhythms of luteinizing hormone, follicle-stimulating hormone, prolactin and testosterone in the serum of male rhesus monkeys. J Endocrinol 83:131-39

Bex FJ, Bartke A, 1977. Testicular LH binding in the hamster: modification by photoperiod and prolactin. Endocrinology 100:1223-26

Bex F, Bartke A, Goldman BD, Dalterio S, 1978. Prolactin, growth hormone, luteinizing hormone receptors, and seasonal changes in testicular activity in the golden hamster. Endocrinology 103:2069-80

Bhanot R, Wilkinson R, 1984. The inhibitory effect of opiates on gonadotrophin secretion is dependent upon gonadal steroids. J Endocinol 102:133-41

Buttle HL, 1974. Seasonal variation of prolactin in plasma of male goats. J Reprod Fertil 37:95-99

Callard GV, Mak P, Solomon DJ, 1986. Effects of short days on aromatization and accumulation of nuclear estrogen receptors in the hamster brain. Biol Reprod 35:282-91

Campbell CS, Finkelstein JS, Turek FW, 1978. The interaction of photo-period and testosterone on the development of copulatory behavior in castrated male hamsters. Physiol Behav 21:409-15

Carr WR, Land RB, 1982. Seasonal variation in plasma concentrations of prolactin in castrated rams of breeds of sheep with different seasonality of reproduction. J Reprod Fertil 66:231-35

Carter JN, Tyson JE, Tolis G, Van Vliet S, Faiman C, Friesen HG, 1978. Prolactin-secreting tumors and hypogonadism in 22 men. N Engl J Med 299:847-52

Chandrashekar V, Bartke A, Sellers K, 1987. Prolactin modulates the gonadotropin response to the negative feedback effect of testos-terone in immature male rats. Endocrinology 120:758-63

Chen CL, Meites J, 1970. Effect of estrogen and progesterone on serum and pituitary prolactin levels in ovariectomized rats. Endocri-nology 86:503-508

Chen HJ, 1981. Spontaneous and melatonin-induced testicular regression in male golden hamsters: augmented sensitivity of the old male to melatonin inhibition. Neuroendocrinology 33:43-46

Cheng CY, Bardin CW, Musto NA, Gunsalus GL, Cheng SL, Ganguly M, 1983. Radioimmunoassay of testosterone-estradiol-binding globulin in humans: a reassessment of normal values. J Clin Endocrinol Metab 56:68-75

Clarke IJ, Burman JW, Funder JW, Findlay JK, 1981. Estrogen receptors in the neuroendocrine tissues of the ewe in relation to breed, season and stage of the estrous cycle. Biol Reprod 24:323-31

Clayton RN, 1982. Gonadotropin-releasing hormone modulation of its own pituitary receptors: evidence for biphasic regulation. Endocri-nology 111:152-61

Clemens JA, Minaguchi H, Voogt JL, Meites J, 1969. Induction of preco-cious puberty in female rats by prolactin. Neuroendocrinology 4:150-56

Davis GJ, Meyer RK, 1973. Seasonal variation in LH and FSH of bilat-erally castrated snowshoe hares. Gen Comp Endocrinol 20:61-68

Day ML, Imakawa K, Pennel PL, Zalesky DD, Clutter AC, Kittok RJ, Kinder JE, 1986. Influence of season and estradiol on secretion of luteinizing hormone in ovariectomized cows. Biol of Reprod 35:549-53

Desjardins C, Ewing LL, Johnson BH, 1971. Effects of light deprivation upon the spermatogenic and steroidogenic elements of hamster testes. Endocrinology 89:791-800

Dohler KD, Wuttke W, 1974. Serum LH, FSH, prolactin and progesterone from birth to puberty in female and male rats. Endocrinology 94:1003-08.

Duncan MJ, Goldman BD, 1984. Hormonal regulation of the annual pelage color cycle in the Djungarian hamster, Phodopus sungorus. II. Role of prolactin. J Exp Zool 230:97-103

Duncan JA, Barkan A, Hebron L, Marshall JC, 1986. Regulation of pitui-tary gonadotropin-releasing hormone (GnRH) receptors by pulsatile GnRH in female rats: Effects of estradiol and prolactin. Endocrinology 118:320-27

Ellis GB, Turek FW, 1979. Time course of the photoperiod-induced change in sensitivity of the hypothalamic-pituitary axis to testosterone feedback in castrated male hamsters. Endocrinology 104:625-30

Finch CE, Jonec V, Wisner JR, Sinha YN, De Vellis JS, Swerdloff RS, 1977. Hormone production by the pituitary and testes of male C57BL/6J mice during aging. Endocrinology 101:1310-17

Fox SR, Hoefer MT, Bartke A, Smith MS, 1987. Suppression of pulsatile LH secretion, pituitary GnRH receptor content and pituitary responsiveness to GnRH by hyperprolactinemia in the male rat. Neuroendocrinology, in press

Frehn JL, Urry RL, Ellis LC, 1974. Effect of melatonin and short photoperiod on Δ4 reductase activity in liver and hypothalamus of the hamster and rat. J Endocrinol 60:507-13

Garcia MC, Ginther OJ, 1976. Effects of ovariectomy and season on plasma luteinizing hormone in mares. Endocrinology 98:958-62

Gaston S, Menaker M, 1967. Photoperiodic control of hamster testis. Science 158:925-28

Gauthier D, Coulaud C, 1986. Effect of underfeeding on testosterone-LH feedback in the bull. J Endocrinol 110:233-38

Glass DJ, Amann RP, Nett TM, 1984. Effects of season and sex on the distribution of cytosolic estrogen receptors within the brain and the anterior pituitary gland of sheep. Biol Reprod 30:894-902

Goldman BD, Matt KS, Roychoudhury P, Stetson MH, 1981. Prolactin release in golden hamsters: photoperiod and gonadal influences. Biol Reprod 24:287-92

Gray GD, Smith ER, Davidson JM, 1980. Gonadotropin regulation in middle-aged male rats. Endocrinology 107:2021-26

Gray GD, Wexlar BC, 1980. Estrogen and testosterone sensitivity of middle-aged female rats in regulation of LH. Exp Gerontol 15:201-07

Grossman A, Moult PJA, McIntyre H, Evans J, Silverstone T, Rees LH, Besser GM, 1982. Opiate mediation of amenorrhoea in hyperprolactinaemia and in weight-loss related amenorrhoea. Clin Endocrinol 17:379-88

Grumbach MM, Conte FA, 1985. Disorders of sexual differentiation. In: Wilson JD, Foster DW (eds.), Textbook of Endocrinology. Philadelphia: WB Saunders, pp. 312-401

Horst HJ, 1979. Photoperiodic control of androgen metabolism and binding in androgen target organs of hamsters (Phodopus sungorus). J Steroid Biochem 11:945-50

Huseman CA, Kugler JA, Schneider JG, 1980. Mechanism of dopaminergic suppression of gonadotropin secretion in men. J Clin Endocrinol Metab 51:209-14

Kalra SP, Kalra PS, 1984. Opioid-adrenergic-steroid connection in regulation of luteinizing hormone secretion in the rat. Neuroendocrinology 38:418-26

Kamberi IA, Mical RS, Porter JC, 1980. Effect of anterior pituitary perfusion and intraventricular injection of catecholamines and indoleamines on LH release. Endocrinology 87:1-12

Klemcke HG, Bartke A, Borer KT, 1984. Regulation of testicular prolactin and luteinizing hormone receptors in golden hamsters. Endocrinology 114:594-603

Klemcke HG, Bartke A, Steger R, Hodges S, Hogan MP, 1986. Prolactin (PRL), follicle-stimulating hormone, and luteinizing hormone are regulators of testicular PRL receptors in golden hamsters. Endocrinology 118:773-82

Kovacevic R, Krsmanovic L, Stojilkovic S, Simonovic I, Maric D, Andjus RK, 1982. Effects of bromocriptine-induced hypoprolactinaemia on the developmental pattern of androgen and LH levels in the male rat. Int J Androl 5:437-47

Legan SJ, Karsh FJ, Foster DL, 1977. The endocrine control of seasonal reproductive function in the ewe: a marked change in response to negative feedback action of estradiol on luteinizing hormone secretion. Endocrinology 101:818-24

L'Hermite M, Delogne-Desnoeck J, Michaux-Duchene A, Robyn C, 1978. Alteration of feedback mechanism of estrogen on gonadotropin by sulpiride-induced hyperprolactinemia. J Clin Endocrinol Metab 47:1132-36

Lincoln GA, Peet MJ, Cunningham RA, 1977. Seasonal and circadian changes in the episodic release of follicle-stimulating hormone, luteinizing hormone and testosterone in rams exposed to artificial photoperiods. J Endocrinol 72:337-49

Lloyd JW, Thomas JA, Mawhinney MG, 1973. A difference in the in vitro accumulation and metabolism of testosterone-1,2-^{3}H by the rat prostate gland following incubation with ovine or bovine prolactin. Steroids 22:473-83

Lu KH, Huang HH, Chen HT, Kurcz M, Mioduszewski R, Meites J, 1977. Positive feedback by estrogen and progesterone on LH release in old and young rats. Proc Soc Exp Biol Med 154:82-85

Marchetti B, Labrie F, 1982. Prolactin inhibits pituitary luteinizing hormone-releasing hormone receptors in the rat. Endocrinology 111:1209-16

Martini L, Celotti F, Massa R, Motta M, 1978. Studies on the mode of action of androgens in the neuroendocrine tissues. J Steroid Biochem 9:411-17

Matt KS, Bartke A, Soares MJ, Talamantes F, Hebert A, Hogan MP, 1984. Does prolactin modify testosterone feedback in the hamster? Suppression of plasma prolactin inhibits photoperiod-induced decreases in testosterone feedback sensitivity. Endocrinology 115:2098-103

Matt KS, Steger R, Bartke A, 1985. How does prolactin modify testosterone feedback--a hypothalamic site of action? Program Annual Meeting Am Soc Zool, Am Zoologist, 25, Abstract #203, p. 35A

Maurel D, Lacroix A, Boissin J, 1984. Seasonal reproductive endocrine profiles in two wild mammals: the red fox (Vulpes vulpes L.) and the European badger (Meles meles L.) considered as short-day mammals. Acta Endocrinologica 105:130-38

McCann SM, Ojeda S, Negro-Vilar A, 1974. Sex steroid, pituitary and hypothalamic hormones during puberty in experimental animals. In: Grumback, MM, Grave GD, Mayer FE (eds.), Control of the Onset of Puberty. John Wiley and Sons, New York, pp. 1-31

McNeilly AS, Sharpe RM, Fraser HM, 1983. Increased sensitivity to the negative feedback effects of testosterone induced by hyperprolactinemia in the adult male rat. Endocrinology 112:22-28

Meyer SL, Goodman RL, 1986. Separate neural systems mediate the steroid-dependent and steroid-independent suppression of tonic luteinizing hormone secretion in the anestrous ewe. Biol Reprod 35:562-71

Morin LP, Fitzgerald KM, Rusak B, Zucker I, 1977. Circadian organization and neural mediation of hamster reproductive rhythms. Psychoneuroendocrinology 2:73-98

Muduuli DS, Sanford LM, Palmer WM, Howland BE, 1979. Secretory patterns and circadian and seasonal changes in luteinizing hormone, follicle stimulating hormone, prolactin and testosterone in the male pygmy goat. J Anim Sci 49:543-53

Murphy BD, Mead RA, McKibbin PE, 1983. Luteal contribution to the termination of preimplantation delay in the mink. Biol Reprod 28:497-503

Negri-Cesi P, Martini L, 1986. Testosterone metabolism in the neuroendocrine structures of the male golden hamsters. J Steroid Biochem (Supplement) 25:35S

Nicoletti I, Filipponi P, Sfrappini M, Fedeli L, Petrelli S, Gregorini G, Santeusanio F, Brunetti P, 1984. Catecholamines and pituitary function. I. Effects of catecholamine synthesis inhibition and subsequent catecholamine infusion on gonadotropin and prolactin serum levels in normal cycling women and in women with hyperprolactinemic amenorrhea. Horm Res 19:158-70

Nolin JM, Campbell GT, Nansel DD, Bogdanove EM, 1977. Does androgen influence prolactin secretion? Endocrinol Res Commun 4:61-70

Odell WD, Swerdloff RS, 1976. Etiologies of sexual maturation: a model system based on the sexually maturing rat. Recent Prog Horm Res 32:245-288

Ojeda SR, Advis JP, Andrews WW, 1980. Neuroendocrine control of the onset of puberty in the rat. Fed Proc 39:2365-71

Quigley ME, Sheehan KL, Casper RF, Yen SSC, 1980. Evidence for an increased opioid inhibition of luteinizing hormone secretion in hyperprolactinemic patients with pituitary microadenomas. J Clin Endocrinol Metab 50:427-30

Panerai AE, Sawynok J, LaBella FS, Friesen HG, 1980. Prolonged hyperpro-lactinemia influences β-endorphin and met-enkephalin in the brain. Endocrinology 106:1804-08

Parkening TA, Collins TJ, Smith ER, 1980. Plasma and pituitary concen-trations of LH, FSH and prolactin in aged female C57BL/6 mice. J Reprod Fertil 58:377-86

Pelletier J, Caraty A, 1981. Characterization of cytosolic 5α-DHT and 17β-estradiol receptors in the ram hypothalamus. J Steroid Biochem 14:603-11

Peluso JJ, Steger RW, Hafez ESE, 1977. Regulation of LH secretion in aged female rats. Biol Reprod 16:212-15

Perryman RL, Thorner MO, 1981. The effects of hyperprolactinemia on sexual and reproductive function in men. J Androl 2:233-42

Piacsek BE, Bonier TM, Tan RC, 1986. Altered testosterone feedback in pubertal male rats raised on reduced caloric intake. J Androl 7:292-97

Pieper DR, 1984. Effects of photoperiod, castration, and gonadotropin-releasing hormone (GnRH) on the number of GnRH receptors in male golden hamsters. Endocrinology 115:1857-62

Prins GS, 1986. PRL effect on cytosol and nuclear androgen receptors in ventral, dorsal and lateral rat prostate, Program and Abstracts Endocrine Society, 68th Annual Meeting, Abstract #712, p. 209

Purvis K, Clausen OPF, Olsen A, Haug E, Hansson V, 1979. Prolactin and Leydig cell responsiveness to LH/hCG in the rat. Arch Androl 3:219-30

Rao MR, Bartke A, Parkening TA, Collins TJ, 1984. Effect of treatment with different doses of bromocriptine on plasma profiles of prolac-tin, gonadotrophins and testosterone in mature male rats and mice. Int J Androl 7:258-68

Ravault JP, 1976. Prolactin in the ram: seasonal variations in the concentration of blood plasma from birth until three years old. Acta Endocrinol, Copenh, 83:720-25

Ravault JP, Martinat-Botte F, Mauget R, Martinat N, Locatelli A, Bariteau F, 1982. Influence of the duration of daylight on prolactin secretion in the pig: hourly rhythm in ovariectomized females, monthly variation in domestic (male and female) and wild strains during the year. Biol Reprod 27:1084-89

Reiter RJ, 1980. The pineal and its hormones in the control of reproduc-tion in mammals. Endocr Rev 1:109-31

Roberts AC, Hastings MH, Martensz ND, Herbert J, 1985. Naloxone-induced secretion of LH in the male Syrian hamster: modulation by photo-period and gonadal steroids. J Endocr 106:243-48

Rotsztejn WH, Charli JL, Pattou E, Kordon C, 1977. Stimulation by dopamine of luteinizing hormone-releasing hormone (LHRH) release from the mediobasal hypothalamus in male rats. Endocrinology 101:1475-83

Schams D, Barth D, 1982. Annual profiles of reproductive hormones in peripheral plasma of the male roe deer (Capreolus capreolus). J Reprod Fertil 66:463-68

Schams D, Reinhardt V, 1974. Influences of the season on plasma prolac-tin in cattle from birth to maturity. Horm Res 5:217-26

Schulte BA, Seal US, Plotka ED, Letellier MA, Verme LJ, Ozoga JJ, Parsons JA, 1981. The effect of pinealectomy on seasonal changes in prolac-tin secretion in the white-tailed deer (Odocoileus virginianus borealis) Endocrinology 108:173-78

Shaar CJ, Euker JS, Riegle GD, Meites J, 1975. Effects of castration and gonadal steroids on serum LH and prolactin in old and young rats. J Endocrinol 66:45-51

Shull JD, Gorski J, 1984. Estrogen stimulates prolactin gene transcription by a mechanism independent of pituitary protein synthesis. Endocrinology 114:1550-57

Stallings MH, Matt KS, Amador A, Bartke A, Siler-Khodr TM, Soares MJ, Talamantes F, 1985. Regulation of testicular LH/hCG receptors in golden hamsters (Mesocricetus auratus) during development. J Reprod Fertil 75:663-70

Steger RW, Peluso JJ, 1979. Hypothalamic-pituitary function in the old irregularly cycling rat. Exp Aging Res 5:303-17

Steger RW, Huang HH, Chamberlain D, Meites J, 1980. Changes in the control of gonadotropin secretion in the transition period between regular cycles and constant estrus in the old female rat. Biol Reprod 22:595-603

Steger RW, Matt KS, Klemcke HG, Bartke A, 1985. Interactions of photoperiod and ectopic pituitary grafts on hypothalamic and pituitary function in male hamsters. Neuroendocrinology 41:89-96

Steger RW, Matt K, 1986. Prolactin modifies the response of the hamster neuroendocrine system to testosterone negative feedback. Biol Reprod 34, Supplement 1, Abstract #268, p. 183

Steger RW, Matt KS, Bartke A, 1986. Interactions of testosterone and short-photoperiod exposure on the neuroendocrine axis of the male Syrian hamster. Neuroendocrinology 43:69-74

Steinberger E, Ficher M, 1969. Differentiation of steroid biosynthetic pathways in developing testes. Biol Reprod Suppl 1:119-33

Styne DM, Grumbach MM, 1978. Puberty in the male and female: its physiology and disorders. In: Yen SSC, Jaffe RB, (eds.), Reproductive Endocrinology. Philadelphia: W.B. Saunders, pp. 189-240

Sweeney CA, Morgan WW, Smith MS, Bartke A, 1985. Altered sensitivity to an opiate antagonist, naloxone, in hyperprolactinemic male rats. Neuroendocrinology 42:1-6

Tahka KM, Teravainen T, Wallgren H, 1982. Effect of photoperiod on the testicular steroidogenesis of the bank vole (Clethrionomys glareolus, Schreber). An in vitro study. Gen Comp Endocrinol 47:377-84

Tamarkin L, Hutchison JS, Goldman BD, 1976. Regulation of serum gonadotropins by photoperiod and testicular hormone in the Syrian hamster. Endocrinology 99:1528-33

Tate-Ostroff B, Stetson MH, 1981. Correlative changes in response to castration and the onset of refractoriness in male golden hamsters. Neuroscience 32:325-29

Thorner MO, Besser GM, 1978. Bromocriptine treatment of hyperprolactinemic hypogonadism. Acta Endocrinol 88:131-46

Thorner MO, Evans WS, MacLeod RM, Nunley WC Jr, Rogol AD, Morris JI, Besser GM, 1980. Hyperprolactinemia: Current concepts of management including medical therapy with bromocriptine. In: Goldstein M, Lieberman A, Cahre DB, Thorner MO, (eds.), Ergot Compounds and Brain Function. Neuroendocrine and Neuropsychiatric Aspects. New York: Raven Press, pp. 165-89

Tucker HA, Koprowski JA, Britt JA, Oxender WD, 1974. Serum prolactin and growth hormone in Holstein bulls. J Dairy Sci 57:1092-94

Turek FW, Elliott JA, Alvis JD Menaker M, 1975. Effect of prolonged exposure to nonstimulatory photoperiods on the activity of the neuroendocrine-testicular axis of golden hamsters. Biol Reprod 13:475-81

Turek FW, 1977. The interaction of photoperiod and testosterone in regulating serum gonadotropin levels in castrated male hamsters. Endocrinology 101:1210-15

Vomachka AJ, Greenwald GS, 1979. The development of gonadotropin and steroid hormone patterns in male and female hamsters from birth to puberty. Endocrinology 105:960-66

Voogt JL, Clemens JA, Meites J, 1969. Stimulation of pituitary FSH release in immature female rats by prolactin implant in median eminence. Neuroendocrinology 4:157-163.

Winters SJ, Troen P, 1984. Altered pulsatile secretion of luteinizing hormone in hypogonadal men with hyperprolactinemia. Clin Endocrinol 21:257-63

Wise PM, Payne AH, Karsch FJ, Jaffe RB, 1975. Cytoplasmic oestrogen receptor complex of female ovine pituitary: changes associated with the reproductive state and oestradiol treatment. J Endocrinol 67:447-52

Whitehead PE, West NO, 1977. Metabolic clearance and production rates of testosterone at different times of the year in the male caribou and reindeer. Can J Zool 55:1692-97

Yellon SM, Goldman BD, 1984. Photoperiod control of reproductive development in the male Djungarian hamster (Phodopus sungorus). Endocrinology 114:664-70

Young LS, Speight A, Charlton HM, Clayton RN, 1983. Pituitary gonadotropin-releasing hormone receptor regulation in the hypogonadotrophic hypogonadal (hpg) mouse. Endocrinology 113:55-61

PART II

REGULATION OF FOLLICULOGENESIS AND OVULATION

STUDIES OF THE PERIOVULATORY INTERVAL IN THE IN VITRO PERFUSED OVARY

Edward E. Wallach and Susan J. Atlas

The Johns Hopkins University School of Medicine
600 N. Wolfe Street, Houck Building Room 264
Baltimore, Maryland 21205

INTRODUCTION

The periovulatory interval refers to that period bracketed by the gonadotropin surge and the development of an effective corpus luteum (Fig. 1). Events taking place within its limits include: growth of the preovulatory follicle and its contents, meiotic maturation and cytoplasmic changes in the oocyte necessary for fertilization and early embryonic development, ovulation (follicle wall disruption), and transformation of the ruptured follicle into a corpus luteum. To maximize the probability of conception, ovulation must be synchronized with oocyte meiotic and cytoplasmic maturation

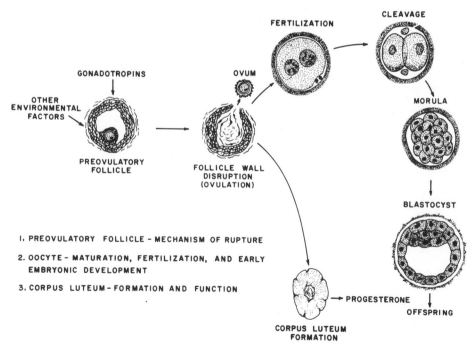

Fig. 1. The periovulatory interval.

179

and the postovulatory follicle must then undergo rapid transformation to a corpus luteum. An effective corpus luteum produces sufficient progesterone to ensure the implantation of a conceptus and
maintenance of a pregnancy. The maturational status of the oocyte
and its embryonic potential, as well as establishment and ultimate
function of the corpus luteum, are dependent upon the quality of
antecedent follicular development. Local events within the ovary
during the periovulatory interval influence: (1) the occurrence and
timing of follicle wall disruption; (2) the maturational status of
the oocyte and its embryonic potential; and (3) the establishment of
an effective corpus luteum.

IN VITRO PERFUSED OVARY

Perfusion of an isolated ovary provides an opportunity to
conduct detailed studies of the intact organ under carefully
regulated conditions independent of systemic influences. Romanoff
and Pincus (1962) studied steroidogenesis in an isolated bovine
ovary perfused with blood. Metabolic pathways including steroidogenesis have been investigated in the human ovary perfused in vitro
with an artificial medium (Kennedy et al., 1975). Synthesis of
progesterone and estradiol from acetate was demonstrated in this
model in response to gonadotropin. In our laboratory Lambertsen and
associates (1976) developed a method of consistently inducing
ovulation in rabbit ovaries perfused in vitro. Intact animals were
stimulated with human chorionic gonadotropin (hCG), and 7 to 8 h
later, one ovary of each animal was removed and perfused. The
timing and rate of ovulation were similar in perfused and
contralateral ovaries which remained in situ. The perfused ovaries
metabolized glucose and lactate continuously throughout the 6 h
experiments. Changes in rates of glucose consumption and lactate
production were comparable to those reported during ovulation in
human ovaries in vitro (Stahler et al., 1974).

It was found that hCG (100 IU) administered directly to the
perfusion medium would stimulate ovulation in ovaries removed from
rabbits that were not pretreated with hCG. Nine of 10 ovaries
ovulated, yielding 3.55 ± 0.22 ovulations per ovulating ovary
(Wallach et al., 1978). These results are similar to the number of
ovulations observed in previous experiments in which hCG was
administered to the intact animal. This observation allows in vitro
exposure of the ovary to a substance of interest before administering hCG. Thus the in vitro perfused rabbit ovary has been
established as a model for studying the periovulatory interval which
enables: (1) isolation of the ovary from extragonadal systemic
influences; (2) direct serial observations of follicle development
and ovulation; (3) addition of substances to the perfusing medium
that may have a direct influence on ovarian function; (4) determination of the composition of the effluent; (5) retrieval of the
ovulated ovum for cytologic and physiologic study; and (6) assessment of corpus luteum formation and function.

The details of the system have been modified over the years,
but the basic components have remained constant. Sexually mature
New Zealand white rabbits weighing approximately 3.5 kilograms are
used. The ovary is isolated and all anastomotic connections are
carefully ligated. A fine glass cannula is inserted approximately 5
mm into the ovarian artery and secured in place with 4-0 silk. Thin
teflon tubing is used to cannulate the ovarian vein (Fig. 2). Each
ovary with artery, vein, and supporting adipose tissue is then

removed from the rabbit and placed in an individual perfusion chamber. The double cannulation makes it possible to collect ovarian arterial and venous samples and to calculate ovarian secretion rates for steroids and other substances. The perfusion apparatus is located in a constant temperature room maintained at 37° C. The system contains three components: oxygenator, peristaltic pump, and perfusion chamber (Fig. 3). The perfusion medium consists of Medium 199 supplemented with insulin, heparin and antibiotics. The total volume (150 ml) approximates the blood volume of a 3.5 kg rabbit, and the flow rate (1.5 ml/minute) that of blood flow through the ovary. The medium is oxygenated by passage over a capillary membrane oxygenator gassed with 5% CO_2/95% O_2. The perfusate is introduced through the ovarian artery cannula and is collected from the ovarian vein cannula and recycled.

The rabbit serves as an ideal species for studying the periovulatory interval by virtue of the relatively short time between gonadotropin stimulus and ovulation in this animal, and because of its consistent ovulatory response to exogenous gonadotropin. Throughout 12 h of perfusion, the rabbit ovary remains morphologically and functionally intact, although edema (Lambertsen et al., 1976; Bjersing et al., 1981) and dissociation of collagen fibrils (Wallach et al., 1984) have been described. Edema has been reduced by incorporating osmotically active particles such as dextran or bovine serum albumin into perfusion solutions. The patterns of progesterone and estradiol release into the recirculated perfusion medium over a 12 h period after exposure to LH or hCG are consistent with those found in ovarian vein blood _in vivo_, (Jansen et al., 1982; Yoshimura et al., 1986a).

OVULATION - FOLLICLE WALL DISRUPTION

Ovulation is the mechanical process that expels the oocyte from the ovarian follicle and transfers the female gamete to a new environment in which fertilization may take place. Ovulation and

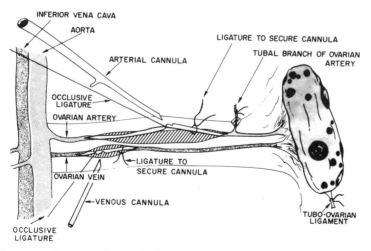

Fig. 2. Cannulation of the rabbit ovarian artery and vein.

oocyte maturation are normally synchronized and coordinated by specific gonadotropin stimuli. A number of local processes intercede between the preovulatory gonadotropin stimulus and ovulation. In a wide range of mammalian species, hormones, enzymes, smooth muscle cells, prostaglandins (PGs), histamine, and catecholamines, participate locally to prepare the follicle for ovulation (Yoshimura and Wallach, 1987a).

Morphology

Cumulus cells in the developing follicle are intimately attached to the oocyte by long microvilli which penetrate the vitelline membrane into the oocyte (Zamboni, 1974). Granulosa cells are also physically associated with one another and are in contact with the antral fluid and cumulus cells. The presence of numerous gap junctions suggest the possibility for exchange of molecules between cumulus cells and the oocyte during follicular development (Gilula et al., 1978). Prior to ovulation, the intimacy of the cumulus/oocyte relationship diminishes (Motlik et al., 1986). The mechanisms responsible for the breakdown of intracellular communication are complex and as yet largely unknown, but probably involve loss of gap junctions (Larsen et al., 1986).

Perifollicular vascular changes in rabbit ovaries during the periovulatory interval have been examined by scanning electron microscopy (SEM) of microcorrosion casts (Kitai et al., 1985a). These casts are prepared by injecting Mercox resin directly into the cannulated ovarian artery. After the resin has hardened, the ovarian tissue is digested with KOH, leaving a cast of the follicular microvasculature. Vascular casts have been prepared from: (1) in situ unstimulated ovaries; (2) in situ ovaries stimulated with hCG; (3) in vitro perfused unstimulated ovaries; (4) in vitro perfused ovaries stimulated with hCG; and (5) in vitro

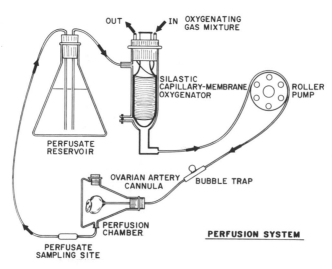

Fig. 3. In vitro rabbit ovary perfusion system.

perfused ovaries following ovulation-inducing doses of non-gonado-tropic substances, including prostaglandin $F_{2\alpha}$ ($PGF_{2\alpha}$), histamine, or norepinephrine. Dilated vessels, extravasation of resin from weakened vessels, and filling defects are observed at the apex of follicles in in situ ovaries 9 to 12 h and in in vitro perfused ovaries 4 to 6 h after stimulation with hCG. In vitro perfused ovaries stimulated with $PGF_{2\alpha}$ or histamine demonstrate dilated capillaries with extravasation of the resin as well as filling defects at the apical region of large follicles. Norepinephrine-stimulated ovaries reveal incomplete filling of vessels, although some large follicles have shown extravasation of resin. Dilated vessels, extravasation of resin, and filling defects characterize preovulatory vascular changes in in situ and in vitro perfused ovaries regardless of the ovulatory stimulus. Extravasation of resin is associated with increased vascular permeability. The filling defects suggest avascularity at the apex, which may weaken the follicle wall and facilitate its rupture.

Progressive degeneration and decomposition of surface epithelial cells have been suggested as significant among the events leading to disruption of underlying connective tissue (Bjersing and Cajander, 1974a). Morphologic studies (Wallach et al., 1984) have demonstrated disappearance of the surface epithelium at the apex of rabbit ovarian follicles perfused with hCG in vitro similar to the in vivo observations of Bjersing and Cajander (1974a). Degradation of collagen fibrils in the theca and tunica albuginea is not as prominent in vitro as in vivo, nor are the multivesicular structures as apparent (Wallach et al., 1984). Changes in the surface epithelium and subsurface connective tissue just prior to ovulation have been compared in in vitro perfused rabbit ovaries and in situ ovaries (Yoshimura et al., 1986b). The subsurface connective tissue was visualized by treatment with sodium dodecyl sulfate (SDS) which removes epithelial cells from the ovary, thereby exposing collagen fibrils and basal lamina. In one experiment, both ovaries were removed for SEM study at time 0 or at 4, 6, 9, or 12 h after hCG administration to enable baseline observations. In a second experiment, both ovaries were removed and perfused for 4-6 h after exposure to hCG. Since ovulation in this perfusion system usually begins at approximately 6 h following hCG administration, observations were initiated at 4 h after hCG exposure. At that time surface epithelial cells had become detached at the follicular apex. Microvilli of the remaining surface epithelial cells were decreased in length and in number. Ovoidal blebs or droplets were observed arising from the underlying tissue. SDS treatment revealed compact collagen fibrils in the center of the apex. The lateral wall of the follicle demonstrated the honeycomb-like architecture characteristic of basal lamina. Neither dissociation nor fragmentation of the collagen fibrils was prominent. These morphologic features are consistent with those observed in ovaries in situ 6 to 9 h after hCG administration. Six h after exposure to hCG in vitro, the surface epithelial cells of the apical region were detached, revealing the connective tissue of the tunica albuginea. Both the degenerative changes of the basal lamina and the appearance of ovoidal blebs or droplets were more prominent just prior to ovulation in the in vitro preparations than they were in in situ ovaries. Following SDS treatment, a defect in the sub-surface connective tissue due to degenerative changes in the basal lamina was visible. Collagen fibrillar networks were more prominent in situ than in vitro; however, the morphologic features in the in vitro ovary at 6 h after hCG administration resembled those of the in situ ovary at 12 h. These observations indicate that during ovulation, degeneration of apical epithelial cells and sub-epithelial follicular tissue occurs synchronously both in situ and in vitro.

Ovarian Smooth Muscle Activity

The identification of autonomic nerves in the ovaries of various mammalian species has prompted the suggestion that the autonomic nervous system may also participate in ovulation by a direct ovarian action (Jacobowitz and Wallach, 1967; Owman et al., 1967). The detection of ovarian smooth muscle tissue and demonstration of its contractility (Okamura et al., 1972; Rocereto et al., 1969; Virutamasen et al., 1972) form the basis for the hypothesis that autonomic nerves within the ovary may influence ovarian smooth muscle contraction at the time of ovulation and assist in achieving follicle rupture and ovum expulsion. Although ovarian contractions accompany the process of ovulation and many agents which suppress ovarian contractions also inhibit ovulation (Wallach et al., 1978), it is not yet clear as to whether ovarian contractions are essential for the occurrence of ovulation. Weiner et al. (1975a) selectively denervated one rabbit ovary and left the contralateral ovary to serve as a control. Ovulation from the denervated ovary and subsequent pregnancy followed mating in this series (Weiner et al., 1975b). Problems inherent in a completely denervated organ (e.g. denervation supersensitivity of receptors to circulating catechol-amines) make such negative findings difficult to evaluate. The occurrence of ovulation from a denervated ovary suggests either that ovarian smooth muscle activity is unnecessary for ovulation or that alternate means of stimulating muscle contractility occur in the absence of normal innervation.

Enzymatic Activity

Proteolytic enzymes appear to play a significant role in the process of mammalian ovulation. Lytic substances have been shown to originate from the surface epithelium at the apical region of the follicle (Rondell, 1970). Collagenolytic enzymes have been detected in ovarian follicles of various species (Ichikawa et al., 1983). Proteases have been shown to promote disruption of the follicle wall in the rabbit (Espey and Lipner, 1965). Espey and Coons (1976) identified two collagenolytic enzymes in rabbit ovarian follicles. Ultrastructural studies of the tunica albuginea and theca externa of the apical region of rabbit Graafian follicles prior to ovulation demonstrate progressive edema, fragmentation of collagen fibrils in the tunica and dissociation of fibroblasts in the theca interna (Bjersing and Cajander, 1974b). Structures identified as lysosomes appear in the surface epithelium in the hours immediately prior to follicle rupture (Bjersing and Cajander, 1974c). These dense bodies in the apical surface epithelium, also noted in in vitro perfused ovaries, are thought to represent a source of lysosomal enzymes which weaken the underlying follicle wall. Their presence is also associated with progressive dissociation of cumulus cells from one another as well as from the oocyte, in response to gonadotropin stimulation. Although PGs can labilize lysosomal membranes, enabling release of lytic enzymes, Curry et al. (1986) have demonstrated that hCG-stimulated collagenase is not dependent upon PG synthesis. Treatment of ovarian follicles with SDS removes the surface epithelium and reveals the collagen fibril network beneath. Such treatment of hamster ovarian follicles induced to ovulate in vitro demonstrates that the collagen network is weakened prior to ovulation (Martin and Miller-Walker, 1983). Collagenase involvement

in ovulation is supported by the observation that pectin, a micro-
bial collagenase inhibitor, blocks ovulation in explanted ovaries
(Ichikawa et al., 1983). Fukumoto et al. (1981) found that in
preovulatory human follicles the highest production of ovarian
collagenase was in granulosa cells and the lowest in the follicle
wall. They concluded that collagenase is produced in granulosa
cells and rapidly transported to the site of collagen degradation in
the apical wall of the follicle.

Rat ovarian granulosa cells produce plasminogen activator in a
fashion which correlates with ovulation. Plasminogen is present in
follicular fluid, and plasmin, the product of the interaction of
plasminogen activator and plasminogen, can weaken the wall of the
follicle in vitro (Strickland and Beers, 1976). In addition to
gonadotropins, certain PGs and adenosine 3'5'-cyclic phosphoric acid
(cAMP) effectively stimulate granulosa cells to produce plasminogen
activator. Under the influence of plasminogen activator, the ovum
is dislodged from the granulosa cell layer and floats free in the
follicular fluid; ultimately the tunica and theca are degraded.
These changes follow the increase in gonadotropin and PG levels
within the follicle. Although Beers et al. (1975) demonstrated that
cultured rat granulosa cells can be stimulated to release plasminogen activator by exposure to LH, Martinat and Combarnous (1983)
reported that the secretion of plasminogen activator is specifically
responsive to FSH. Reich et al. (1985a) examined plasminogen
activator in intact follicles and confirmed that the preovulatory LH
surge stimulates plasminogen activator activity. This group also
demonstrated that intrabursal injections of the cyclooxygenase
inhibitor indomethacin at ovulation-inhibiting doses does not affect
LH-stimulated plasminogen activator activity. Subsequently Canipari
and Strickland (1986) reported that although FSH and LH can both
induce plasminogen activator secretion by granulosa cells, the
response to FSH is more immediate. Granulosa and theca cells have
recently been shown to produce different forms of plasminogen
activator (Canipari and Strickland, 1985); granulosa cells produce

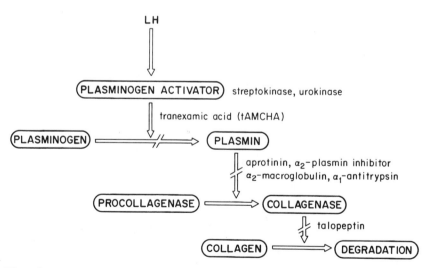

Fig. 4. Proposed scheme for involvement of proteolytic enzymes
in follicle wall disruption.

primarily tissue-type plasminogen activator (tPA) while theca cells produce mostly urokinase-type plasminogen activator (uPA). Both enzymes are under gonadotropin control.

The relationship between PGs and proteolytic enzymes in the process of ovulation is not clear. Although PGs may be involved in the release or activation of collagenolytic enzymes within the follicle wall, blockade of PG synthesis does not inhibit plasminogen activator secretion (Reich et al., 1985a; Espey et al., 1985). Nonetheless, Reich et al. (1985a) demonstrated that indomethacin prevented ovarian collagen breakdown associated with ovulation in rats treated with pregnant mare's serum gonadotropin (PMSG). Espey et al. (1985) on the other hand, conclude that although plasminogen activator may play a mediating role in the preovulatory inflammatory reaction in the follicle wall, it does not directly contribute to follicle rupture.

Studies of ovulation _in vitro_ have helped to elucidate the role of enzymes in the ovulatory process. In the _in vitro_ perfused rabbit ovary system, streptokinase (a plasminogen activator of bacterial origin) induces ovulation in the rabbit ovary in the absence of gonadotropin. Exposure to trans-4-(aminomethyl)-cylohexane-carboxylic acid (tAMCHA), an inhibitor of the conversion of plasminogen to plasmin, significantly reduces ovulatory response to hCG (Yoshimura et al., 1987b). Inclusion of aprotinin, a potent inhibitor of plasmin, in the perfusate also inhibits hCG-induced ovulation. SEM studies of the denuded surface of SDS-treated ovaries perfused with streptokinase reveal prominent loosening and decomposition of collagen in the tunica albuginea (Yoshimura et al., 1987b). These observations further substantiate the role of proteolytic enzymes in the ovulation process. That streptokinase, a plasminogen activator, can accomplish ovulation in the absence of gonadotropin suggests that plasminogen is involved in the dissolution of the follicle wall. A hypothetical enzyme cascade leading to follicle rupture is presented in Fig. 4.

Prostaglandins

Prostaglandins have been implicated in the mechanism of ovulation in several species. Indomethacin delays or inhibits ovulation in the rat, rabbit and monkey (Armstrong and Grinwich, 1972; Tsafriri et al., 1972; Orczyk and Behrman, 1972; Diaz-Infante et al., 1974; Wallach et al., 1975). Levels of PGE and PGF rise within developing rabbit ovarian follicles in response to LH or hCG (LeMaire et al., 1973). PGE and PGF metabolites increase in the effluent of _in vitro_ perfused rabbit ovaries prior to ovulation (Schlaff et al., 1983). The increases in follicular PG content following gonadotropin are blocked by administration of indomethacin (Yang et al., 1973). Osman and Dullaart (1976) have reported that indomethacin alters ovulation by inducing the retrograde release of eggs from follicles through "misplaced" stigmata. In the rabbit, intrafollicular injection of either indomethacin or antiserum specific for F series PGs has also been shown to prevent ovulation in injected follicles without inhibiting ovulation in adjacent, untreated follicles (Armstrong et al., 1974). While these observations suggest that PGs influence ovulation locally at the ovarian level, the precise nature of their role in the process of ovum expulsion has yet to be defined. $PGF_{2\alpha}$ can induce ovulation in the perfused rabbit ovary in the absence of gonadotropin (Kitai et al., 1985b). Administration of indomethacin to the intact rabbit prior to ovarian removal inhibits hCG-induced ovulation both in the _in_ vitro perfused as well as in the in situ contralateral control ovary. That blockade of ovulation can be reversed by addition of $PGF_{2\alpha}$ to the perfusate, further strengthens the significance of $PGF_{2\alpha}$ in the ovulatory mechanism (Hamada et al., 1978). Similar findings were reported by Holmes et al. (1983) using an isolated perfused rabbit ovary. Indomethacin, added directly to the perfusate together with hCG, significantly reduces ovulatory efficiency. Indomethacin does not appear to interfere with the ability of ovulated ova or follicular oocytes to achieve release of the first polar body (i.e. undergo nuclear maturation). Indomethacin treatment, however, does delay the occurrence of follicle rupture (Kobayashi et al., 1981) and leads to degeneration of a high percentage of ova. These findings suggest that $PGF_{2\alpha}$ protects ova from premature degeneration. Although several of the periovulatory events within the follicle, including resumption of ovum meiosis, steroidogenesis, and luteinization, are correlated with local levels of PG, these activities can proceed undisturbed despite blockade of PG synthesis. Follicle rupture, however, is the exception, and ovulation can be prevented by administration of indomethacin or PG antibody. Espey (1980) has likened ovulation to an inflammatory process with PG serving as a mediator. $PGF_{2\alpha}$ and histamine are each individually capable of leading to follicle rupture. No sequence of action of these two agents (histamine, $PGF_{2\alpha}$) could be detected in the perfused rabbit ovary when studied using antihistamines and indomethacin (Kitai et al., 1985b).

The mechanism of PG action on the follicle remains to be elucidated. Inhibition of PG synthesis by the intrabursal administration of indomethacin inhibits collagenolysis, as determined by the ovarian turnover of hydroxyproline (Reich et al., 1985b). Likewise, the usual preovulatory increase in collagenase activity in the rabbit ovary is not present after indomethacin treatment (Kawamura et al., 1984). Curry et al. (1986) however, have demonstrated a preovulatory rise in follicular collagenase dependent upon gonadotropin but not on PG; the source of the collagenase may be the multivesicular structures or dense granules seen in preovulatory follicular and surface epithelial cells.

The work of Espey et al. (1986a) demonstrates two distinct stages of PG production by the ovarian follicle following hCG administration to intact rabbits. Indomethacin significantly reduces PG production within 5 min of administration, but when given 9 h after hCG, ovulation is not suppressed. In achieving ovulation, PGs may mediate a cascade of proteolytic activity in the follicle and/or give rise to vasodilation. Ultrastructural studies of ovarian capillaries in rabbits treated with indomethacin suggest that PGs increase to the permeability of perifollicular capillaries which normally precedes ovulation (Okuda et al., 1983). Vasodilation of the microvasculature of preovulatory follicles following $PGF2\alpha$ perfusion has also been shown by SEM examination of vascular casts (Kitai et al.,1985a).

Bradykinin

Simulating an acute inflammatory reaction, the ovulatory process is preceded by the local release of histamine (Laurent and Bienvenu, 1982; Willoughby et al., 1982), prostaglandins (Vane, 1976), and plasminogen activator (Hamilton et al., 1981). These substances are produced by ovarian follicles in response to gonadotropic hormones. In addition, exposure of ovaries to exogenous histamine, PGs, and serine proteases similar to plasminogen

activator can induce follicle rupture in the absence of gonadotropin. The inflammatory agent bradykinin (BK) may also be involved
in the ovulatory process (Espey, 1980; Smith and Perks, 1979; Smith
and Perks, 1983). This concept is supported by recent evidence of a
significant increase in ovarian kinin-generating activity during the
ovulation process in the rat (Espey et al., 1986b). More recently,
BK has been shown to induce rupture of mature follicles and to
stimulate PG biosynthesis in in vitro perfused rabbit ovaries
(Yoshimura et al., 1986c). Although the PG synthesis inhibitor
indomethacin lowers PG production to basal levels in this system, it
does not block BK-induced ovulation. Moreover, a specific
antagonist of BK significantly inhibits ovulation in hCG-treated
ovaries. An inflammatory reaction in the follicle wall may produce
a cascade of biochemical events which lead to ovulation. BK may
appear in this cascade after the production of PGs. The progression
to ovulation in the absence of appreciable PG biosynthesis is
consistent with Espey's report (Espey et al., 1986a) that the rabbit
follicle can ovulate in vivo despite reduction of PG production by
indomethacin.

OOCYTE MEIOTIC AND CYTOPLASMIC MATURATION

The preovulatory gonadotropin surge initiates a series of
events leading not only to ovulation but also to maturation of one
or more oocytes. Maturation confers upon the oocyte the capacity to
undergo normal fertilization and subsequent embryonic development
and includes changes in both the nucleus and the cytoplasm.

Meiotic Maturation

Mammalian oocytes enter prophase of the first meiotic division
early in fetal life. The process is arrested at the diplotene stage
of prophase, usually around the time of birth, and does not normally
resume until puberty (reviewed by Franchi and Baker, 1973). The
midcycle gonadotropin surge normally signals resumption of meiosis,
morphologically visible as dissolution of the nuclear membrane or
germinal vesicle breakdown (GVBD), in a cohort of oocytes. Following the release of the first polar body, the process is again
arrested at metaphase of the second meiotic division; and in most
mammals meiosis is only completed following fertilization. Meiotic
or nuclear maturation is the process that begins with GVBD and
proceeds to formation of the first polar body and arrest in metaphase II.

Oocyte Maturation Inhibitor (OMI). The capacity of oocytes to
resume meiosis and undergo nuclear maturation appears to be age
related and is acquired in steps (Szybek, 1972; Sorensen and
Wassarman, 1976). Oocytes isolated from fetal mice and cultured in
vitro acquire the ability to resume meiosis to metaphase I on the
same day of development as those that remain in situ (Canipari et
al., 1984). The intrafollicular environment inhibits resumption of
meiosis after the oocytes have gained meiotic competence. Pincus
and Enzmann (1935) demonstrated that oocytes liberated from their
follicles before the gonadotropin surge will undergo spontaneous
meiotic maturation in vitro. Addition of follicular fluid (Chang,
1952) or follicular wall including granulosa cells (Foote and
Thibault, 1969) to in vitro cultured oocytes or coculture of oocytes
with granulosa cells (Tsafriri and Channing, 1975) prevents meiotic
maturation. Tsafriri and Channing concluded that granulosa cells
secrete an inhibitory substance that they named oocyte maturation
inhibitor (OMI).

OMI activity is dependent upon an intact cumulus oophorus (Hillensjo et al., 1979) and its effect is reversed by LH (Tsafriri et al., 1976). While there is evidence that porcine follicular fluid OMI is a small peptide with a molecular weight of approximately 1,000 (Stone et al., 1978), it has not yet been isolated or sequenced. Several investigators have failed to confirm meiotic inhibition with porcine follicular fluid (Liebfried and First, 1980; Racowsky and McGaughey, 1982; Fleming et al., 1983). Other investigators have presented evidence that the active low molecular weight components of follicular fluid may be hypoxanthine (Downs et al., 1985) and adenosine (Eppig et al., 1985).

Cyclic AMP. Gap junction-mediated transmission of follicle cell cAMP to the oocyte may inhibit oocyte maturation (Dekel and Beers, 1978). This hypothesis is supported by the observation that cAMP analogs and phosphodiesterase inhibitors reversibly inhibit maturation in both cumulus-enclosed and denuded oocytes (Cho et al., 1974; Wasserman et al., 1976; Magnusson and Hillensjo, 1977). Gonadotropin stimulation provokes progressive dissociation of cumulus cells from one another as well as from the oocyte, and it has been reported that a striking loss of cumulus cell gap junctions accompanies GVBD in the mouse (Larsen et al., 1986). Disruption of cumulus-oocyte communication could interrupt the direct transfer of cAMP to the oocyte.

The effects of cAMP on nuclear maturation and on ovulation in the in vitro perfused rabbit ovary have recently been reported (Yoshimura et al., 1985a). In these studies, continuous exposure of the perfused rabbit ovary to dibutyryl cAMP [(Bu)$_2$cAMP] (10^{-3} M) for 12 h did not affect GVBD, while exposure to (Bu)$_2$cAMP (10^{-5} to 10^{-3}M) for the first 2 h of perfusion, followed by perfusion without the drug significantly increased the percentage of follicular oocytes achieving GVBD. Forskolin, an activator of adenylate cyclase, postponed resumption of meiotic maturation in in vitro cultured cumulus-enclosed rabbit oocytes (Yoshimura et al., unpublished observations). GVBD was inhibited by incubation with forskolin for 3 h, but by 6 h there was no significant difference between oocytes cultured in the presence or absence of forskolin. In the in vitro perfused ovary, doses of forskolin which accelerate resumption of meiosis also stimulate intrafollicular cAMP synthesis, but the concentration of follicular cAMP returns to basal levels within 6 h of forskolin administration. We have concluded that a transiently elevated level of intrafollicular cAMP may be an essential prerequisite for meiotic resumption during the pre-ovulatory interval.

Cytoplasmic Maturation

Steroids. Morphologic alterations in the nucleus, which have been accepted as evidence of oocyte maturation, may not necessarily be indicative of normal fertilizability of the oocyte (Thibault, 1977). Assessment of full maturation necessitates examining not only the fertilizability of ova, but also their potential for embryonic development.

Oocytes require a specific intrafollicular steroid environment for the completion of the full maturation process (Moor et al.,

1980; Thibault et al., 1975a; Ainsworth et al., 1980). For example, steroids exert a significant influence on the synthesis of cytoplasmic factors which induce normal decondensation of the sperm head and formation of the male pronucleus (Thibault et al., 1975b).

Inhibition of steroid synthesis in the in vitro perfused rabbit ovary with aminoglutethamide phosphate (AGP) does not affect hCG-induced ovulation or nuclear maturation of the ovulated ova (Yoshimura et al., 1986a). After insemination in vitro, a high penetration rate was observed in ova ovulated from hCG-treated ovaries with or without AGP. However, a greater percentage of ovulated ova from control (hCG-treated) ovaries showed evidence of normal fertilization (69.8%) than from hCG-treated ovaries perfused with AGP (24.5%). Interference with ovarian steroidogenesis during the process of follicular development may influence specific cytoplasmic changes in the oocyte, resulting in a reduced probability of normal fertilization. Similar results have been obtained in ovaries perfused with cyanoketone, an inhibitor of β-hydroxysteroid dehydrogenase, using the same model system (Yoshimura et al., 1987c). The addition of estradiol to the perfusate reversed the adverse effects of cyanoketone on fertilizability. The evidence is accumulating that estrogen may be the major steroidal signal during the early inductive phase of cytoplasmic maturation (Osborn and Moor, 1983). Studies of human oocytes obtained for in vitro fertilization (IVF) reveal a close correlation between steroid content of follicular fluid and ability of the oocytes to be fertilized in vitro (Wramsby et al., 1981; Carson et al., 1982; Laufer et al., 1984).

A disparity has been consistently observed between rates of ovulation and establishment of pregnancy in women treated with the estrogen agonist/antagonist clomiphene citrate (CC). Schmidt et al. (1986a) and Laufer et al. (1983) described decreased fertilization and embryonic development rates for mouse oocytes incubated with clomiphene. The perfused rabbit ovary has been used as a means of exposing intrafollicular oocytes to clomiphene prior to ovulation, simulating the period of exposure to clomiphene when administered clinically. Using this model, we have found that clomiphene treatment did not affect fertilization and degeneration rates of ova observed 12 h after insemination (Yoshimura et al., 1985b). However, exposure to clomiphene significantly reduced progression to the morula stage at 60 h following insemination (15.4% for CC plus hCG vs. 48.9% for hCG alone). Although exposure of follicle-enclosed oocytes to clomiphene during follicular maturation in this model does not affect fertilizability of ova ovulated in response to hCG, developmental capacity after fertilization is compromised. The addition of estradiol to the perfusate reverses the deleterious effect of clomiphene on the oocytes and enables normal embryonic development (Yoshimura et al., 1986c). Thus the antiestrogenic effect of clomiphene may influence cytoplasmic maturation of the oocyte, leading to restriction of subsequent developmental capacity, similar to that observed with inhibitors of steroid synthesis (AGP, cyanoketone).

Other drugs that have been shown to affect reproduction may also be influencing estradiol-induced cytoplasmic maturation. Nicotine and anabasine, low molecular weight components of cigarette smoke, inhibit granulosa cell aromatase in a dose dependent manner (Barbieri et al., 1986). Previously, nicotine had been shown to delay ovum cleavage in the rat, presumably through disturbance of endocrine effects on the reproductive tract (Yoshinaga et al., 1979). Cannabinoids have been thought to exert a direct gonadal effect, but their influence on fertilizability of ova and subsequent embryonic development have not been studied (Adashi et al., 1983; Almirez et al., 1983). Likewise danazol inhibits estradiol secretion in cultured human granulosa cells (Olsson et al., 1986). In light of the above mentioned experiments using cyanoketone, AGP, and clomiphene, inhibition of granulosa cell aromatase activity and estradiol production has important implications regarding possible drug-induced and environmental effects on estrogen-dependent processes.

Peptide hormones. As a rule, extrafollicular oocytes which have been allowed to undergo maturation in vitro do not exhibit normal development (reviewed by Thibault, 1977). Shalgi et al. (1979) demonstrated that addition of LH to the in vitro culture of cumulus-enclosed rat oocytes increased their developmental competence. Similar results were found with addition of hCG (Fleming et al., 1985). Addition of gonadotropins, estradiol, and supplementary follicle cells to cultures of cumulus-enclosed oocytes greatly increased development of sheep oocytes to the blastocyst stage (Staigmiller and Moor, 1984). We have attempted to define more precisely hormonal conditions for in vitro maturation which support subsequent fertilization and embryonic development. For this purpose, immature follicular oocytes were recovered from non-stimulated rabbit ovaries and cultured for 12 to 14 h in Ham's F-10 medium supplemented with 20% fetal calf serum and exogenous hormones (Hosoi et al., 1985). Cultured ova were inseminated in vitro and transferred 12 h later to fresh culture medium. The embryos were examined at 120 h following insemination, a time when normally fertilized eggs have usually achieved the morula or blastocyst stage. In each experiment, hormones (LH, FSH, estradiol, prolactin, testosterone) were added in a dose of 1 ug/ml. The results are summarized in Table 1.

These data suggest that hormonal composition in the environment of the oocyte is critical for acquisition of developmental ability. Estradiol appears to be an important constituent in this environment, promoting post-fertilization embryonic development. Of particular additional interest is the possible contribution of prolactin to the maturation process. It is apparent that the preovulatory environment within the follicle may influence nuclear and cytoplasmic maturation of the oocyte, affect its fertilizability, and govern its potential for full and normal development.

CORPUS LUTEUM FORMATION

Shortly after rupture of the Graafian follicle, theca and granulosa cells normally become luteinized and produce progesterone in concentrations capable of preparing the endometrium for implantation of the conceptus. Postovulatory progesterone production by the corpus luteum has been thought to be in part contingent upon an adequate preovulatory gonadotropin surge. Maintenance of the corpus luteum, on the other hand, has been attributed to small amounts of LH produced during the postovulatory interval.

The in vitro ovarian perfusion model can also be applied to studies of the developing corpus luteum. In preliminary experiments, rabbits received hCG (100 IU) intravenously and ovaries were removed for perfusion 7.5 h later (2.5 to 4.5 h prior to anticipated ovulation) (A.M. Dharmarajan, unpublished results). Ovarian perfusion was continued for 8 h during which time samples of the

Table 1. Post-Fertilization Embryonic Development at 120
          hours after insemination

Hormones added	Arrest at 2-4 Cell Stage (%)	Morula or Blastocyst (%)
LH, FSH	4	0
LH, FSH, E_2	4	16
LH, FSH, E_2, PRL	7	56
LH, FSH, E_2, Testosterone	0	0

perfusate were withdrawn hourly both from the cannulated ovarian
vein and from a point in the tubing just proximal to the site of
entry of the cannula into the ovarian artery. Blood samples were
also obtained from the cannulated femoral vein of the anesthetized
animal. The ovulated oocytes were recovered during the 8 h of
perfusion. At the conclusion of perfusion, both the in vitro
perfused ovary and the contralateral ovary which served as an in
situ control were perfusion-fixed, and corpora lutea were dissected
and prepared for histologic study. Progesterone concentrations were
determined in samples of perfusate and blood. Progesterone secretion rates were calculated for each perfused ovary. Results are
presented in Table 2.

Progesterone secretion in the in vitro perfused ovary almost
doubled following ovulation. The postovulatory rise in progesterone
secretion, although rapid, lasted for only 4 h and then gradually
diminished. Progesterone concentrations in the peripheral plasma of
the anesthetized animal rose and remained elevated. Eight h
following cannulation (16 h post hCG), follicular cells of both in
situ and in vitro perfused ovaries exhibited varying degrees of
luteinization. A large cavity was observed and a small percentage
of the cells were luteinized, as evidenced by the presence of lipid
droplets and characteristic changes in smooth endoplasmic reticulum.
These preliminary data support the potential usefulness of the in
vitro perfused rabbit ovary system for the study of corpus luteum
formation.

El-Fouly et al. (1970) have provided evidence that expulsion or
removal of ova from rabbit follicles was a prerequisite for luteinization and postulated that the oocyte plays an inhibitory role in
the luteinization process. The concept of oocyte entrapment in a
luteinized unruptured follicle was initially proposed (Townsend et
al., 1966) as one explanation for the discrepancy between occurrence
of ovulation and pregnancy rate in infertile humans receiving
clomiphene citrate for ovulation induction. Koninckx et al. (1978)
have supported this concept through their clinical correlations of
delayed rise in serum progesterone, abnormal basal body temperature
readings, and lack of visual evidence of ovulatory stigmata in a
group of women with unexplained infertility. Inhibition of ovulation has been observed in rabbits during indomethacin administration
without impairment of the LH-induced pattern of peripheral progesterone levels or alterations in the normal histology of luteinized
granulosa cells, in contrast to the conclusions of El-Fouly et al.
(Armstrong and Grinwich, 1972; Grinwich et al., 1972).

Table 2.   Progesterone Secretion by the In Vitro Perfused Ovary
and Peripheral Plasma Concentrations in In Situ Controls

| Time (h) | Progesterone (mean ± SEM) | |
	In vitro (ug/h/ovary)	In situ (ng/ml)
2^a $(10)^b$	1.40 ± 0.34	7.6 ± 1.8
4 (12)	3.40 ± 0.87	6.3 ± 1.4
6 (14)	1.04 ± 0.21	6.5 ± 1.0
8 (16)	0.59 ± 0.12	3.7 ± 1.1

[a]Time in hours following ovarian removal and perfusion.
[b]Time in hours following hCG administration.

These observations, subsequently confirmed by Phi et al. (1977) in experiments using systemic and intrafollicular injections of LH, PGs, and indomethacin, have led to the additional conclusion that PGs do not act as obligatory mediators of LH in the luteinization of rabbit Graafian follicles.   In the rhesus monkey ovulation can be blocked when administration of ovulation-inducing doses of hMG-hCG is accompanied by indomethacin treatment (10 mg/kg); however, in these experiments progesterone levels in peripheral blood did not differ from those following normal ovulation (Wallach et al., 1975). Indomethacin treatment suppresses ovulation in the pig, but promotes morphologic and histologic changes in unovulated follicles charac-teristic of corpora lutea (Ainsworth et al., 1979).   In these studies the evolution of luteal organization was unaltered by indomethacin treatment.   That resumption of meiosis and oocyte maturation were not impaired by indomethacin treatment provides evidence that oocyte maturation is not PG-mediated. These observa-tions parallel those of Kobayashi et al. (1981) in the isolated perfused rabbit ovary, in which indomethacin treatment (0.5 ug/ml) significantly reduced the occurrence of ovulation in gonadotropin-treated ovaries.   Indomethacin, while inhibiting ovulation in the

LH-treated perfused rabbit ovary, does not interfere with progest-erone production (Schmidt et al., 1986b).   Intraovarian release of oocytes in indomethacin-treated proestrus rats suggests that weakening of the follicular apex may be PG dependent (Osman and Dullaart, 1976).

In the human, clinical criteria used for establishing the luteinized unruptured follicle (LUF) as an entity have been indirect.   The criteria have included: (1) failure to identify a stigma via laparoscopy during the luteal phase of apparently ovulatory cycles;   and (2) serial ultrasonographic findings consistent with luteinization (loss of a clearly demarcated follicle wall and presence of intrafollicular echoes) without rapid decrease in follicle size or evidence of free fluid in the cul-de-sac (Coulam et al., 1982).   In order to substantiate the diagnosis of LUF, histologic studies demonstrating follicular luteinization and an entrapped oocyte would have to be correlated with peripheral plasma progesterone levels indicative of a normally functioning corpus luteum.   Such documentation is obviously not feasible in the human. Schenken et al. (1986) have provided biochemical and histologic evidence for this entity in a spontaneous cycle in a rhesus monkey in which the midcycle surge of LH and FSH was blunted.   Experimental models for studying LUF are currently confined to indomethacin-treated animals and LH administered in increasing doses on diestrus day 2 to 4-day cycling rats (Plas-Roser et al., 1984). Other models for LUF would be valuable in helping to determine the components essential for early establishment of the corpus luteum and maintenance of normal luteal activity.

Obviously, number and status of granulosa cells just prior to ovulation, presence of luteotrophic factors, availability of progesterone precursors, and efficient vascularization of the ruptured follicle, are among the components responsible for normal corpus luteum function. Many of these components reflect follicular status just prior to ovulation. Thus the level of corpus luteum function depends to a great extent upon earlier events during the preovulatory interval.

In conclusion, the periovulatory interval has been identified as the period between the preovulatory gonadotropin surge and establishment of the corpus luteum. Factors influencing follicular development and function are critical in determining reproductive potential in any given cycle. The status of the preovulatory follicle and its environment determine the basic steps leading to reproduction. These include: (1) the timing and efficiency of follicle wall disruption; (2) the ability of the intrafollicular oocyte to resume meiosis, undergo fertilization, and proceed with normal preimplantation development; and (3) the establishment of a corpus luteum capable of promoting implantation and pregnancy maintenance. The perfused mammalian ovary enables us to investigate the periovulatory interval through a model which is isolated from many uncontrollable variables and is amenable to experimental manipulation.

ACKNOWLEDGMENTS

We wish to acknowledge the research efforts of the following scientists for the work herein presented: Alfred Bongiovanni, M.D., A.M. Dharmarajan, Ph.D., Augusto Diaz-Infante, M.D., Yasuo Hamada, M.D., Yoshihiko Hosoi, Ph.D., Hirokatsu Kitai, M.D., Yoshimune Kobayashi, M.D., Christian Lambertsen, M.D., Rosemary Santulli, B.S., Karen H. Wright, M.A., and Yasunori Yoshimura, M.D. We would also like to thank Frances Karas for her expert assistance in preparation of the manuscript.

This research was supported by NIH grant HD19034, the Connelly Foundation, the Rockefeller Foundation, and the Mitchell and Lillian Duberstein Foundation.

REFERENCES

Adashi EY, Jones PBC, Hsueh AJW, 1983. Direct antigonadal activity of cannabinoids: suppression of rat granulosa cell functions. Am J Physiol 244:177-85

Ainsworth L, Tsang BK, Downey BR, Baker RD, Marcus GJ, Armstrong DT, 1979. Effects of indomethacin on ovulation and luteal function in gilts. Biol Reprod 21:401-11

Ainsworth L, Tsang BK, Downey BR, Marcus GJ, Armstrong DT, 1980. Interrelationships between follicular fluid steroid levels, gonadotropic stimuli, and oocyte maturation during preovulatory development of porcine follicles. Biol Reprod 23:621-27

Almirez R, Smith CG, Asch R, 1983. The effects of marijuana extract and delta-9-tetrahydrocannabinol on luteal function in the rhesus monkey. Fertil Steril 39:212-17

Armstrong DT, Grinwich DL, 1972. Blockade of spontaneous and LH-induced ovulation in rats by indomethacin, an inhibitor of prostaglandin biosynthesis. Prostaglandins 1:21-26

Armstrong DT, Grinwich DL, Moon YS, Zamecnik J, 1974. Inhibition of ovulation in rabbits by intrafollicular injection of indomethacin and prostaglandin F antiserum. Life Sci 14:129-40

Barbieri RL, McShane PM, Ryan KJ, 1986. Constituents of cigarette smoke inhibit human granulosa cell aromatase. Fertil Steril 46:232-36

Beers WH, Strickland S, Reich E, 1975. Ovarian plasminogen activator: relationship to ovulation and hormonal regulation. Cell 6:387-94

Bjersing L, Cajander S, 1974a. Ovulation and the mechanism of follicle rupture. II. Scanning electron microscopy of rabbit germinal epithelium prior to induced ovulation. Cell Tissue Res 149:301-12

Bjersing L, Cajander S, 1974b. Ovulation and the mechanism of follicle rupture. V. Ultrastructure of tunica albugenia and theca externa of rabbit Graafian follicles prior to induced ovulation. Cell Tiss Res 153:15-30

Bjersing L, Cajander S, 1974c. Ovulation and the mechanism of follicle rupture. III. Transmission electron microscopy of rabbit germinal epithelium prior to induced ovulation. Cell Tiss Res 149:313-27

Bjersing L, Cajander S, Damber J-E, Janson PO, Kallfelt B, 1981. The isolated perfused rabbit ovary - a model for studies of ovarian function. Cell Tiss Res 216:471-79

Canipari R, Palombi F, Riminucci M, Mangia F, 1984. Early programming of maturation competence in mouse oogenesis. Dev Biol 102:519-24

Canipari R, Strickland S, 1985. Plasminogen activator in the rat ovary. Production and gonadotropin regulation of the enzyme in granulosa and theca cells. J Biol Chem 260:5121-25

Canipari R and Strickland S, 1986. Studies on the hormonal regulation of plasminogen activator production in the rat ovary. Endocrinology 118:1652-59

Carson RS, Trounson AO, Findlay JK, 1982. Successful fertilisation of human oocytes in vitro: concentration of estradiol-17 , progesterone and androstenedione in the antral fluid of donor follicles. J Clin Endocrinol Metab 55:798-800

Chang MC, 1952. The maturation of rabbit oocytes in culture and their maturation, activation, fertilization and subsequent devleopment in the fallopian tubes. J Exp Zool 128:379-405

Cho WK, Stern S, Biggers JD, 1974. Inhibitory effect of dibutyryl cAMP on mouse oocyte maturation in vitro. J Exp Zool 187:383-86

Coulam CB, Hill LM, Breckle R, 1982. Ultrasonic evidence for luteinization of unruptured preovulatory follicles. Fertil Steril 37:524-29

Curry TE Jr, Clark MR, Dean DD, Woessner JF Jr, LeMaire WJ, 1986. The preovulatory increase in ovarian collagenase activity in the rat is independent of prostaglandin production. Endocrinology 118:1823-28

Dekel N, Beers WH, 1978. Rat oocyte maturation in vitro. Proc Nat Acad Sci USA 75:4369-73

Diaz-Infante A, Wright KH, Wallach EE, 1974. Effects of indomethacin and prostaglandin $F2_a$ on ovulation and ovarian contractility in the rabbit. Prostaglandins 5:567-81

Downs SM, Coleman DL, Ward-Bailey PF, Eppig JJ, 1985. Hypoxanthine is the principal inhibitor of murine oocyte maturation in a low molecular weight fraction of porcine follicular fluid. Proc Nat Acad Sci USA 82:454-58

El-Fouly MA, Cook B, Nekola M, Nalbandov AV, 1970. Role of the ovum in follicular luteinization. Endocrinology 87:288-93

Eppig JJ, Ward-Bailey PF, Coleman DL, 1985. Hypoxanthine and adenosine in murine ovarian follicular fluid: concentrations and activity in maintaining oocyte meiotic arrest. Biol Reprod 33:1041-49

Espey LL, 1980. Ovulation as an inflammatory reaction - a hypothesis. Biol Reprod 22:73-106

Espey LL, Coons PJ, 1976. Factors which influence ovulatory degradation of rabbit ovarian follicles. Biol Reprod 14:233-45

Espey LL, Lipner H, 1965. Enzyme-induced rupture of rabbit Graafian follicles. Am J Physiol 208:208-13

Espey LL, Shimada H, Okamura H, Mori T, 1985. Effect of various agents on ovarian plasmonogen activator activity during ovulation in pregnant mare's serum gonadotropin-primed immature rats. Biol Reprod 32:1087-94

Espey LL, Norris C, Saphire D, 1986a. Effect of time and dose of indomethacin on follicular prostaglandins and ovulation in the rabbit. Endocrinology 119:746-54

Espey LL, Miller DH, Margolius HS, 1986b. Ovarian increase in kinin generating capacity in the PMSG/hCG-primed immature rat. Am J Physiol 251:E362-65

Fleming AD, Kalil W, Armstrong DT, 1983. Porcine follicular fluid does not inhibit maturation of rat oocytes in vitro. J Reprod Fertil 69:665-70

Fleming AD, Evans G, Walton EA, Armstrong DT, 1985. Developmental capability of rat oocytes matured in vitro in defined medium. Gamete Res 12:255-63

Foote WD, Thibault C, 1969. Recherches experimentales sur la maturation in vivo des oocytes de truie et de veau. Ann Biol Anim Biochem Biophys 9:329-49

Franchi LL, Baker TG, 1973. Oogenesis and follicular growth. In: Hafez ESE and Evans TN (eds.), Human Reproduction. New York: Harper and Row, pp 53-83

Fukumoto M, Yajima Y, Okamura H, Midorikawa O, 1981. Collagenolytic enzyme activity in human ovary: an ovulatory enzyme system. Fertil Steril 36:746-50

Gilula NB, Epstein ML, Beers WH, 1978. Cell-to-cell communication and ovulation: a study of the cumulus-oocyte complex. J Cell Biol 78:58-75

Grinwich DL, Kennedy TG, Armstrong DT, 1972. Dissociation of ovulatory and steroidogenic actions of luteinizing hormone in rabbit with indomethacin, an inhibitor of prostaglandin biosynthesis. Prostaglandins 1:89-95

Hamada Y, Wright KH, Wallach EE, 1978. In vitro reversal of indomethacin-blocked ovulation by prostaglandin F_2 . Fertil Steril 30:702-06

Hamilton JLA, Bootes A, Phillips PE, Slywka J, 1981. Human synovial fibroblast plasminogen activator: modulation of enzyme activity by antiinflammatory steroids. Arthritis Rheum 24:1296-303

Hillensjo T, Kripner AS, Pomerantz SH, Channing CP, 1979. Action of porcine follicular fluid oocyte maturation inhibitor in vitro: Possible role of cumulus cells. Adv Exp Med Biol 112:283-91

Holmes PV, Janson PO, Sogn J, Kallfelt B, LeMaire WJ, Ahren KB, Cajander S, Bjersing L, 1983. Effects of $PGF_{2\alpha}$ and indomethacin on ovulation and steroid production in the isolated perfused rabbit ovary. Acta Endocrinol 104:233-39

Hosoi Y, Yoshimura Y, Santulli R, Wallach EE, 1985. Influence of hormones on rabbit oocyte maturation in vitro. 32nd Annual Meeting of the Society for Gynecologic Investigation, Phoenix, AZ, (Abstract No. 175P) p.101

Ichikawa S, Ohta M, Morioka H , Murao S, 1983. Blockage of ovulation in the explanted hamster ovary by a collagenase inhibitor. J Reprod Fert 68:17-19

Jacobowitz D, Wallach EE, 1967. Histochemical and chemical studies of the autonomic innervation of the ovary. Endocrinology 81:1132-39

Jansen PO, LeMaire WJ, Kallfelt B, Holmes PV, Cajander S, Bjersing L, Wiqvist N, Ahren K, 1982. The study of ovulation in the isolated perfused rabbit ovary. I. Methodology and pattern of steroidogenesis. Biol Reprod 26:456-65

Kawamura N, Himeno N, Okamura H, Mori T, Fukomato M, Midorikawa O, 1984. Effect of indomethacin on collagenolytic enzyme activities in rabbit ovary. Nippon Sanka Fujinka, Gakkai Zasshi 36:2099

Kennedy JF, Schreiber JR, Andreassen BK, 1975. Perfusion of human ovaries in vitro with artificial medium: metabolism and steroidogenesis. Am J Obstet Gynecol 122:863-71

Kitai H, Yoshimura Y, Wright KH, Santulli R, Wallach EE, 1985a. Microvasculature of preovulatory follicles: comparison of in situ and in vitro perfused rabbit ovaries following stimulation of ovulation. Am J Obstet Gynecol 152:889-95

Kitai H, Kobayashi Y, Santulli R, Wright KH, Wallach EE, 1985b. The relationship between prostaglandins and histamine in the ovulatory process as determined with the in vitro perfused rabbit ovary. Fertil Steril 43:646-51

Kobayashi Y, Santulli R, Wright KH, Wallach EE, 1981. The effect of prostaglandin synthesis inhibition by indomethacin on ovulation and ovum maturation in the in vitro perfused rabbit ovary. Am J Obstet Gynecol 141:53-57

Koninckx PR, Heyns WJ, Corvelyn PA, Brosens IA, 1978. Delayed onset of luteinization as a cause of infertility. Fertil Steril 29:266-69

Lambertsen CJ Jr, Greenbaum DF, Wright KH, Wallach EE, 1976. In vitro studies of ovulation in the perfused rabbit ovary. Fertil Steril 27:178-87

Larsen WJ, Wert SE, Brunner GD, 1986. A dramatic loss of cumulus cell gap junctions is correlated with germinal vesicle breakdown in rat oocytes. Dev Biol 113:517-21

Laufer N, Pratt BM, DeCherney AH, Naftolin F, Merino M, Markett CL, 1983. The in vivo and in vitro effects of clomiphene citrate on ovulation, fertilization, and development of cultured mouse oocytes. Am J Obstet Gynecol 147:633-39

Laufer N, DeCherney AH, Haseltine FP, Behrman HR, 1984. Steroid secretion by the human egg-corona-cumulus complex in culture. J Clin Endocrinol Metab 58:1153-57

Laurent P, Bienvenu J, 1982. Acute inflammatory process. In: Allen RC, Bienvenu J, Laurent P, Suskind RM (eds.), Marker Proteins in Inflammation. New York: Walter de Gruyter, pp. 33-43

LeMaire WJ, Yang NST, Behrman HR, Marsh JM, 1973. Preovulatory changes in the concentration of prostaglandins in rabbit Graafian follicles. Prostaglandins 3:367-76

Liebfried L, First NL, 1980. Effect of bovine and porcine follicular fluid and granulosa cells on maturation of oocytes *in vitro.* Biol Reprod 23:699-704

Magnusson C, Hillensjo T, 1977. Inhibition of maturation and metabolism in rat oocytes by cyclic AMP. J Exp Zool 201:139-47

Martin GG, Miller-Walker C, 1983. Visualization of the three-dimensional distribution of collagen fibrils over preovulatory follicles in the hamster. J Exp Zool 225:311-19

Martinat N and Combarnous Y, 1983. The release of plasminogen activator by rat granulosa cells is hightly specific for FSH activity. Endocrinology 113:433-35

Moor RM, Polge C, Willadsen SM, 1980. Effect of follicular steroids on the maturation and fertilization of mammalian oocytes. J Embryol Exp Morphol 56:319-35

Motlik J, Fulka J, Flechon J-E, 1986. Changes in intercellular coupling between pig oocytes and cumulus cells during maturation *in vivo* and *in vitro.* J Reprod Fertil 76:31-37

Okamura H, Virutamasen P, Wright KH, Wallach EE, 1972. Ovarian smooth muscle in the human being, rabbit, and cat: histochemical and electron microscopic study. Am J Obstet Gynecol 112:183-91

Okuda Y, Okamura H, Kanzaki H, Fujii S, Takenaka A, Wallach EE, 1983. An ultrastructural study of ovarian perifollicular capillaries in the indomethacin treated rabbit. Fertil Steril 39:85-92

Olsson J-H, Hillensjo T, Nilsson L, 1986. Inhibitory effects of danazol on steroidogenesis in cultured human granulosa cells. Fertil Steril 46:237-42

Orczyk GP, Behrman HR, 1972. Ovulation blockade by aspirin or indomethacin - *in vivo* evidence for a role of prostaglandins in gonadotropin secretion. Prostaglandins 1:3-20

Osborn JC, Moor RM, 1983. The role of steroid signals in the maturation of mammalian oocytes. J Steroid Biochem 19: 133-37

Osman P, Dullaart J, 1976. Intraovarian release of eggs in the rat after indomethacin treatment at pro-oestrus. J Reprod Fertil 47:101-3

Owman Ch, Rosengren E, Sjoberg, N-O, 1967. Adrenergic innervation of the human female reproductive organs: a histochemical and chemical investigation. Obstet Gynecol 30:763-73

Phi LT, Moon YS, Armstrong DT, 1977. Effects of systemic and intrafollicular injections of LH, prostaglandins, and indomethacin on the luteinization of rabbit Graafian follicles. Prostaglandins 13:543-52

Pincus G, Enzmann EV, 1935. The comparitive behavior of mammalian eggs *in vivo* and *in vitro.* I. The activation of ovarian eggs. J Exp Med 62:665-75

Plas-Roser S, Kauffmann MT, Aron C, 1984. Progesterone secretion by luteinized unfuptured follicles in mature female rats. J Steroid Biochem 20:441-44

Racowsky C, McGaughey RW, 1982. Further studies of the effects of follicular fluid and membrane granulosa cells on the spontaneous maturation of pig oocytes. J Reprod Fertil 66:505-12

Reich R, Miskin R, Tsafriri A, 1985a. Follicular plasminogen activator: involvement in ovulation. Endocrinology 116:516-21

Reich R, Tsafriri A, Mechanic GL, 1985b. The involvement of collagenolysis in ovulation in the rat. Endocrinology 116:52227

Rocereto T, Jacobowitz D, Wallach EE, 1969. Observations of spontaneous contractions of the cat ovary *in vitro.* Endocrinology 84:1336-41

Romanoff EB, Pincus G, 1962. Studies of the isolated perfused ovary: Methods and examples of application. Endocrinology 71:752-55

Rondell P, 1970. Biophysical aspects of ovulation. Biol Reprod 2: (suppl 2) 64-89

Schenken RS, Werlin LB, Williams RF, Prihoda TJ, Hodgen GB, 1986. Histologic and hormonal documentation of the luteinized unruptured follicle syndrome. Am J Obstet Gynecol 154:839-47

Schlaff S, Kobayashi Y, Wright KH, Santulli R, Wallach EE, 1983. Prostaglandin F$_{2\alpha}$, an ovulatory intermediate in the in vitro perfused rabbit ovary model. Prostaglandins 26:111-21

Schmidt GE, Kim MH, Mansour R, Torello L, Friedman CI, 1986a. The effects of enclomiphene and zuclomiphene citrates on mouse embryos fertilized in vitro and in vivo. Am J Obstet Gynecol 154:727-36

Schmidt G, Holmes PV, Owman Ch, Sjoberg NO, Walles B, 1986b. The influence of indomethacin, prostaglandin E_2 and progesterone production and ovulation in the rabbit ovary perfused in vitro. Biol Reprod 35:815-21

Shalgi R, Dekel N, Kraicer PF, 1979. The effect of LH on the fertilizability and developmental capacity of rat oocytes matured in vitro. J Reprod Fertil 55:429-35

Smith C, Perks AM, 1979. Plasma bradykinogen levels before and after ovulation: studies in women and guinea pigs, with observation on oral contraceptives and menopause. Am J Obstet Gynecol 133:868-76

Smith C, Perks AM, 1983. The kinin system and ovulation: changes in plasma kininogens and in kinin-forming enzymes in the ovaries and blood of rats with 4-day estrous cycles. Can J Physiol Pharmacol 61:736-42

Sorensen RA, Wassarman PM, 1976. Relationship between growth and meiotic maturation of the mouse oocyte. Devel Biol 50:531-36

Stahler E, Spatling L, Bethge HD, Daume E, Buchholz R, 1974. Induction of ovulation in human ovaries perfused in vitro. Arch Gynak 217:1-15

Staigmiller RB, Moor RM, 1984. Effect of follicle cells on the maturation and developmental competence of ovine oocytes matured outside the follicle. Gamete Res 9:221-29

Stone SL, Pomerantz SH, Schwartz-Kripner A, Channing CP, 1978. Inhibitor of oocyte maturation from porcine follicular fluid: further purification and evidence for reversible action. Biol Reprod 19:585-92

Strickland S, Beers WH, 1976. Studies on the role of plasminogen activator in ovulation. J Biol Chem 251:5694-702

Szybek K, 1972. In vitro maturation of oocytes from sexually immature mice. J Endocrinol 54:527-28

Thibault C, 1977. Are follicular maturation and oocyte maturation independent processes? J Reprod Fertil 51:1-15

Thibault C, Gerard M, Menezo Y, 1975a. Preovulatory and ovulatory mechanisms in oocyte maturation. J Reprod Fertil 45:606-10

Thibault C, Gerard M, Menezo Y, 1975b. Acquisition par l'ovocyte de lapine et de veau du facteur de decondensation du noyau du spermatozoide fecondant (MPGF). Ann Biol Anim Biochim Biophys 15:705-14

Townsend SL, Brown JW, Adey FD, Evans JH, Taft HP, 1966. Induction of ovulation. J Obstet Gynecol Brit Comm 73:529-43

Tsafriri A, Channing CP, 1975. An inhibitory influence of granulosa cells and follicular fluid upon porcine oocyte meiosis in vitro. Endocrinology 96:922-27

Tsafriri A, Pomerantz SH, Channing CP, 1976. Inhibition of oocyte maturation by porcine follicular fluid: partial characterization of the inhibitor. Biol Reprod 14:511-16

Tsafriri A, Lindner HR, Zor U, Lamprecht SA, 1972. Physiological role of prostaglandins in the induction of ovulation.

Prostaglandins 2:1-10

Vane JLR, 1976. Prostaglandins as mediators of inflammation. Adv Prostaglandin Thromboxane Res 2:791-99

Virutamasen P, Wright KH, Wallach EE, 1972. Effects of catecholamines on ovarian contractility in the rabbit. Obstet Gynecol 39:225-36

Wallach EE, de la Cruz A, Hunt J, Wright KH, Stevens VC, 1975. The effect of indomethacin on hMG-hCG induced ovulation in the rhesus monkey. Prostaglandins 9:645-58

Wallach EE, Wright KH, Hamada Y, 1978. Investigation of mammalian ovulation with an in vitro perfused rabbit ovary preparation. Am J Obstet Gynecol 132:728-38

Wallach EE, Okuda Y, Kanzaki H, Kobayashi Y, Okamura H, Santulli R, 1984. Ultrastructure of ovarian follicles in in vitro perfused rabbit ovaries: response to human chorionic gonadotropin and comparison with in vivo observations. Fertil Steril 42:127-33

Wassarman PM, Josefowicz WJ, Letourneau GE, 1976. Meiotic maturation of mouse oocytes in vitro: inhibition of maturation at specific stages of nuclear progression. J Cell Sci 22:531-45

Weiner S, Wright KH, Wallach EE, 1975a. Selective ovarian sympathectomy in the rabbit. Fertil Steril 26:253-62

Weiner S, Wright KH, Wallach EE, 1975b. Lack of effect of ovarian denervation on ovulation and pregnancy in the rabbit. Fertil Steril 26:1083-87

Willoughby DA, Sedgewick A, Edwards J, 1982. The inflammatory process. In: Allen RC, Bienvenu J, Laurent P, Suskind RM (eds.), Marker Proteins in Inflammation, New York: Walter de Gruyter, pp. 45-48

Wramsby H, Kullander S, Liedholm P, Rannevik G, Sundstrom P, Thorell J, 1981. The success rate of in vitro fertilization of human oocytes in relation to the concentrations of different hormones in follicular fluid and peripheral plasma. Fertil Steril 36:448-54

Yang NST, Marsh JM, LeMaire WJ, 1973. Prostaglandin changes induced by ovulatory stimuli in rabbit graafian follicles. The effect of indomethacin. Prostaglandins 4:395-404

Yoshimura Y, Hosoi Y, Atlas SJ, Wallach EE, 1985a. Effects of cyclic AMP on rabbit oocyte maturation. Abstracts of the 41st Annual Meeting of the American Fertility Society, Chicago IL, p 78

Yoshimura Y, Kitai H, Santulli R, Wright K, Wallach EE, 1985b. Direct ovarian effect of clomiphene citrate in the rabbit. Fertil Steril 43:471-76

Yoshimura Y, Hosoi Y, Atlas SJ, Bongiovanni AM, Wallach EE, 1986a. The effects of ovarian steroidogenesis on ovulation and fertilizability in the in vitro perfused rabbit ovary. Biol Reprod 35:943-48

Yoshimura Y, Kitai H, Wright KW, Atlas SJ and Wallach EE, 1986b. Surface morphology of ovarian follicles in the in vitro perfused rabbit ovary. Sixth Biennial Ovarian Workshop, Ithaca, NY, July 12-14 (Abstract # 1) p 45

Yoshimura Y, Hosoi Y, Adachi T, Atlas SJ, Ghodgaonkar RB, Dubin NH, Wallach EE, 1986c. The effect of bradykinin on ovulation and prostaglandin production by the perfused rabbit ovary. 68th Annual Meeting of the Endocrine Society, Anaheim, CA, (Abstract No.47) p 77

Yoshimura Y, Hosoi Y, Atlas SJ, Adachi T and Wallach EE, 1986d. Estradiol (E_2) reverses the limiting effects of clomiphene citrate (CC) on early embryonic development in the in vitro perfused rabbit ovary. 42nd Annual Meeting of the American Fertility Society, Toronto, Canada, Sept 27 - Oct 2 (Abstract #

212) p 74

Yoshimura Y, Wallach EE, 1987a. Studies of the mechanism(s) of mammalian ovulation. Fertil Steril 47:22-34

Yoshimura Y, Santulli R, Atlas SJ, Fujii S, Wallach EE, 1987b. The effects of proteolytic enzymes on in vitro ovulation in the rabbit. Am J Obstet Gynecol (in press)

Yoshimura Y, Hosoi Y, Bongiovanni AM, Santulli R, Atlas SJ, Wallach EE, 1987c. Are ovarian steroids required for ovum maturation and fertilization? Endocrinology (in press)

Yoshinaga K, Rice C, Krenn J, Pilot RL, 1979. Effects of nicotine on early pregnancy in the rat. Biol Reprod 20:294-303

Zamboni L, 1974. Fine morphology of the follicle wall and follicle cell-oocyte association. Biol Reprod 10:125-49

SIGNIFICANCE OF ANGIOGENIC AND GROWTH FACTORS

IN OVARIAN FOLLICULAR DEVELOPMENT

Kenneth J. Ryan and Anastasia Makris

Laboratory for Human Reproduction
and Reproductive Biology
Harvard Medical School
Boston, MA 02115

INTRODUCTION

There are many events taking place during follicular development which cannot be explained by the presently known hormones and their inter-relationships as now understood (di Zerga and Hodgen, 1981). With the two-cell model of ovarian steroid secretion, luteinizing hormone (LH) stimulates androgen production by the theca and follicle stimulating hormone (FSH) induces increased aromatase conversion of that androgen to estrogen by the granulosa cells. The FSH and estrogen locally induce LH receptors in the granulosa cell in anticipation of ovulation and formation of the corpus luteum (Hsueh et al., 1984). After much searching for the elusive follicular inhibin it has now been shown that FSH also stimulates granulosa production of inhibin (McLachlan et al., 1986) which along with estrogen provide the feedback from the follicle to the hypothalamic- pituitary unit (Tsonis and Sharpe, 1986). In spontaneous ovulators, estrogen feedback can trigger the LH surge which ultimately results in ovulation. The follicle destined to ovulate can be characterized by a microenvironent of characteristic follicular fluid steroid and gonadotropin concentrations and by an optimal granulosa cell number which distinguish the ripe follicles and their oocytes from those destined for atresia (McNatty et al., 1979). This model of gonadotropin-steroid interaction cannot explain a good deal that is going on during folliculogenesis (Tsafiri et al., 1986). It is likely that other intermediary agents will have to be sought to explain the following:

1. oogenesis and resumption of meiosis after a long period of meiotic arrest

2. granulosa cell proliferation during folliculogenesis

3. the development of the vascular wreath surrounding the follicle and the vascular invasion of the developing corpus luteum soon after ovulation

4. the selection of the cohort of follicles that progress in each cycle to either ovulation or atresia

5. selection of the dominant follicle and the etiology of the passive or active dominance exerted by that dominant follicle over other follicles that could potentially develop in both ovaries

6. the biochemical chain of events leading up to follicle rupture

7. the control of the dynamics and composition of follicular fluid especially with respect to glycosaminoglycans and other nonsteroidal factors.

The search for other factors started with observations that follicular fluid cleared of steroid hormones and gonadotropins had biological properties that might help explain some of the foregoing phenomena (di Zerga et al., 1983, Frederick et al., 1985, Koos, 1986, Tsafiri et al., 1986). The factors which were sought included:

1. oocyte maturation inhibitor

2. inhibitor of FSH stimulated aromatase and LH receptor induction

3. angiogenic factor

4. granulosa (ovarian) growth factor

5. luteinization inhibitor and stimulator

6. LH and FSH binding inhibitors

7. folliculostatin or ovarian inhibin

These putative factors can now be matched against the properties of several peptides newly identified in the ovary or elsewhere. For example inhibin has now been identified, isolated from follicular fluid and the gene cloned (Tsonis and Sharpe, 1986). It has been shown that inhibin is produced by the granulosa cell, and its secretion stimulated by FSH (McLachlan et al., 1986). Follicular inhibin appears to have dynamic changes and properties which fulfill the prophecy of its role in feedback effects that could not be explained by steroids alone. Mullerian inhibiting substance (MIS) has also been identified in ovarian granulosa cells (Takahashi et al., 1986a) and has been proposed as an oocyte maturation inhibitor (Takahashi et al., 1986b). MIS was originally isolated from newborn bovine testes and the genome for both the human and bovine species described (Cate et al., 1986). The amino acid sequence of MIS has interesting homologies to the beta chain of inhibin and also to beta-transforming growth factor (TGFβ) and shares with the latter factor, growth inhibiting properties (Cate et al., 1986). TGFβ also has effects on granulosa cells when tested in vitro, but there is no evidence that it is also made in the ovary under normal circumstances (Adashi and Resnick, 1986, Ying et al., 1986). Fibroblast growth factors (FGFs) have been isolated from the ovary and could be responsible for angiogenesis and granulosa proliferation in the follicle (Gospodarowicz et al., 1985). Insulin like growth factors and insulin proper can enhance both basal and gonadotropin stimulated steroid secretion by granulosa and thecal cells (Barbieri et al., 1986, Hsueh et al., 1984). Other growth factors that have been shown to have biological effects on ovarian cells include epidermal growth factor (EGF), platelet derived growth factor (PDGF) (Hsueh et al., 1984) and alpha-transforming growth factor (TGFα) (Adashi and Resnick, 1986). There are receptors to EGF on granulosa cells that are modulated by gonadotropins and EGF has biological effects on granulosa cells which are similar to those of FGF but in contrast to the latter, there is no

Table 1. Growth Factors Active in the Ovary In Vitro and Their Homologies

Factor	Homologies
Epidermal Growth Factor (EGF)	TGFα
Fibroblast Growth Factors (FGF)	Interleukin 1
αTransforming Growth Factor (TGFα)	EGF
βTransforming Growth Factor (TGFβ)	Inhibin, MIS
Inhibins	MIS, TGFβ
Mullerian Inhibiting Substance (MIS)	Inhibin, TGFβ
Insulin-like Growth Factor 1 (IGF-1)	Insulin, Relaxin
Gonadotropin Releasing Hormone (GnRH)	
Platelet Derived Growth Factor (PDGF)	

evidence that EGF is made in the ovary (Hsueh et al., 1984, Gospodarowicz et al., 1977a,b). Gonadotropin releasing hormones (GnRH) are known to bind to cells in the ovary and produce both inhibitory and stimulating effects. There are, however, species differences in the response observed, and while it is unlikely that hypothalamic GnRH is actively reaching the ovary in vivo, the true ligand for the ovarian receptors to GnRH is unknown (Hsueh et al., 1984). The homologies among many of these growth factors are of interest and make the determination of the physiologically important factors difficult (Table 1.).

We are close to defining a new more complex microenvironment for the follicle which includes peptide growth factors in addition to steroids and gonadotropins. I will restrict the remaining discussion to the phenomena of granulosa cell growth, angiogenesis and the role of fibroblast growth factors as the possible mediators of these events.

GRANULOSA CELL PROLIFERATION

The most dramatic example of granulosa cell proliferation is the hypophysectomized estrogen treated immature rat model which is the source of cells for most in vitro studies. Although in this model, estrogen appears to stimulate granulosa cell division in vivo, there is no evidence that the steroid has this effect in vitro. In the hypophysectomized animal, a pituitary intermediary cannot be invoked and gonadotropins cannot induce granulosa cell proliferation in vitro (Hsueh et al., 1984).

During the follicular growth of a preovulatory human follicle, granulosa cells increase from about one million to over 50 million in number (McNatty et al., 1979). Gospodarowicz demonstrated that both FGF derived from brain and pituitary and EGF derived from the submaxillary gland could induce granulosa cell proliferation in vitro. In these studies the mitogenic effect of both growth factors was additive to that of sera even with a serum concentration of 10%. In addition both factors acted at concentrations 300 to 3000 fold lower than that required for other cell types such as fibroblasts or lens epithelial cells (Gospodarowicz et al., 1977a,b). Makris et al. demonstrated that extracts of whole porcine ovary, thecal cytosol and conditioned media from theca but not granulosa cell cultures, stimulated granulosa cell growth (Makris et al., 1983). FGF has since been isolated from the ovary and it is likely the growth factor is closely related to some form of FGF. Partial purification of the growth factors from the ovary reveals a family of heparin binding mitogens which have microhetero- geneity in amino acid composition (Makris, unpublished). There are obviously a whole array of potential mitogens and growth factors such as EGF, PDGF, TGFs that could also be involved in granulosa cell proliferation but for the moment, FGF is clearly a likely prospect for a physiological role.

ANGIOGENESIS

The cause of the intense proliferation of capillary blood vessels around the developing follicles and the growth of vessels into the newly formed corpus luteum has intrigued biologists for a long time. The concept of an angiogenic factor especially from tumors and rapidly growing tissues was advanced by Folkman (Folkman, 1982). Gospodarowicz isolated an angiogenic factor from the corpus lutuem (Gospodarowicz et al., 1985, 1986a,b) and more recently at least three investigators described angiogenic factors in the follicle. Koos described angiogenic activity derived from granulosa cells (Koos, 1986), di Zerga reported on the presence of angiogenic activity in human and porcine follicular fluid (Frederick et al., 1985) while Makris found such activity in extracts from porcine ovarian stroma and theca (Makris et al., 1984). It is by no means certain that the fractions being described all contain the same active substances. FGF has been shown to be angiogenic and has been isolated from corpora lutea (Gospodarowicz et al., 1985, Montesano et al., 1986). Peptides similar to FGF have been partially purified from nonluteal ovarian tissue including extracts of theca and follicles. The assays used to indicate angiogenesis included _in vivo_ studies with the chick chorioallantoic membrane and the rabbit cornea, and _in vitro_ studies involving endothelial cell migration and proliferation. Although there are several substances known to induce angiogenesis (Schreiber et al., 1986), FGF appears to be the most likely prospect for a physiological role in the ovary (Table 2).

Table 2. Properties of Fibroblast Growth Factor in Ovarian Systems

1. Mitogenic for granulosa cells _in vitro_

2. Mitogenic for capillary endothelial cells _in vitro_

3. Angiogenic in chorioallantoic membrane assay

4. Inhibits FSH induction of LH receptors in granulosa cells

5. Inhibits FSH augmentation of aromatase in granulosa cells

6. Augments progesterone synthesis in granulosa cells

FIBROBLAST GROWTH FACTOR(S)

Although originally derived from brain and pituitary sources, FGFs have now been isolated from many different tissues. It is likely that the twenty or so growth factors and angiogenic factors described for the eye, kidney, tumors, testis, placenta, ovary and prostate will ultimately turn out to be one of the forms of basic or acidic FGF (Gosprodowicz et al., 1986a,b). The two forms of FGF differ with respect to isoelectric point but have a 50-55% homology in their amino acid composition. Acidic FGF has been assigned to the long arm of human chromosome 5 and basic FGF to human chromosome 4. Aside from the presumption of a common evolutionary ancestor, the two FGFs are derived from separate genes (Abraham et al., 1986, Burgess et al., 1986, Gimenez-Gallego et al., 1986, Jaye et al., 1986, Thomas et al., 1985). Basic FGF is supposedly some 30 to 100 times more active than acidic FGF but binding to heparin increases the activity of acidic FGF by that amount without affecting the activity of basic FGF. FGFs bind to cell surface receptors of size 125 and 145 KD and although they compete with one another for receptor sites there is a preference by each for one of the two receptors. FGFs are not ordinarily degraded by target cells and there is no tyrosine phosphorylation of the FGF receptor. FGF acts like EGF and PDGF in inducing diacylgycerol formation, protein kinase C activation and calcium mobilization (Gospodarowicz et al., 1986a,b). FGF can induce transformation in cells which is reversible upon its removal (Huang et al., 1986). Effects on cell proliferation by FGF occur at concentrations 20 to 60 times lower than required for TGF, EGF or PDGF. "FGF-like" proteins have been isolated from the follicle, corpora lutea and stroma of the ovary. This localization with its angiogenic and mitogenic properties, make FGF a contender as an important controlling factor in ovarian function. What remains to be done is to establish the pattern of FGF production, distribution, target cell receptors and action in the ovary in relationship to the cycle and various reproductive states.

REFERENCES

Abraham JA, Mergia A, Whang JL, Tumolo A, Friedman J, Hierrild KA, Gospodarowicz D, Fiddes JC, 1986. Nucleotide sequence of a bovine clone encoding the angiogenic protein, basic fibroblast growth factor. Science 233:545-548

Adashi EY, Resnick CE, 1986. Antagonistic interactions of transforming growth factors in the regulation of granulosa cell differentiation. Endocrinology 119:1879-1881

Barbieri RL, Makris A, Randall RW, Daniels G, Kistner RW, Ryan KJ, 1986. Insulin stimulates androgen accumulation in incubations of ovarian stroma obtained from women with hyperandro- genism. J Clin Endocrinol Metab, 62:904-910

Burgess WH, Mehlman T, Marshak DR, Fraser BA, Maciag T, 1986. Structural evidence that endothelial cell growth factor a is the precursor of both endothelial cell growth factor-a and acid fibroblast growth factor. Proc Natl Acad Sci 83:7216-7220

Cate RL, Mattaliano RJ, Hession C, Tizard R, Farber NM, Cheung A, Ninfa EG, Frey AZ, Gash DJ, Chow EP, Fisher RA, Bertonis JM, Torres G, Wallner BP, Ramachandran KL, Ragin RC, Manganaro TF, MacLaughlin DT, Donahoe PK, Isolation of the bovine and human genes for mullerian inhibiting substance and expression of the human gene in animal cells. Cell 45:685-698

di Zerga GS, Hodgen GD, 1981. Folliculogenesis in the primate ovarian cycle. Endocr Rev 2:27-49 di Zerga GS, Campeau JD, Nakamura RN, Ujita EL, Lobo R, Marrs RP, 1983. Activity of a human follicular fluid protein(s) in spontaneous and induced ovarian cycles. J Clin Endocrinol Metab 57:838-846

Folkman J, 1982. Angiogenesis: Initiation and Control. Ann NY Acad Sci 401:212-227

Frederick JL, Hoa N, Preston DS, Frederick JJ, Campeau JD, Ono T, diZerega GS, 1985. Initiation of angiogenesis by porcine follicular fluid. Am J Obst Gynecol 152:1073-1078

Gimenez-Gallego G, Conn G, Hatcher VB, Thomas KA, 1986. The complete amnio acid sequence of human brain-derived acidic fibroblast growth factor. Biochem Biophys Res Commun 138:611-617

Gospodarowicz D, Ill CR, Birdwell CR, 1977a. Effects of fibroblast and epidermal growth factors in ovarian cell response in vitro. I. Characterization of the response of granulosa cells to FGF and EGF. Endocrinology 100:1108-1120

Gospodarowicz D, Ill CR, Birdwell CR, 1977b. Effects of fibroblast and epidermal growth factors on ovarian cell proliferation in vitro. II. Proliferative response of luteal cells to FGF but not EGF. Endocrinology 100:1121-1128

Gospodarowicz D, Cheng J, Lui GM, Baird A, Esch F, Bohlen P, 1985. Corpus luteum angiogenic factor is related to fibroblast growth factor. Endocrinology. 117:2283-2391

Gospodarowicz D, Neufeld G, Schweigerer L, 1986a. Fibroblast growth factor. Mol Cell Endocr 46:187-204

Gospodarowicz D, Neufeld G, Schweigerer L, 1986b. Molecular and biological characterization of fibroblast growth factor, an angiogenic factor which also controls the proliferation and differentiation of mesoderm and neuroectoderm derived cells. Cell Diff 19:1-17

Hsueh AJW, Adashi EY, Jones PBC, Welsh TH Jr. Hormonal regulation of the differentiation of cultured ovarian granulosa cells. Endocr Rev 5:76-127

Huang SS, Kuo MD, Huang JS, 1986. Transforming growth factor activity of bovine brain-derived growth factor. Biochem Biophys Res Commun 139:619-625

Jaye M, Howk R, Burgess W, Ricca GA, Ing-Ming C, Ravera MW, O'Brien SJ, Modi WS, Maciag T, Drohan WN, 1986. Human endothelial cell growth factor: cloning, nucleotide sequence and chromosome localization. Science 233:541-545

Koos RD, 1986. Stimulation of endothelial cell proliferation by rat granulosa cell-conditioned medium. Endocrinology 119:481-489

Makris A, Klagsbrun MA, Yasumizu T, Ryan KJ, 1983. An endogenous ovarian growth factor which stimulates BALB/3T3 and granulosa cell proliferation. Biol Reprod 19:1135-1141

Makris A, Ryan KJ, Yasumizu T, Hill CL, Zetter BR, 1984. The nonluteal porcine ovary as a source of angiogenic activity. Endocrinology 15:1672-1677

McLachlan RI, Robertson DM, Healy DL, deKretser DM, Burger HG, 1986. Plasma inhibin levels during gonadotropin induced ovarian hyperstimulation for IVF. Lancet 1:1233-1234

McNatty KP, Smith DM, Makris A, Osathanondh R, Ryan KJ, 1979. The microenvironment of the human antral follicle. J Clin Endocrinol Metab 49:851-860

Montesano R, Vassalli J-D, Baird A, Guillemin R, Orci L, 1986. Basic fibroblast growth factor induces angiogenesis in vitro. Proc Natl Acad Sci 83:7297-7301

Schreiber AB, Winkler ME, Derynk R, 1986. Transforming growth factor-a: A more potent angiogenic mediator than epidermal growth factor. Science 232:1250-1253

Takahashi M, Hayashi M, Manganaro TF, Donahoe PK, 1986a. The ontogeny of mullerian inhibiting substance in granulosa cells of the bovine ovarian follicle. Biol Reprod 35:447-454

Takahashi M, Koide SSm Donanoe PK, 1986b.   Mullerian inhibiting substance as ooyte meiosis inhibitor.   Mol Cell Endocrinol 47:225-234

Thomas KA, Rios-Candelore M, Gimenez-Gallego G, DiSalvo J, Bennett C, Rodkey J, Fitzpatrick S, 1985. Pure brain derived acidic fibroblast growth factor is a potent angiogenic vascular endothelial cell mitogen with sequence homology to interleukin. Proc Natl Acad Sci 82:6409-6413

Tsafiri A, Braw RH, Reich R, 1986. Follicular development and the mechanism of ovulation. Infertility: Male and Female. Ed. V. Insler & B. Lunenfeld, Churchill Livingstone, NY, Chap. 4 pp. 73-100

Tsonis CG, Sharpe RM, 1986. Dual gonadol control of follicle stimulating hormone. Nature 321:724-725

Ying SY, Becker A, Ling N, Ueno N, Guillemin R, 1986. Inhibin and beta type transforming factor (TGFB) have opposite modulating effects on follicle stimulating hormone (FSH) - induced aromatase activity of cultured rat granulosa cells.   Biochem Biophys Res Commun 136:969-975

KINETIC ASPECTS OF FOLLICULAR DEVELOPMENT IN THE RAT

Anne N. Hirshfield[1], and Waldemar A. Schmidt[2]

[1]Department of Anatomy
University of Maryland
School of Medicine
Baltimore, MD 21201

[2]Department of Pathology and Laboratory
Medicine
University of Texas Medical School
Houston, TX 77225

INTRODUCTION

The ovary has many similarities to other "renewal tissues" in adults. Renewal tissues are those in which differentiated, functional cells are continuously being replaced by proliferation of more primitive cells. These tissues are composed of a hierarchy of cells: at one end of the hierarchy are stem cells which are less differentiated and can divide without limit; at the other end are mature cells which are highly differentiated and have no capacity for proliferation (Mackillop et al., 1983). When a stem cell divides, each daughter cell has a choice: it can either remain a stem cell, or it can embark on a course of "clonal expansion" leading irreversibly to terminal differentiation (Fig 1). Daughter cells which embark on the second course are known as "transitional cells (Selby et al., 1983) or "committed progenitor cells" (Fitchen et al., 1981). Transitional cells have a limited capacity for cell division. They exhibit a continuous gradient of properties along a unidirectional vector; as cells move down the hierarchy, they acquire the differentiated features associated with specific tissue function, and they progressively lose the potential to divide (Mackillop et al., 1983). The more highly differentiated progeny greatly outnumber the less differentiated progenitor cells within the tissue.

In normal renewal tissues, cells are therefore arrayed along a maturation gradient. As cells progress down this gradient, they become more and more restricted in their developmental program. Although differentiated characteristics may not yet be expressed, transitional cells that are far removed from the parent stem cell are already committed to a particular differentiated fate. As cells progress down this gradient, their daughters multiply more and more rapidly (shorter generation time) but give rise to fewer and fewer subsequent generations (Lajtha, 1983).

The processes of growth and differentiation in transitional cells are regulated by a plethora of molecules. These regulatory molecules are stage-specific: each regulator acts on cells at a different stage along the maturation gradient (Sachs, 1986). It is likely that some of these regulatory molecules are directive whereas others are permissive. Directive inductors convey information to a target cell, eliciting a specific developmental program from a choice of possibilities. Permissive inductors simply permit a particular tissue to continue its normal course of development (Cunha et al., 1983).

Key questions in understanding the processes of growth and differentiation in renewal tissues are:
1) What are the characteristics of transitional cells at various stages in the maturation gradient?
2) At what stages along this maturation gradient do transitional cells become progressively restricted in their developmental possibilities? When do they become irreversibly committed to a particular path of differentiation to the exclusion of all other possible outcomes?
3) What factors control the outcome at each of these branching points in the maturation pathway?
4) What induces stem cell progeny to begin a course of clonal expansion leading irreversibly towards terminal differentiation?
5) What regulates stem cell proliferation?

As in other renewal tissues, follicular development begins with a few, slowly dividing, less differentiated cells. As these cells increase in number, they begin to divide more rapidly; ultimately, cell division ceases and functional signs of differentiation appear. Normal follicles give rise to two possible end stage structures: secondary interstitium and corpora lutea. Some follicles become atretic, their granulosa cells undergoing dissolution and their theca cells differentiating into secondary interstitial gland (Erickson et al., 1985; Guraya and Greenwald, 1964, 1968). The interstitial gland persists for some time, but ultimately involutes. Other follicles ovulate, their granulosa and theca cells both differentiating into luteal cells. The corpus luteum functions for variable lengths of time and then undergoes regression. Thus, in either case, the follicle matures to form an end stage structure which performs its destined function and ultimately disintegrates to be replaced by proliferation of less differentiated cells.

The object of this paper is to characterize the maturation gradient in follicular development, to identify the branching points in the pathway, and to review what is known about the mechanisms which determine the outcome at these branching points. We conclude that most of our knowledge of follicular development is restricted to the penultimate and ultimate stages of this lengthy and complex process. The granulosa cells of large preovulatory follicles are about 10 generations removed from the pregranulosa cells of the primordial follicles from which they are descended. Nearly all studies of follicular development, in vivo and in vitro, focus on events that occur during the 8th, 9th and 10th generations. In contrast, the first 7 generations of follicular development remain largely unexplored.

For the most part, we will restrict our remarks to studies of the rat. The complex regulatory process which selects the appropriate number of follicles for ovulation is most fully expressed in the reproductively competent, adult, cycling rat. Many studies have been performed in prepubertal, infant, pregnant, or lactating rats. Recruitment and selection of ovulatory follicles does not take place in these animals, suggesting that certain regulatory mechanisms may not be operative. Therefore, while studies of other animal models have been extremely enlightening, caution must be used in extending conclusions to the cycling rat.

CHARACTERISTICS OF FOLLICLES AT DIFFERENT STAGES ALONG THE MATURATION
GRADIENT

Nomenclature for Describing Stages of Follicular Growth

In renewal tissues, committed progenitor cells are distributed along
a gradient with an origin at the stem cell and a terminus at the fully
differentiated, end stage cell (Zajicek et al., 1979). The degree of
maturation and the remaining proliferative potential of any given cell are
a function of the number of generations that separate it from its stem
cell parent. In some tissues, such as intestinal epithelium, this maturation gradient is mirrored in the physical organization of the tissue: the
farther a cell is from the stem cell parents (which lie in the intestinal
crypts), the older it is (Zajicek et al., 1979).

The maturational position of granulosa cells in a follicle is mirrored in the size of the follicle; the larger the follicle, the more
generations its granulosa cells are removed from their stem cell parents.

A typical primordial follicle contains approximately 4 granulosa
cells in the largest cross section (LCS). When all four original granulosa cells have undergone division, the follicle would contain 8 granulosa
cells in the LCS. When these daughter cells have all undergone division
(doubled) the follicle would have 16 granulosa cells. If the original
follicular cells are considered to be the first generation, then the 16
cells in this growing follicle would be of the 3rd generation.

A full grown preovulatory follicle contains 2000-2500 granulosa cells
in the LCS (Hirshfield, 1984). Therefore the granulosa cells which make
up this follicle may be considered 10th generation descendants of the
original 4 granulosa cells that comprised the primordial follicle from
which it arose. In other words, the population of granulosa cells must
have doubled 10 times to produce a follicle with 2000 cells. One can thus
characterize follicular maturation on the basis of the number of granulosa
cell generations that must have passed given a starting point of 4 granulosa cells in the original primordial follicle. For example, a follicle
with 32 granulosa cells would have undergone three doublings and would be
in the fourth generation. A follicle with 256 granulosa cells would be in
the 7th generation (Table 1).

Table 1: Classification of follicles by generations

Generation	# of granulosa cells	Diameter	P & P
1	4	--[*]	3a
2	8	--	3a
3	16	39	3b
4	32	·57	4
5	64	82	5a
6	128	119	5b
7	256	172	6
8	512	250	7
9	1024	362	8
10	2048	525	8

[*] Variability in size of the growing oocyte makes diameter
an unreliable measurement of maturity in very small
follicles

213

The concept of follicular generations is meant to be taken as an abstraction rather than a concrete reality. Granulosa cells are probably heterogenous with respect to their cell cycle times. Fast-cycling granulosa cells may ultimately give rise to more progeny than slow cycling cells. Some of the granulosa cells in a preovulatory follicle may be 15 generations removed from the parent stem cell while others may be only 5 generations removed. Despite this caveat, we feel that the generational construct most accurately represents the exponential nature of follicular growth. By describing follicular development in terms of generations of granulosa cells, as much emphasis is given to the progression from 8 cells to 16 cells as to the progression from 1000 cells to 2000 cells.

Application of the Nomenclature to Published Literature on Follicular Development

Few papers have been published which include both biochemical and detailed morphological data. Rather, some papers describe exclusively histological studies (usually on physiologically normal animals), whereas others are confined to biochemical studies (usually on isolated cells derived from hormonally manipulated animal models).

Histological papers are the primary source of data concerning cell proliferation and cell death (atresia). These data are very easy to correlate with cell generations because follicular size is described. Follicular size is usually expressed in terms of follicular diameter, or by means of a classification of follicles originated by Pedersen and Peters (1968) which is based upon the number of granulosa cells in the largest cross section. The conversion chart that we have used to translate these measures of follicular size into cell generations is given in Table 1.

Biochemical papers are the primary source of data concerning differentiation of follicle cells. These data are often difficult to correlate with cell generations because descriptions of follicular size are usually omitted from these papers, or are expressed in vague terms ("small antral follicle"). It is often necessary, therefore, to speculate about what size follicles the cells used in biochemical studies have been obtained from.

Most of the studies of follicular differentiation in the rat have relied upon a few well-established animal models for _in vivo_ studies, and as a source of cells for studies _in vitro_. Granulosa cells are usually obtained for these studies by applying pressure on the entire ovary, which ruptures the follicles and extrudes their cells into collecting medium. An alternative approach is to manually dissect out the largest follicles from the ovary, and then pool the granulosa cells (or thecal shells). When granulosa cells are harvested from ovaries with either method, the majority of cells in the resulting pool are probably derived from the largest follicles because they, though few in number, contain the greatest number of cells. For this reason, it must be assumed that biochemical data, derived from _in vitro_ experiments, applies primarily to the largest follicles in the ovaries of the animal model used.

Figure 1 shows the median number of granulosa cells in the 5 largest follicles of a single ovary from several commonly used animal models. Data were obtained by morphometric analysis of ovaries from at least 5 animals for each treatment; the values indicated are the means of the 5 medians. Although the number of these follicles can be described in single digits, their mass of granulosa cells significantly overshadows the small follicles in the ovary.

In an attempt to obtain pools of cells from smaller follicles, some investigators have used ovaries of infant rats which lack large follicles. The largest follicles in ovaries of 7 day old rats contain only about 40 cells and the largest follicles of 15 day old rats contain only 75 cells. Ovaries of infant rats differ considerably from those of cycling rats in terms of their steroidogenic capabilities (Carson and Smith, 1986), their growth rates (Pedersen, 1969), their responsiveness to gonadotropins (Hunzicker-Dunn et al., 1984), and their susceptibility to atresia (Byskov, 1978). As indicated below, however, data from infant rats are often the only source of information about very small follicles.

Sources of Information

<u>Indicators of cell maturation</u>   Studies of antigenic characteristics of hematopoietic stem cells have demonstrated that some antigens are expressed and others are lost during the maturation process (Fitchen et al, 1981). Thus, some characteristics are unique to progenitor cells at specific stages along the maturation gradient; others are found in end stage, functionally mature cells as well.

Luteal cells and interstitial gland cells are the end stage of the follicular maturation gradient. These functionally mature cells express several tissue-specific characteristics, including LH receptors and the ability to synthesize and secrete large quantities of steroids.

Below, we will attempt to determine in which progenitor generation each of these mature characteristics is first expressed. In addition, we will attempt to define transient functional characteristics which are expressed by progenitor cells but lost before end stage maturation is complete.

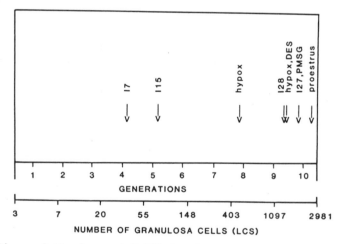

Fig. 1. Size of the largest follicles in commonly-used animal models. Abbreviations: hypox=hypophysectomized, DES=diethylstylbesterol, I=immature (+day of age), PMSG=Pregnant Mare's Serum Gonadotropin, LCS=Largest cross section.

Indicators of proliferative potential   Fast cell production is characteristic of transitional cells approaching terminal differentiation.   In
contrast, stem cells proliferate very slowly, and spend long periods of
time in G0 (non-proliferating state; Lajtha, 1983).   Nevertheless, stem
cells are capable of an infinite number of cell divisions while transit
cells have limited potential divisions.   The maximum generation number in
erythropoietic tissues is 10-17, while in epithelial tissues it is 3-6
(Mackillop et al., 1983).

Granulosa cells obtained from large follicles of the rat proliferate
little, if at all, in vitro, suggesting that their proliferative potential
is exhausted (Sanders and Midgley, 1982; pig: May and Schomberg, 1981).
This is to be expected from committed progenitor cells, far removed from
their stem-cell ancestors.

Unfortunately, the clonogenic capacity of granulosa and theca cells
from normal small and medium follicles has not been studied.   Clonogeni-
city (the ability of a single cell to divide an infinite number of times
and to replace the entire tissue) is a defining characteristic of stem
cells.   This essential information on the proliferative potential of
granulosa and theca cells is not available.

Temporal duration of stages along the maturation gradient   Below, we will
summarize data describing the rate of cell production during follicular
development.   This information will permit us to speculate on the duration
of each generation, providing a temporal framework for the maturation
gradient.

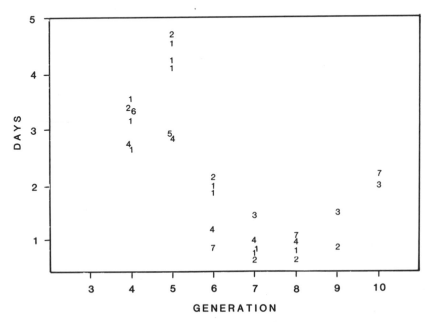

Fig. 2. Published estimates of follicular doubling times. Numbers indicate reference source: 1=Pedersen, 1969; 2=Pedersen,1970a; 3=Hirshfield,
1984; 4=Hage et al, 1978; 5=Hirshfield, 1985; 6=Pedersen and Peters, 1971;
7=Hirshfield, 1986.

Published Data: Fig. 2 illustrates the published estimates of doubling times (generation times) for follicles of different sizes. These data have been culled from several references, and are derived from both adult and immature animals of different ages, and from rats as well as mice. Estimates of generation times vary markedly from study to study.   The variability can, in part, be ascribed to the use of animals of different ages, different techniques, and species differences.

These results are based upon a large number of assumptions about the population of cells being studied.   The replicative behavior of a small subset of cells is used to predict the behavior of the entire population. The accuracy of the resulting growth estimates is strictly dependent upon the degree to which the selected subsample is representative of the rest of the population.   If some cells in the population are not actively dividing, if cell division is somewhat synchronized or subject to diurnal variation, or if the generation time varies widely among cells in the population, estimations of the rate of follicular growth will tend to be overestimated.   This is particularly true with the flash-labeling, frac- tion-labeled mitosis method of cell cycle analysis (Maurer, 1981; Wright and Appleton, 1980).

The rate of granulosa cell proliferation appears to vary with the estrous cycle and with the time of day (Hirshfield, 1984).   Generation time may vary considerably among cells in the same follicle.   Therefore, estimation of the duration of specific growth stages during follicular development is very imprecise at best.   These problems with underlying assumptions may explain the wide discrepancy between estimates of the time required for the entire course of follicular development (in the cyclic mouse, Pedersen (1970a) has estimated this to be 2.7 weeks while Oakberg (1979)   maintains that it requires over 6 weeks).   The reader is cau- tioned that estimates of generation times derived from data in Fig. 3 are speculations based upon shaky foundations. The numbers given are meant to be only rough approximations.

Unpublished Data: Detailed published data on the growth of very small follicles (<20 cells in the LCS) are scarce.   To fill this deficit, we undertook a study of the growth dynamics of small follicles using the method of continuous 3H-TdR infusion.   The procedure and rationale behind this methodology are as follows:

Rats are given continuous infusions of 3H-TdR by means of Alzet Osmotic minipumps. At the time the pumps are inserted in the rat (T_0) the granulosa cells are presumably distributed throughout the 4 phases of the cell cycle: G1,S,G2,M. All cells in S phase at t_0 will take up label when the pumps are inserted.   As time passes, more and more cells will enter S phase, becoming labeled. The cells which have just entered G2 at t_0 will have to pass through the entire length of G2,M, and G1 before they can return to S and take up label.   Under ideal experimental conditions cells in all stages of S phase would be instantaneously labeled as soon as the infusion is begun. However, in actuality, it appears that cells which are in the later portion of S phase at t_0 fail to incorporate enough label to be detected by the autoradiographic procedure.   Thus it is reasonable to assume that cells in mid S phase at t_0 will remain unlabeled until they have traversed the entire cell cycle and returned to mid S phase.   There- fore, if all granulosa cells are actively dividing (growth fraction = 100%), all cells will be labeled when the 3H-TdR infusion period ap- proaches the generation time.   Results of this study are presented in Table 2.

Table 2: Labeling index of small follicles
after continuous infusion with 3H-TdR

Generation	Hours of Infusion				
	24	48	72	144	336
1	31	29	38	45	36
2	19	32	50	56	54
3	33	62	77	95	--
4	60	71	94	98	--
5	58	79	86	99	--
6	50	78	92	97	--

GENERATION BY GENERATION CHARACTERIZATION OF FOLLICULAR DEVELOPMENT

Tenth Generation Follicles

These follicles have more than 2000 granulosa cells in the LCS (>525 um diameter). Follicles of this size are found in ovaries of diestrous and proestrous rats, and in prepubertal intact or hypophysectomized rats pretreated with PMSG or low doses of hCG. Granulosa cells of these follicles are considered the most intensively studied endocrine cell in the body (Research in Reproduction, 1986).

Cell differentiation Tenth generation follicles are highly differentiated and express many mature cell functions. Their granulosa layer is subdivided into regions of specialization (Zoller and Weisz, 1979; Zoller and Weisz, 1978). The theca layer also contains several morphologically distinct types of cells, and is divided into two zones: theca interna and externa (O'Shea, 1970). Theca interna cells are specialized for steroid synthesis; theca externa cells contain actin and myosin (Amsterdam et al., 1976). These follicles contain all the steroidogenic enzymes necessary to produce significant quantities of progesterone, androgens, and estradiol. Moreover, they synthesize and secrete a number of proteins and other substances, including inhibin (Sander et al., 1984), proteoglycans (Yanagishita et al., 1984) relaxin (pig: Bryant-Greenwood et al., 1980), plasminogen activator (Canipari and Strickland, 1986; Wang and Leung, 1983), angiogenic factors (Koos and Le Maire, 1983), and prostaglandin E (Zor et al., 1983). Tenth generation follicles possess receptors for: estradiol (Richards, 1975), FSH (Zeleznik et al., 1974), LH (Bortolussi et al., 1977, 1979), prolactin (Costlow and McGuire, 1977), EGF (Chabot et al., 1986), and possibly progesterone (ewe: McLeod and Haresign, 1984), as well as B-adrenergic receptors (Aguado and Ojeda, 1984; Adashi and Hsueh, 1981). These follicles can be stimulated to ovulate within 24 h by administration of hCG or LH.

Duration of the tenth generation of follicular development Granulosa cells of preovulatory follicles proliferate very slowly (Hirshfield, 1985). Preovulatory follicles of rats with 5 day estrous cycles, are no larger than those of rats with 4 day cycles (Van der Schoot and DeGreef, 1976), despite the extra day of development. Thus, the proliferative potential of preovulatory follicles is apparently exhausted.

The 10th generation follicle, capable of secreting estradiol, and ready to respond to the LH surge, is an ephemeral structure, existing for only about 24 h in the ovary of the cycling rat. Almost as soon as it is formed, the preovulatory follicle begins to degenerate. Serum estradiol concentrations begin to decline on the afternoon of proestrus just before (Brown-Grant et al., 1970) or at the same time as the LH surge begins. If ovulation is experimentally delayed by means of a barbiturate blockade, estradiol falls in the treated group at the same time as the control group even though the LH surge does not occur until the following day (Butcher et al., 1974). Follicles from barbiturate-blocked animals show signs of impaired function within a few hours after the LH surge should have occurred (Braw and Tsafriri, 1980).

Ninth Generation Follicles

These follicles have approximately 1000 granulosa cells in the LCS (360 um diam). Ninth generation follicles are the largest in ovaries of estrous rats, in hypophysectomized prepubertal rats treated with estradiol or DES (Erickson and Hsueh, 1978; Louvet et al., 1975), and in untreated, intact prepubertal rats. These follicles are therefore, the "control" follicles of many in vitro experiments. Follicles of this size always have well-developed antral cavities in intact rats. However, in the total absence of FSH (as in hypophysectomized rats) they lack this feature, but to describe them as "preantral" is misleading. This appellation has the connotation of small, or immature follicles which 9th generation follicles definitely are not.

Cell differentiation   Ninth generation follicles possess receptors for FSH (Nimrod et al., 1976), estradiol (Richards, 1974), androgens (Hillier and DeZwart, 1981) and have LH receptors on their theca cells (Uilenbroek and VanderLinden, 1983; Bortolussi et al., 1977), but they lack LH receptors on granulosa cells (Zeleznik et al., 1974). These follicles lack the aromatase enzymes necessary to synthesize estrogens (Daniel and Armstrong, 1984; Erickson and Hsueh, 1978; Kent and Ryle, 1975), as well as granulosa cell 3B-OHSD (if not exposed to FSH: Zeleznik et al., 1984). They also lack cholesterol side chain cleavage cytochrome P-450 (Zlotkin et al., 1986). However, all of these enzymes can be induced within 24 h by exposure to FSH. Ninth generation follicles can be stimulated by basal concentrations of FSH to synthesize proteoglycans (Mueller et al., 1978). Granulosa cells from "unstimulated" follicles produce fibronectin in culture (Skinner and Dorrington, 1984); exposure to gonadotropins reduces fibronectin synthesis.

In cycling rats, follicles of this size cannot be stimulated to ovulate within 24 h by exposure to hCG (Ying and Greep, 1971); however, untreated prepubertal rats can ovulate in response to hCG (Hirshfield, 1986; Sugawara et al., 1969) beginning at approximately 21 days of age.

Atresia   Atresia of 9th generation follicles is common and appears to play a central role in regulating the number of follicles which ripen in preparation for ovulation (Hirshfield, 1986). A wave of atresia is normally seen in follicles of this size on metestrus (Hirshfield, 1982). This appears to be triggered by the decline of FSH during the morning of estrus, and can be prevented by prolonging the surge of FSH with administration of PMSG (Hirshfield, 1986), resulting in a supernormal number of large, healthy follicles.

Duration of the ninth generation of follicular development   Cell proliferation slows as follicles traverse the 9th generation and approach the preovulatory stage. While granulosa cells abutting the antrum continue to synthesize DNA, mural granulosa cells close to the basement membrane slow down or cease DNA synthesis. The withdrawal of mural cells from the growth fraction becomes more marked as the follicle increases in size (Hirshfield, 1986). A very rough estimate is that the transition from the 9th generation (1000 cells) to the 10th generation (2000 cells) requires 1.5 to 2 days (Fig. 2) of growth.

Eighth Generation Follicles

These follicles contain about 500 granulosa cells in the largest cross section (250 um diameter). Eighth generation follicles are the largest in ovaries of untreated hypophysectomized prepubertal rats, when observed within 2 weeks of surgery (length of the postsurgical period is a critical factor in the condition of the ovary--as time passes the entire reproductive tract becomes progressively more atrophic). These follicles have fully-formed antral cavities in normal cycling rats (Hirshfield and Midgley, 1978); in the absence of FSH an antrum is lacking.

Cell differentiation   Few, if, any of the observations describing functional capabilities of follicles can be ascribed with certainty to 8th generation or smaller follicles. Nearly all of the commonly used models for the study of follicular development contain numerous follicles larger than this size. Pooled granulosa cells isolated for study therefore are probably predominately from the larger follicles. Eighth generation follicles cannot be isolated by manual dissection. Small size and paucity of granulosa cells have prevented the in vitro study of these follicles.

It is likely that follicles from untreated, hypophysectomized, rats are not representative of normal 8th generation follicles in cycling animals. The vast majority of the largest follicles in hypophysectomized rats show extensive evidence of atresia. Moreover lack of exposure to all hypophyseal hormones (LH, FSH, GH in particular), accompanied by low adrenal and thyroid hormone output, is an artificial situation, never encountered under normal physiological conditions.

Cell differentiation   Studies of untreated, hypophysectomized immature rats indicate that granulosa cells of 8th generation follicles possess the enzyme 17-keto-steroid reductase in their granulosa cells (Bogovich and Richards, 1984), however, their theca cells lack the ability to synthesize androgens even when exposed to cAMP (Bogovich and Richards, 1984). Follicles of untreated, hypophysectomized rats have very low, but detectable concentrations of estradiol receptor (Richards, 1975).

Atresia   In cycling rats, the overwhelming majority of follicles become atretic during the 8th generation. While 8th generation follicles comprise a very small proportion of follicles in the ovary, they represent the vast preponderance of atretic follicles in the cycling rat.

Duration of the eighth generation of follicular development   The most rapid follicular growth occurs during the 6th through 8th generations of development. Based on published estimates, 8th generation follicles appear to have a doubling time of approximately 1 to 1.5 days (Fig. 2). Approximately 85-90% of 8th generation follicles were labeled during a 24-h infusion of 3H-TdR (Table 2), indicating that the generation time of these follicles is slightly longer than 24 h. Projections of generation time based on mitotic index following a 2 h metaphase blockade (Hirshfield, 1985), estimate a 17 h generation time. Taken together, these studies suggest that a follicle can grow from 500 to nearly 1000 cells in approximately one day.

Seventh Generation Follicles

These follicles have 256 cells in the LCS (172 um in diameter). Antral spaces begin to appear in follicles of this size in mice (Pedersen and Peters, 1968). Antral spaces are common, but not always present in 7th generation follicles of rats.

Cell differentiation Follicles of this size probably lack LH/hCG receptor on theca cells ("small antral follicles" of intact 20 day old rats; Uilenbroek and VanderLinden, 1983). Their steroidogenic capabilities are not known.

Atresia Atresia (as defined by pyknotic granulosa cells) is rare in 7th generation and smaller follicles in the cycling rat.

Duration of the seventh generation of follicular development Seventh generation granulosa cells have short generation times. A reasonable estimate is that the 7th generation is traversed in 1 to 1.5 days.

Sixth Generation Follicles

These follicles have 128 cells in the LCS (119 um diameter). Sixth generation follicles of mice and hamsters lack antra, but antral spaces are common in follicles of this size in the rat (Hirshfield and Midgley, 1978). The oocyte reaches full size during the 6th generation in the rat (Hirshfield, 1983; Hirshfield and Midgley, 1978) and mouse (Pedersen and Peters, 1968). Uridine incorporation into the nucleus of the oocyte reaches a peak in these follicles (Moore et al., 1974). In the mouse, follicles of this size have three (Pedersen and Peters, 1968) to five (Nicosia and Tojo, 1979) layers of granulosa cells, and 1 to 3 layers of theca cells (Nicosia and Tojo, 1979). Sixth generation follicles are reported to lack a theca layer in the hamster (Roy and Greenwald, 1985).

Cell differentiation The steroidogenic capacity of 6th generation follicles in adult rats is unknown. In infant rats, follicles of this size are capable of producing progesterone even in the absence of exogenous substrate, and they produce androgens (14 day old rat:Carson and Smith, 1986).

Sixth generation follicles (100-200 um) possess more FSH receptor per granulosa cell than at any other stage of development (Presl et al., 1974). Administration of exogenous gonadotropins, especially FSH induces FSH receptor in follicles of this size in infant rats (12 day old rats; Smith and Ojeda, 1986).

Duration of the sixth generation of development Seventy-eight percent of granulosa cells in 6th generation follicles were labeled by a 48 h infusion of 3H-TdR (Table 2). After 3 days, 92% of the granulosa cells were labeled. Therefore, 6th generation granulosa cells are proliferating more slowly than those of the 8th generation. More than 3 days appear to be required for a follicle to grow from 125 to 250 cells.

Fifth Generation Follicles

Fifth generation follicles contain 64-127 granulosa cells in the LCS. In the mouse, follicles of this size are bilaminar; the granulosa cells are arranged in two layers around a growing oocyte (Pedersen and Peters, 1968). These are the largest follicles in the ovaries of the 7 day old mouse (Lunenfeld et al., 1975). The first evidence of an independent blood supply becomes apparent under the light microscope during the 5th generation: follicles 80-160 um are each supplied by 1-2 arterioles (Bassett, 1943).

Cell differentiation   Bortolussi et al. (1979) reported that follicles of this size (100 um) have LH receptors on their theca cells; however, Uilenbroek and Van der Linden (1983) found that no thecal LH receptors in much larger 7th generation follicles. Other indications of granulosa or theca cell differentiation in 5th generation follicles are lacking.

Duration of the fifth generation of follicular development   Eighty-six percent of granulosa cells were labeled after 3 days of continuous 3H-TdR infusion, while 99% were labeled after 6 days (Table 2). Based on the literature, it is estimated that the 5th generation has a duration of approximately 4 days (Fig. 2). Together, these data suggest that at least 4 days are necessary for a follicle to grow from 64 to 125 cells.

Fourth Generation Follicles

These follicles have 32 cells in the LCS (approximately 60 um diameter). Fourth generation follicles have been variously reported to have "one complete ring of follicle cells" (Pedersen and Peters, 1968) or "1-3 layers of cells" (Nicosia and Tojo, 1979). Some authors maintain that 4th generation follicles lack an organized theca layer (Nicosia and Tojo, 1979). Others suggest that a theca layer already exists at this stage (82 um follicles, 8 day old rats; Goldenberg et al., 1973).

Cell differentiation   Data from cycling or prepubertal rats--It is widely believed that granulosa cells of these follicles contain FSH receptor. However data to support this assumption in adult rats are somewhat sketchy. Presl et al. (1974) observed negligible 125I-FSH binding to frozen sections of 23 day old rat ovaries in follicles with a single layer of granulosa cells ("stage 3"--21-60 cells?) and a "significant increase in the uptake of radioiodinated FSH coinciding with the multiplication of granulosa cell layer" ("stage 4; two layers"--61-100 cells?). However, it must be noted that specific binding in this study was only 4 grains above background (18.6 grains vs. 14.6 grains in the presence of excess cold FSH). Other investigators have reported that FSH receptor can be found even in granulosa cells of 1-2 layered follicles (Richards and Midgley, 1976; Eshkol and Lunenfeld, 1972). Very small follicles are not visible in the low power photomicrographs illustrating these papers.

Fourth generation follicles, in the hamster, are capable of synthesizing progesterone (Roy and Greenwald, 1987). The steroidogenic capacity of follicles from the cycling rat has not been evaluated.

Data from infant rats--Follicles with 1-3 layers of granulosa cells have FSH receptors in 10 day old rats (Uilenbroek and Van der Linden, 1983). Fourth generation follicles (the largest in 10 day old rats) cannot produce progesterone in the absence of exogenous substrate, but are capable of converting exogenous pregnenolone to progesterone (Carson and Smith, 1986). Aromatase enzyme first appears in follicles of this size (the largest follicles in 7-8 day old rats) in infant rats (Carson and Smith, 1986). This is unique to the infant ovary as aromatase enzyme is clearly lacking in much larger follicles of older rats.

Granulosa cells from follicles of this size (60-90 um) luteinize spontaneously when cultured in vitro (mouse; Nicosia and Tojo, 1979).

Duration of the fourth generation of follicular development. Ninety-four percent of granulosa cells were labeled after 72 h of continuous 3H-TdR infusion, and 98% were labeled after 144 h (Table 2). Published estimates of cell proliferation in 4th generation follicles vary widely (Fig. 2). It is likely that 3 to 5 days are required to traverse the 4th generation.

First, Second, and Third Generation Follicles

Little, if any, detailed published data are available for follicles with 2-20 granulosa cells in the LCS. Follicles of this size are usually classified as a single group ("type 3a"; Pedersen and Peters, 1968). However, this one category encompasses 3 or 4 generations of granulosa cells. Included in the type 3a classification are both growing follicles and quiescent, primordial follicles.

The zona pellucida first becomes visible in 3rd generation follicles (16 cells; mouse: Oakberg and Tyrell, 1975) whereas second generation follicles lack a zona pellucida (Kang, 1974). Most sources maintain that follicles of this size lack theca cells. However, Familiari et al. (1981) state that these follicles ("primordial and primary") are invested with a theca consisting of a "few layers of concentric fibroblasts and bundles of collagen (mouse)."

Cell differentiation   Little, if anything, is known about follicles of this size in the adult animal. No information is available concerning their steroidogenic capabilities or receptor content. Detailed electron microscopic studies of primordial and primary follicles indicate a lack of morphological specialization in the granulosa cells (Zamboni, 1974).

Data from infant rats--Granulosa cells in follicles of this size ("growing follicles of the newborn mouse") appear to have some 3B-hydroxy-steroid dehydrogenase activity (Rahamim et al., 1976). Again, this appears to be unique to the newborn ovary as much larger follicles of older animals lack 3B-hydroxysteroid dehydrogenase (Zeleznik et al., 1974).

Duration of the first three generations of follicular development   The overwhelming majority of very small follicles in the ovary are quiescent. Follicles in the earliest stages of growth are difficult to distinguish from quiescent, primordial follicles. Several sources suggest that the onset of granulosa cell proliferation in primordial follicles is preceded by a morphological change in the squamous pregranulosa cells: they assume a cuboidal shape (Lintern-Moore and Moore, 1979). The zona pellucida appears to arise concomitantly with this change in the shape of the granulosa cells. We have not been able to confirm these observations; both squamous and cuboidal granulosa cells were labeled following a 24 h infusion of 3H-TdR.

There may, in fact, be no sharp delineation between quiescent, primordial follicles and very small growing follicles. Granulosa cell replication may be sporadic in follicles with less than 20 granulosa cells in the LCS; and cell doubling times may be extremely protracted. Other indicators of early follicular activity also suggest a gradual activation of primordial follicles rather than an abrupt "go/no go" decision: even oocytes of primordial follicles are actively synthesizing RNA; although the rate of RNA synthesis increases as oocyte size increases, the change is so gradual that it is nearly undetectable (Moore et al., 1974).

After 24 h of continuous 3H-TdR infusion, 7% of follicles with 4 granulosa cells in the LCS had at least one labeled cell, while 100% of follicles with 15 or more granulosa cells contained at least 1 labeled

cell. Of these "growing" follicles, those with less than 8 granulosa
cells (1st generation) had a labeling index of 31% after 24 h of continuous infusion (note that this "labeling index" could not have been any
lower as those follicles which lacked labeled cells were excluded from
consideration, and 1 labeled cell out of 3 = 33%;1 out of 4 cells = 25%).
Increasing the infusion period to 3 days did not increase the labeling
index (Table 2) After 6 days of continuous infusion the labeling index of
first generation follicles averaged only 45%. Follicles with 8-16 granulosa cells (2nd generation) showed a similar pattern of 3H-TdR incorporation: a labeling index of 19% after 24 h, 50% after 3 days and only 56%
labeled after 6 days. After 14 days of continuous 3H-TdR infusion, 1st
generation follicles had an average labeling index of only 36%, and 2nd
generation follicles had an average LI of 54% (Table 2).

These observations suggest that the generation time of first and
second generation follicles is considerably longer than 14 days (336
hours) or that some of the granulosa cells in primordial follicles do not
proliferate (the growth fraction is less than 100%).

AT WHAT STAGES ALONG THE FOLLICULAR MATURATION GRADIENT DO TRANSITIONAL
CELLS BECOME PROGRESSIVELY RESTRICTED IN THEIR DEVELOPMENTAL POSSIBILI-
TIES? WHERE ARE THE BRANCHING POINTS IN THE DIFFERENTIATION PROCESS?

A Critical Branching Point in Late Follicular Development

Granulosa and theca cells of large preovulatory follicles are highly
differentiated with respect to their biosynthetic capabilities, hormonal
responsiveness, and morphological appearance. These characteristics of
maturation appear to be acquired during the 9th and 10th generations of
follicular growth. Most follicles undergo atresia and die before development is complete. Although follicles of all sizes can become atretic,
most atresia occurs during the 8th and 9th generations of follicular
development.

The most rapid period of follicular growth occurs during the 8th
generation. Cell number approximately doubles in a single day. Granulosa
cell proliferation and differentiation appear to be reciprocally linked:
as maturational characteristics appear, the rate of granulosa cell proliferation slows dramatically.

Therefore, the 8th and 9th generations represent a crucial turning
point in follicular development. At this stage, the proliferative potential of the granulosa cells becomes exhausted, and the follicle differen-
tiates along one of two paths: into a preovulatory follicle (poised to
become a corpus luteum) or into an atretic follicle (destined to give rise
to secondary interstitium). Most of the signs of cell differentiation
appear suddenly, in rapid succession, during the final stages of development over a few days' time.

Some of the initial signs of differentiation are transient and are
lost before terminal maturation is achieved. In the pig, for example,
elevated estradiol concentrations are not found in the blood until the
last 3 days of follicular maturation, and estradiol secretion diminishes
abruptly during luteinization (Van de Wiel et al., 1981). Similarly,
inhibin production is restricted to the final stages of follicular development (Sander et al., 1984) and diminishes as follicles approach ovulation (DePaolo et al., 1979).

Other functional characteristics are retained by the end-stage, fully mature cells which express them in greatly amplified form (e.g., progesterone production, lipid accumulation, LH receptor content).

An Uneventful Intermediate Period of Follicular Development

Fourth through 7th generation follicles are passing through an uneventful period of development that occupies weeks rather than days; a reasonable guess would be about 12-15 days from the beginning of the fourth generation to the end of the seventh generation.

These stages of development are not distinguished by signs of cell differentiation. The specialized, tissue specific features possessed by this size follicle (FSH receptor, zona pellucida) arose during an earlier stage of development. No new differentiated properties appear to be induced during this lengthy period of growth. Atresia is rare in follicles of this size in adult cycling rats. Thus, this portion of the developmental pathway appears to contain no branching points.

The Earliest Stages of Follicular Growth: Other Critical Branching Points?

The intermediate stages of follicular growth are simply a continuation of what was put in motion at an earlier time. The later stages of follicular development merely bring to fruition the commitments that were made at a much earlier stage of development. The commitment to clonal expansion and the assignment of progenitor cells to the granulosa or theca cell compartment have already occurred.

When do these earlier commitments get made? When do granulosa cells acquire their distinctive FSH receptors which functionally distinguish them from all other cells in the ovary? When does the commitment to a thecal path of differentiation occur in stromal cells that surround the very small follicle? When do the ancestral stem cells of granulosa and theca embark on their course of differentiation? In fact, where are the stem cells, if any, in the adult rat ovary? What alternative maturation pathways are open to their progeny?

Despite their enormous importance for the understanding of ovarian function, the answers to these questions remain unknown. To seek the answers we must study not only the earliest stages of follicular development in the adult rat, but also the organization of the ovary that occurs in embryonic and early postnatal life.

At birth the ovary is a solid mass of tissue composed of thousands of oocytes in meiotic arrest, embedded in the gonadal stroma. The non-germ cell component of the ovary is derived from three embryological sources: the mesenchyme of the gonadal ridge, the coelomic epithelium which covers the gonad, and the rete ovarrii (a component of the Mesonephric [Wolffian] duct; Peters, 1978). It is generally agreed that the mesonephros gives rise to granulosa cells (Upadhyay et al., 1979; Zamboni et al., 1979; Byskov, 1978; Lintern-Moore and Everitt, 1978; Byskov and Lintern-Moore, 1973), although the coelomic epithelium may also be involved (Merchant-Larios, 1979; Peters, 1978; Merchant, 1975). The mesonephros may also be the source of theca cells and primary interstitial tissue of the adult (mouse: Byskov et al., 1985).

Whatever their embryonic origin, ovarian stromal cells proliferate extensively during the perinatal period, insinuating themselves among the oocytes. The generation time of these cells is on the order of 24 h;

nearly 100% of these cells incorporated 3H-TdR during a 24 h infusion period (Hirshfield, unpublished). During the first few days after birth, epithelioid stromal cells encircle individual oocytes and lay down a basement membrane which sequesters them and the oocyte from adjacent ovarian tissue (Byskov et al, 1985). The oocyte, surrounded by a few "pre-granulosa cells" constitutes the primordial follicle. The ovarian stromal cells remaining outside the basement membrane constitute the primary interstitium.

It is believed that all oocytes are encased within a circlet of pregranulosa cells during a short time interval during early postnatal life; oocytes which lack surrounding pregranulosa cells are lost through degeneration within a few days of birth (cow: Ohno and Smith, 1964). Do all progenitor cells for the granulosa compartment become irreversibly committed to their course of differentiation during this perinatal period?

Once they are formed, primordial follicles remain in a quiescent state for varying periods of time. Pregranulosa cells do not divide, in marked contrast to the rapid proliferation of their progenitors. These cells are are often clustered at one pole of the oocyte, leaving other areas of the oocyte exposed save for the thin basement membrane which surrounds it (Zamboni, 1974).

At some point during development of the follicle, the primary interstitial cells outside the developing follicle differentiate into theca cells. These cells begin to proliferate to form a shell outside the basement membrane, enclosing granulosa cells and oocyte. The stage of follicular growth at which the theca cells appear is under dispute. At the source of this dispute is the problem of distinguishing young theca cells from other stromal tissue. Theca cells are usually recognized by histological appearance and their position in concentric rings surrounding follicles. While distinct thecal layers may not appear until follicular growth is well underway, commitment to thecal cell differentiation may occur long before morphological characteristics are evident.

The following points about this commonly-held view of early folliculogenesis are particularly important: 1) that commitment of all granulosa progenitor cells occurs during the few days of the perinatal period, 2) that all of the pregranulosa cells comprising the primordial follicle contribute equally to the descendant populations of granulosa cells (i.e., the growth fraction is 100%), 3) that unlike granulosa cells, commitment of theca progenitor cells occurs continuously throughout reproductive life, and 4) that granulosa cells provide the stimulus that induces commitment of theca cells.

This widely-accepted scenario of early follicular development is based upon careful qualitative descriptive studies. However, to our knowledge, no experimental evidence exists which characterizes the proliferative or differentiative potential of granulosa and thecal cells in very small follicles.

WHAT FACTORS CONTROL THE OUTCOME AT EACH OF THE BRANCHING POINTS IN THE FOLLICULAR MATURATION PATHWAY?

Eighth, Ninth, and Tenth Generation Follicles

An abundance of evidence suggests that FSH and LH (Bogovich et al., 1981; Carson et al., 1981; Richards and Kersey, 1980) are responsible for directing the course of follicular development at the branching point that occurs during the 8th generation. They determine how many follicles enter each of the two developmental paths, and in this way, regulate the number of follicles ripening for ovulation (Hirshfield, 1981; Hirshfield and DePaolo, 1981; Hoak and Schwartz, 1980).

Follicles destined to ovulate during the subsequent cycle, probably traverse the 8th generation on the day of estrus (Hirshfield, 1984). Granulosa cell differentiation appears to be triggered by the prolonged portion of the FSH surge which occurs on estrus morning (Hoak and Schwartz, 1980), although full expression of differentiated function takes 48 h to develop. By 0900 h on proestrus, the maturing follicle is secreting maximal quantities of estradiol (DePaolo et al., 1979).

However, by the 8th generation of follicular development, granulosa and theca cells have already become severely restricted in their developmental possibilities. Only two alternative paths exist: CL vs Interstitium. Neither gonadotropins, nor any of the host of other regulatory agents that affect follicles, can direct granulosa or theca cells down any other developmental routes, nor can they extend the proliferative activity of the granulosa cells. Expression of many maturational characteristics can be enhanced or diminished by manipulation of hormonal conditions, but the ultimate outcome cannot be changed.

Fourth through Seventh Generation Follicles

Numerous factors influence the rate of cell proliferation in these medium sized follicles including stage of the estrous cycle, and time of day. Gonadotropins (Hirshfield, 1985; mouse: Pedersen, 1970b), steroids (Rao et al., 1978) and growth hormones (Hirshfield, unpublished) alter the rate of granulosa cell proliferation. However, the transient vacillations in growth rates which probably occur continuously during the several weeks of development, should have little impact on ovarian function. The magnitude of change in growth rates is so small, and the duration is so short that the overall effect on follicular growth is probably negligible. For example, the mitotic index of 6th generation follicles is approximately 1.6 times higher on estrus than on proestrus (Hirshfield, 1984). Administration of 10 IU PMSG to prepubertal rats only increased mitotic indices 1.4-fold compared with controls (Hirshfield, 1985). Administration of FSH to 21 day old mice resulted in an average 1.3-fold increase in the rate of follicular growth (Pedersen, 1970b). In all cases, the stimulatory effect on cell proliferation is transient, lasting little more than 24 h. By 48 h, indices of cell division decline (Hirshfield, unpublished).

If a 1.4-fold increase in the rate of cell proliferation were maintained for several weeks and applied to all sizes of follicles, the impact on the overall duration of follicular development would be substantial. If a follicle of 32 cells required 13.5 days to grow to preovulatory size in an unstimulated rat, a follicle of 32 cells would require only 9.5 days to grow to preovulatory size in a stimulated rat. A sustained, slight increase in the overall rate of granulosa cell proliferation, resulting in a shortened time frame for follicular growth, is characteristic of immature rats. Non-reproductive factors are most likely responsible for the increased growth rates compared with adults, as proliferation of all tissues is more rapid in immature animals.

If a 1.4-fold increase in the rate of cell proliferation were sustained for only 48 h, the impact on the course of development in small follicles would be minor. A follicle of 64 cells would grow to 103 cells within 48 h in an unstimulated animal whereas it would have only 19 more cells in a stimulated animal. Therefore, most of the extra follicles recruited by PMSG for superovulation probably were rescued from atresia, rather than recruited from the smaller size classes. The 48-56 h usually allowed for PMSG to act are too brief to permit growth of small follicles to an ovulable size. However, the modest increase in cell number following PMSG treatment might alter the fate of 7th generation follicles--a final size of 1591 cells in the stimulated animal versus 910 cells in the unstimulated animal might be sufficient to move the follicle into the group that is competent to ovulate.

First through third generation follicles

The factors that trigger clonal expansion of granulosa cells (activation of primordial follicles) are unknown. Some investigators have postulated that the signal for initiation of follicular development comes from the oocyte (Chang et al., 1978). Some studies suggest that increased gonadotropins may play a role in the activation of primordial follicles (Arendsen de Wolff-Exalto, 1982; mouse:Lintern-Moore and Pantelouris, 1976), while other studies suggest that gonadotropins do not affect the rate of follicular activation (Arendsen, 1982). Studies of aging rats suggest that pregnancy, or prolonged, high concentrations of progesterone may retard the activation of follicles, conserving the follicular reserve, and extending the reproductive lifespan (Lu et al., 1985).

Because the transition from quiescent, primordial follicle to actively growing, primary follicle is so difficult to recognize, little is known about this crucial event. The factors that influence the activation of follicles are ultimately responsible for the number of developing follicles in the ovary at any given time. The regularity of the estrous cycle is closely linked to the number of developing follicles: the more developing follicles, the more regular the estrous cycle (Lacker et al., 1987). Moreover, the duration of the reproductive lifespan is determined by the factors that influence the activation of follicles. When the reserve of oocytes, established before birth, is depleted menopause ensues. Most of the major hypothalamo-pituitary changes associated with the menopausal transition, occur in response to ovarian failure. Exhaustion of the primordial pool of follicles signals the end of reproductive life (Parkening et al., 1985).

Pedersen (1969) has estimated that 1.7 follicles begin to grow per h in 7 day old mice while 0.8 follicles begin to grow per h in 35 day old mice. He has calculated that 80 follicles begin to grow during a 4 day estrous cycle in the adult mouse (Pedersen, 1970a), and that 11 follicles begin to grow during the 24 h of day 2 of pregnancy (Pedersen and Peters, 1971). These estimates are not based upon empirical observation of the behaviour of very small follicles. Rather they are extrapolations from elegant cell kinetic studies of larger follicles. The calculations are predicated upon the assumptions that all granulosa cells in a given follicle are proliferating, that they are proliferating at the same rate, and that the growth rates of all follicles in a given size category are constant. It is likely that many of these assumptions are not true. Therefore, the precise numbers given by Pedersen must be questioned.

Factors that induce commitment of theca cells are also unknown. A "theca cell organizer" has been postulated to emanate from granulosa cells of medium sized follicles (Dubreuil, 1948; Hisaw, 1947). Follicles of 2-3 layers have been reported to release a signal that causes migration of theca cells to the follicle (Eshkol and Lunenfeld, 1972; Peters, 1969). However, the existence of such a substance has not been verified experimentally.

Even less is known about factors that regulate the organization of ovarian stromal elements during the embryonic period and early postnatal life. The arrangement of the granulosa cells within the primordial follicle may be under the direction of the oocyte (Hisaw, 1947). Germ cells which lose their way and end up in the adrenal gland, differentiate into oocytes, and cause the surrounding cells to assume an orientation characteristic of pregranulosa cells (Upahdyay and Zamboni, 1982).

AN ALTERNATIVE SCENARIO FOR EARLY FOLLICULAR DEVELOPMENT

Given the paucity of firm evidence to support the traditional scenario of early follicular development, we would like to suggest an alternative hypothesis concerning the origin and developmental potential of granulosa and theca cells. This hypothesis is based upon principles derived from studies of other renewal tissues.

Like the follicle, other renewal tissues contain more than one cell type which must grow and differentiate as a coordinated unit. The intestine, for example, contains parenchymal cells, supporting connective tissue cells, vascular cells, and nerve fibers. The progenitors of each of these tissue types are hypothesized to be assembled into a single complex unit, the "proliferon" at the outset of clonal expansion (Zajicek, 1977). The proliferon grows, matures, and is ultimately shed as a unit. The intimate association of these different cell types provides a mechanism for coordination of proliferation and differentiation.

In other renewal tissues, and in the developing embryo, the coordination of cell proliferation and differentiation among different tissue types is often controlled by the underlying mesenchyme. Many examples have been found where connective tissue cells direct the growth of overlying tissue by providing essential growth factors and anchoring matrices (Cunha et al., 1983).

We propose that granulosa and theca cells are derived from a single stem cell population. We suspect that this stem cell population continues to reside in the the hilar and medullary regions of the ovary in the adult rat, where most of the primordial follicles are clustered in the adult. We base this hypothesis upon the functional similarity of mature granulosa and theca cells, and upon the observation that ovarian tumors often display characteristics that are intermediate between granulosa and theca cells (particularly in animal models), suggesting a bipotentency in their clonal ancestor (Fox, 1985). Recent studies also point to a common origin of granulosa and theca in mice (Byskov et al., 1985).

Descendants of these stem cells migrate through the ovary towards quiescent oocytes. The earliest arrivals enclose the oocyte, lay down a basement membrane, and differentiate into granulosa cells; those that arrive later are excluded by the basement membrane and differentiate into theca cells. This process of stem cell proliferation, migration and differentiation may occur throughout life, not just in the perinatal period, however most of the cells within the basement membrane probably arrive at their destination during the prenatal and early postnatal period simply as a result of the architecture of the neonatal ovary.

Primordial theca cells, although not morphologically distinguishable from fibroblasts, are thus present at the outset of follicular growth. We suspect that the theca progenitor cells may provide the signal which triggers the onset of granulosa cell proliferation: follicular growth begins when a critical mass of cells is reached, completing the organization of the "proliferon." In addition to granulosa and theca, the progeni-

229

tors of the vascular and neural elements of the follicle are established at this time. We suggest that, as in other tissues, it is the supporting mesenchyme (theca) that directs coordination of cell proliferation in the epithelial (granulosa) compartment of the proliferon.

CONCLUSIONS

In the rat, the 8th and 9th generations of follicular growth represent a crucial turning point in development. It is at this stage that global feedback regulatory control of follicular development is established. Feedback between the follicles and gonadotropin secretion regulates the number of follicles ripening for ovulation.

Gonadotropins appear to play a permissive, not a directive, role in regulating follicular development during the 8th and 9th generations. By this penultimate stage of growth, granulosa and theca cells have become severely restricted in their developmental possibilities. While expression of many maturational characteristics can be advanced or retarded by manipulation of hormonal conditions, the ultimate developmental outcome cannot be changed.

Although it is only the final generations of granulosa and theca cells that express mature functional characteristics, their progenitors probably become committed to a single developmental course early in folliculogenesis. Neither the stages when progenitor cells become irrevocably committed to differentiation into granulosa and theca cells nor the regulatory factors that determine the outcome at these stages have been identified.

We suggest that progenitor cells become committed to the thecal tissue compartment while the follicle is still in the primordial stage of follicular development. We postulate that progenitors of all constituents of the follicle may already be assembled into a unit as follicular growth begins. The accumulation of a sufficient number of theca progenitor cells around the primordial follicle may be the signal for initiation of follicular growth.

We believe that understanding post-natal follicular growth and development rests upon understanding the origin, fates, and capacities of the cells which give rise to the ovary and its various compartments. The first generations of follicular growth are a most promising avenue for future research.

REFERENCES

Adashi EY, Hsueh AJW, 1981. Stimulation of B-adrenergic responsiveness by follicle-stimulating hormone in rat granulosa cells in vitro and in vivo. Endocrinology 108:2170-2178

Aguado LI, Ojeda SR, 1984. Prepubertal ovarian function is finely regulated by direct adrenergic influences: role of noradrenergic innervation Endocrinology 114:1845-1853

Amsterdam A, Lindner HR, Groschel-Stewart U, 1976. Localization of actin and myosin in the rat oocyte and follicular wall by immunofluorescence Anat Rec 187:311-328

Arendsen de Wolff-Exalto E, 1982. Influence of gonadotrophins on early follicle cell development and early oocyte growth in the immature rat J Reprod Fertil 66:537-542

Bassett DL, 1943. The changes in the vascular pattern of the ovary of the albino rat during the estrous cycle. Am J Anat 73:251-291

Bogovich K, Richards JS, 1984. Androgen synthesis during follicular development: evidence that rat granulosa cell 17-ketosteroid reductase is independent of hormonal regulation. Biol Reprod 31:122-131

Bogovich K, Richards JS Reichert LE, 1981. Obligatory role of luteinizing hormone (LH) in the initiation of preovulatory follicular growth in the pregnant rat: specific effects of hCG and FSH on LH receptors and steroidogenesis in theca and granulosa cells. Endocrinology 109:860-867

Bortolussi M, Marini G, Dal Lago A, 1977. Autoradiographic study of the distribution of LH(hCG) receptors in the ovary of untreated and gonadotrophin-primed immature rats. Cell Tissue Res 183:329-342

Bortolussi M, Marini G, Reolon ML, 1979. A histochemical study of the binding of 125I-hCG to the rat ovary throughout the estrous cycle Cell Tissue Res 197:213-226

Braw RH, Tsafriri A, 1980. Follicles explanted from pentobarbitone-treated rats provide a model for atresia. J Reprod Fertil 59:259-265

Brown G, Bunce CM, Guy GR, 1985. Sequential determination of lineage potentials during haemopoiesis Br J Cancer 52:681-686

Brown-Grant K, Exley D, Naftolin F, 1970. Peripheral plasma oestradiol and luteinizing hormone concentrations during the oestrous cycle of the rat. J Endocrinol 48:295-296

Bryant-Greenwood GD, Jeffrey R, Ralph MM, Seamark RF, 1980. Relaxin production by the porcine ovarian Graafian follicle in vitro. Biol Reprod 23:792-800

Butcher RL, Collins WE, Fugo NW, 1974. Plasma concentration of LH, FSH, prolactin, progesterone and estradiol-17B throughout the four-day estrous cycle of the rat. Endocrinology 94:1704-1708

Byskov AGS, 1978. "Follicular Atresia". In: Jones RE (ed.), The Vertebrate Ovary NY Plenum Press pp. 533-562

Byskov AGS, Lintern-Moore S, 1973. Follicle formation in the immature mouse ovary: the role of the rete ovarii. J Anat 116:207-217

Byskov AG, Hoyer PE, Westergaard L, 1985. Origin and differentiation of the endocrine cells of the ovary. J Reprod Fertil 75: 299-306

Canipari R, Strickland S, 1986. Studies on the hormonal regulation of plasminogen activator production in the rat ovary. Endocrinology 118: 1652-1659

Carson R, Smith J, 1986. Development and steroidogenic activity of preantral follicles in the neonatal rat ovary. J Endocrinol 110:87-92

Carson RS, Richards JS, Kahn LE, 1981. Functional and morphological differentiation of theca and granulosa cells during pregnancy in the rat: dependence on increased basal luteinizing hormone. Endocrinology 109:1433-1441

Chabot JG, St. Arnaud R, Walker P, Pelletier G, 1986. Distribution of epidermal growth factor receptors in the rat ovary. Mol Cell Endocrinol 44:99-108

Chang SCS, Ryan RJ, Kang YH, Anderson WA, 1978. Some observations on the development and function of ovarian follicles. In Sreenan JR (ed.), Control of Reproduction in the Cow. The Hauge, Martinus Nijhoff, pp. 3-33

Costlow ME, McGuire WL, 1977. Autoradiographic localization of the binding of 125I-labelled prolactin to rat tissues in vitro. J Endocrinol 75: 221-226

Cunha GR, Chung LWK, Shannon JM, Taguchi O, Fuji H, 1983. Hormone-induced morphogenesis and growth: role of mesenchymal-epithelial interactions Rec Prog Hor Res 39:559-598

Daniel SAJ, Armstrong DT, 1984. Site of action of androgens on FSH-induced aromatase activity in cultured rat granulosa cells. Endocrinology 114:1975-1982

DePaolo LV, Wise PM, Anderson LD, Barraclough CA, Channing CP, 1979. Suppression of the pituitary follicle-stimulating hormone secretion during proestrus and estrus in rats by porcine follicular fluid: possible site. Endocrinology 104:402-408

DePaolo LV, Shander D, Wise PM, Barraclough CA, Channing CP, 1979. Identification of inhibin-like activity in ovarian venous plasma of rats during the estrous cycle. Endocrinology 105:647-658

Dubreuil G, 1948. Sur l'existence d'une substance inductrice a action limitee et locale pour la metaplasie thecale des cellules du stroma cortical ovarien. Ann Endocrinol 9:434-442

Erickson GF, Hsueh AJW, 1978. Induction of aromatase activity by follicle stimulating hormone in rat granulosa cells in vivo and in vitro Endocrinology 102:1275-1282

Erickson GF, Magoffin DA, Dyer CA, Hofeditz C, 1985. The ovarian androgen producing cells: a review of structure/function relationships Endocrinol Rev 6:371-399

Eshkol A, Lunenfeld B, 1972. Gonadotropic regulation of ovarian development in mice during infancy. In: Saxena BB, Beling CG, Gandy HM, (eds.), The Gonadotropins. New York, John Wiley and Sons, pp. 335-346

Familiari G, Correr S, Motta PM, 1981. Gap junctions in theca interna cells of developing and atretic follicles Adv Morph Cells and Tissue 11th Intl Congress, pp. 337-348

Fitchen JH, Foon KA, Cline MJ, 1981. The antigenic characteristics of hematopoietic stem cells. New Eng J Med 305:17-25

Fox H 1985. Sex cord-stromal tumours of the ovary. J Path 145:127-148

Goldenberg RL, Reiter EO, Ross GT, 1973. Follicle response to exogenous gonadotropins: an estrogen-mediated phenomenon. Fertil Steril 24: 121-127

Guraya SS, Greenwald GS, 1964. Histochemical studies on the interstitial gland in the rabbit ovary. Am J Anat 114:495-520

Guraya SS, Greenwald GS, 1964. A comparative histochemical study of interstitial tissue and follicular atresia in the mammalian ovary. Anat Rec 149:411-434

Hage AJ, Groen-Klevant AG, and Welschen R, 1978. Follicle growth in the immature rat ovary. Acta Endocrinol 88:375-382

Hillier SG, DeZwart FA, 1981. Evidence that granulosa cell aromatase induction/activation by FSH is an androgen receptor-regulated process in vitro. Endocrinology 109:1303-1305

Hirshfield AN, 1982. Follicular recruitment in long-term hemicastrate rats. Biol Reprod 27:48-53

Hirshfield AN, 1983. Compensatory ovarian hypertrophy in the long-term hemicastrate rat: size distribution of growing and atretic follicles Biol Reprod 28:271-277

Hirshfield AN, 1984. Stathmokinetic analysis of granulosa cell proliferation in antral follicles of cyclic rats. Biol Reprod 31: 52-58

Hirshfield AN, 1985. Patterns of Cell Proliferation in Follicles Approaching Ovulation: In Toft DO, Ryan RJ (eds.), Proceedings of the Fifth Ovarian Workshop. IL Ovarian Workshops,Inc., pp.249-253

Hirshfield AN, 1985. Comparison of granulosa cell proliferation in small follicles of hypophysectomized, prepubertal and mature rats. Biol Reprod 32:979-987

Hirshfield AN, 1986. Effect of a low dose of pregnant mare's serum gonadotropin on follicular recruitment and atresia in cycling rats Biol Reprod 35:113-118

Hirshfield AN, Midgley AR, 1978. Morphometric analysis of follicular development in the rat. Biol Reprod 19:597-605

Hirshfield AN, DePaolo LV, 1981. Effect of suppression of the surge of follicle stimulating hormone with porcine follicular fluid on follicular development in the rat. J Endocrinol 88:67-71

Hisaw Fl 1947. Development of the Graafian follicle and ovulation Physiol Rev 27:95-119

Hoak DC, Schwartz NAB, 1980. Blockade of recruitment of ovarian follicles by suppression of the secondary surge of follicle-stimulating hormone with porcine follicular fluid. PNAS(USA) 77:4953-4956

Hunzicker-Dunn M, Jungmann RA, Evely L, Hadawi GL, Maizels ET, West, 1984. Modulation of soluble ovarian adenosine 3,5-monophosphate-dependent protein kinase activity during prepubertal development of the rat. Endocrinol 115:302-311

Kang Y, 1974. Development of the zona pellucida in the rat oocyte. Am J Anat 139:535-566

Kent J, Ryle M, 1975. Histochemical studies on three gonadotrophin-responsive enzymes in the infantile mouse ovary. J Reprod Fertil 42: 519-536

Koos RD, Le Maire WD, 1983. Evidence for an angiogenic factor in rat follicles: In Greenwald GS, Terranova PF (eds.), Factors Regulating Ovarian Function. NY Raven Press, pp. 191-196

LaPolt SPS, Matt DW, Shryne JE, Lu JHK, 1985. Analysis of ovarian follicular dynamics in aged and young female rats using continuous 3H-TdR infusion. Endocrine Society Abstract #924.

Lacker HM, Beers WH, Meuli LE, Atkin E, 1987. A theory of follicle selection. Biol Reprod (in press)

Lajtha LG, 1983. Stem cell concepts: In: Potten CS (ed.), Stem Cells: Their Identification and Characterization. Edinburgh Churchill, Livingstone, pp. 1-11

Lintern-Moore S, Everitt AV, 1978. The effect of restricted food intake on the size and composition of the ovarian follicle population in the Wistar rat. Biol Reprod 19:688-691

Lintern-Moore S, Moore GPM, 1979. The initiation of follicle and oocyte growth in the mouse ovary. Biol Reprod 20:773-778

Lintern-Moore S, Pantelouris EM, 1976. Ovarian development in athymic nude mice IV. the effect of PMSG and oestradiol on the growth of the oocyte and follicle. Mech Aging and Devel 5:155-162

Louvet J-P, Harman SM Schreiber JR Ross GT, 1975. Evidence for a role of androgens in follicular maturation. Endocrinology 97:366-372

Lu JHK, LaPolt PS, Nass TC, Matt DW, Judd HL, 1985. Relation of circulating estradiol and progesterone to gonadotropin secretion and estrous cyclicity in aging female rats. Endocrinology 116:1953-1959

Lunenfeld B, Kraiem Z, Eshkol A, 1975. The function of the growing follicle J Reprod Fertil 45:567-574

Mackillop WJ, Ciampi A, Till JE, Buick RN, 1983. A stem cell model of human tumor growth: implications for tumor cell clonogenic assays JNCI 70: 9-16

Maurer HR, 1981. Potential pitfalls of [3H] thymidine techniques to measure cell proliferation. Cell Tissue Kinet 14:111-120

May JV, Schomberg DW, 1981. Granulosa cell differentiation in vitro: the effect of insulin on growth and final integrity Biol Reprod 25: 421-431

McLeod BJ, Haresign W, 1984. Evidence that progesterone may influence subsequent luteal function in the ewe by modulating preovulatory follicle development. J Reprod Fertil 71:381-386

Merchant H, 1975. Rat gonadal and ovarian organogenesis with and without germ cells. an ultrastructural study. Develop Biol 44:1-21

Merchant-Larios H, 1979. Origin of the somatic cells in the rat gonad: an autoradiographic approach. Ann Biol anim Bioch Biophys 19:1219-1229

Moore GPM, Lintern-Moore S, Peters H, and Faber M, 1974. RNA synthesis in the mouse oocyte. Cell Biol 60:416-422

Mueller PL, Schreiber JR, Lucky AW, Schulman JD, Rodbard D, Ross GT, 1978. Follicle stimulating hormone stimulates ovarian synthesis of proteoglycans in the estrogen-stimulated hypophysectomized immature female rat. Endocrinology 102:824-831

Nicosia SV, Tojo R, 1979. Morphogenetic reaggregation and luteinization of mouse preantral follicle cells Am J Anat 156:401

Nimrod A, Erickson GF, Ryan KJ, 1976. A specific FSH receptor in rat granulosa cells: properties of binding in vitro. Endocrinology 98: 56–64

O'Shea JD, 1970. An ultrastructural study of smooth muscle-like cells in the theca externa of ovarian follicles in the rat. Anat Rec 167: 127–131

Oakberg EF, 1979. Timing of oocyte maturation in the mouse and its relevance to radiation-induced cell killing and mutational sensitivity. Mutat Res 59:39–48

Oakberg EF, Tyrrell PD, 1975. Labelling of the zona pellucida of the mouse oocyte. Biol Reprod 12:477–482

Ohno S, Smith JB, 1964. Role of fetal follicular cells in meiosis of mammalian oocytes. Cytogenetics 3:324–333

Parkening TA, Collins TJ, Elder FFB, 1985. Othotopic ovarian transplantation in young and aged C5BL/6J mice. Biol Reprod 32: 989–997

Pedersen T, 1969. Follicle growth in the immature mouse ovary Acta Endocrinol 62:117–132

Pedersen T, 1970a. Follicle kinetics in the ovary of the cyclic mouse Acta Endocrinol 64:304–323

Pedersen T, 1970b. Cell population kinetics of the ovary of the immature mouse after FSH stimulation. In: Butt WR, Crooke AC, Ryle M (eds.), Gonadotropins and Ovarian Development. Edinburgh, E and S Livingstone pp. 312–324

Pedersen T, Peters H, 1968. Proposal for a classification of oocytes and follicles in the mouse ovary. J Reprod Fertil 17:555–557

Pedersen T, Peters H, 1971. Follicle growth and cell dynamics in the mouse ovary during pregnancy. Fertil Steril 22:42–52

Peters H, 1969. The development of the mouse ovary from birth to maturity Acta Endocrinol 62:98–116

Peters H, 1978. Folliculogenesis in mammals. In: Jones RE (eds.), The Vertebrate Ovary. NY Plenum Press, pp.121–140

Presl J, Pospisil J, Figarova V, Krabec Z, 1974. Stage-dependent changes in binding of iodinated FSH during ovarian follicle maturation in rats Endocrinol Exp 8:291–298

Rahamim E, Eshkol A, Lunenfeld B, 1976. Histochemical demonstration of delta5-3betal-hydroxysteroid dehydrogenase activity in ovaries of intact infant mice and mice treated with anti-gonadotropin. Fertil Steril 27:328–34.

Rao MC, Midgley AR Richards JS, 1978. Hormonal regulation of ovarian cellular proliferation. Cell 14:71–78

Research in Reproduction, 1986. More studies on biosynthesis by granulosa cells Research in Reproduction 18: 2.

Richards JS, 1974. Estradiol binding to rat corpora lutea during pregnancy. Endocrinology 95:1046–1053

Richards JS, 1975. Estradiol receptor content in rat granulosa cells during follicular development: modification by estradiol and gonadotropins. Endocrinology 97:1174–1184

Richards JS, Midgley AR Jr, 1976. Protein hormone action: a key to understanding ovarian follicular and luteal cell development. Biol of Reprod 14:82–94

Richards JS, Kersey KA, 1980. Changes in theca and granulosa cell function in antral follicles developing during pregnancy in the rat: gonadotropin receptors, cyclic AMP, and estradiol-17B. Biol Reprod 21:1185–1201

Roy SK, Greenwald GS, 1985. An enzymatic method for dissociation of intact follicles from the hamster ovary: histological and quantitative aspects. Biol Reprod 32:203–215

Roy SK, Greenwald GS, 1987. In vitro steroidogenesis by primary to antral follicles in the hamster during the periovulatory period: effects of FSH, LH, and prolactin. Biol Reprod (in Press).

Sachs L, 1986. Growth, differentiation and the reversal of malignancy Sci Am 254:40-47

Sander HJ, Van Leeuwen ECM, de Jong FH, 1984. Inhibin-like activity in media from cultured rat granulosa cells collected throughout the oestrous cycle. J Endocrin 103:77-84

Sanders MM, Midgley AR, 1982. Rat granulosa cell differentiation: an in vitro model. Endocrinology 111:614-624

Selby P, Buick RN, Tannock I, 1983. A critical appraisal of the "human tumor stem-cell assay". NEJM 308:129-134

Skinner MK, Dorrington JH, 1984. Control of fibronectin synthesis by rat granulosa cells in culture. Endocrinology 115:2029-2031

Smith SS, Ojeda SR, 1986. Neonatal release of gonadotropins is essential for development of ovarian follicle-stimulating hormone (FSH) receptors. Biol Reprod 219-227

Sugawara S, Umezu M, Takeuchi S, 1969. Effect of a single dose of human chorionic gonadotrophin on the ovulatory response of the immature rat J Reprod Fertil 20:333-335

Talbert GB, Meyer RK McShan WH, 1951. Effect of hypophysectomy at the beginning of proestrus on maturing follicles in the ovary of the rat. Endocrinology 49:687-694

Uilenbroek Jth J, Van der Linden R, 1983. Changes in gonadotrophin binding to rat ovaries during sexual maturation Acta Endocrinol 103:413-419

Upadhyay S, Luciani JM Zamboni L, 1979. The role of the mesonephros in the development of indifferent gonads and ovaries of the mouse. Ann Biol Anim Biochim Biophys 19:1179-1196

Upahdyay S, Zamboni L, 1982. Ectopic germ cells: natural model for the study of germ cell sexual differentiation. PNAS 79:6584-6586

Van de Wiel DFM, Erkens J, Koops W, Vos E, Van Landeghem AAJ, 1981. Periestrous and midluteal time courses of circulating LH, FSH, prolactin, estradiol-17B and progesterone in the domestic pig. Biol Reprod 24:223-233

Van der Schoot P, DeGreef WJ, 1976. Dioestrous progesterone and pro-oestrous LH in 4- and 5-day cycles of female rats. J Endocrinol 70: 61-68

Wang C, Leung A, 1983. Gonadotropins regulate plasminogen activator production by rat granulosa cells Endocrinology 112:1201-1207

Wright NA, Appleton DR, 1980. The metaphase arrest technique: a critical review. Cell Tissue Kinet 13:643-663

Yanagishita M, Hascall VC, 1984. Metabolism of proteoglycans in rat ovarian granulosa cell culture. J Biol Chem 259 16:10260-10269

Ying SY, Greep RO, 1971. Responsiveness of follicles to gonadotropins during the estrous cycle of the rat. Endocrinology 89:294-297

Zajicek G, 1977. The intestinal proliferon J. Theor. Biol. 67:515-521

Zajicek G, 1979. Proliferon: the functional unit of rapidly proliferating organs. Med Hypotheses 5:161-174

Zajicek G, Michaeli Y, Regev J, 1979. On the progenitor cell migration velocity. Cell Tiss Kinet 12:453-460

Zamboni L, 1974. Fine morphology of the follicle wall and follicle cell-oocyte association. Biol Reprod 10:125-149

Zamboni L, Bezard L, Mauleon P, 1979. The role of mesonephros in the development of the sheep fetal ovary. Ann Biol Anim Biochim Biophys 19:1153-1178

Zeleznik AJ, Midgley AJ, Reichert LE, 1974. Granulosa cell maturation in the rat: increasing binding of human chorionic gonadotropin following treatment with follicle-stimulating hormone in vivo. Endocrinology 95:818-825

Zlotkin T, Farkash Y, Orly J, 1986. Cell specific expression of immuno-reactive cholesterol side chain cleavage cytochrome P-450 during follicular development in the rat ovary. Endocrinology 119:2809-2820

Zoller LC, Weisz J, 1978. Structure and acid phosphatase activity in granulosa cells of preovulatory follicles in rat ovary. Anat Rec 190:592

Zoller LC, Weisz J, 1979. A quantitative cytochemical study of glucose-6-phosphate dehydrogenase and 3B-hydroxysteroid dehydrogenase activity in the membrana granulosa of the ovulable type of follicle of the rat. Histochemistry 62:125-135

Zor U, Strulovici B, Braw R, Lindner HR, Tsafriri, A, 1983. Follicle stimulating hormone-induced prostaglandin E formation in isolated rat ovarian theca. J Endocrinol 97:43-49

ANIMAL MODELS FOR STUDY OF POLYCYSTIC OVARIES AND OVARIAN ATRESIA

V.B. Mahesh, T.M. Mills, C.A. Bagnell amd B.A. Conway

Department of Physiology and Endocrinology
Medical College of Georgia
Augusta, GA 30912

INTRODUCTION

In 1935, Stein and Leventhal described a syndrome in which there was an association between large, pale polycystic ovaries, menstrual irregularities, ovulatory failure, infertility, hirsutism and obesity. Isolated reports of polycystic ovaries in the human and the management of the associated infertility by partial wedge-resection of the ovaries had been reported in Europe almost a century earlier but a syndrome had not been described. Since 1935 extensive investigations have been undertaken to establish the pathophysiology of the polycystic ovary syndrome and several excellent reviews have been published on the subject (Mahesh et al., 1962; Goldzieher and Axelrod, 1963; Mahesh and Greenblatt, 1964a; Shearman and Cox, 1966; Kirschner and Bardin, 1972; Greenblatt and Mahesh, 1976; Parker and Mahesh, 1976a; Rebar et al., 1976; Mahesh, 1980, 1983, 1984; Yen, 1980, Goldzieher, 1981). In spite of the extensive research on the polycystic ovary syndrome, there is no concensus as to its pathophysiology because the so called syndrome is not a distinct entity in itself. An analysis of 1079 cases published in 1962 showed that 57% of the cases had amenorrhea, 69% hirsutism, 41% obesity, 74% infertility, 21% virilization, and 12% had cyclic menses (Goldzieher and Green, 1962). Variations were also found in the size of the ovary ranging from small to normal to large; in some cases, the tunica was thickened while in others it was not (Smith et al., 1965).

The concept of the polycystic ovary syndrome being a genetic disorder has been advanced based on the occurrence of the syndrome in sisters, identical twins and a mother and a daughter. In addition, isolated examples of the presence of chromosomal abnormalities, and x-chromosome linked or autosomal dominant mode of transmission of the disorder have been reported (reviewed by Mahesh, 1984). In spite of the well documented evidence of the genetic basis of inheritance of the polycystic ovary syndrome, a genetic link is found in a relatively small number of cases and is not a generalized finding.

The presence of elevated androgen secretion with or without an elevation in pregnanetriol and Δ^5-pregnenetriol has been well documented in the polycystic ovary syndrome. In the early 1960s, the adrenal was considered the primary source of the disorder (Perloff et al., 1957; Gallagher et al., 1958; Brooks and Prunty, 1960; Lipsett and Riter, 1960). The concept of the adrenals as a source of excessive androgens was confirmed by the presence of polycystic ovaries in patients with virilizing adrenal tumors and congenital adrenal hyperplasia. By using small doses of ACTH to stimulate steroidogenesis and measuring the ratios of the secretory products, several investigators have reported the presence of mild degrees of abnormality in the 21-hydroxylase, 11β-hydroxylase and 3β-hydroxysteroid dehydrogenase activities in patients with polycystic ovaries (Given et al., 1975; Ikkos and Kellia-Sfikaki, 1975; Lobo and Gobelsmann, 1980, 1981, 1982).

In 1953 Greenblatt observed a fall in urinary 17-ketosteroids after wedge-resection of the ovary in untreated and cortisone treated women with polycystic ovaries, and suggested that the ovary may be a source of excessive androgens in the syndrome. The concept that polycystic ovaries could secrete excessive androgens was confirmed by Mahesh and coworkers in studies in which urinary steroids were measured before and after ovarian wedge-resection and adrenal and ovarian suppression (Mahesh and Greenblatt, 1961, 1964b; Mahesh et al., 1964). Large quantities of dehydroepiandrosterone and Δ^4-androstenedione have also been isolated from ovarian tissue and ovarian venous blood of patients with the polycystic ovary syndrome (Mahesh et al., 1962; Mahesh and Greenblatt, 1962, 1964a). The isolation of dehydroepiandrosterone and Δ^4-androstenedione from ovarian tissue and follicular fluid from polycystic ovaries has subsequently been confirmed by other investigators (Starka et al., 1962; Baulieu et al., 1963; Short and London, 1961; Mahajan et al., 1963).

Early reports on the measurement of urinary gonadotropins by bioassay in patients with the polycystic ovary syndrome indicated a hypersecretion of LH (Ingersoll and McDermott, 1950; McArthur et al., 1958; Taymor and Barnard, 1962). This was confirmed by radioimmunoassays of LH and FSH. Although the serum LH levels were either normal or elevated, the FSH levels were normal or depressed (Mahesh et al., 1970; Yen et al., 1970a; Gambrell et al., 1971, 1973; Berger et al., 1975; DeVane et al., 1975; Duignan et al., 1975; Baird et al., 1977). The high LH concentrations found in patients with the polycystic ovary appear to be the result of increased frequency and/or increased amplitude of pulsatile LH release (Rebar et al., 1976; Baird et al., 1977). Furthermore, when stimulated with LHRH, patients with the polycystic ovary syndrome showed an exaggerated response in the release of LH (Rebar et al., 1976; Berger et al., 1975; Devane et al., 1975; Duignan et al., 1975; Baird et al., 1977; Patton et al., 1975; Taymor et al., 1974; Oettinger et al., 1975; Moltz et al., 1979). These observations have contributed to the concept of an abnormal hypothalamic-pituitary axis in the polycystic ovary syndrome. However, such an abnormal hypothalamic-pituitary axis in the polycystic ovary syndrome is unlikely because a) the immediate lowering of LH after estradiol administration in patients with polycystic ovaries appeared to be comparable to that found in the normal cycle (Yen et al., 1970a; Rebar et al., 1976; Baird et al., 1977), b) the estrogen induced positive LH surge was similar in normal women and patients with polycystic ovaries (Baird et al., 1977), c) treatment of patients having polycystic ovary syndrome with clomiphene citrate resulted in an estradiol surge followed by an ovulatory gonadotropin surge similar to that found in the normal cycle (Mahesh and Greenblatt, 1964a; Yen et al., 1970b; Gambrell et al., 1971; Baird et al., 1977), d) the augmentation of the anterior pituitary gland's responsiveness to LHRH in the release of LH by estrogens in the human is well documented (Jaffe and Keye, 1974; Yen et al., 1974, 1975). In the polycystic ovary syndrome there is a correlation between the level of circulating estrogens and circulating LH (DeVane et al., 1975; Rebar et al., 1976; Baird et al., 1977; Kandeel et al., 1978). It is therefore likely that the elevated circulating LH is due to high circulating estrogens rather than an abnormal feedback. This was confirmed by the finding that the fractional increase in LH after LHRH administration correlates well with the level of circulating estrogens (Rebar et al., 1976) and e) the administration of estrone benzoate to patients with the polycystic ovary syndrome further suppressed serum FSH without altering serum LH and thus produced a pattern of gonadotropins similar to what is seen in patients with the syndrome (Chang et al., 1982).

In summary, there appears to be no agreement on the pathophysiology of the polycystic ovary syndrome because of the variations and diversity of signs and symptoms found to be associated with the presence of polycystic ovaries. The concept of a hypothalamic-pituitary disorder causing the syndrome is debatable due to the demonstration of a normal steroid-gonadotropin feedback system. In the majority of the cases studied, the overriding observation is the presence of excessive androgen secretion of adrenal or ovarian origin associated with ovulatory failure. Suppression of excessive androgens either by the removal of a virilizing tumor, by the wedge-resection of the ovaries or by the administration of glucocorticoids usually results in resumption of ovulatory cycles. This chapter will review our work in experimental animal models to study the relationships between excessive androgen secretion and ovulatory failure.

ANIMAL MODELS FOR THE STUDY OF POLYCYSTIC OVARIES

Considerable species differences exist between the human and experimental animals and thus the use of animal models for the study of the human is limited. Nevertheless, the similarities in key steps in mammalian reproduction make animal models attractive for studies that cannot be carried out in the human.

An animal model used for the study of polycystic ovaries was based on the observation of Browman (1937) and Hemmingsen et al. (1937) that exposure to constant light disrupts the estrous cycle of the rat. Studies conducted to determine the cause of the ovulatory failure showed that the pituitary LH content was decreased (Maric et al., 1965; Bradshaw and Critchlow, 1965). An examination of the morphology of the ovaries revealed the existence of polycystic ovaries typically found associated with ovulatory failure (Singh, 1969). This model has not been used extensively for the study of polycystic ovaries because the steroid-gonadotropin feedback system is modified. Unlike the normal animal, there was an absence of the proestrous type LH surges (Daane and Parlow, 1971). The administration of estradiol did not elicit an LH surge in animals under conditions of constant light (Mennin and Gorski, 1975).

The nymphomanic cow has also been used as an animal model for the study of the pathophysiology of polycystic ovarian disease. The nymphomanic cow has signs of androgen excess associated with large polycystic ovaries and the disorder is x-linked transmitted (Garm, 1949; Roberts, 1955; Short, 1962). The administration of progestins or the mechanical rupture of the ovarian cysts produces estrus in the cow. Because of the size of the animals and the need for large numbers, the nymphomanic cow has not been studied extensively as a model for polycystic ovarian disease.

In 1958 Leathem observed that when female rats were rendered hypothyroid by the administration of thiouracil and then injected with human chorionic gonadotropin, large ovarian cysts were found. Although the use of this model for the study of the pathophysiology of the polycystic ovary syndrome has been suggested repeatedly, it has not been used extensively due to the presence of induced hypothyroidism.

The administration of testosterone propionate during the first 5 days of life to newborn female rats results in ovulatory failure and the formation of polycystic ovaries (Barraclough, 1961). This effect was considered to be due to masculinization of the hypothalamus resulting in a male pattern of tonic, rather than the female pattern of cyclic gonadotropin secretion (Barraclough and Gorski, 1961). The neonatally androgen-sterilized rat has been considered to be a model for the study of polycystic ovarian disease (Goldzieher and Axelrod, 1963). Important differences, however, are present between the human and the rat with respect to exposure to androgens during fetal life and the neonatal period. Patients with virilizing forms of congenital adrenal hyperplasia, who experience massive amounts of androgens in utero as well as during neonatal life, undergo puberty and ovulate normally once the excessive secretion of androgens in suppressed by corticosteroids (Greenblatt, 1958). This major difference raises the question of whether the neonatally androgen-sterilized rat is a suitable model for study.

The observations of Brawer et al. (1978, 1980) that a single injection of a long-acting estrogen, estradiol valerate, to young cycling rats caused persistent estrus, the presence of bilateral polycystic ovaries and multifocal lesions in the hypothalamic arcuate nucleus has yielded another animal model for the study of polycystic ovaries. The estradiol valerate treated animals have a decreased pituitary content and circulating levels of LH accompanied by a tonic elevation in plasma FSH (Schulster et al., 1984). Ovulation can be induced by the administration of LHRH although the amount of LH released is decreased (Hemmings et al., 1983). Unilateral ovariectomy results in the restoration of ovulatory cyclicity even though basal LH levels and the response to LHRH are still attenuated (Farooki et al., 1985). The response of the animals to unilateral ovariectomy is similar to what is observed in the human (Greenblatt, 1961).

The animal model which we have developed for the study of polycystic ovaries is based on the injection of weak androgens such as Δ^4-androstenedione and dehydroepiandrosterone (DHA). Both of these steroids are secreted in large quantities in the polycystic ovary syndrome.

DEHYDROEPIANDROSTERONE TREATED IMMATURE RAT

Mahesh and collaborators used injection of the weak androgen dehydroepiandrosterone (DHA) to immature rats to study the inter-relationship between excessive androgens and ovulatory failure. Since abnormalities of adrenal, ovarian or a hypothalamic-pituitary function have been considered among the causes of the polycystic ovary syndrome, it was important to utilize an animal model in which the organ functions were normal and the feedback mechanisms were undisturbed. The preliminary work of Roy et al. (1962) in the young adult rat showed that Δ^4-androstenedione and dehydroepiandrosterone administration could cause polycystic ovaries and ovulatory failure. Dehydroepiandrosterone was selected because (a) it is a quantitatively important secretory product of the adrenal and has also been isolated from ovarian tissue and ovarian vein blood in the polycystic ovary syndrome, (b) it is one of the first adrenal steroids which increases in the blood prior to the onset of puberty and has been reported to be elevated during the pubertal period in women who eventually develop polycystic ovaries, and (c) because of its weak androgenic activity, it provided flexibility in the dose ranges that could be employed.

After several doses of DHA were tried by Black and Mahesh (1969), the 60 mg/kg BW dose was selected because it gave the most consistent results. Administration of 60 mg/kg BW to 27-day-old female rats for 3 days resulted in a premature ovulation by day 31 in more than 90% of the animals treated whereas untreated animals failed to ovulate until day 36-38 of age (Fig. 1). The ovulation on day 31 of age was brought about by

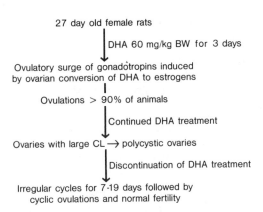

27 day old female rats

DHA 60 mg/kg BW for 3 days

Ovulatory surge of gonadotropins induced
by ovarian conversion of DHA to estrogens

Ovulations > 90% of animals

Continued DHA treatment

Ovaries with large CL ⟶ polycystic ovaries

Discontinuation of DHA treatment

Irregular cycles for 7-19 days followed by
cyclic ovulations and normal fertility

Fig. 1. An immature rat model for androgen induced ovulatory failure and
polycystic ovary formation.

an endogenous ovulatory surge of FSH and LH on day 30 (Knudsen and Mahesh,
1975; Knudsen et al., 1975).   This gonadotropin surge appeared to be
mediated by the ovarian conversion of DHA to estrogens because if such a
conversion was blocked by the administration of a 3β —hydroxysteroid
dehydrogenase antagonist, cyanoketone, the surge did not occur (Knudsen
and Mahesh, 1975). Furthermore, 5α—dihydrotestosterone, an androgen which
cannot be aromatized to estrogen, could not induce precocious ovulation
and vaginal patency.   The gonadotropin surge on day 30 was preceeded by a
rise in estradiol, followed by depletion of the cytoplasmic receptors of
the anterior pituitary and the hypothalamus due to the nuclear translocation (Parker and Mahesh, 1977).   These events were similar to those
found in the normal ovulatory cycle and during the natural onset of
puberty (Greeley et al., 1975; Parker and Mahesh, 1976b). Continued treatment with DHA resulted in slightly elevated levels of serum FSH, depressed
levels of serum LH and elevated levels of serum prolactin (Knudsen et al.,
1975).   The ovaries of DHA treated animals that initially showed large
corpora lutea became polycystic. Discontinuation of DHA treatment
resulted in irregular cycles for 7-19 days followed by the return of
cyclic ovulation and normal fertility.   During the period of DHA
treatment, the cytoplasmic estradiol receptors of the anterior pituitary
and the hypothalamus remained down-regulated. A gradual recovery of these
cytoplasmic receptors occurred after the discontinuation of DHA
administration and this recovery correlated well with the return of serum
LH levels to their normal values (Parker and Mahesh, 1976a).   Serum FSH
and LH levels were also normal during subsequent ovulatory surges after
DHA withdrawal (Knudsen et al., 1975).

Despite the fact that it would be difficult to compare the human and
the rat because of obvious species differences, there were many attractive
features in this animal model. The administration of androgens to the rat
with otherwise normal adrenal, ovarian, and hypothalamic-pituitary
function resulted in inappropriate gonadotropin secretion, ovulatory
failure, and the formation of polycystic ovaries. The effects of the
androgen insult were reversible because normal ovulation and cyclicity
were restored after androgen withdrawal. The animal model differed
from the human because in the rat model, FSH was slightly elevated and LH
was suppressed, as contrasted to elevated LH and low FSH in the human with
the polycystic ovary syndrome. However, it is difficult to equate LH in
the human, a hormone considered important for ovulation as well as the
maintenance of the corpus luteum, to LH in the rat where both LH and
prolactin are required for these functions. The animal model was
therefore of interest for further study of the interrelationships between

241

androgen insult, inappropriate gonadotropin secretion, and ovulatory failure.

DEHYDROEPIANDROSTERONE TREATED ADULT RAT

The DHA treated immature rat remains an interesting model of study of polycystic ovaries. However, two major questions can be raised regarding its use: First, the effective dose of DHA appeared to be large. Second, the effect of DHA in the immature rat may be related to the prepubertal state of the rat when treatment is started and may not be applicable to the adult rat. In an attempt to answer the first question, various doses of DHA were administered for 2 days to the ovariectomized adult rat, starting on the day of ovariectomy (Ward et al., 1978). Inadequate suppression of serum FSH and LH occurred when doses of 7.5 and 15 mg/kg BW of DHA were administered as compared to the gonadotropin levels found in diestrous controls (Fig. 2). The 30 mg/kg BW dose was the lowest dose that suppressed serum LH levels. Even at this dose serum FSH levels were slightly elevated.

Doses of DHA varying from 1.87 to 60 mg/kg BW were administered to adult cycling rats to determine whether DHA administration could induce constant estrus and polycystic ovaries (Ward et al., 1978). The results in Table 1 show that doses of DHA as low as 1.87 mg/kg BW were sufficient to disrupt the ovulatory cycle even through this small dose should not have any significant effect on gonadotropin secretion itself (see 7.5 mg/kg BW dose in Fig. 2). Polycystic ovaries were also found with all doses of DHA used. The possibility of ovarian effects of DHA in the disruption of cyclicity and the formation of polycystic ovaries is thus raised.

Table 1. Induction of Estrous Vaginal Smears by Administration of Various Doses of DHA to Female Rats

Treatment	Days of Estrus/Total Days of Treatment	Estrous Smears (%)
Control 4-day cycles	30/120	25.0
Control 5-day cycles	50/120	41.7
Vehicle-treated controls	29/60	48.3
DHA		
1.87 mg/kg BW	49/60	81.7
3.75 mg/kg BW	54/60	90.0
7.5 mg/kg BW	50/60	83.3
15 mg/kg BW	106/120	88.3
30 mg/kg BW	111/120	92.5
45 mg/kg BW	111/120	92.5
60 mg/kg BW	101/120	84.2

Further indirect evidence suggestive of ovarian effect of DHA comes from the fact that treatment of the adult rat with DHA doses ranging from 15 to 60 mg/kg BW yielded similar levels of circulating serum FSH, LH and prolactin (Fig. 3). Fig. 4 shows that using the 30 mg/kg BW dose of DHA, serum levels of FSH and LH were unchanged if the steroid was administered for 20 days, or 15 days (Mahesh, 1980). If ovariectomy was carried out 15 days after DHA treatment, serum FSH and LH levels increased dramatically within 5 days. This increase could not be suppressed

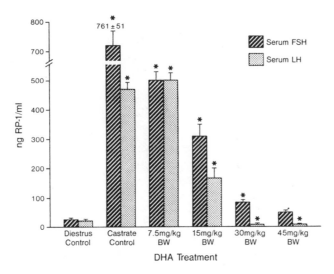

Fig. 2. Effect of DHA treatment on the suppression of the postcastration rise of gonadotropins. Adult cycling rats were ovariectomized and administered various doses of DHA for 20 days. Note lack of suppression of serum FSH and LH levels as compared to diestrus controls with 7.5 and 15 mg/kg BW of DHA. With the 30 mg/kg BW dose, serum FSH levels were elevated and serum LH levels were suppressed. Serum FSH levels were comparable to diestrus controls and serum LH levels were suppressed when the 45 mg/kg BW dose of DHA was used. *Represents $P < 0.05$ as compared to diestrus controls. (Data from Ward et al., 1978).

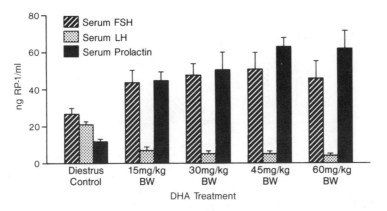

Fig. 3. Effect of various doses of DHA on serum FSH, LH and prolactin levels. Note similar values regardless of the dose of DHA used. (Data from Ward et al., 1978).

adequately even if DHA administration was continued after ovariectomy (Fig. 4). The active involvement of the ovary in steroidogenesis was confirmed by administering 30 mg/kg BW of DHA for 15 days to adult rats followed by ovariectomy in some of the animals. Five days after the discontinuation of DHA administration, the intact controls showed significantly higher levels of serum progesterone, 17α-hydroxyprogesterone, DHA, Δ^4-androstenedione, testosterone, 5α-dihydrotestosterone, and estradiol as compared to the ovariectomized rat (Fig. 5). These results support an active role of the ovary in steroid secretion in the DHA treated rats and raise the possibility of direct androgen effects on the ovary leading to

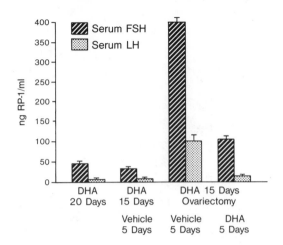

Fig. 4.   Effect of 30 mg/kg BW of DHA on serum FSH and LH after adminis-
          tration for 20 days or 15 days to adult rats.   The serum gonado-
          tropins that were unchanged by withdrawal of DHA administration
          for 5 days showed a dramatic increase if ovariectomy was carried
          out.   The serum FSH and LH could not be suppressed to the levels
          found in DHA treated intact rats even if DHA administration was
          continued after ovariectomy.

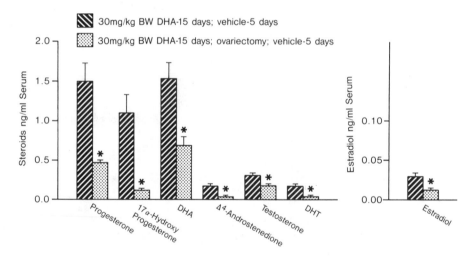

Fig. 5.   Evidence of ovarian production of a variety of steroids in DHA
          induced polycystic ovaries is shown by a remarkable decrease in 7
          blood steroids, 5 days after ovariectomy and withdrawal of DHA
          treatment in ovariectomized rats as compared to intact controls.
          *Represents $P < 0.01$ as compared to non-ovariectomized controls.

altered steroidogenesis which in turn results in altered pituitary gonadotropin secretion and the formation of polycystic ovaries.

DIRECT ANDROGEN EFFECTS ON THE OVARY

     A survey of literature reveals that androgens can both inhibit and
stimulate the growth and development of follicles and certain granulosa
cell functions.   Androgens have been shown to cause follicular atresia in

intact (Payne et al., 1956; Beyer et al., 1974) and hypophysectomized rats (Hillier et al., 1979). An androgen receptor in the granulosa cells of estrogen treated hypophysectomized rats has also been identified (Schrieber et al., 1976; Schrieber and Ross, 1976). The increase in atresia which follows treatment with low doses of LH or hCG in estrogen primed hypophysectomized rats is considered to be due to increased androgen synthesis by the ovary due to gonadotropin stimulation (Louvett et al., 1975a,b). The possible mechanisms of androgen effects on follicular atresia could be the inhibition of ovarian aromatase (Hillier et al., 1980) or an inhibition of the induction of LH receptors by FSH (Farooki, 1980, 1985).

The facilitative effects of androgens on ovarian steroidogenesis and ovulation have also been reported. The synthesis of estrogens by granulosa cells in culture was enhanced by androgens and inhibited by antiandrogens (Hillier and Zwart, 1981). Androgens enhanced granulosa cell progesterone secretion (Quick and Fortune, 1983) and the FSH stimulation of cAMP synthesis (Daniel, 1983). Furthermore, testosterone or 5α–dihydrotestosterone injected 1 h before the injection of hCG to induce ovulation, overcame the inhibition of ovulation caused by anti-bodies to progesterone (Mori et al., 1977). Besides the facilitative effects on steroidogenesis and ovulation, androgens serve as an important source of precursors for ovarian steroidogenesis. Therefore, a detailed study of androgen effects on the ovary was undertaken in vivo.

MODELS FOR THE STUDY OF ANDROGEN EFFECTS ON THE OVARY

Since studies of the effect of androgens on the ovary in vivo are difficult because androgen treatment can alter gonadotropin secretion and thereby alter ovarian steroidogenesis, hypophysectomized rats were used by Bagnell et al. (1982). Furthermore, to eliminate any possible effects of degenerating corpora lutea on follicular growth, immature rats were used. The model consisted of female rats hypophysectomized on day 24 of age which were given an injection of 32 IU of pregnant mare's serum gonado-tropins (PMSG) on day 30 of age. An injection of 10 IU of hCG, given 54 h after PMSG, caused ovulation and resulted in an average of 14.9 ± 1.2 ova per ovulating rat as compared to 9.6 ± 1.4 ova per ovulating rat in intact animals. Thus this model, described in Fig. 6, lends itself to the study of the effects of the non-aromatizable androgen 5 α –dihydro-testosterone (DHT) on follicular atresia and ovulation. The administration of 0.25 to 4 mg/kg BW of DHT as a single injection 30 h after PMSG administration resulted in a decrease in the number of primary, secondary and tertiary follicles and an increase in atretic follicles (Fig. 7). The ovulation rate after 1 injection of DHT (1 mg/kg BW, 30 h after PMSG) or 2

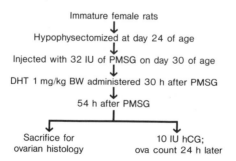

Fig. 6. A hypophysectomized immature female rat model to study the effect of 5α–DHT, a non-aromatizable androgen on follicular histology and ovulation.

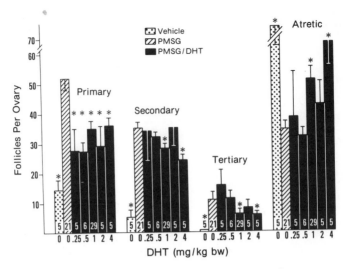

Fig. 7. Effect of DHT on numbers of primary, secondary, tertiary, and atretic ovarian follicles from PMSG primed hypophysectomized immature female rats. Rats hypophysectomized on day 24 were treated with PMSG and DHT according to the scheme shown in Fig. 6. Numbers inside each bar represent the number of animals from which a single ovary was examined. *Represents significant difference from PMSG control (P < 0.05) (From Bagnell et al., 1982).

injections of DHT (given at the time of PMSG administration and 30 h later) resulted in a marked decrease in the ovulation rate when 10 IU of hCG were administered (Table 2). It is, however, important to note that DHT treatment did not influence all the follicles to the same degree and several preantral follicles were able to grow and ovulate in spite of the androgen insult.

Estrogens have a protective effect on follicles from undergoing atresia and diethylstilbestrol treatment has been used frequently to induce follicular growth (Louvet et al., 1975a, 1985b; Lucky et al., 1981; Hillier and DeZwart, 1981). The measurement of androgens and estrogens in normal and atretic follicles shows that atretic follicles commonly have a lowered estrogen content (McNatty et al., 1979a,b). Atretogenic effects of estrogens on follicular atresia after prolonged treatment have also been reported (Hutz et al., 1986; Sadrkhanloo et al., 1987).

In the PMSG-primed immature rat treated with a single injection of DHT, it was thought that a high androgen to estrogen (A/E) ratio could be responsible for follicular atresia. In order to increase follicular estrogen, estradiol was injected in doses of 0.01 to 2 mg/kg BW to PMSG-primed hypophysectomized immature rats. There were no effects on follicular histology (Bagnell et al., 1982). However, the ovulation rate after hCG administration was increased in animals treated with the 2 mg/kg BW dose (Table 2). The absence of an effect of 1 mg/kg BW of estradiol on follicular growth and ovulation was in all probability due to high follicular estrogen caused by the PMSG-treatment, and the fact that only a small amount of injected estrogen enters the follicle. When 1 and 2 mg/kg BW doses of estradiol were administered to the PMSG-primed hypophys-ectomized rats also given a single injection of DHT, the DHT induced decline in healthy follicles, and increase in the number of atretic

246

Table 2. Effects of DHT and Estradiol on Ovulation Induced by hCG in PMSG-Primed Hypophysectomized Immature Rats

Treatment	# of ova per ovulating rat
PMSG (32 IU)	14.9 + 1.2
PMSG/DHT	7.8 + 1.2*
PMSG/DHT (2 injections)	3.2 + 1.4*
PMSG/E$_2$ (1 mg/kg BW)	13.4 + 2.9
PMSG/E$_2$ (2 mg/kg BW)	28.0 + 7.7*
PMSG/DHT/E$_2$ (1 mg/kg BW)	20.5 + 3.3**
PMSG/DHT/E$_2$ (2 mg/kg BW)	23.4 + 2.6**

*Significantly differs from PMSG control
**Significantly differs from PMSG/DHT

follicles was arrested (Fig. 8). The DHT induced decline in the number of ova shed was also reversed by estradiol administration (Table 2). These experiments suggest that by increasing follicular estrogen and/or by decreasing the A/E ratio, the atresia inducing effects of androgens can be prevented.

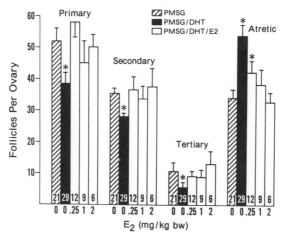

Fig. 8. Effect of estradiol on numbers of primary, secondary, tertiary, and atretic follicles from PMSG/DHT treated hypophysectomized immature female rats. Rats were given PMSG (32 IU) on day 30 and the designated dose of estradiol or vehicle. A second injection of estradiol was given 30 h later at the time of DHT (1 mg/kg BW) treatment. Animals were sacrificed 24 h after the injection of of DHT. *Represents significant difference from PMSG control (P < 0.05). (From Bagnell et al., 1982).

As the supplies of the PMSG used in the above experiments was depleted, a second batch of PMSG, referred to as PMSG II, was obtained for further studies. When PMSG II was administered to hypophysectomized immature female rats, a lower number of secondary follicles was found in

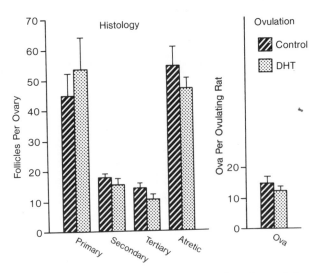

Fig. 9. Effect of a second preparation of PMSG referred to as PMSG II on follicular histology and on ova shed after hCG with or without the administration of DHT. Note the absence of any effects of DHT on follicular histology and ovulation.

the ovaries and DHT no longer decreased the number of ovarian follicles or increased the rate of atresia (Fig. 9). The number of ova shed was also not altered by DHT when 10 IU of hCG was administered. When PMSG-II was used in conjunction with either estradiol (2 mg/kg BW) or two injections of the antiandrogen flutamide (10 mg/kg BW) 6 and 30 h after PMSG-II administration, there was a remarkable increase in the number of primary and secondary follicles (Fig. 10). The increase was similar irrespective of whether the antiandrogen flutamide or estradiol was used. The experiments with PMSG II indicated that this preparation of PMSG probably contained a greater amount of LH-like activity, thus producing a greater quantity of endogenous androgens. Increasing the amount of estrogen in the follicles by the injection of estradiol or decreasing the androgen effects by the use of the flutamide protected against the androgen effect and permitted follicular growth.

The study of follicular growth and follicular atresia is rendered difficult by the variety of follicles present in the ovary at any one time. The atretogenic effects of androgens can be best studied if the ovary contains a large number of follicles sensitive to androgens. Although the hypophysectomized rat is an excellent model for study of ovarian folliculogenesis, its usefulness is limited by the availability of large quantities of FSH and FSH-like preparations with very low or absent LH contaminants that can be used from experiment to experiment without introduction of variability associated with the batch of FSH used. It was, therefore, considered desirable to identify the type of follicle most susceptable to androgen induced atresia and design an animal model for producing such follicles.

It is generally accepted that in the mammalian ovary, the vast majority of follicles which enter into the pool of growing follicles are ultimately lost to atresia and that FSH is able to rescue a few follicles which then go on to ovulation (Tsifriri and Braw, 1984). Furthermore, the atresia rates are not the same for follicles of different sizes based on

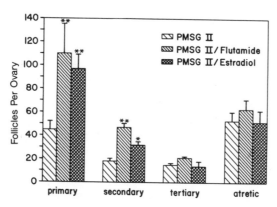

Fig. 10. Effects of flutamide and estradiol on the number of primary, secondary, tertiary and atretic follicles from PMSG II primed hypophysectomized immature female rats. Rats hypophysectomized on day 24 were given PMSG II (20 IU) and either propylene glycol or estradiol (2 mg/kg BW) on day 30. Flutamide (10 mg/kg BW) was given 6 h after PMSG II to a non estrogen treated group. Thirty h after PMSG II, a second injection of flutamide or estradiol or vehicle was given. Animals were sacrificed 24 h after the last injection. *Represents P < 0.05; **Represents P < 0.01.

the classification of Hirshfield and Midgley (1978a,b). Follicles > 500 mμ in diameter are classified as large and are likely preovulatory; these follicles show a low rate of atresia. Follicles of diameters of 200-500 mμ are medium sized, and if < 200 mμ, the follicles are small. Although atresia is not common among the large follicles, the frequency of atresia among the follicles of 200-400 mμ diameter is high (50%). Hirshfield and Midgley (1978a,b) also reported that during the cycle, follicles advanced through the sizes in a very orderly manner with large antral follicles present in the ovary only late in the cycle (on days of diestrus II or proestrus). They further reported that there are no large follicles present on the day of estrus and that medium follicles grow to large size over the 5 days of the cycle.

As healthy, nonatretic follicles grow, their estrogen synthetic capabilities become greater such that the large follicles present in the ovaries are the likely source of the majority of the estrogen produced. Smaller follicles contribute proportionally smaller amounts of estrogen. The high levels of estrogen produced by the follicles likely reflects high activity of the aromatase enzyme system in these follicles. The aromatase system has been reported by several investigators to be highly sensitive to the inhibitory effects of androgens (Hillier et al., 1980a,b).

Taken together, these results can be used to formulate a proposed mechanism for the induction of atresia by treatment with androgens. We postulate that the A/E ratio ultimately dictates the fate of follicles. When the A/E ratio is high, the granulosa cell aromatase activity is inhibited and the follicle will likely degenerate. A low A/E ratio would indicate high aromatase activity in a nonatretic or "healthy" follicle. This explanation can account for the observation that when exogenous testosterone or DHT is administered to a rat, the small follicles are most affected by the androgen insult. In large follicles with the high endogenous estrogen content and synthetic capabilities, androgen treatment would not elevate the A/E ratio above the level which leads to atresia. However, in small follicles, with their limited estrogen synthetic capacity, even small increases in follicular androgen will elevate the ratio above the critical level at which the follicle becomes atretic. This hypothesis also can be used to explain why estrogen treatment protects follicles from the adverse effects of androgen treatment. When estrogen in the follicles is elevated by exogenous administration, the A/E ratio falls and fewer follicles are lost to degenerative changes. Anti-androgens such as flutamide favor follicular growth by decreasing androgen action on ovarian tissue and effectively lowering the A/E ratio.

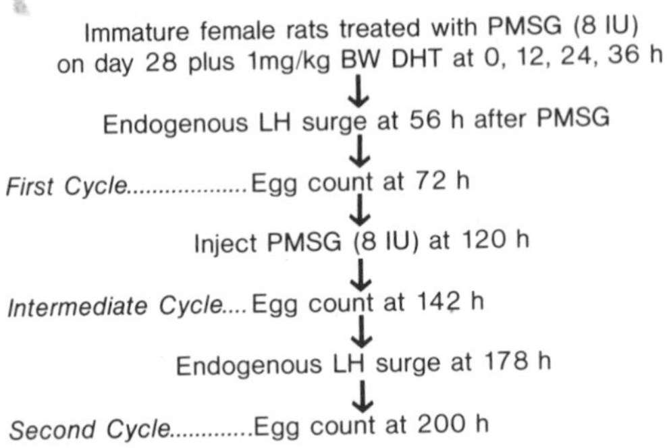

Immature female rats treated with PMSG (8 IU)
on day 28 plus 1mg/kg BW DHT at 0, 12, 24, 36 h
↓
Endogenous LH surge at 56 h after PMSG
↓
First Cycle.................Egg count at 72 h
↓
Inject PMSG (8 IU) at 120 h
↓
Intermediate Cycle....Egg count at 142 h
↓
Endogenous LH surge at 178 h
↓
Second Cycle............Egg count at 200 h

Fig. 11.   Schematic design to evaluate the effect of DHT on two rat cycles, each of which is synchronized by the administration of 8 IU of PMSG.

The results of the study with the two batches of PMSG, along with the proposed mechanism, indicate that it is the small and medium follicles that are most susceptible to androgen induced atresia. This hypothesis was verified by setting up a new animal model. In this model, folliculogenesis was stimulated in immature rats by a single injection of PMSG (8 IU) on day 28 of age, and 4 injections of DHT (1 mg/kg BW) were administered at 0, 12, 24 and 36 h after PMSG (Fig. 10). An endogenous gonadotropin surge occurred 56 h after PMSG. In this surge, the serum FSH levels were comparable in the DHT treated and control groups. The serum LH levels showed a small decrease in the DHT treated group that was not significant. The number of ova per ovulating rat, however, were significantly lowered by the DHT treatment (Cycle 1; Fig. 12). The administration of a second injection of 8 IU of PMSG at 120 h after the first injection brought about the release of ova from follicles that could be ovulated in the absence of an endogenous gonadotropin surge (Intermediate cycle; Fig. 11). A reduced number of ova were released in the control rats while there were no ovulations in the androgen treated rats. The second injection of PMSG induced an endogenous surge of gonado-tropins at 178 h (Cycle 2; Fig. 11). The number of ova shed in the DHT treated animals was likewise reduced as compared to nonandrogen treated controls (Fig. 12).

It is clear from the preceeding results that DHT must exert an inhibitory influence at three different sites in the folliculogenic process. First, DHT apparently inhibits the final developmental stages of follicles which will ovulate less than 36 h after the last DHT injection (Cycle 1). Also affected by the androgen are follicles which are likely preantral at the time of the DHT treatment and would ovulate only in the Intermediate cycle. Based on the kinetic studies of Hirshfield and Midgley (1978a), these follicles must be in the range of 200–400 mµ at the time of the 4 DHT injections. These authors have reported that follicles in that range show a high rate of atresia and may therefore be highly sensitive to an atretogenic factor such as DHT. The third group of

250

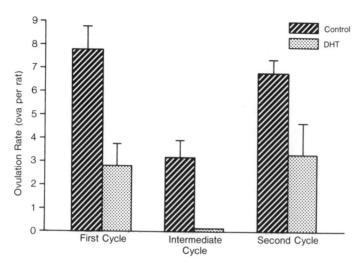

Fig. 12. Ovulation rate in PMSG treated and PMSG/DHT treated rats during the first, intermediate and second cycle as defined in Fig. 11. Note the decrease in ova in DHT treated rats in the first and second cycle and a complete absence in the intermediate cycle.

follicles which is adversely affected by the DHT treatment are probably less than 200 mμ during the androgen treatment. These are the follicles which will grow over the next 8 days to ovulate in Cycle 2.

SUMMARY

In the human, polycystic ovaries are generally accompanied by normal or elevated levels of serum LH, normal or slightly depressed FSH and by high levels of circulating estrogens and androgens. If the excess androgen secretion is reduced by one of several methods, ovulatory cycles are usually restored. Several animal model systems have been proposed for the study of the pathophysiology of the polycystic ovarian syndrome. These include neonatal androgenization, hCG administration to hypothyroid rats, injection of estradiol valerate and maintaining animals in constant light. In a model developed in this laboratory, pubertal or adult rats were treated with the weak androgen, dehydroepiandrosterone (DHA), to induce polycystic ovaries. This treatment also altered the blood levels of LH and FSH but the effect on gonadotropins and on the formation of the degenerative follicles was fully reversed following discontinuation of the androgen injections. The polycystic ovaries of the DHA-treated animals were steroidogenically more active than controls raising the possibility that the DHA was acting directly on the ovary in addition to an action on the pituitary-hypothalamus axis. In order to study the direct effect of androgens on the ovary, another animal model was developed in which immature,, hypophysectomized rats were injected with pregnant mare serum gonadotropin (PMSG) to initiate follicular growth followed by a single injection of dihydrotestosterone (DHT). The androgen caused follicular atresia and decreased the number of ova shed in response to ovulation induction with hCG. The suppressive effects of DHT were entirely prevented by concomitant treatment with estradiol.

The studies with DHT were continued using another batch of PMSG, but the DHT-induced increase in the rate of atresia and suppression of induced ovulation were no longer seen. However, when this same batch of PMSG was given with estrogen or with the antiandrogen flutamide, there was less atresia and the growth of follicles was actually enhanced. Based on these studies, it was postulated that the second batch of PMSG had greater LH activity than the first preparation and that the LH had stimulated endogenous androgen production. The ovarian follicles which appeared to be most susceptible to this DHT effect were small to medium in size and had a low capacity to synthesize estrogen. This possibility was confirmed in another animal model system in which immature rats were injected with PMSG and 4 separate injections of DHT and then sacrificed at several time points over the next 8 days. While the number of ova shed in the first ovulation after the androgen treatment was suppressed, a second gonadotropin-induced ovulation was completely blocked in the DHT injected animals and a subsequent, spontaneous ovulation was attenuated in the androgen-treated rats as well.

ACKNOWLEDGEMENTS

This investigation was supported by research grant HD 16396 from the National Institute of Child Health and Human Development, National Institutes of Health, Bethesda, MD 20205.

REFERENCES

Bagnell CA, Mills TM, Costoff A, Mahesh VB, 1982. A model for the study of androgen effects on follicular atresia and ovulation. Biol Reprod 27:903-914

Baird DT, Corker DT, Davidson DW, Hunter HM, Michie EA and Van Look PFA, 1977. Pituitary ovarian relationships in polycystic ovary syndrome. J Clin Endocrinol Metab 45:798-809

Barraclough CA, 1961. Production of anovulatory, sterile rats by single injections of testosterone propionate. Endocrinology 68:62-67

Barraclough CA, Gorski RA, 1961. Evidence that the hypothalamus is responsible for androgen induced sterility in the female rat. Endocrinology 68:68-79

Baulieu EE, Mauvais-Jarvia P, Corpechot C, 1963. Steroid studies in a case of Stein-Leventhal syndrome with hirsutism. J Clin Endocrinol Metab 23:374-382

Berger MJ, Taymor ML, Patton WC, 1975. Gonadotropin levels and secretory patterns with typical and atypical polycystic ovarian disease. Fertil Steril 26:619-626

Beyer C, Cruz M, Gay V, Jaffe R, 1974. Effects of testosterone and dHT on FSH serum concentration and follicular growth in female rats. Endocrinology 95:722-727

Black JB, Mahesh VB, 1969. Effects of dehydroepiandrosterone on pituitary gonadotropins and ovulation in the rat. Fed Proc 28:506

Bradshaw M, Critchlow V, 1965. Pituitary concentration of LH in three types of constant estrous rats. Endocrinology 78:1007-1014

Brawer JR, Naftolin F, Martin J, Sonnenschein C, 1978. Effects of a single injection of estradiol valerate on the hypothalamic arcuate nucleus and on reproductive function in the female rat. Endocrinology 103:501-512

Brawer JR, Schipper H, Naftolin F, 1980. Ovary-dependent degeneration in the hypothalamic arcuate nucleus. Endocrinology 107:274-279

Brooks RV, Prunity FTG, 1960. Patterns of steroid excretion in three types of post-pubertal hirsutism. J Endocrinol 21:263-276

Browman LG, 1937. Light in its relation to activity and estrous rhythms in the albino rat. J Exp Zool 75(S):375-388

Chang RJ, Mander FP, Lu JKH, Judd HL, 1982. Enhanced disparity of gonadotropin secretion by estrone in women with polycystic ovarian disease. J Clin Endocrinol Metab 54:490-494

Daane TA, Parlow AF, 1971. Serum FSH and LH in constant light-induced persistent estrus: short-term and long-term studies. Endocrinology 88:964-968

Daniel S, 1983. Androgens enhance responsiveness of cultured rat granulosa cells to follicle-stimulating hormone. Abstract 103, 16th Annual Meeting of the Society for the Study of Reproduction, Cleveland.

De Vane GW, Czekala NM, Judd HL, Yen SSC, 1975. Circulating gonadotropins, estrogens and androgens in polycystic ovarian disease. Am J Obstet Gynecol 121:496-500

Duignan NM, Shaw RW, Rudd BT, Holder G, Williams JW, Butt WR, Logan Edwards R, London DR, 1975. Sex hormone levels and gonadotropin release in the polycystic ovary syndrome. Clin Endocrinol 4:287-295

Farookhi R, 1980. Effect of androgens on induction of gonadotropin receptors and gonadotropin stimulated adenosine 3',5'-monophosphate production in rat ovarian granulosa cells. Endocrinology 106:1216-1223

Farookhi R, 1985. Effects of aromatizable and nonaromatizable androgen treatment on luteinizing hormone receptors and ovulation induction in immature rats. Biol Reprod 33:363-369

Farookhi R, Hemmings R, Brawer JR, 1985. Unilateral ovariectomy restores ovulatory cyclicity in rats with a polycystic ovary condition. Biol Reprod 32:530-540

Gallagher TF, Kappas A, Hellman L, Lipsett MB, Pearson OH, West CD, 1958. Adrenal hyperfunction in idiopathic hirsutism and Stein-Leventhal syndrome. J Clin Invest 37:794-799

Gambrell RD, Greenblatt RB, Mahesh VB, 1971. Serum gonadotropin levels and ancillary studies in Stein-Leventhal syndrome treated with clomiphene citrate. Obstet Gynecol 38:850-862

Gambrell RD, Greenblatt RB, Mahesh VB, 1973. Inappropriate secretion of LH in the Stein-Leventhal syndrome. Obstet Gynecol 42:429-440

Garm O, 1949. A study on bovine nymphomania with special reference to etiology and pathogenesis. Acta Endocrinol Suppl 3:1

Given JR, Anderson RN, Ragland JB, Wiser WL, Umstot ES, 1975. Adrenal function in hirsutism. I. Diurnal changes and response of plasma androstenedione, testosterone, 17α-hydroxyprogesterone, cortisol, LH and FSH to dexamethasone and ACTH. J Clin Endocrinol Metab 40:988-1000

Goldzieher JW, 1981. Polycystic ovarian disease. Fertil Steril 25:371-394

Goldzieher JW, Axelrod LR, 1963. Clinical and biochemical features of polycystic ovarian disease. Fertil Steril 14:631-653

Goldzieher JW, Green JA, 1962. The polycystic ovary. I. Clinical and histological features. J Clin Endocrinol Metab 22:325-338

Greeley GH, Muldoon TG, Mahesh VB, 1975. Correlative aspects of LHRH sensitivity and cytoplasmic estrogen receptor concentration in the anterior pituitary and the hypothalamus of the cycling rat. Biol Reprod 13:505-512

Greenblatt RB, 1953. Cortisone in the treatment of the hirsute woman. Am J Obstet Gynecol 66:700-710

Greenblatt RB, 1958. Clinical aspects of sexual abnormalities in man. Recent Prog Horm Res 14:335-404

Greenblatt RB, 1961. The polycystic ovary syndrome. Md State Med J 10:120

Greenblatt RB, Mahesh VB, 1976. The androgenic polycystic ovary. Am J Obstet Gynecol 125:712-726

Hemmings R, Farookhi R, Brawer JR, 1983. Pituitary and ovarian responses to LHRH in a rat with polycystic ovaries. Biol Reprod 29:329-248

Hemmingsen AM, Krarup NB, Axel M, 1937. Rhythmic diurnal variations in the oestrous phenomena of the rat and their susceptibility to light and dark. Kgl Danske Videnskab Selskab Biol Medd 13:1-16

Hillier S, DeZwart F, 1981. Evidence that granulosa cell aromatase induction/activation by FSH is an androgen receptor-regulated process in vitro. Endocrinology 109:1303-1305

Hillier SG, Ross GT, 1979. Effects of exogenous testosterone on ovarian weight, follicular morphology and intraovarian progesterone concentration in estrogen-primed hypophysectomized immature female rats. Biol of Reprod 20:261-268

Hillier SG, van den Boogaard A, Reichert LE, van Hall EV, 1980. Alterations in granulosa cell aromatase activity accompanying preovulatory follicular development in the rat ovary with evidence that 5α reduced androgenic steroids inhibit the aromatase reaction in vitro. J Endocrinology 84:409-419

Hirshfield AN, Midgley AR, 1978a. Morphometric analysis of follicular development in the rat. Biol Reprod 19:597-605

Hirshfield AN, Midgley AR, 1978b. The role of FSH in the selection of large ovarian follicles in the rat. Biol Reprod 19:606-611

Hutz RJ, Dierschke DJ, Wolf RC, 1986. Markers of atresia in ovarian follicular components from Rhesus monkeys treated with estradiol-17β. Biol Reprod 34:65-70

Ikkos DG, Dellia-Sfikaki A, 1975. Plasma steroids in hirsutism. Obstet Gynecol 46:114 (Letter to editor)

Ingersoll RM, McDermott WB, 1950. Bilateral polycystic ovaries, Stein-Leventhal syndrome. Am J Obstet Gynecol 60:117-125

Jaffe RB, Keye WR, 1974. Estradiol augmentation of pituitary responsiveness to gonadotropin releasing hormone in women. J Clin Endocrinol Metab 39:850-855

Kandeel FR, Butt WR, London DR, Lynch SS, Logan Edwards R, Rudd BT, 1978. Oestrogen amplification of LHRH response to clomiphene. Clin Endocrinol 9:429-441

Kirschner MA, Bardin CW, 1972. Androgen production and metabolism in normal and virilized women. Metabolism 21:667-688

Knudsen JF, Mahesh VB, 1975. Initiation of precocious sexual maturation in the immature rat treated with dehydroepiandrosterone. Endocrinology 97:458-468

Knudsen JR, Costoff A, Mahesh VB, 1975. Dehydroepiandrosterone induced polycystic ovaries and acyclicity in the rat. Fertil Steril 26:807-817

Leathem JH, 1958. Hormonal influences on the gonadotropin sensitive hypothyroid rat ovary. Anat Rec 131:487-499

Lipsett MB, Riter B, 1960. Urinary ketosteroids and pregnanetriol in hirsutism. J Clin Endocrinol Metab 20:180-186

Lobo RA, Goebelsmann U, 1980. Adult manifestation of congenital adrenal hyperplasia due to incomplete 21-hydroxylase deficiency minicking polycystic ovarian disease. Am J Obstet Gynecol 138:720-726

Lobo RA, Goebelsmann U, 1981. Evidence for reduced 3β-ol hydroxysteroid dehydrogenase activity in some hirsute women thought to have polycystic ovary syndrome. J Clin Endocrinol Metab 53:394-400

Lobo RA, Goebelsmann U, 1982. Effect of androgen excess on inappropriate gonadotropin secretion as found in the polycystic ovary syndrome. Am J Obstet Gynecol 142:394-401

Louvet JP, Harman SM, Ross GT, 1975a. Effects of hCG, hICSH and hFSH on ovarian weights in estrogen primed hypophysectomized immature female rats. Endocrinology 96:1179-1186

Louvet JP, Harman SM, Schreiber JR, Ross GT, 1975b. Evidence for a role of androgens in follicular maturation. Endocrinology 97:366-372

Lucky A, Schreiber J, Hillier S, Schulman J, Ross G, 1977. Progesterone production by cultured preantral rat granulosa cells: stimulation by androgens. Endocrinology 100:128-133

Mahajan DK, Shah PN, Eik-Nes KB, 1963. Steroids in cyst fluid obtained from polycystic ovaries in hirsute women. J Obstet Gynecol Br Commonw 70:8-12

Mahesh VB, 1980. Current concepts of the pathophysiology of the poly-cystic ovary syndrome. In: Tozzine RI, Reeves G, Pineda RL (eds.), Endocrine Physiopathology of the Ovary. Amsterdam: Elsevier/North-Holland Biomedical Press, pp. 275-295

Mahesh VB, 1983. Various concepts of pathogenesis of polycystic ovarian disease. In: Mahesh VB, Greenblatt RB (eds.), Hirsutism and Virilism. Boston: John Wright, PSG, Inc., pp. 247-276

Mahesh VB, 1984. Hormone secretion and steroidogenesis in the polycystic ovary syndrome. In: Fotherby K, Pal SB (eds.), Steroid Converting Enzymes and Diseases. Berlin-New York: Walter de Gruyter & Co., pp. 97-145

Mahesh VB, Greenblatt RB, 1961. Physiology and pathogenesis of the Stein-Leventhal syndrome. Nature 191:888-890

Mahesh VB, Greenblatt RB, 1962. Isolation of dehydroepiandrosterone and 17α -hydroxy-Δ^5-pregnenolone from the polycystic ovaries of Stein-Leventhal syndrome. J Clin Endocrinol Metab 22:441-448

Mahesh VB, Greenblatt RB, 1964a. Steroid secretions in the normal and polycystic ovary. Recent Prog Horm Res 20:341-394

Mahesh VB, Greenblatt RB, 1964b. Urinary steroid excretion patterns in hirsutism. II. J Clin Endocrinol Metab 24:1293-1302

Mahesh VB, Greenblatt RB, Scholer HFL, Ellegood JO, 1970. Steroid and gonadotropin secretion in the polycystic ovary syndrome. Excerpta Med Int Congr Ser 238:160-167

Mahesh VB, Greenblatt RB, Aydar CK, Roy S, 1962. Secretion of androgens by the polycystic ovary and its significance. Fertil Steril 13:513-530

Mahesh VB, Greenblatt RB, Aydar CK, Roy S, Puebla RA, Ellegood JO, 1964. Urinary steroid patterns in hirsutism. I. J Clin Endocrinol Metab 24:1283-1292

Maric DK, Matsuyama E, Lloyd CW, 1965. Gonadotropin content of pituitaries of rats in constant estrus induced by continuous illumination. Endocrinology 77:529-536

McArthur JW, Ingersoll RM, Worcester J, 1958. The urinary excretion of ICSH and FSH by women with diseases of the reproductive system. J Clin Endocrinol Metab 18:1202-1215

McNatty KP, Markis A, DeGrazia C, Osathanondh R, Ryan K, 1979a. The pro-duction of progesterone, androgens and estrogens by granulosa cells, thecal tissue and stromal tissue from human ovaries in vitro. J Clin Endocrinol Metab 49:687-699

McNatty KP, Makris A, DeGrazia C, Osathanondh R, Ryan K, 1979b. The pro-duction of progesterone, androgens and estrogens by human granulosa cells in vitro and in vivo. J Steroid Biochem 11:775-779

Mennin SP, Gorski RA, 1975. Effects of ovarian steroids on plasma LH in normal and persistent estrous adult female rats. Endocrinology 96:488-491

Moltz L, Rommler A, Schwartz V, Diblngmaier F, Hammerstein J, 1979. Peri-pheral steroid-gonadotropin interactions and diagnostic significance of double stimulation tests with LHRH in polycystic ovarian disease. Am J Obstet Gynecol 134:813-818

Mori T, Suzuki A, Nishimura T, Kambegawa A, 1977. Evidence for androgen participation in induced ovulation in immature rats. Endocrinology 101:623-626

Oettinger V, Paroudakis A, Mahesh VB, Greenblatt RB, 1975. Effect of LHRH stimulation in anovulatory women. Obstet Gynecol 45:614-618

Parker CR, Mahesh VB, 1976a. Interrelationship between excessive levels of circulating androgens in blood and ovulatory failure. J Reprod Med 17:75-90

Parker CR, Mahesh VB, 1976b. Hormonal events surrounding the natural onset of puberty in female rats. Biol Reprod 14:347-353

Parker CR, Mahesh VB, 1977. Dehydroepiandrosterone induced precocious ovulation: Correlative changes in blood steroids, gonadotropins and cytosol estradiol receptors of the anterior pituitary gland and the hypothalamus. J Steroid Biochem 8:173–177

Patton WC, Berger MJ, Thompson IE, Cheng AP, Grimes EM, Taymore ML, 1975. Pituitary gonadotropin responses to synthetic LHRH in patients with typical and atypical polycystic ovarian disease. Am J Obstet Gynecol 121:382–386

Payne RW, Hellbaum AA, Owens JN Jr, 1956. The effects of androgens on the ovaries and uterus of estrogen treated hypophysectomized immature female rat. Endocrinology 59:306–316

Perloff WH, Hadd HE, Channick BJ, Nodine JH, 1957. Hirsutism. Arch Intern Med 100:981–985

Quick S, Fortune J, 1983. Progesterone secretion by rat granulosa cells: The development of response to FSH, LH and testosterone. Abstract 114, 16th Annual Meeting of the Society for the Study of Reproduction, Cleveland

Rebar R, Judd HL, Yen SSC, Rakoff J, Vandenberg G, Naftolin F, 1976. Characterization of the inappropriate gonadotropin secretion in polycystic ovary syndrome. J Clin Invest 57:1320–1329

Roberts SJ, 1955. Clinical observations on cystic ovaries in diary cattle. Cornell Veterinarian 45:497–513

Roy S, Mahesh VB, Greenblatt RB, 1962. Effect of dehydroepiandrosterone and Δ^4-androstenedione on the reproductive organs of female rats. Production of cystic changes in the ovary. Nature 196:42–43

Sadrkhanloo R, Hofeditz C, Erickson GF, 1987. Evidence for wide-spread atresia in the hypophysectomized estrogen-treated rat. Endocrinol 120:146–155

Schreiber JR, Reid R, Ross GT, 1976. A receptor-like testosterone binding protein in ovaries from estrogen stimulated hypophysectomized immature female rats. Endocrinology 98:1206–1213

Schreiber JR, Ross GT, 1976. Further characterization of a rat ovarian testosterone receptor with evidence for nuclear translocation. Endocrinology 99:590–596

Schulster A, Farookhi R, Brawer JR, 1984. Polycystic ovarian condition in estradiol valerate treated rats: spontaneous changes in characteristic endocrine features. Biol Reprod 31:587–593

Shearman RP, Cox RI, 1966. The enigmatic polycystic ovary. Obstet Gynecol Surv 24:1–33

Short RV, 1962. Steroid concentrations in normal follicular fluid and ovarian cyst fluid from cows. J Reprod Fertil 4:27–45

Short RV, London DR, 1961. Defective biosynthesis of ovarian steroids in the Stein-Leventhal syndrome. Br Med J 1:1724–1727

Singh KB, 1969. Induction of polycystic ovarian disease in rats by continuous light. Am J Obstet Gynecol 103:1078–1083

Smith KD, Steinberger E, Perloff WH, 1965. Polycystic ovarian disease. A report of 301 patients. Am J Obstet Gynecol 93:994–1001

Starka L, Matys Z, Janata J, 1962. Der nacheveis von dehydroepiandrosterone in menschlichen sklerocystischen ovarien. Clin Chim Acta 7:776–779

Stein IF, Leventhal ML, 1935. Ammenorrhea associated with bilateral polycystic ovaries. Am J Obstet Gynecol 29:181–186

Taymor ML, Bernard R, 1962. Luteinizing hormone excretion in the polycystic ovary syndrome. Fertil Steril 13:501–512

Taymor ML, Thompson IE, Berger MJ, Patton WC, 1974. LHRH as a diagnostic and research tool in gynecological endocrinology. Am J Obstet Gynecol 120:721–732

Tsafriri A, Braw RH, 1984. Experimental approach to atresia in mammals. Oxf Rev Reprod Biol 6:226–265

Ward RC, Costoff A, Mahesh VB, 1978. The induction of polycystic ovaries in mature cycling rats by the administration of dehydroepi-androsterone. Biol Reprod 18:614–623

Yen SSC, 1980. The polycystic ovary syndrome. Clin Endocrinol 12:177–208

Yen SSC, Lasley BL, Wang CF, Leblanc H, Siler TM, 1975. The operating characteristics of the hypothalamic–pituitary system during the menstrual cycle and observations of biological actions of somatostatin. Recent Prog Horm Res 31:321–363

Yen SSC, Vandenberg G, Siler TM, 1974. Modulation of pituitary responsiveness to LRF by estrogen. J Clin Endocrinol Metab 39:170–177

Yen SSC, Vela P, Rankin J, 1970a. Inappropriate secretion of follicle stimulating hormone and luteinizing hormone in polycystic ovarian disease. J Clin Endocrinol Metab 30:435–442

Yen SSC, Vela P, Ryan KJ, 1970b. Effect of clomiphene citrate in the polycystic ovary syndrome: Relationship between serum gonadotropin and corpus luteum function. J Clin Endocrinol Metab 31:7–13

ANDROGEN-INDUCED CHANGES IN OVARIAN GRANULOSA

CELLS FROM IMMATURE RATS IN VITRO

Everett Anderson, Martin Selig, Gloria Lee and Brian Little

Department of Anatomy and Cellular Biology and Laboratory
for Human Reproduction and Reproductive Biology
Harvard Medical School, Boston, MA 02115
Department of Obstetrics and Gynecology
McGill University, Montreal, P.Q., Canada

INTRODUCTION

Since the classical description of the polycystic ovarian syndrome
(PCO) by Stein and Leventhal (1935) investigators continue to be
intrigued by this complex endrocrinologic disorder. Despite the
continued interest, a satisfactory description of its etiology and
pathogenesis in the ovary remains unexplained. In certain mammalian
models, cystic ovaries have been induced by numerous manipulations.
These treatments include estradiol valerate (Hemmings et al., 1978),
estradiol benzoate, (Kawakami and Visessuvan, 1979),
dehydroepiandrosterone (DHEA) (Roy et al., 1962), treatment with
antibodies to luteinizing hormone releasing hormone, (Fraser and Baker,
1978; Popkin et al., 1983), treatment of hypothyroid rats with human
chorionic gonadotropin (Copmann and Adams, 1981), treatment with
testosterone propionate (Cortes et al., 1971) and constant illumination
(Lawton and Schwartz, 1977) (see Singh, 1969 for various methods of
producing cystic ovaries). In addition to the aforementioned treatments
which produce cysts, atretic follicles are also induced. In humans with
PCO syndrome (and in other species), it has been shown that there is an
associated increase in circulating androgenic steroids, and Greenblatt
and Mahesh (1976) have actually called PCO "the androgenic ovary."
Moreover, it has been observed that the steroid microenvironment of the
follicle affects granulosa cell differentiation, including changes
leading to atresia or luteinization (Payne et al., 1986: Lovett, 1975).
Roy et al. (1962) used DHEA and androstenedione (Δ^4-A) to induce
cystic as well as atretic changes in the ovaries of rats. (Also see
Billiar et al., 1985.) In this paper, we focus attention on the
structural and functional changes of granulosa cells from small antral
follicles cultured with ovine follicular stimulating hormone (O-FHS) and
in the presence of DHEA, and Δ^4-A. DHEA and Δ^4-A were selected
since these androgens have been implicated in atresia and cyst
formation.

Within the antral follicle, the mural granulosa cells rest on a
thick basement membrane. They are columnar in shape, while the rest of
the granulosa cell population assume a polyhedral configuration.

Non-mural granulosa cells can be expressed from the antral follicle producing a group of cells suitable for culturing. We cultured such a group of cells in McCoy's 5A medium with 25 mM Hepes (GIBCO, Grand Island, NY), 1% penicillin-streptomycin, and 50 ng/ml of OFSH (Anderson et al., 1986). The experimental medium was the basic medium plus 10^{-5}M DHEA or Δ^4-A, which had previously been dissolved in 100% ethanol. We also used 10^7 M DHEA or Δ^4-A, and our reason for selecting these concentrations was the known circulating and tissue concentrations of the steroid (Fortune and Armstrong, 1979; Goff and Henderson, 1979; Page and Butcher, 1982). Control experiments were performed with granulosa cells grown in basic medium alone. The control and experimental cultures were incubated for 1,2,4, and 6 days. Viability of the cells, as measured by erythrosin-B exclusion method was approximately 60% (Phillips, 1973).

Cells from control and experimental cultures were prepared for light microscopy and fixed for transmission electron microscopy at the end of each experiment (Anderson and Batten, 1983).

RESULTS AND DISCUSSION

When granulosa cells are initially seeded on the plastic substratum, they are small spherical cells and after two hours, they attach to the substratum. As the cells become attached, they commence spreading. By six days the cells are rather flattened and take on an epitheloid configuration. Some of these cells accumulate small lipid droplets within their cytoplasm.

Two hours after attaching to the substratum, the androgens DHEA or Δ^4-A (10^{-5} and 10^7M) were added. The appearance of cells grown in 10^7M concentration of steroid were the same as those exposed to higher concentrations, but less marked, the results of which will be recorded here.

All cells, whether grown in DHEA or Δ^4-A, accumulated a large number of larger lipid droplets in their cytoplasm. These cells also showed blebs on their surfaces. Frequently they developed blebs, not only on their main body, but on their cytoplasmic processes.

CONTROL

When control cells are examined with the electron microscope, their submicroscopic structure is like that which has already been described (Anderson and Little, 1985). Briefly, the large nucleus is surrounded by a porous nuclear envelope (Fig. 1). In the juxtanuclear position is the Golgi complex associated with centrioles. Within the cytoplasmic matrix are many ribosomes, and rough endoplasmic reticulum with no preferred orientation. There is no endoplasmic reticulum of the smooth variety. Mitochondria are found throughout the cytoplasmic matrix. These organelles are membrane bounded and contain lamellar cristae, which are oriented transverse to the longitudinal axis of the organelle. Coursing through the cytoplasm are microtubules, 6 nm microfilaments and 10 nm filaments. By use of immunofluorescence we have demonstrated that the microfilaments are F-actin (see below) (Anderson et al., 1986) and the 10 nm filaments are vimentin filaments (unpublished). A few small lipid droplets are occasionally seen in the cytoplasm.

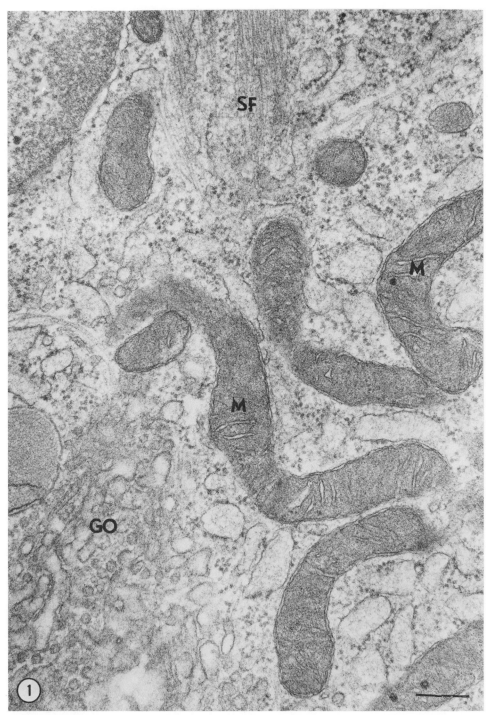

Fig. 1. The cytoplasm of a granulosa cell cultured for four days as a control. N, nucleus; SF, Stress fiber; M, mitochrondria; GO, Golgi complex. Bar = 0.33 μ

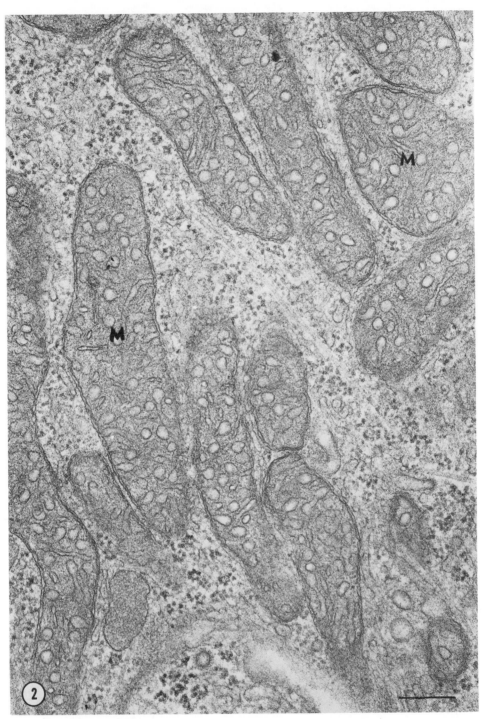

Fig. 2. The cytoplasm of a granulosa cell cultured for four days in 10^{-5} M andiosteindione. MT, mitochrondria with tubular cristae. Bar = 0.25 μ

EXPERIMENTAL

When granulosa cells from the experimental cultures were examined with the electron microscope, dramatic cytological changes were noted, particularly in the mitochondria and appearance and accumulation of large amounts of smooth endoplasmic reticulum. After two days in culture, the mitochondria became rather large and many had a branched configuration. During the two-day period the internal structure of mitochondria have changed; they went from possessing lamella cristae to organelles with villous or tubular cristae (Fig. 2). This change in the cristae pattern may signal ongoing differentiation to a steroidogenic capability. Dimino et al., (1971) have shown that changes in mitochondrial internal structure like those shown in our study are early indicators of follicular luteinization. As indicated above, there developed, in the experimental cells, an abundance of endoplasmic reticulum of the smooth variety (SER) (Fig. 3). This SER is usually associated with lipid droplets. The changes observed in the mitochondria and the appearance of large amounts of SER are consistent with maturation of steroid secreting cells as they differentiate for the secretion of progesterone (Christensen and Gillin, 1969). Granulosa cells in our cultures do secrete progesterone in the medium (Anderson et al., 1987). During steroidogenesis the conversion of cholesterol to progesterone catalyzed by cholesterol side chain cleavage is located in mitochondria (Omura et al., 1966). The acquisition of villous cristae may signal the organelles' final differentiated capacity to carry out the rate limiting step in steroidogenesis. Enzymes for example, which are necessary to convert pregeneolone to progestetone are located in the microsomal fraction of which the SER is a component (Christensen and Gillin, 1969).

By day four there appears in the cytoplasm some acid phosphatase positive heterolysosomes (Fig. 4; also see Anderson et al., 1987). Many of these cells contain large vacuoles. We believe that cellular degeneration was initiated in part by primary lysosomes produced by the Golgi complex. Granulosa cells from the control cultures survived up to beyond six days. We have performed similar experiments on granulosa cells grown in a serum-free medium and have obtained similar results (Anderson et al., 1985).

As indicated in our introduction, DHEA and Δ^4-A have been androgens implicated in atresia and cyst formation. How might these observations in vitro be related to what happens in situ? This is an important issue which is not easy to address. However, in some unpublished observations E.A. made with Dr. Marcus Walker, we used DHEA, as recommended by Roy et al. (1962), to induce cyst formation in rats. When DHEA was administered to immature 27-day old female rats it induced precocious ovulation, chronic acyclicity, anovulation and the formation of ovarian follicular cysts. In the granulosa cells of antral follicles destined to become cysts, a dramatic increase was noted in the amounts of smooth endoplasmic reticulum and mitochondria with tubular cristae. These changes are exactly those seen in the present in vitro cultures and are consistent with the structural profile of active steroidogenic cells. The mural granulosa cells "transform" into epithelial cells while the rest of the population degenerate and are destroyed by macrophages. Perhaps the reason we do not observe the kinds of in situ "transformations" of mural granulosa cells in vitro is because few, if any, mural cells are expressed during our collecting procedure. Similar changes have been noted in the rhesus monkey under stimulation with Δ^4-A (unpublished observation, Anderson, E., Little, B., Billiar, R.B. and Lee, G.).

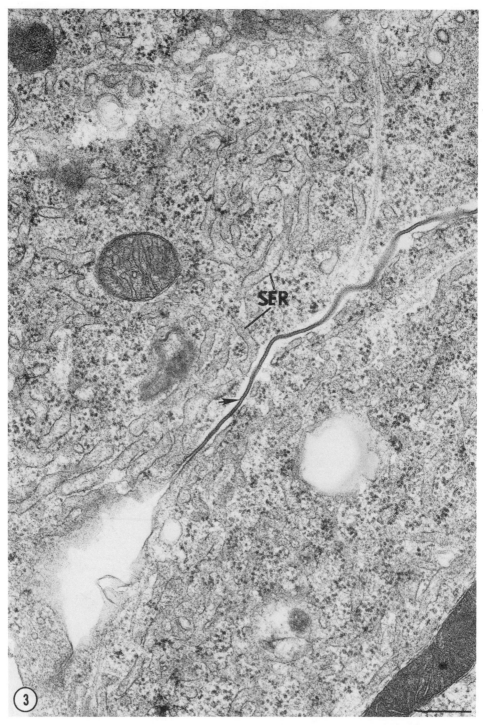

Fig. 3. The cytoplasm of a granulosa cell cultured for two days in 10^{-5}M androstenedione. Arrow, gap junction; SER smooth endoplasmic reticulum. Bar = .25 μ

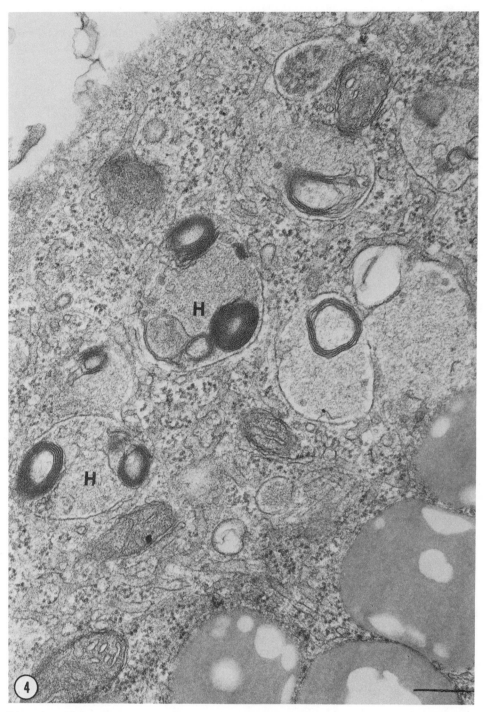

Fig. 4. The cytoplasm of a granulosa cell cultured for four days in 10^{-5}M androstendione. H, heterolysosomes. Bar = .33 μ

In situ, all of the granulosa cells are associated with each other and by both adhering and gap junctions (Albertini and Anderson, 1974; Anderson and Albertini, 1976). Gap junctions are membrane specializations that have been shown to permit cell-to-cell communication and presumably allow regulatory and informational molecules to flow between cells (Revel, 1986). After we punctured the follicle, we dissociated the granulosa cells with ethyleneglycol diamino-tetraacetic acid (EGTA) and hypertonic sucrose as recommended by Campbell (1979). Under these conditions the gap junctions are lost. However, when cells are plated and permitted to spread on the plastic substratum previously treated with type IV collagen and laminin they re-established their gap junctional contacts as shown in Figure 3. (See Anderson et al., 1986.) Granulosa cells grown in vitro have been shown to be electrically coupled (Lawrence et al., 1979).

Granulosa cells cultured in DHEA or Δ^4-A do not remain epitheloid but grow somewhat stellate. In such cells the cytoskeletal system, particularly F-actin appears disorganized. Figures 5-7 show a series of photomicrographs from both control and experimental cultures over a period of 6 days. The cells grow to confluency (Fig. 7A). When we processed similar cells the fine filaments throughout the cell and large parallel oriented regions that run in cords stain intensely with rhodamine phalloidin (Figs. 5, 6, 7B, D & F) indicating the presence of F-actin. The large cord-like areas are prominent in cells from control cultures. When observed with the transmission electron microscope (Fig. 8), the fine filaments are about 6 nm (F-actin) and the larger ones are microfilament bundles (stress fibers, see below). In the EM, the large microfilament bundles have some periodic densities along the longitudinal axis. These densities have been shown to be the location of α-actinin, which is an actin-binding protein (Batten and Anderson, 1981). As can be seen in the experimental cultures, there appears to be little organization of the fine actin filaments and the microfilament bundles (Figs. 5, 6, 7 D and F). In the EM most of the actin filaments appear to be lost and what remains appears disorganized (Fig. 9). Compounds such as dimethylsulfoxide and cytochalasin B are known to destroy cytoplasmic actin filaments and bundles (Sanger et al., 1980); MacLean-Fletcher and Pollard, 1980). Just what mechanism DHEA and Δ^4-A and the other compounds employ to prevent the full compliment of actin filaments is unknown. MacLean-Fletcher and Pollard (1980) believe that cytochalasin B's primary effect is to attach to the barbed ends of F-actin filaments. Hartwig and Stossel (1979) have indicated that such binding eventually leads to a clipping of the actin filaments. It is unknown whether or not myosin is associated with the microfilament bundles in granulosa cells. However, in other cell types it is (Sanger et al., 1986).

As we indicated before, the cells respond to DHEA and Δ^4-A by changing shape. This change in cell shape is also observed in situ, of animals treated with DHEA. It is obvious that the cells lose their confluency and perhaps become uncoupled by a disappearance of the gap junctions and eventually die in four days. The controls maintain their confluency for up to beyond six days. We have no data to indicate that microfilaments (F-actin) are involved in steroidogenesis. However, Hall et al., (1979) believe that microfilaments of adrenal tumor cells are involved in the transport of cholesterol to mitochondria.

Lewis and Lewis (1923) observed fibrous-like regions in the cytoplasm of a variety of cultured cells and referred to them as stress fibers, which are thought to be induced by the culturing process. Nearly all culture cells grown on a substratum develop stress fibers. We now know, as indicated above, that stress fibers are microfilament

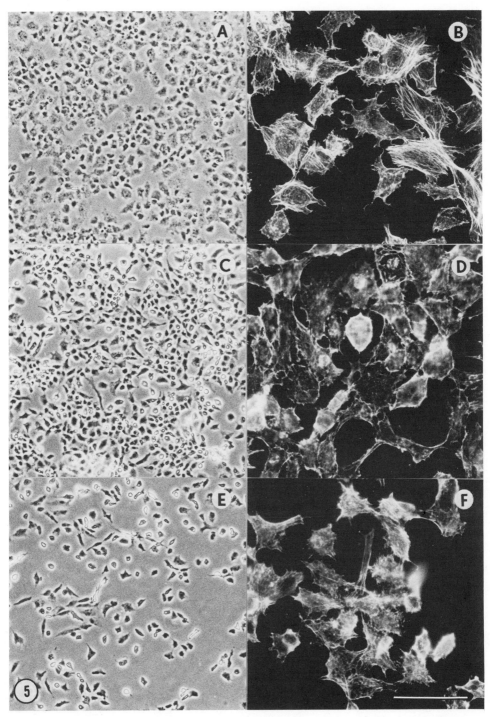

Fig. 5.   Day 2, A-B Controls; C - D, DHEA; E - F, Δ^4-A.   Bar = 40 μ

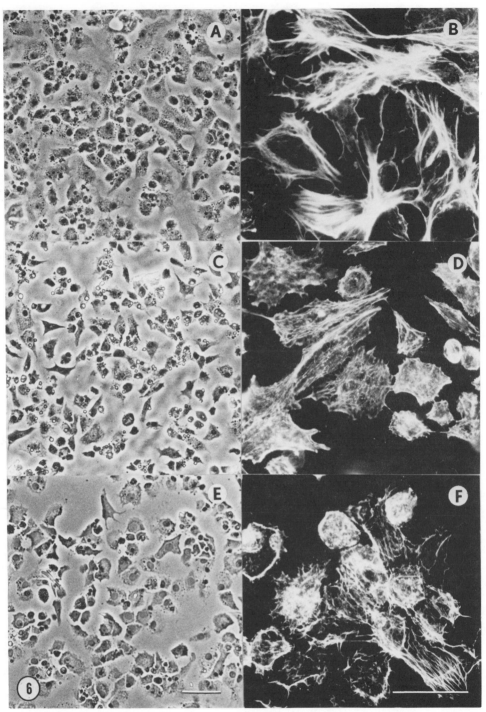

Fig. 6. Day 4, A-B, controls; C-D, DHEA; E and F, Δ^4-A. Bar = 40 μ

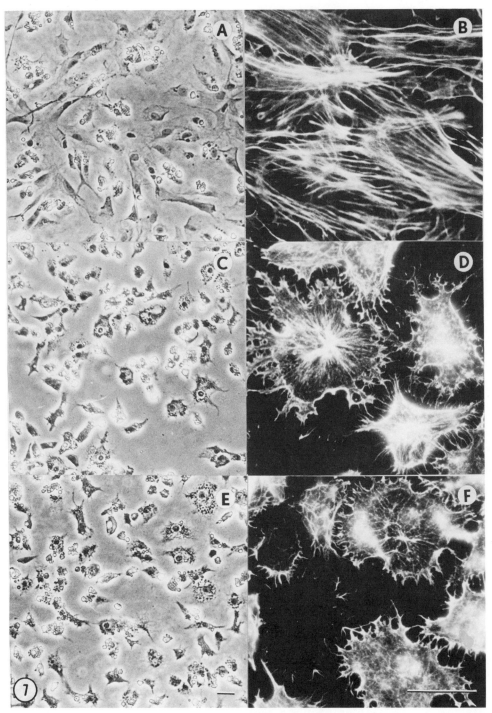

Fig. 7. Day 6, A-B, controls; C-D, DHEA; E and F, Δ^4-A. Bar = 20 μ

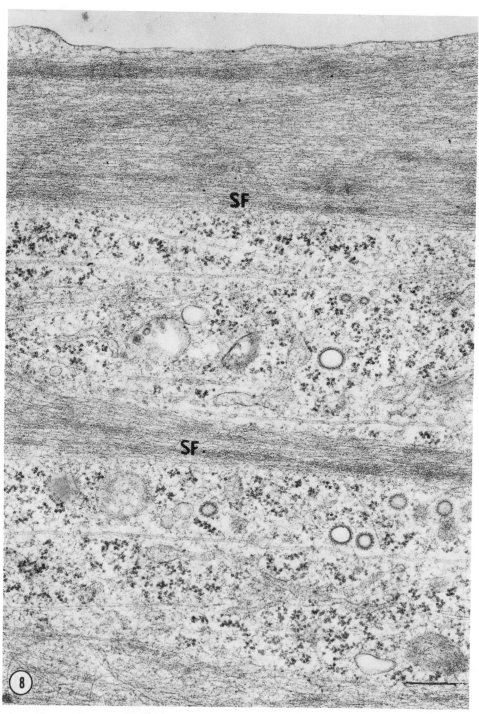

Fig. 8. The cytoplasm of a granulosa cell cultured for four days as a
control. SF stress fibers. Bar = .25 μ

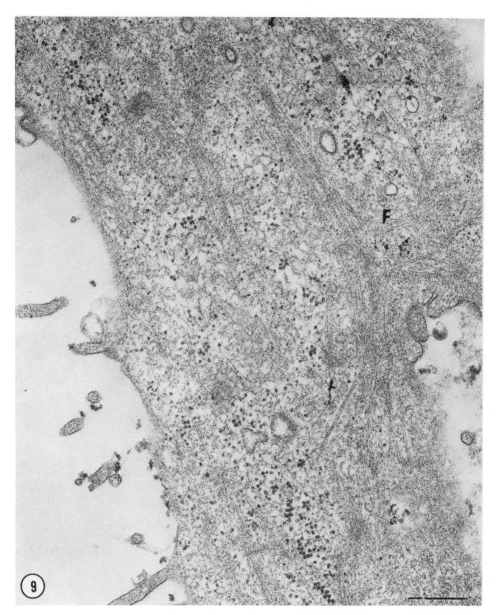

Fig. 9. The cytoplasm of a granulosa cell cultured for six days in 10^{-5}M dehydroepiandrosterone F-filamentous actin. Bar = 0.33 μ

bundles composed of F-actin filaments, with associated proteins and myosin (Weber and Groeschel-Stewart, 1974). We have seen that by rhodamine-labelled phalloidin that actin filaments are located in the cytoplasm of granulosa cells. The question: is there a comparable structure or stress fibers in granulosa cells of follicles that are destined to become a cyst during the DHEA treatment? We have observed a thick band composed of fine 6 nm filaments which is found in both the apical and basal cytoplasmic region of the cells. These bands of filaments appear to exclude organelles from these areas. It should be pointed out that stress-fibers are not only found in cultured cells, but also in cells of tissues, in situ, for example endothelial cells of myometral arterioles and aorta (Rohlick and Olah, 1967; White et al., 1986), and scleroblasts (a modified fibroblast) (Byers and Fijuwara, 1982). We believe that these bands of 6 nm filaments observed in situ may be stress fiber-like structures resulting from some unknown signal during the DHEA treatment.

Why granulosa cells die or transform under the DHEA or Δ^4-A treatment is unknown. For the death of granulosa cells, Schreiber et al., (1976) believe that the androgen enters the granulosa cell where it binds to a receptor and subsequently activates a cell death mechanism.

The results of our studies suggest that androgens alter organelle configuration and cell shape, which may be related to their role in the atretic process and cyst formation associated with the polycystic ovarian condition.

ACKNOWLEDGMENTS

This investigation was supported by the following NIH Grants: HD06645 (Center Grant of Laboratory of Human Reproduction and Reproductive Biology), HD14574 (Dr. Everett Anderson), NIH Grant P50NICHHD 07650, and MRC Grant MA-8678 (Dr. Brian Little). The authors wish to gratefully acknowledge the invaluable secretarial assistance of Hugh Stringer.

REFERENCES

Albertini DF, Anderson E, 1974. The appearance and structure of the intercellular connections during the ontogeny of the rabbit ovarian follicle with particular reference to gap junctions. J Cell Biol 63:234-250

Anderson E, Albertini DF, 1976, Gap junctions between the oocyte and the companion follicle cells in the mammalian ovary. J Cell Biol 71:680-686

Anderson E, and Batten BE, 1983, Surface binding uptake and fate of cationic ferritin in a steroid producing ovarian cell. Tissue and Cell 15:853-871

Anderson E, Little B, 1985, The ontogeny of the rat granulosa cell. In: ToFtDO, Ryan RJ (eds.), Proceedings of the Fifth Ovarian Workshop Champaign, Ill: Ovarian Workshops, pp. 203-226

Anderson E, Selig M, Lee G, 1986, Androgen-induced cellular changes in rat ovarian granulosa cells grown in a serum-free medium. J Cell Biol 103:263a

Anderson E, Little B, Lee GS, 1987. Androgen-induced changes in rat ovarian granulosa cells in vitro tissue and Cell (in press)

Batten BE, Anderson E, 1981. Effects of Ca+2 and Mg+2 deprivation on cell shape in cultured ovarian granulosa cells. Amer J Anat 161:104-114

Billiar RB, Richardson D, Anderson E, Mahajan D, Little B, 1985. The effect of chronic and cyclic elimination of circulating androstene-dione or estrone concentrations on ovarian function in the rhesus monkey. Endocrinol 116:2209-2220

Byers HR, Fujiwara K, 1982. Stress fibers in cells in situ: immunofluorescence visualization with antiactin, antimyosin and anti-alpha-actinin. J Cell Biol 93:804-811

Campbell KL, 1979, Ovarian granulosa cells isolated with EGTA and hypertonic sucrose: Cellular integrity and function. Biol Reprod 21:773-786

Christensen K, Gillim SW, 1969. The correlation of fine structure and function in steroid-secreting cells with emphasis on those of the gonads. In: McKerns KW (ed) The Gonads. New York: Appleton-Century-Crofts pp. 415-488

Copmann TL, Adams WC, 1981. Relationship of polycystic ovary induction to prolactin secretion: prevention of cyst formation by bromocriptine in the rat. Endocrinol 108:1095-1097

Cortes V, McCrocken JA, Lloyd CW, Weisz F, 1971. Progestin production by the ovary of the testosterone-sterilized rat treated with an oculatory dose of LH and the normal proestrus rat. Endocrinal 89:878-885

Dimino MJ, Elfont EA, Berman SK, 1979. Changes in ovarian mitochondria: Early indicators of follicular luteinization. Adv Exp Med Biol 112:505-510

Fortune JE, Armstrong DT, 1979. Androgen production by isolated components of rat ovarian follicles: In: Midgley AR, Sadler WA, (eds), Follicular Development and Function New York: Raven Press, pp. 193-198

Fraser HM, Baker TG, 1978. Changes in the ovaries of rats after immunization against luteinizing hormone releasing hormone. J Endocrinol 77:85-93

Goff AK, Henderson KM, 1979. Changes in follicular fluid and serum concentrations of steroids in PMS treated immature rats following LH administration. Biol Reprod 27:383-392

Greenblatt RB, Mahesh VB, 1976, The androgenic polycystic ovary. Am J Obstet Gynecol 125:712-726

Hall PF, Charpponier C, Nakamura M, Gabbiani C, 1979. The role of microfilaments in the response of adrenal tumor cells to adrenocorticotropic hormone. J Biol Chem 254:9080-9084

Hartwig JH, Stossel TP, 1979. Cytochalasin B and the structure of actin gels. J Mol Biol 134:539-554

Hemmings R, Farookhi R, Brawer JR, 1978. A single subcutaneous injection of estradiol valerate 2 mg administered to young cycling female rats chronic anovulatory acyclicity persistent vaginal cornification. Endocrinol 103:501-502

Karuakami M , Visessuvan S, 1979. The differing responsiveness of the anterial--and middle anterior hypothalamic area to estradiol benzoats implant: inhibition of compensatory ovarian hypertrophy. Endokrinologie 74:271-286

Lawrence TS, Ginzberg RD, Gilula NB, and Beers WH, 1979. Hormonally induced cell shape changes in cultured rat ovarian granulosa cells. J Cell Biol 80:21-36

Lawton IE, Schwartz NB, 1967. Pituitary-ovarian function in rats exposed to constant light: A chronological study. Endocrinol 81:497-508

Lewis WH, Lewis MR, 1924. Behavior of cells in tissue cultures. In Cowdry EV (ed) General Cytology. Chicago: Univ. Chicago Press, pp. 385-447

Lovert JP, Harman SM, Schreiber JR, Ross TG, 1975. Evidence for a role of androgens in follicular maturation. Endocrinol 97:366-372

Maclean-Fletcher A, Pollard, TD, 1980. Mechanism of action of cytochalasin B on actin. Cell 20:329-341

Omura T, Sanders E, Estabrook RW, Cooper DV, Rosenthal O, 1966. Isolation from adrenal cortex of a non-heme iron protein and a flavoprotein functional as a reduced triphospho-pyridine nucleotide-cytochrome P-450 reductase. Arch Biochem Biophys 117:660-673

Page RD, Butcher RL, 1982. Follicular and plasma patterns of steroids in young and old rats during normal and prolonged estrous cycle. Biol Reprod 27:383-392

Payne RW, Hellbaum AA, Owans JN, Jr., 1956. The effects of androgens on the ovaries and uterus of the estrogen treated hypophysectomized immature rat. Endocrinol 59:306-316

Phillips JJ, 1973. Dye exclusion test for cell viability. In: Kruse PF, Patterson MK, (eds). Tissue Culture Methods and Application New York: Academic Press, pp. 406-408

Popkin RM, Fraser HM, Gosalen RG, 1983. Effects of LH-RH immunoneutralization on pituitary and ovarian LH-RH receptors in female rats. J Reprod Fertil 69:245-252

Revel JP, 1986. Gap junctions in development. In: Steinberg MS (ed) Developmental Biology a Comprehensive Synthesis. The Cell Surface in Development and Cancer. New York: Plenum Press Vol. 3:191-204

Rohlick P, Olah I, 1967. Cross-striated fibrils in the endothelium of the rat myometral arterioles. J Ultrastruct Res 18:667-676

Roy S, Mahesh VB, Greenblatt, RB, 1962. Effect of dehydroepiandrosterone and Δ^4-androstenedione on the reproductive organs of female rats: Production of cystic changes in the ovary. Nature 196:42-43

Sanger JW, Gwinn J, Sanger JM, 1980. Dissolution of cytoplasmic bundles and the induction of nuclear actin bundles by dimethyl sulfoxide. J Expt'l Zool 213:227-230

Sanger JW, Balray M, Pochapin M, Sanger JW, 1986. Observations of microfilament bundles in living cells microinjected with fluorescently labeled contractile proteins. J Cell Sci Suppl 5:17-44

Schreiber JR, Reid R, Ross GT, 1976. A receptor-like testosterone binding protein in ovaries from estrogen stimulated hypophysectomized immature female rats. Endocrinol 98:1206-1213

Singh KB, 1969. Persistent estrus: An experimental model of the polycystic ovary syndrome. Obstet Gynecol Survey 24:2-17

Stein IF, Leventhal ML, 1935. Amenorrhea associated with bilateral polycystic ovaries. Am J Obstet Gynecol 29:181-191

Weber K, Groeschel-Stewart U, 1974. Antibody to myosin: the specific visualization of myosin-containing filaments in non-muscle cells. Proc Nat'l Acad Sci U.S.A. 71:4561-4564

White GE, Fujiwara K, 1986. Expression and intracellular distribution of stress fibers in aortic endothelium. J Cell Biol 103:63-70

ESTROGEN-PRODUCING OVARIAN GRANULOSA CELLS: USE OF THE GRANULOSA CELL

AROMATASE BIOASSAY (GAB) TO MONITOR FSH LEVELS IN BODY FLUIDS

Kristine D. Dahl, Nancy M. Czekala, and Aaron J.W. Hsueh

Department of Reproductive Medicine, M-025, University of
California, San Diego, La Jolla, California 92093

INTRODUCTION

Pituitary follicle-stimulating hormone (FSH) is essential for the maintenance of testicular spermatogenesis and ovarian follicle development. Smith (1926,) and Zondek and Aschheim (1925) independently discovered that implants of fresh anterior pituitary into hypophysectomized animals induced an immediate and remarkable growth and maturation of the ovaries. Two apparently separate gonad-stimulating hormones were subsequently detected in menopausal women's urine and human pituitary samples by Ascheim and Zondek (1928) and Fevold et al. (1931), respectively. The primary action of FSH is the development of ovarian follicles, whereas LH is responsible for the transformation of follicles into the corpus luteum. However, precise measurements of FSH in body fluids were not possible prior to the development of a sensitive radioimmunoassay (RIA) for FSH in 1967.

MEASUREMENT OF FSH BY RADIOIMMUNOASSAY VERSUS BIOASSAY

Following the development of a sensitive RIA for insulin (Berson et al., 1964), the technique was adapted for the measurement of FSH in humans (Midgley, 1967; Faiman and Ryan, 1967) and rats (Daane and Parlow, 1971; Bogdanove et al., 1971). Several investigators were able to characterize changes in FSH in women during spontaneous and induced ovulatory cycles (Midgley and Jaffe, 1968; Ross et al., 1970). Since that time, serum levels of immunoreactive FSH have been characterized in humans during puberty, the menstrual cycle, menopause, and many endocrine states (Ross et al., 1970; Crowley et al., 1985; Styne and Grumbach, 1986).

The RIAs are convenient, specific and sensitive; but they have several major deficiencies. Since these assays are based upon the immunoreactivity of the assayed material, discrepancies among assays using different antibodies have been reported (Taymore and Miyata, 1969; Diebel et al., 1973). The difficulty in the choice of appropriate RIA standard adds additional complications. Due to the non-parallelism of competition curves for FSH samples derived from different sources (e.g., serum vs urine) and the fact that most of the serum FSH measurements were assayed against either pituitary or urinary standards, the quantitative aspect of the published FSH data should be interpreted with caution (Albert et al.,

1968; Ryan and Fayman, 1968). Although a radioligand receptor assay has been designed to measure FSH levels (Reichert and Bhalla, 1974; Cheng, 1975), the method has not gained popular use due to interference by serum inhibitors and the difficulty in differentiating agonistic and antagonistic FSH activities.

Although essentially all the available data on FSH levels in body fluids have been obtained by RIA, these data do not necessarily reflect bioactivity. Some modified forms of FSH may not cross-react with the antibody but still retain their bioactivity, whereas some biologically inactive FSH (e.g., deglycosylated forms) may still bind the FSH antibody. Marked disparities between bio- and immuno- estimates of both crude and purified hormone preparations have been observed (Albert et al., 1968; Ryan, 1969; Rosenberg et al., 1971). Data from Bogdanove's lab have suggested that heterogenous populations of FSH are released by the pituitary in castrated rats and the RIA data do not correspond with bioassay results (Diebel et al., 1973; Bogdanove et al., 1974). The bio/immuno ratio of test preparations changes with various physiological conditions; androgens increase this ratio whereas estrogens decrease it. Also, serum FSH from castrated, androgen-treated male rats have a longer in vivo half-life than androgen-deprived rats (Bogdanove et al., 1974). In addition, variations in pituitary and serum FSH have been reported in rhesus monkeys after castration and hormonal replacement (Peckham et al., 1973; Peckham and Knobil, 1976a, 1976b). In direct contrast to results derived from rats, serum and pituitary FSH preparations from both ovariectomized and orchidectomized rhesus monkeys are characterized by a larger apparent molecular size, higher bio/immuno ratio, and a lower rate of disappearance from the circulation when injected into test rats. The role of estrogen in decreasing the bio/immuno ratio was also demonstrated. The observed heterogeneity in bio/immuno ratio of FSH in these studies is believed to be due to variation in the sialic acid content of the hormone; however, disparity between rat and monkey experiments requires further elucidation.

STUDIES ON FSH MICRO-HETEROGENEITY

Two types of FSH molecules with differential ability to bind the lectin Concanavalin A have been found in the pituitary of hamsters (Chappel, 1981). Recent experiments further indicate the presence of six species of immuno-FSH in the pituitary of hamsters based on differences in their isoelectric points (Chappel et al., 1983). It appears that species of FSH with the lowest isoelectric point have the greatest amount of sialic acid (Sherins et al., 1973). In addition, these acidic species of FSH exhibit lower binding in a radioreceptor assay and have lower bioactivity in an in vitro FSH bioassay using plasminogen activator induction as the end point (Chappel et al., 1983). The relative proportions of the most acidic form increase within the pituitary of hamsters prior to the estradiol-induced FSH surge and decrease as the surge occurred (Chappel et al., 1982). Because the proportion of the bioactive and more basic forms of FSH increases with the onset of puberty in rats, Chappel et al. (1983) suggested that the pituitary gland acquires the capacity of transforming FSH species with lower bioactivity to their more potent counterparts, probably by a pituitary neuraminidase (Reichert et al., 1971). In addition, treatment with GnRH increases the relative proportion of the more basic form of pituitary FSH in the hamster (Galle et al., 1983). Studies on human pituitary FSH heterogeneity further suggested that the percentage of acidic forms of FSH increase with age, presumably due to changes in sialic acid content (Wide, 1985). Also, estrogens are believed to induce the formation of more basic forms of FSH (Wide, 1982). However, the physiological significance of this FSH pleomorphism is not clear due, in part, to the lack of a sensitive bioassay to measure circulating, instead of pituitary, FSH levels.

EARLY STUDIES ON FSH BIOASSAYS

The classical FSH bioassays were based upon FSH-induced increases in ovarian weight or follicle size in immature or hypophysectomized rodents (Evan et al., 1939; Steelman and Pohley, 1953; Brown and Wells, 1966). Some investigators also measured increases in testis weight of hypophysectomized rats (Greep et al., 1940; Simpson et al., 1950) or changes in testicular morphology of intact chicks (Siegal and Siegal, 1964). Also, "secondary" FSH responses, such as the increase in uterine weight of immature mice, have been employed (Igarashi and McCann, 1966; Uberoi and Meyer, 1967). Among these assays, the Steelman-Pohley test is the most frequently used. In this assay, bioactivity of FSH is determined by treating immature female rats (21-22 days old) with 20 IU hCG plus test samples containing unknown quantities of FSH for three days and the ovarian weight is measured 72 h after the first injection (Steelman and Pohley, 1953). This assay is superior to other methods because the addition of hCG in the injection protocol not only augments the ovarian weight increase (Simpson et al., 1950), but also minimizes the effect of contaminating LH-like materials in the assay samples; however, the mechanism by which hCG augments the FSH effect was not clear until decades later.

The sensitivity of Steelman-Pohley assay was determined to be 100 ^{m}g for NIH FSH-S1 and 2 IU for 2nd IRP-hMG (Christiansen, 1972a) with a narrow assay range. The index of precision (Lambda: standard deviation of the test samples/slope of the dose-response curve) was determined to be 0.1 for human urinary samples. Also, the addition of LH, TSH, prolactin, ACTH and growth hormone is ineffective in increasing ovarian weight in this assay system (Christiansen, 1972b). The use of this method for assaying FSH of several animal origins was also tested; the regression line for the dose-response curve was shown to be parallel for FSH of ovine, equine, rat and human origins (Parlow and Reichert, 1963). Thus, this bioassay is not species-specific. However, the sensitivity of this assay is low, the procedure is tedious, and the assay variation is large.

With recent development of several in vitro bioassays for pituitary hormones, circulating levels of bioactive LH (Van Damme et al., 1974; Dufau et al., 1976) and lactogenic hormones (Tanaka et al., 1980; Subramanian and Gala, 1986) have been measured for comparison with RIA results. It is clear that disparity between the two types of assays can be found in different physiologic and pathologic conditions.

Although several recent attempts have been made to improve the in vivo FSH bioassay by using in vitro methods, most of these assays were inadequate. Beers and Strickland (1978) proposed an FSH bioassay based on the stimulation of plasminogen activator activity secreted by cultured granulosa cells from PMSG-treated rats. However, subsequent studies have indicated that plasminogen activator activity is controlled by both FSH and LH (Wang and Leung, 1983; Ny et al., 1985). Also, the only application of this method was the measurement of FSH content in conditioned media obtained from GnRH-treated pituitary cell cultures (Beers and Strickland, 1978). Since GnRH treatment alone increases plasminogen activator activity (Wang, 1983), one is concerned about the validity of bio-FSH measurement in samples containing both GnRH and pituitary hormones. Measurement of serum FSH levels using the plasminogen activator assay is further complicated by the presence of serum plasminogen activators, plasmin and alpha $_2$-antiplasmin. Van Damme et al.

(1979) also developing an _in vitro_ bioassay by measuring FSH-stimulated estrogen production in cultured Sertoli cells. Although the sensitivity of this assay was reported to be 0.5 mIU for human FSH, no data on the measurement of serum FSH levels have been presented, probably because of inadequate sensitivity of the cells to FSH. Although these bioassays are applicable to pituitary preparations, there is clearly a need for a sensitive, specific FSH bioassay useful to measure physiologic concentrations of FSH in serum samples.

Based on recently acquired knowledge on the mechanism of FSH action, one can attain a better understanding of the cellular basis of classical FSH bioassays. Estrogens have been shown to stimulate the proliferation of granulosa cells and follicular growth (Smith, 1961; Smith and Bradbury, 1961). Also, the FSH-induced ovarian weight increase used as the end point for the Steelman-Pohley assay was shown to be prevented, at least partially, by the administration of antiserum to estradiol (Reiter et al., 1972). Since FSH acts exclusively on granulosa cells to increase estrogen production, various bioassays employing ovarian weight increases probably represent indirect measurements secondary to the FSH-induced estrogen production by granulosa cells. Based on the 2-cell, 2-gonadotropin theory (reviewed in Hsueh et al., 1984), the mechanism by which FSH increases ovarian weight can be explained (Fig. 1, upper panel). First, injections of hCG increase androgen production by theca interna cells, thus providing androgen precursors for FSH-induced aromatizing enzymes in granulosa cells resulting in an increase in estrogen production. The estrogens secreted by the granulosa cells, in turn, stimulate ovarian growth. Therefore, the increase in ovarian weight, previously believed to be a direct physiologic response of FSH is probably an indirect result of FSH-induced aromatizing enzymes and subsequent estrogenic stimulation of ovarian growth. Thus, a more direct and sensitive bioassay for FSH may be developed if estrogen production (i.e., activity of aromatases in granulosa cells) is monitored directly (Fig. 1, lower panel).

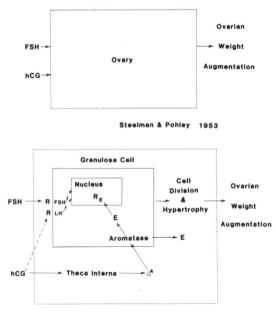

Fig. 1. Hypothetical scheme of the Steelman-Pohley method (upper panel), and the basis for the development of the Granulosa Cell Aromatase Bioassay (lower panel).

DEVELOPMENT OF AN IN VITRO FSH BIOASSAY

As discussed earlier, prior attempts at the development of a specific in vitro bioassay for FSH were often hampered by a lack of sensitivity, or could not be applied to serum samples due to inhibitory serum factors. Successful development of a useful FSH bioassay was therefore dependent upon increasing granulosa cell sensitivity to FSH, as well as pre-treating serum samples to remove inhibitory substances.

Specific FSH binding is found exclusively in the granulosa cell of the follicular complex. FSH specifically induces aromatases in serum-free rat granulosa cell cultures (Dorrington et al., 1975). This effect is FSH dose-dependent and believed to be of physiologic importance (Erickson and Hsueh, 1978). The numerous studies from this laboratory examining the paracrine modulators of FSH actions in granulosa cells led to the discovery of a number of hormones and factors that enhance FSH induction of aromatase activity.

Effect of Estrogens and Androgens

Estrogen receptors are present in granulosa cells and estrogen treatment stimulates granulosa cell proliferation. In vivo studies showed that estrogens enhance FSH-induced follicle growth and ovarian weight gain (Richards and Midgley, 1976). We further demonstrated that estrogens enhance the FSH stimulation of aromatase activity (Adashi and Hsueh, 1982; Zhuang et al., 1982). As shown in Fig. 2, granulosa cells were cultured with increasing concentrations of FSH in the presence or absence of diethystilbestrol (DES). After 3 days, the cells were washed and further incubated in media containing androstenedione for 8 h, and estrogen accumulation measured. FSH stimulated aromatase activity in a dose-dependent manner, and concomitant treatment with DES further augmented estrogen production, as reflected by a decrease in the ED_{50} value for FSH. Thus, within the micro-environment of the ovarian follicle, estrogen serves as an end-product amplifier of its own production. In addition to estrogens, androgens also augment gonadotropin-stimulated aromatase activity in cultured rat granulosa cells (Hillier and DeZwart, 1981). The observation that a non-aromatizable androgen augments FSH action supports the notion that androgens augment FSH-stimulated aromatase activity, both by acting as a substrate for aromatases and by acting as a bona fide hormone through regulation of the aromatase enzyme activity. In the present culture, a saturating concentration of androstenedione is included in all experiments.

Action of Insulin and IGF-I

Since experimentally-induced diabetes in female rats is associated with decreased ovarian function, we have investigated the ovarian actions of insulin (Davoren and Hsueh, 1984). Our result indicates that high doses of insulin exert a specific action on granulosa cells to increase the FSH stimulation of estrogen and progesterone production. Since pharmacological doses of insulin are required to augment FSH action, we further investigated the possibility that insulin may be acting through the receptor of a similar peptide, insulin-like growth factor-I (IGF-I) (Davoren et al., 1985). As shown in Fig. 3, treatment with a submaximal dose of FSH increases estrogen production by cultured granulosa cells. Co-incubation with IGF-I, IGF-II or insulin enhances the FSH action in a dose-dependent manner. Furthermore, radioligand binding studies using labeled IGF-I indicate that high doses of insulin indeed bind to the specific IGF-I receptor found in the granulosa cells (Davoren et al., 1986). Since our recent findings also suggest that growth hormone treatment in hypophysectomized rats increases intraovarian content of

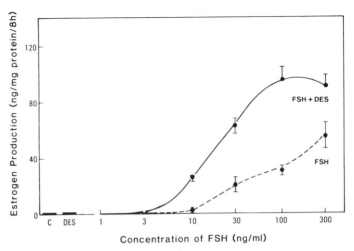

Fig. 2. Estrogen enhancement of FSH–stimulated aromatase activity in cultured granulosa cells. (From Zhuang et al., 1982).

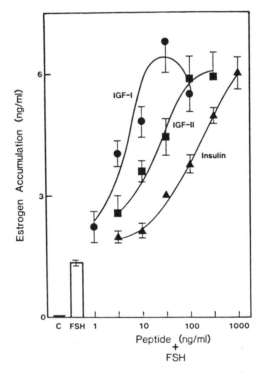

Fig. 3. Dose-dependent enhancement of FSH–stimulated estrogen production by insulin and insulin–like peptides. (From Davoren et al., 1986).

immunoreactive IGF–I (Davoren and Hsueh, 1986), it is possible that growth hormone may act on the ovarian cells to stimulate ovarian production of IGF–I. This, in turn, results in an augmentative action through intraovarian paracrine action of IGF–I on the granulosa cells. Because high concentrations of insulin probably interact with the granulosa cell IGF-I receptor, resulting in similar augmentation of FSH action, insulin has been used in subsequent experiments to replace IGF-I.

Effect of Phosphodiesterase Inhibitors and hCG

Since FSH action at the granulosa cell is believed to be mediated through the cAMP pathway, we tested the potentiating effect of a phosphodiesterase inhibitor, methyl-isobutyl xanthine (MIX), on FSH action (Welsh et al., 1984; Jia and Hsueh, 1986). Upon examining the effect of different doses of MIX, 0.125 mM was shown to be the optimal concentration to augment the FSH stimulation of aromatase activity (Fig. 4).

We further tested the combined actions of these augmenting agents in order to establish a sensitive in vitro FSH bioassay. All cultures were treated with optimal doses of DES and androstenedione. To examine the combined action of MIX and insulin, granulosa cells were cultured in the presence of increasing doses of FSH with or without 0.125 mM MIX, 1 µg/ml insulin, or their combinations. Combined treatment with MIX and insulin resulted in a synergistic augmentation of estrogen production by decreasing the minimum effective dose of FSH from 1 ng to 0.25 ng per culture.

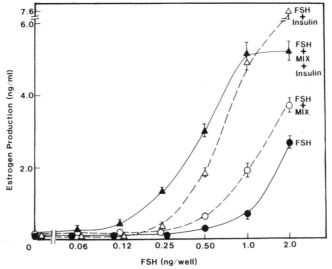

Fig. 4. Synergistic effect of treatment with MIX and insulin on FSH-stimulated estrogen production. (From Jia and Hsueh, 1986).

Although only low concentrations of LH receptors are present in granulosa cells obtained from pre-antral follicles of immature rats, our preliminary results suggest that LH/hCG treatment is capable of enhancing FSH action in the presence of MIX. As shown in Fig. 5, concomitant treatment with increasing concentrations of hCG dose-dependently increased FSH-stimulated estrogen production by granulosa cells incubated with MIX and insulin. The saturating dose of hCG (30 ng/ml) was used for the bioassay.

Elimination of Serum Inhibitors

The combination of hormones and factors in the granulosa cell culture as described above provided a highly sensitive in vitro bioassay for the measurement of FSH activity; however, the addition of 4% gonadotropin-free serum, obtained from hypophysectomized rats, to the cultures substantially decreased estrogen production (Fig. 6, HS = hypophysectomized serum; Jia and Hsueh, 1986). Therefore, separation of FSH and inhibitory serum

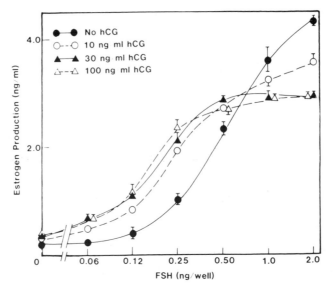

Fig. 5. Augmenting effect of hCG on FSH-stimulated estrogen production.
(From Jia and Hsueh, 1986).

extracts with 40% ethanol precipitates the majority of pituitary proteins
without removing FSH.   Since high concentrations of ethanol are toxic to
cultured cells, we tested the possibility of using polyethylene glycol
(PEG), which is non-toxic to cells at low doses, to eliminate the inhibitory serum factors. PEG has a similar dehydrating property as ethanol and
has been utilized to separate free hormones from hormone-receptor complexes in radioreceptor assays (Roche et al., 1985; Pandian and Bahl,
1977).

PEG treatment of serum samples attenuates the inhibitory action of
serum inhibitors.   Treatment with increasing concentrations of PEG (10-
14%) dose-dependently increases the sensitivity of granulosa cells to FSH,
with a minimum effective dose of 0.12 ng/culture in the presence of 12%
PEG-pretreated gonadotropin-free serum (Fig. 6).   Since PEG pretreatment
may also precipitate serum FSH, recovery of exogenously added I^{125}-FSH or
immunoreactive FSH has been determined in gonadotropin-free serum samples
before and after treatment with different concentrations of PEG. At 12%
PEG, the recovery of FSH determined by counting radioactivities or measuring immunoreactive hormones is between 94 to 98% (Jia and Hsueh, 1986).
Thus, 12% PEG has been used for the pretreatment of serum samples.

Beta-TGF Increases Assay Sensitivity

With the aim of further improving the sensitivity of the GAB assay, we
recently tested the augmenting action of beta-transforming growth factor
(beta-TGF) in the granulosa cells.   Treatment with beta-TGF further
augments the sensitivity of the present system by at least 2-fold (Fig.
7). Therefore, beta-TGF can be added in assays where extremely low levels
of bio-FSH are to be measured.

HORMONE- AND SPECIES-SPECIFICITY OF THE GAB ASSAY

We term the present method as the Granulosa Cell Aromatase Bioassay
(GAB). Hormonal specificity of the present assay was tested in the
presence of MIX, insulin, hCG and PEG-pretreated hypophysectomized rat

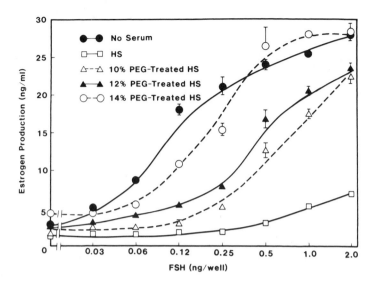

Fig. 6. Serum interference of FSH-stimulated estrogen production: effect of pretreatment with polyethylene glycol. (From Jia and Hsueh, 1986).

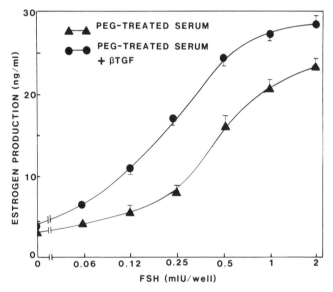

Fig. 7. Dose-dependent enhancement of FSH-stimulated estrogen production by beta-TGF.

serum. As shown in Fig. 8, treatment with increasing concentrations of ovine FSH results in a dose-dependent increase in estrogen production, with a minimum effective dose of 0.12 ng/culture. In contrast, treatment with ovine GH (100 and 300 ng/culture), ovine PRL (100 and 300 ng/culture), rat TSH (30 mIU/culture) and human ACTH (30 mIU/culture) do not stimulate estrogen production. Furthermore, hCG and LH stimulate estrogen production only at extremely high concentrations, probably due to FSH contamination or intrinsic FSH activity in the hormone preparations.

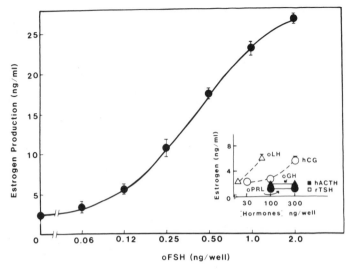

Fig. 8. Hormonal specificity of the granulosa cell aromatase bioassay. (From Jia and Hsueh, 1986).

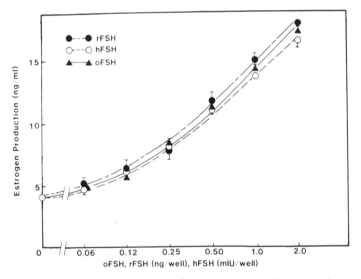

Fig. 9. Dose-dependent stimulation of estrogen production by rat, human and ovine FSH preparations. Granulosa cells were treated with increasing doses of rat FSH (NIH-I-6), human FSH (NIH-hFSH-3), or ovine FSH (NIH-5-15) for 3 days. (From Jia and Hsueh, 1986).

To extend our earlier findings showing that the FSH responsiveness of rat granulosa cells is not species-specific (Hsueh et al., 1983b), the effects of ovine, rat and human FSH on aromatase activity were compared (Jia and Hsueh, 1986). As shown in Fig. 9, the standard curves obtained from ovine, rat and human FSH preparations are parallel to each other. Indeed, our subsequent studies indicate that pituitary FSH preparations from all mammalian species studied show parallel dose-response curves in the GAB assay. Furthermore, FSH preparations from pituitary, serum and urinary sources also result in parallel dose-response curves.

ADVANTAGES AND LIMITATIONS OF THE GAB ASSAY

As compared to other in vivo and in vitro FSH bioassays, the GAB assay offers the following advantages.

1) Sensitivity   The extreme sensitivity of the assay is demonstrated by our ability to measure bio-FSH levels in almost all the serum samples tested.   Furthermore, the high sensitivity of the assay enables one to measure bio-FSH levels in small volumes of urine samples, thereby avoiding potential inhibition by urinary interference substances.

2) Hormone specificity   Although one cannot rule out the presence of unknown factors in body fluids which may regulate granulosa cell aromatase activity, the present assay does not measure circulating levels of multiple pituitary hormones (e.g., growth hormone, ACTH, prolactin and TSH).   Furthermore, the inclusion of a saturating concentration of hCG rules out possibile interference by LH or hCG in the assay.   Due to the inclusion of high concentrations of gonadal steroids (estrogen and androgen) as well as insulin, one also can rule out any possible action of these hormones in the assay.

3) Precision   Calculation of the index of precision was based on standard deviation of the test samples divided by the slope of the dose-response curve.   The index of precision of the GAB assay was shown to be 0.04 which is comparable to other in vitro assays and substantially better than the in vivo Steelman-Pohley assay (index of precision: greater than 0.1).   The inter- and intra-assay errors expressed as the coefficients of variation of a pooled serum sample are 18 and 12%, respectively.

4) Species nonspecificity   The GAB assay has been shown to be useful for measuring FSH bioactivity in all the mammalian species tested.   In general, pituitary FSH preparations from different species show parallel dose-response curves in the present assay.

5) Parallelism for FSH from different sources   In contrast to the nonparallelism observed in the RIA measurement of FSH from different sources, the GAB assay shows unique parallel dose-response curves for FSH activity present in urine, pituitary and serum samples.

Limitations of the GAB assay include two main areas.

1) Due to the use of in vitro cultures, the present methodology does not take into account potential changes in in vivo metabolism of FSH.   It is conceivable that certain potent FSH preparations in the in vitro assay may have a short half life in vivo, therefore exerting transient actions inside the whole organism.

2) Although pre-treatment with PEG eliminates serum inhibitory substances in the GAB assay, one cannot rule out the possibility that abberant forms of bioactive FSH molecules present in selected samples may be removed by the PEG treatment.   Indeed, a recent study by Skaf et al. (1985) suggests the presence of serum factors that are immunologically distinct from FSH but stimulate (or inhibit) granulosa cell steroidogenesis.   This limitation may be minimized by further increasing the sensitivity of the GAB assay in order to delete the PEG treatment step.

Several precautions should also be noted before the application of the GAB assay:

1) <u>Optimization of conditions for PEG pre-treatment</u>   Although the PEG pre-treatment procedure has been optimized for measurement of serum FSH levels in both human beings and rats, it is important to test the concentrations of PEG necessary to precipitate serum inhibitors when applying the assay to other animal species. Likewise, the recovery of FSH following PEG treatment should be monitored.

2) <u>Interference during cell cultures</u>   Multiple hormones and drugs may directly affect granulosa cell functions resulting in erroneous measurement of FSH bioactivity. Although body fluid samples are usually applied in very small amounts in the present bioassay, one has to rule out possible interfering effect of pharmacological agents such as GnRH agonists in serum following drug treatment. Direct inhibitory effects of GnRH agonists on the granulosa cells (Hsueh and Jones, 1981) can be minimized by the inclusion of GnRH antagonists in all the culture samples (Hsueh and Ling, 1979). Similar monitoring of other drugs potentially effective on the granulosa cells is required. Furthermore, high concentrations (greater than 10^{-6}M) of reduced androgens present in selected samples may interfere with the aromatization reaction and these samples may require charcoal pretreatment to remove the steroids.

3) <u>Interference of estrogen RIA</u>   Because granulosa cell aromatase activity is determined by estrogen production using RIA, it is important to rule out possible artifacts during the determination of aromatase activity. Our preliminary results indicate that the measurement of serum bio-FSH in neonatal rats is interfered with by high levels of serum alpha-fetoprotein (Germain et al., 1978). To minimize the interference of estrogen RIA by serum steroid binders (e.g., alpha-fetoprotein or human sex steroid binding protein) during selected conditions, we have extracted steroids from the granulosa cell culture medium before estrogen RIA. During most physiologic conditions, the amount of serum used is too low to contain enough interfering binding proteins.

It is also important to monitor the concentration of estrogens in body fluids (such as follicular fluid) which may contain high levels of estrogens and result in an overestimate of granulosa cell estrogen production. It may be necessary to pretreat these samples with dextran-coated charcoal to strip endogenous estrogens or to correct for granulosa cell estrogen production by assaying samples before and after cell culture.

SERUM BIOACTIVE FSH LEVELS IN WOMEN

The GAB method has been successfully applied to the measurement of bioactive FSH levels in serum and urine samples from humans. To demonstrate the validity of the GAB assay in measuring bio-FSH levels in human serum, granulosa cells were cultured with increasing concentrations of highly purified human pituitary FSH-3 or a crude pituitary gonadotropin preparation (LER-907) to obtain standard curves. Furthermore, increasing aliquots of human serum pretreated with 12% PEG were also added (Fig. 10). All cultures, including the standard curves, were balanced with PEG-pretreated gonadotropin-free serum (from contraceptive pill users) to obtain a final serum concentration of 4%. Human FSH preparation (LER-907) have a working range in the assay between 0.25 to 2.0 mIU per culture. Our subsequent data are presented in units based on the biological potency of LER-907 for comparison with the RIA results.

Treatment of granulosa cells with increasing aliquots of PEG-pretreated serum from women in various physiologic or pathophysiologic states result in dose-response curves parallel to the standard curve

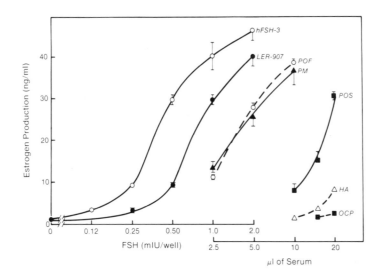

Fig. 10. Dose-dependent stimulation of estrogen production by FSH preparations and by serum from women in various physiologic and pathophysiologic states. Oral contraceptive pill users (OCP), hypothalamic amenorrhea (HA), preovulatory surge (POS), postmenopause (PM) and premature ovarian failure (POF). (From Jia et al., 1986).

(Fig. 10). In contrast, serum from oral contraceptive pill users (OCP) does not stimulate estrogen production at the highest aliquots used. In patients with hypothalamic amenorrhea (HA), the largest aliquot is within the sensitivity range of the assay, and in women during the preovulatory surge (POS), aliquots of 10 to 20 µl result in substantial estrogen production. Furthermore, in postmenopausal women (PM) or patients with premature ovarian failure (POF), aliquots of 2.5 to 10 µl are within the working range of the standard curve. Mean levels of FSH bioactivity in these clinical states are consistent with the clinical presentation: undetectable in oral contraceptive pill users, intermediate in patients with hypothalamic amenorrhea, and highest in postmenopausal women and patients with premature ovarian failure.

Daily serum samples from 7 women throughout ovulatory cycles were assayed for FSH bioactivity or immunoreactivity (Jia et al., 1986). For the purpose of data consolidation and analysis, the day of the LH peak is designated day 0. The mean serum immuno-FSH levels in the 7 women are plotted around the mid-cycle LH peak and revealed a profile in keeping with that found in previous reports (Midgley and Jaffe, 1968; Ross et al., 1970). The serum bio-FSH profile from the same 7 women is very similar to the immuno-FSH profile (Fig. 11). The corresponding B:I ratios range from 1.4 to 3.4, with a mean of 2.5 ± 0.1. The correlation coefficient of the mean values throughout the menstrual cycle obtained by the bioassay and RIA is 0.91 (P < 0.001). Thus, the GAB assay detects FSH bioactivity in all women tested and, in general, the results parallel levels measured by RIA.

URINE BIOACTIVE FSH LEVELS IN WOMEN

Measurement of urinary bio-FSH levels may provide the basis for future noninvasive monitoring of FSH bioactivity in situations, such as puberty onset, when venipuncture is difficult or when prolonged longitudinal

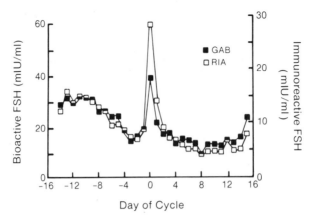

Fig. 11. Mean levels of bioactive and immunoreactive FSH throughout the
menstrual cycle in 7 regularly cycling women, with data centered
around the midcycle LH surge. (From Jia et al., 1986).

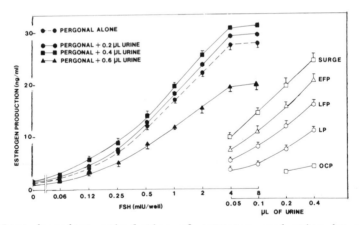

FIG. 12. Dose-dependent stimulation of estrogen production by a urinary
FSH preparation (Pergonal) and human urine from women during the
menstrual cycle, oral contraceptive pill user (OCP), luteal phase
(LP), late follicular phase (LFP), early follicular phase (EFP),
and surge. (From Dahl et al., 1987).

studies are desired. As shown later, the urinary bioassay is also of
great use in animal species without adequate RIAs.

Granulosa cells were cultured with increasing concentrations of
Pergonal, a urinary preparation of FSH or with increasing doses of human
urine with unknown bioactivity (Fig. 12). Pergonal stimulates estrogen
production in a dose-dependent manner with a minimum effective dose of
0.12 mIU per culture. When 0.2 or 0.4 µl/well of gonadotropin-free urine
from contraceptive pill users was added to the standard curve, there was a

slight enhancement of estrogen production as compared to cultures without urine. In contrast, inhibition of Pergonal-stimulated estrogen production was detected when 0.6 µl urine was included. To ensure a constant volume of urine in the total incubation volume of 500 µl all assay samples and standard curves were balanced with the inclusion of 0.4 µl of gonadotropin-free urine from oral contraceptive pill users. Treatment of cells with urine from women during different stages (surge; early follicular phase: EFP; late follicular phase: LFP; luteal phase: LP) of the menstrual cycle resulted in dose response curves parallel to the standard curves; whereas urine from oral contraceptive pill (OCP) users was unable to stimulate estrogen production.

Daily first morning void samples were collected during the menstrual cycle of 6 normal women. The day of the urinary immuno-LH peak is designated as day 0 and the mean urine hormone levels in six women have been plotted around the midcycle LH peak (Fig. 13). All cycles exhibit marked midcycle elevations in urinary bio-FSH levels which coincide with the LH peak. Also, bio-FSH levels increase during the early follicular phase. Although the levels of urine FSH are almost 100-fold higher than those in serum, correlation coefficient analysis of the mean bio-FSH values throughout the menstrual cycle in serum and urine samples indicates that bio-FSH in urine reflects serum levels (P < 0.001).

Urinary estrone conjugates (EC) and pregnanediol-3-glucuronide (PdG) levels were also measured by RIA during the human menstrual cycle. As shown in Fig. 11, urinary estrone conjugate concentration, in general, reflects serum estradiol levels during the menstrual cycle. Furthermore,

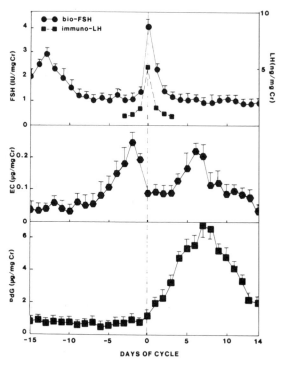

Fig. 13. Mean levels of bioactive FSH, estrone conjugates (EC), and pregnanediol-3-glucuronide (PdG) throughout the menstrual cycle in 6 regularly cycling women, with data centered around the midcycle LH surge. (From Dahl et al., 1987).

the urinary levels of progestin metabolites reflect serum progestrone levels. In conclusion, this study provides the basis for future monitoring reproductive cycles in human and other species using a non-invasive, convenient urinary bioassay method.

AGING STUDIES IN MEN

In healthy men, testis function declines with age, as reflected by decreases in serum testosterone levels and lower sperm production. Also, early studies indicate a decrease in the B/I ratio of LH in older men as compared to young men (Marrama et al., 1984; Warner et al., 1985). In collaboration with Dr. William Bremner and his associates (University of Washington), we have measured serum bio-FSH levels in young and old men. Twenty-three young men (age ranged from 23 to 35 years: mean 27.3 years) and 16 older men (age ranged from 65 to 84 years: mean 71.5 years) were studied. As shown in Fig. 14, immuno-FSH levels are elevated in older men as compared to the young men whereas the bio-FSH levels are comparable for the two populations (Tenover et al., 1987). Consequently, the B/I ratio decreases in older men suggesting microheterogeneity of circulating FSH in the older population. This study suggests the possibility of performing further studies to elucidate the biochemical basis of decreases in the B/I ratio of FSH during aging.

GNRH ANTAGONIST TREATMENT IN NORMAL MEN

Administration of a potent GnRH antagonist in vivo has been shown to suppress gonadotropin levels and spermatogenesis in several animal species (Adams et al., 1986; Weinbauer et al., 1984; Asch et al., 1984). These observations suggest that GnRH antagonists might provide a new approach to fertility regulation in men. Although GnRH antagonists clearly suppress circulating LH levels, only minor decreases in FSH levels were detected

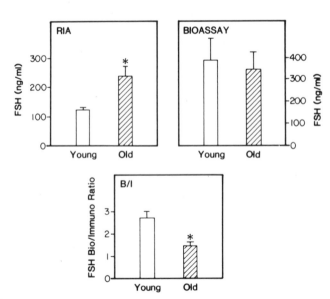

Fig. 14. Serum FSH values by RIA and bioassay and the calculated FSH bioactivity/immunoreactivity ratio in 12 normal young and 13 normal old men. (From Tenover et al., 1987).

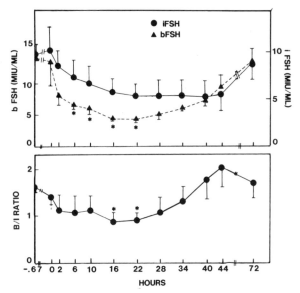

Fig. 15. Effect of the GnRH antagonist on bioactive (b) and immunoreactive
(i) FSH (a), and on the B/I ratio (b) in 4 normal men. (From
Dahl et al., 1986).

using RIA. In men, immuno-FSH levels in serum decrease by only 20 to 30%
(Pavlou et al., 1986). We measured serum bio-FSH levels in 5 normal men
after a single injection (20 mg) of a potent GnRH antagonist (N-Ac-D-Nal-
(2)^1,D-pCl-Phe^2,D-Trp^3,D-h(Arg)Et$_2$^{6},D-Ala^{10}[GnRH]). Although only minimal
suppression of immuno-FSH was detected, pronounced inhibition (79-89%) of
bio-FSH levels occurred (Fig, 15; Dahl et al., 1986). Concomitantly, the
ratio of bio- to immuno-FSH levels decreased drastically after the antagonist administration. These data reinforce earlier expectations that GnRH
antagonists might be potential male contraceptives and provide the finding
of changes in circulating bio- to immuno-FSH levels. These studies also
indicate that the RIA does not consistently provide a good estimate of the
changes in serum bio-FSH levels and the GAB assay is a more discriminating
measure of serum FSH levels than RIA. Measurement of bio-FSH may serve as
a better index for the evaluation of GnRH antagonist action for future
formulation of contraceptive treatment protocols.

The circulating half life of this GnRH antagonist is about 48 h.
Early studies demonstrate that GnRH antagonists have no effect on
granulosa cell aromatase induction by FSH (Hsueh et al., 1983a). To
discount possible interference of the antagonist in the bioassay, a
concentration (10^{-6} M) greater than the estimated circulating levels of
this peptide was added in our cultures. Therefore, the decrease of
bio-FSH levels was not due to the effects of circulating GnRH antagonist
in the bioassay.

The lower B/I ratio, and thus decreasing bio-potency of FSH, after
GnRH antagonist injection may reflect the presence of different molecular
species of FSH. Since FSH is a glycoprotein, changes in FSH activity
could be due to alterations in the degree of glycosylation (Peckham and
Knobil, 1976; Chappel et al., 1983). The deglycosylated FSH has been
shown to behave like antagonists in FSH bioassays (Sairam and Bhargari,
1985). It is possible that treatment with GnRH antagonists may result in

the secretion of deglycosylated or partially glycosylated FSH which decreases the action of intact FSH in the same sample. The rapid decline in both bio- and immuno-FSH levels suggest that the GnRH antagonist may decrease the GnRH-induced FSH biosynthesis and/or release, inhibit the processing of FSH in the late stages of glycosylation, or affect the metabolism of FSH. It is also possible that the GnRH antagonist changes the tertiary structure of the FSH molecule, thereby changing its ability to bind to the receptor but not to the antibodies. However, the variation in the B/I ratio was not due to differential changes in the secretion of alpha and beta subunits of FSH because the RIA only measures intact FSH.

URINE BIOACTIVE FSH LEVELS IN FEMALE LOWLAND GORILLAS

Since pituitary FSH preparations from all mammalian species tested resulted in dose-dependent parallel stimulation of rat granulosa cell estrogen production, the bioassay was further applied to samples from lowland gorillas.

Due to their taxonomic position intermediate between man and monkey, the great apes (including the gorillas, <u>Gorilla gorilla</u>) are of considerable comparative value for assessing the phylogeny of the regulation of the menstrual cycle. The lack of relevant studies is partially attributed to the lack of adequate RIA for the measurement of FSH in these species. With the adaptation of the GAB assay for the measurement of urinary FSH in human samples, we further used this assay to measure bio-FSH in gorilla urine samples. In 4 lowland gorillas studied, menstrual cycle length (28, 29, 31, 38 days) varied considerably. Centering the data around the day of the midcycle immuno-LH peak, the differences between the cycle lengths can be attributed to changes in the length of the follicular phase while the length of the luteal phase remains relatively constant. Due to the difficulty in obtaining gonadotropin-free urine, all gorilla samples were balanced with urine from contraceptive pill users, and the resultant dose response curve was found to be parallel to that induced by Pergonal.

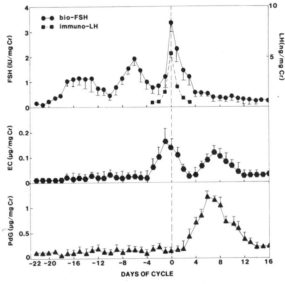

Fig. 16. Levels of bioactive FSH, immunoreactive LH, estrone conjugates (EC), and pregnanediol-3-glucuronide (PdG) throughout the menstrual cycle of 4 lowland gorillas. (From Dahl et al., 1987).

Mean values for 4 cycles were plotted around the mid-cycle LH surge (Fig. 16). In contrast to humans, there are 2 follicular phase elevations in bio-FSH for all the gorilla cycles. The first FSH peak appears 2 to 6 days after the onset of menses whereas the second FSH peak during the late follicular phase occurs 5 to 7 days before the day of the LH peak. The FSH concentration also increases at mid-cycle and then decreases during the luteal phase of the cycle. It is of interest that the second follicular phase elevation of bio-FSH levels precedes the midcycle FSH surge by a fixed duration while the onset of the first follicular phase elevation is variable in accordance with changes in the length of individual follicular phases. Similar to humans, gorillas exhibit conspicuous mid-cycle elevation of urinary estrone conjugates followed by luteal phase elevation of both estrone conjugates and the progestin metabolite. The observed luteal phase increases in estrone conjugates have been detected in human but not monkey cycles, reinforcing the notion that the gorilla is a better model for understanding human cyclicity. This result demonstrates the usefulness of applying the present bioassay to urinary samples in exotic species. The findings may provide better timing for the breeding or artifical insemination of this endangered species as well as the basis for understanding the role of FSH in follicle selection.

SUMMARY

FSH is a key inducer of ovarian follicular and testicular tubule developments. During reproductive life, FSH is released in an orchestrated fashion subject to endocrine control. However, the microheterogeneity of the FSH molecule suggests that studies on circulating FSH bioactivity are essential to understanding gonadal development and function. The study of potential paracrine modulators (estrogens, androgens and growth factors) that could enhance FSH action in granulosa cells and therefore have potential physiologic roles in the development of the dominant follicle led to the development of an in vitro bioassay for FSH. The combined action of enhancing hormones and factors resulted in a granulosa cell assay highly sensitive to the aromatase-inducing action of FSH. This assay is hormone-specific and sensitive, and has been applied to the measurement of FSH bioactivity in serum samples through the use of pre-treatment of serum with polyethylene glycol to remove substances inhibitory in the granulosa cell culture. Recent applications of this assay to human and animal serum and urine samples reveal that physiologic levels of FSH can be measured by this assay. Although bioactive FSH levels in serum and urine samples throughout the normal menstrual cycle generally reflect changes in immunoreactive FSH in serum, substantial decreases in B/I ratio of serum FSH were detected in aging men and normal males treated with a GnRH antagonist. Further applications of the Granulosa Cell Aromatase Bioassay (GAB) in different animals species as well as to other physiologic and pathophysiologic states in humans should be rewarding.

ACKNOWLEDGEMENTS

This work was supported by NIH Research Grants HD-12303 and HD-14084. K.D.D. is the recipient of NIH Postdoctoral Fellowship HD-06875. N.M.C. is currently with the San Diego Zoological Society. AJWH is to whom reprint request and correspondence should be addressed.

REFERENCES

Adams LA, Bremner WJ, Nestor Jr JJ, Vickery BH, Steiner RA, 1986. Suppression of plasma gonadotropin and testosterone in adult male monkeys (Macaca fascicularis) by a potent inhibitory analog of gonadotropin-releasing hormone. J Clin Endocrinol Metab 62:58-63

Adashi EY, Hsueh AJW, 1982. Estrogens augment the stimulation of ovarian aromatase activity by follicle-stimulating hormone in cultured granulosa cells. J Biol Chem 257:6077-6083

Albert A, Rosemberg E, Ross GT, Paulsen CA, Ryan RJ, 1968. Report of the National Pituitary Agency collaborative study on the radioimmunoassay of FSH and LH. J Clin Endocrinol Metab 28:1214-1219

Asch R, Balmaceda JP, Borghi M, 1984. LHRH antagonists in rhesus and cynomolgus monkeys. In: Vickery H, Nestor Jr JJ, Hafex ES (eds). LHRH and its Aanalogs: Contraceptive and Therapeutic Applications. Massachusetts: MTP Press, pp. 107-124.

Beers WH, Strickland S, 1978. A cell culture assay for follicle-stimulating hormone. J Biol Chem: 253:3877-3884

Beers WH, Strickland S, Reich E, 1975. Follicular plasminogen and plasminogen activator and the effect of plasmin on ovarian follicle wall. Cell 6:379-386

Bogdanove EM, Campbell GT, Blair ED, Mula ME, Miller AE, Grossman GH, 1974. Gonad-pituitary feedback involves qualitative change: androgens alter the type of FSH secreted by the rat pituitary. Endocrinology 95:219-228

Bogdanove EM, Schwartz NB, Reichert LE, Midgley AR, 1971. Comparisons of pituitary serum luteinizing hormone (LH) ratios in the castrated rat by radioimmunoassay and OAAD bioassay. Endocrinology 88:644-652

Brown PS, Wells M, 1966. Observations on the assay of human urinary follicle-stimulating hormone by the augmentation test in mice. J Endocrinol 35:199-206

Chappel SC, 1981. The presence of two species of FSH within hamster anterior pituitary glands as disclosed by concanavalin A chromatography. Endocrinology 109:935-942

Chappel SC, Coutifaris C, Jacobs SC, 1982. Studies on the microheterogeneity of FSH present within the anterior pituitary gland of ovariectomized hamsters. Endocrinology 110:847-854

Chappel SC, Ulloa-Aguirre A, Coutifaris C, 1983. Biosynthesis and secretion of follicle-stimulating hormone. Endocrine Rev 4:179-211

Cheng KW, 1975. A radioreceptor assay for follicle stimulating hormone. J Clin Endocrinol Metab 41:581-589

Christiansen P, 1972a. Studies on the rat ovarian augmentation method for follicle-stimulating hormone. Acta Endocrinol 70:636-646

Christiansen P, 1972b. The rat ovarian augmentation method for follicle-stimulating hormone. Acta Endocrinol 70:647-664

Daane TA, Parlow AF, 1971. Serum FSH and LH in constant light-induced persistent estrus: short-term and long-term studies. Endocrinology 88:964-968

Dahl KD, Czekala NM, Lim P, Hsueh AJW, 1987. Monitoring the menstrual cycle of humans and lowland gorillas based on urinary profiles of bioactive follicle-stimulating hormone and steroid profiles. J Clin Endocrinol Metab 64:486-493

Dahl KD, Pavlou SN, Kovacs WJ, Hsueh AJW, 1986. The changing ratio of serum bioactive to immunoreactive follicle-stimulating hormone in normal men following treatment with a potent gonadotropin releasing hormone antagonist. J Clin Endocrinol Metab 63:792-794

Davoren JB, Hsueh AJW, 1986. Growth hormone increases ovarian levels of immunoreactive somatomedin C/insulin-like growth factor I in vivo. Endocrinology 118:888-890

Davoren JB, Hsueh AJW, 1984. Insulin enhances FSH-stimulated steroidogenesis by cultured rat granulosa cells. Mol Cell Endocrinol 35:97-105

Davoren JB, Hsueh AJW, Li LH, 1985. Somatomedin-C augments FSH-induced differentiation of cultured rat granulosa cells. Am J Physiol Endo Metab 249:E26-E33

Diebel ND, Yamamoto M, Bogdanove EM, 1973. Discrepancies between radioimmunoassays and bioassay for rat FSH. Evidence that androgen treatment and withdrawal can alter bioassay-immunoassay ratios. Endocrinology 92:1065-1078

Dorrington JH, Moon YS, Armstrong DT, 1975. Estradiol-17 beta biosynthesis in cultured granulosa cells from hypophysectomized immature rats: stimulation by follicle-stimulating hormone. Endocrinology 97:1328-1331

Dufau ML, Pock R, Neubauer A, Catt KJ, 1976. In vitro bioassay of LH in human serum: The rat interstitial cell testosterone (RICT) assay. J Clin Endocrinol Metab 42:958-969

Erickson GF, Hsueh AJW, 1978. Stimulation of aromatase activity by follicle stimulating hormone in rat granulosa cells in vivo and in vitro. Endocrinology 102:1275-1282

Evans HM, Simpson ME, Tolksdorf S, Jensen H, 1939. Biological studies of the gonadotropic principles in sheep pituitary substance. Endocrinology 25:529-546

Faiman C, Ryan KJ, 1967. Radioimmunoassay for human follicle stimulating hormone. J Clin Endocrinol Metab 77:444-447

Fevold HL, Hisaw FL, Leonard SSL, 1931. The gonad stimulating and the luteinizing hormones of the anterior lobe of the hypophesis. Am J Physiol 97:291-301

Galle PC, Ulloa-Aguirre A, Chappel SC, 1983. Effects of oestradiol, phenobarbitone and luteinizing hormone releasing hormone upon the isoelectric profile of pituitary follicle-stimulating hormone in ovariectomized hamsters. J Endocrinol 99:31-39

Germain BJ, Campbell PS, Anderson JN, 1978. Role of the serum estrogen binding protein in the control of tissue estradiol levels during postnatal development of the female rat. Endocrinology 103:1401-1410

Greep RO, VanDyke HB, Chow BF, 1940. The effect of pituitary gonadotrophins on the testicles of hypophysectomized immature rats. Anat Record 78:88-93

Hillier SG, De Zwart FA, 1981. Evidence that granulosa cell aromatase induction/ activation by follicle-stimulating hormone is an androgen receptor-regulated process in vitro. Endocrinology 109:1303-1305

Hsueh AJW, Adashi EY, Jones PBC, Welsh Jr TH, 1984. Hormonal regulation of the differentiation of cultured ovarian granulosa cells. Endocrine Rev 5:76-127

Hsueh AJW, Adashi EY, Tucker E, Valk C, Ling NC, 1983. Relative potencies of gonadotropin-releasing hormone agonists and antagonists on ovarian and pituitary functions. Endocrinology 112:689-695

Hsueh AJW, Erickson GF, Papkoff H, 1983. Effect of diverse mammalian gonadotropins on estrogen and progesterone production by cultured rat granulosa cells. Arch Biochem Biophys 225:505-511

Hsueh AJW, Jones PBC, 1981. Axim-pituitary actions of gonadotropin releasing hormone. Endocrine Rev 2:437-461

Hsueh AJW, Ling NC, 1979. Effect of an antagonistic analog of gonadotropin release hormone upon ovarian granulosa cell function. Life Science 25:1223-1229

Igarshi M, McCann SM, 1966. A new sensitive bio-assay for follicle-stimulating hormone (FSH). Endocrinology 74:440-445

Jia X-C, Hsueh AJW, 1986. Granulosa cell aromatase bioassay (GAB) for follicle stimulating hormone: validation and application of the method. Endocrinology 119:1570-1577

Jia X-C, Hsueh AJW, 1985. Sensitive in vitro bioassay for the measurement of serum follicle stimulating hormone. Neuroendocrinology 41:445-447

Koenig VL, King E, 1950. Extraction studies of sheep pituitary gonadotrop and lactogenic hormones in alcholic acetate buffers. Arch Biochem 26:219-229

Marrama P, Montanini V, Celani MF, Carani C, Cioni K, Bazzani M, Cavani D, Baraghini GF, 1984. Decrease in luteinizing hormone biological activity/immunoreactivity ratio in elderly men. Matruitas 5:223-231

Midgley RA, 1967. Radioimmunoassay for human follicle-stimulating hormone. J Clin Endocrinol Metab 27:295-299

Midgley Jr AR, Jaff RB, 1968. Regulation of Human Gonadotropins IV. Correlation and serum concentrations of follicle stimulating and luteinizing hormones during the menstrual cycle. J Clin Endocrinol Metab 28:1699-1703

Ny T, Bjersing L, Hsueh AJW, Loskutoff DJ, 1985. Cultured granulosa cells produce two plasminogen activators, each regulated differently by gonadotropins. Endocrinology 116:1666-1668

Pandian MR, Bahl OP, 1977. Labeling of bovine corpus luteal plasma membrane human chorionic gonadotropin or luteinizing hormone (hCG/LH) receptor and its purification and properties. Arch Biochem Biophys 182:420-436

Parlow AF, Reichert LE, 1963. Species differences in follicle-stimulating hormone as revealed by the slope in the Steelman-Pohley assay. Endocrinology 73:740-743

Pavlou SN, Debold CR, Island DP, Wakefield G, Rivier J, Vale W, Rabin D, 1986. Single subcutaneous doses of a luteinizing hormone-releasing hormone antagonist suppress serum gonadotropin and testosterone levels in normal men. J Clin Endocrinol Metab 62:303-308

Peckham WD, Knobil E, 1976a. Qualitative changes in the pituitary gonadotropins of the male Rhesus monkey following castration. Endocrinology 98:1061-1064

Peckham WD, Knobil E, 1976b. The effects of ovariectomy, estrogen replacement, and neuraminidase treatment on the properties of the adenohypophysial glycoprotein hormones of the Rhesus monkey. Endocrinology 98:1054-1060

Peckham WD, Yamaji T, Dierschke DJ, Knobil E, 1973. Gonadal function and the biological and physiochemical properties of follicle-stimulating hormone. Endocrinology 92:1660-1666

Reichert LE, Bhalla VK, 1974. Development of a radioligand tissue receptor assay for human follicle stimulating hormone. Endocrinology 94:483-491

Reiter EO, Goldenberg RL, Vaitukaitis JL, Ross GT, 1972. Evidence for a role of estrogen in the ovarian augmentation reaction. Endocrinology 91:1518-1522

Richards JS, Midgley Jr AR, 1976. Protein hormone action: a key to understanding ovarian follicular and luteal cell development. Biol Reprod 14:82-94

Roche PC, Bergert ER, Ryan RJ, 1985. A simple and rapid method using polyethyl-treated filters for assay of solubilized LH/hCG receptors. Endocrinology 117:790-792

Rosenberg E, Solod EA, Albert A, 1971. Luteinizing hormone activity of human pituitary gonadotropin as determined by the ventral prostrate weight and the ovarian ascorbic acid depletion methods of assay. J Clin Endocrinol Metab 24:714-728

Ross GT, Cargille GM, Lipsalt MB, Rayford PL, Marshall JR, Strott CA, Rodband D, 1970. Pituitary and gonadal hormones in women during spontaneous and induced ovulatory cycles. Rec Prog Horm Res 26:1-62

Ryan RJ, Faiman C, 1968. Proceedings of the informational symposium on protein and polypeptide hormones, Excerpta Medica. International Congress Series No. 161, Part I, pp. 129-146

Ryan KJ, 1969. Chemistry of pituitary gonadotrophins. In: E. Diezfalusy (ed.) Karolinska Symposa on Research Methods in Reproductive Endocrinology, Acta Endocrinol Suppl 12:300-324

Sairam MR, Bhargari GN, 1985. A role for glycosylation of the alpha subunit in transinduction of biological signals in glycoprotein hormones. Science 229:65-67

Sherins RJ, Vaitukaitis JL, Chrambach A, 1973. Physical characterization of hFSH and its desialylation products by isoelectric focusing and electrophoresis in polyacrylamide gel. Endocrinology 92:1135-1141

Siegal HS, Siegal PS, 1964. Genetic variation in chick bioassays for gonadotropins. Virginia J of Science 15:204-217

Simpson ME, Evans HM, Li CH, 1950. Effect of pure FSH alone or in combination with chorionic gonadotropin in hypophysectomized rats of either sex. Anat Rec 106:247

Skaf RA, MacDonald GJ, Shelden RM, Moyle WR, 1985. Use of antisera to FSH to detect non-FSH factors in human serum which modulate rat granulosa cell steroidogenesis. Endocrinology 117:106-113

Smith BD, 1961. The effect of diethylstilbestrol on the immature rat ovary. Endocrinology 69:238-245

Smith BD, Bradbury JT, 1961. Ovarian weight response to varying doses of estrogens in intact and hypophysectomized rats. Proc Soc Exp Biol Med 107:946-952

Smith PE, 1926. Ablation and transplantation of the hypophysis in the rat. Anat Rec 32:221-226

Steelman SL, Pohley FM, 1953. Asssay of the follicle-stimulating hormone based on the augmentation with human chorionic gonadotropin. Endocrinology 53:604-616

Subramanian MG, Gala RL, 1986. Do prolactin levels measured by RIA reflect biologically active prolactin. J Clin Immuno 9:42-46

Tanaka T, Shiu RPC, Gout PW, Beer CT, Noble RL, Friesen HG, 1980. A new sensitive and specific bioassay for lactogenic hormones: measurement of prolactin and growth hormone in human serum. J Clin Endocrinol Metab 51:1058-1063

Taymore ML, and Miyata J, 1969. Discrepancies and similarities of serum FSH and LH patterns as evaluated by different assay methods. Acta Endocrinol Suppl 142:324-357

Uberoi NK, Meyer RK, 1967. Uterine weight of the immature rat as a measure of augmentation of pituitary gonadotrophins by human chorionic gonadotrophin (hCG). Fertil Steril 18:420-428

Van Damme M-P, Robertson DM, Marana R, Ritzen EM, and Diczfalusy E 1979 A sensitive and specific in vitro bioassay method for the measurement of follicle-stimulating hormone activity. Acta Endocrinol 91:224-237

Wang C, 1983. Luteinizing hormone releasing hormone stimulates plasminogen activator production by rat granulosa cells. Endocrinology 112:1130-1132

Wang C, Leung A, 1983. Gonadotropins regulate plasminogen activator production by rat granulosa cells. Endocrinology 112:1201-1207

Warner BA, Dufau ML, Santen RJ, 1985. Effects of aging and illness on the pituitary testicular axis in men: qualitative as well as quantitative changes in luteinizing hormone. J Clin Endocrinol Metab 60:263-268

Weinbauer GF, Surmann FY, Akhtar FB, Shah GV, Vickery BH, Nieschlag E, 1984. Reversible inhibition of testicular function by a gonadotropin releasing hormone antagonist in monkeys (Macaca fasicularis). Fertil Steril 42:906-914

Welsh Jr TH, Jia X-C, Hsueh AJW, 1984. Forskolin and phosphodiesterase inhibitors stimulate rat granulosa cell differentiation. Mol Cell Endocrinol 37:51-60

Wide L, 1982. Male and female forms of human follicle-stimulating hormone in serum. J Clin Endocrinol Metab 55:682-688

Wide L, 1985. Median charge and heterogeneity of human pituitary FSH, LH and follicle-stimulating hormone. II. Relationship to sex and age. Acta Endocrinol 109:190-197

Zondek B, Ascheim S, 1925. Experiments on function and hormone of ovary. Arch G Gynaekol 127:250-292

Zhuang LZ, Adashi EY, Hsueh AJW, 1982. Direct enhancement of gonadotropin-stimulated ovarian estrogen biosynthesis by estrogen and
clomiphene citrate. Endocrinology 110:2219-2221

RECEPTOR-MEDIATED ACTION WITHOUT RECEPTOR OCCUPANCY:

A FUNCTION FOR CELL-CELL COMMUNICATION IN OVARIAN FOLLICLES

Willaim H. Fletcher[1], Craig V. Byus[2], and Donal A. Walsh[3]

[1]Loma Linda University, School of Medicine and J.L. Pettis
Memorial VA Hospital, Loma Linda, CA 92357
[2]Div. of Biomedical Sciences and Dept. of Biochemistry
University of California, Riverside CA 92521
[3]Dept. of Biological Chemistry, University of California
School of Medicine, Davis, CA 95616

INTRODUCTION

In 1959 Furshpan and Potter reported the presence of electrical transmission at the crayfish giant motor synapse. Soon thereafter a similar means of cell-cell communication was described for Mauthner neurons of the goldfish brain (Furshpan and Furakawa, 1962) and for mammalian smooth muscle (Dewey and Barr, 1962). In each instance contacting cells, when impaled with microelectrodes, were found to exchange ions, (perhaps K^+, Na^+, Cl^-) through low resistance pathways connecting adjacent cytoplasms without using extracellular routes. This ionic (or electrotonic) coupling was bidirectional and occurred with an impedance 3-4 orders of magnitude less than that of the transmembrane potential. Since these observations were made in excitable tissues (nerve and muscle) their functional significance was easily reconciled (Bennett et al., 1963).

However, when Kanno and Loewenstein (1966) found that fluorescein ($M_r 332$) when microinjected into one epithelial cell passed within seconds into contacting epithelial cells, it became apparent that the phenomenon of ionic coupling was probably only indicative of a process that was neither confined to excitable tissues nor to ions alone. Thus was born the concept that intercellular communication could be broadly relevant to cellular regulation (Loewenstein and Kanno, 1966; Loewenstein, 1968; Furshpan and Potter, 1968; Sheridan, 1968; Bennett and Trinkhaus, 1970; Gilula et al., 1972).

Added support for this notion came from work by Pitts and co-workers (Subak-Sharpe et al., 1969; see Pitts, 1980; Pitts and Finbow, 1986) who demonstrated that intimately contacting cells could exchange products of metabolism. That is, mutant cells having a defective purine salvage pathway which prevented them from utlizing a specific nucleotide precursor could, nevertheless, incorporate a metabolite of that molecule into their nucleic acid as long as they contacted cells with a functional metabolic pathway. Whether this process of "metabolic cooperation" really reflects a physiologic mechanism is difficult to establish because the assay is based on the uptake of radiolabeled precursors by metabolically competent donor cells and the appearance of a labeled compound/s in nuclei of contacting, recipient mutant cells. Unfortunately, the identity of material in recipient cell nuclei is unknown. More importantly, while the assay is considered to measure metabolic cooperation, no change in precise metabolic events is demonstrated to occur as a result of metabolite transfer.

A recent variation of the metabolic cooperation assay (see Pitts and Finbow,

1986) does suggest that the phenomenon has relevance to <u>in vivo</u> metabolic events. In this study (Yotti et al., 1979; Yancey et al., 1982) Chinese hamster V79 cells that are resistant to 6-thioguanine (6-TG) toxicity grow normally when cultured alone in the presence of 6-TG. When cocultured with V79 cells that are sensitive to 6-TG, resistant V79 cells that are in contact with the sensitive cells are killed by the 6-TG. However, under conditions that prevent metabolic cooperation between resistant and sensitive V79 cells the resistant cells are no longer killed by 6-TG while the sensitive cells are. Clearly, an important, if lethal, metabolite can pass intact from sensitive to resistant cells that are able to cooperate metabolically suggesting that a similar pathway could be used for the <u>in vivo</u> exchange of normal metabolites. The fact that dansylated amino acids, which are fluorescent, can pass from one cell of a contacting pair to its partner (Johnson and Sheridan, 1971) lends support to the physiologic potential for this process.

The clearest evidence that "metabolic cooperation" assays probably are reflective of normal metabolic events was reported by Sheridan et al. (1979). They established that the degree of cooperation between mutant (enzyme-deficient) cells and wild-type partners (the nucleotide donors) was not dependent on the number of wild-type cells in the coculture but was due to the mutant cells stimulating their donor partners to produce more nucleotides. As described by Sheridan et al. (1979) this is true metabolic cooperation and not simply metabolite transfer without metabolic effects.

MORPHOLOGIC SUBSTRATE FOR CELL-CELL COMMUNICATION

Robertson (1963), using high resolution electron microscopy, examined Mauthner cell lateral dendrites and observed specialized regions of contact between membranes of adjacent cells that could serve as the site for the electrotonic coupling Furshpan and Furakawa (1962) found in the same preparation. Robertson's images showed that the membranes were differentiated into a honeycomb of 9nm diameter facets when viewed parallel to the plane of the membrane. These subunits were spaced 10-11nm center-to-center and were aggregated into a nearly hexagonal lattice. When seen in cross section the specialized junction consisted of plasma membranes in such intimate contact that the intercellular space seemed non-existent. Robertson's findings were confirmed and extended to include muscle by Dewey and Barr (1964) who dubbed the junctional specialization the "nexus" (L. to bind; connection). Subsequently, Revel and Karnovsky (1967) found that when the extracellular space of hepatic tissue, was filled with the electron dense tracer colloidal lanthanum hydroxide, junctions unmistakably like those described by Robertson (1963) were crisply detailed. The novelty of Revel and Karnovsky's (1967) finding relative to that of Robertson (1963) was that the lanthanum revealed the existence of a thin, 2-3nm, extracellular cleft separating the membranes of neighbor hepatocytes rather than the near fusion of membranes Robertson (1963) had seen. Because of the markedly narrow but regular intercellular space at the junctional region, Revel and Karnovsky christened the structure a "gap junction", the name most commonly used since.

The freeze-fracture technique was used to give an added dimension to gap junction structure, specialization of the cells membrane interior. With this procedure the junctions of liver (Chalcroft and Bullivant, 1970) and cardiac muscle (Somer and Johnson, 1970; McNutt and Weinstein, 1970) were examined and shown to be essentially similar. In both tissues the cytoplasmic leaflet (P-face) of the plasma membrane was specialized by aggregates of 7-8nm diameter particles that were arranged in a polygonal, sometimes hexagonal, lattice with a center-to-center spacing of 10-11nm. This arrangement of subunits was precisely like that demonstrated by Robertson (1963) and Revel and Karnovsky (1967) in en face views of the gap junction membrane in thin sections.

In freeze-fracture replicas the extracellular leaflet (E face) of gap junctions contained 2-3nm diameter depressions, also arranged in a polygonal lattice with a spacing similar to that of particles in the P-face half membrane. Chalcroft and Bullivant (1970), using the complementary freeze-fracture replica technique, demonstrated a near-perfect register of the P-face particles and E-face pits, indicating that the particles (gap junction subunits) most likely spanned the entire membrane width. In subsequent studies it became evident that, while the thin-section appearance of gap junctions was similar in most vertebrate tissues, there was a great deal of pleiomorphism of gap junction structure in freeze-fracture replicas (reviewed by Gilula, 1975; Bennett and Goodenough, 1978; Zampighi et al., 1980; Hertzberg et al., 1981; Larsen and Risinger, 1985).

For example, in reaggregating Novikoff Hepatoma cells electrotonic coupling develops within 5-15 min after cells contact one another and freeze-fracture reveals that diminutive gap junctions begin to appear in the same time frame (Johnson et al., 1974). These assembling junctions consist of 10 or so P-face particles surrounded by a particle-poor zone of "non-junctional" membrane and a rim of large (ca. 10nm diameter) particles that are distinct from the 8nm particles of the junction itself. This ensemble of particle-poor zone and large particles was called a "formation-plaque" (Johnson et al., 1974) that has been recognized in other tissue/cell preparations wherein gap junction manufacture may be in progress (reviewed by Gilula, 1975; Bennett and Goodenough, 1978; Hertzberg et al., 1981; Larsen and Risinger, 1985).

In rabbit eye the gap junctions of ciliary epithelial cells consist of P-face particles polygonally aggregated into rows that are separated by particle-free aisles (Kogon and Pappas, 1975). Similar arrangements of gap junctional particles have been seen uniting granulosa cells of rabbit (Albertini et al., 1975) and rat ovarian follicles (Fletcher, 1973). It has been suggested that these structural variations in the aggregation of gap junctional particles may reflect differences in the functional status of the junctions as regards their role in intercellular communication (Perrachia, 1973; Perrachia and Perrachia, 1980a,b; Zampighi and Unwin, 1979; Zampighi et al., 1980). While there is evidence both to support and refute this view the observations to date are inconsistent and vary depending on the tissue/cell preparation used. It is, nevertheless, clear that the channels of communication between connected cells can fluctuate from an open to a closed state (Bennett and Goodenough, 1978; Perrachia and Perrachia 1980a,b). Whether this is due to gross changes in the particle packing pattern of gap junctions or more subtle shifts in the molecular configuration of the junction proteins, as some evidence suggests (Zampighi and Unwin, 1979), or to some permutation of both processes remains to be established more conclusively.

The most elegant morphologic studies of gap junctions have been those using x-ray and electron optical diffraction techniques with three-dimensional electron-density mapping and image reconstruction. From these studies the molecular architecture of gap junctions has been revealed to 1.8nm resolution (Makowski et al., 1977; Casper et al., 1977; Baker et al., 1983; Zampighi and Unwin, 1979; Unwin and Zampighi, 1980; Unwin and Ennis, 1984). It is inappropriate here to discuss the full details of these superb studies but a summary of the results is useful. Basically, gap junctions are composed of protein subunits of M_r27kDa (Hertzberg and Gilula, 1979; Hertzberg, 1984) that pack as hexamers forming the essential junctional subunit, the connexon (Makowski et al., 1977; Caspar et al., 1977; Baker et al., 1983). In the center of each connexon is a 1.5nm diameter polar (hydrophilic) channel that apparently spans the entire membrane from cytoplasm to extracellular spaces (Zampighi and Unwin, 1979: Zampighi et al., 1980, 1985 ; Makowski et al., 1977; Makowski, 1985). When connexons of adjacent cell membranes come into register the cytoplasms of connected cells are effectively in continuity. This then is the physical substrate for ionic coupling, fluorescent dye transfer and metabolic cooperation (Gilula et al., 1972) as described above. While there is substantial disagreement as to how the communicating channels are opened and closed (see Makowski, 1985; and Zampighi and Simon, 1985 for back-to-back arguments) results of these studies do show the existence of a transmembrane aqueous channel, of presumable low resistance, that could previously only be speculated on.

Cumulatively all of the studies discussed thus far make it clear that when gap junctions are present cells can communicate ions, fluorescent dyes and amino acids and metabolites or their degradation products in a direct cell-to-cell fashion. The question is, are gap junctions the exclusive mediators of these phenomena? If so, what biologic processes are regulated by intercellular communication? In vertebrate cells and tissues the answer to the first part of the question is almost certainly *yes!* as determined by independent and completely different approaches.

In the first (Zampighi et al., 1985) gap junction membrane was purified to near homogenity and individual connexons obtained by detergent solubilization. When the connexons were then inserted into planar lipid bilayers (artificial membranes) it was found that transbilayer electrical conductance was significantly increased over that of naked lipid bilayers and conductance rose in approximate proportion to the number of connexons inserted per bilayer. Freeze-fracture was used to visually estimate the number of junction subunits that had been inserted into the artificial membranes. Importantly, the progressive rise in conductance with incremental amounts of connexons inserted into the bilayers paralleled in amplitude the quantum jumps (single channel increases) of conductance found in rat lacrimal glands (Neyton and Trautmen, 1985) and cardiac myocytes (Veenstra and DeHaan, 1986). Clearly then, the observations of Zampighi et al. (1985) most likely are reflective of gap junctional communication in vivo. Studies using purified connexons inserted into liposomes (see Peracchia and Girsh, 1985a,c) have come to similar conclusions.

The only caveat to the studies by Zampighi et al. (1985) and Peracchia and Girsh, (1985a,b,c) is that the connexons they used were obtained from eye lens fiber gap junctions which are immunologically distinct from the junctional proteins of other tissues (Hertzberg et al., 1982; Paul and Goodenough, 1983; Hertzberg and Skibbens, 1984) although here to there is some controversy (Traub and Willecke, 1982; Janssen-Timmen et al., 1986).

Additional evidence that gap junctions are the mediators of intercellular communication comes from recent studies employing affinity column purified polyclonal antibody to the M_r27kDa liver gap junction protein. When injected into hepatocytes, myocytes or neurons known to be ionically coupled, the antibody eliminates coupling in one minute or less and prevents the cell-cell passage of fluorescent dyes (Hertzberg et al., 1985). Preimmune serum or non-specific antibodies had no such effect. The antibody also was used to establish what had long been suspected: that the junctional protein was a substrate for phosphorylation by cAMP-dependent protein kinase (Saez et al., 1986).

In elegant experiments using normal rat kidney (NRK) cells that bear a proviral insert from a temperature-sensitive mutant of Rous sarcoma virus (LA-25), Atkinson et al. (1981, 1986) demonstrated an unequivocal relationship between gap junctions and cell-cell communication. In these studies the NRK-LA25 cells are fully communication competent at 39C which is non-permissive for the viral gene. However, upon 30 min. of shifting to 33C, the virus permissive temperature, channels of communication are eliminated as are morphologically identifiable gap junctions. The viral gene product is a pp60^{src} tyrosine protein kinase which Azarnia and Lowenstein (1984) have confirmed as being a probable mediator leading to a reduction of cell-cell communication.

Recently, the gap junction protein has survived the rites of passage through molecular biology. Human and rat liver cDNAs have been cloned and shown to code for the gap junction protein (Paul, 1986; Kumar and Gilula, 1986). Though done independently the cDNAs prepared by these two laboratories differ by only one base in a 1.5 kB cDNA probe. Using Northern blot analysis Paul (1986) found that his rat liver cDNA recognized homologous sequences in rat brain, stomach and kidney RNAs but did not hybridize with RNA from rat heart or bovine lens. The collective results from Paul (1986) and Kumar and Gilula (1986) indicate there is fairly broad sequence homology in the gap junctions of various tissues but there also exist explicit distinctions between junctions of liver and those of the lens and heart. Thus, there is more than one structural form of gap junction. Whether or not this means the different forms mediate different functions has not been shown.

The possibility of physiologically different forms of gap junctions has, however, been shown for the eye lens epithelial system (Miller and Goodenough, 1986), and was found to correlate with structurally distinctive appearances of the junctions in freeze-fracture replicas. In this system lens epithelial cells form gap junctions with each other and with lens fiber cells (a special epithelial cell). Increasing intracellular pH, by exposure of tissues to high CO_2 levels, caused a reduction in communication between epithelial cells but not between them and the lens fiber cells. While these results suggest a potential physiologic diversity of intercellular communication in a single tissue, the effect of that diversity on defined processes is unknown.

THE FUNCTION OF INTERCELLULAR COMMUNICATION

In contrast to the electrophysiological and biochemical phenomena associated with gap junctions and cell-to-cell communication the physiological relevance of this process has been difficult to prove. In electrogenic tissues (nerve and muscle) the junctions probably serve as electrical synapses (Furshpan and Potter, 1959; Dewey and Barr, 1962; Bennett and Goodenough, 1978; Hertzberg et al., 1985) and possibly mediate functional ionic and metabolic coupling (Trachtenberg and Pollen, 1970; Bennett and Goodenough, 1978). However, in non excitable cells, which are the most numerous and functionally diverse sites of intercellular communication, little is known of a role for gap junctions. They are hypothesized to allow the cell-cell exchange of growth controlling molecules (Loewenstein, 1968, 1979, 1981) and there is persuasive evidence supporting this contention (Azarnia and Loewenstein, 1984).

During embryonic development gap junctions unite most cells and there are changes in intercellular communication that coincide with critical events in the differentiation process. For example, in the mouse, electrical coupling, fluorescent dye transfer and gap junctions appear at the beginning of compaction in 8 cell embryos, but not before that time (Lo and Gilula, 1979a,b). Communication between cells of the trophoblast and other layers also is found early in development but is lost in the later stages of the preimplantation embryo. As germ layers begin to be defined there is an even more extensive separation, wherein cells of some regions communicate with each other but not with cells of other developmental compartments (Lo & Gilula, 1979b).

In addition to the existence of developmental compartments the junctions of embryonic cells can be selectively permeable to ions, relative to larger fluorescent tracer dyes (Bennett, 1973), suggesting there may be a heirarchy for the size of molecules that pass through junctions and regulate growth and differentiation. If so, the results of Verselis et al. (1986) indicate that the basis for this selectivity is a change in the number of open communicating channels of gap junctions and not an alteration in diameter of the channel pores. An application of these findings to the developing ovarian follicle is considered later.

That gap junctions most likely have a critical role in embryogenesis can be inferred from the results obtained by Warner et al. (1984). Affinity column purified polyclonal antibody to the Mr27kDa gap junction protein was microinjected into one cell of an 8 cell Xenopus embryo. When the fluorescent dye lucifer yellow was microinjected into the same cell or one of its daughter cells at the 32 cell stage there was no dye transfer to any adjacent cells. However, cells that had been injected with preimmune serum or that recieved no injection maintained normal channels of communication. Antibody injected embryos that were allowed to develop well into the organogenic stages manifested dramatic asymetric teratologies such as lack of an eye and underdevelopment of the brain on one side but not the other (Warner et al., 1984). Although, as the authors point out, these results are not conclusive, they are consistent with a regulatory role for intercellular communication in embryonic development (see Furshpan and Potter, 1968; Sheridan, 1968; Bennett, 1973). More recent studies, however, have shown that, under some circumstances, the preimmune

IgG used by Warner et al. (1984) can reduce junctional communication by 80% relative to that of uninjected controls (Warner and Gurdon, 1987). It appears that in this system more studies must be done before conclusions are made about the relevance of gap junctions to embryogenesis.

In adult tissues, whose component cells unite via gap junctions, a variety of roles for intercellular communication have been postulated. The most long-held of these is its involvement in growth control, including repair and neoplastic transformation. Loewenstein (1966, 1968, 1977, 1979, 1981) has consistently championed the concept that intercellular communication has developed as a means for controlling the appropriate growth and differentiation of tissues (for a comprehensive review of his models, methods and rationale see Loewenstein, 1979, 1981). Early work established that there was a correlation between controlled cell growth and the presence of gap junctions whereas uncontrolled growth occurred in the absence of cell junctions and communication (Loewenstein, 1968). However, subsequent studies revealed that the correlation was not a strict one. Borek et al. (1969) reported that transformed epithelial cells had lost their gap junctions and were communication-incompetent whereas most fibroblastic cells retained junctional communication whether they were normal or neoplastically transformed. Somewhat at contrast with this are the observations of Sheridan (1970) and Johnson and Sheridan (1971) that many tumorogenic cells remain coupled ionically and have gap junctions. This includes epithelial cells such as Novikoff hepatoma (albeit grown in suspension culture) suggesting there is no predictable relationship between neoplastic transformation of epithelial or fibroblastic cells and the loss or retention of gap junctions. Perhaps the best example of this is the observation that when BALB/c3T3 cells undergo methylchloanthrene induced transformation the transformed cells were found as capable of fluorescent dye transfer amongst themselves as were surrounding non-transformed cells. However, transformed cells did not communicate with the normal cells (Enomoto and Yamasaki, 1984). Although no morphological assessments of gap junctions were done the results indicate that upon neoplastic transformation the cells were still able to communicate with each other, but the junctional channels must be altered because transformed cells could not communicate with the normal cells.

Obviously then, a loss of junctional communication is not an obligatory property of the fully transformed state, although this says nothing about the status of gap junctions during the transformation process. Even though there is no information about cell-cell communication throughout the many stages of transformation all evidence to date indicate there is no invariant relationship between gap junctions and neoplastic growth. While there have been numerous studies on this topic (for reviews see Azarnia and Larsen, 1976; Loewenstein, 1979, 1981; Pitts and Finbow, 1986) this conclusion appears to hold true to the present time. This does not mean there is no relevance of cell-cell communication to the control of normal growth and differentiation, as will be discussed there clearly is a relationship. However, it is not yet possible to say that a given treatment will lead to loss of gap junctional communication and a concommitant expression of neoplastic growth. Nor as just cited above, can it be said that the presence of gap junctions assures normal growth control. These points have been often explained previously (Loewenstein, 1979, 1981) but continue to be confused in the literature.

A major reservation about most studies examining the involvement of intercellular communication in any biologic phenomenon, such as growth control and neoplastic transformation is that they are done using cultured cells and the data may not apply fully to in vivo transformation. Studies where attempts have been made to use an in vivo response as a bioassay for the physiologic appropriateness of intercellular communication lend credence to the idea that communication is important to an intact tissue and not just a phenomenon found in vitro.

Goshima (1969, 1971, 1974, 1976) appears to be the first to use cocultures for examining ionic coupling and the contact-dependent passage of bioactive molecules from cell-to-cell. In his preparations fetal mouse myocardial cells, which beat with an intrinsic frequency upon isolation, were cultured with FL cells which form gap junctions with the muscle cells (Goshima, 1971). When an FL cell bridged two myocytes their beat frequency was synchronized, usually at the rate of the cell with the highest frequency, which was also reflected in the FL cells by depolarizing currents of identical rhythmicity. However, when FL cells were pretreated with trypsin then cultured with muscle cells they no longer formed an ionic bridge and each myocyte beat at its intrinsic frequency. The trypsin effect is most likely due to removal of an extracellular piece of the junctional connexons (Makowski et al., 1977; Zampighi and Unwin, 1979) thereby abolishing the intercellular channels. Clearly, these findings indicate that contact-dependent cell-cell communication can regulate a biologic process, the beat frequency of cardiac myocytes.

Lawrence et al. (1978) used an elegant adaptation of Goshima's preparations by employing ovarian granulosa cells cocultured with cardiac myocytes. The cells readily formed gap junctions with each other and were ionically coupled, capable of exchanging fluorescent dyes and cooperating metabolically. When cocultures were exposed to norepinephrine which heart cells, but not ovarian cells, had high affinity receptors for, the ovarian cells secreted plasminogen activator. Ovarian cells cultured alone did not respond to norepinephrine, unless it was used at much higher concentrations. Conversely, treatment of cocultures with FSH caused an increased beat frequency of heart cells in contact with granulosa cells, but had no such effect on heart cells cultured alone. Lawrence et al. (1978) concluded that this two-way intercellular communication coincided only with conditions wherein the heart and ovarian cells contacted via gap junctions, making it likely they were responsible for establishing the channels of communication.

While the studies by Goshima (1969, 1971, 1974, 1976) and Lawrence et al. (1978) make it clear that cell-cell communication may have a physiologic role, they provided no evidence for the mechanism through which cellular regulatory processes could occur. To examine this, the developing ovarian follicle and its constituent epithelium of granulosa cells proved to be an excellent system for study.

CELL-CELL COMMUNICATION IN OVARIAN FOLLICLES: EXPERIMENTAL OBSERVATIONS

Anderson (1971) and Merk (1971) coincidentally reported preliminary observations of gap junctions uniting granulosa cells of, respectively, mouse and rat ovaries. In both cases it was found that cells of secondary or preovulatory follicles were joined by gap junctions whereas prior to development of a thecal layer and beginning antrum formation junctions were not observed.

The morphology of ovarian junctions has been described on numerous occasions (Merk et al., 1972, 1973; Espey and Stutts, 1972; Albertini and Anderson, 1974, 1975; Bjersing and Cajander, 1974; Albertini et al., 1975; Anderson and Albertini, 1976; Amsterdam et al., 1976; Fletcher, 1979; Burghardt and Anderson, 1979; Amsterdam et al., 1981; Burghardt and Anderson, 1981; Campbell and Albertini, 1981; Larsen et al., 1981; Burghardt and Matheson, 1982). However, there are surprisingly few instances where true intercellular communication between granulosa cells has been studied (Gilula et al., 1978; Lawrence et al., 1978; Heller et al., 1981; Dekel et al., 1981; Schultz et al., 1983; Schultz, 1985). In these reports ionic coupling, fluorescent dye transfer and metabolic cooperation were used to assess the potential function of the junctions. As will be described shortly it has only recently been possible to demonstrate that granulosa cells communciate bioactive signals that carry out a defined molecular function.

As shown in Fig. 1 the gap junctions uniting granulosa cells conform to the descriptions reported by Robertson (1963) and Revel and Karnovsky (1967), with the exception that the ovarian junctions are both numerous and large (distance between arrows in panel a) and exceed the extent of junctional membrane in most tissues. On close inspection of thin section preparations the junctions (panel b, between black arrowheads) consist of two cell membranes that are tightly apposed but separated by a 2-3 nm intercellular space. When viewed en face in tissue treated with the colloidal

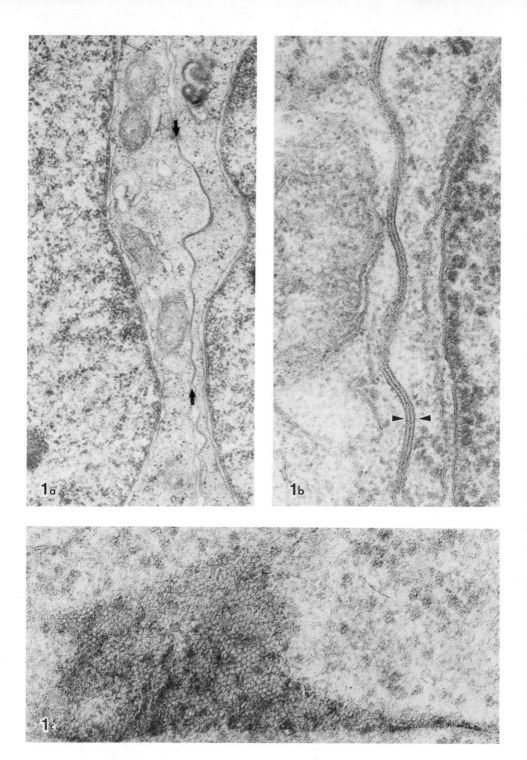

Fig. 1. Gap junctions between granulosa cells of preovulatory follicles from ovaries of 4-day cycling rats killed at 1000h on proestrus. The junctions can be extremely large, extending for 3um or more in cross sections (distance between arrows of panel a). When the extracellular space is filled with colloidal lanthanum (panel b) the tracer occupies a 2-3nm intercellular cleft that separates the closely apposed membranes of adjacent cells (arrowheads). Thus, the entire junctional width is about 16-17nm (distance between arrowheads). When viewed <u>en face</u> (panel c) the junctional region consists of polygonally arranged 8nm facets, in the center of which are occasionally seen an electron density that may correspond to the aqueous channel.

lanthanum technique (Revel and Karnovsky, 1967) the junctional membrane is composed of polygonally aggregated 8-9 nm subunits (panel c) that usually do not have a consistent center-to-center spacing. As discussed previously these subunits are the basic functional units (Zampighi et al., 1985) of gap junctions, the connexons (Makowski et al., 1977).

In freeze-fracture preparations (Fig. 2) the gap junctions joining granulosa cells in preovulatory follicles of four-day cycling rat present at least three distinctive morphologic appearances. The most common is a macular aggregate of 8-8.5 nm diameter P face particles that have no regular packing pattern (Fig. 2a). The E face

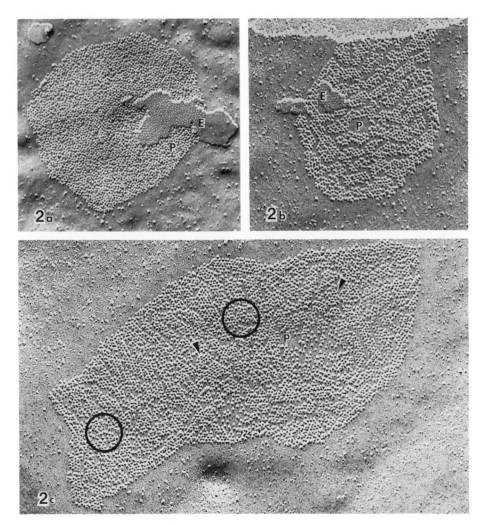

Fig. 2. Freeze-fracture appearance of gap junctions uniting granulosa cells in preovulatory follicles of 4-day cycling rats killed at 1000h on proestrus. The E fracture face represents the outer leaflet of the membrane of the cell facing the reader, while the P face replica derives from the inner leaflet of the membrane of the cell on the other side of the page. Junctions having 3 different arrangements of P face particles (presumably the connexons) are encountered in mature ovarian follicles. In one the particles are randomly arranged (panel a). In other cases the particles are in quasi-hexagonal aggregates that are separated from each other by particle-free aisles (panel b). The third type is a mixture of the first two, having some regions with a geometric particle array (arrowheads), but other areas where particles seem randomly arranged (circles).

of the conjoined cell membrane possesses 3-4 nm pits within the junctional region but is otherwise essentially unembellished. A particle-poor halo surrounding the junctional plaque on the P fracture faces encompasses this and the other formats of ovarian gap junctions.

A less common junctional phenotype is shown in Fig. 2b. Here, the P face particles are aggregated into quasi-hexagonal arrays and there are particle-free aisles that separate these aggregates as demonstrated also by Albertini and Anderson (1974) in the ovary and Kogon and Pappas (1975) in the eye. Again, some portion of the E fracture face with its distinctive pits is present as is a particle-poor perijunctional zone.

The least frequently encountered type of ovarian gap junction is shown in Fig. 2c. These junctions have an appearance that combines features of the junctions shown in panels a and b. That is, the P-face particles are in some regions loosely packed (circle) while elsewhere the particles appear to be in a quasi-hexagonal array (arrowheads). A feature that is most unique for this junction phenotype is the absence of a conjoined E fracture face, even when the junctions are quite large, as shown here. Even so, the particle-poor perijunctional zone is present as usual.

As indicated previously, there is some evidence that structural differences in gap junctions could represent functional differences such as the closed or open state of the communicating channels (Perrachia, 1973; Perrachia and Perrachia, 1980; Zampighi and Unwin, 1979; Miller and Goodenough, 1986). According to these concepts the junction of panel a would be open while that of panel b could be closed and the junction in panel c partly open. However, because of their low frequency and absence of an adjoining E fracture face, we suggest that junctions like that of panel c may be hemi-junctions consisting of the specialized membrane of one granulosa cell which is not intimately apposed to a neighbor cell. To be sure this is speculative but it would be a more conservative mechanism for decreasing the number of communicating channels than is afforded by interiorization and degradation of complete gap junctions (reviewed by Larsen and Risinger, 1985) as has been proposed. Further, this mechanism, disjunction without interiorization, might allow a rapid uncoupling and recoupling of cells such as might be needed in follicles responding to abrupt hormonal changes. Additional experimentation should be done to determine, whether or not, and how ovarian gap junctions can be maintained in an open (communication competent) or closed (no communication) status.

Merk et al. (1972) found that in 21-day-old hypophysectomized rats daily estrogen treatment led to an increase in the number of gap junctions joining granulosa cells of antral follicles whereas without such treatment the incidence of junctions declined significantly within 7 days after pituitary ablation. Whether these results were due to a direct effect of the estrogen or to its influence on gonadotropin release from residual pituitary fragments is not clear. Subsequent studies of rabbit preovulatory follicles by Bjersing and Cajander (1974) confirmed most of the results of Merk et al. (1972) but they found that hCG injection caused a decline in the incidence of gap junctions uniting adjacent granulosa cells within 6-8 hours. This reduction was accompanied by an increase in the number of annular gap junctions which according to Espey and Stutts (1972) are spherical inclusions of gap junctional membrane that has been removed in response to hormonal stimuli prior to ovulation.

It seems that gonadotropin support is not needed for the maintenance of gap junctions which persist uniting granulosa cells up to 90 days after pituitary removal (Burghardt and Anderson, 1981). Under these conditions gap junctional membrane can, nevertheless, be amplified by treating hypophysectomized rats with hCG and estrogen (Burghardt and Anderson, 1981) or with FSH (Burghardt and Matheson, 1982). Cumulatively, these results demonstrate that the ovarian gap junctions respond to exogenous hormones (Merk et al., 1972; Bjersing and Cajander, 1974; Burghardt and Anderson, 1981; Burghardt and Matheson, 1982) and to cyclic changes in the natural hormonal milieu (Espey and Stutts, 1972), all of which are consistent with their having a role in follicular development (Anderson, 1971; Merk, 1971).

However, these observations are more of a correlative nature and do not reveal the mechanism/s by which gap junctional communication could regulate any biologic process, although they do suggest such an involvement.

In support of this we found (Fletcher and Everett, 1973) that intercellular contact via gap junctions appeared to have a role in regulating the preovulatory phase of luteinization in rats with 4-day estrus cycles. At 1000 h on proestrus ripe follicles contained a well-developed stratum granulosum composed of one to two layers of basal columnar cells and the remaining muralia of polygonal cells. In thin section and freeze-fracture preparations it appeared that all granulosa cells joined neighbors via gap junctions. By 2300 h of proestrus, about 9 h after the gonadotropin surge, basal granulosa cells had hypertrophied significantly and now contained numerous lipid droplets that had not been seen previously. The polygonal granulosa cells showed similar morphologic reorganization but to a lesser degree, particularly as regards lipid accumulation. In all cases, granulosa cells retained gap junctional contacts. However, granulosa cells that had lost contact with the follicle wall and were free in the fluid filled antrum (antral cells) had undergone none of these morphological alterations. By 0300 h of estrus structural reorganization was advanced and evident throughout the mural granulosa while the antral cells were unchanged from their appearance at 2300 h. These observations were interpreted to suggest that intercellular communication through gap junctions could allow the passage of hormone-induced "signals", generated in basal granulosa cells, throughout the stratum granulosum. Thus, cells deeper in the layer would be retarded in their responses and those not in contact with the muralia would not receive the signals and, therefore, not differentiate.

Alternatively, there could be a hormone gradient with cells nearest the blood supply binding most of the stimuli thereby denying granulosa cells nearer the antrum access to hormone. This seems unlikely based on electron microscopic observations which show that, except for the numerous close intercellular appositions at gap junctions, the interstitial space of the stratum granulosum is quite large and should furnish no diffusion barrier (Albertini and Anderson, 1974). The later point has been directly established by Payer (1975) who demonstrated that intravenously injected molecules as large as 400 kDa rapidly (5-15 min) appear in the antral space of developing follicles. Given this, the slow spread of preovulatory luteinization from basal to periantral granulosa cells (Fletcher and Everett, 1973) may well be due to the intercellular communication of hormone-induced signals from receptor-bearing basal cells to contacting neighbors. If so, it implies that only a subpopulation of granulosa cells may express hormone receptors.

There appears to be general agreement that LH receptors of preovulatory follicles are predominately confined to granulosa cells nearest the basement lamina while periantral cells bind little of the hormone (cf. Richards and Midgley, 1976; Amsterdam et al., 1981). In contrast, the distribution of receptors for FSH is less clear. When I^{125}-FSH is topically applied to cryotome sections of whole ovaries granulosa cells of nearly all sized follicles seem to uniformly bind the hormone (Richards and Midgley, 1976). But when I^{125}-FSH is injected intravenously and rats killed at time intervals thereafter the hormone binds preferentially to the basal cells of the stratum granulosum of mature follicles (Fig. 3a in Richards and Midgley, 1976). Studies that agree with this are those finding that only a third or so of granulosa cells harvested from preovulatory follicles of cycling swine possess receptors for FSH (May et al., 1980; Murray and Fletcher, 1984; Fletcher and Greenan, 1985). Although the precise source of these cells is unknown their numbers are consistent with their being from the basal stratum granulosum. Clearly, additional efforts need to be made to determine the location of FSH receptors in mature follicles and their time of appearance and distribution throughout follicle development. Perhaps more important will be assessing whether or not, or when, the receptors couple to a definable biologic mechanism. Based on our earlier observations (Fletcher and Everett, 1973) and more recent ones (Murray and Fletcher, 1984; Fletcher and Greenan, 1985) it is probable that, in preovulatory follicles, FSH receptors will be found to predominate in the basal stratum granulosum.

The question is, what advantage to the developing follicle is there in confining LH and FSH sensitivity to a portion of granulosa cells? Likewise, do the remaining cells of a given follicle receive hormone-induced signals from those receptor bearing cells and, if so, do gap junctions mediate this process? Most importantly, are there definable events that are regulated by hormone-initiated cell-cell signal transfer?

The functional importance of gap junctions uniting granulosa cells has been the subject of various hypotheses which differ in details from each other but all have their origins in the postulates originally proposed by Loewenstein (1966, 1968, 1977, 1979, 1981) and Sheridan (1968) (also see Sheridan and Atkinson, 1985).

It has been suggested that the junctions act to coordinate granulosa cell differentiation during the hormone-dependent stages of follicle development (Anderson, 1971; Merk et al., 1972; Albertini and Anderson, 1974). Initially this was thought to be development from beginning antrum formation onwards but more recent observations have shown that in immature rats, gap junctions unite presumptive granulosa cells as soon as they contact each other and surround an oocyte (Fletcher, 1979). This distinction has profound implications because it means that the junctions may be involved in the initiation of follicle development and are likely to help orchestrate that process long before sexual maturation and cyclic changes in gonadotropin levels commence. If so, and the presence of the junctions coincides with senstivity of young follicles to hormonal influence, primary follicles should be able to respond to pituitary or intraovarian hormones at early stages in their development. Funkenstein et al. (1980) found that the ovaries of four- day-old rats, which consisted almost exclusively of small, preantral follicles, respond to FSH by secreting eştrogen using testosterone as a precursor. The authors observed that nearly all follicles had only one or two layers of granulosa cells which are known (Presl et al., 1974) to have a small number of FSH receptors. Combined with our observation that the granulosa cells of these preantral follicles are probably united by gap junctions (Fletcher, 1979) there is ample, if circumstantial, evidence that hormone responsiveness and the capacity for cell-cell communication accrue coincidentally and may be interrelated in regulating the development and economy of ovarian follicles.

While this type of paradigm offers some interesting possibilities for regulating follicle development it proved to be a difficult one to test. Predominantly this is because most hormones that are known to affect follicular recruitment and maturation require cAMP mediation in order to exert their effects (reviewed by Channing et al., 1982). If intercellular communication is involved in follicle regulatory processes it follows that the communicating channels should be able to transmit hormone-initiated signals that control cAMP-dependent processes. Insofar as is known, all actions requiring cAMP-mediation are actually carried out by the cAMP-dependent protein kinases (Walsh and Cooper, 1979; Lohmann and Walters, 1984; Beebe and Corbin, 1986). The cAMP-dependent protein kinases are unique amongst regulatory enzymes because, in order to become active, the holoenzyme must dissociate, giving rise to regulatory subunit (R) and free catalytic subunits (C) according to Equation 1 (Walsh and Cooper, 1979).

$$\text{(Eq. 1.)} \qquad R_2C_2 + 4cAMP = R_2cAMP_4 + 2C$$

The R subunit exists as a dimer and is the cAMP-binding protein while the two monomeric C subunits are cAMP-independent but catalytically active and able to phosphorylate substrate proteins which, in the simplest case, with their function altered, effect the response appropriate to the initial stimulus (Walsh and Cooper, 1979). This reaction and its potential for regulation by intercellular communication are shown in Fig. 3. As the cartoon depicts cells that are in gap junctional contact would be able to share soluble ions, metabolites and other molecules up to $M_r 1000$ (Simpson et al., 1977). This may include, but not be limited to, cyclic nucleotides, phospholipids, Ca^{++}, prostaglandins, polyamines, steroids, and other molecules that contribute to regulation of the protein kinases. Our studies to date have been confined to the cAMP-dependent protein kinases but a role for the calcium-sensitive

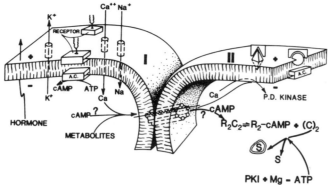

Fig. 3. Diagramatic view of two cells (I and II) united by a gap junction (here represented by a single connexon with its aqueous channel) to depict the events that may be involved in gap junction mediated cell-to-cell transfer of hormone-induced signals. Here, cell I binds hormone whose receptor is coupled to adenylate cyclase, leading to cAMP generation and the transfer of signals to cell II wherein protein kinase dissociation occurs, just as it does in cell I, which generated the hormone-induced signal.

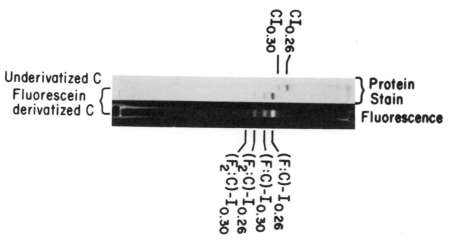

Fig. 4. In vitro formation of a complex between the F:C probe and the protein kinase inhibitor protein. 9.43ug of native catalytic subunits and 2.45ug of inhibitor protein were added and electrophoresed in the left lane and detected by staining with Coomassie (brilliant) blue. The middle lane shows complex formation when 12ug of the F:C probe and 3.06ug of inhibitor protein are added, electrophoresed and detected by protein staining. This same band was photographed using fluorescein optics prior to protein staining (right hand lane). The F:C binds to inhibitor, but to a large extent (80%), the complex consists of C subunit derivitized with a single FITC molecule (F:C) while in the remainder (20%) the C subunit bears two fluorescent molecules (F_2:C). (From Van Patten et al., 1986)

phospholipid-dependent protein kinase C (P.D. kinase) needs to be investigated.

At the outset our working hypothesis was that if gap junctional communication had a role in regulation of ovarian granulosa cells it was probable that activation (dissociation) of the cAMP-dependent protein kinases was involved. The hypothesis predicted (Fig. 3) that upon binding hormone, responsive granulosa cells would dissociate protein kinase and generate "signals" that would be transferred through gap junctions leading to protein kinase dissociation in recipient cells. Clearly, to test this requires an assessment of protein kinase dissociation in individual cells.

It was evident from the work of Steiner and his colleagues (Steiner et al., 1978; Browne et al., 1980; Murtaugh et al., 1982) that immunocytochemical methods would be of limited use in detecting the active (free) forms of the C and R subunits. This is due to the fact shown in Equation 1 that the cAMP-dependent protein kinases are regulated by a dissociation-reassociation reaction rather than by ligand-binding allosteric control (Walsh and Cooper, 1979; Beebe and Corbin, 1986). However, all reported antibodies to the C or R subunit bind to those antigens whether they are in the holoenzyme (inactive) or free (active) form, with equivalent avidity (Murtaugh et al., 1982; Kuettel et al., 1984, 1985). Therefore, upon stimulation there is no net increase in antibody binding over that seen in unstimulated preparations. To overcome this problem direct cytochemical probes were developed that bind exclusively to C subunit (Fletcher and Byus, 1982; Byus and Fletcher, 1982) or R subunit (Fletcher et al., 1986; Van Patten et al., 1986) only when they are in the dissociated, active state.

To localize free C subunit the protein kinase inhibitor protein (PKI) is purified to homogenity using the catalytic subunit affinity column described by Cheng et al. (1986) as a final step. The PKI is then fluoresceinated (Fletcher and Byus, 1982) yielding F:PKI which retains the full catalytic subunit inhibitory activity of native inhibitor protein (Ashby and Walsh, 1972). The F:PKI binds free C subunit with an affinity $(Ki{\sim}2x10^{-10}M)$ log orders greater than that of any substrate $(Ka{\sim}10^{-6}M)$ but does not bind C when it is complexed with R nor does F:PKI bind to the cGMP-dependent protein kinase. Since originally described (Fletcher and Byus, 1982; Byus and Fletcher, 1982) the F:PKI has been used on many cell and tissue systems (Murray et al., 1985; Murray and Fletcher, 1986; and others discussed below) for detecting the spatial and temporal kinetics of protein kinase dissociation and the intracellular distribution of free C subunits. While the F:PKI probe is extremely sensitive and reliable for locating free C subunits it is unable to complex with C bound to R subunit and cannot therefore localize the cAMP-dependent protein kinase holoenzyme, which antibody procedures can do (Murtaugh et al., 1982).

The procedure for identifying the intracellular sites of free R subunit has only recently been described (Fletcher et al., 1986; Van Patten et al., 1986). For this, homogenously pure C and R subunit are recombined to form holoenzyme which is fluoresceinated and then dissociated, by addition of cAMP, to yield F:C. After rechromatography the F:C probe is homogenous, retains full biologic activity, and is exquisitely sensitive for locating intracellular sites of R subunit (Fletcher et al., 1986). Under appropriate conditions the F:C probe also appears to specifically complex with the endogenous PKI (Fletcher et al., 1986; Van Patten et al., 1986). This is shown in Fig. 4 from a protocol wherein F:C was mixed with purified PKI yielding [F:C]I complexes of two types, designated 0.26 or 0.30. The two differ in their apparent isoelectric points and are thought to represent complexes of one form of PKI with either of two forms of C subunit, both of which are fluoresceinated with one (80%) or two (20%) mol/mol of fluoresceinisothiocyanate (Van Patten et al., 1986). Both $[F:C]I_{0.26}$ and $[F:C]I_{0.30}$ have equivalent catalytic activity. As a cytochemical probe F:C binds to free R subunit but not to R in the holoenzyme form, nor to the cGMP-dependent protein kinase or any identifiable substrate (Fletcher et al., 1986; Van Patten et al., 1986). However, when the F:C is used with excess cAMP it is prevented from complexing with R subunit (see equation 1) and it appears instead to bind to endogenous PKI (Fletcher et al., 1986; Van Patten et al., 1986). The F:C probe will not bind R subunit in the holoenzyme form, nor can it distinguish R^{I}

from R^{II} subunits which highly specific antibodies can do (Steiner et al., 1978; Kuettel et al., 1984) On the other hand F:C can differentiate free R subunit from the holoenzyme, which antibodies cannot.

The cytochemical methods are done using cells or tissues that have been exposed to hormone, cyclic nucleotide analogs or other agents then fixed in anhydrous acetone at $-30°C$, rehydrated and stained with one of the fluoroprobes. Prepared in this way cellular C and R subunits retain their biologic potency which is necessary in order for the F:PKI or F:C probes, respectively, to complex with them. Beginning with the development of the F:PKI procedure for localizing free C subunit it became possible to address the question of the functional significance, if any, of intercellular communication to follicular development.

To do so granulosa cells from preovulatory follicles of cycling swine were cocultured with an ACTH-sensitive clonal adrenocortical tumor cell (Y-1). The theory of those studies is the basis for the cartoon of Fig. 3. That is, the two cell types in culture each have specific and unique receptors for peptide hormone, that the other cell type lacks (FSH for granulosa cells and ACTH for their Y-1 partners). In both cases hormone action requires cAMP-mediation and both cell types are known to form gap junctions. Thus, if one member of a mixed cell pair is stimulated with hormone specific for it, then cAMP levels should rise in that cell causing protein kinase dissociation and, in time, similar events should occur in its contacting partner, providing gap junctions actually can transmit bioactive signals. When those studies were done, the observations made essentially followed this scenario but with some interesting variations.

Upon exposure to 8BrcAMP ovarian granulosa cells and Y-1 cells in coculture both dissociated cAMP-dependent protein kinase in a time and dose-dependent fashion (Murray and Fletcher, 1984). When stimulated with FSH however, only granulosa cells and Y-1 cells contacting ovarian cells dissociated protein kinase whereas Y-1 cells not in heterologous contact failed to activate the enzymes. Likewise, when ACTH was used as a stimulus the Y-1 cells dissociated protein kinase as did contacting, but never non-contacting, granulosa cells. This bidirectional exchange of hormone-induced signals clearly had a biologic effect because in FSH stimulated cocultures the Y-1 cells secreted fluorogenic steroids, which are not produced by granulosa cells, whereas FSH had no such effect on Y-1 or granulosa cells cultured alone (Murray and Fletcher, 1984). Control experiments were done to assure that there was no microfocal fusion of Y-1 and granulosa cell membranes and that no anomalous binding of hormone occurred. That is, FSH bound only to ovarian cells and ACTH only to the Y-1 adrenocortical tumor cells.

These data strongly support the contention that hormone-induced intercellular communication can serve a biologic role (Goshima, 1969, 1971, 1974, 1976; Lawrence et al., 1978) and demonstrate that the molecular basis for this is the regulation of cAMP-dependent protein kinases. Given that the protein kinases are important mediators of the protein phosphorylation reactions that regulate metabolic processes as well as proliferation and growth (Walsh and Cooper, 1979; Boynton and Whitfield 1983; Beebe and Corbin 1986) it is likely that this is a major means by which intercellular communication can modulate follicle development (Hunziker-Dunn and Jungmann, 1978a,b,c; Hunziker-Dunn 1982, 1986).

In the context of the basic hypothesis as schematized in Fig. 3 this would mean that granulosa cells need not possess hormone receptors in order to respond to specific stimuli as long as those cells lacking receptors are in gap junctional communication with cells having receptors. As already indicated it does appear that granulosa cells of a given follicle may not uniformly express receptors for LH or FSH (Richards and Midgley, 1976; May et al., 1980; Amsterdam et al., 1981; Murray and Fletcher, 1984). However, what is the relationship between hormone receptor occupancy and hormone-induced cell-cell communication that regulates the cAMP-dependent protein kinases? The coculture preparations using Y-1 tumor cells and granulosa cells do not answer that question because they can not be used to examine signal transfer amongst granulosa cells alone.

To do so preovulatory phase granulosa cells were cocultured with CL-1D cells which do not form gap junctions (Azarnia et al., 1981). This was done to limit the incidence of contact between granulosa cells which even in low confluence cultures send out long processes that connect them with granulosa partners that are 50-100uM distant. In the presence of a 3-fold or greater excess of CL-1D cells granulosa cells usually occur isolated or in clusters of 2-4 cells surrounded by CL-1D cells. The latter do not establish communicating junctions amongst themselves or with nearby granulosa cells while ovarian cells make the junctions with each other as usual.

Cocultures were exposed to hCG and hormone binding sites identified with monoclonal primary antibody (a gift of Dr. Robert Lundak, Techniclone Inc., Santa Ana, CA) and secondary antibody conjugated to rhodamine while protein kinase dissociation was shown using the F:PKI probe for localizing free C subunit. It was found that granulosa cells which bound hCG also dissociated cAMP-dependent protein kinase. Ovarian cells that did not possess hormone receptors nevertheless dissociated protein kinase when, and only when, they contacted reponsive, hormone binding cells. In no case did CL-1D cells bind hormone or contain active enzyme (Fletcher and Greenan, 1985).

Interestingly, there were instances where granulosa cells contacting hormone binding responsive partners failed to dissociate protein kinase (Fletcher and Greenan, 1985), an observation similar to that previously made using Y-1:granulosa cell cocultures (Murray and Fletcher, 1984). In both cases this was interpreted to indicate that either there were no junctions at the point of contact or that if junctions were present, the communicating channels were closed. Clearly, the latter case could have important implications for regulating hormone action in developing follicles but further studies are needed just to verify the initial observation.

Overall, these studies (Murray and Fletcher, 1984; Fletcher and Greenan, 1985) allow the conclusion that gap junctions can furnish the channels of communication for signals that regulate the dissociation of cAMP-dependent protein kinase. Further, in tissues (such as the ovarian follicle) whose cells unite via gap junctions, hormone effects requiring cAMP mediation can be manifest in cells that lack hormone receptors.

An example of this is shown in Figure 5. Cells obtained from a pleural aspirate of a patient with ovarian carcinoma were subcloned twice, each time selecting for their ability to bind hCG. The phase contrast image (panel a) shows the general morphological characteristics of these cells. In panel b all cells excepting one (asterisk) express hormone receptors as determined immunocytochemically (Fletcher and Greenan, 1985); although, the amount of hormone bound varies appreciably amongst cells. Even so, protein kinase dissociation (panel c) measured using F:PKI to locate free C subunit occurs as a consequence either of direct hormone binding (compare cells of panel b to those of panel c) or of hormone-induced intercellular transfer of signal to the receptorless cell (asterisk in panels a-c).

It could be that cells such as the one in Fig. 5 actually possess ample receptors that upon binding hormone initiate protein kinase dissociation, but are insufficient in number to be detected by the immunocytochemical procedure. This is unlikely because as disscussed previously only one-third or fewer of preovulatory phase granulosa cells bind hCG (LH) or FSH as determined by autoradiography (Presl et al., 1974; May et al., 1980; Amsterdam et al., 1981; Murray and Fletcher, 1984) or immunocytochemistry (Fletcher and Greenan, 1985). Thus, if our methods for receptor detection are insufficently sensitive there should be many instances where isolated granulosa cells (as in the cocultures with excess CL-1D cells) dissociate protein kinase but lack appreciable receptors. Given that only the cells that bind detectable amounts of hormone, or are in contact with those cells, dissociate protein kinase it seems that the cytological approaches are amply sensitive to reveal receptor occupancy which causes, by one mechanism or the other, enzyme activation.

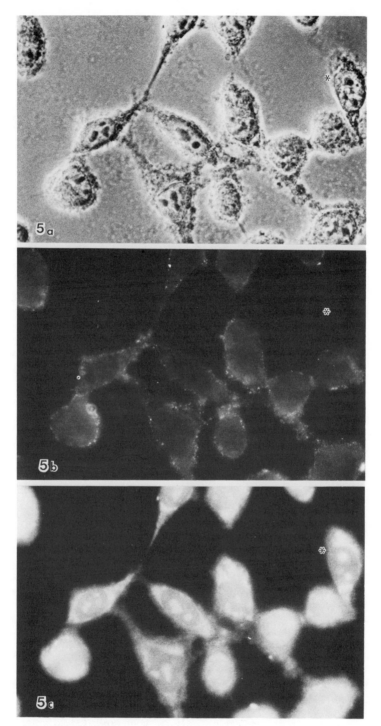

Fig. 5. Clonal cells from a patient with ovarian carcinoma. The cells have an essentially fibroblastic morphology (panel a). Upon exposure to 1 ug/ml hCG and location of hCG bound to receptors by immunocytochemistry (panel b) all cells but 1 (asterisk) bind hormone, (spots of fluorescence at cell periphery), in varying amounts. Even so, that cell dissociated protein kinase and free C subunits are detectable in it as well as in its hormone binding neighbors (compare panels b and c). Electron microscopy has shown that these cells unite by small gap junctions.

How then, do these observations relate to our understanding of the hormone regulation of follicle differentiation?

CELL-CELL COMMUNICATION IN OVARIAN FOLLICLES: PHYSIOLOGIC SIGNIFICANCE

Given that gap junctions unite granulosa cells at the earliest stages of follicle development (Fletcher, 1979) and remain a consistent feature at least until the time of ovulation (Gilula et al., 1978) they must be considered to be of potential functional importance throughout the entirety of follicle development. It should be noted that while only junctional communication between granulosa cells can be considered here cummulus granulosa cells (the corona radiata) establish gap junctions with the oocyte (Anderson and Albertini, 1976) and the junctions allow the bidirectional transfer of ions and fluorescent dyes (Gilula et al., 1978). This avenue of communication may well be involved in regulating the resumption of oocyte meiosis (Maller and Krebs, 1977) and in metabolic and other responses of the oocyte to hormone (Dekel et al., 1981; Heller et al., 1981; Schultz et al., 1983a,b). At the moment we have no data that directly bear on this potentially important aspect of follicle control about which there is very little known concerning the molecular mechanisms of hormone-initiated regulation (Maller and Krebs, 1977; reviewed by Schultz, 1985).

In contrast, as described above, there is convincing evidence that communication amongst granulosa cells allows the exchange of hormone-induced bioactive molecules that can regulate the cAMP-dependent protein kinases and, thereby specific physiologic events (Murray and Fletcher, 1984; Fletcher and Greenan, 1985; Murray and Fletcher, 1986). Classically, the action of peptide hormones upon granulosa cells has been thought to necessitate their binding to target cells within which effects are manifested (Richards and Midgley, 1976; reviewed by Channing et al., 1982), primarily by cAMP-mediated mechanisms (Hunziker-Dunn and Jungmann, 1978 a-c; Hunziker-Dunn 1982, 1986).

The ability of granulosa cells to respond to hormone even though they lack detectable hormone receptors is clearly a new aspect of the post-recognitive mechanisms of hormone action. It can be argued that the data leading to this conclusion are derived from in vitro preparations which may not reflect accurately the circumstances occurring in vivo. A number of observations strongly diminish this possibility. To begin with, in the granulosa:Y-1 cell cocultures, stimulation with FSH initiated the production of steroids by Y-1 cells (Murray and Fletcher, 1984). This is an appropriate physiologic response of Y-1 cells to increasing cAMP levels that, given the experimental design, could only have occurred as a result of signals communicated to Y-1 cells from FSH responsive granulosa cells. In other studies (Fletcher et al., in preparation) we found that when 4-day cycling rats were injected with FSH at 1000 h of proestrus the hormone caused a rapid (within 15 min) rise in protein kinase dissociation in basal granulosa cells but not in deeper layers of the follicle wall or in the theca interna. By 60 min after hormone injection all mural granulosa cells showed activation of the enzyme as did the cummulus granulosa cells, but to a lesser extent. This progressive activation of protein kinase beginning in the basal granulosa cells and moving from them to the more periantral granulosa cells is consistent with our previous observations of progressive cellular differentiation in preovulatory follicles (Fletcher and Everett, 1973) and with the fact that hormone binding appears to be confined to the basal cells (Amsterdam et al., 1974; Richards and Midgley, 1976). A most reasonable interpretation of these results is that signals that are generated in receptor-bearing cells are communicated to neighbor cells and therein regulate dissociation of cAMP-dependent protein kinase. This is fully in accord with the data from in vitro preparations (Murray and Fletcher, 1984; Fletcher and Greenan, 1985) indicating that in both instances the underlying mechanisms are probably similar.

It appears then that intercellular communication as a means of regulating the cAMP-dependent protein kinases does have a potentially influential role in follicle development. Given the results to date it is evident that one effect of gap junction mediated signal transfer is to allow greater numbers of cells to respond to hormones than are able to bind those agents. This would have the advantage of keeping the stratum granulosum under tightly coordinated hormonal control relative to that which would occur if all granulosa cells equivalently expressed hormone receptors. In the latter case a response to hormone exposure would be a cacophony of protein kinase activation whose kinetics, amplitude and duration of response represent a compendium of the response by each cell. In contrast, if only a subpopulation of cells bind hormone and respond by transmitting signals to neighbors the resultant activation of protein kinase would be well-coordinated temporally as would the amplitude and duration of response.

Another feature of a developing follicle having subsets of cells each with specific and unique hormone receptors is that the follicle should be capable of concurrently responding to more than one stimulus. It is easy to envision that in the face of multiple hormone exposure (eg. FSH, LH and oocyte maturation inhibitor or inhibin) each subset of cells may be capable of mounting a response that is additive, antagonistic or independent of the response by other cell ensembles within the same follicle. This is obviously speculative but it is consistent with the observations of heterogenous hormone binding and of protein kinase dissociation by preovulatory follicles as discussed above. This postulated capacity of receiving numerous stimuli would have the effect of allowing the follicle to act as an integrator with its ultimate response to the hormone milieu being the result of interactions via cell-cell communication between the various hormone sensitive subsets of granulosa cells.

AREAS FOR FUTURE STUDY

The effect of receptor occupancy and protein kinase dissociation by individual cells needs to be correlated with a biologic endpoint, such as steroid production, by the responding cell. This could be done using an additional in situ assay like the reverse hemolytic plaque technique.

The studies thus far have revealed that cell-cell communication does transmit hormone-induced signals from receptor-bearing cells to receptorless partners. However, the identity of the signal/s actually being transmitted is unknown. This extremely important question needs to be answered.

The question of cell subsets with distinctive populations of hormone receptors residing in a given follicle should be looked at. Flow cytometric analysis of granulosa cells whose hormone occupied receptors have been labeled with specific antibodies seems a reasonable approach that should answer the basic question.

Hormone-induced signal transfer that regulates the cAMP-dependent protein kinases should be tested in viable cells. By this means classical assays of intercellular communication like ionic coupling could be done in concert with an examination of enzyme activation thereby further assuring that gap junctions indeed mediate the transmission of the hormone-induced signals.

Support: USPHS grants HD 13704, HD 21318, AM 21019: AM 07310, CA 23743

REFERENCES

Albertini DF, Anderson E, 1974. The appearance and structure of intercellular connections during the ontogeny of the rabbit ovarian follicle with particular reference to gap junctions. J Cell Biol 63:234-250
Albertini DF, Fawcett DW, Olds PJ, 1975. Morphological variations in gap junctions of ovarian granulosa cells. Tissue & Cell 7:389-405
Albertini DF, Anderson E, 1975. Structural modifications of lutein cell gap junctions during pregnancy in the rat and mouse. Anat Rec 181:171-194

317

Amsterdam A, Koch Y, Liberman ME, Lindner HR, 1975. Distribution of binding sites for human chorionic gonadotropin in the preovulatory follicle of the rat. J Cell Biol 67:894

Amsterdam A, Knecht M, Catt KJ, 1981. Hormonal regulation of cytodifferentiation and intercellular communication in cultured granulosa cells. Proc Natl Acad Sci USA 78:3000-3004

Anderson E, 1971. Intercellular junctions in the differentiating graafian follicle of the mouse. Anat Res 169:473 (abstract)

Anderson E, Albertini DF, 1976. Gap junctions between the oocyte and companion follicle cells in the mammalian ovary. J Cell Biol 71:680-686

Atkinson MM, Menko AS, Johnson RG, Sheppard JR, Sheridan JD, 1981. Rapid and reversible reduction of junctional permeability in cells infected with a temperature-sensitive mutant of avian sarcoma virus. J Cell Biol 91:573-578

Atkinson MM, Anderson SK, Sheridan JD, 1986. Modification of gap junctions in cells transformed by a temperature-sensitive mutant of rous sarcoma virus. J. Membrane Biol. 91:53-64

Ashby CD, Walsh CD, 1972. Characterization of the interaction of a protein inhibitor with adenosine 3':5'-monophosphate-dependent protein kinases. J Biol Chem 247:6637-6642

Azarnia R, Larsen WJ, 1976. Intercellular communication and cancer. In: DeMello (ed.), Intercellular Communication. New York: Plenum Publishing, pp. 145-172

Azarnia R, Dahl G, Loewenstein WR, 1981. Cell junction and cyclic AMP: III. Promotion of junctional membrane permeability and junctional membrane particles in a junction-deficient cell type. J Membrane Biol 63:133-146

Azarnia R, Loewenstein WR, 1984. Intercellular communication and the control of growth: X. Alteration of junctional permeability by the src gene. A study with temperature-sensitive mutant Rous sarcoma virus. J Membrane Biol 82:191-205

Baker TS, Caspar DLD, Hollingshead CJ, Goodenough DA, 1983. Gap junction structures. IV. Asymmetric features revealed by low-irradiation microscopy. J Cell Biol 96:204-216

Beebe SJ, Corbin JD, 1986. Cyclic Nucleotide-dependent protein kinases. In: The Enzymes, vol. 17, pp. 43-111

Bennett MVL, Aljure E, Nakajima Y, Pappas GD, 1963. Electrotonic junctions between teleost spinal neurons: electrophysiology and ultrastructure. Science 141:262

Bennett MVL, 1966. Physiology of electrotonic junctions. Ann NY Acad Sci 137:509-539

Bennett MVL, Trinkaus JP, 1970. Electrical coupling between embryonic cells by way of extracellular space and specialized junctions. J Cell Biol 44:592-606

Bennett MVL, 1973. Function of electrotonic junctions in embryonic and adult tissues. Fed Proc 32:65-75

Bennett MVL, Goodenough DA, 1978. Gap junctions, electrotonic coupling and intercellular communication. Neuro Res Prog Bull 16:373-486

Bjersing L, Cajander S, 1974. Ovulation and the mechanism of follicle rupture. IV. Ultrastructure of membrana granulosa of rabbit graffian follicles prior to induced ovulation. Cell Tiss Res 153:1-14

Borek C, Higashino S, Loewenstein WR, 1969. Intercellular communication and tissue growth. IV. Conductance of membrane junctions of normal and cancerous cells in culture. J Membrane Biol 222:78-86

Boynton AL, Whitfield JF, 1983. The role of cyclic AMP in cell proliferation: a critical assessment of the evidence. Adv Cyclic Nucleotide Res 15:193-294

Browne CL, Lockwood AH, Su J-L, Beavo JA, Steiner AL, 1980. Immunofluorescent localization of cyclic nucleotide-dependent protein kinase on the mitotic apparatus of cultured cells. J Cell Biol 87:336-345

Burghardt RC, Anderson E, 1979. Hormonal modulation of ovarian interstitial cells with particular reference to gap junctions. J Cell Biol 81:104-114

Burghardt RC, Anderson E, 1981. Hormonal modulation of gap junctions in rat ovarian follicles. Cell Tiss Res 214:181-193

Burghardt RD, Matheson RL, 1982. Gap junction amplification in rat ovarian granulosa cells. Dev Biol 94:206-215

Byus CV, Fletcher WH, 1982. Direct cytochemical localization of catalytic subunits dissociated from cAMP-dependent protein kinase in Reuber H-35 hepatoma cells. II. Temporal and spatial kinetics. J Cell Biol 93:727-734

Campbell KL, Albertini DF, 1981. Freeze-fracture analysis of gap junction disruption in rat ovarian granulosa cells. Tissue & Cell 13:651-668

Caspar DLD, Goodenough DA, Makowski L, Phillips WC, 1977. Gap junction structures. I. Correlated electron microscopy and X-ray diffraction. J Cell Biol 74:605-628

Channing CP, Anderson LD, Hoover DJ, Kolena J, Osteen KG, Pomerantz SH, Tanabe K, 1982. The tole of nonsteroidal regulators in control of oocyte and follicular maturation. In: Recent Progress in Hormone Research, vol. 38. New York: Academic Press, pp. 331-408

Chalcroft JP, Bullivant S, 1970. An interpretation of liver cell membrane and junction structure based on observations of freeze-fracture replicas of both sides of the fracture. J Cell Biol 47:49-60

Cheng H-C, Kemp BE, Pearson RB, Smith AJ, Misconi L, Van Patten SM, Walsh DA, 1986. A potent synthetic peptide inhibitor of the cAMP-dependent protein kinase. J Biol Chem 261:989-992

Dekel N, Lawrence TS, Gilula NB, Beers WH, 1981. Modulation of cell-to-cell communication in the cumulus-oocyte complex and the regulation of oocyte maturation by LH. Dev Biol 86:356-362

Dewey MM, Barr L, 1962. Intercellular connection between smooth muscle cells: The nexus. Science 137:670-672

Dewey MM, Barr L, 1964. A study of the structure and distribution of the nexus. J Cell Biol 23:553-585

Enomoto T, Yamasaki H, 1984. Lack of intercellular communication between chemically transformed and surrounding nontransformed BALBc/3T3 cells. Cancer Res 44:5200-5203

Epsey LL, Stutts RH, 1972. Exchange of cytoplasm between cells of the membrana granulosa in rabbit ovarian follicles. Biol Reprod 6:168-175

Fletcher WH, 1973. Diversity of intercellular contacts in the rat ovary. J Cell Biol 59:101a

Fletcher WH, Everett JW, 1973. Ultrastructural reorganization of rat granulosa cells on the day of proestrus. Anat Rec 175:320

Fletcher WH, Anderson NC, Everett JW, 1975. Intercellular communication in the rat anterior pituitary gland. An in vivo and in vitro study. J Cell Biol 67:469-476

Fletcher WH, 1979. Intercellular junctions in ovarian follicles: a possible functional role in follicle development. In: Midgley AR, Sadler WA (eds.), Ovarian Follicular Development and Function. New York: Raven Press, pp. 113-120

Fletcher WH, Byus CV, 1982. Direct cytochemical localization of catalytic subunits dissociated from cAMP-dependent protein kinase in Reuber H-35 hepatoma cells. I. Development and validation of fluoresceinated inhibitor. J Cell Biol 93:719-726

Fletcher WH, Greenan JRT, 1985. Receptor mediated action without receptor occupancy. Endocrinology 116:1660-1662

Fletcher WH, Van Patten SM, Cheng H-C, Walsh DA, 1986. Cytochemical identification of the regulatory subunit of the cAMP-dependent protein kinase by use of fluorescently labeled catalytic subunit. J Biol Chem 261:5504-5513

Funkenstein B, Nimrod A, Lindner HR, 1980. The development of steroidogenic capability and responsiveness to gonadotropins in cultured neonatal rat ovaries. Endocrinology 106:98

Furshpan EJ, Potter DD, 1959. Transmission at the giant motor synapses of the crayfish. J Physiol (London) 145:289-325

Furshpan EJ, Furakawa T, 1962. Intracellular and extracellular responses of several regions of the Mauthner cell of the goldfish. J Neurophysiol 25:732-771

Furshpan EJ, Potter DD, 1968. Low resistance junctions between cells in embryos and tissue culture. Curr Top Dev Biol 3:95-127

Gilula NB, Reeves OR, Steinbach A, 1972. Metabolic coupling, ionic coupling and cell contacts. Nature (London) 235:262-265

Gilula NB, 1975. Junctional membrane structure. In: Tower DB (ed.), The Nervous System, vol. 1: The Basic Neurosciences. New York: Raven Press, pp. 1-11

Gilula NB, Epstein ML, Beers WH, 1978. Cell-to-cell communication and ovulation.

A study of the cummulus-oocyte complex. J Cell Biol 78:58-75

Goshima K, 1969. Synchronized beating of and electrotonic transmission between myocardial cells mediated by heterotypic strain cells in monolayer culture. Exp Cell Res 58:420-426

Goshima K, 1971. Synchronized beating of myocardial cells mediated by FL cells in monolayer culture and its inhibition by trypsin-treated FL cells. Exp Cell Res 65:161-169

Goshima K, 1974. Initiation of beating in quiescent myocardial cells by norepinephrine, by contact with beating cells and by electrical stimulation of adjacent FL cells. Exp Cell Res 84:223-234

Goshima K, 1976. Antagonistic influences of dibutyryl cyclic AMP and dibutyryl cyclic GMP on the beating rate of cultured mouse myocardial cells. J Mol Cell Cardiol 8:713-725

Heller DT, Cahill DM, Schultz RM, 1981. Biochemical studies of mammalian oogenesis: Metabolic cooperativity between granulosa cells and growing mouse oocytes. Dev Biol 84:455-464

Hertzberg EL, Gilula NB, 1979. Isolation and characterization of gap junctions from rat liver. J Biol Chem 254:2138-2147

Hertzberg EL, Lawrence TS, Gilula NB, 1981. Gap junctional communication. Ann Rev Physiol 43:479-491

Hertzberg EL, Anderson DJ, Friedlander M, Gilula NB, 1982. Comparative analysis of the major polypeptides from liver gap junctions and lens fiber junctions. J Cell Biol 92:53-59

Hertzberg EL, 1984. A detergent-independent procedure for the isolation of gap junctions from rat liver. J Biol Chem 259:9936-9943

Hertzberg EL, Skibbens RV, 1984. A protein homologous to the 27,000 dalton liver gap junction protein is present in a wide variety of species and tissues. Cell 39:61-69

Hertzberg EL, Spray DC, Bennett MVL, 1985. Reduction of gap junctional conductance by microinjection of antibodies against the 27-kDa liver gap junction polypeptide. Proc Natl Acad Sci USA 82:2412-2416

Hunzicker-Dunn M, Jungmann RA, 1978a. Rabbit ovarian protein kinases. I. Effect of an ovulatory dose of human chorionic gonadotropin or luteinizing hormone on the subcellular distribution of follicular and luteal protein kinases. Endocrinology 103:420-430

Hunzicker-Dunn M, Jungmann RA, 1978b. Rabbit ovarian protein kinases. II. Effect of an ovulatory dose of human chorionic gonadotropin or luteinizing hormone on the multiplicity of follicular and luteal protein kinases. Endocrinology 103:431-440

Hunzicker-Dunn M, Jungmann RA, 1978c. Rabbit ovarian protein kinases. III. Gonadotrophin-induced activation of soluble adenosine 3',5'-monophosphate-dependent protein kinases. Endocrinology 103:441-450

Hunzicker-Dunn M, 1982. Rat ovarian nuclear protein kinases. Biochim et Biophys 714:395-406

Hunzicker-Dunn M, 1986. Unique properties of the follicle-stimulating hormone- and cholera toxin-sensitive adenylyl cyclase of immature granulosa cells. Endocrinology 118:302-311

Janssen-Timmen U, Traub O, Dermietzel R, Rabes HM, Willecke K, 1986. Reduced number of gap junctions in rat hepatocarcinomas detected by monoclonal antibody. Carcinogenesis 7:1475-1482

Johnson RG, Hammer M, Sheridan JD, Revel JP, 1974. Gap junction formation between reaggregated Novikoff hepatoma cells. Proc Natl Acad Sci USA 71:4536-4540

Johnson RG, Sheridan JD, 1971. Junctions between cancer cells in culture: ultrastructure and permeability. Science 174:717-719

Kanno Y, Loewenstein WR, 1966. Cell-to-cell passage of large molecules. Nature (London) 212:629-631

Kogon M, Pappas GD, 1975. Atypical gap junctions in the ciliary epithelium of the albino rabbit eye. J Cell Biol 66:671-676

Kuettel MR, Schwoch G, Jungmann RA, 1984. Localization of cyclic AMP-dependent protein kinase subunits in rat hepatocyte nuclei by immunogold electron microscopy. Cell Biology International Reports 8:949-957

Kuettel MR, Squinto SP, Kwast-Welfeld J, Schwoch G, Schweppe JS, Jungmann RA, 1985. Localization of nuclear subunits of cyclic AMP-dependent protein kinase by the immuncolloidal gold method. J Cell Biol 101:965-975

Kumar NM, Gilula NB, 1986. Cloning and characterization of human and rat liver cDNAs coding for a gap junction protein. J Cell Biol 103:767-776

Larsen WJ, Tung HN, Polking C, 1981. Response of granulosa cell gap junctions to human chorionic gonadotropin (HCG) at ovulation. Biol Reprod 25:1119-1134

Larsen WJ, Risinger MA, 1985. The dynamic life histories of intercellular membrane junctions. In: Satir B (ed.), Reviews in Cell Biology, vol. 4. New York: Alan R. Liss, pp. 151-216

Lawrence TS, Beers WH, Gilula NB, 1978. Transmission of hormonal stimulation by cell-to-cell communication. Nature 272:501-506

Lo CW, Gilula NB, 1979a. Gap junctional communication in the preimplantation mouse embryo. Cell 18:399-409

Lo CW, Gilula NB, 1979b. Gap junctional communication in the post-implantation mouse embryo. Cell 18:411-422

Loewenstein WR, Kanno Y, 1966. Intercellular communication and the control of tissue growth. Lack of communication between cancer cells. Nature (London) 209:1248-1250

Lowenstein WR, 1968. Communication through cell junctions. Implications in growth control and differentiation. Dev Biol 2:151

Loewenstein WR, 1977. Permeability of membrane junctions. Ann NY Acad Sci 137:441-472

Loewenstein WR, 1979. Junctional intercellular communication and the control of growth. Biochim et Biophys Acta 560:1-65

Loewenstein WR, 1981. Junctional intercellular communication: the cell-to-cell membrane channel. Phys Rev 61:829-913

Lohmann SM, Walter U, 1984. Regulation of the cellular and subcellular concentrations and distribution of cyclic nucleotide-dependent protein kinases. In: Greengard P, Robison GA (eds.), Advances in Cyclic Nucleotide and Protein Phosphorylation Research vol. 18. New York: Raven Press, pp. 63-117

Makowski L, Caspar DLD, Phillips WC, Goodenough DA, 1977. Gap junction structures. II. Analysis of the X-ray diffraction data. J Cell Biol 74:629-645

Makowski L, 1985. Structural domains in gap junctions: implications for the control of intercellular communication. In: Bennett MVL, Spray DC (eds.), Gap Junctions. New York: Cold Spring Harbor Laboratory, pp. 5-12

Maller JH, Krebs EG, 1977. Progesterone-stimulated meiotic cell division in *Xenopus* oocytes: induction by regulatory subunit and inhibition by catalytic subunit of adenosine 3':5'-monophosphate-dependent protein kinase. J Biol Chem 252:1712-1718

May JV, McCarty K, Reichert LE, Schomberg DW, 1980. Follicle-stimulating hormone-mediated induction of functional luteinizing hormone/human chorionic gonadotropin receptors during monolayer culture of porcine granulosa cells. Endocrinology 107:1041-1049

McNutt NS, Weinstein RS, 1970. The ultrastructure of the nexus. A correlated thin-section and freeze-cleave study. J Cell Biol 47:666-687

Merk FB, 1971. In: Arceneaux CJ (ed.), Proceedings 29th Annual Meeting Electron Microscopy Society of America. Baton Rouge: Claitor's Publishing Division, p. 554

Merk FB, Botticelli CR, Albright JT, 1972. An intercellular response to estrogen by granulosa cells in the rat ovary; an electron microscope study. Endocrinology 90:992-1007.

Merk FB, Albright JT, Botticelli CR, 1973. The fine structure of granulosa cell nexuses in rat ovarian follicles. Anat Rec 175:107-126

Miller TM, Goodenough DA, 1986. Evidence for two physiologically distinct gap junctions expressed by the chick lens epithelial cell. J Cell Biol 102:194-199

Murray SA, Fletcher WH, 1984. Hormone-induced intercellular signal transfer dissociates cyclic AMP-dependent protein kinase. J Cell Biol 98:1710

Murray SA, Byus CV, Fletcher WH, 1985. Intracellular kinetics of free catalytic units dissociated from adenosine 3',5'-monophosphate-dependent protein kinase in adrenocortical tumor cells (Y-1). Endocrinology 116:364-374

Murray SA, Fletcher WH, 1986. Cyclic AMP-dependent protein kinase-mediated desensitisation of adrenal tumour cells. Mol Cell Endocrinol 47:153-161

Murtaugh MP, Steiner AL, Davies PJA, 1982. Localization of the catalytic subunit of cyclic AMP-dependent protein kinase in cultured cells using a specific antibody. J Cell Biol 95:64-72

Neyton J, Trautman A, 1985. Single-channel currents of an intercellular junction. Nature 317:331-335

Paul DL, Goodenough DA, 1983. In vitro synthesis and membrane insertion of bovine MP26, an integral protein from lens fiber plasma membrane. J Cell Biol 96:633-638

Paul DL, 1986. Molecular cloning of cDNA for rat liver gap junction protein. J Cell Biol 103:123-134

Payer AF, 1975. Permeability of ovarian follicles and capillaries in mice. Am J Anat 142:295-318

Peracchia C, 1973. Low resistance junctions in crayfish. J Cell Biol 57:66-76

Peracchia C, Peracchia, LL, 1980a. Gap junction dynamics: reversible effects of divalent cations. J Cell Biol 87:708-718

Peracchia C, Peracchia LL, 1980b. Gap junction dynamics: reversible effects of hydrogen ions. J Cell Biol 87:719-727

Peracchia C, Girsch SJ, 1985a. Permeability and gating of lens gap junction channels incorporated into liposomes. Current Eye Res 4:431-439

Peracchia C, Girsch SJ, 1985b. Functional modulation of cell coupling: evidence for a calmodulin-driven channel gate. Am J Physiol 248:765-782

Peracchia C, Girsch SJ, 1985c. An in vitro approach to cell coupling: permeability and gating of gap junction channels incorporated into liposomes. In: Bennett MVL, Spray DC (eds.), Gap Junctions. New York: Cold Spring Harbor Laboratory, pp. 191-203

Pitts JD, 1980. The role of junctional communication in animal tissues. In Vitro 16:1049-1056

Pitts JD, Finbow ME, 1986. The gap junction. J Cell Sci Suppl 4:239-266

Presl J, Pospisil J, Figarova V, Krabed Z, 1974. Stage-dependent changes in binding of iodinated FSH during ovarian follicle maturation in rats. Endocrinol Exp (Bratisl) 8:2

Revel JP, Karnovsky MJ, 1967. Hexagonal array of subunits in intercellular junctions of the mouse heart and liver. J Cell Biol 33:C7-C12

Richards JS, Midgley AR, 1976. Protein hormone action: a key to understanding ovarian follicular and luteal cell development. Biol Reprod 14:82-94

Robertson JD, 1963. The occurrence of a subunit pattern in the unit membranes of club endings in Mauthner cell synapses in goldfish brains. J Cell Biol 19:201-221

Saez JC, Spray DC, Nairn AC, Hertzberg E, Greengard P, Bennett MVL, 1986. cAMP increases junctional conductance and stimulates phosphorylation of the 2kDa principal gap junction polypeptide. Proc Natl Acad Sci USA 83:2473-2477

Schultz RM, Montgomery RR, Ward-Bailey PF, Eppig JJ, 1983a. Regulation of oocyte maturation in the mouse: possible roles of intercellular communication, cAMP, and testosterone. Dev Biol 95:294-304

Schultz RM, Montgomery RR, Belanoff JR, 1983b. Regulation of mouse oocyte meiotic maturation: implication of a decrease in oocyte cAMP and protein dephosphorylation in commitment to resume meiosis. Dev Biol 97:264-273

Schultz RM, 1985. Roles of cell-to-cell communication in development. Biol Reprod 32: 27-42

Sheridan JD, 1968. Electrophysiological evidence for low-resistance intercellular junctions in the early chick embryo. J Cell biol 37:650-666

Sheridan JD, 1970. Low-resistance junctions between cancer cells in various solid tumors. J Cell Biol 45:91-99

Sheridan JD, Finbow ME, Pitts JD, 1979. Metabolic interactions between animal cells through permeable intercellular junctions. Exp Cell Res 123:111-117

Sheridan JD, Atkinson MM, 1985. Physiological roles of permeable junctions: some possibilities. Ann Rev Physiol 47:337-353

Simpson J, Rose B, and Loewenstein WR, 1977. Size limit of molecules permeating the junctional membrane channels. Science 195:294-296

Somer JR, Johnson EA, 1970. Comparative ultrastructure of cardiac cell membrane specializations. Amer J Cardiol 25:184-194

Steiner AL, Koide Y, Earp HS, Bechtel PJ, Beavo JA, 1978. Compartmentalization of cyclic nucleotides and cyclic AMP-dependent protein kinases in rat liver. Immunocytochemical demonstration. Adv Cyclic Nucleotide Res 9:691-705

Subak-Sharpe H, Burk RR, Pitts JC, 1969. Metabolic cooperation between biochemically marked mammalian cells in tissue culture. J Cell Sci 4:353-367

Trachtenberg MC, Pollen DA, 1970. Neuroglia: biophysical properties and physiological function. Science 167:1248-1252

Traub O, Willecke K, 1982. Cross-reaction of antibodies against liver gap junction protein (26K) with lens fiber junction protein (MIP) suggests structural homology between these tissue specific gene products. Biochem Biophys Res Commun 109:895-901

Unwin PMT, Zampighi G, 1980. Structure of the junction between communicating cells. Nature 283:545-549

Unwin PMT, Ennis PD, 1984. Two configurations of a channel-forming membrane protein. Nature 307:609-613

Van Patten SM, Fletcher WH, Walsh DA, 1986. The inhibitor protein of the cAMP-dependent protein kinase-catalytic subunit interaction. J Biol Chem 261:5514-5523

Veenstra RD, DeHaan RL, 1986. Measurement of single channel currents from cardiac gap junctions. Science 233:972-974

Verselis V, White RL, Spray DC, Bennett MVL, 1986. Gap junctional conductance and permeability are linearly related. Science 234:461-464

Walsh DA, Cooper RH, 1979. The physiological regulation and function of cAMP-dependent protein kinases. In: Biochemical Actions of Hormones, vol. 6. Academic Press, pp. 1-75

Warner AE, Gurdon JB, 1987. Functional gap junctions are not required for muscle gene activation by inductionin Xenopus embryos. J Cell Biol 104:557-564

Warner AE, Guthrie SC, Gilula NB, 1984. Antibodies to gap-junctional protein selectively disrupt junctional communication in the early amphibian embryo. Nature 311:127-131

Yancey SB, Edens JE, Trosko JE, Chang C-C, Revel J-P, 1982. Decreased incidence of gap junctions between chinese hamster V-79 cells upon exposure to the tumor promoter 12-0-tetradecanoyl phorbol-13-acetate. Exp Cell Res 139:329-340

Yotti LP, Chang CC, Trosko JE, 1979. Elimination of metabolic cooperation in chinese hamster cells by a tumor promoter. Science 206:1089-1091

Zampighi G, Unwin PNT, 1979. Two forms of isolated gap junctions. J Mol Biol 135: 451-464

Zampighi G, Corless JM, Robertson JD, 1980. On gap junction structure. J Cell Biol 86:190-198

Zampighi GA, Hall JE, Kreman M, 1985. Purified lens junctional protein forms channels in planar lipid films. Proc Natl Acad Sci USA 82:8468-8472

Zampighi GA, Simon SA, 1985. The structure of gap junctions as revealed by electron microscopy. In: Bennett MVL, Spray DC (eds.), Gap Junctions. New York: Cold Spring Harbor Laboratory, pp. 13-22

PART III

REGULATION OF CORPUS LUTEUM FUNCTION IN PREGNANCY

COMPARATIVE ASPECTS OF THE REGULATION OF CORPUS LUTEUM

FUNCTION IN VARIOUS SPECIES

Fredrick Stormshak, Mary B. Zelinski-Wooten, and
Salah E. Abdelgadir

Departments of Animal Science and
Biochemistry and Biophysics
Oregon State University, Corvallis, Oregon 97331

INTRODUCTION

Central to understanding the pivotal role of the corpus luteum in governing reproductive cycles of mammals has been the study of those factors that control the function of this gland. Research on control of luteal function has encompassed a broad spectrum of mammalian taxa and has evolved from early studies to identify the source and nature of controlling factors to present day attempts to resolve their action at the level of the luteal cell . From this research data have emerged leading to the realization that among mammals a diversity of factors regulate the function of the corpus luteum. The fact that function of the corpus luteum, in most mammals, is influenced by hormones from several sources, namely the pituitary gland, uterus and placenta, makes this organ truly unique among endocrine glands. As the student of the corpus luteum is well aware, some of these hormones are involved in promoting steroidogenesis and(or) prolonging the life span of the corpus luteum while others serve to provoke its demise. In recent years several excellent reviews have appeared that discuss differences among mammals in regard to the hormonal regulation of luteal function (Rothchild, 1981; Keyes et al., 1983; Niswender et al., 1985; Khan-Dawood and Dawood, 1986). Consequently, certain portions of the present treatise on luteal function may appear to be redundant. Although old ground may be trod upon once again, an effort will be made to present new information.

Because of extensive literature on luteal function in various mammals, some selection of research to be covered has been necessary. This presentation will, therefore, focus on aspects of luteal function during the estrous cycle (domestic and laboratory animals), menstrual cycle (primates), pseudopregnancy (laboratory animals) and early gestation in the aforementioned species. In addition, some attention will be given to peculiarities of luteal function in those species exhibiting obligate delayed implantation (European badger, skunk, mink).

MORPHOLOGICAL AND STEROIDOGENIC CHARACTERISTICS OF THE LUTEAL CELL

Before discussing the role of various hormones in regulating the function of the corpus luteum, it is deemed appropriate to briefly review morphological and steroidogenic properties of the luteal cell. The fine structure of luteal cells from a number of species has been described by Enders (1973). Although the luteal cell undergoes morphological changes during the course of its life span, distinguishing characteristics of this cell are similar among species. In general, the luteal cell, besides possessing a nucleus, is characterized by presence of an abundant, smooth endoplasmic reticulum, comparatively minor stretches of rough endoplasmic reticulum, mitochondria with mostly tubular cristae and a prominent Golgi apparatus. The intracellular matrix is also characterized by presence of membrane-bound lipid droplets, granules and lysosomes. Adjacent luteal cells are in close apposition and a variety of junctions existing between cells have been described. One of the more unique associations between cells is a 20 nm "septate-like zone of adhesion" (Friend and Gilula, 1972) that is granular in nature. The significance of this granular zone of adhesion remains unknown.

The importance of the plasma membrane of the luteal cell as a signal transducer has been recognized for a number of years. From this standpoint the plasma membrane has been investigated primarily for its concentration of various hormone receptors during different reproductive states. With the exception of proteins, including those contributing to the structures of receptors, the bulk of the plasma membrane is composed of cholesterol and phospholipids. However, relatively little is known about the lipid composition of the luteal cell plasma membrane and much less is known about the changes in its composition that occur in association with altered cellular function. Data have been reported for the lipid composition of membranes or whole ovaries of the rabbit (Morin, 1968) and rat (Strauss and Flickinger, 1977) and corpora lutea of the cow (Scott et al., 1968), ewe (Waterman, 1980) and sow (Holman and Hofstetter, 1965). It is emphasized, however, that these data represent the lipids present in not only the plasma membrane but all other subcellular organelle membranes as well. Recently, Zelinski and Stormshak (1983) and Zelinski (1986) reported the lipid composition of a membranous fraction isolated from ovine corpora lutea, which, because of its enrichment with 5'-nucleotidase, was assumed to consist primarily of plasma membranes. There appears to be considerable agreement among the earlier data reported for heterogeneous membranes isolated from luteal homogenates and those data obtained for enriched membrane preparations of ovine corpora lutea (Zelinski, 1986) with respect to lipid composition. In terms of total percentage, phospholipids present in greatest quantities in luteal plasma membranes are phosphatidylcholine (about 50%), phosphatidylethanolamine (14-33% depending upon the species), phosphatidylserine (4-6%), phosphatidylinositol (4-6%), sphingomyelin (6-10%) and cardiolipin (< 4%). Principal fatty acids present in phospholipids of ovarian and luteal tissue are palmitic (16:0), stearic (18:0), oleic (18:1), linoleic (18:2) and arachidonic (20:4) acid, and there are minor quantities of unsaturated fatty acids of chain length greater than 20 carbons. Lipids play a critical role in regulating membrane fluidity, which, in the case of the luteal cell, is essential for ensuring coupling of the hormone with its receptor, internalization of the hormone-receptor complexes and exocytosis of secretory products. In general terms, membrane fluidity encompasses both rate and extent of movement of lipids as well as integral membrane proteins. It might be anticipated, therefore, that major changes in membrane composition and fluidity would be most evident during marked changes in luteal function. Indeed it has been shown by use of a variety of techniques that natural and induced luteolysis in the rat and cow is associated with a reduction in membrane fluidity (Buhr et al., 1979; Carlson et al., 1981, 1982, 1984; Goodsaid-Zalduondo et al., 1982). In the rat, decreases in membrane fluidity during luteolysis did not appear to result from major changes in the fatty acid composition or cholesterol:phospholipid ratio (Carlson et al., 1981). In contrast, membranes of regressing bovine corpora lutea had increased levels of sphingomyelin, which may have promoted the observed reduction in membrane fluidity (Goodsaid-Zalduondo et al., 1982). Luteolysis in the ewe was not associated with changes in any phospholipid, but rather significant changes were detected in arachidonic acid levels of phosphatidylcholine and phosphatidylinositol. The phosphatidylcholine content of arachidonic acid was increased while that of phosphatidylinositol was decreased (Zelinski, 1986). In addition, the phosphatidylinositol content of docasatetraenoic acid (22:4), a metabolite of arachidonate, was also markedly increased in membranes of regressing corpora lutea of ewes. Collectively, results of these studies have provided some information about luteal membranes; however, during only one stage of the life span of the corpus luteum. More reports on luteal plasma membranes will likely appear in the near future, particularly in view of recent interest in phosphatidylinositol and its relationship to activation of protein kinase C.

A discussion of the corpus luteum would not be complete without at least mentioning the primary biological reason for its existence. The major secretory product of the corpus luteum is, of course, progesterone, whose time-honored function is to ensure the maintenance of pregnancy. In addition to progesterone, the corpus luteum produces other steroids, the nature and quantity of which vary among mammalian species (Savard, 1973). This variation in steroid secretory products indicates that while gross morphological characteristics of the luteal cell may be similar among species, differences do exist in the complexity of its steroidogenic pathway. Extremes in steroidogenesis are represented on the one hand by the bovine corpus luteum which secretes progesterone, 20β-hydroxy-4-pregnen-3-one and pregnenolone, and on the other by the human corpus luteum which secretes progesterone, 20α- hydroxy-4-pregnen-3-one, pregnenolone, 17β-hydroxyprogesterone, 4-androstenedione, estrone and 17β-estradiol. Besides progesterone and the estrogens, the types of steroids secreted by corpora lutea of other species lie between these two extremes. Conversion of cholesterol to pregnenolone is the rate limiting step in the synthesis of progesterone. While corpora lutea of some species possess enzymes to synthesize cholesterol from acetate many derive this sterol from high or low density lipoproteins (Gwynne and Strauss, 1982).

The corpus luteum also synthesizes protein hormones, and this aspect of luteal function in recent years has been the focus of considerable research. The primary peptides synthesized by the luteal cell are relaxin and oxytocin. Compared with our knowledge about control of luteal steroidogenesis, little is known about the factors that regulate synthesis of these peptides. Because luteal relaxin is the subject of another chapter in this book, it will not be discussed in detail in this presentation. However, a brief discussion of the comparative aspects of luteal synthesis of oxytocin is deemed appropriate and will appear in a following section.

Until recently it was simply taken for granted that the corpus luteum developed from granulosa and theca interna cells as reported by Corner (1919). During the intervening years reports appeared that the mature corpus luteum consisted of two cell types identifiable by light microscopy, but the significance of these observations was essentially not recognized. Neither the methodology nor state of knowledge was at hand to answer questions about the origin of these cells and their functional attributes. However, advances in technology have now permitted the dispersion and isolation of relatively pure populations of luteal cells. Based upon size, two cell types referred to simply as large and small have been isolated from bovine (Uresley and Leymarie, 1979; Koos and Hansel, 1981), ovine (Rodgers and O'Shea, 1982; Niswender et al., 1985), porcine (Lemon and Loir, 1977), primate (Hild-Patito et al, 1986) and rabbit (Hoyer et al., 1986) corpora lutea. The morphological and biochemical characteristics of these cells have been recently reviewed by Niswender et al. (1985). Briefly, small luteal cells range from 10 to about 20 µm in diameter and large cells range from 20 to greater than 35 µm in diameter. Although fewer in number, the large cells have been reported to account for approximately 30% of the ovine corpus luteum on a volume basis, compared with 16% for small cells. Both cell types possess the fine structure characteristic of luteal cells and secrete progesterone. In addition, the cytoplasm of large cells is endowed with rough endoplasmic reticulum and membrane-bound secretory granules, while such features are essentially absent in small cells. Thus, the large cells most likely synthesize relaxin and oxytocin. While both cell types secrete progesterone, the cells apparently differ in the mechanisms by which synthesis of this steroid is promoted. Plasma membranes of large and small cells are inversely related with respect to their concentrations of receptors for luteinizing hormone (LH) and prostaglandins $F_2\alpha$ ($PGF_2\alpha$) and E_2 (PGE_2). Plasma membranes of isolated small cells have been shown to be richly populated with LH receptors while plasma membranes of large cells contain predominantly receptors for $PGF_2\alpha$ and PGE_2. Consistent with these concentrations of receptors, small but not large cells have been shown to respond to LH, dbcAMP, cholera toxin and forskolin with enhanced synthesis of progesterone. Large cells do, however, respond to such agents as cholera toxin and forskolin with increased intracellular levels of adenosine-3',5'-cyclic monophosphate (cAMP). It has been suggested that failure of large cells to repond to LH or increased levels of cAMP may be due to the fact that synthesis of progesterone is already at a maximum. Unstimulated, large cells secrete about 20 times as much progesterone as do small cells. The mechanisms governing progesterone synthesis in large cells are intriguing and warrant further study.

Because of the origin of the corpus luteum, it would be logical to assume that these two cell types are representatives of the parent population of granulosa and theca cells. Alila and Hansel (1984), using monoclonal antibodies developed against theca-specific and granulosa-specific antigens, monitored the fate of large and small cells of the bovine corpus luteum during the estrous cycle and gestation. From days 4-6 of the cycle, 77% of the large luteal cells bound the granulosa antibody while 70% of the small cells bound the theca antibody. However, by days 16-18 of the cycle the percentage of large cells binding to granulosa antibody decreased to about 30% while a greater percentage (about 40%) of large cells now bound the theca antibody. The percentage of small cells binding granulosa antibody on days 16-18 of the cycle was negligible but a comparatively large percentage (about 60%) of these cells bound the theca antibody. Although no large cells bound the granulosa antibody after day 100 of gestation, decreasing percentages of large cells did bind the theca antibody, at least up to day 200 of gestation. By the end of gestation only a low percentage of small cells bound theca antibody. To account for differences in binding of granulosa and theca antibody, the possibility was advanced that small cells differentiated into large cells, a premise consistent with the hypothesis previously proposed by Donaldson and Hansel (1965a). In fact, these investigators originally suggested that LH stimulated transformation of small into large cells. This possibility is supported by recent data of Niswender et al. (1985) who showed that treatment of ewes with human chorionic gonadotropin (hCG) during midcycle increased the number of large cells. Alila and Hansel (1984) also reported that there were cells that failed to bind to either antibody at all stages of the cycle and gestation studied. Thus it is possible that these cells arise from some cell (stem cell) other than granulosa or theca cells.

Because large and small luteal cells differ in biochemical characteristics, changes in relative proportions of these cell types during the life span of the corpus luteum could be important in terms of understanding factors that govern this gland's function. Schwall et al. (1986) have

330

quantitated the numbers of small and large dissociated cells that stained positively for 3β-hydroxysteroid dehydrogenase during various stages of the ovine estrous cycle. Growth of the corpus luteum was promoted by an increase in numbers of small and large cells, with the population of small cells vastly outnumbering that of large cells up to midcycle. Thereafter, there occurred a marked reduction in number of small cells so that at the time of luteal regression the ratio of large to small cells was greater than during earlier stages of the cycle. Similarly, morphometric analyses of ovine luteal slices have also revealed an increase in number of small cells as the cycle advanced, but no decrease in the number of small cells was detected during luteolysis (O'Shea et al., 1986; Farin et al., 1986). In addition, these investigators found the number of large luteal cells to remain constant but to increase in size from the mid to late phases of the cycle. In contrast to the situation in the ewe, the fully developed corpus luteum of the pseudopregnant rabbit consists predominantly of large cells (Hoyer et al., 1986). However, large and small cells of the rabbit corpus luteum are similar to those found in ruminants from the standpoint of their response to LH _in vitro_.

The fact that extrinsic factors regulate luteal function and life span in a majority of species cannot be denied. However, coexistence of two luteal cell types, which differ in their response to secretagogues, raises questions regarding the functional relationship of these cells to each other. Indeed, paracrine has suddenly emerged as a popular "buzz" word in discussions related to the corpus luteum.

CONTROL OF LUTEAL FUNCTION BY PITUITARY GONADOTROPINS AND ESTROGENS

Luteotropins, by definition, are hormones usually of extrinsic origin that prolong the life span of the corpus luteum and(or) stimulate luteal synthesis and secretion of progesterone _in vivo_ or _in vitro_. This definition has its shortcomings because in some instances no physiological basis exists for a hormone that is luteotropic _in vitro_. For example, epinephrine can stimulate adenylate cyclase activity and progesterone synthesis by luteal tissue of the cow (Condon and Black, 1976), ewe (Jordan et al., 1978), and rabbit (Hunzicker-Dunn, 1982) but there is no evidence to suggest that this hormone has the same effect _in vivo_. The term luteotropin also implies a direct action on the luteal cell but this may not always be the situation. Further, it has generally, but perhaps erroneously, been assumed that a hormone which stimulates progesterone synthesis is the same hormone that maintains the structural integrity of the corpus luteum.

In the following sections an attempt will be made to discuss the luteotropic effects of pituitary gonadotropins and estrogens. A discussion of "embryonic" luteotropins will appear elsewhere in this chapter.

Luteinizing Hormone

It is perhaps appropriate to first discuss the role of LH, because this gonadotropin has been shown to be luteotropic in virtually all mammalian species. Research that established LH as being a luteotropin in various domestic and laboratory animals has been reviewed previously (Hansel et al., 1973; Rothchild, 1981; Niswender et al., 1985); therefore, brief mention of only salient experimental results will be included. In the cow and ewe LH is the primary luteotropin. Although some might argue otherwise, the report by Mason et al. (1962) demonstrating that LH-stimulated steroidogenesis by bovine luteal slices _in vitro_ served as an impetus for further examining the luteotropic properties of this gonadotropin, not only in the cow but other species as well. Subsequent to this report, Donaldson and Hansel (1965b) found that exogenous LH could prolong the life span of the bovine corpus luteum during the estrous cycle and prevent the luteolytic effects of oxytocin administered early in the cycle (Donaldson et al., 1965). Evidence that LH is luteotropic in the ewe was provided by Kaltenbach et al. (1968) who showed that constant infusions of crude pituitary extracts containing LH and FSH activity maintained luteal function in hypophysectomized pregnant and nonpregnant animals. This observation was later confirmed by Karsch et al. (1971) who reported that constant infusions of LH prolonged the life span and function of the ovine corpus luteum during the estrous cycle. Whether LH is luteotropic in the sow is debatable because in this species hypophysectomy shortly after onset of estrus allows ovulation and development of corpora lutea that are slightly smaller by days 13-14 of the estrous cycle but otherwise fully functional (du Mesnil du Buisson and Leglise, 1963). These data suggest that only the ovulatory surge of LH is required for formation of corpora lutea in the sow and that, thereafter, structure and function are autonomous almost for the duration of the cycle. However, in previously hysterectomized gilts, hypophysectomy caused luteal regression within about one week (du Mesnil du Buisson and Leglise, 1963; Anderson et al., 1965b) but daily injections of hCG, LH or crude pituitary preparations caused luteal maintenance. More recently, Guthrie and Rexroad (1981) found that a single injection of hCG given to sows on day 12 of the estrous cycle prolonged the functional life span of corpora lutea. Thus, in the sow, LH appears to be required for luteal maintenance beyond the normal duration of the estrous cycle. Surprisingly, in vitro LH or hCG has little effect in stimulating progesterone synthesis by porcine corpora lutea (Duncan et al., 1961). During the estrous cycle, the corpus luteum of the mare is sustained by pituitary gonadotropins because treatment of pony mares with antiserum to an equine pituitary preparation caused luteal regression (Pineda et al., 1973). Dependence of the equine corpus luteum on hypophyseal gonadotropin during the estrous cycle is also supported by the ability of daily injections of pituitary extracts to prolong luteal life span (Ginther, 1979). Initial studies by VandeWeile et al. (1970) involving hypophysectomy of women suggested that LH was luteotropic during the menstrual cycle. More recently, the essential role of LH in maintaining luteal function during the menstrual cycle of women was confirmed by Mais et al. (1986). These investigators showed that blocking LH secretion during the menstrual cycle by administration of a gonadotropin-releasing hormone (GnRH) antagonist led to luteal regression. In the monkey, as in the human, it has been demonstrated that luteal function during the menstrual cycle is dependent upon LH (Moudgal et al., 1971; Groff et al., 1984). Dependence of the monkey corpus luteum on LH has most vividly been demonstrated by Hutchison and Zeleznik (1984) using "hypothalamic-clamped" monkeys. In studies by these investigators monkeys were subjected to bilateral lesions of the arcuate region in the hypothalamus, causing them to become anovulatory. Subsequently, chronic pulsatile infusions of GnRH into the monkeys reestablished gonadotropin secretion and ovulatory cycles. However, halting the infusion of GnRH during the early or mid-luteal phase of the cycle abruptly ended secretion of LH with eventual reduction in luteal function and, finally, premature regression of the corpus luteum. Surprisingly, temporary cessation of LH secretion for 3 days did not result in irreversible loss of luteal function. After reinitiating infusions of GnRH and, hence, LH secretion, the corpus luteum was rejuvenated in its ability to synthesize and secrete progesterone, and its life span was of normal duration. These latter data suggest that the primary role of LH in the primate is to maintain steroidogenesis, while the structural integrity of the corpus luteum may be dependent upon other factors. In addition to the foregoing in vivo studies, it has been amply demonstrated that luteal tissue of the human and monkey respond in vitro to LH with increased synthesis of progesterone and estrogen (Rice et al.,

1964; Marsh and LeMaire, 1974; Stouffer et al., 1980; Eyster and Stouffer, 1985).

Mechanism of Action of LH. The plasma membrane is the initial site of action of LH in triggering a cascade of biological events that ultimately result in expression of a functional corpus luteum. The mechanism of action of LH on the luteal cell has been shown to conform to the "second messenger" model originally proposed by Sutherland and reviewed about a decade ago by Marsh (1976). Briefly, LH binds to its receptor in the plasma membrane of the luteal cell wherein adenylate cyclase is activated to convert adenosine triphosphate (ATP) to cAMP, the second messenger, which, in turn, activates cAMP-dependent protein kinase(s). Protein kinase activity results in phosphorylation of steroidogenic enzymes and(or) other proteins necessary for the synthesis and secretion of pro-gesterone. Due to their structural (Pierce and Parsons, 1981; Sairam, 1983) and functional similarities, LH and hCG bind to the same receptor site in gonadal tissues, hereafter referred to as the "LH receptor". In recent years progress has been made to obtain relatively pure preparations of luteal cell LH receptor, permitting insight regarding its structure and chemical properties. One of the most comprehensive studies has involved the isolation and purification of the LH receptor of bovine corpora lutea (Dattatreyamurty et al., 1983). These investigators concluded that the LH receptor was an oligomeric glycoprotein consisting of two identical sub-units of M_r = 120,000 - 140,000 which, in turn, were composed of two disulfide-bonded chains of M_r = 85,000 and 38,000. Using a photoaffinity labeled hCG preparation, Rapoport et al. (1984) isolated rat ovarian LH receptor under reducing conditions without protease inhibitors and repor-ted three components with M_r = 106,000, 85,000 and 80,000. However, only the largest component was able to bind hCG with high affinity in a satura-ble manner. This component, after subtraction of the molecular weight of the covalently-linked subunit of hCG, had a M_r = 86,000. Ascoli and Segaloff (1986), using ^{125}I-labelled derivatives of hCG and ovine LH cross-linked to the LH receptor of porcine granulosa cells, concluded that the receptor consisted of two disulfide-bonded subunits of M_r = 83,000 and 23,000, but that the hormone bound only to the M_r = 83,000 subunit. Collectively, these data suggest that the LH receptor is a noncovalently bonded oligomer with the minimal binding unit consisting probably of two different chains joined by disulfide bonds.

The relationship between luteal concentrations of LH receptors and progesterone secretion has been monitored in a number of species. In pregnant rats the luteal concentrations of LH and prolactin receptors appear to be inversely related throughout gestation, with a major increase in LH receptors occurring between days 8-11 of gestation (Richards and Midgley, 1976). From days 12-18 of gestation luteal concentrations of LH receptor plateau and increases in serum progesterone that occur during this interval are not accompanied by concomitant increases in the levels of receptor. However, changes in luteal concentrations of LH receptors have been shown to be highly correlated with changes in luteal function during various stages of the estrous cycle of the ewe (Diekman et al., 1978a), sow (Ziecik et al., 1980), cow (Spicer et al., 1981), mare (Roser and Evans, 1983), and during the menstrual cycle of the human (Rao et al., 1977) and monkey (Cameron and Stouffer, 1982). Diekman et al. (1978a) also investigated the possibility that maintenance of the ovine corpus luteum during early gestation involved changes in the number and(or) affinity of luteal LH receptors. These investigators found that the number of unoccupied and occupied LH receptors in corpora lutea obtained from ewes on days 12, 16 and 20 of gestation were identical to those quantified on day 12 of the estrous cycle. Thus, the presence of the embryo, which ensures luteal maintenance, does so through some mechanism other than alteration of luteal concentration of LH receptor. Our under-standing of the basic cellular mechanisms underlying luteal regression is as yet incomplete. Thus, it is perhaps appropriate to mention the results of studies conducted to examine the relationship of changes in concentrations of LH receptor to changes in progesterone secretion during induced luteal regression. Spicer et al. (1981) found that injection of $PGF_2\alpha$ into cows during the midluteal phase of the cycle to induce luteal regression was followed in 12 hours by a 75% reduction in serum concentrations of progesterone. However, specific binding of labeled hCG to luteal homogenates did not change by 12 hours but decreased 97% by 24 hours after the injection of $PGF_2\alpha$. Similarly, injection of $PGF_2\alpha$ into ewes during the midluteal phase of the cycle caused serum concentrations of progesterone to decrease before significant changes were detected in luteal concentrations of unoccupied and occupied LH receptors (Diekman et al., 1978b). Thus, while increased concentrations of LH receptors are positively associated with increased luteal steroidogenesis, a decrease in LH receptors is not the primary determinant of reduction in luteal function, at least at the end of the estrous cycle. This is not surprising if one recalls that small luteal cells apparently are endowed with the majority of LH receptors, while large luteal cells, which presumably synthesize more progesterone than small cells, possess the majority of receptors for $PGF_2\alpha$ and, thus, are the likely targets for $PGF_2\alpha$ during induced as well as natural luteolysis. It should be emphasized, however, that large luteal cells are not totally devoid of LH receptors, possessing about 10% of the number of LH receptor sites/cell found in small cells (Niswender et al., 1985). Although large cells are unresponsive to LH in vitro, no studies have yet been conducted to determine the degree of receptor occupation in these cells post-isolation, whether their capacity to secrete progesterone at a maximal rate might require a lower than expected occupancy of LH receptors or whether these cells stripped of any endogenous LH retain their maximal steroidogenic capabilities. It is also possible that in the intact corpus luteum the plasma membranes of large cells are populated with a greater number of LH receptors than is apparent and that these receptors or their binding properties are destroyed by the methods of cell dispersion employed. Hopefully, further research will provide answers for these unknown aspects of the large luteal cell.

Prolactin

Prolactin (PRL) appears to play no role in regulating luteal function during the estrous cycles or menstrual cycles of domestic animals and primates, respectively. Administration of PRL to cows (Smith et al., 1957) or injections of bromocryptine into ewes (Niswender et al., 1976) to reduce serum concentrations of PRL do not alter luteal function and life span during the estrous cycle. Similarly, exogenous PRL is not luteotropic during the estrous cycle of the sow (Duncan et al., 1961) or in hypophysectomized-hysterectomized gilts (du Mesnil du Buisson et al., 1964). It is noteworthy, however, that PRL may be luteotropic during the terminal stages of gestation in the gilt. Administration of purified porcine PRL to hysterectomized or hysterectomized-hypophysectomized gilts during the time period equivalent to days 110-120 after ovulation enhanced or maintained luteal secretion of progesterone and markedly stimulated secretion of relaxin (L.L. Anderson, Iowa State University, personal communication). Neither increased nor decreased secretion of PRL during the menstrual cycles of women or monkeys has been shown to alter normal luteal function (del Pozo et al., 1975; Richardson et al., 1985). Prolactin has also been shown to be without effect in stimulating steroidogenesis by luteal tissue of the cow (Mason et al., 1962), sow (Duncan et al., 1961), woman (Rice et al., 1964) and monkey (Stouffer et al., 1980). In contrast to its apparent unimportance as a luteotropin in the above mentioned species, PRL alone or in combination with other pituitary gonadotropins is critical for maintaining luteal function in the rat, mouse and hamster. These latter species are characterized by short estrous cycles during which time luteal function is not dependent upon or, at best, is minimally dependent upon hypophyseal gonadotropins. Hypophysectomy of the rat (Smith, 1930) or hamster (Greenwald and Rothchild, 1968) shortly after ovulation does not alter the normal life span of the corpora lutea. Cervical stimulation by mating or similar stimulation in the rat causes daily surges of PRL (Smith and Neill, 1976) that support luteal maintenance up to day 8 or 9 of pregnancy or pseudopregnancy, at which time the corpora lutea become dependent upon LH (Morishige and Rothchild, 1974). After day 8 or 9 corpora lutea of the rat remain dependent upon LH for the duration of pseudopregnancy. However, it is evident that in the event of pregnancy, luteal function is maintained after day 12 by placental luteotropins because hypophysectomy at this time does not terminate pregnancy (Pencharz and Long, 1933). Requirements for maintenance of luteal function during the first one-half of gestation in the mouse appear to be similar to those in the rat. Both pseudopregnant (Sinha et al., 1978) and pregnant (Barkley et al., 1978) mice exhibit recurring PRL surges that are initiated on the day of cervical stimulation. In the pregnant mouse hypophysectomized after implantation, but before day 10, the only gonadotropins reported to maintain pregnancy and luteal function were a combination of follicle stimulating hormone (FSH) and PRL (Choudary and Greenwald, 1969). Later research by Mednick et al. (1980), however, demonstrated that luteal function in the mouse was dependent upon PRL prior to implantation and upon LH after implantation and until day 10 of gestation. Thereafter, as in the rat, placental luteotropins become the major regulators of luteal function because neither hypophysectomy (Newton and Beck, 1939) nor administration of antiserum to LH (Mednick et al., 1980) on day 10 of gestation interferes with continuation of pregnancy. For the pregnant hamster, Greenwald (1967, 1973) reported that maintenance of luteal function depended upon a minimal luteotropic complex of FSH and PRL, with the latter hormone being primarily responsible for maintaining structural integrity of the corpus luteum. Subsequently, rather convincing evidence was acquired by Mukku and Moudgal (1975) and even by Terranova and Greenwald (1979) that LH was an important luteotropin in the pregnant hamster. In fact, the latter investigators demonstrated that corpora lutea of the hamster became LH-dependent on day 4 of gestation and remained so until at least day 12. Thus, the hamster and mouse differ from the rat in that luteal function in the former two species becomes dependent upon LH at or near the time of implantation, whereas in the rat it occurs as many as 4 or 5 days after this event. The hamster also differs from both the rat and mouse from the standpoint of the necessity of the pituitary after midgestation; removal of the pituitary on day 12 of gestation causes corpora lutea to remain as normal appearing structures but apparently without significant function, because pregnancy is terminated (Greenwald, 1967). In the guinea pig, with a comparatively longer cycle than that of the rat, mouse, and hamster, hypophyseal stalk-section soon after ovulation has been shown not to impair the growth of corpora lutea or their ability to secrete progesterone (Illingsworth and Perry, 1971). In addition, these same researchers demonstrated that daily administration of PRL to animals hypophysectomized early during the cycle promoted luteal maintenance. Thus, it appears that PRL is luteotropic in the nonpregnant guinea pig. However, the guinea pig corpus luteum may also be responsive to LH because in hysterectomized animals hCG blocked the luteolytic actions of exogenous $PGF_2\alpha$ while exogenous FSH and PRL were ineffective (Tam et al., 1982).

Limited information exists about the mechanisms by which PRL can affect luteal function. Besides somehow ensuring cellular integrity of the corpus luteum, PRL may serve to prevent reduction of progesterone to 20α-hydroxy-4-pregnen-3-one in rat luteal cells (Armstrong et al., 1970). Prolactin also appears to be essential for maintaining luteal cell concentrations of LH receptors (Grinwich et al., 1976; Gibori and Richards, 1978).

Estrogens

Estrogen is luteotropic in the rat, an observation made by Merckel and Nelson (1940) almost 50 years ago. Subsequently, Takayama and Greenwald (1973) reported that regression of corpora lutea in rats hypophysectomized and hysterectomized at midgestation could be curtailed by daily injections of estrogen. In rats hypophysectomized and hysterectomized on day 12 of gestation, daily injections of estradiol or an implant of testosterone maintained luteal progesterone synthesis, increased luteal nuclear concentrations of estrogen receptor and luteal concentrations of estradiol (Gibori and Keyes, 1978). These investigators suggested that intraluteal estrogen formed by aromatization of testosterone played an important role in regulating luteal function during pregnancy in the rat. Gibori et al. (1979) demonstrated that serum contrations of androgen in the rat markedly increased during the second one-half of gestation in the rat and, subsequently, showed the placenta to be the source of this steroid (Gibori and Sridaran, 1981). Thus, in this species, estrogen derived from placental androgens plays a significant role in maintaining luteal function.

Daily injections of estrogen initiated on day 11 of the estrous cycle have been shown to prolong the life span of corpora lutea in the sow (Kidder et al., 1955; Gardner et al., 1963). Pig blastocysts begin to synthesize estrogen in increasing quantitites from about day 9-12 of gestation. A decrease in synthesis follows between days 13-14 and then increased production begins after day 15 (Pope et al., 1982; Bazer et al., 1986). Current evidence suggests that in the sow estrogens produced by the blastocyst convert the uterus from an endocrine to an exocrine organ with respect to its secretion of $PGF_2\alpha$ (Bazer and Thatcher, 1977; Bazer et al., 1986). Thus, in the sow, the luteotropic action of estrogen is indirect, through blockage or suppression of the secretion of uterine $PGF_2\alpha$ which would otherwise cause luteal regression.

The rabbit is the only species in which estrogen has been shown to be the primary luteotropin. Robson (1937) was first to report that exogenous estrogens maintained corpora lutea in hypophysectomized pseudopregnant rabbits. Because the actions of estrogen in regulating luteal function in the rabbit are the subject of another chapter, this subject will not be pursued further. It is interesting, however, that corpora lutea of rabbits do possess receptors for LH (Hunzicker-Dunn and Birnbaumer, 1976). The role of this gonadotropin in regulating luteal function during pseudopregnancy and pregnancy in this species is an enigma.

EMBRYONIC LUTEOTROPINS

Mammalian species exhibit a wide variety of requirements for luteal maintenance during pregnancy. The extension of luteal life span can be achieved by one or more mechanisms which include luteotropic stimulation by anterior pituitary hormones, inhibition of the luteolytic effects of $PGF_2\alpha$ by conceptus-derived products acting as antiluteolysins, or the action of placental luteotropins (Heap and Flint, 1984). In general, luteal function is regulated largely by the continued secretion of luteotropins. It is evident that the anterior pituitary provides the main luteotropic stimuli (LH, PRL) throughout pregnancy in rabbits, pigs and goats because hypophysectomy (and ovariectomy) always result in abortion. As mentioned previously, rats, mice and hamsters require mainly PRL and LH (and possibly FSH) as luteotropins during the first one-half of gestation, after which placental proteins maintain the corpus luteum, whose presence is required for the duration of gestation. Hypophysectomized ewes infused with exogenous LH maintain luteal function during the first one-half of gestation. Although hypophysectomy in combination with hormone replacement studies has not been performed in cows and mares, neutralization of circulating LH during early pregnancy results in abortion, suggesting the necessity of this pituitary hormone for luteal maintenance. The placenta acquires the ability to synthesize progesterone during the second one-half of gestation in the ewe, cow and mare, precluding the need for the corpus luteum.

An exception to anterior pituitary regulation of luteal life span during pregnancy is observed in women, nonhuman primates, guinea pigs and the tammar wallaby. Although these species require the presence of the corpus luteum throughout gestation, hypophysectomy has no detrimental effect on the duration of pregnancy because the placenta produces the luteotropic stimulus. The precise nature of the luteotropin in the guinea pig and wallaby is not known, but has been well characterized in the woman and nonhuman primate. Chorionic gonadotropin (hCG, mCG), an LH-like glycoprotein produced by the developing syncytiotrophoblast, first appears in the maternal circulation 9-12 days postovulation but may be secreted even earlier by the preimplantation blastocyst (Hearn, 1986). Chorionic gonadotropin acts directly on the corpus luteum, replacing LH as the luteotropic stimulus for progesterone synthesis. By virtue of its long half-life in the circulation, hCG can act systemically to rescue the corpus luteum from the luteolytic effects of $PGF_2\alpha$. The action of chorionic gonadotropin as the sole luteotropin in human and nonhuman primates represents the most efficient mechanism known in mammalian species whereby the conceptus ensures the maintenance of pregnancy.

Rat decidual tissue has recently been demonstrated to contain and secrete a distinct PRL-like protein (M_r = 23,500) between days 6-12 of gestation (Jayatilak et al., 1985; Herz et al., 1986) which appears to assume the luteotropic role of pituitary PRL in sensitizing luteal cells to the tropic effects of LH and estradiol. At midgestation, placental lactogen (rPL) produced by the trophoblast appears (Kelly et al., 1975), coincident with decline in decidual luteotropin, and assumes the PRL-like luteotropic role for the duration of gestation. In fact, two molecular species of rPL have been characterized, one that peaks on day 12 of gestation and declines on day 14, and the other occurring between days 17-21 of pregnancy. Androgen secretion from the placenta begins at midpregnancy (day 12) and continues until parturition, while the corpus luteum also continues to produce testosterone (Warshaw et al., 1986). Ovarian androgen secretion appears to be under placental control from day 12-18 of pregnancy and placental androgen secretion is independent of the pituitary (Sridaran and Gibori, 1983). These observations suggest a role for a chorionic gonadotropin in regulating placental and luteal androgen production. A chorionic gonadotropin-like material identified in mouse placentae has a secretory pattern resembling that of testosterone in the maternal circulation (Wide and Wide, 1979). Attempts to identify a rat chorionic gonadotropin by hybridization analysis using cDNA probes encoding the α- and β- subunits of rat pituitary LH revealed the absence of LH-like material in rat placenta (Carr and Chin, 1985). These data suggest that rodent placental molecules containing LH-like activity may differ in structure from LH or hCG or may not exist; however, there is the possibility that hybridization analysis using LH probes may not be sensitive enough to detect mRNA's for a putative chorionic gonadotropin that may exist in low abundance.

The maintenance of luteal function during early pregnancy in domestic animals (ewe, cow, sow, mare) appears to be regulated largely by conceptus-derived steroids and proteins that function primarily as antiluteolysins within the uterus. Conceptus-derived products (proteins, PGE_2) have also been implicated as luteotropins during the first one-half of gestation. Comparison of the roles of conceptus secretory products in the maintenance of early pregnancy has been recently reviewed by Bazer et al. (1986), and salient features of this work will be provided below.

Ewe

The ovine conceptus signals its presence to the maternal system between days 12-13 postmating by maintaining luteal function in a local, rather than systemic, manner. Ability of proteins from homogenates of day 13 conceptuses to extend luteal life span when introduced into the uterus of nonpregnant ewes demonstrated that these biochemicals were somehow involved in promoting luteal maintenance. However, these proteins were unable to stimulate progesterone synthesis in vitro, nor did they contain any PRL or LH activity (Ellinwood et al., 1979). Subsequent characterization of proteins secreted by the day 13 ovine conceptus revealed three closely related isoelectric species having molecular weights of approximately 17,000 that were named ovine trophoblast protein 1 (oTP-1; Godkin et al., 1982b). Secretion of oTP-1 occurred transiently between days 12-21 of gestation and appeared to originate from the trophoblast. Infusion of oTP-1, as well as total conceptus proteins, into the uterine lumen of nonpregnant ewes between days 12-18 postestrus has been shown to prolong luteal life span. Ovine trophoblast protein-1 appears to exert an anti-luteolytic effect by altering the function of the surface and upper glandular epithelium of the uterine endometrium, and not by direct luteotropic action on the corpus luteum. Incubation of endometrial explants with oTP-1 induced the synthesis of six additional peptides that remain to be characterized. Daily uterine infusion of total conceptus proteins, obtained from day 16 blastocysts cultured in vitro, into nonpregnant ewes treated with either estradiol or oxytocin suppressed the total quantity, amplitude and frequency of pulsatile $PGF_2\alpha$ release (Fincher et al., 1986). These latter observations support previous reports that pulsatile release of $PGF_2\alpha$ observed during luteolysis is reduced to only a single episode on similar days of pregnancy (McCracken et al., 1984; Zarco et al., 1984). The mechanisms whereby conceptus-secretory proteins could act to reduce $PGF_2\alpha$ synthesis are not known. Levels of oxytocin receptors observed during early pregnancy were lower than those observed during luteolysis (McCracken et al., 1984); however, oTP-1 did not appear to block oxytocin-induced $PGF_2\alpha$ release by binding directly to oxytocin receptors (Bazer et al., 1986) nor did it appear to affect tonic secretion of PGF by the conceptus and endometrium. An effect of these proteins on estrogen receptor-mediated events also appears unlikely because some estrogen-induced proteins are required for normal pregnancy.

It is also possible that ovine conceptus proteins could favor the endometrial conversion of arachidonic acid to PGE_2, another proposed antiluteolysin/luteotropin acting during early pregnancy in the ewe. Evidence for the local delivery of a substance from the gravid uterine horn directly to the corpus luteum-bearing ovary has been provided (Mapletoft et al., 1976), and by virtue of its structural similarities to $PGF_2\alpha$, PGE_2 could pass from the uterine vein into the ovarian artery. Natural and estradiol-induced luteal regression are prevented by intra-uterine infusion of PGE_2, and the ovine conceptus can synthesize PGE_2 (see Inskeep and Murdoch, 1980). Although utero-ovarian levels of both PGE_2 and $PGF_2\alpha$ increase as pregnancy and the estrous cycle advance to day 14, the PGE_2:$PGF_2\alpha$ ratio is maximal on day 13 of pregnancy (Silvia et al., 1984). Prostaglandin E_2 could maintain luteal function through direct action on the luteal cell by binding to specific receptors and stimulating adenylate cyclase activity and progesterone synthesis (Fletcher and

Niswender, 1982) or by antagonizing the luteolytic actions of $PGF_2\alpha$ (Fitz et al, 1984).

Detectable levels of placental lactogen (PL) are present in the ovine and bovine trophoblast as early as day 14 and 17 of gestation, respectively (Martal and Djiane, 1977; Flint et al., 1979). The physical properties of oPL (M_r = 22,000; isoelectric point between 7.7 and 8.4) distinguish it from oTP-1 (Chan et al., 1976). Specific binding of oPL to ovine corpora lutea was demonstrated (Chan et al., 1978), but its possible role in regulating luteal function had not been tested until recently. Infusion of oPL alone or in combination with $PGF_2\alpha$ into the autotransplanted ovary in ewes on day 12 of an induced estrous cycle did not stimulate progesterone secretion or prevent $PGF_2\alpha$-induced luteolysis (Schramm et al., 1984). The effectiveness of oPL in suppressing endometrial $PGF_2\alpha$ secretion has not been investigated. Thus, a role for PL as an embryonic luteotropin in the ewe, as well as the cow, remains unresolved. Unlike the conceptus of the cow or ewe, the porcine conceptus apparently produces no PL (Kelly et al., 1976).

Cow

Bovine conceptuses have also been shown to secrete a group of acidic proteins (M_r = 22,000 - 26,000), referred to as bovine trophoblast protein 1 (bTP-1), between days 16-24 of gestation (Bartol et al., 1985). An additional group of conceptus proteins (M_r = 16,000 - 18,000) are produced between days 21-38 of gestation (Godkin et al., 1986). Similar to the effect of oTP-1 in the ewe, the introduction of bTP-1 into the uterine lumen prolonged the interestrous interval of nonpregnant cows and suppressed estradiol-stimulated $PGF_2\alpha$ secretion by the uterus (Thatcher et al., 1985). Presence of the bovine conceptus was also associated with attenuation of the episodic release of $PGF_2\alpha$ from the uterus that normally occurs during luteolysis (Thatcher et al., 1985), and with inhibition of oxytocin-induced $PGF_2\alpha$ secretion (Lafrance and Goff, 1985). Similarities among the antiluteolytic effects elicited by ovine and bovine conceptus-derived proteins, in addition to the ability of trophoblastic vesicles to prolong luteal maintenance after interspecies transfer to recipient cows and ewes, has led to the elegant demonstration of the immunological homology that exists between oTP-1 and bTP-1 (Helmer et al., 1987). Further characterizaion of these conceptus proteins will likely enhance our understanding of their underlying mechanism of action in promoting luteal maintenance during early gestation in ewes and cows.

Sow

As mentioned previously, biphasic estrogen production by the porcine conceptus appears to be indirectly luteotropic by causing $PGF_2\alpha$ to be sequestered within the uterine lumen; however, the mechanism underlying this event is unknown (Bazer et al., 1986). Evidence regarding a direct effect of estrogen in regulating luteal function in the sow is equivocal. Exogenous estradiol has been reported to increase concentrations of unoccupied LH receptors in corpora lutea of intact and hysterectomized gilts (Garverick et al, 1982). Ball and Day (1982) also found that extracts of day 16-25 porcine embryos, presumably containing estrogens, maintained progesterone synthesis in the presence of $PGF_2\alpha$ when implanted into corpora lutea of pregnant sows. In contrast, Watson and Maule Walker (1978) reported that estradiol did not enhance in vitro progesterone synthesis of corpora lutea removed from sows on days 18-22 of gestation nor did it inhibit the luteolytic action of $PGF_2\alpha$. In addition to estrogens, porcine conceptuses secrete a class of acidic proteins (M_r = 20,000 - 25,000) and a class of basic proteins (M_r = 35,000 - 50,000) between days 10-18 of gestation (Godkin et al., 1982a). The possible luteotropic/

antiluteolytic function of these proteins has not been investigated. Porcine conceptuses also synthesize PGE_2 (Davis et al., 1983; Lewis et al.,
1983) which is reflected by an increased ratio of PGE_2 to $PGF_2\alpha$ in uterine
flushings of gilts on days 11-14 of gestation compared with the ratio of
these protaglandins in flushings of nonpregnant gilts (Geisert et al.,
1982). However, luteal maintenance in nonpregnant sows is not altered by
intrauterine infusion of PGE_2 (Bazer et al., 1982). Recently, Marengo et
al. (1986) provided convincing evidence that neither PGE_2 nor PGE_1 are
involved in estrogen-mediated exocrine secretion of $PGF_2\alpha$ into the uterine
lumen of the sow.

Mare

Maintenance of luteal function during early gestation in the mare
appears to be regulated by antiluteolytic effects of the conceptus. It
has been suggested that the equine conceptus causes inhibition of uterine
PGF synthesis because concentrations of PGF in uterine fluid and uterine
venous plasma are reduced on days 11-16 of gestation relative to those
measured in nonpregnant mares during luteolysis (for review, see Sharp et
al., 1984). In fact, PGF production was found to be greatly suppressed
during coincubation of day 13-14 equine conceptus with endometrium (Sharp
et al., 1984). Presence of the conceptus also has been shown to inhibit
oxytocin-induced stimulation of uterine PGF synthesis in the mare (Goff
and Pontriand, 1985). The nature of the antiluteolytic factor(s) produced
by the equine conceptus is not known with certainty. Administration of
estrogen to mares during the late luteal phase of the cycle has been
effective in maintaining luteal function in some studies, but not in
others (Sharp et al., 1984). Although the equine conceptus synthesizes
estrogen between days 8-20 of gestation (Zavy et al., 1979), the role of
this steroid as an antiluteolysin remains in question.

Five proteins (M_r = 50,000 - 400,000) are secreted transiently by
equine conceptuses between days 12-14 of gestation and several acidic
proteins of low molecular weight are produced by the chorioallantois on
days 15-16 of gestation (Fazleabas and McDowell, 1983). However, the
significance of these proteins with respect to luteal maintenance in the
mare has not been determined.

Subsequent to the initial antiluteolytic actions of the conceptus
that promote luteal maintenance, the function of the corpus luteum of the
mare is markedly affected by equine chorionic gonadotropin (eCG, previously referred to as PMSG). Discovered by Cole and Hart (1930), eCG
production by the endometrial cups (derived from chorionic girdle cells
that invade the maternal endometrium) is first detectable in maternal
systemic blood on approximately day 37, increases to maximal concentrations on days 55-65 and declines to low levels on days 120-150 of gestation (Ginther, 1979). A transient reduction in progesterone secretion by
the primary corpus luteum of the mare, which occurs between days 8-28 of
gestation, is followed by a marked increase in progesterone secretion
between days 28-60 that is attributable to the luteotropic effects of eCG.
Secondary corpora lutea, structures unique to the equine ovary, begin to
form about day 40 of gestation and are stimulated by eCG to produce progesterone until their demise which occurs, along with that of the primary
corpus luteum, on about day 180 (Ginther, 1979). Although the bioactivity
of eCG is predominantly FSH-like, its action in vivo resembles that of LH.
Consequently, eCG ensures continued luteal function until the placenta
acquires the synthetic capacity to become the major source of progestins
during the remainder of gestation.

LUTEAL FUNCTION DURING OBLIGATE DELAYED IMPLANTATION

Obligate delayed implantation occurs as a regular event in the reproductive cycle of most pinnipeds, mustelids, bears, armadillos, an equatorial fruit bat and the Roe deer (Daniel, 1970). In what follows, discussion of luteal function during delayed implantation will be for the most part restricted to the situation in the European badger, mink and spotted skunk.

The length of time embryos remain in a state of dormancy or diapause varies both within and among species that exhibit this biological phenomenon. In the European badger delayed implantation usually occurs from February or March to December or January, a period of 10-11 months (Bonnin et al., 1978). Delayed implantation in mink, which mate from late February to April, varies anywhere from 5-55 days (Martinet et al., 1981). In the western variety of spotted skunk, which mates in September, delayed implantation lasts approximately 7 months (Mead, 1975). It is obvious that in these three species, and probably others mentioned above, time of mating as well as implantation is cued by seasonal changes in photoperiod. In mink and spotted skunks, implantation is associated with exposure of the animal to increasing daylength, while in the European badger it is associated with decreasing daylength (Mead, 1981).

During delayed implantation, corpora lutea of the mink, spotted skunk and European badger remain as distinct but relatively inactive structures. Inactivity of corpora lutea in these species is exemplified by histological appearance, low plasma progesterone levels, lack of a progestational endometrium and absence of implantation (Hansson, 1947; Enders, 1952; Møller, 1973; Bonnin et al., 1978; Mead, 1981; Sarker and Canivenc, 1982). Hysterectomy of pseudopregnant mink (Duby et al., 1972) and pregnant skunks (Mead and Swannack, 1978) does not deter luteal activation or time (near the vernal equinox) that it occurs. Thus, hormones of uterine or embryonic origin do not appear to be involved in activating corpora lutea of these species. However, hypophysectomy of mink or spotted skunks during the period of embryonic diapause prevents luteal activation and blastocyst implantation (Mead, 1975; Murphy et al., 1981). Further, hypophysectomy of mink after the preimplantation rise in progesterone leads to an abrupt reduction in progesterone concentration as well as degeneration of embryos (Murphy et al., 1980). Results of these studies indicate that the pituitary gland is required for luteal activation and blastocyst implantation. It is also noteworthy that Mead (1975) could detect no change in the histological appearance of corpora lutea in the hypophysectomized skunk. This observation suggests that during delayed implantation, corpora lutea of the skunk, and most likely those of the mink and European badger, remain independent of support from the pituitary gland. Delayed implantation appears to be a period of luteostasis in these species during which time corpora lutea may be totally autonomous or maintained by a "luteostasin", which, if such a factor exists, is not of pituitary origin.

It has been demonstrated that PRL is luteotropic and required for promoting transition of corpora lutea from an inactive to an active state in mink (Allais and Martinet, 1978; Papke et al., 1980). These data are consistent with the increase in systemic concentrations of PRL that occurs shortly before implantation in this species (Martinet et al., 1981). Further, Rose et al. (1986) found that unoccupied prolactin receptors in the ovaries of mated mink decreased as serum concentrations of PRL increased after the vernal equinox. In contrast, administration of ovine PRL to spotted skunks for 7 days during delayed implantation had no effect on plasma progesterone levels or duration of the preimplantation period (Mead, 1981). While PRL is obviously important for promoting luteal progesterone synthesis in mink, it is unknown whether this effect is by activation of steroidogenic enzymes by itself or via stimulation of increased numbers of LH receptors as occurs in the rat. There is some evidence that LH and(or) FSH may be luteotropic in mink after implantation occurs. Murphy et al. (1984) demonstrated that administration of monoclonal antibodies against GnRH to mink during gestation significantly reduced systemic levels of progesterone.

Although progesterone is required for implantation and maintenance of pregnancy in mink and skunks, all attempts to induce implantation in these species with exogenous progesterone or estrogen alone or in combination have failed (Cochrane and Shackelford, 1962; Holcomb, 1967; Mead, 1981). Mead (1981) concluded that other ovarian factors in addition to progesterone were required to initiate implantation in the skunk, and possibly in other mustelids.

LUTEOLYSIS BY PROSTAGLANDIN $F_2\alpha$

Exogenous estradiol is luteolytic in sheep, cattle, guinea pigs, hamsters, and primates. While primate corpora lutea can synthesize estradiol, a role for endogenous estradiol in luteal regression of the primate has not been demonstrated. In ewes, the function of endogenous estradiol in promoting luteolysis has been studied extensively and reviewed by McCracken et al. (1984). In fact, the endocrine regulation of $PGF_2\alpha$ secretion in ewes appears to be controlled primarily by estradiol and oxytocin. Pulsatile surges of oxytocin originating from the corpus luteum occur during luteolysis in the ewe (Flint and Sheldrick, 1986). Increases in utero-ovarian venous levels of $PGF_2\alpha$ are observed prior to oxytocin pulses on day 15 of the cycle (Moore et al., 1986), suggesting that uterine $PGF_2\alpha$ initiates oxytocin release from the corpus luteum during the later stages of luteolysis. McCracken et al. (1984) postulated the following sequence of events regulating luteolysis in the ewe. Because the uterotropic actions of progesterone appear to decline as the luteal phase progresses, endogenous estradiol is able to stimulate oxytocin receptor synthesis in the endometrium. Endogenous luteal oxytocin interacts with its receptor to cause secretion of $PGF_2\alpha$ from the endometrium. Luteal regression is initiated as a result of the countercurrent transfer of $PGF_2\alpha$ from the uterine vein to the ovarian artery. Further release of oxytocin from the corpus luteum may be caused by $PGF_2\alpha$, and oxytocin binding to the endometrium further reinforces $PGF_2\alpha$ release in a positive feedback manner (also suggested by Flint and Sheldrick, 1983). This latter release of luteal oxytocin appears to cause the pulsatile secretion of $PGF_2\alpha$ on days 14-15 of the estrous cycle. Because oxytocin receptors may be desensitized for a period of time subsequent to oxytocin binding, hour-long pulses of endometrial $PGF_2\alpha$ release occur every 6 hours, which is the time necessary for estradiol to induce the synthesis of new oxytocin receptors. Five, but not four, pulses of $PGF_2\alpha$ administered over a 24 hour period were found to cause permanent luteal regression (McCracken and Schramm, 1983). However, Hooper et al. (1986) recently demonstrated that while uterine $PGF_2\alpha$ release during luteolysis in the ewe was associated with luteal oxytocin release, secretion of this nonapeptide was not always dependent upon stimulation by uterine $PGF_2\alpha$.

Prostaglandin $F_2\alpha$ induces both functional and structural luteolysis; the former results from the acute actions of the luteolysin that lead to reduced progesterone production. These actions are reversible upon hysterectomy (except in women and primates), neutralization with antibodies, treatment with synthesis inhibitors, or the presence of a viable conceptus. Exposure of the corpus luteum to $PGF_2\alpha$ for an additional 24 to 48 hours beyond functional luteolysis results in irreversible structural and enzymatic changes that ultimately complete luteal regression. A direct action of $PGF_2\alpha$ on luteal cells is supported by the existence of specific receptors located within the plasma membranes of ovine (Powell et al., 1974a), bovine (Powell et al., 1976; Lin and Rao, 1977), equine (Kimball and Wyngarden, 1974), human (Powell et al., 1974b) and rat (Luborsky-Moore et al., 1979) corpora lutea. In view of the fact that luteal development and steroidogenesis in a number of species had been shown to be dependent upon LH-induced accumulation of cAMP, Henderson and McNatty (1975) proposed that $PGF_2\alpha$ initiates functional luteolysis by interfering with LH activation of adenylate cyclase activity within luteal plasma membranes. In support of this hypothesis, exposure of rat (Lahav et al., 1976; Thomas et al., 1978), ovine (Fletcher and Niswender, 1982), bovine (Marsh, 1971), human (Hamberger et al., 1979) and primate (Stouffer et al., 1979) corpora lutea to $PGF_2\alpha$ _in vitro_ was shown to inhibit gonadotropin stimulation of adenylate cyclase activity and cAMP formation. Indeed, natural luteolysis in cows (Garverick et al., 1985), sows (Ritzhaupt et al., 1986) and primates (Eyster et al., 1985) as well as $PGF_2\alpha$-induced luteolysis in ewes (Agudo et al., 1984) and pseudopregnant rats (Khan and Rosberg, 1979) was accompanied by decreases in basal and(or) LH-stimulated adenylate cyclase activity. Progesterone secretion declined within minutes of $PGF_2\alpha$ administration to rats (Grinwich et al., 1976) and ewes (McCracken and Schramm, 1983), while attenuation of adenylate cyclase activity was noted by 15 and 120 minutes in rats (Lahav et al., 1976) and ewes (Agudo et al., 1984), respectively.

The cellular mechanism whereby functional luteolysis is initiated by $PGF_2\alpha$ has only recently been demonstrated to involve phosphoinositide metabolism in rat (Leung et al., 1986) and bovine (West et al., 1986) luteal cells _in vitro_. Hormone-induced hydrolysis of plasma membrane phosphatidylinositol 4,5-bisphosphate (PIP_2), a metabolite of phosphatidylinositol (PI) results in the generation of inositol 1,4,5-triphosphate (IP_3) and diacylglycerol (DG), whose actions as second messengers in target cells elicit calcium mobilization from the endoplasmic reticulum and activation of protein kinase C, respectively (Berridge and Irvine, 1984; Nishizuka et al., 1984). Within seconds of exposure to $PGF_2\alpha$, rat and bovine luteal cells exhibited a rapid decrease in PIP_2 and a concomitant increase in IP_3, which presumably led to increases in intracellular calcium. Effects of calcium on LH-induced adenylate cyclase activity in rat luteal cells _in vitro_ were examined by Dorflinger et al. (1984). The calcium ionophore, A23187, inhibited LH-induced cAMP accumulation only in the presence of extracellular calcium. Ability of $PGF_2\alpha$ to inhibit LH-induced adenylate cyclase activity in intact luteal cells was not affected by a) the absence of extracellular calcium, b) verapamil, a calcium channel blocker, or c) an influx of extracellular calcium. However, incubation of isolated rat luteal membranes in the presence of 5-20 μM calcium produced a dose-dependent decrease in LH-stimulated adenylate cyclase activity. Collectively, these observations suggest that acute increases in intracellular calcium inhibit the activation of adenylate cyclase in a manner similar to the inhibition produced by $PGF_2\alpha$. Hansel and Dowd (1986) also provided evidence to indicate that elevated intracellular calcium levels induced by A23187 and methyl-isoxanthine treatment of small cells from bovine corpora lutea inhibit LH-stimulated progesterone synthesis _in vitro_. These observations in rat and bovine corpora lutea are consistent with current reports of $PGF_2\alpha$-induced increases in IP_3 accumulation in luteal cells of the same species, which may reflect increases in intracellular calcium mobilization. An exhaustive study on the properties of a calcium-magnesium ATPase in rat luteal memranes revealed that it functions as a calcium-extrusion pump (Verma and Penniston, 1981). A dramatic reduction in luteal calcium-ATPase activity occurred within 1 to 2 hours post-$PGF_2\alpha$ treatment of rats containing functional corpora lutea (Albert et al., 1984). Thus, intracellular calcium levels in luteal cells sustained as a result of PGF$_2\alpha$-mediated PIP$_2$ hydrolysis may provide a central mechanism whereby functional luteolysis ensues in many species.

Effects of elevated intracellular calcium could be manifested in many ways. An increase in phosphodiesterase activity, a calcium-dependent enzyme that inactivates cAMP, was noted within 2 hours of PGF$_2\alpha$ administration to ewes (Agudo et al., 1984). Another important consequence of elevated intracellular calcium levels in luteal cells could be the activation of phospholipase A$_2$, an enzyme known to catalyze the release of arachidonic acid from the 2-position of the phospholipid glycerol moiety of phosphatidylcholine, phosphatidylethanolamine and phosphatidylinositol. Riley and Carlson (1985) observed that the rapid rigidification of luteal plasma membranes from PGF$_2\alpha$-treated rats was calcium dependent. Exposure of nonregressed luteal membranes to phospholipase A$_2$ caused virtually identical changes in the polarization properties indicative of decreased membrane fluidity as in membranes obtained from regressing corpora lutea. Further studies suggested the activation of an endogenous phospholipase A$_2$ during luteolysis in the rat (Riley and Carlson, 1986). It is possible, although not yet proven, that calcium-dependent decreases in plasma membrane fluidity observed during luteolysis may provide a physical means of preventing the activation of adenylate cyclase by LH. Lateral movement of LH-receptor complexes within luteal plasma membranes of ewes (Niswender et al., 1985) and rats (Luborsky et al., 1984) remain to be correlated with adenylate cyclase activity, although PGF$_2\alpha$ prevented LH-receptor micro-aggregation observed in functional rat corpora lutea.

Whatever the mechanism, the ability of exogenous agents normally capable of stimulating the N$_s$ regulatory subunit (GTP analogues) and(or) the catalytic subunit (cholera toxin, NaF) of adenylate cyclase was unaltered in murine (Torjesen and Aakvaag, 1986) and ovine (Fletcher and Niswender, 1982; Agudo et al., 1984) corpora lutea exposed to PGF$_2\alpha$, indicating that the luteolysin interferes with the coupling of the LH-receptor and cyclase. Alterations in the physical properties of the luteal plasma membrane could affect other enzymes as well. Kim and Yeoun (1983) reported a decrease in Na+-K+-ATPase activity of rat luteal membranes within 1 hour of incubation with PGF$_2\alpha$. In addition to the possible attenuation of adenylate cyclase and other enzyme activity by PGF$_2\alpha$ during luteal regression, degenerative changes in plasma membranes correlating with increased phospholipase A$_2$ activity were suggested to result from elevations in luteal prostaglandin synthesis. Such changes could accelerate regression in a positive feedback manner and(or) the generation of superoxide anions through the action of lipoxygenases on arachidonic acid (Riley and Carlson, 1985). Rothchild (1981) proposed that PGF$_2\alpha$, whether of uterine or ovarian origin, could stimulate its own production in luteal tissue of all species, thus contributing to the completion of luteolysis in a paracrine fashion. As in human and primates, luteal and(or) stromal tissues of rats, rabbits, sows (see McCracken and Schramm, 1983) and cows (Hansel and Dowd, 1986) are capable of producing PGF$_2\alpha$. A central, physiological role for intraluteal prostaglandin synthesis during luteolysis in nonprimate species remains to be investigated.

Prolonged exposure to PGF$_2\alpha$ eventually leads to the irreversible changes associated with structural luteolysis. These changes could result from a direct action on the luteal cell or, indirectly, from a reduction in blood flow through the luteal vascular bed. Decreases in LH receptor concentrations, membrane fluidity, and steroidogenic enzyme activity, as well as increases in lysosomal enzyme activity, occur during the later stages of luteolysis. Although physiological concentrations of PGF$_2\alpha$ did not affect luteal blood flow in rabbits and ewes, an eventual reduction in blood flow to the corpus luteum, as observed in primates, rats, rabbits and ewes upon exposure to high concentrations of $PGF_{2}\alpha$ and(or) as regression progresses, may contribute to structural luteolysis (McCracken and Schramm, 1983) as evidenced by the resemblance between initial structural changes and those induced by hypoxia in ewes. The combination of these events would culminate in complete luteolysis.

Current dogma assigning $PGF_{2}\alpha$ as the major physiologic regulator of luteolysis may also be revised in the near future. Hansel and Dowd (1986) discuss evidence that strongly suggests a role for lipoxygenase products of arachidonic acid metabolism, such as 5-hydroxyeicosatetraenoic acid (5-HETE) as a uterine luteolysin in cattle. Investigations concerning the involvement of lipoxygenase products as possible endocrine and(or) paracrine regulators of luteolysis in mammalian corpora lutea may contribute greatly to our understanding of luteal regression.

OXYTOCIN IN THE CORPUS LUTEUM

A corpus luteum factor with oxytocic action was suggested as early as the beginning of this century. Ott and Scott (1910) found that an aqueous extract of corpus luteum increased milk flow in the goat, and Schafer and Mackenzie (1911) and Mackenzie (1911) showed that injection of an extract of ovine corpus luteum induced milk let-down in the lactating cat. These interesting observations were not pursued further until this decade when oxytocin was identified in corpora lutea of several species. The corpus luteum of the cow (Fields et al., 1983; Wathes et al., 1983a,b; Abdelgadir et al., 1986) ewe (Wathes and Swann, 1982) monkey (Khan-Dawood et al., 1984) woman (Wathes et al., 1982; Khan-Dawood and Dawood, 1983) and rabbit (Khan-Dawood and Dawood, 1984) has been shown to contain measurable quantities of oxytocin.

Luteal oxytocin is not only immunologically indistinguishable, but is also biologically similar to its pituitary counterpart. Luteal extracts caused a significant increase in contraction of uterine muscle, stimulated uterine strips in vitro and increased intramammary pressure in a manner similar to authentic oxytocin in rats (Wathes and Swann, 1982; Wathes et al., 1983b). Moreover, ovine, bovine and human luteal oxytocin extracts elute at the same position as pituitary oxytocin on Sephadex G50 and reverse-phase high-performance liquid chromatography (HPLC; Wathes and Swann, 1982; Wathes et al., 1982; Wathes et al., 1983a; Fields et al., 1983; Sheldrick and Flint, 1983a; Schaeffer et al., 1984; Dawood and Khan-Dawood, 1986). Dispersed cell cultures of ovine and bovine corpora lutea incorporated labeled cysteine into a peptide that eluted at the same position as oxytocin on HPLC. As occurs in the hypothalamus, the synthesis of luteal oxytocin involved the formation of an approximately 14-K precursor protein that was subsequently cleaved to form neurophysin and oxytocin (Swann et al.,1984). The oxytocin gene is highly transcribed in the bovine corpus luteum. Luteal cDNA sequence analysis as well as cell-free translation studies showed that luteal and hypothalamic mRNA for oxytocin were essentially similar. However, the active corpus luteum produces approximately 250 times more oxytocin mRNA than a single hypothalamus (Ivell and Richter, 1984).

In the ewe and cow oxytocin is first detectable in follicular fluid and granulosa cells during or shortly after the LH surge (Kruip et al., 1985; Wathes et al., 1986). Measurement of luteal oxytocin-specific mRNA throughout the estrous cycle of the cow showed that gene transcription was maximal accompanying ovulation and decreased thereafter. Highest concentrations of mRNA were detected around day 3; these declined sharply around day 7 and reached a basal level by day 11 of the cycle, after which only very low levels were detectable (Ivell et al., 1985). Apparently, changes in luteal concentrations of oxytocin during the bovine estrous cycle do not occur simultaneously with those of mRNA for this peptide. Oxytocin concentrations in both ovine and bovine luteal tissues were maximal during the early luteal phase with peak levels occurring around day 8 of the cycle (Sheldrick and Flint, 1983b; Wathes et al., 1984; Abdelgadir et al., 1986). In the cow, luteal oxytocin concentrations increased from about .4 µg/g (wet weight) on day 4 of the cycle to more than 4 µg/g on day 8 and declined thereafter to about .6 µg/g on day 12 and less than .1 µg/g on day 16 (Abdelgadir et al., 1986). The observed difference in time between occurrence of maximal levels of oxytocin mRNA and of concentrations of this peptide hormone in luteal tissue suggests that during early stages of the cycle only some of the prohormone is processed. This lag period may be controlled by certain endocrine factors that regulate synthesis and processing of the hormone in granulosa cells (Wathes et al., 1986).

In contrast to the high oxytocin concentrations measured in ruminants, extremely low concentations were found in the sow corpus luteum, with maximal levels (10 ng/g) detected on day 5 of the cycle (Pitzel et al., 1984). Similarly, both rat and rabbit ovarian and luteal tissues, respectively, contained low levels of oxytocin (Khan-Dawood, 1984). Oxytocin concentration in primate luteal tissue was also much lower than that of ruminants. Wathes et al. (1982) detected oxytocin levels of about 30 ng/g while Khan-Dawood (1983) reported oxytocin concentrations of 11-53 ng/g in human luteal tissue. In cynomolgus monkeys the concentration ranged from 3.4 to 602.5 ng/g wet weight of tissue, with the highest concentration at the midluteal phase of the cycle (Khan-Dawood et al., 1984).

The first evidence of a luteolytic effect of oxytocin was described by Armstrong and Hansel (1959) and Hansel and Wagner (1960). Injection of oxytocin into heifers between days 3-6 of the cycle resulted in a significant decrease in its duration. These data were later confirmed in the cow (Labhsetwar et al., 1964; Donaldson and Takken, 1968; Harms et al., 1969) and the goat (Cooke and Knifton, 1981; Cooke and Homeida, 1982). Both active and passive immunization against oxytocin extended the length of the cycle in ewes and goats (Sheldrick et al., 1980; Schams et al., 1983; Cooke and Homedia, 1985). In the ewe, rhesus monkey, mare, sow, rat, rabbit and guinea pig injection of oxytocin has thus far proved ineffective in reducing estrous cycle length (Duncan et al., 1961; Donovan, 1961; Brinkley and Nalbandov, 1963; Milne, 1963; Neely et al., 1979; Wilks, 1983).

As discussed in a previous section of this chapter, ovarian secretion of oxytocin is believed to be stimulated by $PGF_2\alpha$. Indeed, treatment of cows and ewes with an analog of $PGF_2\alpha$ caused an immediate increase in luteal oxytocin secretion (Walters et al., 1983; Flint and Sheldrick, 1983; Schallenberger et al., 1984) and degranulation of luteal cells (Heath et al., 1983). Conversely, oxytocin treatment stimulated the release of $PGF_2\alpha$ from the uterine endometrium of the ewe (Roberts and McCracken, 1976), goat, (Cooke and Homeida, 1982) and cow (Newcomb et al., 1979; Milvae and Hansel, 1980). It appears that the luteolytic action of oxytocin is normally mediated by this response, because it can be prevented by hysterectomy in the cow (Armstrong and Hansel, 1959; Anderson et al., 1965a; Ginther et al., 1967) and by simultaneous inhibition of prostaglandin synthesis in the goat (Cooke and Knifton, 1981; Cooke and Homeida, 1983).

Oxytocin may also have a luteotropic effect in the regulation of luteal function. In low concentrations oxytocin stimulated progesterone production by isolated bovine and human luteal cells in the early luteal phase whereas it inhibited both basal and hCG-stimulated progesterone release at high concentrations (Tan et al., 1982a,b).

ACTION OF GnRH ON THE CORPUS LUTEUM

Administration of GnRH or its agonistic analogs to the rat (Kledzik et al., 1978; Harwood et al., 1980; Jones and Hsueh, 1980), human (Koyama et al., 1978; Casper and Yen, 1979), nonhuman primate (Asch et al., 1981) and cow (Rodger and Stormshak, 1986) interfered with luteal function. Treatment with LH can overcome inhibitory effects of GnRH on rat luteal progesterone synthesis in vitro (Behrman et al., 1980) and treatment with hCG can override in vivo inhibitory effects of GnRH in the human (Casper et al., 1980). Exogenous GnRH or GnRH-agonistic analogs may act directly on the ovary or indirectly, via released gonadotropins, to alter luteal steroidogenesis. Receptors for GnRH have been detected in rat luteal cells (Clayton et al., 1979) but none have been found in luteal tissue of the cow, ewe and sow (Brown and Reeves, 1983). In addition, no direct effects of GnRH on human luteal cell function in vitro could be detected (Williams and Behrman, 1983), suggesting an absence of GnRH receptors in this tissue. Casper et al. (1980) suggested that GnRH-induced early luteal regression in the human may be due to a decrease in gonadotropin secretion caused by GnRH-induced pituitary refractoriness. In the cow, injection of GnRH during the early or midluteal phase of the cycle caused release of LH that may have resulted in down-regulation of luteal LH receptors (Rodger and Stormshak, 1986). Such a possible mode of action for GnRH was supported by the observation that luteal concentrations of LH receptors were depressed after treatment with the decapeptide. However, if this is the mechanism by which GnRH acts in vivo to alter luteal pro-gesterone synthesis in the cow, then the bovine luteal cell must be more sensitive to LH receptor down-regulation during the mid than early luteal phase of the cycle. A single intravenous injection of GnRH into cows on day 2 of the cycle was followed by a 6 day lag period before altered luteal function was detected. On the other hand, a similar injection on day 10 of the cycle caused serum concentrations of progesterone to be sig-nificantly depressed after only 48 hours. Thus, the possibility cannot be excluded that in the cow exogenous GnRH may act to alter luteal function by two different mechanisms, depending upon the stage of the cycle when it is administered.

It should be mentioned that other investigators have shown that injection of GnRH or a GnRH agonistic analog into cows during the mid-luteal phase of the cycle increased serum concentrations of progesterone (Kittok et al., 1973; Milvae et al., 1984). Milvae et al. (1984) reported that luteal function of GnRH-treated cows was suppressed during the succeeding cycle. Reasons for the differences in effects of GnRH in the study by Rodger and Stormshak (1986) and GnRH-agonistic analog in the study by Milvae et al. (1984) are not known.

It has been proposed that the inhibitory action of GnRH in the rat luteal cell results from ability of the decapeptide to inhibit calcium extrusion from the cytosol (Williams and Behrman, 1983). Consequently, elevated levels of intracellular calcium prevent activation of adenylate cyclase by LH via a protein kinase C-dependent phosphorylation or calmodulin-dependent process. This mechanism of action of GnRH is suppor-ted by recent data of Leung et al. (1986) who found that GnRH binding to receptors in plasma membranes of rat luteal cells in primary culture acti-vated hydrolysis of phosphoinositides, which ultimately leads to stimula-tion of protein kinase C activity.

CONCLUSION

Among mammals there appear to be some commonalities, and yet some unique differences, with respect to the response of the corpus luteum to secretagogues. For example, LH and PGF$_2\alpha$ have, for most species studied, proven to be the "primary" luteotropin and luteolysin, respectively. However, some species seem to have evolved a corpus luteum whose function is also regulated by "secondary" luteotropins or luteolysins. In fact, for some species, a "secondary" luteotropin regulating luteal function has become more important than LH, even though the luteal cell has retained the capacity to respond to this gonadotropin. It is obvious from the variety of secretagogues that affect its function that the corpus luteum is a more complex endocrine gland than was thought even as late as just a decade ago. It is envisioned that future research will provide answers about whether luteal cells commmunicate with each other and, if so, the nature of the intercellular messenger. Studies will likely continue to elucidate the interrelationships between the phosphoinositide and arachidonic acid cascades within the luteal cell and their significance in regulating the life span of the corpus luteum.

ACKNOWLEDGMENTS

The authors greatly appreciate the efforts of Helen Chesbrough who has spent countless hours editing and typing the manuscript. The authors also thank Anthony Archibong, Carrie Cosola and OV Slayden for their assistance in preparing this chapter.

REFERENCES

Abdelgadir SE, Swanson LV, Oldfield JE, Stormshak F, 1986. In vitro release of oxytocin from bovine corpora lutea by prostaglandin F$_2\alpha$. Biol Reprod 34 (Suppl. 1):135.

Agudo LSp, Zahler WL, Smith MF, 1984. Effect of prostaglandin F$_2\alpha$ on the adenylate cyclase and phosphodiesterase activity of ovine corpora lutea. J Anim Sci 58:955-62.

Albert PJ, Preston SL, Behrman HR, 1984. Prostaglandin-induced luteolysis linked to inhibition of calcium pump activity. Excerpta Medica, Intl Cong Series 656:340.

Alila HW, Hansel W, 1984. Origin of different cell types in the bovine corpus luteum as characterized by specific monoclonal antibodies. Biol Reprod 31:1015-25.

Allais C, Martinet L, 1978. Prolactin, the luteotrophic factor in mink (Mustela vison). In: Gaillard PJ, Boer HH (ed.), Proc 8th Int Symp Comp Endocrinol. Amesterdam: Elsevier, p. 83.

Anderson LL, Bowerman AM, Melampy RM, 1965a. Oxytocin on ovarian function in cycling and hysterectomized heifers. J Anim Sci 24:964-68.

Anderson LL, Leglise PC, du Mesnil du Buisson F, Rombauts P, 1965b. Interaction des hormones gonadotropes et de l'uterus dans le maintien du tissu luteal ovarien chez la truie. C R Acad Sci (Paris) 261:3675-8.

Armstrong DT, Hansel WJ, 1959. Alteration of the bovine estrous cycle with oxytocin. J Dairy Sci 42:533-42.

Armstrong DT, Knudsen KA, Miller LS, 1970. Effects of prolactin upon cholesterol metabolism and progesterone biosynthesis in corpora lutea of rats hypophysectomized during pseudopregnancy. Endocrinology 86:634-41.

Asch RH, Siler-Khodr TM, Smith CG, Schally AV, 1981. Luteolytic effect of D-Trp[6]-luteinizing hormone-releasing hormone in the rhesus monkey (Macaca mulatta). J Clin Endocrinol Metab 52:565-71.

Ascoli M, Segaloff DL, 1986. Effects of collagenase on the structure of the lutropin/choriogonadotropin receptor. J Biol Chem 261:3807-09.

Ball GD, Day BN, 1982. Local effects of $PGF_2\alpha$ and embryonic extracts on luteal function in swine. J Anim Sci 54:150-54.

Barkley MS, Bradford GE, Geschwind II, 1978. The pattern of plasma prolactin concentration during the first half of mouse gestation. Biol Reprod 19:291-96.

Bartol FF, Roberts RM, Bazer FW, Lewis GS, Godkin JD, Thatcher WW, 1985. Characterization of proteins produced in vitro by periattachment bovine conceptuses. Biol Reprod 32:681-93.

Bazer FW, Geisert RD, Thatcher WW, Roberts RM, 1982. The establishment and maintenance of pregnancy. In: Cole JA, Foxcroft GR (ed.), Control of pig reproduction. London: Butterworth Scientific, pp. 227-52.

Bazer FW, Thatcher WW, 1977. Theory of maternal recognition of pregnancy in swine based on estrogen controlled endocrine versus exocrine secretion of prostaglandin F by the uterine endometrium. Prostaglandins 14:397-400.

Bazer FW, Vallet JL, Roberts RM, Sharp DC, Thatcher WW, 1986. Role of the conceptus secretory products in establishment of pregnancy. J Reprod Fertil 76:841-50.

Behrman HR, Preston SL, Hall AK, 1980. Cellular mechanism of the anti-gonadotropic action of luteinizing hormone-releasing hormone in the corpus luteum. Endocrinology 107:656-64.

Berridge MJ, Irvine RF, 1984. Inositol triphosphate, a novel second messenger in cellular signal transduction. Nature 312:315-21.

Bonnin M, Canivenc R, Ribes CL, 1978. Plasma progesterone levels during delayed implantation in European badger (Meles meles) J. Reprod Fertil 52:55-8.

Brinkley HJ, Nalbandov AV, 1963. Effect of oxytocin on ovulation in rabbits and rats. Endocrinology 73:515-17.

Brown JL, Reeves JJ, 1983. Absence of specific luteinizing hormone releasing hormone receptors in ovine, bovine and porcine ovaries. Biol Reprod 29:1179-82.

Buhr MM, Carlson JC, Thompson JE, 1979. A new perspective on the mechanism of corpus luteum regression. Endocrinology 105:1330-35.

Cameron JL, Stouffer RL, 1982. Gonadotropin receptors of the primate corpus luteum. II. Changes in available luteinizing hormone-and chorionic gonadotropin-binding sites in macaque luteal membranes during the nonfertile menstrual cycle. Endocrinology 110:2068-73.

Carlson JC, Buhr MM, Gruber MY, Thompson JE, 1981. Compositional and physical properties of microsomal membrane lipids from regressing rat corpora lutea. Endocrinology 108:2124-28.

Carlson, JC, Buhr MM, Riley JCM, 1984. Alterations in the cellular membranes of regressing rat corpora lutea. Endocrinology 114:521-26.

Carlson JC, Buhr MM, Wentworth R, Hansel W, 1982. Evidence of membrane changes during regression in the bovine corpus luteum. Endocrinology 110:1472-76.

Carr FE, Chin WW, 1985. Absence of detectable chorionic gonadotropin sub-unit messenger ribonucleic acids in the rat placenta throughout gestation. Endocrinology 116:1151-57.

Casper RF, Yen SS, 1979. Induction of luteolysis in the human with a long-acting analog of luteinizing hormone-releasing factor. Science 205:408-10.

Casper RF, Sheehan KL, Yen SS, 1980. Chorionic gonadotropin prevents LRF-agonist-induced luteolysis in the human. Contraception 21:471-78.

Chan JSD, Robertson HA, Friesen HG, 1976. The purification and characterization of ovine placental lactogen. Endocrinology 98:65-76.

Chan JSD, Robertson HA, Friesen HG, 1978. Distribution of binding sites for ovine placental lactogen in sheep. Endocrinology 102:632-40.

Choudary JB, Greenwald GS, 1969. Luteotrophic complex of the mouse. Anat Rec 163:201-07.

Clayton RN, Harwood JP, Catt KJ, 1979. Gonadotropin-releasing hormone analogue binds to luteal cells and inhibits progesterone production. Nature 282:90-92.

Cochrane RL, Shackelford RM, 1962. Effects of exogenous oeştrogen alone or in combination with progesterone on pregnancy in the intact mink. J Endocrinol 25:101-06.

Cole HH, Hart GH,1930. The potency of blood serum of mares in progressive stages of pregnancy in effecting the sexual maturity of the immature rat. Am J Physiol 93:57-68.

Condon WA, Black DL, 1976. Catecholamine-induced stimulation of progesterone by the bovine corpus luteum in vitro. Biol Reprod 15:573-78.

Cooke RG, Homeida AM, 1982. Plasma concentrations of 13,14-dihydro-15-keto prostaglandin $F_2\alpha$ and progesterone during oxytocin-induced estrous in the goat. Theriogenology 18:453-60.

Cooke RG, Homeida AM, 1983. Prevention of the luteolytic action of oxytocin in the goat by inhibition of prostaglandin synthesis. Theriogenology 20:363-65.

Cooke RG, Homeida AM, 1985. Suppression of prostaglandin $F_2\alpha$ release and delay of luteolysis after active immunization against oxytocin in the goat. J Reprod Fertil 75:63-68.

Cooke RG, Knifton A, 1981. Oxytocin-induced estrous in the goat. Theriogenology 16:95-97.

Daniel Jr JC, 1970. Dormant embryos of mammals. Bio Sci 20:411-15.

Dattatreyamurty B, Ratham P, Saxena BB, 1983. Isolation of the luteinizing hormone-chorionic gonadotropin receptor in high yield from bovine corpora lutea. J Biol Chem 258:3140-58.

Dawood MY, Khan-Dawood FS, 1986. Human ovarian oxytocin: Its source and relationship to steroid hormones. Am J Obstet Gynecol 154:756-63.

Davis DL, Pakras PL, Dey SK, 1983. Prostaglandins in swine blastocysts. Biol Reprod 28:1114-18.

del Pozo E, Golstein M, Friesen H, Brun del R, Eppenberger U, 1975. Lack of action of prolactin suppression on the regulation of the human menstrual cycle. Am J Obstet Gynecol 123:719-23.

Diekman MA, O'Callaghan P, Nett TM, Niswender GD, 1978a. Validation of methods and quantification of luteal receptors for LH throughout the estrous cycle and early pregnancy in ewes. Biol Reprod 19:999-1009.

Diekman MA, O'Callaghan P, Nett TM, Niswender GD, 1978b. Effects of prostaglandin $F_2\alpha$ on the number of LH receptors in ovine corpora lutea. Biol Reprod 19:1010-13.

Donaldson LE, Hansel W, 1965a. Histological study of bovine corpora lutea. J Dairy Sci 48:905-09.

Donaldson LE, Hansel W, 1965b. Prolongation of the life span of the bovine corpus luteum by single injections of bovine luteinizing hormone. J Dairy Sci 48:903-04.

Donaldson LE, Hansel W, Van Vleck LD, 1965. The luteotropic properties of luteinizing hormone and the nature of oxytocin induced luteal inhibition in cattle. J Dairy Sci 48:331-37.

Donaldson LE, Takken A, 1968. The effect of exogenous oxytocin on corpus luteum function in the cow. J Reprod Fertil 17:373-83.

Donovan BT, 1961. The role of the uterus in the regulation of the oestrous cycle. J Reprod Fertil 2:508-10.

Dorflinger LJ, Albert PJ, Williams AT, Behrman HR, 1984. Calcium is an inhibitor of luteinizing hormone-sensitive adenylate cyclase in the luteal cell. Endocrinology 114:1208-15.

du Mesnil du Buisson F, Leglise PC, Anderson LL, 1964. Hypophysectomy in pigs. J Anim Sci 23:1226-27.

du Mesnil du Buisson F, Leglise PC, 1963. Effet de l'hypophysectomie sur les corps jaunes de la truie. Resultats preliminaires. C R Hebd Seanc Acad Sci 257:261-63.

Duby RT, Pilbeam T, Travis HF, 1972. The influence of melatonin and hysterectomy on plasma progesterone levels of the mink (Mustela vison). Proc Ann Mtg Soc Study Reprod Abstr 73.

Duncan GW, Bowerman AM, Anderson LL, Hearn WR, Melampy RM, 1961. Factors influencing in vitro synthesis of progesterone. Endocrinology 68:199-207.

Ellinwood WE, Nett TM, Niswender GD, 1979. Maintenance of the corpus luteum of early pregnancy in the ewe. I. Luteotrophic properties of embryonic homogenates. Biol Reprod 21:281-88.

Enders RK, 1952. Reproducion in the mink (Mustela vison). Proc Am Phil Soc 96:691-755.

Enders AC, 1973. Cytology of the corpus luteum. Biol Reprod 8:158-82.

Eyster KM, Stouffer RL, 1985. Adenylate cyclase in the corpus luteum of the rhesus monkey. II Sensitivity to nucleotides, gonodotropins catecholamines, and nonhormonal activators. Endocrinology 116:1552-58.

Eyster, KM, Ottobre JS, Stouffer RL, 1985. Adenylate cyclase in the corpus luteum of the rhesus monkey. III. Changes in basal and gonadotropin-sensitive activities during the luteal phase of the menstrual cycle. Endocrinology 117:1571-77.

Farin CE, Moeller CL, Sawyer HR, Gamboni F, Niswender GD, 1986. Morpho-metric analysis of cell types in the ovine corpus luteum throughout the estrous cycle. Biol Reprod 35:1299-308.

Fazleabas AT, McDowell KJ, 1983. Synthesis and release of polypeptides by horse conceptus tissue and extra-embryonic membranes during early and midgestation. Biol Reprod 28(Suppl 1):138.

Fields PA, Eldridge RK, Fuchs AR, Roberts RF, Fields MJ, 1983. Human placental and bovine corpora luteal oxytocin. Endocrinology 112:1544-46.

Fincher KB, Hansen PJ, Thatcher WW, Roberts RM, Bazer FW, 1986. Ovine conceptus secretory proteins suppress induction of prostaglandin $F_2\alpha$ release by estradiol and oxytocin. J Reprod Fertil 76:425-33.

Fitz TA, Mock EJ, Mayan MH, Niswender GD, 1984. Interactions of prosta-glandins with subpopulations of ovine luteal cells. II. Inhibitory effects of $PGF_2\alpha$ and protection by PGE_2. Prostaglandins 28:127-38.

Fletcher PW, Niswender GD, 1982. Effect of $PGF_2\alpha$ on progesterone secretion and adenylate cyclase activity in ovine luteal tissue. Prostaglandins 23:803-18.

Flint APF, Henville A, Christie WB, 1979. Presence of placental lactogen in bovine conceptuses before attachment. J Reprod Fertil 56:305-08.

Flint APF, Sheldrick EL, 1983. Evidence for a systemic role for ovarian oxytocin in luteal regression in sheep. J Reprod Fertil 67:215-25.

Flint APF, Sheldrick EL, 1986. Ovarian oxytocin and the maternal recog-nition of pregnancy. J Reprod Fertil 76:831-39.

Friend DS, Gilula NB, 1972. A distinctive cell contact in the rat adrenal cortex. J Cell Biol 53:148-63.

Gardner ML, First NL, Casida LE, 1963. Effect of exogenous estrogens on corpus luteum maintenance in gilts. J Anim Sci 22:132-34.

Garverick HA, Polge C, Flint ARF, 1982. Ostradiol administration raises luteal LH receptor levels in intact and hysterectomized pigs. J Reprod Fertil 66:371-77.

Garverick HA, Smith MF, Elmore RG, Morehouse GL, Agudo LSp, Zahler WL, 1985. Changes and interrelationships among luteal LH receptors, adenylate cyclase activity and phosphodiesterase activity during the bovine estrous cycle. J Anim Sci 61:216-23.

Geisert RD, Renegar RH, Thatcher WW, Roberts RM, Bazer FW, 1982. Establishment of pregnancy in the pig: I. Interrelationships between preimplantation development of the pig blastocyst and uterine endometrial secretions. Biol Reprod 27:925-39.

Gibori G, Chatterton Jr RT, Chien JL, 1979. Ovarian and serum concentrations of androgens throughout pregnancy in the rat. Biol Reprod 21:53-56.

Gibori G, Keyes PL, 1978. Role of intraluteal estrogen in the regulation of the rat corpus luteum during pregnancy. Endocrinology 102:1176–82.

Gibori G, Richards JS, 1978. Dissociation of two distinct luteotropic effects of prolactin: regulation of luteinizing hormone–receptor content and progesterone secretion during pregnancy. Endocrinology 102: 767–74.

Gibori G, Sridaran R, 1981. Sites of androgen and estradiol production in the second half of pregnancy in the rat. Biol Reprod 24:249–56.

Ginther OJ, 1979. Reproductive Biology of the Mare. Ann Arbor: McNaughton and Gunn, Inc., pp. 192–358.

Ginther OJ, Woody CO, Mahajan S, Janakiraman K, Casida LE, 1967. Effect of oxytocin administration on the oestrous cycle of unilaterally hysterectomized heifers. J Reprod Fertil 14:225–29.

Godkin JD, Bazer FW, Lewis GS, Geisert RD, Roberts RM, 1982a. Synthesis and release of polypeptides by pig conceptuses during the period of blastocyst elongation and attachment. Biol Reprod 27:977–87.

Godkin JD, Bazer FW, Moffatt J, Sessions F, Roberts RM, 1982b. Purification and properties of a major, low molecular weight protein released by the trophoblast of sheep blastocysts at day 13–21. J Reprod Fertil 65:141–50.

Godkin JD, McGrew S, 1986. Characterization of bovine conceptus protein production during early pregnancy. Biol Repeod 34(Suppl 1):148.

Goff AK, Pontriand, 1985. Effect of pregnancy on the oxytocin stimulation of PGF_α in the mare. Biol Reprod 32(Suppl 1):93.

Goodsaid–Zalduondo F, Rintoul DA, Carlson JC, Hansel W, 1982. Luteolysis-induced changes in phase composition and fluidity of bovine luteal cell membranes. Proc Natl Acad Sci USA 79:4332–36.

Greenwald GS, 1967. Luteotropic complex of the hamster. Endocrinology 80:118–30.

Greenwald GS, 1973. Further evidence for a luteotropic complex in the hamster: progesterone determinations of plasma and corpora lutea. Endocrinology 92:235–42.

Greenwald GS, Rothchild I, 1968. Formation and maintenance of corpora lutea in laboratory animals. J Anim Sci 27(Suppl 1):139–62.

Grinwich DL, Hichens M, Behrman HR, 1976. Control of LH receptor by prolactin and prostaglandin $F_2\alpha$ in rat corpora lutea. Biol Reprod 14:212–18.

Groff TR, Madhwa Raj HG, Talbert LM, Willis DL, 1984. Effects of neutralization of LH on corpus luteum function and cyclicity in Macaca fascicularis. J Clin Endocrinol Metab 59:1054–57.

Guthrie HD, Rexroad CE, 1981. Endometrial prostaglandin F release in vitro and plasma 13,14-dihydro-15-keto-prostaglandin F in pigs with luteolysis blocked by pregnancy, estradiol benzoate or human chorionic gonadotropin. J Anim Sci 52:330–39.

Gwynne JT, Strauss III JF, 1982. The role of lipoproteins in steroidogenesis and cholesterol metabolism in steroidogenic glands. Endo Rev 3:299–329.

Hamberger L, Nilsson L, Dennefors B, Khan I, Sjogren A, 1979. Cyclic AMP formation by isolated human corpora lutea in response to hCG-interference by $PGF_2\alpha$. Prostaglandins 17:615–21.

Hansel W, Concannon PW, Lukaszewska JH, 1973. Corpora lutea of the large domestic animals. Biol Reprod 8:222–45.

Hansel W, Dowd JP, 1986. New concepts of the control of corpus luteum function. J Reprod Fertil 78:755–65.

Hansel W, Wagner WC, 1960. Luteal inhibition in the bovine as a result of oxytocin injection, uterine dilatation, and intrauterine infusion of seminal and preputial fluids. J Dairy Sci 43:796–805.

Hansson A, 1947. The physiology of reproduction in the mink (Mustela vison,schreb.) with special reference to delayed implantation. Acta Zool Stockh 28:1–136.

Harms PG, Niswender GD, Malven PV, 1969. Progesterone and luteinizing hormone secretion during luteal inhibition by exogenous oxytocin. Biol Reprod. 1:228–33.

Harwood JP, Clayton RN, Catt KJ, 1980. Ovarian gonadotropin-releasing hormone receptors. I. Properties and inhibition of luteal cell function. Endocrinology 107:407–13.

Heap RB, Flint APF, 1984. Pregnancy. In: Austin CR, Short RV, (ed.), Reproduction in mammals: 3, Hormonal control of reproduction. New York: Cambridge University Press, pp. 153–94.

Hearn JP, 1986. The embryo-maternal dialogue during early pregnancy in primates. J Reprod Fertil 76:809–19.

Heath E, Weinstein P, Merritt B, Shanks R, Hixon J, 1983. Effects of prostaglandins on the bovine corpus luteum: granules, lipid inclusions and progesterone secretion. Biol Reprod 29:977–85.

Helmer SD, Hansen PJ, Anthony RV, Thatcher WW, Bazer FW, Roberts RM, 1987. Identification of bovine trophoblast protein-1, a secretory protein immunologically related to ovine trophobast protein-1. J Reprod Fertil 79:83–91.

Henderson KM, McNatty KP, 1975. A biochemical hypothesis to explain the mechanism of luteal regression. Prostaglandins 9:779–97.

Herz Z, Khan I, Jayatilak PG, Gibori G, 1986. Evidence for the secretion of decidual luteotropin: a prolactin-like hormone produced by rat decidual cells. Endocrinology 118:2203–09.

Hild-Petito S, Shiigi S, Stouffer RL, 1986. Evidence for multiple cell populations in the primate corpus luteum. Biol Reprod 34(Suppl 1):127.

Holcomb LC, 1967. Effects of progesterone treatment on delayed implantation in mink. Ohio J Sci 67:24–31.

Holman RT, Hofstetter HH, 1965. The fatty acid composition of the lipids from bovine and porcine reproductive tissues. J Am Oil Chem Soc 42:540–44.

Hooper SB, Watkins WB, Thorburn GD, 1986. Oxytocin, oxytocin-associated neurophysin, and prostaglandin $F_2\alpha$ concentrations in the utero-ovarian vein of pregnant and nonpregnant sheep. Endocrinology 119:2590–97.

Hoyer PB, Keyes PL, Niswender GD, 1986. Size distribution and hormonal responsiveness of dispersed rabbit luteal cells during pseudopregnancy. Biol Reprod 34:905–10.

Hunzicker-Dunn M, 1982. Epinephrine-sensitive adenylyl cyclase activity in rabbit ovarian tissues. Endocrinology 110:233–40.

Hunzicker-Dunn M, Birnbaumer L, 1976. Adenylyl cyclase activities in ovarian tissues. II. Regulation of responsiveness to LH, FSH, and PGE in the rabbit. Endocrinology 99:185–97.

Hutchison JS, Zeleznik AJ, 1984. The rhesus monkey corpus luteum is dependent upon pituitary gonadotropin secretion throughout the luteal phase of the menstrual cycle. Endocrinology 115:1780–86.

Illingworth DV, Perry JS, 1971. The effect of hypophysial stalk-section on the corpus luteum of the guinea pig. J Endocrinol 50:625–35.

Inskeep EK, Murdoch WT, 1980. Relation of ovarian functions to uterine and ovarian secretion of prostaglandins during the estrous cycle and early pregnancy in the ewe and cow. In: Greep RO (ed.), Reproductive physiology III: Int Rev Physiol, Vol 22. Baltimore: University Park Press, pp. 325–56.

Ivell R, Brackett KH, Fields MJ, Richter D, 1985. Ovulation triggers oxytocin gene expression in the bovine ovary. FEBS Lett 190:263–67.

Ivell R, Richter D, 1984. The gene for the hypothalamic peptide hormone oxytocin is highly expressed in the bovine corpus luteum: biosynthesis, structure and sequence analysis. EMBO J 3:2351–54.

Jayatilak PG, Glaser LA, Basuray R, Kelly PA, Gibori G, 1985. Identification and partial characterization of a prolactin-like hormone produced by rat decidual tissue. Proc Natl Acad Sci USA 82:217–22.

Jones PBC, Hsueh AJW, 1980. Direct inhibitory effect of gonadotropin-releasing hormone upon luteal luteinizing hormone receptor and steroidogenesis in hypophysectomized rats. Endocrinology 107:1930-36.

Jordan III AW, Caffery JL, Niswender GD, 1978. Catecholamine-induced stimulation of progesterone and adenosine 3', 5',-monophosphate production by dispersed ovine luteal cells. Endocrinology 103:385-92.

Kaltenbach CC, Graber JW, Niswender GD, Nalbandov AV, 1968. Luteotrophic properties of some pituitary hormones in nonpregnant or pregnant hypophysectomized ewes. Endocrinology 32:818-24.

Karsch FJ, Cook B, Ellicott AR, Foster DL, Jackson GL, Nalbandov AV, 1971. Failure of infused prolactin to prolong the life span of the corpus luteum of the ewe. Endocrinology 89:272-75.

Kelly PA, Shiu RPC, Robertson MC, Friesen HG, 1975. Characterization of rat chorionic mammotropin. Endocrinology 96:1187-95.

Kelly PA, Tsushima T, Shiu RPC, Friesen HG, 1976. Lactogenic and growth hormone-like activities in pregnancy determined by radioreceptor assays. Endocrinology 99:765-74.

Keyes PL, Gadsby JE, Yuh K-CM, Bill III CH, 1983. The corpus luteum. In: Greep RO (ed.), Reproductive Physiology IV, Int Rev Physiol, Vol. 27. Baltimore: University Park Press, pp. 57-97.

Khan MI, Rosberg S, 1979. Acute suppression of PGF$_2\alpha$ on LH, epinephrine and fluoride stimulation of adenylate cyclase in rat luteal tissue. J Cyclic Nucl Res 5:55-63.

Khan-Dawood FS, Dawood MY, 1983. Human ovaries contain immunoreactive oxytocin. J Clin Endocrinol Metab 57:1129-32.

Khan-Dawood FS, Dawood MY, 1984. Presence of oxytocin (OT) in steroid producing glands of the cow, rabbit and rat. Scientific Abstracts of the 31st Annual Meeting of the Society for Gynecologic Investigation. p 149. St. Louis: CV Mosby.

Khan-Dawood FS, Dawood MY, 1986. Paracrine regulation of luteal function. J Clin Endocrinol Metab 15:171-84.

Khan-Dawood FS, Marut EL, Dawood MY, 1984. Oxytocin in the corpus luteum of the cynomolgus monkey (Macaca fascicularis). Endocrinology 115:570-74.

Kidder HE, Casida LE, Grummer RH, 1955. Some effects of estrogen injections on the estrual cycle of gilts. J Anim Sci 14:470-74.

Kim I, Yeoun DS, 1983. Effect of prostaglandin F$_2\alpha$ on Na$^+$-K$^+$-ATPase activity in luteal membranes. Biol Reprod 29:48-55.

Kimball FA, Wyngarden LJ, 1977. Prostaglandin Fα specific binding in equine corpora lutea. Prostaglandins 13:553-64.

Kittok RJ, Britt JH, Convey EM, 1973. Endocrine response after GnRH in luteal phase cows and cows with ovarian follicular cysts. J Anim Sci 37:985-89

Kledzik GS, Cusan L, Auclair C, Kelly PA, Labrie F, 1978. Inhibitory effect of a luteinizing hormone (LH)- releasing hormone agonist on rat ovarian LH and follicle-stimulating hormone receptor levels during pregnancy. Fertil Steril 29:560-64.

Koos RD, Hansel W, 1981. The large and small cells of the bovine corpus luteum: ultrastructural and function differences. In: Schwartz NB, Hunzicker-Dunn M (ed.), Dynamics of Ovarian Function. New York: Raven Press, pp. 197-203.

Koyama T, Ohkura T, Kumasaka T, Saito M, 1978. Effect of postovulatory treatment with a luteinizing hormone-releasing hormone analog on the plasma level of progesterone in women. Fertil Steril 30:549-52.

Kruip TAM, Vullings HGB, Schams D, Jonis J, Klarenbeek A, 1985. Immunocytochemical demonstration of oxytocin in bovine ovarian tissues. Acta Endocrinol (Copenh) 109:537-42.

Labhsetwar AP, Collins WE, Tyler WJ, Casida LE, 1964. Effect of progesterone and oxytocin on the pituitary-ovarian relationship in heifers. J Reprod Fertil 8:77-83.

Lafrance M, Goff AK, 1985. An antiluteolytic action of the bovine embryo via the hormonal axis oxytocin-prostaglandin $F_2\alpha$. Biol Reprod 33:1113–19.

Lahav M, Freud A, Lindner HR, 1976. Abrogation by prostaglandin $F_2\alpha$ of LH-stimulated cyclic AMP accumulation in isolated rat corpora lutea of pregnancy. Biochem Biophys Res Comm 68:1294–300.

Lemon M, Loir M, 1977. Steroid release in vitro by two luteal cell types in the corpus luteum of the pregnant sow. J. Endocrinol 72:351–59.

Leung PCK, Minegishi T, Ma F, Zhou F, Ho-Yen B, 1986. Induction of polyphosphoinositide breakdown in rat corpus luteum by prostaglandin $F_2\alpha$. Endocrinology 119:12–18.

Lewis GS, Waterman RA, 1983. Metabolism of arachidonic acid in vitro by porcine blastocysts and endometrium. Prostaglandins 25:871–80.

Lin MT, Rao CV, 1977. {^{3}H} prostaglandins binding to dispersed bovine luteal cells: evidence for discrete prostaglandin receptors. Biochem Biophys Res Comm 78:510–16.

Luborsky JL, Slater WT, Behrman HR, 1984. Luteinizing hormone (LH) receptor aggregation: modification of ferritin-LH binding and aggregation by prostaglandin $F_2\alpha$ and ferritin-LH. Endocrinology 115:2217–26.

Luborsky-Moore JL, Wright K, Behrman HR, 1979. Demonstration of luteal cell membrane receptors for prostaglandin $F_2\alpha$ by ultrastructural and binding analysis. Adv Exp Med Biol 112:633–38.

Mackenzie K, 1911. An experimental investigation of the mechanism of milk secretion with special reference to the action of animal extracts. Q J Exp Physiol 4:305–30.

Mais V, Kazer RR, Cetel NS, Rivier J, Vale W, Yen SSC, 1986. The dependency of folliculogensis and corpus luteum function on pulsatile gonadotropin secretion in cycling women using a gonadotropin-releasing hormone antagonist as a probe. J Clin Endocrinol Metab 62:1250–55.

Mapletoft RJ, Lapin DR, Ginther OJ, 1976. The ovarian artery as the final component of the local luteotropic pathway between a gravid uterine horn and ovary in ewes. Biol Reprod 15:414–21.

Marengo SR, Bazer FW, Thatcher WW, Wilcox CJ, Wetteman RP, 1986. Prostaglandin $F_2\alpha$ as the luteolysin in swine: VI. Hormonal regulation of the movement of exogenous $PGF_2\alpha$ from the uterine lumen into the vasculature. Biol Reprod 34:284–92.

Marsh JM, 1971. The effect of prostaglandins on the adenyl cyclase of the bovine corpus luteum. Ann NY Acad Sci 180:416–25.

Marsh JM, 1976. The role of cyclic AMP in gonadal steroidogenesis. Biol Reprod 14:30–53.

Marsh JM, Le Maire WJ, 1974. Cyclic AMP accumulation and steroidogensis in the human corpus luteum: effect of gonadotropins and prostaglandins. J Clin Endocrinol Metab 38:99–106.

Martal J, Djiane J, 1977. The production of chorionic somatomammotrophin in sheep. J Reprod Fertil 49:285–89.

Martinet L, Allais C, Allain D, 1981. The role of prolactin and LH in luteal function and blastocyst growth in mink (Mustela vison). J Reprod Fertil Suppl 29:119–30.

Mason NR, Marsh JM, Savard K, 1962. An action of gonadotropin in vitro. J Biol Chem 237:1801–06.

McCracken JA, Schramm W, 1983. Prostaglandin and corpus luteum regression. In: Curtis-Prior PB (ed.), A handbook of prostaglandin related compounds. Edinburgh: Churchill-Livingstone, pp. 1–104.

McCracken JA, Schramm W, Okulicz WC, 1984. Hormone receptor control of pulsatile secretion of prostaglandin $F_2\alpha$ from the ovine uterus during luteolysis and its abrogation in early pregnancy. Anim Reprod Sci 7:31–55.

Mead R, 1975. Effects of hypophysectomy on blastocyst survival, progesterone secretion and nidation in the spotted skunk. Biol Reprod 12:526–33.

Mead R, 1981. Delayed implantation in mustelids, with special emphasis on the spotted skunk. J Reprod Fertil Suppl 29:11–24.

Mead R, Swannack A, 1978. Effects of hysterectomy on luteal function in the western spotted skunk (Spilogale putorius latifrons). Biol. Reprod 18:379–83.

Mednick DL, Barkley MS, Geschwind II, 1980. Regulation of progesterone secretion by LH and prolactin during the first half of pregnancy in the mouse. J Reprod Fertil 60:201–07.

Merckel C. Nelson WO, 1940. The relation of the estrogenic hormone to the formation and maintenance of corpora lutea in mature and immature rats. Anat Rec 76:391–409.

Milne JA, 1963. Effects of oxytocin on the oestrous cycle of the ewe. Aust Vet J 39:51–52.

Milvae RA, Hansel W, 1980. Concurrent uterine venous and ovarian arterial prostaglandin F concentrations in heifers treated with oxytocin. J Reprod Fertil 60:7–15.

Milvae RA, Murphy BD, Hansel W, 1984. Prolongation of the bovine estrous cycle with a gonadotropin-releasing hormone analog. Biol Reprod 31:664–70

Møller DM, 1973. The progesterone concentrations in the peripheral plasma of the mink (Mustela vison) during pregnancy. J Endocrinol 56:121–33.

Moore LG, Choy VJ, Elliot RL, Watkins WB, 1986. Evidence for the pulsatile release of prostaglandin $F_2\alpha$ inducing the release of ovarian oxytocin during luteolysis in the ewe. J Reprod Fertil 76:159–66.

Morin RJ, 1968. Ovarian phospholipid composition and incorporation of [I-^{14}C] acetate into the phospholipid fatty acids of ovaries from non-pregnant and pregnant rabbits. J Reprod Fertil 17:111–17.

Morishige WK, Rothchild I, 1974. Temporal aspects of the regulation of corpus luteum function by luteinizing hormone, prolactin and placental luteotrophin during the first half of pregnancy in the rat. Endocrinology 95:260–74.

Moudgal NR, MacDonald GJ, Greep RO, 1971. Effect of hCG antiserum on ovulation and corpus luteum formation in the monkey (Macaca fascicularis).J Clin Endocrinol Metab 32:579–81.

Mukku V, Moudgal NR, 1975. Studies on luteolysis:effect of antiserum on luteinizing hormone on sterols and steroid levels in pregnant hamsters. Endocrinology 97:1455–59.

Murphy BD, Concannon PW, Travis HF, Hansel W, 1981. Prolactin: the hypophyseal factor that terminates embryonic diapause in mink. Biol Reprod 25:487–91.

Murphy BD, Humphrey WD, Shepstone SL, 1980. Luteal function in mink: the effect of hypophysectomy after the preimplantation rise in progesterone. Anim Reprod Sci 3:225–32.

Murphy BD, Rajkumar K, Silversides DW, 1984. Luteotrophic control of the mink corpus luteum during the postimplantation phase of gestation. 3rd Congr Int Sci Prod Anim. Fourrure Versailles (France) Commun No. 32.

Neely DP, Stabenfeldt GH, Sauter CL, 1979. The effect of exogenous oxytocin on luteal function in mares. J Reprod Fertil 55:303–08.

Newcomb R, Booth WD, Rowson LEA, 1979. The effect of oxytocin treatment on the levels of prostaglandin F in the blood of heifers. J Reprod Fertil 49:17–24.

Newton WH, Beck N, 1939. Placental activity in the mouse in the absence of the pituitary gland. J Endocrinol 1:65–75.

Nishizuka Y, Takai Y, Kishimoto A, Kikkawa U, Kaibuchi K, 1984. Phospholipid turnover in hormone action. Rec Prog Horm Res 40:301–45

Niswender GD, Reimers TJ, Diekman MA, Nett TM, 1976. Blood flow: a mediator of ovarian function. Biol Reprod 14:64–81.

Niswender GD, Schwall RH, Fritz TA, Farin CE, Sawyer HR, 1985. Regulation of luteal function in domestic ruminants: new concepts. Rec Prog Horm Res 41:101–51.

O'Shea JD, Rodgers RT, Wright PJ, 1986. Cellular composition of the sheep corpus luteum in the mid- and late luteal phases of the oestrous cycle. J Reprod Fertil 76:685–91.

Ott I, Scott JC, 1910. The galactagogue action of the thymus and corpus luteum. Proc Soc Exp Biol Med 8:49.

Papke RL, Concannon PW, Travis HF, Hansel W, 1980. Control of luteal function and implantation in the mink by prolactin. J Anim Sci 50:1102–07.

Pencharz RI, Long JA, 1933. Hypophysectomy in the pregnant rat. Am J Anat 53:117–39.

Pierce JG, Parsons TF, 1981. Glycoprotein hormones: structure and function. Ann Rev Biochem 50:465–95.

Pineda MH, Garcia MC, Ginther OJ, 1973. Effect of antiserum against an equine pituitary fraction on corpus luteum and follicles in mares during diestrus. Am J Vet Res 34:181–83.

Pitzel L, Konrad W, Wolfgang H, Annemarie K, 1984. Neurohypophyseal hormones in the corpus luteum of the pig. Neuroendocrinol Lett 6:1–6.

Pope WF, Maurer RR, Stormshak F, 1982. Intrauterine migration of the porcine embryo-interaction of embryo, uterine flushings and indomethacin on myometrial function in vitro. J Anim Sci 55:1169–78.

Powell WS, Hammarström S, Samuelsson B, 1974a. Prostaglandin $F_2\alpha$ receptor in ovine corpora lutea. Eur J Biochem 41:103–07.

Powell WS, Hammerström S, Samuelsson B, 1976. Localization of a prostaglandin $F_2\alpha$ receptor in bovine corpus luteum plasma membranes. Eur J Biochem 61:605–11.

Powell WS, Hammarström S, Samuelsson B, Sjöberg B, 1974b. Prostaglandin $F_2\alpha$ receptor in human corpora lutea. The Lancet 1:1120.

Rao CV, Griffin LP, Carman Jr FR, 1977. Gonadotropin receptors in human corpora lutea of the menstrual cycle and pregnancy. Am J Obstet Gynecol 128:146–53.

Rapoport B, Hazum E, Zor U, 1984. Photoaffinity labeling of human chorionic gonadotropin-binding sites in rat ovarian plasma membranes. J Biol Chem 259:4267–71.

Rice BF, Hammerstein J, Savard K, 1964. Steroid hormone formation in the human ovary. II. Action of gonadotropins in vitro in the corpus luteum. J Clin Endocrinol Metab 24:606–15.

Richards JS, Midgley AR, 1976. Protein hormone action: a key to understanding ovarian follicular and luteal cell development. Biol Reprod 14:82–94.

Richardson DW, Goldsmith LT, Pohl CR, Schallenberger E, Knobil E, 1985. The role of prolactin in the regulation of the primate corpus luteum. J Clin Endocrinol Metab 60:501–04.

Riley JCM, Carlson JC, 1985. Calcium-regulated plasma membrane rigidification during corpus luteum regression in the rat. Biol Reprod 32:77–82.

Riley JCM, Carlson JC, 1986. Association of phospholipase A activity with membrane degeneration during luteolysis in the rat. Biol Reprod 34(Suppl 1):135.

Ritzhaupt LK, Nowak RA, Calvo FO, Khan IM, Bahr JM, 1986. Adenylate cyclase activity of the corpus luteum during the oestrous cycle of the pig. J Reprod Fertil 78:361–66.

Roberts JS, McCracken JA, 1976. Does prostaglandin $F_2\alpha$ released from the uterus by oxytocin mediate the oxytocic action of oxytocin? Biol Reprod 15:457–63.

Robson JM, 1937. Maintenance by oestrin of the luteal function in hypophysectomized rabbits. J Physiol 90:435–39.

Rodger LD, Stormshak F, 1986. Gonadotropin-releasing hormone-induced alteration of bovine corpus luteum function. Biol Reprod 35:149–56.

Rodgers RJ, O'Shea JD, 1982. Purification, morphology, and progesterone production and content of three cell types isolated from the corpus luteum of the sheep. Austr J Biol Sci 35:441–55.

Rose J, Oldfield JE, Stormshak F, 1986. Changes in serum prolactin concentrations and ovarian prolactin receptors during embryonic diapause in mink. Biol Reprod 34:101–06.

Roser JF, Evans JW, 1983. Luteal luteinizing hormone receptors during the postovulatory period in the mare. Biol Reprod 29:499–510.

Rothchild I, 1981. The regulation of the mammalian corpus luteum. Rec Prog Horm Res 37:183–298.

Sairam MR, 1983. Gonadotropic hormones: relationship between structure and function with emphasis on antagonists. In: Li CH (ed.), Hormonal proteins and peptides, Vol XI. New York: Academic Press, pp. 1–79.

Sarker NJ, Canivenc R, 1982. Luteal vascularization in the European badger (Meles meles L.) Biol Reprod 26:903–08.

Savard K, 1973. The biochemistry of the corpus luteum. Biol Reprod 8:183–202.

Schams D, Prokopp S, Barth D, 1983. The effect of active and passive immunization against oxytocin on ovarian cyclicity in ewes. Acta Endocrinol (Copenh) 103:337–44.

Schaeffer JM, Liu J, Hsueh AJW, Yen SCC, 1984. Presence of oxytocin and arginine vasopressin in human ovary, oviduct and follicular fluid. J Clin Endocrinol Metab 59:970–73.

Schafer EA, Mackenzie K, 1911. The action of animal extracts on milk secretion. Proc R Soc Lond B 84:16–22.

Schallenberger E, Schams D, Bullermann B, Walters DL, 1984. Pulsatile secretion of gonadotrophins, ovarian steroids and ovarian oxytocin during prostaglandin-induced regression of the corpus luteum in the cow. J Reprod Fertil 71:493–501.

Schramm W, Friesen HG, Robertson HA, McCracken JA, 1984. Effect of exogenous ovine placental lactogen on luteolysis induced by prostaglandin $F_2\alpha$ in sheep. J Reprod Fertil 70:557–65.

Schwall RH, Gamboni F, Mayan MH, Niswender GD, 1986. Changes in the distribution of sizes of ovine luteal cells during the estrous cycle. Biol Reprod 34:911–18.

Scott TW, Hansel W, Donaldson LE, 1968. Metabolism of phospholipids and the characterization of fatty acids in bovine corpus luteum. Biochem J 108:317–23.

Sharp DC, Zavy MT, Vernon MW, Bazer FW, Thatcher WW, Berglund LA, McDowell KJ, 1984. The role of prostaglandins in the maternal recognition of pregnancy. Anim Reprod Sci 7:269–82.

Sheldrick EL, Flint APF, 1983a. Ovarian oxytocin and luteal function in the early pregnant sheep. Anim Reprod Sci 10:101–13.

Sheldrick EL, Flint APF, 1983b. Luteal concentrations of oxytocin decline during early pregnancy in the ewe. J Reprod Fertil 68:477–80.

Sheldrick EL, Mitchell MD, Flint APF, 1980. Delayed luteal regression in ewes immunized against oxytocin. J Reprod Fertil 59:37–42.

Silvia WJ, Ottobre JS, Inskeep EK, 1984. Concentrations of prostaglandins E_2, $F_2\alpha$ and 6-keto-prostaglandin $F_1\alpha$ in the utero-ovarian venous plasma of nonpregnant and early pregnant ewes. Biol Reprod 30:936–44.

Sinha YN, Wickes MA, Baxter SR, 1978. Prolactin and growth hormone secretion and mammary gland growth during pseudopregnancy in the mouse. J Endocrinol 77:203–12.

Smith PE, 1930. Hypophysectomy and replacement therapy in the rat. Am J Anat 45:205–73.

Smith VR, McShan WH, Casida LE, 1957. On maintenance of the corpora lutea of the bovine with lactogen. J Dairy Sci 40:443.

Smith MS, Neill JD, 1976. Termination at midpregnancy of the two daily surges of plasma prolactin initiated by mating in the rat. Endocrinology 98:696–701.

Spicer LJ, Ireland JJ, Roche JF, 1981 Changes in serum LH, progesterone, and specific binding of ^{125}I-hCG to luteal cells during regression and development of bovine corpus lutea. Biol Reprod 25:832–41.

Sridaran R, Gibori G, 1983. Control of placental and ovarian secretion of testosterone in the pregnant rat. In: Greenwald GS, Terranova PF (ed.), Factors regulating ovarian function. New York: Raven Press, pp. 87–91.

Stouffer RL, Coensgen JL, Hodgen GD, 1980. Progesterone production by luteal cells isolated from cynomolgus monkeys: effects of gonadotropin and prolactin during acute incubation and cell culture. Steroids 35:523–32.

Stouffer RL, Nixon WE, Hodgen GD, 1979. Disparate effects of prostaglandins on basal and gonadotropin-stimulated progesterone production by luteal cells isolated from rhesus monkeys during the menstrual cycle and pregnancy. Biol Reprod 20:897–903.

Strauss JF, Flickinger GL, 1977. Phospholipid metabolism in cells from highly luteinized rat ovaries. Endocrinology 101:883–89.

Swann RW, O'Shaughnessy PJ, Birkett SD, Wathes DC, Porter DG, Pickering PT, 1984. Biosynthesis of oxytocin in the corpus luteum. FEBS Lett 174:262–66.

Takayama M, Greenwald GS, 1973. Direct luteotropin action of estrogen in the hypophysectomized-hysterectomized rat. Endocrinology 92:1405–13.

Tam WH, Beveridge WK, Tso EC-F, 1982. Effects of gonadotropin, hypophysectomy and prostaglandin $F_2\alpha$ on corpora lutea and ovarian follicles, and evidence for the presence of $PGF_2\alpha$ receptors in the ovarian follicles and interstitium of non-pregnant guinea pigs. J Reprod Fertil 64:9–17.

Tan GJS, Tweedale R, Biggs JSG, 1982a. Effect of oxytocin on the bovine corpus luteum of early pregnancy. J Reprod Fertil 66:75–78.

Tan GJS, Tweedale R, Biggs JSG, 1982b. Oxytocin may play a role in the control of the human corpus luteum. J Endocrinol 95:65–70.

Terranova PF, Greenwald GS, 1979. Antiluteinizing hormone: chronic influence on steroid and gonadotropin levels and superovulation in the pregnant hamster. Endocrinology 104:1013–19.

Thatcher WW, Knickerbocker JJ, Bartol FF, Bazer FW, Roberts RM, Drost M, 1985. Maternal recognition of pregnancy in relation to the survival of transferred embryos: endocrine aspects. Theriogenology 23:129–44.

Thomas JP, Dorflinger LJ, Behrman HR, 1978. Mechanism of the rapid antigonadotropic action of prostaglandins in cultured luteal cells. Proc Natl Acad Sci USA 75:1344–48.

Torjesen PA, Aakvaag A, 1986. Characterization of the adenylate cyclase of the rat corpus luteum during luteolysis induced by a prostaglandin $F_2\alpha$ analogue. Mol Cell Endocrinol 44:237–42.

Ursely J, Leymarie P, 1979. Varying response to luteinizing hormone of two cell types isolated from bovine corpus luteum. J Endocrinol 83:303–10.

Walters DL, Schams D, Bullermann B, Schallenberger E, 1983. Pulsatile secretion of gonadotropins, ovarian steroids and ovarian oxytocin during luteolysis in the cow. Biol Reprod 28(Suppl 1):142.

Walters DL, Schams D, Schallenberger E, 1984. Pulsatile secretion of gonadotrophins, ovarian steroids and ovarian oxytocin during the luteal phase of the oestrous cycle in the cow. J Reprod Fertil 71:479–91.

Warshaw ML, Johnson DC, Khan I, Eckstein B, Gibori G, 1986. Placental secretion of androgens in the rat. Endocrinology 119:2642–48.

Waterman RA, 1980. Lipid accumulation with sheep corpora lutea. J Anim Sci 53(Suppl 1):374.

Wathes DC, Swann RW, 1982. Is oxytocin an ovarian hormone? Nature 297:225–27.

Wathes DC, Swann RW, Birkett SD, Porter DG, Pickering BT, 1983a. Characterization of oxytocin, vasopressin and neurophysin from the bovine corpus luteum. Endocrinology 113:693–98.

Wathes DC, Swann RW, Hull MGR, Drief JO, Porter DG, Pickering BT, 1983b. Gonadal sources of the posterior pituitary hormones. Prog Brain Res 60:513–20.

Wathes DC, Swann RW, Pickering BT, Porter DG, Hull MGR, Driefe JO, 1982. Neurohypophysial hormones in the human ovary. Lancet II: 410-12.

Wathes DC, Swann RW, Porter DG, Pickering BT, 1986. Oxytocin as an ovarian hormone. In: Ganten D, Pfaff D, (ed.), Current topics in neuroendocrinology, Vol. 6. Berlin Heidelberg: Springer-Verlag, pp 129-52.

Watson J, Maule Walker FM, 1978. Progesterone secretion by the corpus luteum of the early pregnant pig during superfusion in vitro with $PGF_2\alpha$, LH and oestradiol. J Reprod Fertil 52:209-12.

West LA, Weakland LL, Davis JS, 1986. Prostaglandin $F_2\alpha$ stimulates phosphoinositide hydrolysis and inositol 1,4,5-triphosphate (IP_3) synthesis in isolated bovine luteal cells. Biol Reprod 34(Suppl 1):138

Wide L, Wide M, 1979. Chorionic gonadotropin in the mouse from implantation to term. J Reprod Fertil 57:5-9.

Wilks JW, 1983, The effect of oxytocin on the corpus luteum of the monkey. Contraception 28:267-72.

Williams AT, Behrman HR, 1983. Paracrine regulation of the ovary by gonadotropin releasing hormone and other peptides. Sem Reprod Endocrinol 1:269-77.

VandeWeile RL, Bogumil J, Dyrenfurth I, Ferin M, Jewelewicz R, Warren M, Rizkallah T, Mikhail G, 1970. Mechanisms regulating the menstrual cycle in women. Rec Prog Horm Res 26:63-103.

Verma AK, Penniston JT, 1981. A high affinity Ca^{2+}-stimulated and Mg^{2+}-dependent ATPase in rat corpus luteum plasma membrane fractions. J Biol Chem 256:1269-75.

Zarco L, Stabenfeldt GH, Kindahl H, Bradford GE, Basu S, 1984. A detailed study of prostaglandin $F_2\alpha$ release during luteolysis and establishment of pregnancy in the ewe. Biol Reprod 30(Suppl 1):153.

Zavy MT, Mayer R, Vernon MW, Bazer FW, Sharp DC, 1979. An investigation of the uterine luminal environment of nonpregnant and pregnant pony mares. J Reprod Fertil Suppl 27:403-11.

Zelinski MB, 1986. Plasma membrane composition and luteinizing hormone receptors of ovine corpora lutea during early pregnancy. PhD Dissertation, Oregon State Univ, Corvallis.

Zelinski MB, Stormshak F, 1983. Plasma membrane composition of ovine corpora lutea during pregnancy. J Anim Sci 57(Suppl 1):382.

Ziecik A, Shaw HJ, Flint APF, 1980. Luteal LH receptors during the oestrous cycle and early pregnancy in the pig. J Reprod Fertil 60:129-37.

ROLE OF ESTROGEN AND THE PLACENTA IN THE MAINTENANCE

OF THE RABBIT CORPUS LUTEUM

P. Landis Keyes[1] and John E. Gadsby[2]

[1]Department of Physiology and
Reproductive Endocrinology Program
The University of Michigan
Ann Arbor, MI 48109

[2]Department of Anatomy
Physiological Science and Radiology
School of Veterinary Medicine
North Carolina State University
Raleigh, NC 27606

INTRODUCTION

The rabbit is among those species in which the placenta secretes low or physiologically insignificant quantities of progesterone, and therefore, the corpora lutea must remain steroidogenically active through-out gestation (Hilliard, 1973; Thau and Lanman, 1974). If the young embryo is to survive, it must transmit a signal that in some way halts or overrides incipient luteal regression, reflected in declining serum progesterone values by 15 days after ovulation in non-pregnant animals (Keyes et al., 1983a). The subject of this manuscript is an exploration of the mechanisms that are responsible for placental maintenance of luteal function. We begin with a brief summary of the literature, setting the stage for the specific hypotheses and experiments reported herein.

At least two major factors are critical for extension of luteal function through the end of pregnancy which lasts about 32 days: a factor produced by the conceptus and estrogen produced by the ovarian follicles (Keyes et al., 1983a). Evidence that the conceptus or fetal placenta contributes a factor, that either directly or indirectly affects luteal function, comes from numerous investigators (Greep, 1941; Chu et al., 1946; Holt and Ewing, 1974; Lanman and Thau, 1979; Browning and Wolf, 1981; Nowak and Bahr, 1983; Gadsby and Keyes, 1984). The placental factor appears to be produced by the fetal trophoblast cells rather than by maternal decidual cells (Holt and Ewing, 1974). The role of estrogen as an essential luteotropin in the rabbit has been known since the early report of Robson (1939). Estrogen produced by the ovarian follicles is vital for the maintenance of luteal progesterone synthesis throughout pregnancy (Keyes and Nalbandov, 1967; Keyes and Armstrong, 1968; Spies et al., 1968; Rippel and Johnson, 1976; Gadsby et al., 1983). In the experiments to follow, we have examined several hypotheses directed toward a better understanding of the nature of the interaction between the fetal placenta and estrogen in the control of the corpus luteum.

361

HYPOTHESIS 1. THE LUTEOTROPIC EFFECT OF FETAL PLACENTAL FACTOR(S) REQUIRES THE SIMULTANEOUS ACTION OF ESTROGEN

Californian rabbits were mated to fertile males (Day 0) and on Day 4 of pregnancy were hypophysectomized via the parapharyngeal route as described previously (Bill and Keyes, 1983). Hypophysectomy was performed in order to assess the luteotropic activity of a putative fetal placental factor in the absence of pituitary hormones and endogenous ovarian estrogen. At this time a Silastic capsule containing 17β-estradiol was placed subcutaneously; in the presence of this implant, peripheral serum estradiol concentration averages 5 pg/ml, which is in the range of physiological concentrations in pregnancy (Browning et al., 1980). On Day 20, after confirming by palpation that animals were pregnant, groups of animals were given either a single im injection of the synthetic progestin, medroxyprogesterone acetate (Depo-Provera, 5 mg; Upjohn Company, Kalamazoo, MI) or an injection of vehicle only. The rationale for the injection of progestin was that pregnancy and the presence of putative placental factor(s) could be maintained by a compound that did not cross-react in the radioimmunoassay for serum progesterone, which was being used to monitor the activity of the corpora lutea. On Day 21, the estradiol capsule was removed from all animals using local anesthetic. Blood samples were taken at appropriate intervals via the ear vein for measurement of serum progesterone.

The results are shown in Fig. 1 and 2. In animals which were not treated with Depo-Provera (Fig. 1), removal of the estradiol implant caused a striking decline in serum progesterone, from 11 ng/ml on Day 21 to less than 1 ng/ml on Day 22, remaining at this value through Day 30. These animals either aborted between Days 23 and 24 (2 animals), or retained dead fetuses (1 animal). Fig. 2 shows the effect of removal of estradiol in the animals treated with Depo-Provera (MPA). A precipitous decline in progesterone occurred, from 17 ng/ml on Day 21 to less than 1 ng/ml on Days 22 through 30, despite the maintenance of viable fetuses through Day 30 in all 3 animals. Serum estradiol values were 5 pg/ml on Day 20 and 1.2 pg/ml on Day 22 after removal of the estradiol implant, indicating that estrogen withdrawal had occurred within 24 h.

These data provide additional support for the idea that during the second half of pregnancy luteal progesterone synthesis is acutely dependent upon estradiol, and hence upon the steroidogenic activity of ovarian follicles, which are the sources of estradiol in the rabbit (Mills and Savard, 1973; Keyes and Nalbandov, 1967). Further, the data indicate that placental factor(s) is not directly luteotropic or steroidogenic in the absence of estrogen.

We have used another approach to test the efficacy of placental factor(s) in maintaining the activity of corpora lutea in the absence of endogenous estrogen. Dutch rabbits were given a single im injection of 5 mg of Depo-Provera on Day 20 of pregnancy to ensure continued fetal viability. To eliminate, acutely, follicular estrogen synthesis, all animals were given a minimally effective ovulating dose of human chorionic gonadotropin (hCG; 10 IU iv) on Day 21. Ovulation of follicles causes immediate cessation of estrogen synthesis (Mills and Savard, 1973). On Day 22, all animals were anesthetized, and the newly-forming corpora lutea were excised, leaving the original corpora lutea of pregnancy intact. At this time, in some animals, a Silastic capsule containing estradiol was implanted sc, to restore serum estradiol values. Blood samples were taken at appropriate intervals.

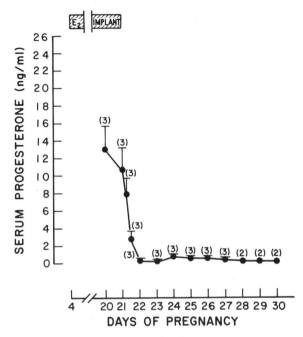

Fig. 1. Effect of estradiol withdrawal on serum progesterone concentrations in rabbit hypophysectomized on Day 4 of pregnancy. An
estradiol (E$_2$) implant was placed sc on Day 4 and removed on Day
21. Retention of dead fetuses or abortion occurred in all
animals. Data expressed as mean ± SEM; ( ) = number of animals.
(From Gadsby et al., 1983).

The results are shown in Fig. 3. An injection of hCG caused a rise
in serum progesterone within 6 h, from 16 ng/ml to 27 ng/ml, followed by a
precipitous decline which continued unabated, to reach a mean value of
2 ng/ml by 24 h after hCG injection. In animals without further treatment,
serum progesterone values remained low and by Day 27 had risen gradually,
to a mean value of 7 ng/ml. In contrast, in the animals which received
estradiol replacement on Day 22, serum progesterone showed a brisk rise to
reach original values of 16 to 22 ng/ml on Days 24 to 27. Serum estradiol
values declined after injection of hCG, from 5 pg/ml on Day 20 to 0.8
pg/ml on Day 22 (24 h after hCG injection). In other rabbits injected
with only vehicle (no hCG injected) serum estradiol values on Days 20 and
22 were 5 and 8 pg/ml, serum progesterone was relatively stable around 17
ng/ml throughout the period of study (Days 20 to 27), and injection of
Depo-Provera had no effect on serum progesterone values.

These results reveal the necessity of follicular estrogen in order
for luteal progesterone synthesis to be maintained. Further, the data
indicate that in the absence of estrogen, placental factor(s) is not
steroidogenic in the corpora lutea, but may maintain responsiveness to
estradiol. The replacement of estradiol clearly shows that the corpora
lutea retain their steroidogenic potential and estrogen responsiveness for
at least 24 h after an ovulating dose of chorionic gonadotropin. This

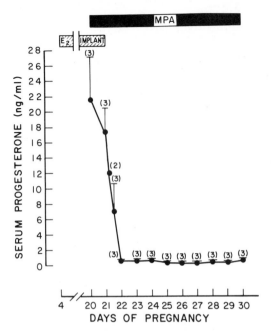

Fig. 2. Effect of estradiol withdrawal on serum progesterone concentrations in rabbits hypophysectomized on Day 4 of pregnancy and
treated with the progestin, medroxyprogesterone acetate (MPA)
beginning on Day 20. An estradiol (E$_2$) implant was placed sc on
Day 4 and removed on Day 21. All animals had viable fetuses on
Day 30. Data expressed as mean ± SEM; ( ) = number of animals.
(From Gadsby et al., 1983).

result is in full accord with previous reports that corpora lutea exposed
to low doses of hCG (10 to 20 IU/rabbit) are not desensitized to the
steroidogenic action of estrogen (Yuh and Keyes, 1981; Keyes et al.,
1983b). If hCG (20 IU) is injected in rabbits pretreated with estrogen,
serum progesterone does not decline below pre-hCG stimulation values
(Keyes et al., 1983b). Therefore, the precipitous decline in serum
progesterone values in animals in Fig. 3 is not attributable to
desensitizing actions of hCG in the luteal tissue, but rather to the loss
of follicular estrogen synthesis. The gradual rise in serum progesterone
by Day 27 in rabbits without estradiol treatment probably reflects a
gradual return of follicular estrogen synthesis, and continued luteal
responsiveness to estrogen in the presence of the conceptuses maintained
by Depo-Provera.

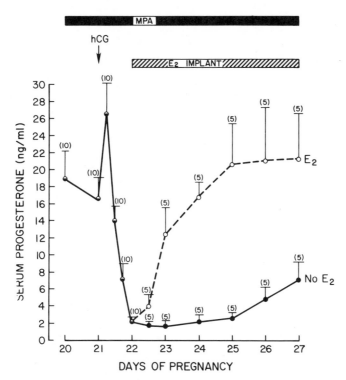

Fig. 3. Effect of hCG-induced ovulation and estradiol treatment (E implant) on serum progesterone concentrations in pregnant rabbits. All animals were treated with the progestin, medroxyprogesterone acetate (MPA) beginning on Day 20 of pregnancy; hCG (10 IU) was injected in all animals on Day 21. Controls received no estradiol (no E). Data expressed as mean ± SEM; ( ) = number of animals. (From Gadsby et al., 1983).

The preceding experiments confirm the hypothesis that the putative fetal placental factor(s) is not effective in maintaining luteal function in the absence of estrogen. Further, it is clear that in the presence of conceptuses, estradiol can stimulate and restore progesterone secretion. However, it is not clear from the above experiments if placental factor(s) is required for the action of estrogen to be expressed in the second half of pregnancy. In the experiments to follow, all conceptuses were removed and estradiol maintained to determine if the conceptus has a vital role in the luteotropic process in the second half of pregnancy.

HYPOTHESIS 2. THE LUTEOTROPIC EFFECT OF ESTROGEN IN PREGNANCY REQUIRES THE SIMULTANEOUS ACTION OF FETAL PLACENTAL FACTOR(S)

Dutch rabbits were anesthetized with xylazine-ketamine and were hysterectomized or sham hysterectomized on Day 21 of pregnancy. Serum progesterone was measured to detect changes in luteal function. The results are shown in Fig. 4. In sham hysterectomized animals, serum progesterone values remained relatively stable through the entire experiment, which was terminated on Day 27. In contrast, hysterectomy caused a marked decline in serum progesterone from a mean value of 13 ng/ml on Day 21 to 5 ng/ml on Day 22, and by Days 24 to 27, the values were 1 ng/ml or less. Serum estradiol values were not different through the experiment for the two groups, with mean values ranging from 4 to 8 pg/ml (Fig. 5).

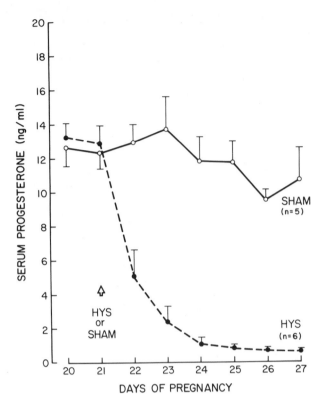

Fig. 4. Effect of hysterectomy (HYS) or sham hysterectomy (SHAM) performed on Day 21 of pregnancy on serum progesterone concentrations. Data expressed as mean ± SEM; (n) = number of animals.
(From Gadsby and Keyes, 1984).

In a second experiment rabbits were anesthetized and hysterectomized
on Day 21 of pregnancy, and treated with estradiol beginning either the
day before hysterectomy or the day after hysterectomy. The estradiol was
administered by a small Silastic capsule placed sc and was designed to
deliver physiological amounts of estradiol. The results are shown in Fig.
6. In animals treated with estradiol beginning on Day 20, hysterectomy
caused a decline in serum progesterone values, from 19 ng/ml on Day 21 to
13 ng/ml on Day 22; by Days 24 to 27 values had fallen to 1 to 2 ng/ml.
Hysterectomy of animals treated with estradiol beginning on Day 22 caused
a sharp decline in progesterone from a mean value of 16 ng/ml on Day 21 to
5 ng/ml on Day 22; thereafter progesterone values remained around 2 to 3
ng/ml. The decline in progesterone in all hysterectomized groups (Fig. 4
and 6) was accompanied by some loss in luteal wet weights, which were
determined at autopsy on Day 27. Luteal weight for sham hysterectomized
animals was 16 mg/corpus luteum, which was significantly (p < 0.01)
greater than weights for corpora lutea in hysterectomized groups (9 to 10
mg/corpus luteum).

The sharp decline in serum progesterone observed within the first 24
h after hysterectomy (Fig. 4 and 6), is interpreted to reflect declining

366

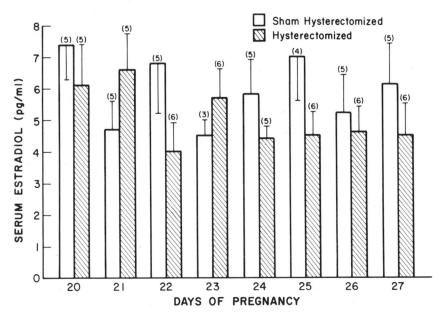

Fig. 5. Effect of hysterectomy or sham hysterectomy performed on Day 21 of pregnancy on serum estradiol concentrations. Data expressed as mean ± SEM; ( ) = number of animals. (From Gadsby and Keyes, 1984).

rates of progesterone secretion. To obtain more direct evidence, a separate experiment was performed in which corpora lutea were removed 24 h after hysterectomy or sham hysterectomy and incubated for determination of in vitro rates of progesterone production. On Day 21 of pregnancy, rabbits were anesthetized with xylazine and ketamine and were hysterectomized or sham hysterectomized. Twenty-four h later the animals were again anesthetized and the corpora lutea, with adherent ovarian tissues removed by dissection, were sliced and incubated (approximately 10 mg per flask) in medium 199 (containing 10 mM Hepes, 50 u/ml penicillin and 50 μg/ml streptomycin) for 12 h at 37C in the presence of 95% oxygen: 5% carbon dioxide. The incubations were performed in triplicate and medium was changed completely at 4 and 8 h. Progesterone was measured by specific radioimmunoassay without extraction of the medium. The data are shown in Fig. 7 and are expressed as the cumulative progesterone production over the 12 h of incubation; means of triplicate incubations were calculated for each animal. Luteal tissue from hysterectomized animals produced/released significantly (p < 0.01) less progesterone than did luteal tissue from sham hysterectomized animals.

These experiments confirm the above hypothesis that estradiol either produced endogenously by the ovarian follicles or administered by Silastic capsule, is not an effective luteotropic stimulus in the absence of conceptuses removed by hysterectomy. The declining values of serum progesterone observed after hysterectomy cannot be easily dismissed as the result of non-specific effects of surgical trauma: sham hysterectomy, which consisted of surgery and manipulation of the uterus, had no effect on serum progesterone (Fig. 4); further, it is well known that

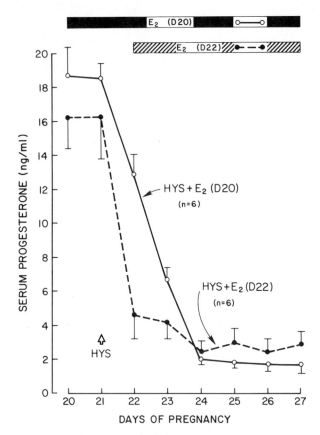

Fig. 6. Effect of estradiol-17β (E₂) on serum progesterone concentrations in rabbits hysterectomized (HYS) on Day 21 of pregnancy (arrow). Data expressed as mean ± SEM; (n) = number of animals. (From Gadsby and Keyes, 1984).

hysterectomy of non-pregnant animals actually extends the period of progesterone secretion for a few days (Hilliard et al., 1974; Miller and Keyes, 1976). The argument could be advanced that hysterectomy caused an interruption of estradiol secretion, long enough to cause loss of luteal function. However, this seems improbable, because serum estradiol values were not different after hysterectomy (Fig. 5), and the administration of estradiol did not prevent the loss of luteal function (Fig. 6). Finally, the direct measurement of progesterone produced in vitro by luteal tissue 24 h after hysterectomy (Fig. 7) confirmed that progesterone secretion had diminished. The diminished amounts of progesterone produced in vitro, most likely reflect reduced progesterone synthesis, since the in vitro production of the major metabolite, 20 α-dihydroprogesterone, was also reduced (data not shown).

The preceding experiments support the idea that the corpora lutea remain functional in the second half of pregnancy through the combined actions of estrogen and factor(s) from the conceptus or fetal placenta. Neither estrogen, which is considered to be the principal luteotropic hormone in this species (Hilliard, 1973; Bill and Keyes, 1983), nor

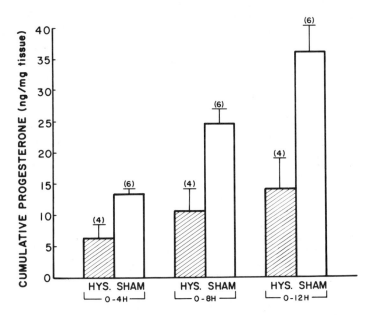

Fig. 7. In vitro progesterone production by rabbit luteal tissue. Corpora
lutea were removed 24 h (Day 22) after hysterectomy (HYS) or sham
hysterectomy (SHAM) performed on Day 21 of pregnancy. Sliced
luteal tissue from each animal was incubated in triplicate in
medium 199 for 12 h at 37C; medium was changed completely at 4 and
8 h. Data are presented as cumulative progesterone values in
medium during 0 to 4 h, 0 to 8 h, and 0 to 12 h; mean $\pm$ SEM; ( ) =
number of animals.

factor(s) of fetal placental origin, have demonstrable luteotropic effects
when acting alone. It should be emphasized that estradiol is an effective
luteotropic hormone in pseudopregnancy; the corpora lutea are maintained
and secrete normal quantities of progesterone through the action of
estradiol, even in the absence of the pituitary (Bill and Keyes, 1983).
The mechanisms that are responsible for the transition from luteal
dependence on estrogen to estrogen and placental factor(s) are not understood. To gain insight into this phenomenon, the hypothesis was tested
that placental factor(s) promotes responsiveness of the corpora lutea to
estrogen by maintaining estrogen receptor in the corpora lutea.

HYPOTHESIS 3. FETAL-PLACENTAL FACTOR(S) PROMOTES LUTEAL FUNCTION IN THE
THIRD WEEK OF PREGNANCY BY MAINTAINING ESTROGEN RECEPTOR IN THE CORPORA
LUTEA

Rabbits were hysterectomized or sham hysterectomized on Day 21 of
pregnancy as described above, and 24 h later (Day 22), the corpora lutea
were removed for estrogen receptor determination according to procedures
published previously (Yuh and Keyes, 1979). Blood samples were also taken
to confirm that serum progesterone had fallen after hysterectomy as
observed above (Fig. 4). The corpora lutea were homogenized in Tris-HCL/
EDTA buffer at 4C. The luteal homogenate was centrifuged at 800 xg for 10
min at 4C to obtain the crude nuclear fraction. The 800 xg supernatant
was centrifuged at 105,000 xg for 1 h at 4C, to obtain the cytoplasmic
receptor preparation (cytosol; 105,000 xg supernatant). Specific tritiated

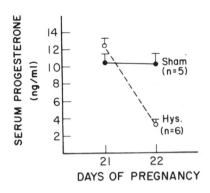

a) Serum progesterone

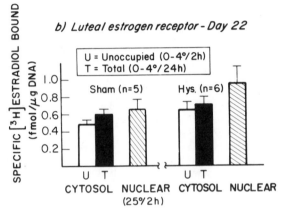

b) Luteal estrogen receptor - Day 22

Fig. 8. Serum progesterone concentrations (a) and luteal estrogen receptor
concentrations (b) in rabbits, hysterectomized (HYS) or sham
hysterectomized (SHAM) on Day 21 of pregnancy. Blood samples were
collected on Days 21 and 22 for measurement of serum progesterone.
On Day 22, corpora lutea were removed, homogenized and cytosol
(105,000 xg supernatant) and crude nuclear (800 xg pellet)
fractions were prepared. Unoccupied cytosol (0–4C for 2 h), total
cytosol (0–4C for 24 h; exchange assay) and total nuclear (25C for
2 h; exchange assay) estrogen receptor concentrations were
measured (see text for further details). Data expressed as mean ±
SEM; (n) = number of animals.

estradiol–17β binding to cytosol and nuclear estrogen receptors was determined by incubating cytosol with 0.6 nM tritiated estradiol and the
nuclear fraction with 1 nM tritiated estradiol, in the presence or absence
of a 100–fold excess diethylstilbestrol (DES), at 0–4C for 2 h (unoccupied
cytosol receptor), 0–4C for 24 h (total cytosol receptor, exchange assay),
and 25C for 2 h (total nuclear receptor, exchange assay). The binding
characteristics (dissociation constant and binding capacity) of cytosol
(0–4C for 24 h) and nuclear receptors (25C for 2 h) were examined by

saturation binding analysis by the method of Scatchard (1949), using a range of tritiated estradiol concentrations (1×10^{-7} M to 2.5×10^{-10}M, cytosol; 2×10^{-7}M to 7.5×10^{-10}M, nuclear) in the presence or absence of 100-fold excess of DES. Separation of tritiated estradiol bound to cytosol receptor, from unbound was achieved by adding dextran-coated charcoal. After mixing and centrifugation at 800 xg for 15 min at 4C, the supernatant was taken for determination of radioactivity by liquid scintillation counting. The nuclear receptor was washed 3 times in ice-cold Tris/EDTA buffer to remove unbound steroid, and the bound tritiated estradiol was extracted in ethanol and radioactivity determined. DNA contents of the crude homogenate (or cytosol equivalent) and nuclear pellets were determined by the method of Burton (1956), and the data were expressed as fmol bound per µg DNA. For saturation binding analysis, bound and unbound tritiated estradiol were calculated per 10 µg DNA and dissociation constant and binding capacity were determined by the method of Scatchard (1949).

The results are shown in Fig. 8 and in Table 1. Following hysterectomy, serum progesterone declined from a mean value of 12 ng/ml on Day 21 to 3 ng/ml on Day 22; whereas, in the sham hysterectomized animals serum progesterone values remained stable, around 10 ng/ml on both days. Measurement of luteal estrogen receptor concentration revealed no differences between the two groups, either with respect to cytosol receptor or nuclear receptor. Further, as indicated in Table 1, the dissociation constants and binding capacities determined by Scatchard analysis were similar for the two groups.

Table 1. Scatchard Analysis of Luteal Estrogen Receptor After Hysterectomy or Sham Hysterectomy of Pregnant Rabbits[a]

		Dissociation Constant (M)	Binding Capacity (M)
Cytosol	SHAM	$5.0 \ (\pm 1.8) \times 10^{-11}$	$2.4 \ (\pm 0.6) \times 10^{-11}$
	HYS	$4.8 \ (\pm 1.8) \times 10^{-11}$	$2.7 \ (\pm 0.7) \times 10^{-11}$
Nuclear	SHAM	$1.2 \ (\pm 0.6) \times 10^{-10}$	$0.8 \ (\pm 0.2) \times 10^{-11}$
	HYS	$1.6 \ (\pm 0.2) \times 10^{-10}$	$1.4 \ (\pm 0.7) \times 10^{-11}$

[a]Corpora lutea were removed 24 h after hysterectomy or sham hysterectomy performed on Day 21 of pregnancy (see text for details of receptor analysis). Data are mean $\pm$ SEM of 3 independent experiments (pools of tissue); each pool has luteal tissue from 5 animals.

If estrogen receptor is not affected, quantitatively, within 24 h after hysterectomy, is luteal estradiol content altered in the same period of time? Dutch rabbits were anesthetized with xylazine and ketamine and were hysterectomized or sham hysterectomized on Day 21 of pregnancy. Twenty-four h later the animals were again anesthetized, adherent interstitial and follicular tissues were removed from the corpora lutea by dissection using a dissecting microscope, and the corpora lutea were weighed and then frozen. In preparation for estradiol radioimmunoassay, the corpora lutea were thawed, homogenized in ice-cold phosphate buffered saline, extracted with benzene, and the extracts chromatographed on Sephadex LH-20 columns as described previously (Holt et al., 1975). The results are shown in Fig. 9. Twenty-four h after sham hysterectomy, luteal estradiol concentration was 0.8 pg/mg of luteal tissue, and after hysterectomy, 0.6 pg/mg; these values were not different, statistically.

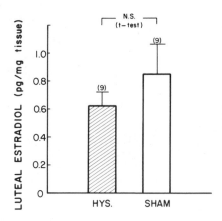

Fig. 9. Luteal estradiol concentrations, 24 h (Day 22) following sham
hysterectomy or hysterectomy performed on Day 21 of pregnancy.
Data expressed as mean ± SEM; ( ) = number of animals; N.S. = nonsignificant by t – test.

From these experiments, we must reject the hypothesis that
hysterectomy-induced loss of luteal progesterone synthesis is due to
concomitant loss of estrogen receptor. Furthermore, as suggested by the
serum estradiol measurements in Fig. 5, luteal tissue estradiol concentrations have not changed after hysterectomy (Fig. 9), leading to the
conclusion that estradiol availability within the luteal tissue (and
therefore to the receptor) has not changed. In light of the absence of
quantitative changes in estrogen receptor within 24 h after hysterectomy,
is there a precedent for acute changes in luteal estrogen receptor? Yuh
and Keyes (1981) reported that the injection of "desensitizing" doses
(50–100 IU) of human chorionic gonadotropin caused a loss of luteal
steroidogenesis and simultaneous loss of luteal estrogen receptor, both
cytoplasmic and nuclear, within 24 h; associated with these changes was
the loss of steroidogenic response to the luteotropic hormone, estradiol.
In the present experiments, hysterectomy caused a decrease in luteal
steroidogenesis which was not accompanied by detectable changes in luteal
estrogen receptor concentrations or binding characteristics. These results
would appear to exclude the possibility that placental factor(s) can
regulate the estrogen receptor. However, in view of the static nature of
these receptor measurements, it is not possible to rule out effects on the
dynamics of estrogen binding to receptor, retention within the nucleus,
and receptor replenishment in the cytoplasm. Furthermore, there may be
events beyond those associated with receptor interactions, which are
regulated by placental factor(s), and which are responsible for the loss
of responsiveness to estrogen.

In this manuscript attention has been directed to the activity of a putative placental factor that in some way prevents luteal regression and enhances luteal responsiveness to estrogen in the third week of pregnancy. Since implantation normally occurs at the end of the first week after ovulation (Böving, 1961), what constitutes maternal recognition of pregnancy in this species? At least two laboratories have reported the existence of a chorionic gonadotropin-like activity in the preimplantation rabbit blastocyst that is associated with a preimplantation elevation in serum progesterone and therefore inferred to stimulate the corpus luteum (Haour and Saxena, 1974; Fuchs and Beling, 1975). These interesting findings were not supported by subsequent independent investigations from several laboratories. Using three bioassay systems, Holt et al. (1976) failed to detect gonadotropic activity in rabbit blastocysts. Sundaram et al. (1975) using another bioassay also failed to detect LH-like activity in rabbit blastocysts. Frequent sampling of peripheral blood through the period of implantation did not reveal differences in serum progesterone between pregnant and pseudopregnant rabbits, suggesting that the signal for maternal recognition of pregnancy is transmitted some time after implantation (Fig. 10; Browning et al., 1980). From these profiles of serum progesterone, and our knowledge of the requirement of a placental factor in the third week of pregnancy, it appears plausible that maternal recognition of pregnancy consists of preventing the onset of the luteolytic process, that normally ensues two weeks after ovulation in non-pregnant rabbits.

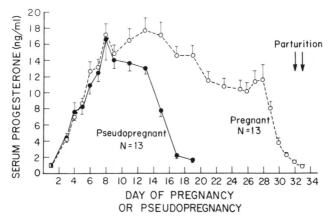

Fig. 10. Serum progesterone concentrations in pregnant and pseudopregnant Dutch rabbits. Pseudopregnancy was induced by mating to a vasectomized male; the day following fertile or sterile mating is designated Day 1. Data expressed as mean ± SEM; N = number of animals. (From Keyes et al., 1983a).

DISCUSSION AND CONCLUDING REMARKS

From the experiments reported here, we have gained additional evidence for dual regulation of the corpora lutea during the latter half of pregnancy. As is characteristic of the pseudopregnant rabbit, follicular estrogen is also a necessary luteotropic hormone in pregnancy. However, in contrast to pseudopregnancy, estrogen is not a sufficient luteotropic hormone in pregnancy, requiring one or more factors, presumably produced by the fetal placenta. The present experiments show convincingly, that neither estrogen, produced by the follicles or administered exogenously, nor factor(s) from the conceptuses maintained artifically by Depo-Provera, have luteotropic activity when acting alone. In order to gain an understanding of the luteotropic process in pregnancy, we must know the identity of the placental factor(s), and develop a reliable bioassay for its activity, and we must gain insight into the cellular mechanisms for the coordinate actions of estrogen and placental factor(s).

For purposes of this discussion, it is useful to review briefly what is known about the process of luteal regression in the rabbit. It seems reasonable to assert that the placental factor must inhibit the synthesis and/or action of a luteolytic substance, thereby enabling the corpora lutea to continue to respond to estrogen for two weeks longer than the normal duration of pseudopregnancy. Unlike some other species, in which the uterus has a major role in killing the corpus luteum through the secretion of prostaglandins (Horton and Poyser, 1976), the non-gravid uterus of the pseudopregnant rabbit does not appear to have a major role in determining the onset of luteolysis. Hysterectomy does not prevent the major decline in serum progesterone concentration at the end of pseudopregnancy (Miller and Keyes, 1976). Neither the pituitary, nor the non-luteal portion of the ovary, nor local uterine factors appear to have critical roles in luteal regression in the rabbit: regression of the corpora lutea occurs at about the normal time (approximately two weeks after ovulation) in estradiol-treated hypophysectomized rabbits (Bill and Keyes, 1983), and in rabbits with corpora lutea transplanted beneath the kidney capsule (Miller and Keyes, 1976). Furthermore, since the administration of estradiol either in non-physiological (Miller and Keyes, 1976) or physiological (Bill and Keyes, 1983) amounts does not alter the course of luteal regression, and since serum estradiol values rise toward the end of pseudopregnancy (Browning, et al., 1980), it is clear that the normal course of luteal regression is not caused by estrogen withdrawal.

In view of the physiological luteolytic role of uterine prostaglandin $F_{2\alpha}$ in sheep and guinea-pig (Horton and Poyser, 1976), and the luteolytic activity of prostaglandin $F_{2\alpha}$ when administered in rabbits (Carlson and Gole, 1978; Laudanski et al., 1979; Kehl and Carlson, 1981), it has been suggested that prostaglandins, produced by the corpus luteum itself, might be luteolytic (Hoffman, 1974; Miller et al., 1983). A similar "autolytic" mechanism involving luteal prostaglandin has been implicated for the human female and the Rhesus monkey, two other species in which the uterus does not control luteal regression at the end of the non-fertile cycle (Neill et al., 1969; Beling et al., 1970; Challis et al., 1977; Swanston et al., 1977). However appealing the idea might be that a luteal prostaglandin is the luteolytic substance in the rabbit, evidence is not at hand to support this idea. Luteal tissue prostaglandins do not increase toward the end of pseudopregnancy (Miller et al., 1983). Furthermore, the administration of indomethacin in pseudopregnant rabbits does not prevent luteal regression (O'Grady et al., 1972). Attention has been directed recently to the role of luteal oxytocin in the luteolytic process in a number of species (Wathes, 1984), but its role in the rabbit is unknown. Thus, no clear picture emerges as to the precise mechanism for luteal regression in this species; only generalities can be proposed, until new insights are gained into this obscure area.

Having reviewed briefly our current knowledge of potential luteolytic factors in the rabbit, what is known about the identity of placental factor(s) that might influence luteal function? Greep (1941) showed that the rabbit fetal placenta contained little or no estrogen activity, although it has been shown to possess aromatase activity (George et al., 1978). The putative placental factor affecting the corpora lutea does not appear to be estrogen, since the corpora lutea are not maintained in the presence of viable conceptuses when follicular estrogen synthesis is eliminated either by ovarian x-irradiation (Keyes and Armstrong, 1968), by injection of hCG (see Fig. 3), or by hypophysectomy (see Fig. 2). There is no convincing evidence for the existence of a chorionic gonadotropin (CG) in the rabbit. Fetal placental extracts or medium conditioned with incubated fetal placental tissue showed little or no chorionic gonadotropin-like activity in _in vitro_ or _in vivo_ bioassays (Browning et al., 1982). A rabbit placental lactogen has been isolated and purified; this lactogen has a low affinity for rabbit mammary prolactin receptors (Bolander and Fellows, 1976). However, the role, if any, of this placental lactogen in the corpus luteum is unknown. Prostaglandin E_2 concentrations are elevated in uterine vein blood in pregnant rabbits after Day 11 (Lytton and Poyser, 1982). This might be considered a potential luteotropic substance, since it has been shown that rabbit luteal tissue possesses a prostaglandin E-stimulable adenylate cyclase (Hunzicker-Dunn and Birnbaumer, 1976; Abramowitz and Birnbaumer, 1979). The rabbit placenta contains a GnRH-like peptide (Nowak et al., 1984). However, this peptide seems an unlikely candidate as luteotropin, because rabbit luteal tissue does not respond to GnRH agonist, nor does it possess receptor for GnRH (Thorson et al., 1985).

The current state of knowledge about the identity of the placental factor that is suspected to affect luteal function is far from gratifying. However, from the experiments presented here, we have a clearer understanding of the requirement for this factor, in that it appears to be necessary to allow the continued action of estrogen. The identity of this factor and elucidation of its mode of action have the potential for opening new avenues of research. Assuming that this factor acts directly in the corpus luteum, it might be viewed as a molecule that inhibits the production or action of a luteolytic agent; alternatively, it might in some way enhance the basic action of estrogen in the luteal cell.

ACKNOWLEDGEMENTS

This research was supported by grants NIH-HD07127, HD13645, HD-11311, and by Biomedical Research Support Grant. J.E.G. supported by a NATO Fellowship and the Mellon Foundation.

REFERENCES

Abramowitz J, Birnbaumer L, 1979. Prostacyclin activation of adenylyl cyclase in rabbit corpus luteum membranes: comparison with 6-keto prostaglandin $F_{1\alpha}$ and prostaglandin E_1. Biol Reprod 21:609-616

Beling CG, Marcus SL, Markham SM, 1970. Functional activity of the corpus luteum following hysterectomy. J Clin Endocrinol Metab 30:30-39

Bill CH, Keyes PL, 1983. 17β-estradiol maintains normal function of corpora lutea throughout pseudopregnancy in hypophysectomized rabbits. Biol Reprod 28:608-617

Bolander Jr FF, Fellows RE, 1976. The purification and characterization of rabbit placental lactogen. Biochem J 159:775-782

Böving BG, 1961. Anatomical analyses of rabbit trophoblast invasion. Contrib Embryol Carnegie Instit Washing 37:33

Browning JY, Amis MM, Meller PA, Bridson WE, Wolf RC, 1982. Luteotropic and antiluteolytic activities of the rabbit conceptus. Biol Reprod 27:665-672

Browning JY, Keyes PL, Wolf RC, 1980. Comparison of serum progesterone, 20α –dihydroprogesterone, and estradiol-17β in pregnant and pseudo-pregnant rabbits: evidence for postimplantation recognition of pregnancy. Biol Reprod 23:1014-1019

Browning JY, Wolf RC, 1981. Maternal recognition of pregnancy in the rabbit: effect of conceptus removal. Biol Reprod 24:293-297

Burton K, 1956. A study of the conditions and mechanism of the diphenylamine reaction for the colorimetric estimation of deoxyribo-nucliec acid. Biochem J 62:315-323

Carlson JC, Gole JWD, 1978. CL regression in the pseudopregnant rabbit and the effects of treatment with prostaglandin $F_{2\alpha}$ and arachidonic acid. J Reprod Fertil 53:381-387

Challis JRG, Calder AA, Dilley S, Forster CS, Hillier K, Hunter DJS, Mackenzie IZ, Thorburn GD, 1977. Production of prostaglandins E and F_α by corpora lutea, corpora albicantes and stroma from the human ovary. J Endocrinol 68:401-408

Chu JP, Lee CC, You SS, 1946. Functional relation between the uterus and the corpus luteum. J Endocrinol 4:392-398

Fuchs A-R, Beling C, 1975. Evidence for early ovarian recognition of blastocysts in rabbits. Endocrinology 95:1054-1058

Gadsby JE, Keyes PL, 1984. Control of corpus luteum function in the pregnant rabbit: role of the placenta ("placental luteotropin") in regulating responsiveness of corpora lutea to estrogen. Biol Reprod 31:16-24

Gadsby JE, Keyes PL, Bill CH, 1983, Control of corpus luteum function in the pregnant rabbit: role of estrogen and lack of a direct luteo-tropic role of the placenta. Endocrinology 113:2255-2262

George FW, Tobleman WT, Milewich L, Wilson JD, 1978. Aromatase activity in developing rabbit brain. Endocrinology 102:86-91

Greep RO, 1941. Effects of hysterectomy and of estrogen treatment on volume changes in the corpora lutea of pregnant rabbits. Anat Rec 80:465-477

Haour F, Saxena BB, 1974. Detection of a gonadotropin in rabbit blasto-cyst before implantation. Science 185:444-445

Hilliard J, 1973. Corpus luteum function in guinea pigs, hamsters, rats, mice and rabbits. Biol Reprod 8:203-221

Hoffman LH, 1974. Luteal regression induced by arachidonic acid in the pseudopregnant rabbit. J Reprod Fertil 36:401-404

Holt JA, Ewing LL, 1974. Acute dependence of ovarian progesterone output on the presence of placentas in 21-day pregnant rabbits. Endocrinology 94:1438-1444

Holt JA, Heise WF, Wilson SM, Keyes PL, 1976. Lack of gonadotropic activity in the rabbit blastocyst prior to implantation. Endocrinology 98:904-909

Holt JA, Keyes PL, Brown JM, Miller JB, 1975. Premature regression of corpora lutea in pseudopregnant rabbits following the removal of polydimethylsiloxane capsules containing 17β–estradiol. Endocrinology 97:76-82

Horton EW, Poyser NL, 1976. Uterine luteolytic hormone: a physiological role for prostaglandin $F_{2\alpha}$. Physiol Rev 56:595-651

Hunzicker-Dunn M, Birnbaumer L, 1976. Adenylyl cyclase activities in ovarian tissues. II. Regulation of responsiveness to LH, FSH, and PGE, in the rabbit. Endocrinology 99:185-197

Kehl SJ, Carlson JC, 1981. Assessment of the luteolytic potency of various prostaglandins in the pseudopregnant rabbit. J Reprod Fertil 62:117-122

Keyes PL, Armstrong DT, 1968. Endocrine role of follicles in the regulation of corpus luteum function in the rabbit. Endocrinology 83:509-515

Keyes PL, Gadsby JE, Yuh K-CM, Bill CH, 1983a. The corpus luteum. In: Greep RO (ed.), Reproductive Physiology IV, International Review of Physiology. Baltimore: University Park Press, pp 57-97

Keyes PL, Nalbandov AV, 1967. Maintenance and function of corpora lutea in rabbits depend on estrogen. Endocrinology 80:938-946

Keyes PL, Possley RM, Yuh K-CM, 1983b. Contrasting effects of estradiol-17β and human chorionic gonadotrophin on steroidogenesis in the rabbit corpus luteum. J Reprod Fertil 69:579-586

Lanman JT, Thau RB, 1979. Effect of the fetal placenta and of a rabbit pituitary extract on plasma progesterone in fetectomized rabbits. J Reprod Fertil 57:341-344

Laudanski T, Batra S, Akerlund M, 1979. Prostaglandin-induced luteolysis in pregnant and pseudopregnant rabbits and the resultant effects on the myometrial activity. J Reprod Fertil 56:141-148

Lytton FDC, Poyser NL, 1982. Concentrations of $PGF_{2\alpha}$ and PGE_2 in the uterine venous blood of rabbits during pseudopregnancy and pregnancy. J Reprod Fertil 64:421-429

Miller JB, Jarosik C, Stanisic D, Wilson Jr L, 1983. Alterations in plasma and tissue prostaglandin levels in rabbits during luteal regression. Biol Reprod 29:824-832

Miller JB, Keyes PL, 1976. A mechanism for regression of the rabbit corpus luteum: uterine-induced loss of luteal responsiveness to 17β-estradiol. Biol Reprod 15:511-518

Mills TM, Savard K, 1973. Steroidogenesis in ovarian follicles isolated from rabbits before and after mating. Endocrinology 92:788-791

Neill JD, Johansson EDB, Knobil E, 1969. Failure of hysterectomy to influence the normal pattern of cyclic progesterone secretion in the Rhesus monkey. Endocrinology 84:464-465

Nowak RA, Bahr JM, 1983. Maternal recognition of pregnancy in the rabbit. J Reprod Fertil 69:623-627

Nowak RA, Wiseman BS, Bahr JM, 1984. Identification of a gonadotropin-releasing hormone-like factor in the rabbit fetal placenta. Biol Reprod 31:67-75

O'Grady JP, Caldwell BV, Auletta FJ, Speroff L, 1972. The effects of an inhibitor of prostaglandin synthesis (indomethacin) on ovulation, pregnancy, and pseudopregnancy in the rabbit. Prostaglandins 1:97-106

Rippel RH, Johnson ES, 1976. Regression of corpora lutea in the rabbit after injection of a gonadotropin-releasing peptide (39320). Proc Soc Exp Biol Med 152:29-32

Robson JM, 1939. Maintenance of pregnancy in the hypophysectomized rabbit by the administration of oestrin. J Physiol 95:83-91

Scatchard G, 1949. The attractions of proteins for small molecules and ions. Ann NY Acad Sci 51:660-672

Spies HG, Hilliard J, Sawyer CH, 1968. Maintenance of corpora lutea and pregnancy in hypophysectomized rabbits. Endocrinology 83:354-367

Sundaram K, Connell KG, Passantino T, 1975. Implication of absence of hCG-like gonadotrophin in the blastocyst for control of corpus luteum function in the pregnant rabbit. Nature 256:739-741

Swanston IA, McNatty KP, Baird DT, 1977. Concentration of prostaglandin $F_{2\alpha}$ and steroids in the human corpus luteum. J Endocrinol 73:115-122

Thau R, Lanman JT, 1974. Evaluation of progesterone synthesis in rabbit placentas. Endocrinology 94:925-926

Thorson JA, Marshall JC, Bill CH, Keyes PL, 1985. D-Ala$_6$-des-Gly$_{10}$ - gonadotropin-releasing hormone ethylamide: absence of binding sites and lack of a direct effect in rabbit corpora lutea. Biol Reprod 32:226-231

Wathes DC, 1984. Possible actions of gonadal oxytocin and vasopressin. J Reprod Fertil 71:315-345

Yuh K-CM, Keyes PL, 1979. Properties of nuclear and cytoplasmic estrogen receptor in the rabbit corpus luteum: evidence for translocation. Endocrinology 105:690–696

Yuh K-CM, Keyes PL, 1981. Effects of human chorionic gonadotropin in the rabbit corpus luteum: loss of estrogen receptor and decreased steroidogenic response to estradiol. Endocrinology 108:1321–1327

DECIDUAL LUTEOTROPIN SECRETION AND ACTION: ITS ROLE

IN PREGNANCY MAINTENANCE IN THE RAT

G. Gibori, P.G. Jayatilak, I. Khan, B. Rigby,
T. Puryear, S. Nelson, and Z. Herz

Department of Physiology and Biophysics
College of Medicine, University of Illinois
Chicago, IL 60612

INTRODUCTION

The decidual tissue is the result of proliferation, hypertrophy and differentiation of endometrial cells. Decidualization of the endometrial stroma accompanies implantation in several mammalian species. Man and other primate species exhibit this phenomenon, and several rodent species, of which the rat is a fine example, exhibit it to marked degree (De Feo, 1967). In humans, decidualization occurs normally with each menstrual cycle. In other species, including the rat, decidual tissue does not develop unless a stimulus is applied to a sensitized uterus. This stimulus may be either the contact of the blastocyst with the endometrium or artificial stimulation of the uterus (Lobel et al., 1965). The ability of the rat uterus to respond, however, depends on the completion of a basic hormonal sequence: exposure to progesterone for at least 48 h and to a minute amount of estrogen at the end of this period (Psychoyos, 1973; Glasser and McCormack, 1980). A state of peak uterine sensitivity then results 20-24 h following the completion of this sequence. In normal pregnancy and pseudopregnancy this occurs on day 5 (De Feo, 1967; Psychoyos, 1973), at which time the uterus is capable of responding to a variety of stimuli which elicit the decidual response. Stromal cells proliferate, hypertrophy and differentiate into decidual cells between the 5th and 11th day. Following this period, no evidence for further growth is seen. Regression of the decidual tissue begins on day 12 and necrotic changes are extensive by day 15.

A complex feedback mechanism exists between the decidua and the corpus luteum. Secretion of progesterone by the corpus luteum is crucial to the maintenance of the decidua (Deanesley, 1973; Butterstein and Hirst, 1977). If the source of this hormone is removed, the decidua collapses. However, decidual degeneration occurs even in the presence of continuous exogenous progesterone administration (Finn and Porter, 1975). The decrease in steroid receptors that occurs as decidualization proceeds (Armstrong et al., 1977; Talley et al., 1977; Martel et al., 1984) may explain why exogenous hormone treatment fails to delay the regression of the decidua. The presence of the decidua

is, in turn, necessary for the prolongation of the life span of the corpus luteum and the maintenance of progesterone synthesis.
Decidualization of the rat uterus prolongs pseudopregnancy from 13 to 20 days (De Feo, 1967). Interestingly, hysterectomy also produces the same effect. A widely accepted explanation for this similarity is that the luteolytic action of the uterus is abolished by hysterectomy and decidualization (Melampy et al., 1964; Finn and Porter, 1975). It has been suggested that decidualization prevents the production of the luteolytic substance prostaglandin $F_{2\alpha}$ which reaches the ovaries without being released in the peripheral circulation and thus accounts for a prolonged pseudopregnancy (Wilson et al., 1970; Pharriss et al., 1972). Although $PGF_{2\alpha}$ levels increase in uterine venous plasma at the end of pseudopregnancy, the presence of the decidual tissue does not reduce its production or release by the uterus (Weems et al., 1975; Castracane and Shaikh, 1976; Weems et al., 1979) nor does it decrease the content of prostaglandins in the corpora lutea (Weems et al., 1979). Recent evidence suggests that hysterectomy and decidualization of the uterus may prolong luteal function by completely different mechanisms. Freeman and collaborators (Freeman, 1979; Gorospe et al., 1981; Gorospe and Freeman, 1982) reported that the uterus secretes a substance which restrains the secretion of pituitary prolactin. Thus, removal of the uterus may prolong prolactin secretion and pseudopregnancy because of the withdrawal of a uterine prolactin inhibitory factor.

It has been well documented that the rat placenta secretes a luteotropic hormone as early as day 7 of pregnancy (Alloiteau, 1957; Morishige and Rothchild, 1974; Tabarelli et al., 1982). This luteotropic hormone is considered to have prolactin-like activity because it sustains progesterone production when prolactin secretion is selectively suppressed. Administration of a dopamine agonist prior to implantation causes a dramatic decrease in progesterone synthesis and abortion. Once implantation occurs, prolactin is no longer required and its withdrawal from the circulation does not affect luteal progesterone production. Several investigators (Yoshinaga and Adams, 1967; Yoshinaga et al., 1973; Morishige and Rothchild, 1974) suggested that the luteotropic hormone produced by the young placenta may be rat placental luteotropin. Rat placental luteotropin, a 40-50,000 MW prolactin-like hormone, is secreted by the fetal trophoblastic tissue in large amounts between days 10-13 (Linkie and Niswender, 1973; Kelly et al., 1975; De Greef et al., 1977; Robertson et al., 1982). However, this large molecular weight luteotropin is not detectable earlier in pregnancy . Using indirect evidence, Alloiteau (1957) was the first to hypothesize the production of a luteotropic hormone by the young placenta. Aware of the possibility that this hormone might be different from rat placental luteotropin, he called it facteur placentaire precoce. It has been subsequently determined that, unlike rat placental luteotropin, this hormone originates not from the trophoblast but from the decidua (Basuray and Gibori, 1980). In pregnant rats, trophoblastic tissue from young placentas (days 7-10) has no luteotropic activity, yet decidual tissue maintained in situ after removal of the trophoblast sustains progesterone synthesis after prolactin withdrawal.

Our studies have demonstrated that the decidual tissue of the rat produces a prolactin-like hormone which modulates luteal progesterone and follicular steroid synthesis (Gibori et al., 1974, 1981, 1984; Rothchild and Gibori, 1975; Basuray and Gibori, 1980; Basuray et al., 1983; Jayatilak et al., 1984, 1985; Gibori et al., 1985; Herz et al., 1986). We have investigated the role of decidual luteotropin in

ovarian function by suppressing prolactin secretion in pseudopregnant
rats with or without decidual tissue and by comparing luteal and
follicular steroid-ogenesis in these animal models. The results of
these investigations suggest that decidual luteotropin possesses
several physiological properties of prolactin. Decidual luteotropin
prevents the involution of the corpus luteum that follows prolactin
withdrawal and sustains luteal cell progesterone production (Gibori et
al., 1974; Castracane and Rothchild, 1976; Basuray and Gibori, 1980).
It maintains the luteal cell content of LH receptors and LH responsive
adenylyl cyclase (Gibori et al., 1984) and sustains the capacity of
luteal cells to secrete estradiol when stimulated by LH (Gibori et
al., 1985). Decidual luteotropin appears also to synergize with
estradiol to sustain progesterone biosynthesis (Gibori et al., 1981).
In contrast to its effect on the corpus luteum, decidual luteotropin
prevents the follicle from producing large amounts of estradiol
(Gibori et al., 1985). Because follicles are the principal source of
this hormone, ovarian secretion of estradiol in the ovarian vein is
greatly reduced by the presence of the decidual tissue. This
inhibitory action of decidual luteotropin may be of importance for
fetal survival since high levels of estradiol in the circulation cause
abortion.

The production of a prolactin-like molecule by decidual cells is
not unique to the rat. A few years after we proposed that the rat
decidual tissue secretes a prolactin-like hormone, several laboratories
reported production of prolactin by human decidualized endometrium
(Golander et al., 1978; Riddick et al., 1978). Antisera developed
against human pituitary prolactin recognize human decidual prolactin,
making quantitation of the hormone relatively easy. Decidual prolactin
is synthesized during normal gestation (Golander et al., 1978; Riddick
et al., 1978), during pregnancies with ectopic implantation sites
(Maslar et al., 1980) and in the luteal phase of the menstrual cycle
(Maslar and Riddick, 1979). Decidual prolactin is also synthesized by
endometrial cells which undergo decidualization in vitro (Daly et al.,
1983). Human decidual prolactin is identical to pituitary prolactin
(Golander et al., 1978; Tomita et al., 1982), and there is good
homology between the nucleotide sequence of human decidual prolactin
and pituitary prolactin messenger RNA (Takahashi et al., 1984).
Furthermore, the amino acid sequence of decidual prolactin, determined
from the nucleotide sequence of decidual prolactin cDNA, is identical
to that of pituitary prolactin (Takahashi et al., 1984). In contrast
to humans, the decidual luteotropin of the rat is not immunologically
similar to rat pituitary prolactin (Gibori et al., 1974) but competes
with prolactin for the same receptor site on luteal membranes
(Jayatilak et al., 1985). Characterization of decidual luteotropin
has revealed that this hormone is a protein with a molecular weight of
approximately 23,500, is heat labile, digestible by trypsin, and
contains disulfide bridges but no carbohydrate moieties (Jayatilak et
al., 1985).

Decidual luteotropin is detectable in the decidual tissue as early
as day 6 (i.e., 24 hours after the induction of decidualization). Its
highest tissue concentration is reached on day 9. By day 11 decidual
luteotropin levels begin to decline and become undetectable by day
15. This decline also corresponds to the disappearance of the luteotropic effect of the decidual tissue. After day 11, the decidual
tissue of either pregnant or pseudopregnant rats loses its luteotropic
activity and becomes incapable of sustaining luteal cell progesterone
production (Rothchild and Gibori, 1975; Gibori et al., 1981; Jayatilak
et al., 1984).

If prolactin and decidual luteotropin are both produced between days 6-11 and both affect ovarian function in a similar fashion, what, then, is the physiological role of decidual luteotropin? Furthermore, if the luteotropic activity of the decidua disappears from day 12, how does decidual luteotropin prolong pseudopregnancy? The ability of decidual luteotropin to maintain progesterone synthesis could be regarded as a physiological "fail safe mechanism" which would ensure the continuation of progesterone production if prolactin production is suboptimal. However, some of our data (Gibori et al., 1981) indicate that decidual luteotropin prevents prolactin from acting on the corpus luteum. Corpora lutea of hypophysectomized pseudopregnant rats will respond to prolactin by secreting progesterone only if no decidual tissue is present in the uterus. This inhibitory effect disappears when the decidual tissue is removed, suggesting that although the decidual tissue and the pituitary are both sources of luteotropin in rats with decidualized uteri, only the decidual tissue is responsible for sustaining progesterone synthesis. Our finding that the luteo-tropic activity of the decidual tissue disappears on day 12 of pseudo-pregnancy was puzzling and contradicted the conclusion that this hormone caused prolongation of corpus luteum function until day 21. However, our recent evidence that decidual luteotropin inhibits ovarian estradiol secretion (Gibori et al., 1985) may provide an explanation for such an effect. By reducing estradiol production, decidual luteotropin may prevent the LH surge that occurs normally around day 10 of pseudopregnancy (De Greef and Zeilmaker, 1979) and thus prevent desensitization of the luteal cell by LH and inhibition of progesterone production. The sustained progesterone production may, in turn, maintain prolactin surges (De Greef and Zeilmaker, 1978; Murakami et al., 1980; Gorospe and Freeman, 1981) and result in prolongation of luteal cell function and pseudopregnancy.

The objective of this investigation was to determine whether decidual RNA directs the synthesis of decidual luteotropin and to study the secretion and action of decidual luteotropin in vitro.

TRANSLATION AND HYBRIDIZATION OF DECIDUAL RNA

To determine whether poly(A)+ mRNA isolated from decidual tissue directs the synthesis of decidual luteotropin in cell-free systems, RNA was extracted from 4 g of tissue using standard guanidinium isothiocyanate/cesium chloride methods (Glisin et al., 1974; Ullrich et al., 1977). Poly(A)$^+$ mRNA was isolated by oligo (dT)-cellulose chromatography as described by Maniatis et al. (1982). Aliquots of the poly(A)$^+$ mRNA were translated in a cell-free system derived from rabbit reticulocyte lysates pretreated with micrococcal nuclease (Phelham and Jackson, 1976). The in vitro protein synthesis was performed in the presence of [^{35}S]-methionine, and the labeled peptides that were synthesized were analyzed by sodium dodecyl sulfate-polyacrylamide gel electrophoresis (SDS-PAGE) (Laemmli, 1970). One major cell-free translation product had an apparent molecular weight of approximately 28,000 (Fig. 1, line A and B).

To determine which of the proteins bind to prolactin receptors, translated proteins from tRNA were allowed to bind to rat luteal prolactin receptors in the presence or absence of excess ovine prolactin. The resultant bound proteins were eluted from the receptor with MgCl$_2$, precipitated and analyzed by SDS-PAGE. One major translated protein of approximately 28,000 MW whose binding was markedly inhibited by cold prolactin, was eluted from the receptor (Fig. 1, line B). This is slightly greater than the estimated MW for the native hormone and appears to represent a prohormone for decidual luteotropin.

Since decidual luteotropin appears to be part of the prolactin family, we attempted to select mRNA coding for decidual luteotropin from total RNA using cDNA probes for mouse prolactin, GH, proliferin, and proliferin-related protein kindly provided by Dr. Daniel I. Linzer (Linzer and Talamantes, 1985; Linzer et al., 1985). As shown in Fig. 2, there were no detectable hybridizable bands in the decidual or in the placental RNA samples, whereas single bands were observed in both the mouse pituitary and placental RNA indicating low homologies among prolactin, proliferin, proliferin-related protein, and either decidual or placental prolactin-like hormones.

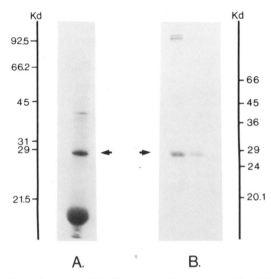

A.   B.

Fig. 1.   Translation, Binding and Elution of Decidual RNA.

Line A: RNA was extracted from day 9 decidual tissue using guanidinium isothiocyanate/cesium chloride, and poly(A)$^+$ mRNA was isolated by oligo (dt)-cellulose chromatography. Aliquots of the poly(A)$^+$ mRNA were translated in the presence of ^{35}S-methionine in a cell-free system derived from a rabbit reticulocyte lysate pretreated with micrococal nuclease.   The synthesized labeled peptides were analyzed by SDS-PAGE.
Line B: Total RNA was translated as described above. Two million cpm of TCA precipitable protein was added to 200 ul of rat ovarian membrane rich in prolactin receptors in a total volume of 1 ml 50 mM Tris, 20 mM $MgCl_2$ and were incubated for 16 hours at 22° C in the presence or absence of 5 ug ovine prolactin. Membranes were pelleted, washed thoroughly and resuspended in 5 M $MgCl_2$ for 1 hour to elute the bound protein.   After centrifugation, the supernatant was diluted, dialyzed, lyophilized and reconstituted.   The eluted proteins were analyzed by SDS-PAGE. The left lane represent total binding and the right lane represent binding in the presence of cold prolactin.

The second goal of this investigation was to establish a direct role for decidual luteotropin in luteal secretion of progesterone. Decidual luteotropin was extracted from a cytosolic extract of day 9 decidual tissue as previously described (Jayatilak et al., 1985), lyophilized and reconstituted. Partially purified decidual luteotropin increases progesterone production by luteal cells in a dose-related manner (Fig. 4). Although these experiments were performed with crude preparations, they represent the first direct evidence that the decidua contains a substance which stimulates luteal cell progesterone production. The results further suggest that this hormone is highly potent in vitro since as little as 14 ng/ml significantly stimulates progesterone synthesis.

To further investigate the role of the decidual tissue on luteal cell function, we cocultured luteal and decidual cells. Decidual and luteal cells were first cultured independently (Figs. 4 and 5). Binucleated cells and clusters of nuclei were often found in the decidual cell culture (Figs. 4 and 5).

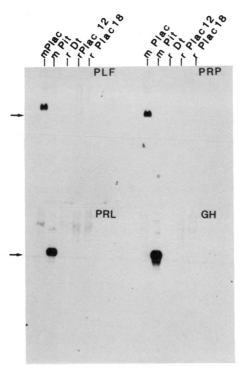

Fig. 2. Hybridization of mouse prolactin, GH, proliferin and proliferin-related protein cDNA to RNA derived from mouse pituitary and placenta and rat decidual and placental tissues. RNA was also isolated from day 12 and day 18 rat placentas, tissues known to produce two different prolactin-like hormones which bind to prolactin receptors. Mouse placenta and pituitary were used as tissue control. Total RNA (10 ug/lane) was denatured, transferred to nitrocellulose filters and hybridized under low stringency conditions with labeled cDNA probes.

Luteal cells have a very different appearance in culture. They are extremely rich in lipid droplets which become stained intensely with the Red O dye (Fig. 6). Since decidual cells in culture do not take up the Red O dye, luteal cells could be easily distinguished from decidual cells in subsequent co-culture experiments.

Progesterone releases into the media by either 50,000, 100,000 or 250,000 decidual cells was negligible. In contrast, luteal cells secreted significant amounts of progesterone (approximately 15 ng/250,000 cells/24 hours). However, as depicted in Fig. 8, luteal cell progesterone secretion did not increase in the presence of decidual cells at any cell concentration used.

To determine whether decidual luteotropin was secreted in culture levels of this hormone were measured in both the decidual cells and the culture media. After dispersion, decidual cells contained very small decidual luteotropin and secreted no hormone into the media. Amounts of these negative results suggested that either the decidual cells did not produce decidual luteotropin in culture because of sub-optimal conditions or that other secretory products accumulated in the static culture and inhibited the production of this hormone. To test the latter possibility, we studied the decidual cells using a perifusion system with continuous media changes. Cells were placed in the chamber over a bedding of Bio Gel. The chambers were fit into a water bath maintained at 37⁰ C. The medium, also maintained at 37⁰ C, was constantly aerated and delivered to the cell at a rate of 1 ml/h. Because decidual luteotropin is temperature sensitive, the spent media was collected at 4⁰ C.

Decidual cells $(2, 10$ or $30 \times 10^6)$ were cultured under either static conditions without change of media or in the perifusion system. The mechanical dispersion of decidual cells caused a dramatic decrease in the cell content of decidual luteotropin. Thus, after dispersion and before culture, decidual cells contained no detectable hormone. Under static conditions, no hormone was released in the media at any

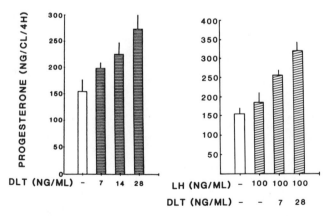

Fig. 3. Effect of decidual luteotropin on luteal progesterone production. Decidual luteotropin was extracted
from day 9 decidual tissue as described previously
(Jayatilak et al., 1985). Corpora lutea obtained from
day 15 pregnant rats were incubated with different doses
of decidual luteotropin in the presence or absence of LH.

time or by any decidual cell concentration. Perifused cells also did not secrete any hormone for the first 4 hours. However, by the 5th hour of culture cells began to secrete decidual luteotropin (Fig. 9). This production increased with time, reached peak values between 7-15 h of perifusion and declined thereafter.

When the net accumulation of hormone was calculated per x 10^6 cells, as presented in Fig. 10 (lower panel), it was found that the best yield of hormone was obtained with 10 x 10^6 cells. Thus, although hormone secretion was dependent on cell increasing the number, cell density in the perifusion chamber reduced the overall capacity of the cells to release decidual luteotropin.

To determine whether decidual explants also release decidual luteotropin in the perifusion system, decidual tissue obtained from day 9 pseudopregnant rats was minced into small pieces. One gram of tissue was transferred into each chamber and cultured for 40 h. The results presented in Fig. 11 indicate that decidual explants secrete decidual luteotropin as effectively as decidual cells cultured under similar conditions. The profile of hormone release by tissue explants was similar to that obtained with decidual cells. However, no delay in hormone secretion was observed and explants began to release decidual

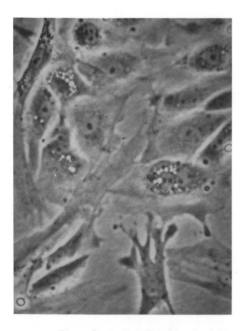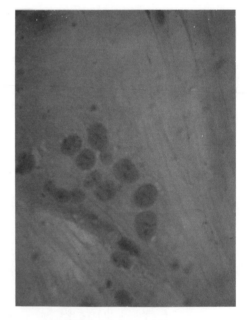

Fig. 4. & 5. Decidual cell culture (X100). Cluster of nuclei in cultured decidual cells (X250). Decidual tissue was obtained from day 9 pseudopregnant rats. Cells were dispersed with collagenase (2,000 ug/g) and DNase (3,000 u/g). Cell viability was checked by trypan blue exclusion. 250,000 living cells were cultured in 1 ml RPMI-1640 containing 5% fetal calf serum supplemented with penicillin (100 u/ml), streptomycin sulfate (100 ug/ml), and fungizone (0.2 ug/ml) (X250).

luteotropin from the first hour of culture. This difference is probably due to the presence of decidual luteotropin in the explants and its absence in the dispersed decidual cells before perifusion. The amount of hormone secreted by the explants in the first two hours of perifusion was equivalent to the initial tissue content of decidual luteotropin. Levels of hormone released in a 40-hour period the initial tissue content exceeded by almost 200-fold . A significant inhibition in hormone production was obtained with cycloheximide treatment. In a 25-hour perifusion, cycloheximide caused a 75% inhibition in hormone release (Fig. 12).

The finding that decidual cells secrete decidual luteotropin only with continuous medium change suggested to us that decidual cells may secrete an inhibitor of decidual luteotropin which accumulates under static culture conditions and prevents decidual luteotropin synthesis and/or release. A number of indirect findings strongly suggest that the inhibitor, if it exists, differs from decidual luteotropin. Dispersed decidual cells contain negligible amounts of decidual luteo-tropin and do not produce any hormone in a static culture. To be capable of inhibiting its own production, decidual luteotropin must first be present. If decidual luteotropin does not inhibit its own secretion, the inhibiting factor may have been produced by non-decidual luteotropin secreting cells. To investigate this possibility, 40-60 x 10^6 decidual cells were layered on a 10-60% discontinuous Percoll gradient. Five bands were obtained. The sixth band contained red blood cells. Band 1 contained primarily cell debris. The largest viable decidual cells were in Band 2 and the smallest cells were in Band 4. A mixed cell population and cells obtained from Bands 2-4 were cultured independently, and their capacity to produce decidual luteotropin was investigated.

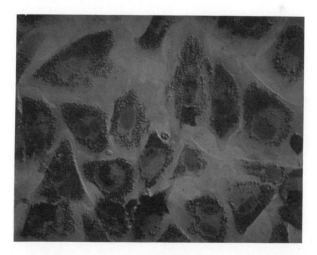

Fig. 6. Corpora lutea obtained from day 9 pseudopregnant rats were dispersed with collagenase (50 u/ml), dispase (2.4 u/ml) and DNase (200 u/ml). Cell viability was checked with either trypan blue or diacylfluorescin which fluoresce only in living cells. Cells were cultured in 1 ml of medium (McCoy's 5a/Ham's F_{12}; 1:1 ratio) with 10% fetal calf serum supplemented with penicillin (100 u/ml), streptomycin sulfate (100 ug/ml), and fungizone (0.2 ug/ml) (X250).

For the first 24 hours, cells were cultured in medium (McCoy's 5a/Ham's F_{12}; 1:1) containing 10% fetal calf serum (FCS) and antibiotics. Thereafter the medium was changed to medium containing no FCS and was removed every 24 hours. As expected, a mixed cell population did not produce decidual luteotropin when cultured under static conditions. However, the large decidual cells obtained from Band 2 secreted significant amounts of decidual luteotropin under the same conditions. Thus, when large decidual cells are isolated from smaller cells, they become capable of secreting decidual luteotropin. These results suggest that an inhibitor is removed with the removal of the smaller decidual cell population and that luteotropin secretion and/or synthesis is locally inhibited by neighboring cells.

The rat decidual tissue is composed of mesometrial and antimesometrial zones. The antimesometrial cells are generally large, closely packed and polyploid. These cells show morphological evidence of protein synthetic activity. Cells of the mesometrial zone are smaller and loosely packed (Krehbiel, 1937; O'Shea et al., 1983). Since decidual luteotropin is secreted by large decidual cells and since only the large decidual cells of the antimesometrium appear to have the ability to secrete proteins (O'Shea et al., 1983), it became of interest to determine whether decidual luteotropin is secreted only by cells of the antimesometrial zone of the decidua. Reduction of intracellular space and tight attachment between cells in the antimesometrial region allows for a visual separation of the darker, more abundant antimesometrial region from the mesometrial area. Thus, these two zones were dissected again and levels of hormones were measured in both tissue preparations. Decidual luteotropin was found almost totally in the antimesometrial tissue (mesometrium, 21 $\pm$ 81 ng/g; antimesometrium 275 $\pm$ 81 ng/g).

Once we demonstrated that the decidual tissue secretes decidual luteotropin and determined the site of its secretion, we became

Fig. 7.   Co-culture of luteal and decidual cells.   Cells were cultured as described for Fig. 6.   Arrow shows luteal cells (X250).

interested in determining whether the decidual tissue is also a site of action for decidual luteotropin. We found that rat decidual tissue possesses high affinity binding sites for prolactin (Jayatilak and Gibori, 1986) and that the locally produced decidual luteotropin binds specifically to these receptors (Fig. 13).

The study of the ontogeny of prolactin receptors in the decidual tissue has revealed that no binding sites can be detected for 48 hours after the induction of decidual tissue formation. In contrast,

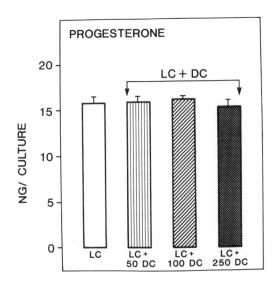

Fig. 8. Luteal cells (250 x 10³) were cultured either alone or in the presence of 50 x 10³, 100 x 10³ or 250 x 10³ decidual cells in 2 ml culture medium prepared as described in Fig. legend 6. LC—luteal cells; DC—decidual cells.

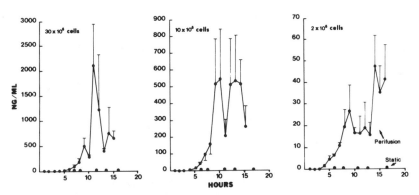

Fig. 9. Hourly secretory patterns of decidual luteotropin by dispersed decidual cells maintained in either a perifusion system or static culture. Different concentrations of decidual cells (2, 10, 30 x 10⁶) were either perifused with medium 199 containing L-glutamine and 25 mM HEPES in chambers over a bedding of 0.5 ml Biogel 2 or were cultured without change of media. (Reprinted in part from Herz et al., 1986.)

389

decidual luteotropin becomes detectable within 24 hours of decidualization (Fig. 14).

After its induction, decidualization occurs in the antimesometrial area during the first 48 hours. There is little decidualization of the mesometrium at this stage (Krehbiel, 1937). Decidual tissue that was obtained 24 and 48 hours after the induction of decidualization

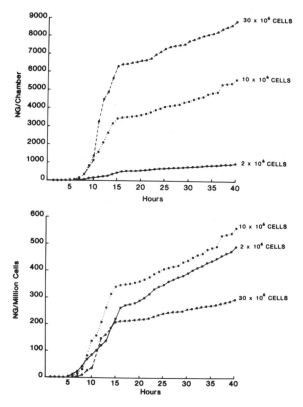

Fig. 10. Hourly accumulation of decidual luteotropin released by 2, 10 or 30 x 10^6 decidual cells during a 40-hour perifusion period. Cumulative levels of decidual luteotropin released per hour (upper panel). The yield of decidual luteotropin produced by different concentrations of cells as calculated for 10^6 cells (lower panel).

Table 1. Density, Size and Decidual Luteotropin Secretion of Decidual Cells Separated on Discontinuous Percoll Gradient

Band	Gradient (%)	Density (g/ml)	Cell Size (microns)	Decidual Luteotropin (ug/10^6 cells/24 h)
1	10-20	1.0273	73.2 + 7.4	--
2	20-30	1.0396	53.1 + 5.65	4 + 0.37
3	30-40	1.0516	29.8 + 4.52	ND
4	40-50	1.0636	27.1 + 2.49	ND
5	50-60	1.0756	20.34 + 2.00	--

secretes decidual luteotropin (Fig. 14) but does not contain receptor site for prolactin/decidual luteotropin. At 72 hours, decidualization of the mesometrial zone occurs and binding sites for prolactin/decidual luteotropin become detectable (Fig. 14; Jayatilak and Gibori, 1986). This temporal relationship suggested that decidual luteotropin production and decidual luteotropin binding sites may be compartmentalized in two different zones of the decidual tissue. To investigate this possibility, mesometrial and antimesometrial tissues were dissected out 96 hours after decidualization was induced. The results of this experiment indicate that this hormone is found almost totally in the antimesometrial tissue, and twice as many receptors were found in the mesometrial as in the antimesometrial area. However, since tissue contamination during dissection cannot be excluded, we isolated the large cells of the antimesometrial tissue from the small cells of the mesometrium by density gradient and then determined which cell type contains binding sites for prolactin/decidual luteotropin.

Binding sites were localized to the mixed cell population before separation and the small cells obtained from Band 4 of the gradient. These results are presented in Table 2 together with the cellular localization of decidual luteotropin secretion. It appears from these results that the site of decidual luteotropin production and the site of receptors for this hormone are compartmentalized in two different cell types. Decidual luteotropin appears to be secreted by the large

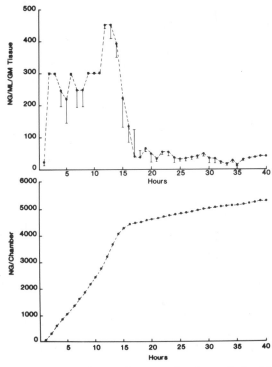

Fig. 11. Hourly secretion and accumulation of decidual luteotropin by decidual explants. Decidual tissue of eleven day 9 pseudopregnant rats were coursely minced into small pieces. One gram of tissue was transferred into each perifusion chamber and perifused with culture medium at a rate of 1 ml/hour. The upper panel represents hourly release; the lower panel depicts the accumulation of hormone at each time indicated.

cells of the antimesometrium and to bind to the small cells of the mesometrial zone. The small mesometrial cells are known to be rich in glycogen which may serve as a nutritive reservoir for the growing embryo. Recently, Freemark and Handwerger (1984) have shown that prolactin—like hormones are capable of increasing glucose uptake and glycogen synthesis. It is therefore possible that decidual luteotropin secreted by the antimesometrium acts on the mesometrial cell to stimulate glycogen deposition.

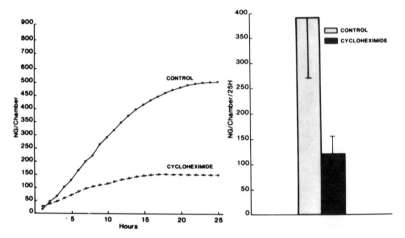

Fig. 12. Effect of cycloheximide (0.05 mM) on the secretion of decidual luteotropin by decidual explants. The right panel represents total levels of hormone obtained in 25 hours of perifusion. (Reprinted from Herz et al., 1986.)

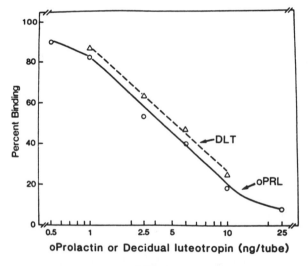

Fig. 13. Displacement of [^{125}I]–ovine prolactin from decidual membranes by decidual extract. A cytosolic fraction of day 9 decidual tissue was obtained and various aliquots of this extract were assayed for their ability to compete with prolactin for binding sites in in the decidual tissue. (Reprinted from Jayatilak and Gibori, 1986.)

392

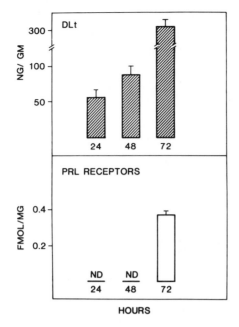

Fig. 14. Comparison between the decidual cell content of decidual luteotropin and prolactin/decidual luteotropin receptors at different times after the induction of decidualization.

Table 2. Cellular Localization of Decidual Luteotropin Secretion and Decidual Luteotropin Binding in the Decidual Tissue.

Band	Prolactin Receptors (fmol/mg protein)	Decidual Luteotropin (ug/10^6 cells/24 h)
Mixed cells	2.2 + 0.2	ND
Small cells	ND	4 + 0.37
Large cells	2.9 + 0.2	ND

SUMMARY

Studies of rat decidual luteotropin production and action have revealed that decidual mRNA directs the synthesis of a 28,000 MW protein in a cell-free system which binds to prolactin receptors in luteal cells and appears to represent a prohormone for decidual luteotropin. Hybridization studies indicate that although rat decidual and prolactin-like placental hormones bind to prolactin receptors, they possess little homology to other members of the prolactin family. In addition, results of this investigation have revealed a possible physiological relationship between the mesometrial and the antimesometrial cells of the decidual tissue. The large antimesometrial cells produce decidual luteotropin in which secretion and/or synthesis is inhibited by the neighboring mesometrial cells. Since mesometrial cells possess binding sites for decidual luteotropin, it is possible that decidual luteotropin acts on the mesometrial cell

393

to affect the formation of its own inhibitor. Mesometrial cells are rich in glycogen, whose synthesis is stimulated by prolactin-like hormones in other tissues. Therefore, decidual luteotropin may also act on these cells to enhance glycogen formation. In summary, decidual luteotropin appears to have at least two sites of action—the luteal cell, where it can substitute for prolactin in maintaining progesterone production, and the mesometrial cell, where its role remains to be investigated.

ACKNOWLEDGEMENTS

The authors gratefully acknowledge Dr. Daniel Linzer for the cDNA probes and for his help in the hybridization studies. This work was supported by grants NIH HD 12356 and BRSG RR 05369-24.

REFERENCES

Alloiteau JJ, 1957. Evolution normale des corps jaunes gestatifs chez la ratte hypophysectomisee an moment de la nidation. C R Soc Biol 151:2009-2011

Armstrong EG Jr, Tobert JA, Talley DJ, Villee CA, 1977. Changes in progesterone receptor levels during deciduomata development in the pseudopregnant rat. Endocrinology 101:1545-1551

Basuray R, Gibori G, 1980. Luteotropic action of the decidual tissue in the pregnant rat. Biol Reprod 23:507-512

Basuray R, Jaffe RC, Gibori G, 1983. Role of decidual luteotropin and prolactin in the control of luteal cell receptors for estradiol. Biol Reprod 28:551-556

Butterstein GM, Hirst JA, 1977. Serum progesterone and fetal morphology following ovariectomy and adrenalectomy in the pregnant rat. Biol Reprod 16:654-660

Castracane VD, Rothchild I, 1976. Luteotropic action of decidual tissue in the rat: comparison of jugular and uterine vein progesterone level and the effect of ligation of the utero-ovarian connections. Biol Reprod 15:497-503

Castracane VD, Shaikh AA, 1976. Effect of decidual tissue on the uterine production of prostoglandins in pseudopregnant rats. J Reprod Fert 46:101-104

Daley DC, Maslar IA, Riddick DH, 1983. Prolactin production during in vitro decidualization of proliferative endometrium. Am J Obstet Gynecol 145:672-678

Deanesley R, 1973. Termination of early pregnancy in rats after ovariectomy is due to immediate collapse of the progesterone-dependent decidua. J Reprod Fertil 35:183-186

De Feo VJ, 1967. Decidualization. In: Wynn RM (Ed.), Cellular biology of the uterus. New York: Appleton-Century Crofts Press, pp. 191-290

De Greef WJ, Dullaart J, Zeilmaker GH, 1977. Serum concentration of progesterone, luteinizing hormone, follicle stimulating hormone and prolactin in pseudopregnant rats: effect of decidualization. Endocrinology 101:1054-1063

De Greef WJ, Zeilmaker GH, 1978. Regulation of prolactin secretion during the luteal phase of the rat. Endocrinology 102:1190-1198

De Greef WJ, Zeilmaker GH, 1979. Serum prolactin concentrations during hormonally induced pseudopregnancy in the rat. Endocrinology 105:195-199

Finn CA, Porter DG, 1975. The decidual cell reaction. In: Finn CA, Porter DG (eds.), The Uterus. Massachusetts: Publishing Sciences Group, Inc, pp 74-85

Freeman ME, 1979. A direct effect of the uterus on the surges of prolactin induced by cervical stimulation in the rat. Endocrinology 105:387-390

Freemark M, Handwerger S, 1984. Ovine placental lactogen stimulates glycogen synthesis in fetal rat hepatocytes. Am J Physiol 246:E21-E24

Gibori G, Rothchild I, Pepe GJ, Morishige WK, Lam P, 1974. Luteotrophic action of the decidual tissue in the rat. Endocrinology 95:1113-1118

Gibori G, Basuray R, McReynolds B, 1981. Luteotropic role of the decidual tissue in the rat: dependency on intraluteal estradiol. Endocrino logy 108:2060-2066

Gibori G, Kalison B, Basuray R, Rao MC, Hunzicker-Dunn M, 1984. Endocrine role of the decidual tissue: decidual luteotropin regulation of luteal adrenyl cyclase activity, luteinizing hormone receptors, and steroidogenesis. Endocrinology 115:1157-1163

Gibori G, Kalison B, Warshaw ML, Basuray R, Glaser LA, 1985. Differential action of decidual luteotropin on luteal and follicular production of testosterone and estradiol. Endocrinology 116:1784-1791

Glasser SR, McCormack SA, 1979. Functional development of rat trophoblast and decidual cells during establishment of the hemochorial placenta. Adv Biosci 25:165-197

Glisin V, Crkvenjakov R, Byus C, 1974. Ribonucleic acid isolated by cesium chloride centrifugation. Biochemistry 13:2633-2637

Golander A, Hurley T, Barrett J, Hizi A, Handwerger S, 1978. Prolactin synthesis by human chorion-decidual tissue: a possible source of amniotic fluid prolactin. Science 202:311-312

Gorospe WC, Freeman ME, 1981. An ovarian role in prolonging and terminating the two surges of prolactin in pseudopregnant rats. Endocrinology 108:1293-1298

Gorospe WC, Mick C, Freeman, ME, 1981. The obligatory role of the uterus for termination of prolactin surges initiated by cervical stimulation. Endocrinology 109:1-4

Gorospe WC, Freeman ME, 1982. The imprint provided by cervical stimulation for the initiation and maintenance of daily prolactin surges: Modulation by the uterus and ovaries. Endocrinology 110:1866-1870

Herz Z, Khan I, Jayatilak PG, Gibori G, 1985. Evidence for the synthesis and secretion of decidual luteotropin; a prolactin-like hormone produced by rat decidual cells. Endocrinology 118:2203-2209

Jayatilak PG, Glaser LA, Warshaw ML, Herz Z, Grueber JR, Gibori G, 1984. Relationship between luteinizing hormone and decidual luteotropin in the maintenance of luteal steroidogenesis. Biol Reprod 31:556-564

Jayatilak PG, Glaser LA, Basuray R, Kelly PA, Gibori G, 1985. Identification and partial characterization of a prolactin-like hormone produced by the rat decidual tissue. Proc Natl Acad Sci 82:217-221

Jayatilak PG, Gibori G, 1986. Ontogeny of prolactin receptors in rat decidual tissue: binding by locally produced prolactin-like hormone. J Endocrinol 110:115-121

Kelly PA, Shiu RPC, Robertson MC, Friesen HG, 1975. Characterization of rat chorionic mammotropin. Endocrinology 96:1187-1195

Krehbiel RH, 1937. Cytological studies of the decidual reaction in the rat during early pregnancy and in the production of deciduomata. Physiol Zool 10:212-234

Laemmli UK, 1970. Cleavage of structural proteins during the assembly of the head of bacteriophage T4. Nature 227:680-685.

Linkie DM, Niswender GD, 1973. Characterization of rat placental luteotropin. Physiological and physicochemical properties. Biol Reprod 8:48–57

Linzer DI, Talamantes F, 1985. Nucleotide sequence of mouse prolactin and growth hormone mRNAs and expression of these mRNAs during pregnancy. J Biol Chem 260:9574–9579

Linzer DI, Lee SJ, Ogren L, Talamantes F, Nathans D, 1985. Identification of proliferin mRNA and protein in mouse placenta. Proc Natl Acad Sci USA 82:4356–4359

Lobel BL, Tic L, Shelesnyak MC, 1965. Studies on the mechanism of nidation. XVII. Histochemical analysis of decidualization in the rat. Part 1: Framework: Oestrous cycle and pseudopregnancy. Acta Endocrinol Copenh 50:452–468

Maniatis T, Risch EF, Sambrook J, 1982. In: Molecular cloning. A laboratory manual. New York: Cold Spring Harbor Laboratory

Martel D, Monier MN, Psychoyos A, De Feo VJ, 1984. Estrogen and progesterone receptors in the endometrium, myometrium and metrial gland of the rat during the decidualization process. Endocrinology 114:1627– 1634

Maslar IA, Riddick DH, 1979. Prolactin production by human endometrium during the normal menstrual cycle. Am J Obstet Gynecol 135:751–754

Melampy RM, Anderson LL, Kragt CL, 1964. Uterus and life span of rat corpora lutea. Endocrinology 75:501–506

Morishige WK, Rothchild I, 1974. Temporal aspects of the regulation of corpus luteum function by luteinizing hormone, prolactin and placental luteotrophin during the first half of pregnancy in the rat. Endocrin-ology 95:260–274

Murakami N, Takahashi M, Suzuki Y, 1980. Induction of pseudopregnancy and prolactin surges by a single injection of progesterone. Biol Reprod 22:253–258

O'Shea JD, Kleinfeld RG, Morrow HA, 1983. Ultrastructure of decidualization in the pseudopregnant rat. Am J Anat 166:271–298

Pharriss BB, Tillson SA, Erikson RR, 1972. Prostaglandins in luteal function. Recent Progr Horm Res 28:51–73

Phelham HRB, Jackson RJ, 1976. An efficient mRNA-dependent translation system from reticulocyte lysates. Eur J Biochem 67:247–256

Psychoyos A, 1973. Hormonal control of ovo-implantation. Vitam Horm 32:201–255

Riddick DH, Luciano AA, Kusmik WF, Maslar IA, 1978. De novo synthesis of prolactin by human decidua. Life Sci 23:1913–1929

Robertson MC, Gillespie B, Friesen HG, 1982. Characterization of the two forms of rat placental lactogen (rPL): rPL-I and rPL-II. Endocrinology 111:1862–1866.

Rothchild I, Gibori G, 1975. The luteotrophic effect of decidual tissue: the stimulating effect of decidualization on serum progesterone level of pseudopregnant rats. Endocrinology 97:838–842

Takahashi H, Nabeshima Y, Nabeshima YI, Ogata K, Takeuchi S, 1984. Molecular cloning and nucleotide sequence of DNA complementary to human decidual prolactin mRNA. J Biochem 95:1491–1499

Talley DJ, Tobert JA, Armstrong EG Jr, Villee CA, 1977. Changes in estrogen receptor levels during deciduimata development in the pseudopregnant rat. Endocrinology 101:1538–1544

Ullrich A, Shine J, Chirgwim J, Pictet R, Tischer E, Rutter WJ, Goodman HM, 1977. Rat insulin genes: construction of plasmids containing the coding sequences. Science 196:1313–1315

Weems CW, Pexton JE, Butcher RL, Inskeep EK, 1975. Prostaglandin F in uterine tissue and venous plasma of pseudopregnant rats: effects of deciduomata. Biol Reprod 13:282–289

Weems CW, 1979. Prostaglandins in uterine and ovarian compartment and in plasma from the uterine vein, ovarian artery and vein, and abdominal aorta of pseudopregnant rats with or without deciduomata. Prostaglandins 17:873-890

Wilson L, Butcher RL, Inskeep EK, 1970. Studies on the relation of decidual cell response to luteal maintenance in the pseudopregnant rat. Biol Reprod 3:342-346

Yoshinaga K, Adams CE, 1967. Luteotropic activity in the young conceptuses in the rat. J Reprod Fertil 13:505-509

Yoshinaga K, MacDonald GJ, Greep RO, 1973. Influence of various doses of LH on fetal survival in hypophysectomized rats. Proc Soc Exp Biol Med 140:893-895

ROLE OF PLACENTAL LACTOGEN AND PROLACTIN IN HUMAN PREGNANCY

Stuart Handwerger and Michael Freemark

Departments of Pediatrics and Physiology
Duke University Medical Center, Durham, NC

INTRODUCTION

Placental lactogen and prolactin along with growth hormone form a family of protein hormones with striking homologies in chemical and biological properties (Josimovich and McLaren,1962; Sherwood et al., 1971). The three hormones have molecular weights in the range of 22-24 kD and exhibit both lactogenic and somatotrophic properties (Table 1). Human placental lactogen (hPL; human chorionic somatomammotrophin, hCS) is synthesized by the syncytiotrophoblast cells of the placenta, the same cells which synthesize and secrete human chorionic gonadotropin. Human prolactin during pregnancy is synthesized and secreted by both the anterior pituitary gland and the decidua.

Both placental lactogen and prolactin appear to have numerous biologic actions during pregnancy. Investigations of the physiology of placental lactogen performed during the past few years in our and other laboratories strongly suggest a role for placental lactogen in the regulation of fetal somatomedin production and carbohydrate and amino acid metabolism. Recent studies of the physiology of prolactin suggest a role for prolactin in fetal pulmonary surfactant production and suppression of the immune response. In this review, we will focus our discussion primarily on the effects of placental lactogen on maternal and fetal metabolism and on fetal growth and development. In addition, we will discuss some aspects of the physiology of decidual prolactin and the effects of prolactin on fetal metabolism and development. Although placental lactogen and prolactin are potent lactogenic hormones, we will not discuss the mammotrophic and lactogenic effects of these hormones in this review.

Table 1. Some Physical Properties of HPL, HPRL, and HGH

Property	HPL	HPRL	HGH
Molecular weight	21,600	22,500	21,500
Isoelectric point	4.8	4.6	4.6
Number of amino acid residues	191	198	191
Disulfide bonds	2	3	2
Number of identical residues (%)	vs HGH: 85 vs HPRL: 13		

Synthesis and Secretion During Pregnancy

HPL is first detected in maternal plasma at about six weeks of pregnancy (Grumbach and Kaplan, 1964). The concentration then increases linearly until about the thirtieth week of gestation when peak concentrations of 5000 to 7000 ng/ml are attained (Grumbach and Kaplan, 1964). Near term, the secretion rate of hPL is about 1.0 gm/day, a rate considerably greater than that of any other polypeptide hormone (Grumbach et al., 1968). Throughout pregnancy, the plasma concentration of hPL in the mother is positively correlated with placental mass and is greater in multiple than in singleton gestations (Tyson, 1972; Saxena et al., 1969). The plasma concentrations of hPL in aborted fetuses at 12-20 weeks of gestation are in the range of 50-280 ng/ml, and the concentration of hPL in umbilical cord (fetal) plasma immediately after delivery is about 80-125 ng/ml (Kaplan and Grumbach, 1965). A summary of the concentrations of hPL, hPRL, and hGH in maternal and fetal plasma and in amniotic fluid during pregnancy is shown in Table 2. Since radiolabeled hPL does not cross the placenta from the maternal to the fetal circulation, the hPL in fetal blood apparently derives from the direct secretion of the hormone into the fetal circulation (Saaman et al., 1966).

The Role of Placental Lactogen in the Pregnant Mother

HPL has effects on carbohydrate and protein metabolism which are qualitatively similar to those of hGH. Like hGH, hPL enhances insulin secretion, impairs glucose tolerance, and promotes nitrogen retention in children with hypopituitarism and in nonpregnant and pregnant women (Grumbach et al., 1966; Grumbach et al., 1968; McGarry et al., 1972) and stimulates insulin synthesis and secretion in isolated pancreatic islets (Martin and Friesen, 1969). In addition, some studies have demonstrated that hPL, like hGH, stimulates lipolysis in rat and rabbit adipose tissues in vitro and in human subjects in vivo (Felber et al., 1972). However, several studies have failed to detect lipolytic actions of hPL in humans (Beck and Daughaday, 1967; Berle et al., 1974).

Prolactin also appears to impair glucose tolerance and induce a hyperinsulinemic state. Studies of hyperprolactinemic females have demonstrated that these women often have an exaggerated insulin response to oral glucose which is frequently reversed following treatment with bromocriptine. In addition, the administration of human prolactin to normal females has been reported to stimulate an increase in plasma free fatty acids and glycerol (Berle et al., 1974). The demonstration that both placental lactogen and prolactin appear to be diabetogenic and induce a hyperinsulinemic state similar to that observed during pregnancy strongly suggests a role for these hormones in the regulation of carbohydrate metabolism during pregnancy.

Table 2. The concentrations of hPL, hPRL, and hGH in maternal and fetal plasma and in amniotic fluid during pregnancy

| | Maternal Plasma | | Fetal Plasma | Amniotic Fluid |
	1st Trimester	Term	Term	Term
hPL (ng/ml)	300-500	6800 $\pm$ 2100	80-125	55-70
hPRL (ng/ml)	30	130-200	150-250	15
hGH (ng/ml)	4.0-6.0	4.0-6.0	33.5 $\pm$ 4.6	1200

HPL also has somatotrophic effects in postnatal tissues which are qualitatively similar to those of hGH. When administered to hypophysectomized postnatal rats, hPL stimulates epiphyseal growth (Kaplan and Grumbach, 1964), weight gain (Arezzuni et al., 1972), sulfate incorporation into chondroitin sulfate (Kaplan and Grumbach, 1964), thymidine incorporation into cartilage DNA (Marakawa and Raben, 1968; Breuer, 1969), and hepatic ornithine decarboxylase (ODC) activity (Butler et al., 1978) with a potency 1-15% that of growth hormone. In addition, small doses of hPL also potentiate the actions of growth hormone in stimulating epiphyseal growth and thymidine incorporation into cartilage DNA from immature rats (Breuer, 1969).

Since plasma hPL concentrations increase markedly during pregnancy, while plasma hGH concentrations do not change, Grumbach and co-workers presented a cogent argument for a major role of hPL as a maternal "growth hormone" of the second half of pregnancy (Grumbach et al., 1968). They suggested that hPL acts as an antagonist of insulin action, inducing glucose intolerance, lipolysis and proteolysis in the mother and thereby promoting the transfer of glucose and amino acids to the fetus. Though attractive, this hypothesis remains essentially unproven.

A role for hPL in the generation of maternal somatomedin during pregnancy is suggested by the significant correlation between maternal plasma hPL and maternal insulin-like growth factor I (IGF-I) levels (Furlanetto et al., 1978) and by the demonstration that plasma IGF-I levels in GH-deficient and prolactin-deficient hypopituitary women increase dramatically to normal levels during the course of pregnancy and then fall to low baseline levels immediately after delivery (Hall et al., 1984).

To clarify the roles of placental lactogen in the regulation of growth and metabolism during pregnancy, our laboratories nearly 10 years ago initiated studies using the pregnant ewe and fetus as model systems to study the physiologic actions of placental lactogen and the regulation of placental lactogen secretion. We succeeded in purifying ovine placental lactogen (oPL) to homogeneity (Handwerger et al., 1974), characterizing many of its physicochemical properties (Hurley et al., 1977a), developing a specific homologous radioimmunoassay for its detection (Handwerger et al., 1977), measuring its concentration in plasma and other biological fluids (Handwerger et al., 1977) and studying the effects of various factors on oPL secretion in the mother and fetus. These studies, subsequently confirmed by other laboratories (Chan et al., 1976; Martal and Dijane, 1977; Martal, 1978), indicate that oPL has striking homologies in its chemical, biological and immunological properties with ovine growth hormone (oGH).

The intravenous administration of oPL to non-pregnant and pregnant sheep as an intravenous bolus stimulated insulin secretion and decreased plasma glucose and alpha-amino nitrogen concentrations (Handwerger et al., 1976). Plasma free fatty acid concentrations initially decreased by 40-50% and then gradually increased to concentrations 50-100% above preinjection values. While hPL has weak somatotropic effects in postnatal subprimate tissues, oPL bound to growth hormone receptors and stimulated epiphyseal growth (Chan et al., 1976), weight gain (Chan et al., 1976), liver ornithine decarboxylase activity (Butler et al., 1978), somatomedin-C production (Hurley et al., 1977) and amino acid transport (Freemark and Handwerger, 1982) in postnatal rats with a potency equal to (and occasionally exceeding that of) growth hormone. The demonstration that oPL has metabolic and somatotropic effects in postnatal sheep tissues which are similar to those of oGH are therefore consistent with a "growth hormone-like" role for oPL during pregnancy. The potent soma-totropic effects of oPL in postnatal subprimate tissues may be explained by the high affinity of oPL for binding to subprimate GH receptors, while the relatively weak somatotropic effects of hPL in subprimate tissues is due to the low or absent specific binding of hPL to these tissues. A summary of the proposed functions of placental lactogen in the pregnant women is shown in Table 3.

In a recent study, the infusion of oPL antiserum to three pregnant ewes in late pregnancy (131 days) over a 12 hour period has been reported to have no effects on glucose and free acid metabolism and serum somatomedin concentrations (Waters et al., 1985). The significance of these latter experiments, however, is unclear since oPL antiserum was infused for only a relatively short time at a single point during pregnancy and the half-life of somatomedin in the pregnant sheep may be greater that 12 hours.

Role of Placental Lactogen in the Fetus

The demonstration that placental lactogen concentrations in maternal serum greatly exceed the concentrations of placental lactogen in umbili-cal cord blood prompted several investigators to conclude that the effects of placental lactogen on fetal metabolism are mediated indirectly by changes in maternal metabolism and not by direct effects of placental lactogen on the fetus. However, more recent studies have demonstrated that plasma concentrations of placental lactogen in the ovine fetus (Gluckman et al., 1979) and the human fetus (Kaplan et al., 1972; Freemark and Handwerger, unpublished observations) at mid-gestation exceed umbilical cord plasma concentrations at term. In addi-tion, as discussed below, studies from our and other laboratories indicate that placental lactogen has direct effects on fetal metabolism. Together these studies suggest that placental lactogen may act directly on fetal tissues to regulate fetal metabolism and growth.

In our initial studies of the direct effects of placental lactogen on fetal tissues, we examined the effects of oPL on ornithine decar-boxylase (ODC) activity in fetal rat liver in vivo (Hurley et al.,1980). In these experiments, oPL stimulated a dose-dependent increase in ODC activity, and the potency of oPL in stimulating fetal hepatic ODC acti-vity was comparable to that observed in postnatal liver. Since ODC is the rate-limiting enzyme in the synthesis of the polyamines, a group of compounds which play critical roles in the regulation of protein and nucleic acid metabolism, these studies further suggest that placental lactogen might contribute to the regulation of fetal metabolism and growth. While growth hormone and prolactin stimulated ODC activity in postnatal liver with potencies nearly identical to oPL, neither ovine and rat growth hormones nor ovine and rat prolactins stimulated fetal ODC activity, even at doses many fold greater than the half-maximal effective dose of oPL.

Table 3. Proposed functions of PL in the pregnant mother

(1) Antagonist of insulin action, promoting transfer of glucose and amino acids to the fetus
(2) Preparation of the breast for lactation
(3) Stimulation of maternal somatomedin production

Placental lactogen also has direct anabolic effects on amino acid meta-bolism in rat skeletal muscle (Freemark and Handwerger, 1983). In studies using hemidiaphragms from postnatal hypophysectomized female rats, oPL stimulated the intracellular uptake of the non-metabolizable amino acid amino-isobutyric acid (AIB) with a potency identical to those of ovine and rat growth hormones (Freemark and Handwerger, 1982)(Figure 1). In studies using hemidiaphragms from 20 day old rat fetuses, OPL also stimulated a dose-dependent increase in AIB transport, but neither ovine nor rat growth hormone affected AIB uptake in the fetal hemi-diaphragms – even at concentrations 50 times greater than the minimal effective dose for stimulation in adult tissues (Freemark and Handwerger, 1983)(Figure 2). These studies, like our earlier studies of the effects of oPL and oGH on ornithine decarboxylase activity in fetal (Hurley et al., 1980) and postnatal liver (Butler et al., 1978), suggest that oPL acts as a fetal "growth hormone" and that oGH has little or no biologic activity in fetal tissues.

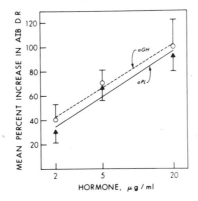

Figure 1.. Comparison of the effects of oPL and oGH on AIB uptake by hemidiaphragms from intact female rats. Values represent the mean ± SEM of the percent stimulation of AIB uptake above paired controls. From Freemark and Handwerger, 1982.

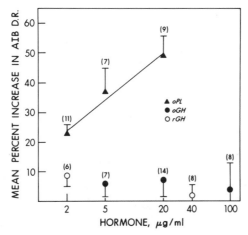

Figure 2. The effects of oPL, oGH, and rGH on AIB transport in fetal rat hemidiaphragms. The number of paired hemidiaphragms in each group is indicated by the numbers in parentheses. Values represent the mean ± SEM of the percent stimulation of AIB uptake above paired controls. From Freemark and Handwerger, 1983.

403

A direct role for placental lactogen in the regulation of fetal carbohydrate metabolism was suggested by our studies of the effects of oPL on glycogen metabolism in cultured fetal hepatocytes (Freemark and Handwerger, 1984a and b; Freemark and Handwerger, 1985)(Figure 3). OPL, at concentrations within the physiologic range (60-250 ng/ml), stimulated dose-dependent increases in ^{14}C-glucose incorporation into glycogen and total cellular glycogen content of hepatocytes from fetal sheep and rats. In addition to stimulating glycogen synthesis, oPL antagonized the glycogenolytic effect of glucagon in cultured fetal rat hepatocytes. OPL alone had no effect on basal phosphorylase (Pa) activity but caused a dose-dependent inhibition of both glucagon and cAMP-stimulated Pa activity, suggesting that the inhibitory effect of oPL on glucagon action is exerted, at least in part, at a site distal to the intracellular accumulation of cAMP. These studies suggest that placental lactogen may contribute to the storage of glycogen in fetal liver by both stimulating glycogen synthesis and inhibiting glycogen degradation. The effect of oPL on glycogen storage has important physiological implications since a marked increase in hepatic glycogen stores occurs in all mammals during late gestation and hepatic glycogenolysis is the major source of circulating glucose for the newborn brain, renal cortex and hematopoietic cells in the early hours after birth.

Mammalian growth hormones and prolactins from several species, including the sheep, were found to have effects on glycogen metabolism in fetal liver which were similar to oPL, but the potencies of ovine growth hormone and ovine prolactin were only one-tenth and one-twentieth that of oPL, respectively (Freemark and Handwerger, 1986). The maximal glycogenic effects of growth hormone and oPL were identical, and the effects of the two hormones at maximal concentrations were not additive, suggesting that the two hormones stimulate glycogen synthesis by similar mechanisms. In contrast, the maximal glycogenic effect of oPL was 2-4 times that of insulin, and oPL and insulin had additive effects at maximal concentrations and synergistic effects at low concentrations. These

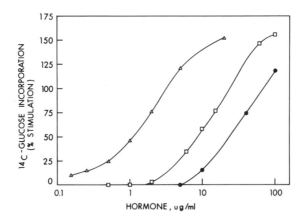

Figure 3. The effects of oPL (△——△), oGH (▢——▢), and oPRL (●——●) on glycogen synthesis in ovine fetal hepatocytes. Hepatocytes were incubated for 4 hours in medium containing ^{14}C-glucose (0.5 uCi/ml) as well as hormones or diluent. Each point represents the mean of three determinations using hepatocytes from three fetal sheep. In all cases the SEM was less than 10% of the mean value. Hepatocytes exposed to diluent alone incorporated 1.55 $\pm$ 0.10 nmol ^{14}C-glucose/mg protein. The glycogen content of diluent-exposed hepatocytes was 89 $\pm$ 7.2 ug glycogen/ mg protein. OPL at concentrations of 150 ng/ml to 20 ug/ml also stimulated a dose-dependent increase in total cellular glycogen content (10-69%, p<0.01). From Freemark and Handwerger, 1986.

findings suggest that oPL and insulin stimulate glycogen synthesis in fetal liver through different mechanisms and that oPL and insulin may act in concert to promote the storage of glycogen in fetal liver.

Although direct growth promoting effects of placental lactogen on the fetus have not been demonstrated in vivo, studies from this and other laboratories also suggest a direct role for placental lactogen in the regulation of fetal growth. As indicated above, oPL stimulates weight gain, tibial epiphyseal growth and somatomedin-C (insulin-like growth factor I) production in postnatal hypophysectomized rats with a potency equal to that of growth hormone. In addition, both oPL and oGH stimulate the production of IGF-I in cultured fibroblasts of postnatal rats.

Studies in the rat and sheep indicate that fetal plasma concentrations of insulin-like growth factor II (IGF-II) greatly exceed fetal plasma concentrations of IGF-I, suggesting a major role for IGF-II in the regulation of growth in the fetal rat and lamb (Gluckman and Butler, 1983). In studies performed in collaboration with Drs. Sally Adams, Peter Nissley, and Matthew Rechler at the National Institutes of Health, we examined the effects of oPL on somatomedin-C/IGF-I and MSA/IGF-II production in fetal rat tissues in vitro (Adams et al., 1983). We observed that cultured rat embryo fibroblasts synthesize large amounts of IGF-II and barely detectable amounts of IGF-I and that fibroblasts cultures from postnatal rats 2-50 days of age synthesize primarily IGF-I, thus mimicking the developmental switch from IGF-II to IGF-I production that is observed in serum. OPL at physiologic concentrations stimulated a significant dose-dependent increase in IGF-II synthesis in fetal fibroblasts, while growth hormone at even much higher concentrations had no effect (Figure 4). In addition, a variety of other polypeptide and steroid hormones and growth factors, including prolactin, insulin, estrogen, testosterone, and epidermal growth factor, had no effect on the production of IGF-II by fetal fibroblasts. By contrast, both oPL and growth hormone stimulated IGF-I synthesis in fibroblasts from older rats but had no effects on IGF-II synthesis (Figure 5). The effects of oPL and growth hormone were completely blocked by cycloheximide. Together these studies provide a mechanism by which placental lactogen may affect fetal growth and suggest possible ontogenic differences in the regulation of somatomedin production and the control of growth. In the fetus, growth may be regulated primarily by IGF-II which is under the control of placental lactogen, while in postnatal life growth appears to be regulated primarily by IGF-I which is under the control of growth hormone. During pregnancy, maternal IGF-I production may be controlled in part by placental lactogen.

Recent studies by Hill and co-workers strongly support a direct role for hPL in the regulation of human fetal metabolism and growth. Hill has demonstrated that hPL stimulates amino acid transport (Hill et al., 1986), DNA synthesis (Hill et al., 1985) and somatomedin (IGF-I) production in human fetal fibroblasts and myoblasts (Hill et al., 1986), but hGH and prolactin are devoid of these activities. These biologic effects of hPL are observed at concentrations (50-250 ng/mL) within the physiologic range for the human fetus at mid-gestation. Since the biological actions of oPL and hPL in the fetus are similar to the actions of growth hormone in postnatal animals, these studies implicate a role for placental lactogen as a "fetal growth hormone". The lack of somatotropic and metabolic activity of growth hormone in the fetus is in accordance with other clinical and experimental observations suggesting that growth hormone does not play a central role in the regulation of fetal growth (Table 4).

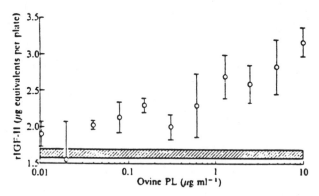

Figure 4. Stimulation of IGF-II production in rat embryo fibroblasts by ovine PL. Third passage fibroblasts derived from 16-day rat embryos were preincubated in serum-free medium for 24 hours and then fresh media containing different concentrations of PL were added. The vertical lines through the data points indicated + 1 s.d. The hatched bar shows the IGF-II level in control cultures (+ 1 s.d.). Significant stimulation (Student's t-test) was achieved at PL concentrations (ug ml^{-1}) of 0.04 (p<0.025), 0.080 (p<0.025), 0.16 (p<0.005), 0.32 (p<0.05), 0.62 (p<0.05), 1.25 (p<0.005), 2.5 (p<0.005), 5.0 (p<0.001, and 10 (p<0.001). From Adams et al., 1983.

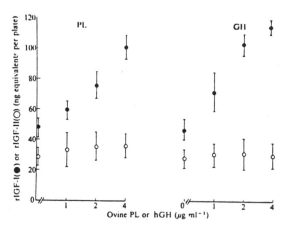

Figure 5. Stimulation of IGF-I production in adult rat fibroblasts cultured with ovine PL or hGH. Adult (25 days) rat lung fibroblasts in serum-free medium were incubated for 48 hours with hGH. The vertical line indicates + 1 s.d. From Adams et al., 1983.

Table 4. Evidence that GH does not play a major role in fetal growth.

(1) Deficiency of GH receptors in many fetal tissues
(2) Little or no biologic activity of GH in many fetal tissues
(3) An excess of fetal GH, as noted in transgenic mice bearing metallothionein-hGH fusion genes, does not accelerate fetal growth
(4) Deficiency or absence of GH in the mammalian fetus does not limit fetal weight gain or linear growth in utero
(5) Fetal GH deficiency is not associated with decreased somatomedin production in the fetus

The absence of growth hormone action in the fetus is not due to low growth hormone concentrations in the fetus. In fact, fetal plasma concentrations of growth hormone (25-200 ng/ml) greatly exceed plasma growth hormone concentrations in postnatal animals (0-10 ng/ml). Studies from a number of laboratories, however, have demonstrated that the binding of radiolabeled growth hormone to tissues of postnatal rats, mice, sheep and humans greatly exceeds the binding of growth hormone to fetal tissues, suggesting that the lack of somatotropic and metabolic activity of GH in the fetus results from a deficiency of specific fetal growth hormone receptors. The marked increase in growth hormone binding to mammalian tissues after birth provides a mechanism whereby growth hormone may acquire somatotropic activity during the postnatal period.

Placental Lactogen and Growth Hormone Receptors

Studies by several laboratories have demonstrated that the mammalian placental lactogens compete with radiolabeled growth hormone and prolactin for binding to growth hormone (somatotropic) and prolactin (lactogenic) receptors in postnatal mammalian tissues (Lesniak et al., 1977; Emane et al., 1977). As a result of these findings, it has been suggested that the metabolic effects of placental lactogen may be mediated through binding to growth hormone and/or prolactin receptors. However, our studies comparing the binding of radiolabeled placental lactogen, growth hormone and prolactin to hepatic membranes of fetal lambs at mid-and late gestation and to postnatal sheep from 1 day to 7 months of age strongly suggest the presence of a placental lactogen binding site in fetal tissues which is distinct from growth hormone and prolactin binding sites in postnatal tissues (Freemark and Handwerger, 1986; Freemark et al., 1986) (Figure 6). Specific ^{125}I-oPL binding sites in fetal liver were detected as early as mid-gestation, and the number of oPL binding sites in fetal liver increased markedly during the latter half of gestation, reaching a peak (11-12 fmoles/mg protein) at 3-7 days prior to parturition. In contrast, there was little or no specific binding of ^{125}I-oGH or ^{125}I-oPRL (less than 1.2 fmoles/mg protein) to ovine fetal liver. In addition, the potency of oPL (Kd 0.1-0.5 nM, 2.2-11 ng/ml) in competing for ^{125}I-oPL binding sites in fetal liver was 20-90 times that of oGH and 1000-2000 times that of oPRL. Thus the relative order of the potencies of the three hormones in competing for binding to ^{125}I-oPL binding sites in fetal liver was similar to the relative order of potencies of the three hormones in stimulating glycogen synthesis in cultured hepatocytes from fetal lambs. These findings suggest that the biologic effects of placental lactogen (and perhaps growth hormone and prolactin) in ovine fetal liver may be mediated through binding to fetal oPL receptors.

During the first week after birth in the sheep, there is a marked increase in the specific binding of both ^{125}I-oGH and ^{125}I-oPL, but the number of ^{125}I-oPL binding sites exceeds the number of ^{125}I-oGH binding sites at all developmental stages (Figure 6). Competitive binding studies indicated that the affinity of oGH (Kd .25-.3 nM) for ^{125}I-oGH binding sites in postnatal liver is identical to that of oPL. On the other hand, the affinity of oGH (Kd 1.5 nM) for ^{125}I-oPL binding sites in postnatal liver is only one-sixth that of oPL (Kd .25-.3 nM). The affinity of ^{125}I-oGH and ^{125}I-oPL binding sites for prolactin is very low (Kd 350 nM). These findings suggest the presence of two separate and distinct receptors in postnatal sheep liver - the placental lactogen receptor, which has high affinity for oPL and low affinity for oGH and oPRL, and the GH receptor, which has high affinity for oGH and oPL and low affinity for oPRL. Specific placental lactogen binding sites predominate in fetal liver, while growth hormone binding sites appear soon after birth.

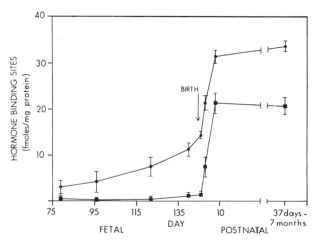

Figure 6. Ontogeny of ovine placental lactogen (oPL) (•——•) and ovine growth hormone (oGH) (■ —— ■) binding sites in sheep liver. Each point represents mean $\pm$ SE of data obtained using hepatic microsomes from 3 or 4 sheep. Similar results were obtained in 4 separate experiments. From Freemark et al., 1986.

To determine whether the marked increase in growth hormone binding after birth reflects a developmental change in the affinity of the placental lactogen receptor for growth hormone or the appearance of a new, structurally distinct growth hormone binding site, we compared the molecular weights of complexes formed by the covalent cross-linking of ^{125}I-oPL and ^{125}I-oGH to hepatic membranes from fetal lambs and from pregnant sheep and postnatal nonpregnant sheep (Freemark et al., in press)(Figure 7). Cross-linking of ^{125}I-oPL to fetal hepatic membranes yielded a major radiographic band with an estimated molecular weight of 60 kD. The intensity of the band was reduced in a specific, dose-dependent fashion by low concentrations of unlabeled OPL (.27-2.7 nM) or by higher concentrations of unlabeled oGH (90 nM) or oPRL (1 uM). In contrast, there was no specific cross-linking of ^{125}I-oGH or ^{125}I-oPRL to fetal hepatic membranes. The low affinity of the fetal placental lactogen binding site for growth hormone or prolactin distinguishes the placental lactogen binding site in the fetus from previously described somatotropic and lactogenic binding sites in postnatal tissues. Since the molecular weight of oPL is approximately 22 kD, the estimated molecular weight of the placental lactogen binding site is approximately 38 kD. Two lines of evidence strongly suggest that the 38 kD binding site represents a distinct fetal placental lactogen receptor or a subunit or component of that receptor: 1) unlabeled oPL specifically competes with ^{125}I-oPL for binding to the 38 kD protein at concentrations comparable to the plasma concentration of oPL in the fetal lamb; and 2) the relative order of potencies of oPL, oGH and oPRL in competing for binding to the 38 kD hepatic protein was similar to the relative order of potencies of the three hormones in stimulating glycogenesis in ovine fetal hepatocytes.

Cross-linking of ^{125}I-oGH to hepatic membranes from postnatal sheep yielded two major radiographic bands with apparent molecular weights of 75 kD and 140 kD which were distinguishable from the 60 kD band formed by the cross-linking of ^{125}I-oPL to fetal liver. The relative potencies of oGH, oPL and oPRL in reducing the intensities of the 75 kD and 140 kD bands were similar to the relative potencies of these hormones in competing for growth hormone binding sites in postnatal sheep liver but

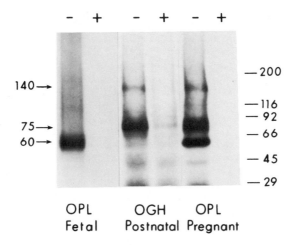

Figure 7. Cross-linking of ^{125}I-oPL to fetal and pregnant sheep liver microsomes and ^{125}I-oGH to postnatal sheep liver microsomes incubated in presence of ^{125}-radiolabeled hormone alone (+) incubated in presence of radiolabeled hormone and 1 ug unlabeled hormone. The numbers refer to molecular weight in kD.

were strikingly different than the relative potencies of the hormones in competing for placental lactogen binding sites in fetal liver. These observations suggest that a new and structurally distinct GH receptor appears soon after birth in the sheep.

Our competitive binding studies of oPL and oGH to hepatic plasma membranes of pregnant and non-pregnant postnatal sheep strongly suggest the presence of distinct placental lactogen and growth hormone receptors in postnatal tissues. The cross-linking of ^{125}I-oPL to hepatic membranes of postnatal nonpregnant and pregnant sheep yielded radiographic bands with molecular weights of 60 kD, 75 kD, and 140 kD. The intensities of all three bands were reduced in a dose-dependent fashion by unlabeled OPL (0.9-9.0 nM). In contrast, unlabeled oGH (18 nM) completely abolished the 75 kD and 140 kD bands but reduced the intensity of the 60 kD band by only 20-30%. These findings suggest that oPL binds with high affinity to two functionally and structurally distinct receptors in pregnant and non-pregnant postnatal sheep liver, the placental lactogen and the growth hormone receptor. OGH, on the other hand, binds with high affinity to the postnatal growth hormone receptor but has low affinity for the hepatic placental lactogen receptor. Thus, while the biologic effects of oGH in postnatal sheep liver appear to be mediated through binding to the growth hormone receptor, the biologic effects of placental lactogen in pregnant sheep liver might be mediated by binding to the placental lactogen receptor, the growth hormone receptor, or both receptors.

The findings of these studies provide insight into the relative roles of the fetal and placental hormones in the regulation of fetal metabolism and growth. The presence of distinct placental lactogen binding sites in ovine fetal tissues provides a mechanism whereby oPL may exert direct anabolic effects of fetal amino acid and carbohydrate metabolism and fetal somatomedin production. Since the biologic effects of oPL in fetal tissues are similar to the effects of oGH in postnatal tissues, the presence of distinct fetal placental lactogen binding sites implicates a role for placental lactogen as a "growth hormone" in the ovine fetus. Whereas the lack of somatotropic and metabolic activity of growth hormone in the fetus appears to result from a relative deficiency of specific growth hormone receptors in fetal tissues, the emergence of new and distinct growth hormone binding sites after birth provides a mechanism whereby growth hormone may assume a critical role in the regulation of postnatal growth and development. A summary of the evidence suggesting a direct role for placental lactogen in fetal growth and development is shown in Table 5.

Studies demonstrating that hPL stimulates amino acid transport, DNA synthesis and somatomedin production in human fetal fibroblasts and myoblasts also suggest a direct role for hPL in the regulation of fetal metabolism and growth. In addition, the demonstration that hPL stimulates somatomedin production in human fetal skeletal muscle while hGH and hPRL are without effect provides indirect evidence for the presence of a distinct fetal hPL receptor. The relative binding of radiolabeled hPL and hGH to homologous human fetal tissues, however, has not yet been examined.

Although placental lactogen has direct metabolic and somatotrophic effects in fetal tissues, it is evident that placental lactogen is not the only hormonal factor involved in the regulation of fetal growth. Women with very low hPL concentrations resulting from a deletion of two of the three genes coding for hPL have normal pregnancies and normal sized babies at birth (Wurzel et al., 1982). Since fetal growth, like postnatal growth, is undoubtedly controlled by many factors, the low hPL concentrations in the fetus may have resulted in compensatory changes in growth hormone and other factors involved in somatomedin production. One possible explanation for normal fetal growth in pregnant women with gene deletions for hPL comes from the recent studies of Frankenne et al. (1986). Using newly developed monoclonal antibodies to hPL and hGH, these investigators assayed the hormone content of the placenta of a woman with a deletion of the hPL-A, hPL-B and hGH-V genes whose pregnancy was characterized by an apparent lack of immunoreactive hPL in blood and placental tissue. The "abnormal" placenta actually contained hPL-like immunoactivity (1.8 ug/gm wet weight) and hGH-like immunoactivity (0.15 0.18 ug/gm wet weight) which was distinct from normal hPL and pituitary hGH and which differed from the recently described placental GH. Frankenne et al. (1986) hypothesized that the "abnormal" placenta may produce hPL-like and hGH-like molecules through the expression of placental genes (for instance, the hPL-L gene or an altered hGH-N gene) which are not expressed under normal conditions. It is possible that these hPL-like or hGH-like gene products may assume hPL-like roles in the mother and/or the fetus, sustaining normal fetal growth and development during those pregnancies complicated by a deficiency or absence of normal hPL production. It is also possible that low hPL concentrations during pregnancy may have caused alterations in the number and/or affinity of hPL, hGH, or hPRL receptors in fetal tissues. Unfortunately the concentrations of hGH, hPRL and other growth factors in mothers with hPL gene deletion and their newborns have not been reported and receptor studies of hormone binding in lymphocytes or other tissues have not been performed.

Table 5. Evidence suggesting a direct role for PL in fetal growth
and development

(1) PL detected in human, ovine and bovine fetal sera
(2) Distinct PL receptors demonstrable in fetal tissues
(3) PL has anabolic effects on fetal amino acid and carbohydrate
metabolism and stimulates somatomedin production in fetal
tissues
(4) Low cord blood hPL may be associated with intrauterine
growth retardation

PROLACTIN

Concentrations During Pregnancy

During pregnancy, human prolactin increases above normal luteal
phase concentrations at about 30 to 35 days (Barberia et al., 1975). The
prolactin concentration then increases at a relatively linear rate
reaching a maximum of 200 to 225 ng/ml at about 38 weeks. The range of
prolactin concentrations, however, may vary considerably among normal
women. Prolactin has been detected in fetal blood as early as 16 weeks
of gestation and reach concentrations equal or slightly higher than
maternal concentrations at or near term (Winters et al., 1975).
Prolactin is present in amniotic fluid at concentrations which greatly
exceed those in either maternal or fetal serum (Tyson et al., 1972;
Schenker et al., 1975: McNeilly et al., 1977; Clements et al., 1977;
Andersen, 1982). A schematic representation of prolactin concentrations
in maternal and fetal serum and in amniotic fluid is shown in Figure 8.

Since amniotic fluid prolactin is indistinguishable from pituitary
prolactin by chemical, biological and immunologic techniques (Schenker
et al., 1975; Hwang et al., 1974), most investigators initially suggested
that amniotic fluid prolactin is derived from either the maternal
(Schenker et al., 1975; Josimovich et al., 1974) or fetal pituitary
(Clements et al., 1977; Fang and Kim, 1975). An extra-pituitary source
for amniotic fluid prolactin, however, was suggested by the demonstra-
tions that the pattern of serial prolactin concentrations in amniotic
fluid during pregnancy is markedly different than the patterns of serial
prolactin concentrations in maternal and fetal sera (Tyson et al., 1972;
Andersen, 1982) and that amniotic fluid prolactin concentrations are not

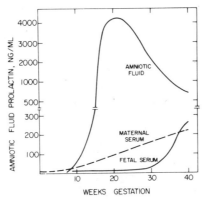

Figure 8. Schematic representation of human prolactin concentrations in
maternal and fetal serum and in amniotic fluid during pregnancy.

411

reduced by either maternal hypophysectomy (Walsh et al., 1977; Riddick et al., 1979), fetal death (Walsh et al., 1977), or bromocriptine therapy (Bigazzi et al., 1979; Lehtovirta and Ranta, 1981).

Several lines of evidence now implicate the decidua as the source of the amniotic fluid prolactin. Experiments from numerous laboratories indicate that decidual tissue in vitro synthesizes and releases a pro-lactin which is identical to pituitary and amniotic fluid prolactin in chemical, biological, and immunological properties (Golander et al., 1978; Riddick et al., 1978; Golander et al., 1979b) and which is transported in vitro across the membranes to the luminal side of the amnion (i.e., towards the amniotic cavity) (Riddick and Maslar, 1981; McCoshen et al., 1982a; Fukamatsu et al., 1984). In addition, studies have shown that the amounts of prolactin released by decidual explants from different times during pregnancy are correlated with the prolactin concentration in matched amniotic fluid samples (Rosenberg et al., 1980). Structural identity between pituitary and decidual prolactin has been established by the finding that the nucleotide sequences of the cDNAs for pituitary and decidual prolactin are identical (Clements et al., 1983; Takahashi et al., 1984). To date, decidual prolactin has not been detected in other species, but a protein has been partially purified from rat decidua which has biological properties similar to prolactin but which does not cross-react immunologically with rat or other prolac-tins (Herz et al., 1986).

Studies from our laboratory indicate that the factors regulating the synthesis and release of decidual prolactin and pituitary prolactin are distinct. Unlike pituitary prolactin, the synthesis and release of decidual prolactin is not affected by thyrotropin releasing hormone, dopamine, and bromocriptine (Golander et al., 1979a). The release of decidual prolactin, unlike that of pituitary prolactin, is inhibited by arachidonic acid (Handwerger et al., 1981) and a protein other than prolactin which is released by the decidua (Markoff et al., 1983). In addi-tion, the synthesis and release of decidual prolactin are stimulated selectively by a protein released by the placenta (Handwerger et al., 1983).

Effects on Ovary

The physiologic role of prolactin in human ovarian physiology is unclear. In the human, the corpus luteum persists throughout pregnancy and is the main source of steroids prior to the ninth week of gestation. Although luteotropic in other species, prolactin in the human does not appear to be important for implantation or maintenance of early pregnancy. In a study of 30 women who had unprotected intercourse during the time of ovulation and who were subsequently treated with bro-mocriptine, five women had detectable hCG concentrations in the urine and three became pregnant, a rate of pregnancy not statistically dif-ferent from that of women not receiving bromocriptine (Ylikorkala et al., 1979). Furthermore, elevation or suppression of prolactin levels in normal women during the sixth and ninth week of pregnancy had no effect on serum progesterone levels. Elevation or suppression of prolactin secretion for one week had no effect on serum hCG concentrations although increased hCG concentrations had been detected in women receiving bromocriptine throughout pregnancy. It therefore does not appear that high prolactin levels during human pregnancy interfere with hormone secretion from the corpus luteum or feto-placental unit.

HPL has been shown to bind to the corpus luteum (Moodbidri et al., 1973) and to be luteotrophic in pseudopregnant rats (Josimovich and Brande, 1964). HPL, however, has not been observed to be luteotrophic in pregnant women (Saxena, 1971), and the role – if any– of hPL in ovarian physiology is unclear.

Effects on Water and Mineral Metabolism in Amniotic Fluid

Since prolactin affects water and ion transport in lower vertebrates (Ensor, 1978) and binds specifically to amniotic membranes with high affinity (Josimovich et al., 1977; McCoshen et al., 1982b), several investigators have suggested a role for amniotic fluid prolactin in the regulation of amniotic fluid volume and ion content (Josimovich et al., 1977; Leotonic and Tyson, 1977; Stray-Pedersen, 1982; Tyson et al., 1984; Johnson et al., 1985; Raabe, 1986). Ovine prolactin (oPRL) was shown to promote water transfer across amniotic membranes of the human, guinea pig, rhesus monkey and rat (Page et al., 1974; Manku et al., 1975; Hold and Perks, 1975; Leontic et al., 1979; McCoshen et al., 1982a). In addition, the injection of oPRL into the amniotic fluid of rhesus monkeys at 115-116 days of gestation has been shown to markedly reduce or prevent rapid changes in the movement of sodium, potassium and water into or from the fetal extracellular fluid following changes in the tonicity of the amniotic fluid (Josimovich et al., 1977). These findings suggest that amniotic fluid prolactin may protect the fetus from sudden changes in the extracellular fluid content of water and electrolytes which might occur as a consequence of altered amniotic fluid tonicities.

Effects on Pulmonary Surfactant Synthesis

Although the data are not conclusive, many studies suggest that prolactin may also be one of the factors involved in the regulation of surfactant synthesis in the fetus. In the rhesus monkey, amniotic fluid prolactin concentrations are correlated with the amniotic fluid lecithin/sphingomyelin (L/S) ratio (an index of lung maturity), the lung phosphatidylcholine concentration and lung stability (Johnson et al., 1985), although such correlations have not been observed in one study in the human (Hatjis et al., 1981). In addition, the intramuscular administration of oPRL to 24 day old rabbit fetuses has been shown to increase the total phospholipid, phosphatidylcholine and dipalmitoylphosphatidylcholine content of lung extracts (Hamosh and Hamosh, 1977), although these findings have not been confirmed (Ballard et al., 1978). In two recent studies (Mendelson et al., 1981; Snyder et al., 1983), the synthesis of surfactant by human fetal lung explants maintained in serum-free medium has been shown to be under multihormonal control by prolactin, cortisol and insulin in a manner analogous to the multihormonal control of casein synthesis by mammary gland explants. A possible role for fetal pituitary prolactin in the regulation of surfactant synthesis has been suggested by the demonstration in some (Hauth et al., 1978; Gluckman et al., 1978) but not all studies (Schober et al., 1982; Yuei et al., 1982) of a positive correlation between fetal prolactin concentrations and the amniotic fluid L/S ratio and an inverse relationship between prolactin concentrations in cord plasma and the incidence of respiratory distress syndrome.

Other Effects of Prolactin

Several studies have demonstrated that prolactin inhibits uterine contractility (Horrobin et al., 1973; Mugambi et al., 1974), suppresses the immune response (Contractor and Davies, 1973; Karmali et al., 1974), and is involved in the maintenance of T-cell immunocompetence (Hiestand et al, 1986). Since prolactin is present in decidual tissue throughout gestation, decidual prolactin may act locally on the uterus to inhibit myometrial contractility and prevent immunologic rejection of the blastocyst and fetus. The observation that human decidual tissue contains specific membrane receptors for prolactin further supports an autocrine/ paracrine role for decidual prolactin (McWey, 1982). Since pituitary prolactin stimulates the synthesis of 1,25-(OH)2 D$_3$ by the kidney (Spanos et al, 1976), decidual prolactin may also be involved in the regulation of the synthesis of 1,25-(OH)2 D$_3$ by decidual and placental tissues.

SUMMARY

In summary, studies from our and other laboratories strongly suggest that placental lactogen has direct effects on fetal growth and metabolism as well as on maternal metabolism. Prolactin may be important in the regulation of water and ion transport across the amnion, the production of surfactant by the fetal lung, and the immune response during pregnancy. A summary of the postulated effects of placental lactogen on maternal and fetal physiology is depicted in Figure 9 and a summary of the postulated effects of prolactin during pregnancy is shown in Table 6. Undoubtedly, future studies of the physiology of placental lactogen and prolactin will uncover new functions for these hormones during gestation.

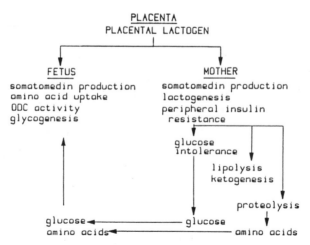

Figure 9. Possible mechanisms by which placental lactogen regulates maternal and fetal metabolism and growth.

Table 6. Proposed Roles of Prolactin in Human Pregnancy

(1) Impairs glucose tolerance
(2) Stimulates lipolysis and insulin secretion
(3) Affects water and ion transport across the amnion
(4) Stimulates surfactant production by fetal lung
(5) Suppresses the immune response
(6) Inhibits uterine contractility

ACKNOWLEDGEMENTS

We thank Gayle Kerr for secretarial assistance. Supported by NIH grants HD 07447, HD 15201 and HD 00656 and a March of Dimes, Basil O'Connor Starter Grant, #5-503.

REFERENCES

Adams SO, Nissley SP, Handwerger S, Rechler MM, 1983. Developmental patterns of insulin-like growth factor I and II synthesis and regulation in rat fibroblasts. Nature 302:150-153

Andersen JR, 1982. Prolactin in amniotic fluid and maternal serum during uncomplicated human pregnancy. Dan Med Bull 29:266-274

Arezzuni C, DeGori V, Tarli P, Neri P, 1972. Weight increase of body and lymphatic tissue in dwarf mice treated with human chorionic somatomammotropin (hCS). Proc Soc Exp Biol Med 141:98-100

Ballard PL, Gluckman PD, Breheir A, Kitterman JA, Kaplan SL, Rudolph AM, Grumbach MM, 1978. Failure to detect an effect of prolactin on pulmonary surfactant and adrenal steroids in fetal sheep and rabbits. J Clin Invest 62:879-883

Barberia JM, Abu-Fadel S, Kletzky OA, Nakamura RM, Mishell DR, 1975. Serum prolactin levels in early human gestation. Am J Obstet Gynecol 121:1107-1110

Beck P, Daughaday WH, 1967. Human placental lactogen: Studies of its acute metabolic effects and disposition in normal man. J Clin Invest 46:103-110

Berle P, Finsterwalder E, Apostolakis M, 1974. Comparitive studies on the effect of human growth hormone, human prolactin and human placental lactogen on lipid metabolism. Horm Metab Res 6:347-350

Bigazzi M, Rongo R, Lancranjan I, Ferraro S, Branconi F, Buzzoni P, Martorana G, Scarselli GF, Del Pozo E, 1979. A pregnancy in an acromegalic women during bromocriptine treatment: Effects of growth hormone and prolactin in the maternal, fetal and amniotic compartments. J Clin Endocrinol Metab 48:9-12

Breuer CB, 1969. Stimulation of DNA synthesis in cartilage of hypophysectomized rats by native and modified placental lactogen and available hormones. Endocrinology 85:989-999

Butler SR, Hurley TW, Schanberg SM, Handwerger S, 1978. Ovine placental lactogen stimulation of ornithine decarboxylase activity in brain and liver of neonatal rats. Life Sci. 22:2073-2078

Chan JSD, Robertson HA, Friesen HG, 1976. Purification and characterization of ovine placental lactogen. Endocrinology 98:65-76

Clements JA, Reyes FI, Winter JSD, Faiman C, 1977. Studies on human sexual development. IV. Fetal pituitary and serum and amniotic fluid concentration of prolactin. J Clin Endocrinol Metab 44:408-413

Clements J, Whitfeld P, Cooke N, Healy D, Matheson B, Shine J, Funde J, 1983. Expression of the prolactin gene in human decidua-chorion. Endocrinology 112:1133-1134

Contractor SF, Davies H, 1973. Effect of human chorionic somatomam-motrophin and human chorionic gonadotropin on phytohaemagglutinin-induced lymphocyte transformation. Nat New Biol 243:284-286

Emane MN, Delouis C, Kelly PA, Dijane J, 1986. Evolution of prolactin and placental lactogen receptors in ewes during pregnancy and lactation. Endocrinology 118:695-700

Ensor DM, 1978. Comparative Endocrinology of Prolactin. London: Chapman and Hall, LTD

Fang VS, Kim MH, 1975. Study on maternal, fetal, and amniotic human prolactin at term. J Clin Endocrinol Metab 41:1030-1034

Felber JP, Zaragoza N, Benuzzi-Badoni M, Genazzani AR, 1972. The double effect of human chorionic somatomammotropin (HCS) and pregnancy on lipogenesis and lipolysis in the isolated rat epididymal fat pad and fat pad cells. Hormone Metab Res 4:293-296

Frankenne F, Hennen G, Parks JS, Nielsen PV, 1986. A gene deletion in the hGH/hCS gene cluster could be responsible for the placental expression of hGH and/or hCS like molecule absent in normal subjects. Proceedings of the 68th Annual Meeting of the Endocrine Society, Anaheim, CA, June, Abstract #388

Freemark M, Handwerger S, 1982. Ovine placental lactogen stimulates amino acid transport in rat diaphragm. Endocrinology 110:2201-2203

Freemark M, Handwerger S, 1983. Ovine placental lactogen, but not growth hormone, stimulates amino acid transport in fetal rat diaphragm. Endocrinology 112:402-404

Freemark M, Handwerger S, 1984a. Ovine placental lactogen stimulates glycogen synthesis in fetal rat hepatocytes. Am J Physiol 246:E21-E24

Freemark M, Handwerger S, 1984b. Synergistic effects of ovine placental lactogen and insulin on glycogen metabolism in fetal rat hepatocytes. Am J Physiol 247:E714-E718

Freemark M, Handwerger S, 1985. Ovine placental lactogen inhibits glucagon-induced glycogenolysis in fetal rat hepatocytes. Endocrinology 116:1275-1280

Freemark M, Handwerger S, 1986. The glycogenic effects of placental lactogen and growth hormone in ovine fetal liver are mediated through binding to specific fetal oPL receptors. Endocrinology 118:613-618

Freemark M, Comer M, Handwerger S, 1986. PL and GH binding sites in sheep liver: Striking differences in ontogeny and function. Am J Physiol 251:E328-333

Freemark M, Comer M, Korner G, Handwerger S, 1987. A unique PL receptor: Implications for fetal growth. Endocrinology, in press

Fukamatsu Y, Tomita K, Fukuta T, 1984. Further evidence of prolactin production from human decidua and its transport across fetal membrane. Gynecol Obstet Invest 17:309-316

Furlanetto RW, Underwood LE, Van Wyk JJ, Handwerger S, 1978. Serum immunoreactive somatomedin-C is elevated late in pregnancy. J Clin Endocrinol Metab 47: 695-698

Gluckman PD, Ballard PL, Kaplan SL, Liggins GC, Grumbach MM, 1978. Prolactin in umbilical cord blood and the respiratory distress syndrome. J Ped 93:1011-1014

Gluckman PD, Butler JH, 1983. Partutition-related changes by insulin-like growth factors I and -II in the perinatal lamb. J Endocrinol 99:223-232

Gluckman PD, Kaplan SL, Rudolph AM, Grumbach MM, 1979. Hormone ontogeny in the ovine fetus. II. Ovine chorionic somatommotropin in mid-and late gestation in the fetal and maternal circulation. Endocrinology 104:1828-1833

Golander A, Barrett J, Hurley T, Barry S, Handwerger S, 1979a. Failure of bromocriptine, dopamine and thyrotropin-releaseing hromone to affect prolactin secretion by decidual tissue in vitro. J Clin Endocrinol Metab 49:787-789

Golander A, Hurley T, Barrett J, Handwerger S, 1979b. Synthesis of prolactin by human decidua in vitro. J Endocrinol 82:263-267

Golander A, Hurley T, Barrett J, Hizi A, Handwerger S, 1978. Prolactin synthesis by human chorion-decidual tissue: A possible source of amniotic fluid prolactin. Science 202:311-313

Grumbach MM, Kaplan SL, 1964. On the placental origin and purification of chorionic "growth hormone-prolactin" and its immunoassay in pregnancy. Trans NY Acad Sci 27:167-188

Grumbach MM, Kaplan SL, Abrams CL, Beel JJ, Conte FA, 1966. Plasma free fatty acid response to the administration of chronic growth hormone-prolactin. J Clin Endocrinol Metab 26:478-482

Grumbach MM, Kaplan SL, Sciarra JJ, Burr IM, 1968. Chorionic growth hormone-prolactin (CGP): Secretion, disposition, biologic activity in man, and postulated function as the "growth hormone" of the second half of pregnancy. Ann NY Acad Sci 148:501-531

Hall K, Enberg G, Hellen E, Lundin G, Ottoson-Seeberger A, Sara V, Trystad O, Ofverholm V, 1984. Somatomedin levels in pregnancy. Longitudinal study in healthy subjects and patients with growth hormone deficiency. J Clin Endocrinol Metab 59:587-594

Hamosh M, Hamosh P, 1977. The effect of prolactin on the lecithin content of fetal rabbit lung. J Clin Invest 59:1002-1005

Handwerger S, Barry S, Barrett J, Markoff E, Zeitler P, Cwikel B, Siegel M, 1981. Inhibition of the synthesis and secretion of decidual prolactin by arachidonic acid. Endocrinology 109:2016-2021

Handwerger S, Barry S, Markoff E, Barrett J, Conn, PM, 1983. Stimulation of the synthesis and release of decidual prolactin by a placental polypeptide. Endocrinology 112:1370-1374

Handwerger S, Crenshaw,Jr, C, Maurer WF, Barrett J, Hurley TW, Golander A, Fellows RE, 1977. Studies of ovine placental lactogen secretion by homologous radioimmunoassay. J Endocrinol 72:27-34

Handwerger S, Fellows RE, Crenshaw MC, Hurley T, Barrett J, Maurer WF, 1976. Ovine placental lactogen: Acute effects on intermediary metabolism in pregnant and nonpregnant sheep. J Endocrinol 69:133-137

Handwerger S, Maurer W, Barrett J, Hurley T, Fellows RE, 1974. Evidence for homology between ovine and human placental lactogens. Endo Res Comm 1:403-413

Hatjis CG, Wu CH, Gabbe SG, 1981. Amniotic fluid prolactin levels and lecithin/spingomyelin ratios during the third trimester of human gestation. Am J Obstet Gynecol 138:435-440

Hauth JC, Parker CR, MacDonald PC, Porter JC, Johnston JM, 1978. A role of fetal prolactin in lung maturation. Obstet Gynecol 51:81-88

Herz Z, Khan I, Jayatilak PG, Gibori G, 1986. Evidence for the secretion of decidual luteotropin: A prolactin-like hormone produced by rat decidual cells. Endocrinology 118:2203-2209

Hiestand PC, Mekler P, Nordmann R, Grieder A, Permmongkol C, 1986. Prolactin as a modulator of lymphocyte responsiveness provides a possible mechanism of action for cyclosporine. Proc Natl Acad Sci USA 83:2599-2603

Hill DJ, Crace CJ, Milner RDG, 1985. Incorporation of ^{3}H-thymidine by isolated fetal myoblasts and fibroblasts in response to human placental lactogen (hPL): Possible mediation of hPL action by release of immunoreactive SM-C. J Cell Physiol 125:337-344

Hill DJ, Crace CJ, Strain AJ, Milner RDG, 1986. Regulation of amino acid uptake and deoxyribonucleic acid synthesis in isolated human fetal fibroblasts and myoblasts: Effect of human placental lactogen, somatomedin-C, multiplication-stimulating activity, and insulin. J Clin Endocrinol Metab 62:753-760

Hold WF, Perks AM, 1975. The effect of prolactin on water movement
through the isolated amniotic membrane of the guinea pig. Gen Comp
Endocrinol 26:153-156

Horrobin DR, Lipton A, Muiruri KL, Manku MS, Bramley PS, Burstyn PG,
1973. An inhibitory effect of prolactin on the response of rat
myometrium to oxytocin. Experientia 29:109-112

Hurley TW, D'Ercole AJ, Handwerger S, Underwood LE, Fulanetto RW,
Fellows RE, 1977. Ovine placental lactogen induces somatomedin: A
possible role in fetal growth. Endocrinology 101:1635-1638

Hurley TW, Handwerger S, Fellows RE, 1977. Isolation and structural
characterization of ovine placental lactogen. Biochemistry 16:
5598-5604

Hurley TW, Kuhn CM, Schanberg SM, Handwerger S, 1980. Differential
effects of placental lactogen, growth hormone and prolactin on rat
liver ornithine decarboxylase activity in the perinatal period. Life
Sci 27:2269-2275

Hwang P, Murray J, Jacobs J, Niall H, Friesen H, 1974. Human amniotic
fluid prolactin, purification by affinity chromatography and amino-
terminal sequence. Biochem 13:2354-2458

Johnson JWC, Tyson JE, Mitzner W, Beck JC, Andressen B, London WT,
Villar J, 1985. Amniotic fluid prolactin and fetal lung maturation.
Am J Obstet Gynecol 153:372-380

Josimovich JB, Brande BL, 1964. Chemical properties in biological
affects of human placental lactogen (HPL). Trans NY Acad Sci 27:161-
166

Josimovich JB, McLaren JA, 1962. Presence in the human placenta and
term serum of a highly lactogenic substance immunologically related
to pituitary growth hormone. Endocrinology 71:209-220

Josimovich JB, Merisko K, Boccella L, 1977. Amniotic prolactin control
over amniotic and fetal extracellular fluid water and electrolytes in
the rhesus monkey. Endocrinology 100:564-570

Josimovich JB, Weiss G, Hutchinson DL, 1974. Sources and disposition of
pituitary prolactin in maternal circulation, amniotic fluid, fetus
and placenta in the pregnant rhesus monkey. Endocrinology 94:1364-
1371

Kaplan SL, Grumbach MM, 1964. Studies of a human and simian placental
hormone with growth hormone-like and prolactin-like activities. J
Clin Endocrinol Metab 25:1370-1374

Kaplan SL, Grumbach MM, 1965. Serum chorionic "growth hormone-prolactin"
and serum pituitary growth hormone in mother and fetus at term. J
Clin Endocrinol Metab 25:1370-1374

Kaplan SL, Grumbach MM, Shepard TH, 1972. The ontogenesis of human
fetal hormones. I. Growth hormone and insulin. J Clin Invest 51:
3080-3093

Karmali RA, Lauder I, Horrobin DF, 1974. Letter: Prolactin and the
immune response. Lancet 2:106-107

Lehtovirta P, Ranta T, 1981. Effect of short-term bromocriptine treatment on amniotic fluid concentrations in the first half of pregnancy.
Acta Endocrinol 97:559-561

Leontic EA, Schruefer JJ, Andreassen B, Pinto H, Tyson JE, 1979.
Further evidence for the role of prolactin on human feto-placental
osmoregulation.. Am J Obstet Gynecol 133:435-438

Leontic EA, Tyson JE, 1977. Possible osmoregulatory role for amniotic
fluid prolactin. In: Crosignani PG, Robyn C,(eds.), Prolactin and
Human Reproduction. London: Academic Press, pp.37-45

Lesniak MA, Gorden P, Roth J, 1977. Reactivity of non-primate growth
hormone and prolactin with growth hormone receptors on cultured human
lymphocytes. J Clin Endocrinol Metab 44:838-849

Manku MS, Mtabaji JP, Horrobin DF, 1975. Effect of cortisol prolactin and ADH on the amniotic membrane. Nature 258:78-79

Marakawa S, Raben MS, 1968. Effect of growth hormone and placental lactogen on DNA synthesis in rat costal cartilage and adipose tissue. Endocrinology 83:645-650

Markoff E, Howell S, Handwerger S, 1983. Inhibition of decidual prolactin by a decidual peptide. J Clin Endocrinol Metab 57:1282-1286

Martal J, Dijane J, 1977. Mammotrophic and growth promoting activities of a placental hormone in sheep. J Steroid Biochem 8:415-417

Martal J, 1978. Placental growth hormone in sheep: Purification, properties and variations. Ann Biol An 18:45-51

Martin JW, Friesen HG, 1969. Effect of human placental lactogen on the isolated islets of Langerhans in vitro. Endocrinology 84:619-621

McCoshen JA, Tagger OY, Wodzicki A, Tyson JE, 1982a. Choriodecidual adhesion promotes decidual prolactin transport by human fetal membranes. Am J Physiol 243:552-557

McCoshen JA, Tomika K, Fernandy C, Tyson JE, 1982b. Specific cells of human amnion selectively localize prolactin. J Clin Endocrinol Metab 55:166-169

McGarry EE, Beck JC, 1972. Biological effects of non-primate prolactin and human placental lactogen. In: Wolstenholme GEW, Knight J, (eds.), Lactogenic Hormones. London: Churchill Livingston, pp. 361-389.

McNeilly AS, Gilmore D, Jeffery D, Dobbie G, Chond T, 1977. The origin of prolactin in amniotic fluid: fetal or maternal? In: Croisignani PG, Robyn C,(eds.), Prolactin and Human Reproduction. London: Academic Press, pp. 21-26

McWey LA, Singhas CA, Rogol AD, 1982. Prolactin binding sites on human chorion-decidua tissue. Am J Obstet Gynecol 144:283-288

Mendelson CR, Johnston JM, MacDonald PC, Snyder JM, 1981. Multihormonal regulation of surfactant synthesis by human fetal lung in vitro: J Clin Endocrinol Metab 53:307-317

Moodbidri SB, Sheth AR, Rao SS, 1973. Binding of radioiodinated human placental lactogen to buffalo corpus luteum in vitro. J Reprod Fert 35:453-460

Mugambi M, Mati JKG, Muriuki PG, Thairu K, 1974. An inhibitory action of prolactin on non-pregnancy human myometrium in vitro. E Afr Med J 51:392-396

Page KR, Abramovich DR, Smith MR, 1974. Water transport across isolated term human amnion. J Mem Biol 18:49-60

Raabe MA, McCoshen JA, 1986. Epithelial regulation of prolactin effect on amnionic permability. Am J Obstet Gynecol 154:130-134

Riddick D, Luciano A, Kusmik W, Maslar I, 1978. De novo synthesis of prolactin by human decidua. Life Sci 23:1913-1922

Riddick D, Luciano A, Kusmik W, Maslar I, 1979. Evidence for a non-pituitary source of amniotic fluid prolactin. Fert Steril 31:35-39

Riddick DH, Maslar IA, 1981. The transport of prolactin by human fetal membranes. J Clin Endocrinol Metab 52:220-224

Rosenberg SM, Maslar IA, Riddick DH, 1980. Decidual production of prolactin in late gestation: Further evidence for a decidual source of amniotic fluid prolactin. Am J Obstet Gynecol 138:681-685

Samaan N, Yen S, Friesen HG, Pearson O, 1966. Serum placental lactogen levels during pregnancy and trophoblastic disease. J Clin Endocrinol Metab 26:1303-1308

Saxena B, Emerson K, Selenkow H, 1969. Serum placental lactogen levels as an index of placental function. New Eng J Med 281:225-231

Saxena BN, 1971. Protein-polypeptide hormones of the human placenta. Vit Horm 29:95-151

Schenker JG, Ben-David M, Polishuk WZ, 1975. Prolactin in normal pregnancy. Relationship of maternal, fetal, and amniotic fluid levels. Am J Obstet Gynecol 123:834-838

Schober E, Simbruner G, Salzer H, Husslein P, Spona J, 1982. The relationship of prolactin in cord blood, gestational age and respiratory compliance after birth in newborn infants. J Perinat Med 10:23-26

Sherwood LS, Handwerger S, McLaurin WE, Lanner M, 1971. Amino acid sequence of human placental lactogen. Nature New Biol 233:59-61

Snyder JM, Longmuir KJ, Johnston JM, Mendelson CR, 1983. Hormonal regulation of the synthesis of lamellar body phosphatidylglycerol and phosphatidylinositol in fetal lung tissue. Endocrinology 112:1012-1018

Spanos E, Colstan KW, Evans IMS, Golante LS, Macauley SJ, MacIntyre I, 1976. Effect of prolactin on vitamin D. metabolism. Mol Cell Endocrinol 5:163-167

Spellacy WN, Buhi WC, 1969. Pituitary growth hormone and placental lactogen levels measured in normal term and at early and late postpartum periods. Am J Obstet Gynecol 105:888-896

Stray-Pedersen S, 1982. The effect of prolactin on fetal membrane transport. J Perinatal Med 10:121-122

Takahashi H, Nabeshima Y, Nabeshima Y-I, Ogata K, Takeuchi S, 1984. Molecular cloning and nucleotide sequence of DNA complementary to human decidual prolactin mRNA. J Biochem 95:1491-1499

Tyson JE, 1972. Human chorionic somatomammotropin. Obstet Gynecol Ann 1:421-452

Tyson JE, Hwang P, Guyda H, Friesen HG, 1972. Studies of prolactin secretion in human pregnancy. Am J Obstet Gynecol 113:14-20

Tyson JE, Mowat GS, McCoshen JA, 1984. Simulation of a probable biologic action of decidual prolactin on fetal membranes. Am J Obstet Gynecol 148:296-300

Walsh SW, Meyer RK, Wolf RC, Friesen HG, 1977. Corpus luteum and-fetoplacental functions in monkeys hypophysectomized during late pregnancy. Endocrinology 100:845-850

Waters MJ, Oddy VH, McCloghry CE, Gluckman PD, Duplock R, Owens PC, Brinsmead MW, 1985. An examination of the proposed roles of placental lactogen in the ewe by means of antibody neutralization. J Endocrinol 106:377-386

Winters AJ, Colston C, MacDonald PC, Porter JC, 1975. Fetal plasma prolactin levels. J Clin Endocrinol Metab 41:626-629

Wurzel JM, Parks JS, Herd JE, Nielsen PV, 1982. A gene deletion is responsible for absence of human chorionic somatomammotropin. DNA 1: 251-257

Ylikorkala O, Huhtaniemi I, Tuimala R, Seppala M, 1979. Subnormal postconceptional levels of prolactin do not interfer with the early events of human pregnancy. Fert Steril 32:286-288

Yuei BH, Phillips WDP, Cannon W, Sy L, Redford D, Burch P, 1982. Prolactin, estradiol, and thyroid hormones in umbilical cord blood of neonates with and without hyaline membrane disease: A study of 405 neonates from mid pregnancy to term. Am J Obstet Gynecol 142:698-703

REGULATION OF RELAXIN SECRETION AND ITS ROLE IN PREGNANCY

L. L. Anderson

Department of Animal Science
Iowa State University
Ames, IA 50011

INTRODUCTION

Relaxin plays a critical role in suppressing uterine motility during pregnancy and in remodeling connective tissue in preparation for imminent parturition. Its biological actions in softening the cervix and symphysis pubis in the guinea pig were first described by Hisaw (1926). Numerous biological studies were conducted with partially purified preparations of relaxin which led to a consensus of its importance in reproductive processes. Following the purification and characterization of porcine relaxin by Sherwood and O'Byrne (1974), there has been intensive research on the chemistry of relaxin and its physiology. Although relaxin is produced primarily during pregnancy in several mammalian species, it also is found in the male reproductive tract and can maintain sperm motility but little is known of its physiological significance (Loumaye et al., 1980; Essig et al., 1982). Recent reviews on relaxin include those by Bryant-Greenwood (1982; 1985), Porter (1984) and Kemp and Niall (1984). This review will briefly describe the current status of the species sources of relaxin, cellular localization of the hormone, and the biological actions of relaxin particularly during pregnancy. Major emphasis will be placed on the regulation of relaxin secretion during pregnancy in the rat and farm animals.

SPECIES SOURCES OF RELAXIN AND MEASURE OF IMMUNOREACTIVITY IN DIFFERENT SPECIES

Relaxin is a peptide hormone of about 6,000 daltons and found primarily in reproductive tissues during pregnancy. The corpus luteum in the pig is a rich source of this hormone. Relaxin, like insulin, consists of A and B chains linked by disulfide bonds and it is derived from a C-chain precursor. The cDNA sequences coding for rat, porcine and human preprorelaxins have been reported (Hudson et al., 1981; Haley et al., 1982; Hudson et al., 1984). On the basis of the similarity of the disulfide links and other structural features between relaxin and insulin, an insulin-like structure has been predicted for relaxin (Schwabe and McDonald, 1976, 1977a,b; Bedarkar et al., 1977). Insulins and relaxins differ, however, by 75% of the amino acid sequence but they retain a similar cross-linking and chain size. Computer graphic modeling of the three-dimensional structure of porcine relaxin was used to develop schematic diagrams showing the chain folding of insulin as derived from the X-ray analysis of insulin crystals and proinsulin, IGF-1, IGF-2 and relaxin (Figure 1, Bedarkar et al., 1977: Blundell and Humbel, 1980). Although the tertiary structures of relaxin and insulin are similar, the surfaces of the two molecules are very different. Furthermore, there are marked differences in the conserved residues on the surface of relaxin molecules from different species that could affect receptor binding and tissue specificity (Dodson et al., 1982). Relaxin and insulin are not immunologically cross-reactive and relaxin shows neither insulin-like biological activity nor the ability to bind insulin receptors (Schwabe et al., 1978). Relaxin does enhance the ability of adipocytes to bind insulin; however, and this is mediated through an increase in insulin receptor affinity (Olefsky et al., 1982; Jarrett et al., 1984). Like insulin, relaxin is similarly conserved during evolution and may be a gene duplication product of insulin, but arguments have been presented to counter such a relation of insulin and relaxin (Schwabe, 1983a). For example, relaxin has been extracted from the protozoan Tetra-hymena pyriformis (Table 1, Schwabe et al., 1983). Although relaxin is a hormone associated with reproduction in mammalian development, its presence in single cellular organisms would predate any biological functions of the hormone about one or two billion years.

A partial list is presented in Table 1 of species sources of relaxin as well as some species in which relaxin immunoreactivity has been measured by heterologous bioassays and radioimmunoassays. The initial studies on relaxin were carried out on extract prepared from pregnant sow ovaries (Hisaw, 1926). Porcine relaxin was the first hormone to be purified and characterized (Sherwood and O'Byrne, 1974) and sequenced for amino acid residues (Schwabe and McDonald, 1976, 1977a,b; James et al., 1977). Chemical procedures for the isolation of naturally occurring porcine relax-

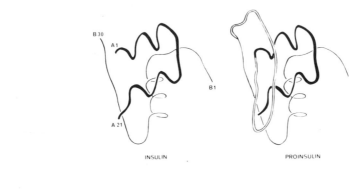

INSULIN   PROINSULIN

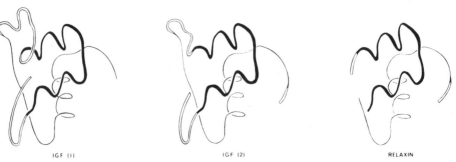

IGF (1)   IGF (2)   RELAXIN

Fig. 1.   A schematic representation of insulin, proinsulin, insulin-like growth factors and relaxin.   (From Blundell and Humbel, 1980.)

Table 1. Relaxin Sources and Immunoactivity Measured in Different Species

Species	Source	Amino Acid Sequence	Hormone for Measure of Immuno-activity
Pig	Ovary	+	porcine
Rat	Ovary	+	rat
Human	Ovary	+	porcine human
Sand tiger shark	Ovary	+	porcine shark
Dogfish shark	Ovary	+	porcine shark
Mouse	Ovary	−	porcine
Cattle	Ovary	−	porcine
Rabbit	Placenta	−	rabbit porcine
Horse	Placenta	−	equine porcine
Cat	Placenta	−	porcine
Guinea pig	Uterus	−	porcine
Sheep	Placenta Uterus Ovary	−	porcine
Ciliated protozoa	Tetrahymena pyriformis	−	porcine

in and large-scale preparation of the B29 hormone were presented by Büllesbach and Schwabe (1985). The primary structure of the entire porcine preprorelaxin coding sequence has been determined from analysis of recombinant cDNA clones (Haley et al., 1982). Rat relaxin was purified and characterized by Sherwood (1979). Its complete amino acid sequence was derived by molecular cloning and characterization of cDNA sequences coding for rat relaxin (Hudson et al., 1981; Niall et al., 1982). Similarly, the amino acid sequence for human relaxin was deduced by recombinant techniques (Hudson et al., 1984). Porcine and human relaxins also have been prepared by solid phase peptide synthesis (Tregear et al., 1983). The amino acid sequences for the sand tiger shark and dogfish shark are of interest for their evolutionary history. Although they differ by only 20% in their amino acids, the biological activities of these two shark relaxins are markedly different in the mouse and guinea pig (Gowan et al., 1981; Reinig et al., 1981; Schwabe, 1983a). Among species, relaxins differ more than insulins and this may account for the poor immunological crossreactivity of relaxins from different species. Bioactive and immunoreactive relaxin was found in the ovary and peripheral blood of the mouse during pregnancy (Schwabe et al., 1978; O'Byrne and Steinetz, 1976). In cattle, low concentrations of immunoreactive relaxin were found in corpora lutea during late pregnancy (Fields et al., 1980). During late pregnancy, relaxin levels in peripheral blood are low (< 200 pg/ml) a week preceding parturition and increase to peak concentrations (> 800 pg/ml) on day of parturition in beef heifers (Anderson et al., 1982b).

CELLULAR LOCALIZATION OF RELAXIN

The ovary is considered a primary source of relaxin in some mammalian species, particularly those in which it is required to maintain pregnancy. The pig, rat and mouse are examples of such species (Schwabe et al., 1978; Anderson, 1982). Extraovarian sources of relaxin include the placenta, endometrial gland cells and decidual tissue adjacent to fetal membranes. Examples of such species are the horse, guinea pig, cat, sheep and human. Particularly useful techniques for the localization of relaxin in reproductive tissues include immunocytochemistry at the light and electron microscopy level. With the use of relaxin antisera homologous to the species being studied, these techniques can be used to provide conclusive evidence for localization of the hormone. The utilization of heterologous relaxin antisera, however, has provided useful information concerning cellular localization of relaxin and binding sites for the hormone. The following is a brief account of the localization of relaxin and related peptides in reproductive tissues of a few selected species.

Porcine Ovary

The initial studies on fine structure of porcine corpora lutea throughout pregnancy and after hysterectomy revealed marked changes in granular endoplasmic reticulum and striking numbers of electron-dense membrane-limited granules (Belt et al., 1970, 1971). These features were virtually absent during the 21-day estrous cycle (Cavazos et al., 1969). By day 28 of pregnancy, (duration of pregnancy is about 114 days in the pig) granular endoplasmic reticulum begins to increase in quantity with clusters of granules (0.1-0.2 μm in diameter) evident for the first time in moderate numbers. Bioassays indicated a parallel rise in relaxin, and elevated for the first time above basal levels found in the estrous cycle (Anderson et al., 1973b). During midgestation there are further increases in granular endoplasmic reticulum and Golgi as well as a continued increase in the population of granules. The granule population continues to increase throughout most of pregnancy which coincided with the steady increase in relaxin bioactivity (Figure 2). By days 100 to 110, the concentrations of granules reach a peak and levels of relaxin in luteal tissue are maximal (Figures 3A, B). The rapid disappearance of the granules coincided with an abrupt decrease in relaxin levels in luteal tissue and a simultaneous increase in concentrations of this hormone in ovarian venous blood just preceding parturition. In the last day before delivery, the level of relaxin in the corpora lutea becomes minimal, the cells are depleted of granules and the level of ovarian vein relaxin is low. Within a few hours of parturition, the luteal cells are nearly free of granules. Exocytosis of granules is rarely observed, but solubilization of them is evident the last days preceding delivery. Sequential bleedings every sixth day revealed a similar increase in serum relaxin concentrations from days 6 to 114 in hysterectomized and pregnant gilts (Anderson et al., 1983). By day 112, luteal tissue relaxin levels and granule populations decrease by about half and their disappearance coincides with relaxin release during late pregnancy. Light microscopy fluorescent and peroxidase-antiperoxidase immunohistochemistry confirmed that the relaxin was localized in luteal cells of late pregnant pigs (Larkin et al., 1977, 1983). Electron microscopy immunoperoxidase demonstrated that relaxin was packaged in small granules found in the luteal cells (Kendall et al., 1978; Larkin et al., 1983). Electron microscopy immunocytochemistry using porcine relaxin antiserum and goat antirabbit immunoglobulin G-colloidal gold demonstrated that relaxin was localized in the small dense granules in luteal cells at day 110 of pregnancy in the pig (Figure 4; Orci, L.,Steinetz, B. G. and Anderson, L. L., unpublished observations). This late stage of pregnancy corresponds to a time when the population of granules and luteal tissue concentrations of relaxin are high (Figure 2;

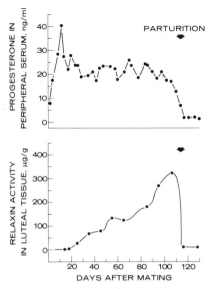

Fig. 2. Mean values of progesterone in peripheral blood serum of six Yorkshire gilts and bioactive relaxin in luteal tissue from 81 gilts throughout pregnancy, parturition and during early lactation. Arrows indicate parturition which occurs about day 114-115. Relaxin was measured by interpubic ligament formation (Steinetz et al., 1960) in estrogen-primed mice in a six-point assay with 20 mice used for each of three dilutions of NIH reference relaxin (lot no. 0148). (From Anderson et al., 1973b; 1982b; Hard and Anderson, 1979).

Anderson et al., 1973b; 1983). Using similar immunoglobulin G-colloidal gold techniques, Larkin et al. (1983) observed gold-labeled granules at day 45 of pregnancy. Fields and Fields (1985) observed increasing numbers of gold-labeled granules from days 17-75 of pregnancy, and maximal numbers of them at day 106 of pregnancy and day 110 of pseudopregnancy in the pig. Although relaxin levels in luteal tissue are low during the porcine estrous cycle (Anderson et al., 1973b; Sherwood and Rutherford, 1981), relaxin was found to be localized in the corpus luteum using the avidin-biotin immuno-peroxidase method and an antiserum to porcine relaxin (Ali et al., 1986). Relaxin immunostaining was undetectable on day 3 and became evident by days 7 and 9. At day 11 staining intensity increased and persisted through day 15 and was absent after day 18. Relaxin immunostaining appeared to be located throughout the cytoplasm of the luteal cell. These changes in immunostaining correspond with relaxin levels as determined by bioassay and radioimmunoassay at early, middle and late stages of the estrous cycle (Anderson et al., 1973b; Sherwood and Rutherford, 1981).

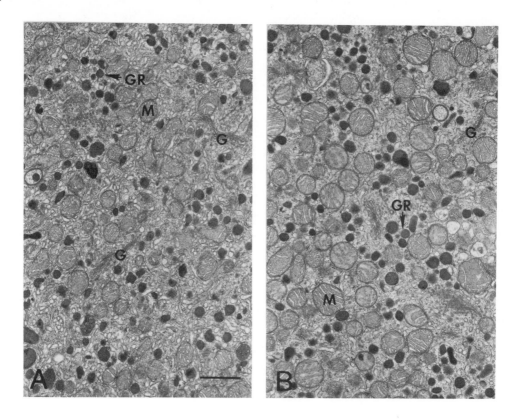

Fig. 3A and B. Electron photomicrographs of porcine luteal tissue at day 100 in (A) late pregnancy and (B) after hysterectomy. The population of electron-dense granules in both reproductive states becomes maximal at this time and coincides with peak concentrations of relaxin in luteal tissue. GR = granule, M = mitochondria, G = Golgi, and bar = 1 μm. (Stromer, M. H. and Anderson, L. L., unpublished observations.)

Although the corpus luteum is the major source of relaxin, the hormone isolated from ovaries of pregnant sows exhibits microheterogeneity (Sherwood and O'Byrne, 1974). An extra-luteal source of relaxin includes ovarian follicles from prepuberal gilts, and pregnant and non-pregnant sows with normal or polycystic ovaries (Anderson et al., 1973b; Afele et al., 1979; Matsumoto and Chamley, 1980). Porcine follicular fluid contains high molecular weight forms of relaxin or prorelaxins (e.g., 42,000; 27,500; 19,000; 10,000) in addition to a molecule of approximately 6,000 (Frieden and Yeh, 1977; Kwok et al., 1978; Matsumoto and Chamley, 1980). Porcine follicular cells can produce relaxin *in vitro* and the preovulatory theca cells have been identified as a source of the hormone (Bryant-Greenwood et al., 1980; Evans et al., 1983).

Rat Ovary

Long (1973) observed the presence of a distinct granule population in luteal cells of the pregnant rat. The duration of pregnancy is about 22 days in this species. The granule population first appears about day 14, increases to peak numbers by day 20, and is absent after parturition. These changes in granule population parallel the increase and decrease of relaxin content in the ovaries of the rat (Anderson et al., 1973a). The granules were suggested as a site of storage of relaxin in this species.

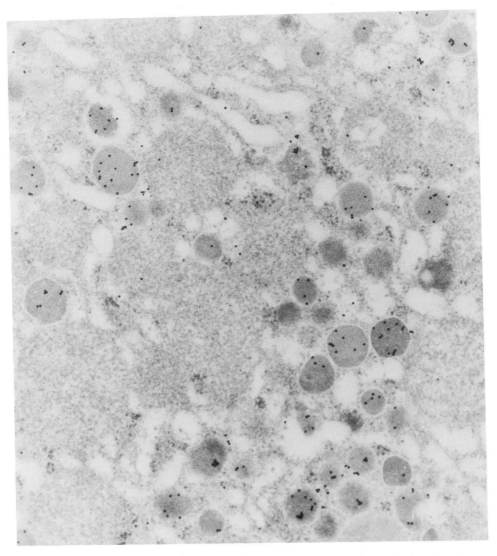

Fig. 4. Electron photomicrograph of a porcine luteal cell at day 110 of pregnancy. Immunocytochemistry using porcine relaxin antiserum (R6) and goat antirabbit immunoglobulin G-colloidal gold indicates the localization of colloidal gold particles (10 nm in diameter) overlaying the electron dense granules (200-600 nm in diameter) and their absence in other parts of the cellular cytoplasm. (Orci, L., Steinetz, B. G., and Anderson, L. L.; unpublished observations.)

Indirect immunofluorescence with rabbit anti-porcine serum suggested the presence of relaxin in the rat corpus luteum during late pregnancy but lacked specificity for the granule population (Anderson et al., 1975). Luteal cell fractionation studies indicated that the greatest relaxin bioactivity is present in the granule-rich fraction (Anderson and Long, 1978). The appearance and disappearance of the granules during pregnancy parallel not only ovarian relaxin bioactivity and immunoactivity but also peripheral serum levels of relaxin immunoactivity (Sherwood et al., 1980; Sherwood and Rutherford, 1981). With the use of light microscopy and a highly specific rabbit antirat-relaxin serum and the peroxidase-antiperoxidase complexing technique, relaxin was localized in the ovaries of rats

on days 12 and 20 of pregnancy (Golos et al., 1984). Ovaries from nonpregnant lactating rats contained very little, if any, relaxin immunostaining.
By using these same immunostaining techniques at the ultrastructural level,
relaxin was localized in electron-dense, membrane-bound granules (maximum
diameter, 270 nm), which are present in the cells during the last third of
gestation (Anderson and Sherwood, 1984). In the granule-rich luteal cells
the granules appear in clusters. These studies provided conclusive
evidence that the electron-dense granules represent the subcellular sites
of relaxin localization within luteal cells of pregnant rats. The utilization of antiserum to porcine relaxin and goat anti-rabbit immunoglobulin G-
colloidal gold revealed gold particles limited to the small granules in
luteal tissue from days 17 and 20 of pregnancy (Fields, 1984). Exocytosis
involving the incorporation of the granule membrane into the cell membrane
followed by release of the product was rarely observed. The results
provide unequivocal evidence that the corpus luteum is the principal source
of relaxin in the pregnant rat. With the use of a radiolabeled cDNA probe
for rat relaxin and histochemical techniques, relaxin mRNA was observed
within the corpus luteum of pregnancy but not in other parts of the ovary
(Hudson et al., 1981). Relaxin was undetectable in rat uteri, placenta and
metrial glands by biological and immunological methods (Bloom et al., 1958;
Kroc et al., 1959; Anderson et al., 1975; Sherwood et al., 1980).

Mouse Ovary

Relaxin bioactivity has been demonstrated in the ovaries of pregnant
mice (Steinetz et al., 1959). Immunocytochemical localization of relaxin
was found in luteal cells in ovaries from mice during the later half of
pregnancy (days 10-18) by use of rabbit antirat relaxin serum with the
unlabeled antibody peroxidase-antiperoxidase technique (Anderson et al.,
1984). Electron-dense membrane-bound granules were first observed on day
12, reached maximum numbers by days 16 and 18 of pregnancy and were absent
by day 2 postpartum. The appearance and disappearance of this granule
population paralleled relaxin immunostaining in mouse luteal cells.

Rabbit Placenta

The placenta is a major source of relaxin in the rabbit. This
initially was suggested by elevated blood levels of bioactive relaxin in
ovariectomized does with their pregnancy maintained by progesterone (Zarrow
and Rosenberg, 1953). Relaxin has been isolated and purified from rabbit
placental tissue and it yielded a molecular weight of approximately 7,200,
but its amino acid sequence is not yet known (Eldridge and Fields, 1985).
At the light microscopy level, immunohistochemical staining with guinea pig
antiporcine relaxin serum indicated that relaxin was located in the
syncytiotrophoblast cells of the placental labyrinth at days 23 and 30 but
not at day 16 of pregnancy. Examination of the fine structure of the
rabbit syncytiotrophoblast revealed membrane-bound granules (150-400 nm in
diameter) found near the Golgi and in close association with the cell
membrane (Eldridge and Fields, 1986). These granules labeled positively
for relaxin after treatment with guinea pig antirabbit relaxin serum and
goat antiguinea pig immunoglobulin G-colloidal gold. These results provide
evidence to suggest that relaxin is synthesized and secreted from the
syncytiotrophoblast of the rabbit placenta. Furthermore the subcellular
site of storage for relaxin is the electron-dense, membrane-bound granules
in this species (Eldridge and Fields, 1986). In contrast, the rabbit ovary
contains little immunohistochemical staining for relaxin.

Bovine Ovary

The ultrastructure of cells from corpora lutea of pregnant cows
reveals distinctive features of protein synthetic capacity (i.e., stacks of

granular endoplasm reticulum, prominent Golgi) and membrane-bound secretory granules (Fields et al., 1985). The duration of gestation is about 283 days in cattle. The granules (150-300 nm in diameter) appear to be packaged in the Golgi, accumulate at a paranuclear region, and migrate as a group to the cell membrane where they exocytose. They were first observed on day 45 and increased to peak numbers about day 200. The granules were found both in the small (10-15 μm in diameter) and large (20-50 μm in diameter) luteal cells, and there was no evidence of a transition of cell morphology as pregnancy advanced. Although there is evidence for relaxin in the bovine corpus luteum its presence in membrane-bound granules has not been confirmed (Fields et al., 1980; 1985). Oxytocin and neurophysin as well as mRNA for oxytocin have been isolated from the bovine corpus luteum, and there is evidence that these peptides are associated with the granules (Fields and Fields, 1986; Wathes et al., 1983; Schams et al., 1983; Ivell et al., 1985).

Ovine Ovary

The corpus luteum of the sheep contains secretory granules (Gemmell et al., 1974; Sawyer et al., 1979). Oxytocin is synthesized and secreted by the large luteal cells and its concentrations parallel those of progesterone during the estrous cycle (Wathes and Swann, 1982; Flint and Sheldrick, 1982; Webb et al., 1981). By using an immunogold complex, secretory granules (200-300 nm in diameter) for oxytocin were identified in large luteal cells (Theodosis et al., 1986). Relaxin was not detected in sections of ovary, endometrium or placenta by the avidin-biotin peroxidase complex or protein-A gold techniques (Renegar and Larkin, 1985). Although small dense granules were occasionally observed in tissues from ewes at days 45 and 140 of gestation, relaxin was not detected. Porcine relaxin and antiserum to porcine relaxin were used to radioimmunoassay relaxin in ovine corpora lutea, placentomes and intercotyledonary endometrium (Wathes et al., 1986). The major source of relaxin immunoreactivity was the placentomes, but there was no correlation of relaxin concentration with stage of pregnancy.

Guinea Pig Uterus

Earlier evidence obtained with bioassays demonstrated that the uterus was a source of relaxin in the guinea pig (Hisaw et al., 1944). Immunocytochemical localization of relaxin in cells of the endometrial glands of pregnant guinea pigs was indicated by light microscopy with peroxidase-antiperoxidase staining and antiserum to porcine relaxin (Pardo et al., 1980). The numbers of glands staining for relaxin increased to peak values just before parturition (Pardo and Larkin, 1982). Extracts of pregnant uteri contained biological activity as based upon the mouse interpubic ligament assay and cross-reactivity with antiserum raised against porcine relaxin in immunodiffusion plate assays. Furthermore, relaxin was localized in granules of endometrial gland cells when sections of uterus were treated with porcine relaxin antiserum and immunoglobulin G-labeled colloidal gold, and examined by electron microscopy (Larkin et al., 1983; Pardo et al., 1984). The uterine cervix and symphysis pubis are the principal target tissues for relaxin (Gates et al., 1981).

Human Ovary, Placenta and Decidua

The presence of bioactive and immunoreactive relaxin has been demonstrated in the human corpus luteum of pregnancy (Weiss et al., 1976; Weiss, 1983; Loumaye et al., 1978; O'Byrne et al., 1978a,b). Immunohistochemical evidence was presented by Mathieu et al. (1981) for relaxin in the human corpus luteum of pregnancy. This was based on light microscopy examination of luteal tissue stained by the peroxidase-antiperoxidase method and the use of antiserum to porcine relaxin. Relaxin has been identified in the basal plate cells of the human placenta with the immunoperoxidase technique and antiserum to porcine relaxin (Fields and Larkin, 1981). These basal plate cells likely arise from the cytotrophoblast and are distinct from decidual cells. The human decidua is considered a source of relaxin as based on three lines of evidence of relaxin bioactivity: 1) isolated from cultures of decidual cells; 2) extracted from decidua that was removed from the maternal surface of fetal membranes; and 3) indirect immunofluorescence of decidual cells with antisera raised against porcine relaxin (Bigazzi et al., 1980, 1982). The immunofluorescent reaction was found within the cytoplasm as well as the cell membrane. The fine structure of the decidual cells from human pregnancies at term revealed morphology typical for protein synthesis but with an absence of secretory granules (Bigazzi et al., 1983).

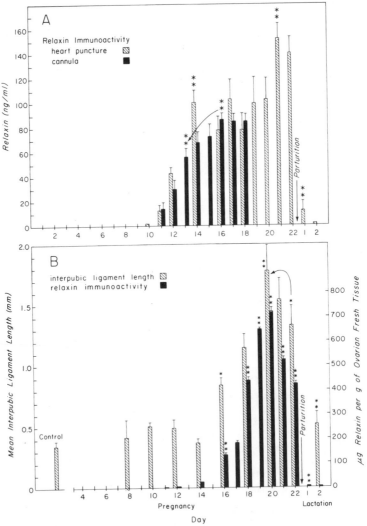

Fig. 5. (A) Relaxin immunoactivity levels in peripheral sera of anesthetized (heart puncture) and unanesthetized (cannula) rats. (B) Relaxin immunological and biological activities, as based on the interpubic ligament bioassay, as well as interpubic ligament length during pregnancy. Values are means ± SE. *P<0.05; **P<0.01. (From Sherwood et al., 1980).

RELAXIN SECRETION AND ITS REGULATION DURING PREGNANCY

Rat

In 1973a, Anderson and colleagues reported the levels of relaxin biological activity in the ovaries of the rat throughout pregnancy and during lactation. Relaxin increased only during the later half of pregnancy (days 12-20), declined just before parturition, and remained ' low during lactation. Relaxin levels remained consistently low between days 6 and 13 of pseudopregnancy. In pseudopregnant, hysterectomized rats or in animals bearing deciduomata, relaxin remained consistently low from days 6-21. The levels of relaxin were not related to the age of the corpus luteum since the diestrous interval is of about the same duration in hysterecto-mized animals or rats with deciduomata as in pregnancy. The results indicated that the conceptuses may stimulate the ovary to increase not only progesterone production (Hashimoto et al., 1968) but also relaxin produc-tion during the later half of pregnancy. With the purification and characterization of a 6,600 dalton rat relaxin, ovarian levels of relaxin were confirmed, and they corresponded to shifts in blood concentrations of the hormone during the later half of pregnancy (Figure 5A and B; Sherwood, 1979; Sherwood and Crnekovic, 1979; Sherwood et al., 1980). Interpubic ligament length increased during this same period of increasing immuno-active relaxin and estrogens in the peripheral blood. A prepartum surge in relaxin immunoactivity was interpreted to be associated with luteolysis. A surge in relaxin also was noted during parturition. In later studies, multiple forms of relaxin, designated C1, C2, and C3 representing apparent molecular weights of 69,000, .14,200, and 6,600, were found to be secreted during late pregnancy (Figure 6, Sherwood et al., 1984). The distribution of relaxin immunoactivity among these components changed in a consistent manner with day of pregnancy, but its physiological significance is unknown. The placenta is not the source of relaxin because the hormone becomes undetectable after ovariectomy (Goldsmith et al., 1981). It regulates ovarian secretion of relaxin, however, as indicated by: 1)

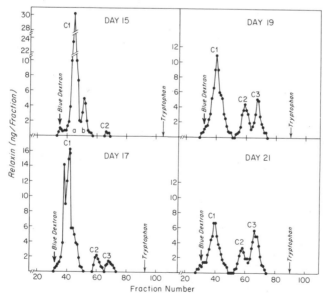

Fig. 6. Relaxin immunoactivity profiles obtained after gel filtration of serum on Sephacryl S-200 on different days of late pregnancy in the rat. Refer to text for molecular weight designations of C1, C2 and C3. (From Sherwood et al., 1984.)

greater relaxin blood levels as the number of conceptuses increases; and 2) continued secretion of high levels of relaxin after fetectomy (Figure 7, Golos and Sherwood, 1982). The maternal pituitary seems to suppress relaxin indirectly during the second half of pregnancy as indicated by greater levels of circulating relaxin as well as placental lactogen and chorionic gonadotropin after hypophysectomy (Blank and Dufau, 1983; Ochiai et al., 1983; Golos and Sherwood, 1984). This suppressive effect of the pituitary on corpus luteum function and its secretion of relaxin is not dependent upon the presence of nonluteal ovarian tissue (Golos and Sherwood, 1986). Furthermore, there are progressive changes in the distribution of the relaxin immunoactive components (C1, C2, and C3) as related to increasing numbers of conceptuses regardless of the presence or absence of the maternal pituitary (Sherwood et al., 1986). Indirect evidence was interpreted to indicate that maternal pituitary luteinizing hormone (LH) plays a role in the prepartum release of relaxin, luteolysis and parturition (Gordon and Sherwood, 1982). This was based on an observed delay in parturition and sustained levels of serum relaxin in hypophysectomized rats or in intact ones given LH-antiserum during late pregnancy. Whether endogenous LH is related to these events is not established. Regardless, the placenta markedly influences the relaxin production by the ovary in this species. This abrupt increase in relaxin secretion in intact rats reflects increased levels of preprorelaxin mRNA in the ovary from days 15 to 22 of pregnancy, and virtually undetectable at parturition (Crish et al., 1986a). These findings indicate that the concentration of relaxin in the rat is directly dependent on preprorelaxin mRNA levels available for translation. These abrupt changes in relaxin gene expression are important in regulating the amount of the peptide in the ovary, but the control of relaxin gene expression is not understood. Reducing the number of conceptuses results in lower levels of preprorelaxin mRNA in the ovaries from days 10 to 20 of pregnancy (Crish et al., 1986b), which correlates with reduced relaxin secretion as shown by Golos and Sherwood (1982). Since

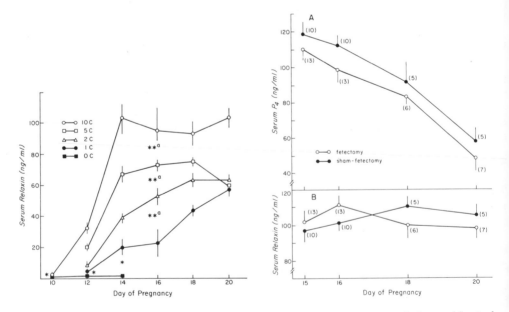

Fig. 7. (Left panel) Serum levels of relaxin in rats containing adjusted numbers of conceptuses (C). (Right panel) Serum progesterone (A) and relaxin (B) levels in fetectomized and sham-fetectomized animals. Number of observations is indicated in parenthesis. Values are means ± SE. *P < 0.05; **P < 0.01. (From Golos and Sherwood, 1982).

relaxin levels remain low in hysterectomized-pseudopregnant rats (Anderson et al., 1973a) and drop after hysterectomy of pregnant rats (Goldsmith et al., 1981), it is of interest to determine whether shifts occur in preprorelaxin mRNA levels in the ovaries after removal of the conceptuses. Hysterectomy of pregnant rats on day 10, before the rise in preprorelaxin and relaxin levels, results in continued low preprorelaxin mRNA concentrations in the ovaries to day 19 (Figure 8; Crish et al., 1986b). When hysterectomy is performed on day 14 after the rise in preprorelaxin and relaxin levels, preprorelaxin mRNA levels drop by day 16 and become undetectable by day 20 (Figure 8). These results clearly indicate that the conceptuses are necessary for both the marked increase in ovarian preprorelaxin that occurs at midgestation and the maintenance of these elevated levels during the second half of pregnancy in the rat. Although the stimuli for the increase and maintenance of ovarian preprorelaxin may depend indirectly on placental hormones (e.g., androgen, estrogens, placental lactogen, chorionic gonadotropin), these seem unrelated to the demise in ovarian preprorelaxin mRNA near the end of pregnancy and the prepartum surge in relaxin before luteolysis. Injections of 17β-estradiol and testosterone attenuate relaxin secretion after hysterectomy of pregnant rats (Goldsmith et al., 1981, 1982) as well as preprorelaxin mRNA levels in the ovary, but a nonaromatizable androgen, 5α-dihydrotestosterone, was in-

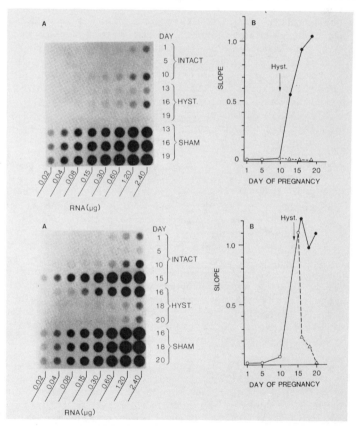

Fig. 8. Preprorelaxin mRNA levels in ovaries of mated (●) and hysterectomized (△) rats at day 10 (upper panel) and day 14 (lower panel) as compared with intact controls. (A) Dot blot analyses of total ovarian RNA filtered on nitrocellulose, hybridized with ^{32}P-labeled preprorelaxin cDNA and exposed to X-ray film. (B) Preprorelaxin mRNA levels were quantitated densitometrically. (From Crish et al., 1986b).

effective (Crish et al., 1986b). Although preprorelaxin mRNA levels are 200-fold less during the estrous cycle as compared with late pregnancy, there are greater concentrations in estrous vs proestrous rats. These albeit low preprorelaxin mRNA levels during estrus and at the time of mating may entrain relaxin gene expression at midgestation and its termination near parturition in the rat.

The uterus and cervix are known to be the primary target tissues for radiolabeled relaxin in the rat (Cheah and Sherwood, 1980; Weiss and Bryant-Greenwood, 1982). High endogenous levels of relaxin during the later half of gestation in the rat attribute to normal fetal development and preparation for parturition. The myometrium remains quiescent from day 20 until a few hours before the onset of labor (Fuchs, 1978). The effects of PGE and PGF are thought to be uterotonic while thromboxane A_2 (TxA_2) and prostacyclin (PGI_2) have vasoconstrictor and vasodilator properties. A series of prostanoids that include PGF, PGE, TxB_2 (a stable metabolite of TxA_2) and 6-keto-$PGF_{2\alpha}$ (a stable metabolite of PGI_2) are produced by the uterus and they are significantly augmented during late pregnancy (Parnham et al., 1975; Wilson and Huang, 1983). The fetoplacental unit markedly increases their net production in adjacent uterine tissue as contrast to that in an empty uterine horn. On the last days of pregnancy (days 20 and 21), however, factors from the systemic circulation impinge upon the uterus that are responsible for the acute increases in uterine prostanoids by tissues adjacent or non-adjacent to the fetoplacental unit (Wilson and Huang, 1985). Relaxin may be important in this and other species during periods of declining progesterone and increasing estrogen domination of the uterus near term for prevention of premature parturition (Porter, 1979a,b). This has been demonstrated indirectly by maintaining pregnancies in ovariectomized animals with progesterone and estrogen and in the presence or absence of relaxin (Kroc et al., 1959; Downing and Sherwood, 1985a,b,c; 1986). Relaxin treatment in such animals reduces the length of gestation, duration of labor and delivery, and maintains fetal survival rates similar to those in intact controls, whereas these events are prolonged with reduced fetal survival in the absence of relaxin. Progesterone inhibits myometrial activity during early and midgestation but later, when progesterone decreases, its inhibitory role may be taken over by relaxin myometrial receptors (Porter, 1979b). Endogenous estrogens in cyclic rats and estrogen injected into ovariectomized animals can increase uterine relaxin receptors (Mercado-Simmen et al., 1982a). Relaxin in conjunction with estrogen plays an important role in reducing myometrial activity during the second half of pregnancy. For example, specific receptors for relaxin have been described in rat uterine membrane-enriched fractions (Mercado-Simmen et al., 1980). During normal pregnancy they increase from approximately 10 pmol/mg protein at day 13 to a peak of 160 pmol/mg protein by day 17, and then drop to < 10 pmol/mg protein 2 days before parturition, a time when endogenous blood levels of relaxin are elevated (Mercado-Simmen et al., 1982a; Sherwood et al., 1980).

Estrogen enhances the binding activity of the uterus for relaxin by increasing the concentration of myometrial relaxin receptors, that is coincident with increasing blood levels of relaxin. The sensitivity of the uterus diminishes, however, by continual exposure to increasing levels of relaxin and results in reduced numbers of available myometrial receptors in late pregnancy. Endogenous relaxin blood levels are basal at the onset of increased intrauterine pressure cycles that signal the onset of labor (Downing and Sherwood, 1985b). Thus, relaxin receptors in the uterus may be initially enhanced by estrogen and later reduced as a result of decreased responsiveness of the uterus by relaxin, but estrogen and relaxin may not be the only modulators of relaxin receptor number. Estrogen can increase myometrial receptors for oxytocin; however, the numbers of them during normal pregnancy are unrelated to blood levels of estrogens (Soloff,

1975). Uterine receptors for oxytocin in pregnant rats remain low throughout pregnancy and increase markedly only at the onset of parturition
(Soloff et al., 1979). The paucity of oxytocin receptors before term may
explain the inability of oxytocin to induce parturition more than a few
hours before delivery. During the late prepartum period the peak blood
levels of relaxin, in the presence of decreasing numbers of relaxin
receptors, may allow endogenous estrogens to maximally stimulate oxytocin
receptors in preparation for the neurogenic release of oxytocin at the
onset of parturition. Relaxin acutely decreases prostaglandin $F_{2\alpha}$-induced
uterine contractions but some desensitization of the uterus occurs during
prolonged infusion of relaxin (Porter et al., 1981; Cheah and Sherwood,
1981). Thus, the uterine contractions necessary for the delivery of the
fetuses may be stimulated by oxytocin or prostaglandin $F_{2\alpha}$, or a combination of them.

Relaxin inhibition of myometrial contractile activity may be mediated
by mechanisms that include: 1) an increase in cAMP; 2) a decrease in myosin
light chain phosphorylation; 3) alteration of the kinetic properties of
myosin light chain kinase; and 4) a rapid calcium efflux from the cells
(Nishikori et al., 1982; Hsu et al., 1985; Hsu and Sanborn, 1986a,b; Rao
and Sanborn, 1986). Contraction and relaxation of smooth muscle seem to
require the phosphorylation and dephosphorylation of 20,000-molecular
weight light chains (LC_{20}) by a specific myosin light chain kinase (MLCK)
(Adelstein and Eisenberg, 1980). Phosphorylation of LC_{20} by MLCK increases
ATPase activity and muscle contraction while phosphatases reverse phosphorylation. Relaxin treatment alters the kinetic properties of myometrial
cell MLCK that correlates with increases in intracellular calcium efflux
within 1 minute (Hsu and Sanborn, 1986a; Rao and Sanborn, 1986). Relaxin
can relax uterine strips within 2 minutes and myometrial cells within 1
minute without a detectable increase in cAMP (Hsu and Sanborn, 1986a). In
the presence of 1-methyl-3-isobutyl xanthine (MIX) or forskolin, relaxin
elevates cAMP concentrations in a time- and dose-dependent manner in
myometrial but not in stromal cells. Prostacyclin (PGI_2) is present in the
myometrium of some species (e.g., rats, sheep) and it is a potent inhibitor
of uterine activity, but a possible inhibitory role during pregnancy is
undefined (Williams et al., 1978; Lye and Challis, 1982). The role of cAMP
in smooth muscle relaxation is unclear, but it is known that a α-adrenergic
mediated uterine relaxation is not required as a mechanism for relaxin's
action (Sanborn et al., 1981). The cAMP elevation induced by relaxin may

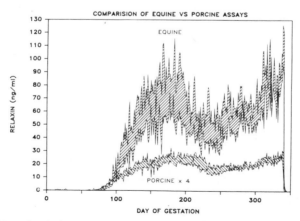

Fig. 9. Relaxin concentrations (standard error of
mean values) in peripheral plasma of pregnant mares throughout gestation were determined by porcine and equine radioimmunoassays. (From Stewart, 1986).

contribute directly to uterine relaxation via regulation of MLCK activity
by cAMP-dependent protein kinase, or possibly indirectly by altering
calcium dynamics in the myometrium.

Cervical distensibility, dilatation and softening are critical for
passage of the fetus at birth. In the rat, these processes begin by day 12
and increase steadily to parturition (Harkness and Harkness, 1959). The
cervical changes include increases in water content, solubility and
reorganization of collagen as well as hyaluronic acid and is characterized
by changes in tensile (creep) properties (Harkness and Harkness, 1959;
Golichowski et al., 1980; Williams et al., 1982). Cervical and uterine
motility seem to be independent during the estrous cycle and pregnancy.
Progesterone and estrogen treatment in rats ovariectomized during late
pregnancy is inadequate for normal cervical dilatation at parturition (Kroc
et al., 1959). The addition of relaxin, however, allows similar cervical
distensibility as found in intact controls (Hollingsworth et al., 1979;
Downing and Sherwood, 1985c); it also rapidly promotes uterine collagen as
well as glycogen and protein synthesis (Vasilenko et al., 1980; Frieden and
Adams, 1985). Relaxin causes the cervix to transform dense collagen fiber
bundles to a loose network of collagen fibers surrounded by extracellular
matrix containing hexosamines and glycosaminoglycans, and this process
begins as early as day 9 (Golichowski et al., 1980; Fosang et al., 1984).
Estrogen blood levels remain low until day 20 of pregnancy and peak near
parturition (Shaikh, 1973). High levels of estrogen in combination with
relaxin stimulate collagen synthesis at term (Downing and Sherwood, 1985c).
During the course of pregnancy and the postpartum period there are marked
differences in the activities of collagenase, proteoglycanase and β-
glucuronidase in the rat uterus and cervix (Too et al., 1986). These
enzymes are major determinants of the integrity of the extracellular matrix
of these tissues. In estrogen primed animals the addition of relaxin can
overcome the inhibiting effect of estrogen on uterine proteoglycanase
secretion without affecting β-glucuronidase levels. In the cervix,
however, relaxin decreases collagenase and proteoglycanase secretion while
not affecting β-glucuronidase levels. Thus relaxin clearly has an effect
on the enzymes involved in collagen and proteoglycan degradation and likely

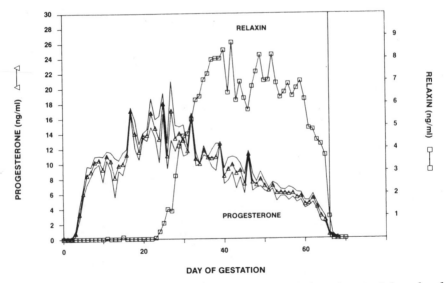

Fig. 10. Relaxin and progesterone immunoreactivity in peripheral plasma
during pregnancy in the cat. Progesterone values are means ± SE.
(From Stewart and Stabenfeldt, 1985.)

plays an important role in the remodeling of the uterus and cervix which occurs during pregnancy and immediately postpartum.

Horse

The equine placenta is the primary source of relaxin (Stewart and Stabenfeldt, 1981; Stewart et al., 1982a,b). Equine relaxin was purified and partially characterized from term placentas and used to develop a homologous radioimmunoassay (Stewart, 1986; Stewart and Papkoff, 1986). Although daily profiles of plasma levels of equine and porcine relaxin immunoreactivity were of different magnitude, they were comparable in mares throughout gestation (Figure 9). Relaxin was observed from about day 75 of gestation through foaling. The physiological significance of placental relaxin and its role in parturition are unknown in this species.

Cat

Relaxin plasma immunoreactivity during gestation in the cat suggests that the placenta is the primary source of the hormone, but it has not been identified in placental tissue (Stewart and Stabenfeldt, 1985). This evidence is based on a porcine relaxin antiserum (R6). During early pregnancy when the corpora lutea are functional, relaxin immunoreactivity was not detected (Figure 10). The onset of relaxin secretion between days 20 and 25 suggests that the placenta either is secreting relaxin or it is stimulating relaxin production by the ovary. The relaxin pattern is similar to that in the mare. Unlike the pig and rat, prepartum surges of relaxin are not evident in the cat or mare (Figures 9 and 10). Proges- terone profiles during gestation in the cat differ markedly from relaxin and implicate the ovaries as the major source of this steroid. During pseudopregnancy in the cat the corpora lutea do not produce detectable amounts of relaxin. The presence of relaxin in peripheral plasma during the later two-thirds of pregnancy in the cat may play a role in uterine quiescence, but little is known concerning the physiological significance of this hormone in this species.

Cattle

Immunoreactive relaxin concentrations in peripheral blood serum of beef heifers, as based on porcine relaxin antiserum, are low during late pregnancy and increase only on the day of parturition (day 283; Anderson et al., 1982b). In heifers bled sequentially during a 30-hour period on day 273 of pregnancy, endogenous relaxin in peripheral plasma ranged from undetectable levels (<180 pg/ml) to 1 ng/ml (Figure 11; Musah et al.,

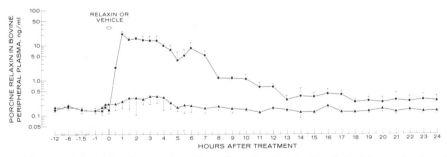

Fig. 11. Peripheral plasma levels of relaxin in beef heifers during late pregnancy after administration of porcine relaxin (3000 U; n = 4 heifers, ●) and vehicle (0.1 M PBS or 0.1 M PBS-gel; n = 4 heifers, ▲) either im or in the cervical os on day 273 of gestation. Values are means ± SE. (From Musah et al., 1987a.)

1987a). Administration of porcine relaxin increases both pelvic area and cervical dilatation in beef heifers during late pregnancy (Perezgrovas and Anderson, 1982; Musah et al., 1986a,b). Intracervical application of either 3000 U relaxin once or twice (two infusions 12 hours apart) on day 277-278 caused a highly significant increase in pelvic area and cervical dilatation within 12 hours after treatment as compared with control heifers given 1 ml 0.01 M PBS-gel vehicle (Figure 12; Musah et al., 1986a). Relaxin treatment causes an increase in both the rate and linear increase as well as absolute increase in pelvic height and width as compared with those parameters in gel-vehicle-treated controls. Heifers of three genetic frame sizes (small, medium and large) were represented in each treatment group in this study, and the growth in pelvic height and width differed markedly among the different frame sizes, but cervical dilatation was unrelated to frame size (Musah et al., 1986b). Thus, these relatively low blood concentrations of relaxin in late pregnancy may inhibit uterine myometrial activity and modulate oxytocin release or action. Furthermore, the low blood levels of relaxin may contribute to calving ease by greater pelvic and cervical development as compared with undetectable immunoreactive relaxin in those heifers experiencing difficult calving (dystocia). The mechanism by which relaxin may mediate pelvic canal expansion in cattle however, is unknown.

The intervals between administration of relaxin or the PBS-gel vehicle and calving were 2.0, 2.5, and 5.3 days for heifers given relaxin-double, relaxin-single, and PBS-gel vehicle, respectively; duration of gestation

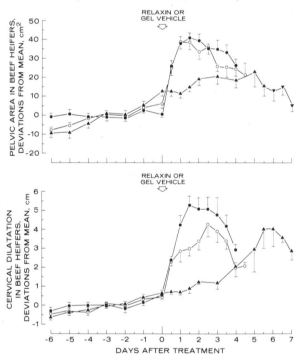

Fig. 12. Sequential changes in pelvic area (upper) and cervical diameter (lower) in beef heifers receiving relaxin-double (● ; n = 17), relaxin-single (O ; n = 14), and gel vehicle (▲; n = 16). Day of treatment is day 0. Values are means ± SE. (From Musah et al., 1986b.)

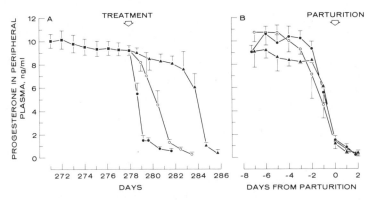

Fig. 13. Progesterone concentrations in peripheral plasma in relation to day of treatment (A) and day of parturition (B) in beef heifers receiving relaxin-double (● ; n = 17), relaxin-single (○ ; n = 14), and gel vehicle (▲ ; n = 16). ■; Sequential profiles of progesterone in plasma in all heifers before treatment. Day of parturition = day 0. Values are mean ± SE. (From Musah et al., 1986b).

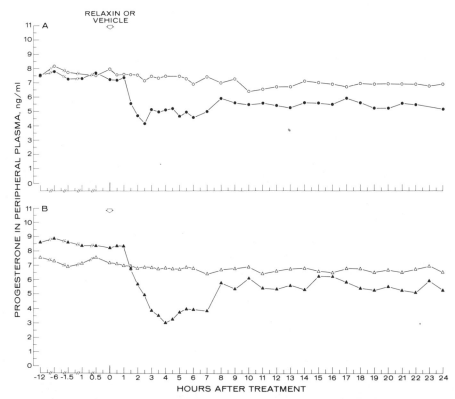

Fig. 14. Mean blood plasma progesterone levels after cervical os (A; n = 4 heifers) and intramuscular (B; n = 4 heifers) administration of porcine relaxin (3000 U; ● and ▲) and vehicle (○ and △) at 0 hours on day 273 of gestation. Pooled SE = 1.28 and 1.77 ng/ml for A and B, respectively. (From Musah et al., 1987a.)

was significantly reduced (P<0.002) by relaxin compared with control heifers (Musah et al., 1986a). The shifts in progesterone, estrone and 17β-estradiol secretion reflect premature parturition induced by relaxin in cattle (Musah et al., 1986a). Progesterone plasma levels decreased soon after relaxin treatment (Figure 13), whereas estrone and 17β-estradiol concentrations peaked earlier than found in gel-vehicle control heifers. There was no incidence of retained placenta beyond 24 hours postpartum (0 of 31 heifers), whereas the placenta was retained in 2 of 16 controls. The intracervical or intramuscular administration of porcine relaxin (3000 U) to primiparous beef heifers during late pregnancy (10 days before expected parturition) caused an acute elevation in peripheral plasma levels of relaxin (P<0.01), a rapid depression of progesterone (P<0.01), and a transient elevation (P<0.05) of estrone and 17β-estradiol (Figures 11, 14, and 15; Musah et al., 1987a). A significant decrease in blood levels of progesterone was evident in relaxin-treated heifers as early as 90 minutes after treatment (Figure 14). The decrease in progesterone secretion may be complete and thereby leading to parturition (Figure 13), or incomplete and followed by a rebound (Figure 14) (Musah et al., 1986b, 1987a). The mechanisms by which exogenous porcine relaxin mediates luteolysis in beef heifers are unknown. Estrogen biosynthesis during pregnancy in cattle is primarily from the placenta, which derives steroids from the maternal compartment. Progesterone production at this time is primarily of luteal origin. The rapid decrease in progesterone secretion caused by exogenous relaxin could make available precursors for the biosynthesis of estrogens by the placenta. The significant rise in estrone and 17β-estradiol plasma levels occurs within 4 hours after the abrupt decrease in progesterone secretion. We speculate that both oxytocin and prostaglandin $F_{2\alpha}$ may be involved in a feedback interaction with relaxin to mediate these changes.

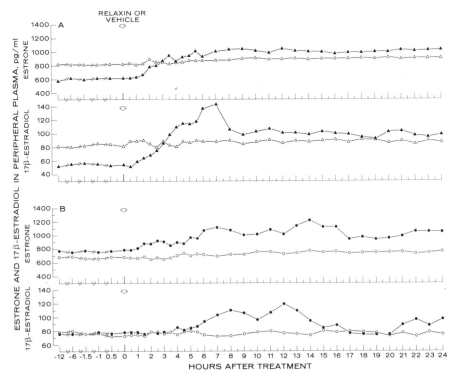

Fig. 15. Mean peripheral plasma estrone and 17β-estradiol levels after intramuscular (A; n = 4 heifers) and cervical os (B; n = 4 heifers) administration of porcine relaxin (3000 U; ▲ and ●) and vehicle (△ and ○). Estrone pooled SE = 173 pg/ml and 17β-estradiol pooled SE = 32.6 pg/ml. (From Musah et al., 1987a.)

Induction of parturition with PGF$_2\alpha$, glucocorticoids or their analogues often results in a high incidence of dystocia, retention of fetal membranes, and a reduction in subsequent fertility (Laster et al., 1973; Paisley et al., 1986). When relaxin treatment is combined with cloprostenol or dexamethasone 10 days before expected parturition, the induction interval is shortened, and the incidence and duration of retained fetal membranes are markedly reduced in beef heifers (Musah et al., 1986c). This may result, at least in part, from the effects of relaxin on the loosening of connective tissues of the placentomes. Furthermore, relaxin combined with cloprostenol or dexamethasone reduces the incidence of dystocia (Musah et al., 1986d, 1987b). Pelvic area and cervical dilatation are increased markedly in cloprostenol-treated heifers given relaxin either intramuscularly or in the cervical os as compared with cloprostenol-treated controls (Figure 16). Similar pelvic development and cervical dilatation were observed in heifers given relaxin in combination with dexamethasone 10 days before expected parturition.

Overall these results provide strong evidence that relaxin causes an early (e.g., beginning within 90 minutes) and acute decrease in progesterone secretion, followed by a significant elevation in estrogen secretion within a few hours in beef heifers during late pregnancy. Exogenous relaxin induces earlier parturition without complications of dystocia and retained placenta. When it is used in combination with a prostaglandin or a glucocorticoid, earlier calving results and problems of dystocia and retained placenta are markedly reduced. The mechanisms by which relaxin affects parturition in cattle require further study.

Pig

Functional corpora lutea are essential to define the interestrous interval whether relating to the nongravid or gravid state in the pig. A feature unique to this species is its dependence on corpora lutea for their production of progesterone throughout pregnancy; ovariectomy at any time results in abortion within 36 hours (du Mesnil du Buisson and Dauzier, 1957; Belt et al., 1971). Pregnancy can be maintained in ovariectomized

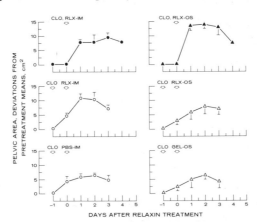

Fig. 16. Mean deviations in pelvic area (cm^2) from pretreatment means in primiparous beef heifers after administering relaxin (RLX: 3000 U) either in the cervical os (OS) or intramuscular (IM) with cloprostenol (CLO; 500 µg) during late pregnancy. There were 5 heifers in each group; values are means ± SE. (From Musah et al., 1986d).

gilts by exogenous progesterone sufficient to maintain plasma progesterone levels of 4-5 ng/ml (Ellicott and Dziuk, 1973). Removal of the corpora lutea (luteectomy) decreases progesterone and relaxin and increases $PGF_{2\alpha}$-metabolite in maternal plasma, and prematurely terminates gestation without causing onset of lactation (Kertiles and Anderson, 1979; Nara et al., 1981). There is a high incidence of stillbirths after ovariectomy or luteectomy in the pig. Removal of follicles, however, does not affect pregnancy or parturition (Nara et al., 1981). During brief 21-day estrous cycles the corpora lutea develop (> 450 mg) and progesterone secretion reaches peak concentrations (e.g., > 30 ng/ml in peripheral blood) from days 8-12; luteolysis occurs rapidly after day 16 (Masuda et al., 1967; Hard and Anderson, 1979). Luteolysis after day 13 in this species is associated with an increase in plasma $PGF_{2\alpha}$ and corresponding decreases in progesterone and the proportion of cholesterol bound to cytochrome P450 side-chain cleavage enzyme ($P450_{scc}$) in luteal tissue (Gleeson et al., 1974; Torday et al., 1980). Prostaglandin $F_{2\alpha}$ administration at this time in hysterectomized gilts causes a similar decrease in progesterone and $P450_{scc}$. These changes in $P450_{scc}$ are consistent with a direct block of LH stimulation by $PGF_{2\alpha}$. In this species hypophysectomy after the preovulatory LH surge is followed by normal development of the corpora lutea to day 10 and normal luteal regression after that time (du Mesnil du Buisson and Léglise, 1963; Anderson et al., 1967). After the initial LH stimulation at estrus, the porcine corpora lutea do not require pituitary support until after day 10, and they are refractory to $PGF_{2\alpha}$ during that time (Diehl et al., 1974). Thus, once formed the porcine corpora lutea do not require luteotropic support up to 10 days. At the time of luteolysis, $PGF_{2\alpha}$, possibly of uterine origin, may cause a decrease in cholesterol available to $P450_{scc}$ and the subsequent decrease in progesterone production. Since corpora lutea in cyclic gilts contain little relaxin, this hormone likely plays no significant role in luteal demise during the estrous cycle. Relaxin levels in luteal tissue remain relatively low during this period of maximal progesterone secretion in cyclic gilts (Figure 2; Anderson et al., 1973b; Sherwood and Rutherford, 1981). Human chorionic gonadotropin (hCG) mimics the preovulatory LH surge and delays luteolysis in cyclic gilts and in hysterectomized pigs following hypophysectomy (Anderson et al., 1967; Guthrie and Rexroad, 1981). Exogenous estrogen also delays luteolysis in cyclic gilts and reduces uterine secretion of $PGF_{2\alpha}$ (Gardner et al., 1963; Frank et al., 1977, Moeljono et al., 1977). Injection of hCG at midcycle delays luteolysis and causes a transitory increase in plasma estrogen concentrations (Guthrie and Bolt, 1983). The absence of elevated $PGF_{2\alpha}$-metabolite plasma levels, typically observed at the termination of a normal luteal phase, may result from an antagonism of hCG on the action of endogenous $PGF_{2\alpha}$ or possibly a reduction in uterine luteolytic activity caused by increased follicular estrogen production. Exogenous estrogen in nongravid gilts extends luteal function beyond 116 days, and these aging corpora lutea produce relaxin in amounts similar to those in pregnant and hysterectomized gilts (Anderson et al., 1973b).

After mating, the corpora lutea remain large (> 450 mg) throughout pregnancy and regress (< 75 mg) soon after parturition which occurs about day 114. Progesterone reaches peak blood levels by days 8-12 and remains elevated (e.g., $\cong$20 ng/ml peripheral blood) until an abrupt decrease to basal levels near parturition (Figure 2; Adair et al., 1982). During pregnancy in the sow two morphologically distinct cell types (large and small) have been isolated (Lemon and Loir, 1977). The large luteal cells contribute the major part of progesterone whereas the smaller cells secrete small quantities of progesterone in the absence of gonadotropic stimulation. When the two cell types are recombined in vitro they secrete more progesterone than released by each cell type alone (Lemon and Mauléon, 1982). Furthermore, superfusate from one cell type causes increased progesterone by the other (large) cell type. By use of a reverse hemolytic plaque assay for porcine relaxin, it was found that a proportion (50%) of the large (25 μm in diameter) porcine luteal cells is capable of secreting relaxin whereas the small ones do not release relaxin (Taylor et al., 1987). The relaxin secretory capacity of cells increases with advancing stages of pregnancy (e.g., days 25-80). Gilts can maintain pregnancy and deliver normal piglets following stepwise reduction of the number of corpora lutea to as few as one to three (Thomford et al., 1984). Progesterone secretion is maintained and the normal prepartum peak plasma levels of relaxin are comparable to normal pregnant gilts. The corpora lutea are the major source of relaxin in this species but seemingly few of them are capable of secreting sufficient amounts of relaxin to be compatible with normal pregnancy and parturition. With such few corpora lutea there results a marked increase in the number of large nonluteinized follicles. The corpora lutea may act normally to suppress follicular development either directly or at the level of the hypothalamus and pituitary gland. Likewise, in hysterectomized gilts with aging corpora lutea (day 174) progesterone serum levels decrease and the number of large cystic follicles increases possibly as a result of insufficient inhibition of a tropic factor(s) (Musah et al., 1984). It seems that an increase to peak blood levels of relaxin must precede the onset of lactation (Sherwood et al., 1978; Nara et al., 1982b). The ovaries are quiescent during lactation and relaxin and progesterone remain at low to undetectable levels. In contrast, relaxin blood levels remain low to day 28 of pregnancy and increase steadily to peak values just preceding parturition (Belt et al., 1971; Sherwood et al., 1975; 1978; Anderson et al., 1983). Estrone and 17β-estradiol are primarily of fetal-placental origin in this species, and circulating levels of these steroids peak just preceding parturition (Fèvre et al., 1968; Anderson et al., 1983). Blood levels of both estrogens at day 12 of pregnancy are about 10 pg/ml and they increase to a peak (estrone, 2,600 pg/ml; 17β-estradiol, 300 pg/ml) just preceding parturition. Thus progesterone and relaxin production and release by the porcine corpora lutea differ markedly throughout pregnancy.

The hysterectomized gilt serves as a useful model to examine the mechanisms regulating the secretion of these hormones. Corpora lutea in gilts hysterectomized during the early part of the estrous cycle are fully developed by day 8 (> 450 mg) and are maintained until day 150 (Anderson et al., 1983). The circulating levels of progesterone in hysterectomized gilts are consistently greater than those during corresponding stages throughout normal pregnancy (Adair et al., 1982). After hysterectomy, peripheral blood levels of estrone and 17β-estradiol remain consistently low (< 20 pg/ml) from days 12-114 and in marked contrast to those found during pregnancy (Anderson et al., 1983). Exogenous 17β-estradiol between days 45-108 decreases progesterone secretion by aging corpora lutea in hysterectomized gilts compared with that in sesame oil-treated controls (Musah et al., 1984). This decreased level of progesterone secretion is similar to that found during the later half of normal pregnancy (Adair et al., 1982). Progesterone secretion in hysterectomized gilts is greater than that during pregnancy at a time when circulating estrone and 17β-estradiol levels are low in the former and peak in the later (Adair et al., 1982; Anderson et al., 1983). Uterine or conceptus metabolism of progesterone to estrogens may partly account for the lower concentrations of progesterone in pregnant compared with hysterectomized gilts. Luteotropic support of young corpora lutea is primarily by luteinizing hormone (LH) (Anderson et al., 1965, 1967), whereas in older corpora lutea, prolactin may play a prominent role (du Mesnil du Buisson, 1973). For example, in hypophysectomized-hysterectomized gilts, exogenous LH maintains luteal function from days 20-30 (Anderson et al., 1965; 1967). The mechanisms by which estrogen decreases progesterone secretion in hysterectomized gilts may relate to the high secretion rates of estrone and 17β-estradiol during

Table 2. Relaxin Concentrations in Fresh Tissue Homogenates of Corpora Lutea Obtained at Sequential Laparotomies of Gilts during Late Pregnancy and Early Lactation compared with Those in Aging Corpora Lutea of Hysterectomized Gilts

Repro-ductive Stage	No. of Gilts	Relaxin in Porcine Luteal Tissue (μg/g)			
		Day 100	Day 112	Day 124	Day 136
Pregnancy and lac-tation	4	2001 ± 288	918 ± 108	0.053 ± 0.006	0.003 ± 0.001
After hys-terec-tomy[a]	4	1825 ± 416	1113 ± 231	700 ± 222[b]	333 ± 39[c]

Values are the mean ± SE.
[a]Corpora lutea in unmated gilts were marked with black silk suture at hysterectomy on day 6 (day 0 = estrus).
[b]$p < 0.025$.
[c]$p < 0.001$.
Anderson et al. (1983).

the later half of gestation. Thus, in the absence of the uterus exogenous estrogen may decrease progesterone secretion by indirectly decreasing adenohypophysial hormone secretion, directly by inhibiting luteal function, or a combination of both.

Relaxin concentrations in peripheral blood serum are low in hysterectomized gilts on day 6 and similar to those in pregnant animals (Anderson et al., 1983). During the period from days 6-114, every sixth day bleeding revealed a steady increase in relaxin concentrations from day 6 to peak levels on day 114 in both groups of hysterectomized and pregnant gilts. From days 18-114 serum relaxin levels in hysterectomized gilts remain consistently higher than those in pregnant gilts. After parturition relaxin is low and remains low from days 120-168 in lactating dams. During this same period relaxin in peripheral serum of hysterectomized gilts remains significantly greater than that in lactating dams. Relaxin concentrations in fresh tissue homogenates of aging corpora lutea show abrupt changes in hysterectomized and pregnant/lactating gilts (Table 2). Relaxin concentrations were highest on day 100 and similar in pregnant and hysterectomized gilts. By day 112 luteal tissue relaxin decreased by half in both groups. The corpora lutea remain large (averaging > 450 mg) on days 124 and 136 in hysterectomized gilts, whereas they regress (averaging < 75 mg) in the lactating dams. Relaxin concentrations in luteal tissue from hysterectomized gilts on days 124 and 136 are markedly greater than in lactating dams (Table 2).

These results indicate that the porcine estrous cycle is controlled by the transitory span of the corpus luteum and its dependence upon the luteolytic action of the uterus. The production of primarily progesterone, but not relaxin, by granulosa lutein cells seems to regulate these relatively brief 21-day cycles. Prolonged extension of the life span of the porcine corpus luteum for not only progesterone but also relaxin production and release requires the presence of the conceptuses or complete absence of the uterus.

Maximal plasma concentrations of relaxin occur about 15 hours before parturition, and they are associated with elevated endogenous plasma $PGF_{2\alpha}$ and termination of luteal function (Belt et al., 1971; Sherwood et al., 1978; 1979; 1981; Nara and First, 1981b). This prepartum relaxin surge also is associated with a decline in circulating progesterone, which seems to be initiated by an increase in $PGF_{2\alpha}$ before parturition. Parturition can be induced in this species by dexamethasone, prostaglandin $F_{2\alpha}$ or prostaglandin E_2 (Diehl et al., 1974; Coggins and First, 1977; Vaje et al., 1980). A nonluteolytic dose (50 μg) of $PGF_{2\alpha}$, as based on maintenance of progesterone blood levels in late pregnant gilts, causes peak relaxin concentrations in peripheral blood within 10 minutes (Nara et al., 1982a). Thus, $PGF_{2\alpha}$ can cause relaxin release from porcine corpora lutea independent of a luteolytic effect. Whether $PGF_{2\alpha}$ acts through separate mechanisms for relaxin release and luteolysis or whether these two events may be affected in a similar manner in the luteal cell was not established. Parturition can be delayed by maintaining high endogenous blood levels of progesterone with injected progesterone or endogenous progesterone from induced corpora lutea (Coggins et al., 1977). Although injected progesterone delays parturition, the prepartum relaxin surge occurs at the normal time (e.g., day 113) (Sherwood et al., 1978). Thus, the surge in relaxin levels in pigs experiencing progesterone-delayed parturition is not sufficient by itself to initiate parturition within 24 hours. Furthermore, relaxin blood levels are markedly increased approximately 1 hour after administration of a luteolytic dosage (5 mg) of $PGF_{2\alpha}$ (Sherwood et al., 1979). The prepartum surge in plasma relaxin occurs coincident with an increase in $PGF_{2\alpha}$ and a decrease in progesterone when parturition is advanced, delayed or occurs at the expected time (Nara and First, 1981b). Prevention of prostaglandin synthesis by indomethacin prevents release of relaxin as well as luteolysis whereas simultaneous treatment with $PGF_{2\alpha}$ induces luteolysis and relaxin release (Vaje et al., 1980; Nara and First, 1981a,b). Although indomethacin prevents relaxin release, after such drug treatment a prepartum relaxin surge occurs just preceding the delayed parturition (Sherwood et al., 1979). These results were interpreted to indicate that release of relaxin depends upon and is a result of luteolysis.

When only a small fragment of uterus remains after partial hysterectomy, proximal connections between the remaining uterine horn and its adjacent ovary are required for luteal regression in the pig (Anderson et al., 1969). Estrous cycles continue, however, after autotransplantation of the uterus to the abdominal wall (Anderson et al., 1963; du Mesnil du Buisson and Rombauts, 1963). Furthermore, pregnancy can proceed normally following the transfer of embryos to the uterus of gilts with ovaries autotransplanted to the uterus or abdominal wall (Martin et al., 1978). Thus, usual vascular and neural connections between the uterus and ovary are not an absolute requirement for maintenance of pregnancy and normal parturition in this species. A signal for maintenance of corpora lutea of pregnancy or their demise at parturition need not require a proximal route between the uterus and ovaries, but apparently can be conveyed by a systemic route. That fetal corticosteroids may initiate synthesis or release of $PGF_{2\alpha}$ and the events leading to parturition in the pig is suggested by premature increases in $PGF_{2\alpha}$-metabolite blood levels caused by dexamethasone and prevented by indomethacin in the presence of dexamethasone (Nara and First, 1981a).

Exogenous porcine relaxin induces marked cervical dilatation during late pregnancy in gilts. Relaxin, given daily from 105 days after mating and before luteectomy on day 110, significantly a) induced premature cervical dilatation, b) decreased the interval from luteectomy to delivery of the first neonate and c) reduced the duration of delivery of all neonates in the litter compared with those parameters in luteectomized controls. Cervical growth and distensibility in this species depend upon both estrogens and relaxin. Exogenous porcine relaxin induced cervical dilatability at a stage in pregnancy when endogenous blood levels of estrone and 17β-estradiol reach peak values and levels of progesterone begin a gradual decline (Kertiles and Anderson, 1979; Adair et al., 1982; Anderson et al., 1983). Relaxin is necessary for a normal duration of parturition and frequency of live births. In the absence of relaxin in peripheral blood of ovariectomized gilts given only progesterone for 10 days before parturition there results prolonged parturition and a high incidence of stillbirths (Nara et al., 1982b). Injections of relaxin at 6-hour intervals into such gilts results in a normal rapid delivery of live piglets. The prolonged lack of relaxin in ovariectomized gilts may be caused by insufficient dilatation of the cervix at parturition which results in dystocia and a high frequency of stillbirths. Thus relaxin facilitates expulsion of the fetuses at birth and contributes to a high incidence of live piglets. Although $PGF_{2\alpha}$ alone can induce parturition in gilts and sows, combining it with relaxin improves synchrony of the onset of delivery (Butler and Boyd, 1983). Vaginally applied porcine relaxin and $PGF_{2\alpha}$ applied separately induce similar histologic changes in the rabbit cervix which are comparable with those seen in the cervix following spontaneous onset of labor (MacLennan et al., 1985). These features include a dissolution of collagen bundles and an apparent increase in the ground substance. A unique giant cell infiltrate, containing mostly neutrophils, is seen in relaxin-treated and control rabbits with spontaneous labor. Relaxin and prostaglandin $F_{2\alpha}$ can induce the normal structural changes in the rabbit cervix associated with spontaneous cervical ripening and parturition.

In the pig the binding presumably of ^{125}I-relaxin (porcine) to both the myometrium and cervix is rapid and specific, but the specificity is not absolute; proinsulin and insulin inhibit but at concentrations far greater than native unlabeled relaxin (Mercado-Simmen et al., 1982b). Estrogen increases the activity of relaxin receptors in the porcine myometrium and cervix. Such differences in receptor concentrations may relate to the biological actions of the hormone since relaxin in early pregnancy promotes uterine quiescence but at prepartum it is needed for cervical dilatation. This could be expressed as differences in relaxin receptor concentrations in the uterus and cervix in response to different hormonal determinants. Myometrial activity in the ovariectomized estrogen-treated pig is abruptly inhibited for a period > 2 hours when relaxin is injected intravenously at a dosage that produces plasma levels comparable to those prepartum animals (Porter and Watts, 1986). Progesterone also is a potent inhibitor of myometrial activity in this species but it requires a much longer period to exert its maximum effect. During relaxin inhibition the uterus remains responsive to oxytocin, but progesterone abolishes the responsiveness of the myometrium to oxytocin. Although estrogen is required to maintain relaxin receptors in ovariectomized animals (Mercado-Simmen et al., 1982b), estradiol benzoate seems to have no effect on myometrial activity in vivo. It is proposed that relaxin plays an important role in the preparation of the myometrium for delivery by providing an inhibitory control during progesterone withdrawal (Porter and Watts, 1986). This view is strengthened by their observations that: a) relaxin is able to inhibit the gravid pig uterus in vivo, and b) the evolution of labor activity in the sow is inversely correlated with plasma relaxin concentrations.

During lactation in the sow luteal regression is rapid and complete, and ovarian and steroidogenic activity are minimal (Anderson et al., 1969, 1983; Adair et al., 1982). Relaxin blood levels were found to be low to non-detectable during this time and seemingly unaffected by suckling (Anderson et al., 1973b, 1983; Sherwood et al., 1981). Afele et al. (1979) reported, however, that suckling and nuzzling of lactating sows caused acute and transient increases in plasma relaxin but their physiological significance remained obscure. Plasma prolactin concentrations increase during suckling but they are not accompanied by detectable changes in relaxin secretion (Kendall et al., 1983). Stimulation of prolactin secretion by injection of haloperidol likewise fails to increase plasma relaxin levels during lactation. Conversely, in prepartum gilts given bromocriptine prolactin plasma levels remain low whereas relaxin concentrations follow a normal rise similar to that found in untreated animals (Taverne et al., 1982). By using intensive bleeding frequencies during periods of suckling, Whitely et al. (1985) observed rapid episodic relaxin secretion (e.g., peaks of 2-4 ng/ml plasma) after the suckling stimulus. Intravenous infusion of oxytocin is known to elicit a rapid sixfold increase in relaxin in ovarian venous plasma during late pregnancy (Anderson et al., 1973b). When oxytocin is injected into lactating sows relaxin is released in a similar episodic and rapid manner as seen in response to a suckling stimulus (Whitely et al., 1985). Therefore oxytocin and relaxin may cause opposing effects of contraction and relaxation of myoepithelial cells of the mammary gland as is found in myometrial cells of the uterus. Since only one milk ejection occurs per suckling period in the pig (Ellendorff et al., 1982), the release of endogenous relaxin could play an antagonistic role to limit milk let-down at each suckling episode.

Although the cause of the well-known relaxin surge by aging corpora lutea a few hours before parturition in the pig is unknown, it is thought that relaxin release may relate to fetal-placental factor(s) and final luteal demise. To investigate critically relaxin and progesterone secretion by aging corpora lutea, nonmated gilts were hysterectomized on day 6 of the estrous cycle to compare hormone secretion profiles with animals during late pregnancy and lactation (Felder et al., 1986). The results indicate that a precisely timed peak release of relaxin and coincident decrease in progesterone secretion occur in unmated hysterectomized gilts at the same time as those occurring a few hours preceding parturition during normal pregnancy (Figure 17). A surge release of relaxin occurred on day 113 in both pregnant and hysterectomized gilts. Furthermore, these results indicate not only higher plasma levels of progesterone after hysterectomy, but also an abrupt decrease to 50% ($\cong$ 16 ng/ml) that coincides with the decrease on pregnant gilts from days 110-114. The marked shifts in relaxin and progesterone secretion in these hysterectomized gilts occur in the complete absence of structural luteal regression, but with an obvious abrupt change in hormone secretion. The corpora lutea not only remain large (> 450 mg), but also retain fine structural features and at least some hormone secretory capacity to days 124 and 136 (Adair et al., 1982; Anderson et al., 1983; Musah et al., 1984). The higher level of relaxin secretion by the aging corpora lutea in these hysterectomized compared with lactating gilts coincides with maintenance of luteal cells that contain electron-dense cytoplasmic granules. It is evident that concentrations of relaxin and progesterone follow similar secretory patterns during pregnancy and lactation and after hysterectomy in the pig. Porcine corpora lutea and endometrium secrete prostaglandin F during the estrous cycle and pregnancy (Guthrie et al., 1978; Watson and Patek, 1979; Patek and Watson, 1983). Because prostaglandin metabolism is rapid (Lands, 1979), its presence in corpora lutea probably means that it is made there (Rothchild, 1981). There is no evidence concerning uterine and luteal secretion of PGF during late pregnancy in this species. Whether luteal prostaglandins play a role in the abrupt decrease in progesterone secretion and peak relaxin release by aging corpora lutea in hysterectomized pigs is not known.

The ephemeral nature of mammalian corpora lutea can be characterized by their autonomy of progesterone and relaxin secretion as shown here, responsiveness to the luteolytic effects of prostaglandins, and their

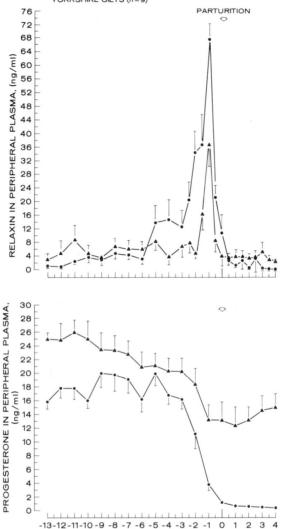

Fig. 17. Relaxin (upper panel) and progesterone
(lower panel) concentrations in peripheral plasma during pregnancy and
lactation (●) compared with those in
unmated gilts hysterectomized (▲) on
day 6 after estrus. Day 0 is the day
of parturition (114 days) in pregnant
animals and the day after peak relaxin
levels (113 days) in hysterectomized
and pregnant animals. Relaxin in
peripheral plasma from 0800 and 2000
hour bleedings are included from days
110-120. The number of gilts in each
group is indicated in parentheses.
Values are the mean ± SE. (From
Felder et al., 1986.)

ability to make prostaglandins (Anderson et al., 1969; Rothchild, 1981; Adair et al., 1982). Extrinsic (uterine) prostaglandins may switch on intrinsic (luteal) prostaglandins to cause luteolysis at the termination of an estrous cycle or pregnancy, whereas after hysterectomy, the rate of luteal regression may be limited to the intrinsic process. Whether the cataclysmic events associated with luteal demise during late pregnancy result primarily from prostaglandins of uterine and luteal cell origin and during late hysterectomy from luteal cells is not known. Nevertheless, a precisely timed signal, possibly of ovarian origin or from the central nervous system and pituitary gland, entrains in hysterectomized and pregnant pigs on days 112-113 for the surge release of relaxin and coincident decrease in progesterone secretion. The aging porcine corpora lutea may be preprogrammed to have an inherent life-span of approximately 113 days and release relaxin as a result of the ending of their life-span regardless of extrinsic factors. During pregnancy, uterine prostaglandins may cause the immediate death of the corpora lutea and, therefore, indirectly cause the immediate release of relaxin. On the other hand, uterine prostaglandins are not present after hysterectomy, the corpora lutea are not immediately killed, but remain large and seemingly healthy, as based on fine structural features to at least day 136, and they release relaxin during a longer period compared with release during pregnancy and lactation. This would also account for the lower peak blood levels of relaxin in hysterectomized animals with those in gilts during late pregnancy. Therefore, the uterus, fetuses, and their associated placental estrogens are not required for a surge release of relaxin and coincident decrease in progesterone secretion normally found at the termination of gestation in the pig. While these results show an effective maternal control system for depressing plasma progesterone and elevating plasma relaxin, no evidence is provided to show that feto-placental signals do not assist in terminating progesterone production and pregnancy.

Luteinizing hormone stimulates relaxin secretion by cultured granulosa cells from preovulatory porcine follicles, whereas follicle stimulating hormone (FSH) increases progesterone but it is a poor stimulus for relaxin secretion (Loeken et al., 1983). Prolactin is ineffective on either relaxin or progesterone secretion by such cells from large follicles. Likewise, several prostaglandins (PGA_1, PGE_1, PGE_2, $PGF_{2\alpha}$) do not stimulate relaxin secretion by these cells. Luteinizing hormone, FSH, PGE_1, as well as relaxin stimulate the release of plasminogen activator by rat granulosa cells in vitro (Too et al., 1982; 1984). The presence of relaxin in follicular fluid and its production by theca and granulosa cells may indicate a role in follicular growth and connective tissue remodeling.

The possible influence of the hypothalamo-hypophysial axis in in vivo regulation of relaxin and progesterone secretion by aging corpora lutea was investigated in hysterectomized and pregnant gilts (Felder, K. J. and Anderson, L. L., unpublished observations). It is known that prolactin secretion is tonically inhibited by the hypothalamus in the pig; prolactin blood concentrations remain elevated after hypophysial stalk transection (Anderson et al., 1982a). Episodic secretion of LH, however, is abolished after stalk transection (Berardinelli et al., 1981). In gilts hypophysectomized soon after mating the blastocysts develop and progesterone secretion is maintained during the first 14 days of gestation (du Mesnil du Buisson et al., 1964). In mated gilts hypophysial stalk-transected at days 30 and 50 the pregnancies fail, whereas in those stalk-transected at days 70 and 90 the pregnancies are maintained to normal term (du Mesnil du Buisson and Denamur, 1969). Exogenous prolactin maintains pregnancy and progesterone secretion at least 10 days in gilts hypophysectomized at day 70, whereas LH is ineffective (du Mesnil du Buisson, 1973). In hypophysectomized-hysterectomized gilts, LH or hCG maintains luteal function and progesterone secretion to day 30, whereas the corpora lutea regress in those similarly treated but with the uterus intact (Anderson et al., 1965, 1967). These results from hypophysectomized gilts indicate that LH is required to maintain progesterone secretion and pregnancy between days 14 and 50 whereas after day 70 prolactin is essential.

During late pregnancy in intact gilts prolactin blood levels increase markedly during the 30 hours preceding parturition (Taverne et al., 1979; 1982; Dusza and Krzymowska, 1981; Kendall et al., 1982; Felder, J. K., Klindt, J., Bolt, D. J., and Anderson, L. L., unpublished observations); however, in hysterectomized gilts prolactin levels remain low from days 112-120. The rise in prolactin secretion in pregnant gilts does not depend on the steady decrease in progesterone that begins 2 to 3 days earlier (Figure 17; Felder et al., 1986). Plasma LH levels at 20-minute intervals remain low (< 0.5 ng/ml) on days 109, 112 and 116 during late pregnancy/ lactation and after hysterectomy (Felder, K. J. and Anderson, L. L., unpublished observation). Parvizi et al., (1976) observed that episodic LH release was followed by an episodic progesterone peak 3 weeks before parturition but progesterone was lower and independent of LH release just before parturition. During the last 3 days of pregnancy, relaxin and prolactin blood levels increase steadily with their peak concentrations occurring about the same time whereas progesterone decreases rapidly. The fall in progesterone, however, is not a prerequisite for increased prolactin and relaxin secretion because these hormonal changes can be dissociated and still maintain the mechanisms of parturition and lactation. The relaxin and prolactin peaks coincide during normal pregnancy but they are independent of a common stimulus. For example, suppression of prolactin secretion by bromocriptine does not interfere with the occurrence of the relaxin peak. Furthermore, the prepartum prolactin peak is independent of progesterone blood levels or the progesterone:estrogen ratio. Endogenous $PGF_{2\alpha}$ blood levels remain low during late pregnancy and peak near the onset of parturition (Silver et al., 1979; First and Bosc, 1979). Oxytocin increases only a few hours before delivery and further rises occur during expulsion of the fetuses (Ellendorff et al., 1979). Fetal plasma concentrations of corticosteroids begin a steady rise 6 days before delivery but in the maternal blood they increase only on the day of parturition (Silver et al., 1979). Thus, it seems that during late pregnancy in the pig control of the precisely timed relaxin surge preceding parturition is independent of blood concentrations of decreasing progesterone, rising estrone and 17β-estradiol, and increasing fetal and maternal corticosteroids. Furthermore, peak levels of $PGF_{2\alpha}$ and oxytocin occur after the relaxin surge. The concept that these ovarian, placental and fetal hormones do not control relaxin secretion is reinforced by the precisely timed relaxin surge on day 113 in hysterectomized gilts. Daily injections of purified porcine LH (3.2 mg/day) and prolactin (2.0 mg/day) from days 110-120 increased circulating levels of these hormones in pregnant and hysterectomized gilts (Felder, K. J., Klindt, J., Bolt, D. J., and Anderson, L. L., unpublished observations). Pregnancy was prolonged about 2 days (to day 116) by exogenous LH and prolactin. Both LH and prolactin increased relaxin secretion by the aging corpora lutea in late pregnancy but plasma levels of relaxin dropped abruptly after parturition (Table 3). The increase in progesterone likely resulted from luteinization of numerous follicles caused by LH treatment. These luteinized follicles, however, were unable to secrete significant quantities of relaxin by day 120. Prolactin treatment increased relaxin and progesterone secretion by aging corpora lutea without luteinization of follicles in pregnant and hysterectomized gilts (Table 3). These results provide clear evidence that porcine prolactin prolongs secretory function by aging corpora lutea with sustained relaxin and progesterone secretion in pregnant and hysterectomized gilts. Furthermore, porcine prolactin maintains aging corpora lutea and progesterone secretion from days 110-120 in hypophysectomized-hysterectomized gilts (Li, Y., Molina, J. R., Klindt, J., Bolt, D. J. and Anderson, L. L., unpublished observations).

Table 3. Effects of Luteinizing Hormone (LH) and Prolactin (PRL) on Relaxin and Progesterone Secretion in Pregnant and Hysterectomized Gilts

Experimental Group and Treatment	No. of Gilts	Mean Plasma Concentrations, ng/ml			
		Relaxin		Progesterone	
		Day 115	Day 120	Day 115	Day 120
Pregnancy/Lactation					
LH	4	39	< 1	11	31
PRL	4	40	< 1	5	< 1
Control	7	14	< 1	3	< 1
After Hysterectomy					
LH	4	15	5	41	59
PRL	4	21	11	23	19
Control	9	4	3	13	11

The control of the precisely timed surge release of relaxin on day 113 in the pig may derive from the hypothalamo-pituitary axis. There is evidence in the rat that oxytocin secretion by the hypothalamus is profoundly altered by relaxin. Intracerebroventricular infusion of relaxin (1 μg per rat) disrupts the pattern of reflex milk ejection without affecting the response of the mammary gland to oxytocin (Summerlee et al., 1984). It is possible that relaxin might also suppress the release of oxytocin during the prepartum period when endogenous levels of relaxin are elevated. It is known that intracerebroventricular infusion of oxytocin (0.4 μg per rat) elicits maternal behavior (e.g., grouping, licking, crouching, nest building and retrieval) in virgin female rats presented with 2- to 7-day old pups (Pedersen and Prange, 1979). It has been clearly shown in the rabbit that endogenous plasma levels of oxytocin are extremely low before birth (e.g., < 60 pg/ml) and increase 50-fold with the onset of the expulsive phase of delivery (O'Byrne et al., 1986). When high blood levels of relaxin are maintained by intravenous infusion of relaxin during late pregnancy in the rat, the gestation is prolonged and the delivery lengthened without causing dystocia or affecting the number of live births (Jones and Summerlee, 1986a,b). Naloxone reverses the inhibitory effects of relaxin, increases oxytocin levels and shortens deliveries (Eltringham et al., 1986; Jones and Summerlee, 1986c). Relaxin suppresses oxytocin release possibly through an opioid system and this may be important in the control of the timing of birth (Jones and Summerlee, 1986c). Other evidence can be considered for possible peripheral regulation of relaxin's effect on myoepithelial cells (O'Byrne and Summerlee, 1985). Relaxin fails to inhibit reflex milk ejection in adrenalectomized rats whereas it is inhibitory in sham-operated controls. This may indicate that relaxin can exert its inhibitory effect on oxytocin release via the adrenal glands possibly by releasing opioids. These results are consistent with the hypothesis that relaxin has a central action suppressing the release of oxytocin as well as the well-known peripheral actions on inhibition of myometrial and myoepithelial activities and cervical distensibility. Clearly, surge releases of relaxin and oxytocin are precisely timed during late pregnancy and their regulation may be dependent on one another. Relaxin peak release at this time may sustain inhibition of myometrial activity during the brief period of falling progesterone and increasing estrogen secretion. The final decline in blood levels of relaxin could signal the neurogenic release of oxytocin in the presence of maximum estrogen levels to switch on intense uterine contractility for expulsion of the fetuses. During lactation relaxin and oxytocin may modulate reflex milk ejection to the suckling stimulus that allows an appropriate interval for milk accumulation before responding positively to the subsequent suckling stimulus.

CONCLUSIONS

Extensive experimental evidence has accumulated for multifaceted biological actions of relaxin during pregnancy and parturition, and postpartum physiology. Conclusive evidence now is available on the site(s) (e.g., corpus luteum, ovarian follicles, placenta, endometrium) for the production of relaxin in different species. Less clear is our understanding of the endogenous regulation of relaxin secretion, how its secretion is affected by other hormones, and specific receptors for relaxin.

In the rat, the ovary is the major site of relaxin production but the placenta and fetuses clearly control its production and release. In the pig, relaxin is produced primarily by the corpora lutea but the placenta, fetuses or even the uterus are not required for its production or the precisely timed surge release of relaxin 113 days after estrus or mating. Thus, the aging corpora lutea may be preprogrammed to release relaxin as a result of an inherent life-span defined through evolutionary development of the reproductive cycle as the duration of gestation of about 114 days. Although these ephemeral structures are capable of an extended life span to 150 days in hysterectomized pigs, they retain this precisely timed signal for relaxin release on day 113. On the other hand, the control of the precisely timed signal for relaxin release could derive from the central nervous system and pituitary gland. Intracerebroventricular application of relaxin causes profound effects on oxytocin secretion by the hypothalamus and neural lobe of the pituitary gland in the rat. Modulation of prolactin secretion by the anterior pituitary gland or prolactin treatment extends not only the life span of aging corpora lutea in the pig but also increases relaxin and progesterone secretion by them. Well designed _in vivo_ and _in vitro_ studies will be required to further our understanding of the relative importance of the hypothalamo-pituitary axis in possibly regulating precisely timed relaxin secretion in pregnant and non-pregnant animals. The placenta has been identified as the major site for relaxin production in the rabbit, horse, guinea pig, cat and sheep, but factors controlling relaxin production and secretion are unknown.

Finally, it is emphasized that in cattle, relaxin administered during late pregnancy induces premature parturition. Furthermore, it induces an acute depression of progesterone secretion beginning within 90 minutes. These early and marked luteolytic effects of relaxin on progesterone secretion could be by direct or indirect actions via mechanisms that are yet unknown.

ACKNOWLEDGMENTS

This work was supported in part by U.S. Department of Agriculture, ARS, CSRS, and OGPS Competitive Grants 85-CRCR-1-1862 and 86-CRCR-1-2130. Journal Paper J-12559 of the Iowa Agriculture and Home Economics Experiment Station, Ames, IA (Projects 2444, 2638, 2754, 2797 and 2273, the last a contributing project to North Central Regional Research Project NC-113).

REFERENCES

Adair V, Anderson LL, Stromer MH, McDonald WG, 1982. Progesterone, estrone, and estradiol-17β secretion and the fine structure of aging corpora lutea in hysterectomized and pregnant gilts. J Anim Sci [Suppl 1] 55:333 (Abstract 534).

Adelstein RS, Eisenberg E, 1980. Regulation and kinetics of the actin-myosin ATP interaction. Annu Rev Biochem 49:921-33.

Afele S, Bryant-Greenwood GD, Chamley WA, Dax EM, 1979. Plasma relaxin immunoactivity in the pig at parturition and during nuzzling and suckling. J Reprod Fert 56:451-57.

Ali MS, McMurtry JP, Bagnell CA, Bryant-Greenwood GD, 1986. Immunocyto-chemical localization of relaxin in corpora lutea of sows throughout the estrous cycle. Biol Reprod 34:139-43.

Anderson LL, 1982. Relaxin localization in porcine and bovine ovaries by assay and morphologic techniques. Adv Exp Med Biol 143:1-67.

Anderson LL, Adair V, Stromer MH, McDonald WG, 1983. Relaxin production and release after hysterectomy in the pig. Endocrinology 113:677-86.

Anderson LL, Bast JD, Melampy RM, 1973a. Relaxin in ovarian tissue during different reproductive stages in the rat. J Endocrinol 59:371-72.

Anderson LL, Berardinelli JG, Malven PV, Ford JJ, 1982a. Prolactin secre-tion after hypophysial stalk transection in pigs. Endocrinology 111:380-84.

Anderson LL, Bland KP, Melampy RM, 1969. Comparative aspects of uterine-luteal relationships. Recent Progr Horm Res 25:57-104.

Anderson LL, Butcher RL, Melampy RM, 1963. Uterus and occurrence of oestrus in pigs. Nature 193:311-12.

Anderson LL, Dyck GW, Mori H, Henricks DM, Melampy RM, 1967. Ovarian function in pigs following hypophysial stalk transection or hypophysec-tomy. Am J Physiol 212:1188-94.

Anderson LL, Ford JJ, Melampy RM, Cox DF, 1973b. Relaxin in porcine corpora lutea during pregnancy and after hysterectomy. Am J Physiol 225:1215-19.

Anderson LL, Léglise PC, du Mesnil du Buisson F, Rombauts P, Courrier R, 1965. Interaction des hormones gonadotropes et de l'utérus dans le maintien du tissu lutéal ovarien chez la Truie. CR Acad Sci Paris 261:3675-78.

Anderson LL, Perezgrovas R, O'Byrne EM, Steinetz BG, 1982b. Biological actions of relaxin in pigs and beef cattle. Ann New York Acad Sci 380:131-50.

Anderson MB, Sherwood OD, 1984. Ultrastructural localization of relaxin immunoreactivity in corpora lutea of pregnant rats. Endocrinology 114:1124-27.

Anderson MB, Vaupel MR, Sherwood OD, 1984. Pregnant mouse corpora lutea: Immunocytochemical localization of relaxin and ultrastructure. Biol Reprod 31:391-97.

Anderson ML, Long JA, 1978. Localization of relaxin in the pregnant rat. Bioassay of tissue extracts and cell fractionation studies. Biol Reprod 18:110-17.

Anderson ML, Long JA, Hayashida T, 1975. Immunofluorescence studies on the localization of relaxin in the corpus luteum of the pregnant rat. Biol Reprod 13:499-504.

Bedarkar S, Turnell WG, Blundell TL, Schwabe C, 1977. Relaxin has confir-mational homology with insulin. Nature 270:449-51.

Belt WD, Anderson LL, Cavazos LF, Melampy RM, 1971. Cytoplasmic granules and relaxin levels in porcine corpora lutea. Endocrinology 89:1-10.

Belt WD, Cavazos LF, Anderson LL, Kraeling RR, 1970. Fine structure and progesterone levels in the corpus luteum of the pig during pregnancy and after hysterectomy. Biol Reprod 2:98-113.

Berardinelli JG, Anderson LL, Ford JJ, Christenson RK, 1981. Neuroendo-crine regulation of luteinizing hormone secretion in estrogen-treated, ovariectomized gilts. J Anim Sci [Suppl 1] 53:296 (Abstract 424).

Bigazzi M, Bruni P, Nardi E, Petrucci F, Pollicino G, Franchini M, Scarselli G, Farnararo M, 1982. Human decidual relaxin. Ann New York Acad Sci 380:87-97.

Bigazzi M, Nardi E, Bruni P, Petrucci F, 1980. Relaxin in human decidua. J Clin Endocrinol Metab 51:939-41.

Bigazzi M, Nardi E, Petrucci F, Scarselli G, 1983. Synthesis of relaxin by human decidua. In: Bigazzi M, Greenwood FC, Gasparri F (ed.), Biology of Relaxin and Its Role in the Human. Amsterdam: Excerpta Medica, pp. 206-12.

Blank MS, Dufau ML, 1983. Rat chorionic gonadotropin: Augmentation of bioactivity in the absence of the pituitary. Endocrinology 112:2200-02.

Bloom G, Paul KG, Wiqvist N, 1958. A uterine-relaxing factor in the pregnant rat. Acta Endocrinol 28:112-18.

Blundell TL, Humbel RE, 1980. Hormone families: Pancreatic hormones and homologous growth factors. Nature 287:781-87.

Bryant-Greenwood GD, 1982. Relaxin as a new hormone. Endocrine Rev 3:62-90.

Bryant-Greenwood GD, 1985. Current concepts on the role of relaxin. Res Reprod 17:1-4.

Bryant-Greenwood GD, Jeffrey R, Ralph MM, Seamark RF, 1980. Relaxin production by the porcine ovarian graafian follicle in vitro. Biol Reprod 23:792-800.

Büllesbach EE, Schwabe C, 1985. Naturally occurring porcine relaxins and large-scale preparation of the B29 hormone. Biochemistry 24:7717-22.

Butler WR, Boyd RD, 1983. Relaxin enhances synchronization of parturition induced with prostaglandin $F_{2\alpha}$ in swine. Biol Reprod 28:1061-65.

Cavazos LF, Anderson LL, Belt WD, Henricks DM, Kraeling RR, Melampy RM, 1969. Fine structure and progesterone levels in the corpus luteum of the pig during the estrous cycle. Biol Reprod 1:83-106.

Cheah SH, Sherwood OD, 1980. Target tissues for relaxin in the rat: Tissue distribution of injected I-labeled relaxin and tissue changes in adenosine 3',5'-monophosphate levels after in vitro relaxin incubation. Endocrinology 106:1203-09.

Cheah SH, Sherwood OD, 1981. Effects of relaxin on in vivo uterine contractions in conscous and unrestrained estrogen-treated and steroid-untreated ovariectomized rats. Endocrinology 109:2076-83.

Coggins EG, First NL, 1977. Effect of dexamethasone, methallibure and fetal decapitation on porcine gestation. J Anim Sci 44:1041-49.

Coggins EG, Van Horn D, First NL, 1977. Influence of prostaglandin $F_{2\alpha}$, dexamethasone, progesterone and induced CL on porcine parturition. J Anim Sci 46:754-62.

Crish JF, Soloff MS, Shaw AR, 1986a. Changes in relaxin precursor mRNA levels in the rat ovary during pregnancy. J Biol Chem 261:1909-13.

Crish JF, Soloff MS, Shaw AR, 1986b. Changes in relaxin precursor messenger ribonucleic acid levels in ovaries of rats after hysterectomy and removal of conceptuses, and during the estrous cycle. Endocrinology 119:1222-28.

Diehl JR, Godki RA, Killian DB, Day BN, 1974. Induction of parturition in swine with prostaglandin $F_{2\alpha}$. J Anim Sci 38:1229-34.

Dodson GG, Eliopoulos EE, Isaac NW, McCall MJ, Niall HD, North ACT, 1982. Rat relaxin: insulin-like fold predicts a likely receptor binding region. Int J Biol Macromol 4:399-405.

Downing SJ, Sherwood OD, 1985a. The physiological role of relaxin in the pregnant rat. I. The influence of relaxin on parturition. Endocrinology 116:1200-05.

Downing SJ, Sherwood OD, 1985b. The physiological role of relaxin in the pregnant rat. II. The influence of relaxin on uterine contractile activity. Endocrinology 116:1206-14.

Downing SJ, Sherwood OD, 1985c. The physiological role of relaxin in the pregnant rat. III. The influence of relaxin on cervical extensibility.

Endocrinology 116:1215-20.

Downing SJ, Sherwood OD, 1986. The physiological role of relaxin in the pregnant rat. IV. The influence of relaxin on cervical collagen and glycosaminoglycans. Endocrinology 118:471-79.

du Mesnil du Buisson F, 1973. Facteurs luteotropes chez la truie. In: Denamur R, Netter A (ed.), Le Corp Jaune. Paris: Masson et Cie, pp. 225-37.

du Mesnil du Buisson F, Dauzier L, 1957. Influence de l'ovariectomie chez la truie pendant la gestation. C R Soc Biol 151:311-13.

du Mesnil du Buisson F, Denamur R, 1969. Mécanismes du contrôle de la fonction lutéale chez la truie, la brébis et la vache. In: Gaul C (ed.), IIIrd Int Congr Endocrinol Amsterdam: Excerpta Medica Int Congr Ser 184, pp. 927-34.

du Mesnil du Buisson F, Léglise PC, 1963. Effet de l'hypophysectomie sur les corps jaunes de la truie. Résultats préliminaires. C R Acad Sci Paris D257:261-63.

du Mesnil du Buisson F, Rombauts P, 1963. Effet d'autotransplants uterins sur le cycle oestrien de la truie. C R Acad Sci Paris 256:4984-86.

du Mesnil du Buisson F, Léglise PC, Anderson LL, Rombauts P, 1964. Maintien des corps jaunes et de la gestation de la truie au cours de la phase preimplantatoire apres hypophysectomie. Vth Int Congr Reprod Artif Insem, Trento. 3:571-75.

Dusza L, Krzymowska H, 1981. Plasma prolactin levels in sows during pregnancy, parturition and early lactation. J Reprod Fert 61:131-34.

Eldridge RK, Fields PA, 1985. Rabbit placental relaxin: Purification and immunohistochemical localization. Endocrinology 117:2512-19.

Eldridge RK, Fields PA, 1986. Rabbit placental relaxin: Ultrastructural localization in secretory granules of the syncytiotrophoblast using rabbit placental relaxin antiserum. Endocrinology 119:606-15.

Ellendorff F, Forsling ML, Poulain DA, 1982. The milk ejection reflex in the pig. J Physiol 333:577-94.

Ellendorff F, Taverne M, Elsaesser F, Forsling M, Parvizi N, Naaktgeboren C, Smidt D, 1979. Endocrinology of parturition in the pig. Anim Reprod Sci 2:323-34.

Ellicott AR, Dziuk PJ, 1973. Minimum daily dose of progesterone and plasma concentration for maintenance of pregnancy in ovariectomized gilts. Biol Reprod 9:300-04.

Eltringham L, O'Byrne KT, Summerlee AJS, 1986. Opioids, from the adrenal glands, may mediate relaxin inhibition of oxytocin release in lactating rats. J Physiol 371:183P.

Essig M, Schoenfeld C, Amelar RD, Dubin L, Weiss G, 1982. Stimulation of human sperm motility by relaxin. Fertil Steril 38:339-43.

Evans G, Wathes DC, King GJ, Armstrong DT, Porter DG, 1983. Changes in relaxin production by the theca during the preovulatory period in the pig. J Reprod Fert 69:677-83.

Felder KJ, Molina JR, Benoit AM, Anderson LL, 1986. Precise timing for peak relaxin and decreased progesterone secretion after hysterectomy in the pig. Endocrinology 119:1502-09.

Fèvre J, Léglise PC, Rombauts P, 1968. Du role de l'hypophyse et des ovaires dan la biosynthèse des oestrogènes au cours de la gestation chez la truie. Ann Biol anim Biochim Biophys 8:225-33.

Fields MJ, Fields PA, 1986. Luteal neurophysin in the nonpregnant cow and ewe: Immunocytochemical localization in membrane-bounded secretory granules of the large luteal cell. Endocrinology 118:1723-25.

Fields PA, 1984. Intracellular localization of relaxin in membrane-bound granules in the pregnant rat luteal cell. Biol Reprod 30:753-62.

Fields PA, Fields MJ, 1985. Ultrastructural localization of relaxin in the corpus luteum of the nonpregnant, pseudopregnant, and pregnant pig. Biol Reprod 32:1169-79.

Fields PA, Larkin LH, 1981. Purification and immunohistochemical localization of relaxin in the human term placenta. J Clin Endocrinol Metab

52:79-85.

Fields MJ, Dubois W, Fields PA, 1985. Dynamic features of luteal secretory granules: Ultrastructural changes during the course of pregnancy in the cow. Endocrinology 117:1675-82.

Fields MJ, Fields PA, Castro-Hernandez A, Larkin LH, 1980. Evidence for relaxin in corpora lutea of late pregnant cows. Endocrinology 197:869-76.

First NL, Bosc MJ, 1979. Proposed mechanisms controlling parturition and the induction of parturition in swine. J Anim Sci 48:1407-21.

Flint APF, Sheldrick EL, 1982. Ovarian secretion of oxytocin is stimulated by prostaglandin. Nature 297:587-88.

Fosang AJ, Handley CJ, Santer V, Lowther DA, Thorburn GD, 1984. Pregnancy related changes in the connective tissue of the ovine cervix. Biol Reprod 30:1223-35.

Frank M, Bazer FW, Thatcher WW, Wilcox CJ, 1977. A study of prostaglandin $F_{2\alpha}$ as a luteolysin in swine: III. Effects of estradiol valerate on prostaglandin F, progestins, estrone and estradiol concentrations in the utero-ovarian vein of non-pregnant gilts. Prostaglandins 14:1183-96.

Frieden EH, Adams WC, 1985. Stimulation of rat uterine collagen synthesis by relaxin. Proc Soc Exp Biol Med 180:39-43.

Frieden EH, Yeh L, 1977. Evidence for a "prorelaxin" in porcine relaxin concentrates. Proc Soc Exp Biol Med 154:407-11.

Fuchs AR, 1978. Hormonal control of myometrial function during pregnancy and parturition. Acta Endocrinol [Suppl] 221:1-71.

Gardner ML, First NL, Casida LE, 1963. Effect of exogenous estrogens on corpus luteum maintenance in gilts. J Anim Sci 22:132-34.

Gates GS, Flynn JJ, Ryan RJ, Sherwood OD, 1981. In vivo uptake of ^{125}I-relaxin in the guinea pig. Biol Reprod 25:549-54.

Gemmell RT, Stacy BD, Thorburn GD, 1974. Ultrastructural study of secretory granules in the corpus luteum of the ewe during the estrous cycle. Biol Reprod 11:447-62.

Gleeson AR, Thorburn GD, Cox RI, 1974. Prostaglandin F concentration in the utero-ovarian venous plasms of the sow during late luteal phase of the estrous cycle. Prostaglandins 5:521-30.

Goldsmith LT, de la Cruz J, Weiss G, Castracane VD, 1982. Steroid effects on relaxin secretion in the rat. Biol Reprod 27:886-90.

Goldsmith LT, Grob HS, Scherer KJ, Surve A, Steinetz BG, Weiss G, 1981. Placental control of ovarian immunoreactive relaxin secretion in the pregnant rat. Endocrinology 109:548-52.

Golichowski AM, King SR, Mascaro M, 1980. Pregnancy-related changes in rat cervical glycosaminoglycans. Biochem J 192:1-8.

Golos TG, Sherwood OD, 1982. Control of corpus luteum function during the second half of pregnancy in the rat: A direct relationship between conceptus number and both serum and ovarian relaxin levels. Endocrinology 111:872-78.

Golos TG, Sherwood OD, 1984. Evidence that the maternal pituitary suppresses the secretion of relaxin in the pregnant rat. Endocrinology 115:1004-10.

Golos TG, Sherwood OD, 1986. The suppressive effect of the maternal pituitary on relaxin secretion during the second half of pregnancy in rats does not require the presence of the nonluteal ovarian tissue. Biol Reprod 34:595-601.

Golos TG, Weyhenmeyer JA, Sherwood OD, 1984. Immunocytochemical localization of relaxin in the ovaries of pregnant rats. Biol Reprod 30:257-61.

Gordon WL, Sherwood OD, 1982. Evidence that luteinizing hormone from the maternal pituitary gland may promote antepartum release of relaxin, luteolysis, and birth in rats. Endocrinology 111:1299-1310.

Gowan LK, Reinig JW, Schwabe C, Bedarkar S, Blundell TL, 1981. On the primary and tertiary structure of relaxin from the sand tiger shark. FEBS Lett 129:80-82.

Guthrie HD, Bolt DJ, 1983. Changes in plasma estrogen, luteinizing hormone, follicle-stimulating hormone and 13,14-dihydro-15-keto-prostaglandin $F_2\alpha$ during blockade of luteolysis in pigs after human chorionic gonadotropin treatment. J Anim Sci 57:993-1000.

Guthrie HD, Rexroad Jr CE, 1981. Endometrial prostaglandin F release in vitro and plasma 13, 14-dihydro-15-keto-prostaglandin $F_2\alpha$ in pigs with luteolysis blocked by pregnancy, estradiol benzoate or human chorionic gonadotropin. J Anim Sci 52:330-39.

Guthrie HD, Rexroad Jr CE, Bolt DJ, 1978. In vitro synthesis of progesterone and prostaglandin F by luteal tissue and prostaglandin F by endometrial tissue from the pig. Prostaglandins 16:433-440.

Haley J, Hudson P, Scanlon D, John M, Cronk M, Shine J, Tregear G, Niall H. 1982. Porcine relaxin: Molecular cloning and cDNA structure. DNA 1:155-62.

Hard DL, Anderson LL, 1979. Maternal starvation on progesterone secretion, litter size, and growth in the pig. Am J Physiol 237:E273-78.

Harkness ML, Harkness RD, 1959. Changes in the properties of the uterine cervix of the rat during pregnancy. J Physiol 148:524-27.

Hashimoto I, Henricks DM, Anderson LL, Melampy RM, 1968. Progesterone and pregn-4-en-20α-ol-3-one in ovarian venous blood various reproductive states in the rat. Endocrinology 82:333-41.

Hisaw FL, 1926. Experimental relaxation of the pubic ligament of the guinea pig. Proc Soc Exp Biol Med 23:661-63.

Hisaw FL, Zarrow MX, Money WL, Talmage RV, Abramowitz AA, 1944. Importance of the female reproductive tract in the formation of relaxin. Endocrinology 34:122-34.

Hollingsworth M, Isherwood CNM, Foster RW, 1979. The effect of oestradiol, progesterone, relaxin and ovariectomy on cervical extensibility in the late pregnant rat. J Reprod Fert 56:471-77.

Hsu CH, Sanborn BM, 1986a. Relaxin affects the shape of rat myometrial cells in culture. Endocrinology 118:495-98.

Hsu CJ, Sanborn BM, 1986b. Relaxin treatment alters the kinetic properties of myosin light chain kinase activity in rat myometrial cells in culture. Endocrinology 118:499-505.

Hsu CJ, McCormack SM, Sanborn BM, 1985. The effect of relaxin on cyclic adenosine 3',5'-monophosphate concentrations in rat myometrial cells in culture. Endocrinology 116:2029-35.

Hudson P, Haley J, Cronk M, Shine J, Niall H, 1981. Molecular cloning and characterization of cDNA sequences coding for rat relaxin. Nature 291:127-31.

Hudson P, John M, Crawford R, Haralambidis J, Scanlon D, Gorman J, Tregear G, Shine J, Niall H, 1984. Relaxin gene expression in human ovaries and the predicted structure of a human preprorelaxin by analysis of cDNA clones. EMBO J 3:2333-39.

Ivell R, Brackett KH, Fields MJ, Dietmar R, 1985. Ovulation triggers oxytocin gene expression in the bovine ovary. FEBS Lett 190:263-67.

James R, Niall H, Kwok S, Bryant-Greenwood G, 1977. Primary structure of porcine relaxin: Homology with insulin and related growth factors. Nature 267:544-46.

Jarrett JC, Ballejo G, Saleem TH, Tsibris JCM, Spellacy WN, 1984. The effect of prolactin and relaxin on insulin binding by adipocytes from pregnant women. Am J Obstet Gynecol 149:250-55.

Jones SA, Summerlee AJS, 1986a. Effects of porcine relaxin on the length of gestation and duration of parturition in the rat. J Endocrinol 109:85-88.

Jones SA, Summerlee AJS, 1986b. Relaxin prolongs gestation and delivery time in the rat. J Physiol 371:181P.

Jones SA, Summerlee AJS, 1986c. Relaxin acts centrally to inhibit oxytocin release during parturition: an effect that is reversed by naloxone. J Endocrinol 111:99-102.

Kemp BE, Niall HD, 1984. Relaxin. Vit Horm 41:79-115.

Kendall JZ, Plopper CG, Bryant-Greenwood GD, 1978. Ultrastructural immunoperoxidase demonstration of relaxin in corpora lutea from a pregnant sow. Biol Reprod 18:94-98.

Kendall JZ, Richards GE, Shih LN, 1983. Effect of haloperidol, suckling, oxytocin, and hand milking on plasma relaxin and prolactin concentrations in cyclic and lactating pigs. J Reprod Fert 69:271-77.

Kendall JZ, Richards GE, Shih LN, Farris TS, 1982. Plasma relaxin concentrations in the pig during the periparturient period: Association with prolactin, estrogen and progesterone concentrations. Theriogenology 17:677-87.

Kertiles LP, Anderson, LL, 1979. Effect of relaxin on cervical dilatation, parturition and lactation in the pig. Biol Reprod 21:57-68.

Kroc RL, Steinetz BG, Beach VL, 1959. The effects of estrogens, progestagens, and relaxin in pregnant and nonpregnant laboratory rodents. Ann New York Acad Sci 75:942-80.

Kwok SCM, Chamley WA, Bryant-Greenwood GD, 1978. High molecular weight forms of relaxin in pregnant sow ovaries. Biochem Biophys Res Commun 82:997-1005.

Lands WEM, 1979. The biosynthesis and metabolism of prostaglandins. Annu Rev Physiol 41:633-52.

Larkin LH, Fields PA, Oliver RM, 1977. Production of antisera against electrophoretically separated relaxin and immunofluorescent localization of relaxin in the porcine corpus luteum. Endocrinology 101:679-85.

Larkin LH, Pardo RJ, Renegar RH, 1983. Sources of relaxin and morphology of relaxin containing cells. In: Bigazzi M, Greenwood FC, Gasparri F (ed.), Biology of Relaxin and Its Role in the Human. Amsterdam: Excerpta Medica, pp. 191-205.

Laster DB, Glimp HA, Cundiff LV, Gregory KE, 1973. Factors affecting dystocia and effects of dystocia on subsequent reproduction in beef cattle. J Anim Sci 36:695-705.

Lemon M, Loir M, 1977. Steroid release in vitro by two cell types in the corpus luteum of the pregnant sow. J Endocrinol 72:351-59.

Lemon M, Mauléon P, 1982. Interaction between two luteal cell types from the corpus luteum of the sow in progesterone synthesis in vitro. J Reprod Fert 64:315-23.

Loeken MR, Channing CP, D'Elletto R, Weiss G, 1983. Stimulatory effect of luteinizing hormone upon relaxin secretion by cultured porcine preovulatory granulosa cells. Endocrinology 112:769-71.

Long JA, 1973. Corpus luteum of pregnancy in the rat-ultrastructural and cytochemical observations. Biol Reprod 8:87-99.

Loumaye E, DeCooman S, Thomas K, 1980. Immunoreactive relaxin-like substances in human seminal plasma. J Clin Endocrinol Metab 50:1142-43.

Loumaye E, Teuwissen B, Thomas K, 1978. Characterization of relaxin radioimmunoassay using Bolton-Hunter reagent. First results in plasma during pregnancy and in placenta, corpora lutea and ovarian cysts in woman. Gynecol Obstet Invest 9:262-67.

Lye SL, Challis JRG, 1982. Inhibition by PGI_2 of myometrial activity in vivo in nonpregnant ovariectomized sheep. J Reprod Fert 66:311-15.

MacLennan AH, Katz M, Creasy R, 1985. The morphologic characteristics of cervical ripening induced by the hormones relaxin and prostaglandin $F_2\alpha$ in a rabbit model. Am J Obstet Gynecol 152:691-96.

Martin PA, Bevier GW, Dziuk PJ, 1978. The effect of disconnecting the uterus and ovary on the length of gestation in the pig. Biol Reprod 18:428-33.

Masuda H, Anderson LL, Henricks DM, Melampy RM, 1967. Progesterone in ovarian venous plasma and corpora lutea of the pig. Endocrinology 80:240-46.

Mathieu P, Rahier J, Thomas K, 1981. Localization of relaxin in human gestational corpus luteum. Cell Tissue Res 219:213-16.

Matsumoto D, Chamley WA, 1980. Identification of relaxins in porcine follicular fluid and in the ovary of the immature sow. J Reprod Fert

58:369-75.

Mercado-Simmen RC, Bryant-Greenwood GD, Greenwood FC, 1980. Characterization of the binding of ^{125}I-relaxin to rat uterus. J Biol Chem 225:3617-23.

Mercado-Simmen RC, Bryant-Greenwood GD, Greenwood FC, 1982a. Relaxin receptor in the rat myometrium: Regulation by estrogen and relaxin. Endocrinology 110:220-26.

Mercado-Simmen RC, Goodwin B, Ueno MS, Yamamoto SY, Bryant-Greenwood GD, 1982b. Relaxin receptors in the myometrium and cervix of the pig. Biol Reprod 26:120-28.

Moeljono MPE, Thatcher WW, Bazer FW, Frank M, Owens LJ, Wilcox CJ, 1977. A study of prostaglandin $F_{2\alpha}$ as the luteolysin in swine: II. Characterization and comparison of prostaglandin F, estrogens and progestin concentrations in utero-ovarian vein plasma of nonpregnant and pregnant gilts. Prostaglandins 14:543-55.

Musah AI, Ford JJ, Anderson LL, 1984. Progesterone secretion as affected by 17β-estradiol after hysterectomy in the pig. Endocrinology 115:1876-82.

Musah AI, Schwabe C, Anderson LL, 1987a. Acute decrease in progesterone and increase in estrogen secretion caused by relaxin during late pregnancy in beef heifers. Endocrinology 120:317-24.

Musah AI, Schwabe C, Willham RL, Anderson LL, 1986a. Pelvic development as affected by relaxin in three genetically selected frame sizes of beef heifers. Biol Reprod 34:363-69.

Musah AI, Schwabe C, Willham RL, Anderson LL, 1986b. Relaxin on induction of parturition in beef heifers. Endocrinology 118:1476-82.

Musah AI, Schwabe C, Willham RL, Anderson LL, 1986c. Induction of parturition in beef heifers with relaxin and cloprostenol or dexamethasone: Pelvic and cervical dilatation and incidence of dystocia. J Anim Sci [Suppl 1] 63:377 (Abstract 530).

Musah AI, Schwabe C, Willham RL, Anderson LL, 1986d. Induction of parturition in cattle with relaxin and cloprostenol or dexamethasone: Effects on progesterone secretion and onset of parturition. J Anim Sci [Suppl 1] 63:377 (Abstract 474).

Musah AI, Schwabe C, Willham RL, Anderson LL, 1987b. Induction of parturition, progesterone secretion and delivery of placenta in beef heifers given relaxin with cloprostenol or dexamethasone. Biol Reprod (In press).

Nara BS, First NL, 1981a. Effect of indomethacin on dexamethasone-induced parturition in swine. J Anim Sci 52:788-93.

Nara BS, First NL, 1981b. Effect of indomethacin and prostaglandin $F_{2\alpha}$ on parturition in swine. J Anim Sci 51:1360-70.

Nara BS, Ball GD, Rutherford JE, Sherwood OD, First NL, 1982a. Release of relaxin by a nonluteolytic dose of prostaglandin $F_{2\alpha}$ in pregnant swine. Biol Reprod 27:1190-95.

Nara BS, Darmadja D, First NL, 1981. Effect of removal of follicles, corpora lutea or ovaries on maintenance of pregnancy in swine. J Anim Sci 52:794-801.

Nara BS, Welk FA, Rutherford JE, Sherwood OD, First NL, 1982b. Effect of relaxin on parturition and frequency of live births in pig. J Reprod Fert 66:359-65.

Niall HD, Hudson P, Haley J, Cronk M, Shine J, 1982. Rat preprorelaxin: Complete amino acid sequence derived from cDNA analysis. Ann New York Acad Sci 380:13-21.

Nishikori K, Weisbrodt NW, Sherwood OD, Sanborn BM, 1982. Relaxin alters rat uterine myosin light chain phosphorylation and related enzymatic activities. Endocrinology 111:1743-45.

O'Byrne EM, Steinetz BG, 1976. Radioimmunoassay (RIA) of relaxin in sera of various species using an antiserum to porcine relaxin. Proc Soc Exp Biol Med 152:272-76.

O'Byrne EM, Carriere BT, Sarensen L, Segaloff A, Schwabe C, Steinetz BG,

1978a. Plasma immunoreactive relaxin levels in pregnant and nonpregnant women. J Clin Endocrinol Metab 47:1106-10.

O'Byrne EM, Flitcraft JF, Sawyer WK, Hochman J, Weiss G, 1978b. Relaxin bioactivity and immunoactivity in human corpora lutea. Endocrinology 102:1641-44.

O'Byrne KT, Summerlee AJS, 1985. Relaxin suppression of oxytocin release occurs at the neurohypophysis in the rat. J Physiol 365:49P.

O'Byrne KT, Ring JPG, Summerlee AJS, 1986. Plasma oxytocin and oxytocin neurone activity during delivery in rabbits. J Physiol 370:501-13.

Ochiai K, Latolt, Kelly PA, Rothchild I, 1983. The importance of a luteolytic effect of the pituitary in understanding the placental control of the rat's corpus luteum. Endocrinology 112:1687-95.

Olefsky JM, Saekow M, Kroc RL, 1982. Potentiation of insulin binding and insulin action by purified porcine relaxin. Ann New York Acad Sci 380:200-16.

Paisley LA, Mickelsen WD, Anderson PB, 1986. Mechanism and therapy for retained fetal membranes and uterine infections of the cow: A review. Theriogenology 25:353-81.

Pardo RJ, Larkin LH, 1982. Localization of relaxin in endometrial gland cells of pregnant, lactating and ovariectomized, hormone treated guinea pigs. Am J Anat 164:79-90.

Pardo RJ, Larkin LH, Fields PA, 1980. Immunocytochemical localization of relaxin in endometrial glands of the pregnant guinea pig. Endocrinology 107:2110-112.

Pardo RJ, Larkin LH, Renegar RH, 1984. Immunoelectron microscopic localization of relaxin in endometrial gland cells of the pregnant guinea pig. Anat Rec 209:373-79.

Parnham MJ, Sneddon JM, Williams KI, 1975. Evidence for a possible foetal control of prostaglandin release from the pregnant rat uterus in vitro. J Endocrinol 65:429-37.

Parvizi N, Elsaesser F, Smidt D, Ellendorff F, 1976. Plasma luteinizing hormone and progesterone in the adult female pig during the oestrous cycle, late pregnancy and lactation, and after ovariectomy and pentobarbitone treatment. J Endocrinol 69:193-203.

Patek CE, Watson J, 1983. Factors affecting steroid and prostaglandin secretion by reproductive tissues of cycling and pregnant sows in vitro. Biochim Biophys Acta 755:17-22.

Pedersen CA, Prange Jr AJ, 1979. Induction of maternal behavior in virgin rats after intracerebroventricular administration of oxytocin. Proc Natl Acad Sci USA 76:6661-65.

Perezgrovas R, Anderson LL, 1982. Effect of porcine relaxin on cervical dilatation, pelvic area and parturition in beef heifers. Biol Reprod 26:765-76.

Porter DG, 1979a. Relaxin: Old hormone, new prospect. Oxford Rev Reprod Biol 1:1-57.

Porter DG, 1979b. The myometrium and the relaxin enigma. Anim Reprod Sci 2:77-96.

Porter DG, 1984. Relaxin: A multipurpose hormone. In: Labrie F, Proulx L (ed.). Endocrinology International Congress Series No. 655. New York, Elsevier Science Publishing Co, pp 522-26.

Porter DG, Watts AD, 1986. Relaxin and progesterone are myometrial inhibitors in the ovariectomized non-pregnant mini-pig. J Reprod Fert 76:205-13.

Porter DG, Downing SJ, Bradshaw JMC, 1981. The inhibition of oxytocin or $PGF_{2\alpha}$ driven activity by relaxin in the rat is estrogen dependent. J Endocrinol 89:399-404.

Rao MR, Sanborn BM, 1986. Relaxin increases calcium efflux from rat myometrial cells in culture. Endocrinology 119:435-37.

Reinig JW, Daniel LN, Schwabe C, Gowan LK, Steinetz BG, O'Byrne EM, 1981. Isolation and characterization of relaxin from the sand tiger shark (Odontaspis taurus). Endocrinology 109:537-43.

Renegar RH, Larkin LH, 1985. Relaxin concentrations in endometrial, placental, and ovarian tissues and in sera from ewes during middle and late pregnancy. Biol Reprod 82:840-47.

Rothchild I, 1981. The regulation of the mammalian corpus luteum. Recent Prog Horm Res 37:183-298.

Sanborn BM, Weisbordt NW, Sherwood OD, 1981. Evidence against an obligatory role for catecholamine release or prostacyclin synthesis in the effects of relaxin on the rat uterus. Biol Reprod 24:987-92.

Sawyer HR, Abel Jr JH, McClellan MC, Schmitz M, Niswender GD, 1979. Secretory granules and progesterone secretion by ovine corpora lutea in vitro. Endocrinology 104:476-86.

Schams D, Walters DL, Schallenberger E, Butterman B, Karg H, 1983. Ovarian oxytocin in the cow. Acta Endocrinol [Suppl] (Copenh) 253:147 (Abstract 164).

Schwabe C, 1983a. Relaxin sequences. In: Bigazzi M, Greenwood FC, Gasparri F (ed.), Biology of Relaxin and Its Role in the Human. Amsterdam: Excerpta Medica, pp. 22-31.

Schwabe C, 1983b. N-α-formyl-tyrosyl-relaxin - A reliable tracer for relaxin radioimmunoassay. Endocrinology 113:814-15.

Schwabe C, McDonald JK, 1976. Primary structure of the A chain of porcine relaxin. Biochem Biophys Res Commun 70:397-405.

Schwabe C, McDonald JK, 1977a. Demonstration of the pyroglutamyl residue at the N terminus of the B-chain of porcine relaxin. Biochem Biophys Res Commun 74:1501-04.

Schwabe C, McDonald JK, 1977b. Relaxin: A disulfide homolog of insulin. Science 197:914-15.

Schwabe C, LeRoith D, Thompson RP, Shiloach J, Roth J, 1983. Relaxin extracted from protozoa (Tetrahymena pyriformis). J Biol Chem 258:2778-81.

Schwabe C, Steinetz B, Weiss G, Segaloff A, McDonald JK, O'Byrne E, Hochman J, Carriere B, Goldsmith L, 1978. Relaxin. Recent Prog Horm Res 34:123-211.

Shaikh AA, 1973. Estrone and estradiol levels in the ovarian venous blood from rats during the estrous cycle and pregnancy. Biol Reprod 5:297-307.

Sherwood OD, 1979. Purification and characterization of rat relaxin. Endocrinology 104:886-92.

Sherwood OD, Crnekovic VE, 1979. Development of a homologous radioimmunoassay for rat relaxin. Endocrinology 104:893-97.

Sherwood OD, O'Byrne EM, 1974. Purification and characterization of porcine relaxin. Arch Biochem Biophys 160:185-96.

Sherwood OD, Rutherford JE, 1981. Relaxin immunoactivity levels in ovarian extracts obtained from rats during various reproductive states and from adult cycling pigs. Endocrinology 108:1171-77.

Sherwood OD, Crnekovic VE, Gordon WL, Rutherford JE, 1980. Radioimmunoassay of relaxin throughout pregnancy and during parturition in the rat. Endocrinology 107:691-98.

Sherwood OD, Golos TG, Key RH, 1986. Influence of the conceptuses and the maternal pituitary on the distribution of multiple components of serum relaxin immunoactivity during pregnancy in the rat. Endocrinology 119:2143-47.

Sherwood OD, Key RH, Tarbell MK, Downing SJ, 1984. Dynamic changes of multiple forms of serum immunoactive relaxin during pregnancy in the rat. Endocrinology 114:806-13.

Sherwood OD, Nara BS, Crnekovic VE, First NL, 1979. Relaxin concentrations in pig plasma after the administration of indomethacin and prostaglandin $F_{2\alpha}$ during late pregnancy. Endocrinology 104:1716-21.

Sherwood OD, Nara BS, Welk FA, First NL, Rutherford JE, 1981. Relaxin levels in the maternal plasma of pigs before, during and after parturition and before, during and after suckling. Biol Reprod 25:65-71.

Sherwood OD, Rosentreter KR, Birkhimer ML, 1975. Development of a radioimmunoassay for porcine relaxin using ^{125}I-labeled polytrosyl-relaxin. Endocrinology 96:1106-13.

Sherwood OD, Wilson ME, Edgerton LA, Chang CC, 1978. Serum relaxin concentrations in pigs with parturition delayed by progesterone administration. Endocrinology 102:471-75.

Silver M, Barnes RJ, Comline RS, Fowden AL, Clover L, Mitchell MD, 1979. Prostaglandins in the foetal pig and prepartum endocrine changes in mother and foetus. Anim Reprod Sci 2:305-22.

Soloff MS, 1975. Uterine receptor for oxytocin: Effects of estrogen. Biochem Biophys Res Commun 65:205-12.

Soloff MS, Alexandrova M, Fernstrom MJ, 1979. Oxytocin receptors: Triggers for parturition and lactation? Science 204:1313-15.

Steinetz BG, Beach VL, Kroc RL, 1959. The physiology of relaxin in laboratory animals. In: Lloyd CH (ed), Recent Progress in Endocrinology of Reproduction. New York: Academic Press, pp. 389-427.

Steinetz BG, Beach VL, Kroc RL, Stasilli NR, Nussbaum RE, Nemith PJ, Dunn RK, 1960. Bioassay of relaxin using a reference standard: A simple and reliable method utilizing direct measurement of interpubic ligament formation in mice. Endocrinology 67:102-15.

Stewart DR, 1986. Development of a homologous equine relaxin radioimmunoassay. Endocrinology 119:1100-04.

Stewart DR, Papkoff H, 1986. Purification and characterization of equine relaxin. Endocrinology 119:1093-99.

Stewart DR, Stabenfeldt GH, 1981. Relaxin activity in the pregnant mare. Biol Reprod 25:281-89.

Stewart DR, Stabenfeldt GH, 1985. Relaxin activity in the pregnant cat. Biol Reprod 32:848-54.

Stewart DR, Stabenfeldt GH, Hughes JP, 1982a. Relaxin activity in foaling mares. J Reprod Fert [Suppl] 32:603-09.

Stewart DR, Stabenfeldt GH, Hughes JP, Meagher DM, 1982b. Determination of the source of equine relaxin. Biol Reprod 27:17-24.

Summerlee AJS, O'Byrne KT, Paisley AC, Breeze MF, Porter DG, 1984. Relaxin affects the central control of oxytocin release. Nature 309:372-74.

Taverne M, Bevers M, Bradshaw JMC, Dieleman SJ, Willemse AH, Porter DG, 1982. Plasma concentrations of prolactin, progesterone, relaxin and oestradiol-17β in sows treated with bromocriptine or indomethacin during late pregnancy. J Reprod Fert 65:85-96.

Taverne M, Willemse AH, Dieleman SJ, Bevers M, 1979. Plasma prolactin, progesterone and oestradiol-17β concentrations around parturition in the pig. Anim Reprod Sci 1:257-63.

Taylor MJ, Clark CL, Frawley LS, 1987. Analysis of relaxin release by reverse hemolytic plaque assay: influence of gestational age and $PGF_{2\alpha}$. Endocrinology (In press).

Theodosis DT, Wooding FBP, Sheldrick EL, Flint APF, 1986. Ultrastructural localization of oxytocin and neurophysin in the ovine corpus luteum. Cell Tissue Res 243:129-35.

Thomford PJ, Sander HKL, Kendall JZ, Sherwood OD, Dziuk PJ, 1984. Maintenance of pregnancy and levels of progesterone and relaxin in the serum of gilts following a stepwise reduction in the number of corpora lutea. Biol Reprod 31:494-98.

Too CK, Bryant-Greenwood GD, Greenwood FC, 1984. Relaxin increases the release of plasminogen activator, collagenase, and proteoglycanase from rat granulosa cells in vitro. Endocrinology 115:1043-50.

Too CK, Kong JK, Greenwood FC, Bryant-Greenwood GD, 1986. The effect of oestrogen and relaxin on uterine and cervical enzymes: Collagenase, proteoglycanase and β-glycuronidase. Acta Endocrinol 111:394-403.

Too CK, Weiss TJ, Bryant-Greenwood GD, 1982. Relaxin stimulates plasminogen activator secretion by rat granulosa cells in vitro. Endocrinology 111:1424-26.

Torday JS, Jefcoate CR, First NL, 1980. Effect of prostaglandin $F_{2\alpha}$ on steroidogenesis by porcine corpora lutea. J Reprod Fert 58:301-10.

Tregear G, Yu-cang Du, Ke-zhen W, Southwell C, Jones P, Johm M, Gorman J, Kemp B, Niall HD, 1983. The chemical synthesis of relaxin. In: Bigazzi M, Greenwood FC, Gasparri F (ed.), Biology of Relaxin and Its Role in the Human. Amsterdam: Excerpta Medica, pp. 42-55.

Vaje S, Elsaesser F, Elger W, Ellendorff F, 1980. Induktion der Geburt beim Schwein mit einem Prostaglandin E. Dtsch Tierarztl Wocheuschr 87:77-79.

Vasilenko P, Frieden EH, Adams WC, 1980. Effect of purified relaxin on uterine glycogen and protein in the rat. Proc Soc Exp Biol Med 163:245-48.

Wathes DC, Swann RW, 1982. Is oxytocin an ovarian hormone? Nature 297:225-27.

Wathes DC, Rees JM, Porter DG, 1986. Identification and measurement of relaxin in the ewe. Soc Study Fertil, Annual Conference, p. 58 (Abstract 88).

Wathes DC, Swann RW, Birkett SD, Porter DG, Pickering BT, 1983. Characterization of oxytocin, vasopressin, and neurophysin from the bovine corpus luteum. Endocrinology 113:693-98.

Watson J, Patek CE, 1979. Steroid and prostaglandin secretion by the corpus luteum, endometrium and embryos of cyclic and pregnant pigs. J Endocrinol 82:425-28.

Webb R, Mitchell MD, Falconer J, Robinson JS, 1981. Temporal relationships between peripheral plasma concentrations of oxytocin, progesterone and 13,14-dihydro-15-keto prostaglandin $F_{2\alpha}$ during the estrous cycle and early pregnancy in the ewe. Prostaglandins 22:443-54.

Weiss G, 1983. The secretion and role of relaxin in pregnant women. In: Bigazzi M, Greenwood FC, Gasparri F (ed.), Biology of Relaxin and Its Role in the Human. Amsterdam: Excerpta Medica, pp. 304-10.

Weiss G, O'Byrne EM, Steinetz BG, 1976. Relaxin: A product of the human corpus luteum of pregnancy. Science 194:948-49.

Weiss TJ, Bryant-Greenwood GD, 1982. Localization of relaxin binding sites in the rat uterus and cervix by autoradiography. Biol Reprod 27:673-79.

Whitely J, Willcox DL, Hartmann PE, Yamamoto SY, Bryant-Greenwood GD, 1985. Plasma relaxin levels during suckling and oxytocin stimulation in the lactating sow. Biol Reprod 33:705-14.

Williams KI, Dembinska-kiec A, Zmuda A, Gryglewski RJ, 1978. Prostacyclin formation by myometrial and decidual fractions of the pregnant rat uterus. Prostaglandins 15:343-50.

Williams LM, Hollingsworth M, Dixon JS, 1982. Changes in the tensile properties and fine structure of the rat cervix in late pregnancy and during parturition. J Reprod Fert 66:203-11.

Wilson Jr L, Huang LS, 1983. Effects of the fetal-placental unit on uterine prostaglandin levels in the pregnant rat. Prostaglandins 25:725-37.

Wilson Jr L, Huang LS, 1985. Contributions of the fetoplacental unit to augmented uterine prostaglandin levels periparturition in the rat. Biol Reprod 33:612-17.

Zarrow MX, Rosenberg B, 1953. Sources of relaxin in the rabbit. Endocrinology 53:593-98.

PART IV

REGULATION OF TESTICULAR CELL FUNCTIONS AND SPERMATOGENESIS

REGULATION OF MITOCHONDRIAL AND MICROSOMAL CYTOCHROME P-450 ENZYMES

IN LEYDIG CELLS

Anita H. Payne, Onyeama O. Anakwe, Dale B. Hales,
Markos Georgiou, Louise M. Perkins and Patrick G. Quinn

Departments of Obstetrics and Gynecology
Biological Chemistry and The Reproductive
Endocrinology Program
The University of Michigan
Ann Arbor, Michigan 48109

INTRODUCTION

Testosterone biosynthesis in Leydig cells is dependent on both acute and chronic stimulation by the anterior pituitary hormone, luteinizing hormone (LH). Binding of LH to specific high-affinity receptors on the surface of Leydig cells (Catt et al., 1972) results in increased production of intracellular cAMP (Dufau et al., 1973). Both the acute and the chronic effects of LH are mediated by increases in cyclic AMP. A schematic representation of LH stimulated testosterone production in Leydig cells is shown in Fig. 1. Although the exact mechanism is not known, acute stimulation by trophic hormones or cAMP in all steroidogenic tissues (testis, ovary and adrenal) results in an increased amount of the endogenous precursor, cholesterol, which associates with the mitochondrial cholesterol side-chain cleavage enzyme (P-450$_{scc}$). Chronic treatment of intact or hypophysectomized rats with LH, or its analog, human chorionic gonadotropin (hCG), results in an increased capacity for LH-stimulated testosterone production (Zipf et al., 1978; Payne et al., 1980) and induces enzymes of the steroidogenic pathway (Purvis et al., 1973a,b; O'Shaughnessy and Payne, 1982). In contrast, treatment of intact animals with a single high dose of LH or hCG causes Leydig cell steroidogenic desensitization which by definition is characterized by a diminished capacity to produce testosterone in response to subsequent, acute stimulation with trophic hormone or cAMP (Zipf et al., 1978; Sharpe, 1977; Hsueh et al., 1977; Saez et al., 1978; Cigorraga et al., 1978; Chasalow et al., 1979) and decreases in the microsomal P-450 activities, 17α-hydroxylase and C$_{17-20}$ lyase (Cigorraga et al., 1978; Chasalow et al., 1979; O'Shaughnessy and Payne, 1982) and a decreased capacity for pregnenolone production (Cigorraga et al., 1978; Nozu et al., 1981a,b) as well as reduced testicular mitochondrial P-450 (Luketich et al., 1983). This steroidogenic desensitization is independent of the loss of LH receptors or the desensitization of the adenylate cyclase system since it cannot be overcome by addition of exogenous cAMP (Tsuruhara et al., 1977; Chasalow et al., 1979; Cigorraga et al., 1978).

This chapter describes studies of the mechanisms that regulate the activities and the amounts of the mitochondrial and microsomal P-450 enzymes of Leydig cells. Enzymes involved in the biosynthesis of

467

testosterone from cholesterol are shown in Fig. 1. This biosynthesis requires the action of three cytochrome P-450 enzyme activities, the mitochondrial cholesterol side-chain cleavage (P-450$_{scc}$) as well as the microsomal 17α-hydroxylase and C$_{17-20}$ lyase activities which are catalyzed by a single protein referred to as P-450$_{17\alpha}$ (Nakajin and Hall, 1981; Nakajin et al., 1981; Suhara et al., 1984; Zuber et al., 1986).

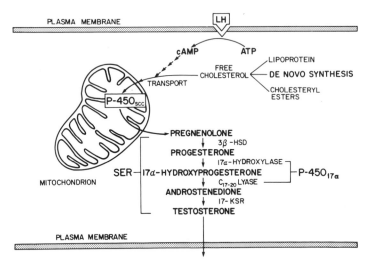

Fig. 1. LH-and cAMP-stimulated testosterone production in Leydig cells.

DOWN-REGULATION OF MITOCHONDRIAL AND MICROSOMAL P-450 ENZYMES

We previously reported that administration of a single high dose of LH or hCG to rats resulted 72 h later in a decrease in the activities of 17α-hydroxylase and C$_{17-20}$lyase and a diminished capacity for testosterone production to subsequent in vitro stimulation by gonadotropin or cAMP (steroidogenic desensitization) (O'Shaughnessy and Payne, 1982). The addition of an exogenous source of cholesterol in the form of high density lipoprotein to the in vitro incubation restored the capacity for testosterone production of the Leydig cells from hCG injected rats (Quinn et al., 1981). These data suggest that depletion of endogenous cholesterol stores rather than the decrease in the microsomal P-450 enzymes is the major cause of the diminished capacity of Leydig cell testosterone production following the administration of a desensitizing dose of hCG.

To investigate the mechanism by which the microsomal P-450 activities are decreased and the relationship of this decrease in enzyme activities to the diminished steroidogenic capacity, Quinn and Payne (1984) developed a primary culture system of purified mouse Leydig cells. Various hypotheses have been proposed for the decrease in microsomal P-450 activities. Studies from our laboratory indicate that the mechanism by which microsomal P-450 activities are decreased in desensitized Leydig cells involves oxygen free radical-initiated damage (Quinn and Payne, 1984, 1985). An alternative mechanism has been proposed by Nozu et al. 1981a,b) and Cigorraga et al. (1978). These investigators suggest that increased production of estradiol, by hCG- or cAMP-stimulated rat Leydig cells causes an estradiol receptor-mediated decline in microsomal P-450 activities and steroidogenic capacity. Such a mechanism, however, has not been verified by other investigators (Benahmed et al., 1982, Brinkmann et al., 1981, 1982; Damber et al., 1983).

Cultured mouse Leydig cells exhibited time dependent decreases in
17α-hydroxylase and C_{17-20} lyase activities when maintained under standard
culture conditions (95% air, 5% CO_2), (Fig. 2). Inclusion of the hydroxyl
radical scavenger dimethylsulfoxide (Me_2SO) in the culture medium or the
reduction of oxygen tension from 19 to 1% O_2 preserved these enzymes activities and the combined effects of low oxygen and Me_2SO were synergistic
(Fig. 2). These data demonstrate that Leydig cell microsomal P-450
activities, 17α-hydroxylase and C_{17-20} lyase are very sensitive to oxygen-
mediated damage. When Leydig cells were stimulated with 1 mM 8-Br-cAMP
which resulted in a 15-fold increase in testosterone production, a marked
decrease in 17α-hydroxylase activities, relative to non-stimulated control
cultures, was observed at 24 and 48 h in cultures incubated under standard
conditions (19% O_2), (Fig. 3). The addition of Me_2SO to the medium of
cells cultured at 19% O_2 had no significant effect on the pattern observed,
but the absolute level of enzyme activities was higher (Fig. 3). As shown
in Fig. 4 when Leydig cells were cultured at a reduced oxygen tension
(1%), the 17α-hydroxylase activities of cAMP-stimulated cells were maintained at higher levels than either at 19% O_2 or at 19% O_2 + Me_2SO and
were not significantly different (p > 0.05) from control values. C_{17-20}
lyase activities responded in an identical pattern as was observed for
17α-hydroxylase activities, (Quinn and Payne, 1984). However, the activity
of 3β-hydroxysteroid dehydrogenase isomerase, a microsomal enzyme which is
not a P-450 enzyme, was stable during 48 h incubation of cultures in both

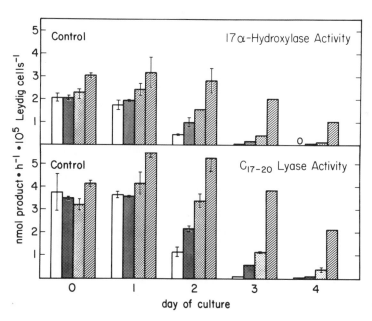

Fig. 2. Effects of oxygen tension and Me_2SO on P-450 activities of
control mouse Leydig cells in primary culture. Enzyme activities
were determined during a 1-hour period of incubation after a 3 h
attachment period on day 0 and then at 24 h intervals. Cells were
maintained at 19% O_2 (□), 19% O_2 + 100 mM dimethylsulfoxide
(Me_2SO, ▨), 1% O_2 (▨), and 1% O_2 + 100 mM Me_2SO (▨).
17α -Hydroxylase activity was determined by measuring the conversion of [³H]progesterone to [³H]17α-progesterone, [³H]andro-
stenedione, and [³H]testosterone. C_{17-20} Lyase activity was
determined by measuring the conversion of [³H]17α-hydroxypro-
gesterone to [³H]androstenedione and [³H]testosterone. Determinations were done in duplicate. Values are the means ± the
range of two experiments. (From Quinn and Payne, 1984).

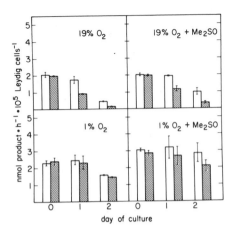

Fig. 3. Effects of oxygen tension and Me$_2$SO on 17α-hydroxylase activity of control and cAMP-stimulated mouse Leydig cells in primary culture. 17α-Hydroxylase activity was determined as described for Fig. 2 after 3 h attachment period on day 0 of culture and then at 24 h and 48 h in control (□) and cAMP-stimulated (▨) Leydig cell cultures, which were treated with 1 mM 8-Br-cAMP during the initial 24 h of culture. Determinations were done in duplicate. Values are the means ± the range of two experiments. (From Quinn and Payne, 1984).

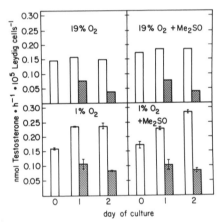

Fig. 4. Maximal testosterone production in response to 8-Br-cAMP of cultured mouse Leydig cells. Cultured Leydig cells were incubated in fresh medium containing 1 mM 8-Br-cAMP for a 3 h period at 0, 24 and 48 h. Control cells (□) had not been previously exposed to 8-Br-cAMP, whereas cAMP treated-cells (▨) had been exposed to 1 mM 8-Br-cAMP during the initial 24 h of culture. Determinations were done in duplicate. Values are the means ± the range of two experiments. (From Quinn and Payne, 1984).

control and cAMP stimulated Leydig cells under all culture conditions (Quinn and Payne, 1984).

When steroidogenic capacity of cAMP-treated cells was assessed at 24 or 48 h it was found that their ability to produce testosterone was diminished 50 and 70%, respectively, in response to subsequent acute stimulation with cAMP, regardless of the culture conditions, (Fig. 4).

Since cAMP treated Leydig cells cultured at 1% O_2 showed no greater loss of enzyme activity than did controls, loss of microsomal P-450 activities cannot account for the diminished testosterone production of these desensitized Leydig cells as has been suggested by Nozu et al. (1981a,b). These observations by Quinn and Payne (1984) are consistent with the hypothesis that the enhanced loss of Leydig cell microsomal P-450 activities in cAMP-treated cultures is caused by the increased production of testosterone which interacts with the P-450 to form a pseudosubstrate· P-450·O_2 complex from which damaging oxygen-derived free radicals are released due to the inability of the steroid product to be hydroxylated. This hypothesis is based on studies in adrenal cell cultures and was proposed by Hornsby (1980) and Hornsby and Crivello (1983a,b). If this hypothesis is correct then the addition of the products of the microsomal P-450 enzyme catalyzed reaction to the Leydig cell cultures should bring about the same oxygen-mediated loss in 17α-hydroxylase and C$_{17-20}$ lyase activities as cAMP stimulation. In a subsequent study we investigated the effect of addition to the culture medium of the products, androstenedione or testosterone, at a concentration of 2 μM, which is equivalent to the concentration of testosterone present in the medium of cultures following stimulation with LH or cAMP for 24 h (Quinn and Payne, 1985). It was observed that in cultures incubated at 19% O_2, androstenedione or testosterone caused a similar decrease in 17α-hydroxylase activity as was observed in cultures treated with LH or cAMP and that this steroid-induced decrease in P-450 activity was prevented by reduction of the oxygen tension (Fig. 5). Additional studies demonstrated that the oxygen-mediated loss in enzyme activities was only observed with steroid analogs

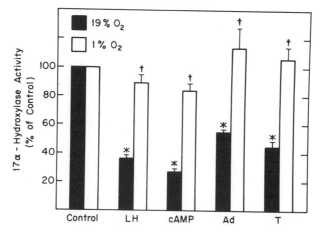

Fig. 5. Effects of steroid products on 17α-hydroxylase activity of mouse Leydig cells. Cultures were maintained at 19% O_2 and treated for 48 h with no addition (control), 100 ng/ml of luteinizing hormone (LH), 1 mM 8-Br-cAMP (cAMP), 2 μM androstenedione (Ad), or 2 μM testosterone (T). Cultures were washed to remove treatment agents and 17α-hydroxylase activity was determined as described for Fig. 2. Results are expressed as a percentage of the respective control activities of cultures maintained at either 19% or 1% O_2. (From Quinn and Payne, 1985).

471

structurally related to testosterone or androstenedione and could not be demonstrated in cultures treated with cortisol, estradiol, or the androgen agonist methyltrienolone (R1881), (Table 1). Treatment of Leydig cell cultures with cyproterone acetate, an androgen receptor antagonist, did not prevent the cAMP- or testosterone-induced decreases in 17α-hydroxylase activity, (Quinn and Payne, 1985). Similar effects on C_{17-20} lyase activity were observed in steroid-product treated cultures, (Quinn and Payne, 1985). These data support the hypothesis that the decline in microsomal P-450 activities of cAMP-desensitized Leydig cells results from an increased concentration of steroid products, which act as pseudosubstrates for the P-450 enzymes to increase generation of oxygen-derived, free-radical species, resulting in inactivation of the P-450 enzymes. The current results do not support any role for estrogen or androgen receptor-mediated actions on the loss of microsomal P-450 activities during steroidogenic desensitization. Since the alternative mechanism involving an estradiol-mediated action for the decrease in microsomal P-450 activities was based on studies in rats (Nozu et al., 1981a,b), our laboratory examined whether the loss in microsomal P-450 activities in rat Leydig cells can also be explained by a steroid product-induced, oxygen-mediated mechanism. As shown in Table 2 a similar steroid product-enhanced, oxygen-mediated mechanism is responsible for the loss in C_{17-20} lyase activity in cAMP desensitized rat Leydig cells. The data presented in Table 2 that support this mechanism include the protection of C_{17-20} lyase activity by aminoglutethimide, which blocks testosterone production, and partial reversal of the loss when Leydig cell cultures were incubated at a reduced oxygen tension and an oxygen-dependent loss in enzyme activity when exogenous testosterone was added to the culture medium. The addition of a high concentration of estradiol (1 µM), a concentration similar to those used by Nozu et al. (1981b) had no effect on C_{17-20} lyase activity of rat Leydig cells. Furthermore, the presence of an estradiol receptor antagonist (LY 156758) did not prevent the loss of C_{17-20} lyase activity in cAMP-treated cultures. These data demonstrate that estradiol is not involved in the loss of microsomal P-450 activities in rat Leydig cells and that the mechanism of loss of microsomal P-450 enzymes is similar in rat and mouse Leydig cells.

Table 1. Effect of Steroids and Androgen Receptor Agonist on 17α-Hydroxylase Activity at Different Oxygen Tensions.[a]

| Treatments | 17α-Hydroxylase Activity | |
| | 19% O_2 | 1% O_2 |
	(Per Cent)	
Control	100	100
Testosterone	44 ± 2	84
Epitestosterone	27 ± 1	54 ± 3[b,c]
17α-Methyltestosterone	56 ± 2	72 ± 3[b,c]
Cortisol	106 ± 4	97 ± 7
Estradiol	111 ± 8	120 ± 8
R1881	84 ± 4	88 ± 6

[a] Mouse Leydig cell cultures were maintained at 19 and 1% O_2 and treated for 48 h with no additions (Control, n = 8), or 2 µM of each of the following: testosterone (n = 8), epitestosterone (n = 8), 17α-Methyltestosterone (n = 5), cortisol (n = 5), estradiol (n = 6), methyltrienolone (R1881), n = 3). Cultures were then washed to remove treatment agents and 17α-hydroxylase activity was determined as described in legend for Fig. 2. Results are expressed as mean $\pm$ range of the indicated number of experiments (n). From Quinn and Payne (1985).

[b] $p < 0.05$ versus control

[c] $p < 0.05$ versus 19% O_2

Table 2. Effect of Treatment with cAMP, Steroid Products, Androgen Receptor Agonist and Estradiol Receptor Antagonist on C_{17-20} Lyase Activity at Different Oxygen Tensions.[a]

Treatments	C_{17-20} Lyase Activity 19% O_2 (Per Cent)	1% O_2
Control	100	100
cAMP	45 $\pm$ 3[b]	71 $\pm$ 2[b,c]
Aminoglutethimide (AG)	108 $\pm$ 4	112 $\pm$ 4
cAMP + AG	84 $\pm$ 3	89 $\pm$ 5
Testosterone	62 $\pm$ 5[b]	95 $\pm$ 7[c]
R1881	84 $\pm$ 7	97 $\pm$ 9
Estradiol	99 $\pm$ 4	99 $\pm$ 3
LY 156758	99 $\pm$ 5	—
cAMP + LY	49 $\pm$ 5	—

[a]Rat Leydig cell cultures were maintained at 19 and 1% O_2 and treated for 48 h with no additions (Control, n = 5), 1 mM 8-Br-cAMP (cAMP, n = 5), 0.3 mM aminoglutethimide (n = 5), 10 µM Testosterone (n = 5), 2 µM methyltrienolone (R1881, n = 4), 1 µM estradiol (n = 4, or 100 nM LY 156758 (n = 4), or 8-Br-cAMP plus LY 156758 (cAMP + LY, n = 4). Cultures were then washed to remove treatment agents and C_{17-20} lyase activity was determined as described in legend for Fig. 3. Results are expressed as mean $\pm$ S.E. of the indicated number of independent experiments, each of which was done in duplicate.
[b]$p < 0.05$ versus control
[c]$p < 0.05$ versus 19% O_2

Although there are numerous studies on the effect of prolonged stimulation of steroidogenesis with LH/hCG or cAMP on Leydig cell microsomal P-450 activities, little is known about the effect of this type of treatment on mitochondrial P-450$_{scc}$ activity. P-450$_{scc}$ plays a key role in regulating the amount of androgen production. It is a mitochondrial P-450 enzyme which cleaves the side-chain of cholesterol to yield pregnenolone, the first committed step in steroid hormone biosynthesis, (Simpson, 1979). It has been reported that maximal hCG-stimulated pregnenolone production of Leydig cells from rats injected with a single high dose of hCG is reduced (Cigorraga et al., 1978; Nozu et al., 1981a,b). In another study, (Luketich et al., 1983) reported that injection of mice with a high dose of hCG resulted in reduced testicular mitochondrial P-450 measured spectrally. The mechanism by which pregnenolone production or mitochondrial P-450 is reduced in rodents injected with high doses of gonadotropin has not previously been investigated. This laboratory used a rat Leydig cell culture system to examine whether the P-450$_{scc}$ activity is decreased by an oxygen-mediated, product-enhanced mechanism as was observed for the microsomal P-450 activities. As illustrated in Table 3, treatment of rat Leydig cell cultures with 1 mM cAMP for 48 h at ambient oxygen tension (19%) caused a marked decrease in the P-450$_{scc}$ activity which was prevented when cultures were incubated at a reduced oxygen tension (1%). Inhibition of cholesterol metabolism by the addition of aminoglutethimide to the cultures prevented the cAMP-induced loss of P-450$_{scc}$ activity observed at 19% oxygen. These data suggest that enhanced steroid production is responsible for the decrease in P-450$_{scc}$ activity in an oxygen-dependent manner as was observed for the microsomal P-450 activities. Pregnenolone is the immediate product of the cholesterol side-chain cleavage reaction. We therefore investigated whether the enhanced loss of P-450$_{scc}$ activity in cAMP-treated cultures was due to the increased production of pregnenolone. Further metabolism of pregnenolone

Table 3. Effect of Steroid Production and Oxygen Tension on $P\text{-}450_{scc}$ Activity.[a]

| Treatments | $P\text{-}450_{scc}$ Activity | |
	19% O_2 (Per Cent)	1% O_2
Control	100	100
cAMP	41 ± 3[b]	90 ± 5[c]
AG	73 ± 5	85 ± 4
cAMP + AG	92 ± 4	121 ± 9

[a]Rat Leydig cell cultures were maintained at 19 and 1% O_2 and treated for 48 h with no addition (Control, n = 5), 1 mM 8-Br-cAMP (cAMP, n = 5), 0.3 mM aminoglutethimide (AG, n = 3), or 1 mM 8-Br-cAMP plus aminoglutethimide (cAMP + AG, n = 4). Cultures were then washed to remove treatment agents and $P\text{-}450_{scc}$ activity was determined by measuring the conversion of side-chain labeled $[26,27\text{-}^3H]25\text{-}$hydroxycholesterol to $[^3H]$ labeled 4-hydroxy-4-methyl-pentanoic acid during a 1-h incubation. Results are expressed as mean $\pm$ S.E. of the indicated number of independent experiments (n), each of which was done in duplicate.
[b]p < 0.05 versus control
[c]p < 0.05 versus 19% O_2

was blocked by incubating Leydig cell cultures in the presence of cyanoketone, an inhibitor of 3β-hydroxysteroid dehydrogenase, and SU-10603, an inhibitor of 17α-hydroxylase. The effectiveness of these inhibitors in blocking pregnenolone metabolism is shown in Table 4. In the absence of inhibitors, testosterone was the major androgen secreted into the medium by either control or cAMP-treated Leydig cell cultures maintained at 19% oxygen. The addition of cyanoketone plus SU-10603 during the 2-day culture period resulted in pregnenolone being the major steroid product with only minimal amounts of progesterone and testosterone being produced. A similar pattern of steroid production was observed in cultures incubated at 1% oxygen. Table 4 illustrates that accumulation of pregnenolone in the presence of inhibitors increases from 0.4 ng to 54 ng in control cultures. This increase in pregnenolone resulted in reduction of $P\text{-}450_{scc}$ activity, by 80%, in culture incubated at ambient oxygen tension, and this decrease was completely prevented in cultures incubated at a reduced

Table 4. Steroid Product Accumulation in Leydig Cell Cultures in the Presence and Absence of Inhibitors of Pregnenolone Production.[a]

Treatments	Pregnenolone	Progesterone (ng)	Testosterone
Control	0.4	2.3	48
SU-10603 (SU)	7.3	31	4
Cyanoketone (CK)	4.1	1.4	9
CK + SU	54	4.1	0.9
cAMP	6.8	16	451
cAMP + CK + SU	271	11.7	9

[a]Rat Leydig cell cultures were maintained at 19% O_2 and were treated for 48 h with no additions (Control), 1 mM 8-Br-cAMP (cAMP), 3 μM cyanoketone, 15 μM SU-10603, cyanoketone plus SU-10603, or 8-Br-cAMP plus cyanoketone plus SU-10603. At the end of the treatment period, culture media were collected and analyzed by specific radioimmunoassays for pregnenolone, progesterone and testosterone. Values are means of duplicate determinations of a representative experiment.

oxygen tension, (Table 5). When pregnenolone was maximally increased in cAMP-treated cultures in the presence of the inhibitors, P-450$_{scc}$ activity was completely eliminated at 19% O$_2$ but was protected to near control values at 1% O$_2$, (Table 5). These data demonstrate that pregnenolone causes oxygen-mediated damage to P-450$_{scc}$ activity and suggest that the enhanced loss of this enzyme activity in cAMP-treated cultures is due to increased mitochondrial concentrations of pregnenolone near the active site of the enzyme. These data also indicate that testosterone or other steroid products of pregnenolone metabolism are not responsible for the cAMP-induced loss in P-450$_{scc}$ activity.

Table 5. Effect of Inhibitors of Pregnenolone Metabolism on P-450$_{scc}$ Activity.[a]

Treatments	P-450$_{scc}$ Activity	
	19% O$_2$ (Per Cent)	1% O$_2$
Control	100	100
cAMP	27 $\pm$ 9	102 $\pm$ 12
Cyanoketone (CK)	104 $\pm$ 7	145 $\pm$ 11
SU-10603 (SU)	75 $\pm$ 6	71 $\pm$ 4
CK + SU	20 $\pm$ 2	97 $\pm$ 5
cAMP + CK + SU	N.D.*	77 $\pm$ 16

[a] Rat Leydig cell cultures were maintained at 19 and 1% O$_2$ and were treated for 48 h with no additions (Control), 1 mM 8-Br-cAMP, 3 μM cyanoketone, 15 μM SU-10603, cyanoketone plus SU-10603, (CK + SU), or 8-Br-cAMP plus cyanoketone plus SU-10603 (cAMP + CK + SU). Cultures were then washed to remove treatment agents and P-450$_{scc}$ activity was determined as described in legend for Table 3. Results are expressed as mean $\pm$ range of two independent experiments, each of which was done in duplicate.
* Non detectable.

To examine if the observed oxygen-dependent, product-enhanced decreases in P-450 activities are due to damage to and concomitant loss of the enzyme protein or to an inactivation of the catalytic activity of the enzyme with no change in the amount of the enzyme, the amount of immunoreactive mouse P-450$_{17\alpha}$ and rat P-450$_{scc}$ was determined by Western blotting. Fig. 6 illustrates that treatment of mouse Leydig cell cultures for 2 days with 1 mM cAMP at 19% oxygen resulted in a marked decrease (30% of control) in the amount of immunoreactive P-450$_{17\alpha}$. This decrease in the amount of the enzyme was prevented when cAMP-treated Leydig cells were incubated at a reduced atmosphere of oxygen. 17α-hydroxylase activity measured in parallel cultures was reduced to 38% in Leydig cells treated with cAMP at 19% O$_2$ relative to control and this cAMP-induced decrease was prevented at 1% O$_2$ (106% of control). To determine if the decrease in the amount of P-450$_{17\alpha}$ that occurs in cAMP-treated cultures at 19% oxygen is mediated by testosterone or structurally related steroids, Leydig cell cultures were incubated at 19% oxygen for two days and treated with testosterone, epitestosterone, cortisol, estradiol or the androgen agonist, methyltrienolone (R1881). Fig. 7 demonstrates that, of the compounds tested, only testosterone and its 17α-epimer, epitestosterone, caused a decrease in the amount of P-450$_{17\alpha}$. These decreases in the amount of the enzyme, specifically caused by testosterone or epitestosterone and not by estradiol, cortisol or the androgen agonist, are similar to the decreases in 17α-hydroxylase activity described above. The observed decreases in amount of P-450$_{17\alpha}$ were not due to decreased de novo synthesis since similar decreases were observed in cultures incubated in the presence of cycloheximide (10 μg/ml) (data not shown).

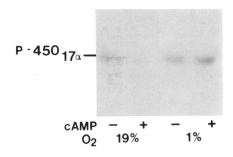

$$P-450_{17\alpha}-$$

| cAMP | — | + | — | + |
| O$_2$ | 19% | | 1% | |

Fig. 6. Effect of oxygen tension and cAMP treatment on the amount of immunoreactive P-450$_{17\alpha}$. Mouse Leydig cell cultures were treated for 48 h at 19% O$_2$ (left two lanes) or 1% O$_2$ (right two lanes) in the presence (+) or absence (−) of 1 mM 8-Br-cAMP (cAMP). Lysates were prepared from treated Leydig cells and immunoreactive P-450$_{scc}$ was determined by immunoblotting. Specifically bound antibody was detected by ^{125}I-protein A.

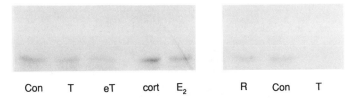

Con   T   eT   cort   E$_2$ R   Con   T

Fig. 7. Effect of steroid hormones and an androgen agonist on the amount of immunoreactive P-450$_{17\alpha}$. Mouse Leydig cell cultures were incubated for 48 h at 19% O$_2$ and treated with medium only (Con) or with 2 μM each of the following agents: testosterone (T), epitestosterone (eT), cortisol (cort), estradiol (E); R1881 (R), Con or T. Immunoreactive P-450$_{17\alpha}$ was determined as described for Fig. 6.

Fig. 8 illustrates the amount of immunoreactive P-450$_{scc}$ and P-450$_{17\alpha}$ from rat Leydig cells incubated for 2 days at 19 and 1% oxygen in the presence of inhibitors of pregnenolone production and in the presence or absence of cAMP. Accumulation of pregnenolone caused a marked decrease in immunoreactive P-450$_{scc}$ and this decrease was prevented in cultures incubated at 1% oxygen. These oxygen-mediated, pregnenolone-induced decreases in the amount of P-450$_{scc}$ are similar to the decreases in P-450$_{scc}$ activity described above. In contrast, no decrease in the amount of P-450$_{17\alpha}$ was observed from these treatments. This finding is consistent with the hypothesis that the oxygen-mediated decrease in P-450$_{17\alpha}$ in cAMP-treated cultures is due to increased testosterone production since addition of cyanoketone and SU-10603 inhibits testosterone production by Leydig cells (Table 4).

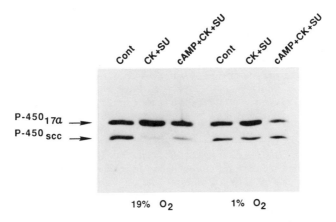

Fig. 8. Effect of cAMP and inhibitors of pregnenolone production on the amount of immunoreactive P-450$_{scc}$ and P-450$_{17\alpha}$. Rat Leydig cell cultures were incubated at 19 and 1% O_2 and were treated for 48 h with no additions (Cont), 3 mM cyanoketone plus 15 mM SU-10603 (CK + SU) or 1 mM 8-Br-cAMP plus CK plus SU. Immunoreactive P-450$_{scc}$ and P-450$_{17\alpha}$ were determined by immunoblotting. Specifically bound P-450$_{scc}$ and P-450$_{17\alpha}$ antibodies were detected by ^{125}I-protein A.

Taken together, these various studies indicate that the enhanced loss of Leydig cell microsomal and mitochondrial cytochrome P-450 activities in cAMP-treated cultures is caused by the increased production of testosterone and pregnenolone, respectively. These products of the enzyme catalyzed reactions act as pseudosubstrates for the proximal P-450 by binding to its P-450 and triggering the release of reactive oxygen species. These oxygen radicals either directly or indirectly, via lipid peroxidation, damage the P-450's leading to increased degradation of the enzymes. The data suggest that the decrease in P-450 activities are due either entirely or to a major extent to decreased amounts of these enzymes and not due to inactivation of the existing enzymes.

REGULATION OF INDUCTION OF MITOCHONDRIAL AND MICROSOMAL P-450 ENZYMES

The above studies describe the mechanisms by which degradation of mitochondrial and microsomal P-450's occur in Leydig cells. Little is known about the regulation of induction of these enzymes in Leydig cells. Purvis et al. (1973a,b) reported that mitochondrial and microsomal P-450, measured spectrally, decreased following hypophysectomy and that daily treatment with hCG increased both of these P-450's. The changes in spectrally measured microsomal P-450 correlated closely to increases in the activities of 17α-hydroxylase and C_{17-20}lyase. These data suggest that LH plays a role in regulating the testicular P-450 enzymes; however, the changes in P-450's could be a reflection of changes in the number of Leydig cells following hypophysectomy and treatment with hCG (Christensen and Peacock, 1980). In a recent study using a specific IgG fraction from antiserum raised against purified bovine adrenocorticortical P-450$_{scc}$, Anderson and Mendelson (1985) reported that treatment for 48 h with either hCG or cAMP of rat Leydig cells, after 8 days in culture, resulted in increased de novo synthesis of P-450$_{scc}$ indicating that LH via cAMP may regulate the amount of P-450$_{scc}$.

Studies from our laboratory demonstrated that chronic treatment of mouse Leydig cells in culture with LH or cAMP caused an increase in 17α-hydroxylase and C_{17-20} lyase activities (Malaska and Payne, 1984). Treatment with cAMP was found to be more effective than treatment with LH in inducing the enzyme activities. In the same study it was shown that the continuous presence of cAMP in the medium was essential for induction of the enzyme activities. In a subsequent study Rani and Payne (1986) showed that inhibition of steroid product formation by the addition of aminoglutethimide to the Leydig cell cultures caused a marked increase in the cAMP induction of the microsomal P-450 activities, 17α-hydroxylase and C_{17-20} lyase (Fig. 9). Aminoglutethimide by itself had no effect on the P-450 activities, indicating that the induction of the enzyme activities was solely due to the action of cAMP.

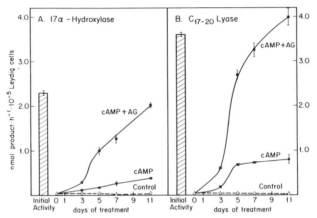

Fig. 9. Effect of aminoglutethimide (AG) on cAMP-mediated induction of 17α-hydroxylase (A) and C_{17-20} lyase (B) activities. Mouse Leydig cell cultures were maintained at 19% O_2 for 6 days. On day 6, cultures were transferred to 1% O_2 and treated for 11 days with 0.5 mM 8-Br-cAMP in the absence or presence of 0.5 mM AG. Initial activity is enzyme activity from freshly isolated Leydig cells. Enzyme activities were determined as described for Fig. 2. (From Rani and Payne, 1986).

We examined whether the marked increase in cAMP induction of microsomal P-450 activities when cholesterol metabolism is inhibited is due to the decreased production of androstenedione and/or testosterone (Rani and Payne, 1986). As can be seen in Fig. 10, the addition of testosterone or androstenedione at a concentration of 5 μM resulted in a significant decrease in enzyme activity compared to that in cells treated with cAMP plus aminoglutethimide. The combination of testosterone plus androstenedione was not more effective than either steroid alone. In addition, the addition of 50 nM estradiol to the Leydig cell cultures did not significantly reduce 17α-hydroxylase activity. When estradiol was added together with testosterone, the observed decrease was no greater than that caused by testosterone alone. None of the treatments with steroid products, either by themselves or in combination, reduced cAMP-induced aminoglutethimide-enhanced enzyme activity to the level of activity in cultures treated only with cAMP. In the studies shown in Fig. 10, a concentration of 0.5 mM 8-Br-cAMP was used. In subsequent studies it was found that treatment with lower concentrations of 8-Br-cAMP (0.05 mM) results in greater and more rapid induction of 17α-hydroxylase activity. The treatment of mouse Leydig cell cultures with 0.05 mM cAMP, cAMP plus aminoglutethimide and with the addition of 5 μM testosterone to the cAMP plus aminoglutethimide treated cultures reduces the 17α-hydroxylase activity to levels observed with cAMP treatment only (Fig. 11).

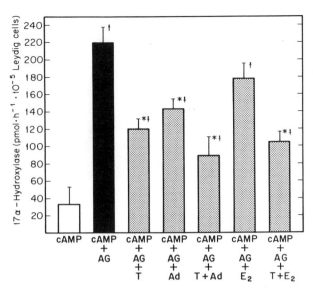

Fig. 10. Effect of steroid products on cAMP-mediated induction of 17α-hydroxylase activity in the presence of aminoglutethimide (AG). Mouse Leydig cell cultures were maintained at 19% O_2 for 6 days. On day 6, cultures were transferred to 1% O_2 and treated for 3 days with 0.5 mM 8-Br-cAMP (cAMP) or 0.5 mM cAMP plus 0.5 mM AG plus the indicated steroids: 5 μM testosterone (T), 5 μM androstenedione (Ad), and 50 nM 17β-estradiol (E_2). 17α-Hydroxylase activity was determined as described for Fig. 2. Data were statistically analyzed by analysis of variance and Duncan's new multiple range test. †, significantly ($p < 0.05$) different from cAMP, *, significantly ($p < 0.05$) different from cAMP plus AG. (From Rani and Payne, 1986).

Changes in 17α-hydroxylase activity are reflective of changes in the amount of the enzyme protein as measured by immunoblotting. Fig. 12 shows that without cAMP, P-450$_{17α}$ is nondetectable. Following treatment with cAMP an increase in the amount of P-450$_{17α}$ is observed at 24 h which continues to increase during the 4 day-treatment period. The addition of aminoglutethimide to the cultures markedly enhances the amount of immunoreactive P-450 at all times and this aminoglutethimide-enhanced increase in amount is completely prevented by the addition of testosterone to cAMP plus aminoglutethimide treated cultures. These data suggest that testosterone produced during cAMP induction of P-450$_{17α}$ either causes increased degradation of the newly synthesized enzyme or inhibits de novo synthesis. If the effect of testosterone is on degradation rather than de novo synthesis, it would be expected that the rate of de novo synthesis in cAMP treated-cultures would be the same in the presence or absence of aminoglutethimide, but the total amount and activity of enzyme would be increased in the presence of the inhibitor of cholesterol metabolism.

De novo synthesis of P-450$_{17α}$ was determined using ^{35}S-methionine incorporation into newly synthesized proteins during a 3 h-period of incubation. Fig. 13 is an autoradiogram of immunoisolated P-450$_{17α}$ from Leydig cell lysates obtained from cultures treated for various periods of time with no additions or with added cAMP or cAMP plus aminoglutethimide.

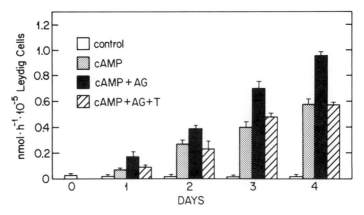

Fig. 11. Effect of steroid products on cAMP-induced 17α-hydroxylase activity. Mouse Leydig cell cultures were incubated for 7 days at 19% oxygen prior to initiation of treatment marked day 0. Cultures were treated for 4 days with medium only (control), 0.05 mM 8-Br-cAMP (cAMP), cAMP plus 0.5 mM aminoglutethimide (AG) or cAMP plus AG plus 5 μM testosterone. 17α-Hydroxylase activity was determined as described for Fig. 2. Values represent the mean and the range of two independent experiments.

Without the addition of cAMP de novo synthesis was negligible. cAMP alone caused a marked time dependent increase in the rate of de novo synthesis up to day 3 of treatment. This cAMP-induced increase was enhanced approximately two fold by the addition of aminoglutethimide (Table 6). The effect that addition of testosterone has on de novo synthesis of cAMP plus aminoglutethimide treated cultures is shown in Fig. 14. It is obvious from this figure that the addition of testosterone decreases de novo synthesis of $P-450_{17\alpha}$ to the rate observed in cultures treated with cAMP, only. These data indicate that testosterone affects de novo synthesis as well as degradation of $P-450_{17\alpha}$, as described above. Preliminary data from our laboratory indicate that the effect of testosterone on de novo synthesis of $P-450_{17\alpha}$ is distinct from the effect of testosterone on increasing degradation of the enzyme. Testosterone modulation of de novo synthesis of $P-450_{17\alpha}$ appears to be an androgen receptor-mediated action occurring at low concentrations of testosterone. The data thus suggest that testosterone produced during chronic cAMP treatment of Leydig cells negatively regulates the amount of $P-450_{17\alpha}$ by two distinct mechanisms: 1) modulation of cAMP-induced de novo synthesis and 2) increased degradation by acting as a pseudosubstrate enhancing oxygen radical-mediated damage to the enzyme.

To investigate whether amount and synthesis of $P-450_{17\alpha}$ and $P-450_{scc}$ in Leydig cells are coordinately regulated, changes in the amount and in the rate of de novo synthesis of these two P-450's were studied in the absence and presence of cAMP treatment. Fig. 15 illustrates an auto-radiogram of a representative immunoblot of $P-450_{17\alpha}$ and $P-450_{scc}$ from the same cultures. Day 0 represents the amount of these two enzymes measured 3 h after plating of the cells. As can be seen from this immunoblot, the amount of $P-450_{scc}$ changes very little during eleven days in culture in the absence of cAMP. Treatment with cAMP initiated on day 7 brought about only a minimal increase in the total amount of $P-450_{scc}$. In sharp contrast, the amount of $P-450_{17\alpha}$ in the absence of cAMP treatment decreases to 50% by day two, relative to day 0, and is essentially

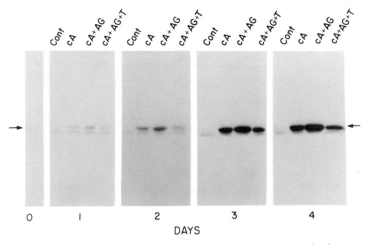

DAYS

Fig. 12. Effect of cAMP, cAMP plus aminoglutethimide (AG) and cAMP plus aminoglutethimide plus testosterone on the amount of immuno-reactive $P-450_{17\alpha}$. Mouse Leydig cell cultures were treated as described for Fig. 11. At the indicated time, immunoreactive $P-450_{17\alpha}$ was determined as described for Fig. 6.

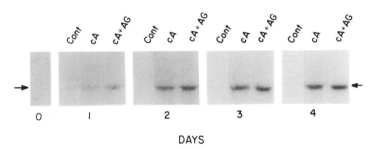

DAYS

Fig. 13. Effect of aminoglutethimide (AG) on the rate of cAMP-induced de novo synthesis of $P-450_{17\alpha}$. Mouse Leydig cell cultures were treated with medium only, with cAMP or cAMP plus AG as described for Fig. 11. At the indicated time cell proteins were radio-labeled by incubating cultures for 3 h with ^{35}S-methionine following a 1.5 h incubation in methionine-free medium. Newly synthesized ^{35}S-labeled $P-450_{17\alpha}$ was immunoisolated from cell lysates and separated by SDS-gel electrophoresis and subjected to autoradiography. Each lane represents immunoprecipitated $P-450_{17\alpha}$ isolated from an equal amount of TCA-precipitable radioactivity.

Table 6. Effect of cAMP and cAMP Plus Aminoglutethimide (AG) on the Rate of De Novo Synthesis of $P-450_{17\alpha}$.

Treatment	Days				
	0	1	2	3	4
		(^{35}S Units)[a]			
Control	0.4	0.4	0.5	0.4	0.3
cAMP	–	1.0	2.4	3.3	3.4
cAMP + AG	–	1.9	4.3	6.7	7.1

[a] Each lane in the autoradiogram illustrated in Fig. 13 was scanned by laser densitometry. The areas of the peak were integrated and the results are expressed as relative units of ^{35}S incorporation into immuno-isolated $P-450_{17\alpha}$.

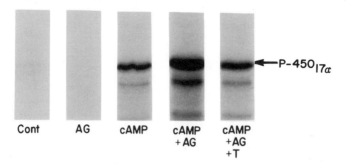

Cont   AG   cAMP   cAMP   cAMP
+AG   +AG
+T

Fig. 14. Effect of testosterone on cAMP-induced, aminoglutethimide (AG) enhanced _de novo_ synthesis of P-450$_{17\alpha}$. Mouse Leydig cell cultures were incubated for 7 days and treatment was initiated on day 7 and continued for 4 days. Cultures were treated with medium only, with 0.5 mM AG, with 0.05 mM 8-Br-cAMP (cAMP) with AG plus cAMP or with cAMP plus AG plus 5 μM testosterone (T). Rate of _de novo_ synthesis of P-450$_{17\alpha}$ was determined as described for Fig. 13.

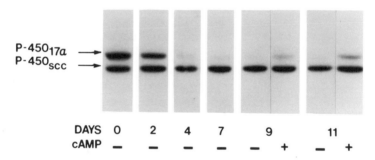

P-450$_{17\alpha}$
P-450$_{scc}$

DAYS 0   2   4   7   9   11
cAMP −   −   −   −   − +   − +

Fig. 15. Amount of immunoreactive P-450$_{17\alpha}$ and P-450$_{scc}$ in mouse Leydig cells during eleven days in culture. Mouse Leydig cells were cultured for 11 days at 19% oxygen. On day 7, some cultures were treated with 0.05 μM 8-Br-cAMP (cAMP) for either 2 days (9 days in culture) or 4 days (11 days in culture). Immunoreactive P-450$_{17\alpha}$ and P-450$_{scc}$ were determined from the same Leydig cell lysates at the indicated time. Day 0 represents the amount of P-450's measured 3 h after plating of the cells. Immunoreactive P-450$_{17\alpha}$ and P-450$_{scc}$ were determined as described for Figs. 6 and 8. Each lane represents equal amounts of protein.

undetectable by day 4. Treatment with cAMP from day 7 to day 11 results in small increases in the amount of P-450$_{17\alpha}$, 10% of control by day 9 and 20% of control by day 11. Fig. 16 is a representative autoradiogram illustrating the rate of _de novo_ synthesis of P-450$_{17\alpha}$ and P-450$_{scc}$ in Leydig cell cultures during a similar 11 day period. The rate of synthesis of P-450$_{scc}$ in the absence of cAMP increases somewhat between day 0 and day 4 and remains constant through day 11. Treatment with cAMP initiated

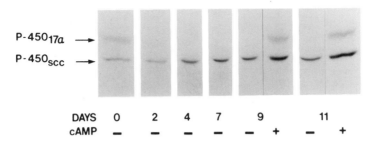

Fig. 16. Rate of de novo synthesis of P-450$_{17\alpha}$ and P-450$_{scc}$ in mouse Leydig cells during eleven days in culture. Culture conditions and treatment schedule as described for Fig. 15. Newly synthesized ^{35}S-labeled P-450$_{17\alpha}$ and P-450$_{scc}$ were immuno-isolated from the same lysates and separated as described for Fig. 13.

on day 7 results in approximately a 2-fold increase in the rate of de novo synthesis relative to nontreated cultures on the same day. In the absence of cAMP no de novo synthesis of P-450$_{17\alpha}$ is observed at any day except day 0. Treatment with cAMP, from day 7 on, increases the rate of de novo synthesis approximately 2-fold over the initial rate of synthesis determined on day 0. These data demonstrate that cAMP is obligatory for de novo synthesis of P-450$_{17\alpha}$ but not for P-450$_{scc}$. In the absence of cAMP, de novo synthesis of P-450$_{17\alpha}$ ceases, while P-450$_{scc}$ exhibits high constitutive synthesis which can be increased by cAMP.

Leydig cell responsiveness to acute stimulation by cAMP was determined in similarly treated cultures. Leydig cell cultures were stimulated with 1 mM cAMP and the amounts of pregnenolone, progesterone, androstenedione and testosterone produced during a two-hour period were measured. Results shown in Fig. 17 illustrate that total steroid production (pregnenolone, progesterone, androstenedione plus testosterone) decreased somewhat between the initial day of culture and day 2, but remained constant over the next 9 days in culture. Chronic treatment with cAMP between days 7 and 11 did not increase total steroid production. Although, total cAMP-stimulated steroid production essentially did not change during the 11 day culture period, the pattern of the major steroid produced in response to acute stimulation by cAMP changed markedly. In control cultures, by day 4, testosterone production decreased while progesterone increased. Between days 7 and 11, in nontreated cultures, testosterone production essentially ceased and total steroid production was attributable to progesterone plus small amounts of pregnenolone (Table 7). Treatment with cAMP initiated on day 7 reversed this pattern of steroid production with testosterone again being the major steroid produced in response to acute cAMP stimulation. Under all conditions, pregnenolone and androstenedione that were produced paralleled the amounts of progesterone and testosterone produced, respectively, (Table 7 and Fig. 17). However, the total amount of pregnenolone and androstenedione produced contributed little to the total steroid production. These data demonstrate that the reduced capacity for testosterone production in nontreated Leydig cell cultures is not due to a decreased capacity for cAMP-stimulated steroid production but reflects changes in the amount of P-450$_{17\alpha}$.

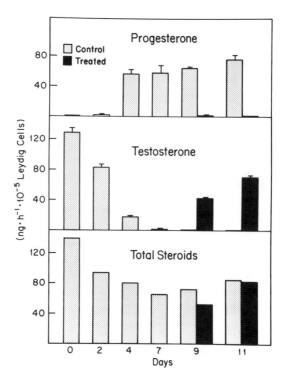

Fig. 17. Acute cAMP-stimulated progesterone, testosterone and total steroid production in mouse Leydig cells during eleven days in culture. Culture conditions and chronic cAMP treatment schedule same as described for Fig. 15. At the indicated time cells were washed for 1 h in medium only prior to incubation for 2 h in medium containing 1 mM 8-Br-cAMP. The media were collected, boiled for 5 min and centrifuged to remove denatured protein. Amounts of steroids were determined by specific radioimmunoassay. Total steroids represents the sum of pregnenolone, progesterone, androstenedione and testosterone.

Table 7. Acute cAMP-Stimulated Pregnenolone and Androstenedione Production in Leydig Cells During Eleven Days in Culture.[a]

| Days | Pregnenolone | | Androstenedione | |
	Control	cAMP $(ng \cdot h^{-1} \cdot 10^{-5}$	Control Leydig cells)	cAMP
0	0.7 + 0.1	–	9.1 + 0.9	–
2	1.0 + 0.1	–	6.9 + 0.3	–
4	4.7 + 0.5	–	0.8 + 0.1	–
7	3.7 + 0.1	–	0.3 + 0.02	–
9	5.2 + 0.4	0.9 + 0.1	0.3 + 0.03	5.6 + 0.4
11	6.7 + 0.3	1.1 + 0.2	0.3 + 0.03	8.1 + 0.5

[a]Experimental description same as for Fig. 17

484

CONCLUSION

The studies presented in this chapter indicate that the regulation of steroidogenic P-450's in Leydig cells is complex. One aspect of this regulation is degradation of both $P-450_{scc}$ and $P-450_{17\alpha}$. Acute stimulation of Leydig cells with high amounts of LH/hCG or cAMP leads to increased production of pregnenolone and testosterone. These steroid products, present at elevated concentrations near the active site of their respective P-450's, bind to the enzyme and in the presence of oxygen form a $P-450\cdot pseudosubstrate\cdot O_2$ complex. Due to the inability of the pseudo-substrate to be hydroxylated, the complex breaks down, giving rise to reactive oxygen-free radical species. These oxygen radicals either directly or indirectly damage the P-450's leading to increased degradation of the enzymes.

In contrast to the mechanism of degradation, regulation of de novo synthesis of the P-450's is different. cAMP is essential for de novo synthesis of $P-450_{17\alpha}$ but not for $P-450_{scc}$ which exhibits high constitutive synthesis that can be further increased by cAMP. In the absence of chronic cAMP stimulation, $P-450_{17\alpha}$ disappears thus leading to the inability of Leydig cells to produce testosterone in response to an acute stimulus by LH or cAMP. However, the capacity of Leydig cells to increase total steroid production in response to an acute stimulus by cAMP is not appreciably affected by long term (11 days) absence of chronic cAMP stimulation. These results imply that in the absence of trophic hormone stimulation of Leydig cells, the amount of $P-450_{17\alpha}$ becomes the limiting factor in determining the amount of testosterone that can be produced. Furthermore, these results suggest that reduced testosterone production in response to acute stimulation by LH in hypophysectomized rats as reported by Hauger et al. (1977) is not due to the inability of the Leydig cells to respond to acute administration of LH or hCG, but most likely reflects the disappearance of $P-450_{17\alpha}$ in the absence of chronic LH stimulation. Future in vivo studies, in which the concentration of other serum steroids are determined and the total amounts of $P-450_{scc}$ and $P-450_{17\alpha}$ is measured in testicular tissue from hypophysectomized animals, should resolve this question. It should be noted that as little as 10% of the amounts of $P-450_{17\alpha}$ that is found in freshly isolated Leydig cells, is an adequate amount of enzyme for the conversion of C_{21} steroids to C_{19} steroids.

The studies, furthermore, provide evidence that testosterone produced during cAMP treatment of Leydig cells negatively regulates the amount of $P-450_{17\alpha}$ by two distinct mechanisms: 1) modulation of cAMP induced de novo synthesis and 2) increased degradation by acting as a pseudosubstrate. Whether de novo synthesis of $P-450_{scc}$ is also modulated by steroid products remains to be determined.

With the recent availability of cDNA probes specific for $P-450_{scc}$ and $P-450_{17\alpha}$, future studies will be directed towards elucidating the regulation of expression of these Leydig cell steroidogenic enzymes at the transcriptional level.

ACKNOWLEDGEMENTS

The authors gratefully acknowledge the following collaborators whose experiments have contributed to the studies presented in this chapter: C.S. Sheela Rani and Tamie Malaska. We are indebted to Linli Sha and M. Margaret Snow for technical assistance and to Rita Lemorie for preparation of the manuscript. Preparation of the manuscript and studies reported in this article were supported by National Institute of Health Grants HD-08358 and HD-17916. Dale B. Hales, Louise M. Perkins and Patrick G. Quinn were supported in part by the National Institutes of Health Training Grant

HD-07048. Patrick G. Quinn's present address is Department of Molecular Physiology and Biophysics, School of Medicine, Vanderbilt University, Nashville, TN 37232.

REFERENCES

Anderson CM, Mendelson CR, 1985. Regulation of steroidogenesis in rat Leydig cells in culture: Effect of human chorionic gonadotropin and dibutyryl cyclic AMP on the synthesis of cholesterol side chain cleavage cytochrome P-450 and adrenodoxin. Arch Biochem Biophys 238:378-387

Benahmed M, Dellamonica C, Haour F, Saez JM, 1981. Specific low density lipoprotein in cultured Leydig cell steroidogenesis. Biochem Biophys Res Commun 99:1123-1130

Brinkmann AO, Leemborg FG, van der Molen HJ, 1981. hCG-induced inhibition of testicular steroidogenesis: an estradiol-mediated process? Molec Cell Endocrinol 24:65-72

Brinkmann AO, Leemborg I, Rommerts F, van der Molen H, 1982. Translocation of the testicular estradiol receptor is not an obligatory step in the gonadotropin-induced inhibition of C_{17-20}lyase. Endocrinology 110:1834-1836

Catt KJ, Tsuruhara T, Dufau ML, 1972. Gonadotropin binding sites of the rat testis. Biochim Biophys Acta 279:194-201

Chasalow F, Marr H, Haour F, Saez JM, 1979. Testicular steroidogenesis after human chorionic gonadotropin desensitization in rats. J Biol Chem 254:5613-5617

Christensen AK, Peacock KC, 1980. Increase in Leydig cell number in testes of adult rats treated chronically with an excess of human chorionic gonadotropin. Biol Reprod 22:383-391

Cigorraga SB, Dufau ML, Catt KJ, 1978. Regulation of luteinizing hormone receptors and steroidogenesis in gonadotropin-desensitized Leydig cells. J Biol Chem 253:4297-4304

Damber JE, Bergh A, Selstam G, Sodergard R, 1983. Estrogen receptor and aromatase activity in the testis of the unilateral cryptorchid rat. Arch Androl 11:259-264

Dufau ML, Watanabe K, Catt KJ, 1973. Stimulation of cyclic AMP production by the rat testis during incubation with hCG in vitro. Endocrinology 92:6-11

Hauger RL, Chen YDI, Kelch RP, Payne AH, 1977. Pituitary regulation of Leydig cell function in the adult male rat. J Endocrinol 74:57-66

Hornsby PJ, 1980. Regulation of cytochrome P-450-supported 11β-hydroxylation of deoxycortisol by steroids, oxygen, and antioxidants in adrenocortical cell cultures. J Biol Chem 255:4020-4027

Hornsby PJ, Crivello JF, 1983a. The role of lipid peroxidation and biological antioxidants in the function of the adrenal cortex. Part 1: A background review. Mol Cell Endocrinol 30:1-20

Hornsby PJ, Crivello JF, 1983b. The role of lipid peroxidation and biological antioxidants in the function of the adrenal cortex. Part 2. Mol Cell Endocrinol 30:123-147

Hsueh AJW, Dufau ML, Catt KJ, 1977. Gonadotropin-induced regulation of luteinizing hormone receptors and densensitization of testicular 3':5' cyclic AMP and testosterone responses. Proc Natl Acad Sci 74:592-595

Luketich JD, Melner MH, Guengerich FP, Puett D, 1983. Effects of human choriogonadotropin on mitochondrial and microsomal cytochrome P-450 levels in mouse testis. Biochem Biophys Res Commun 111:424-429

Malaska T, Payne AH, 1984. Luteinizing hormone and cyclic AMP-mediated induction of microsomal cytochrome P-450 enzymes in cultured mouse Leydig cells. J Biol Chem 259:11654-11657

Nakajin S, Hall PF, 1981. Microsomal cytochrome P-450 from neonatal pig testis. J Biol Chem 256:3871-3876

Nakajin S, Hall PF, Onoda M, 1981. Testicular microsomal cytochrome P-450 for C_{21} steroid side chain cleavage. J Biol Chem 256:6134-6139

Nozu K, Dufau ML, Catt KJ, 1981a. Estradiol receptor-mediated regulation of steroidogenesis in gonadotropin-desensitized Leydig cells. J Biol Chem 256:1915-1922

Nozu K, Deheija A, Zawistowich L, Catt KJ, Dufau ML, 1981b. Gonadotropin-induced receptor regulation and steroidogenic lesions in cultured Leydig cells. Induction of specific protein synthesis by chorionic gonadotropin and estradiol. J Biol Chem 256:12875-12882

O'Shaughnessy PJ, Payne AH, 1982. Differential effects of single and repeated administration of gonadotropins on testosterone production and steroidogenic enzymes in Leydig cell populations. J Biol Chem 257:11503-11509

Payne AH, Wong KL, Vega MM, 1980. Differential effects of single and repeated administrations of gonadotropins on luteinizing hormone receptors and testosterone synthesis in two populations of Leydig cells. J Biol Chem 255:7118-7122

Purvis JL, Canick JA, Rosenbaum JH, Hollogitas J, Latif SA, 1973a. Control of cytochrome P-450 in rat testis mitochondria by human chorionic gonadotropin. Arch Biochem Biophys 159:32-38

Purvis JL, Canick JA, Latif SA, Rosenbaum JH, Hologgitas J, Menard RH, 1973b. Lifetime of microsomal cytochrome P-450 and steroidogenic enzymes in rat testis as influenced by human chorionic gonadotropin. Arch Biochem Biophys 159:39-49

Quinn PG, Dombrausky LJ, Chen YDI, Payne AH, 1981. Serum lipoproteins increase testosterone production in hCG-desensitized Leydig cells. Endocrinology 109:1790-1792

Quinn PG, Payne AH, 1984. Oxygen-mediated damage of microsomal cytochrome P-450 enzymes in cultured Leydig cells. J Biol Chem 259:4130-4135

Quinn PG, Payne AH, 1985. Steroid product-induced, oxygen-mediated damage of microsomal cytochrome P-450 enzymes in Leydig cell cultures. J Biol Chem 260:2092-2099

Rani CSS, Payne AH, 1986. Adenosine 3',5'-monophosphate-mediated induction of 17α-hydroxylase and C_{17-20} lyase activities in cultured mouse Leydig cells is enhanced by inhibition of steroid biosynthesis. Endocrinology 118:1222-1228

Saez JM, Haour F, Cathiard AM, 1978. Early hCG-induced desensitization in Leydig cells. Biochem Biophys Res Commun 81:552-558

Sharpe RM, 1977. Gonadotropin induced reduction in the steroidogenic responsiveness of the immature rat testis. Biochem Biophys Res Commun 76:956-962

Simpson ER, 1979. Cholesterol side-chain cleavage, cytochrome P-450 and the control of steroidogenesis. Mol Cell Endocrinol 13:213-227

Suhara K, Fujimura Y, Shiroo M, Katagiri M, 1984. Multiple catalytic properties of the purified and reconstituted cytochrome P-450 ($P-450_{scII}$) system of pig testis microsomes. J Biol Chem 259:8729-8738

Tsuruhara T, Dufau ML, Cigorraga S, Catt KJ, 1977. Hormonal regulation of testicular luteinizing hormone receptors. J Biol Chem 252:9002-9009

Zipf WB, Payne AH, Kelch RP, 1978. Dissociation of lutropin-induced loss of testicular lutropin receptors and lutropin-induced desensitization of testosterone synthesis. Biochim Biophys Acta 540:330-336

Zuber MX, Simpson ER, Waterman MR, 1986. Expression of bovine 17α-hydroxylase cytochrome P-450 cDNA in nonsteroidogenic (COS 1) cells. Science 234:1258-1261

DEMONSTRATION OF hCG BINDING SITES AND hCG STIMULATED STEROIDOGENESIS

IN DIFFERENT POPULATIONS OF INTERSTITIAL CELLS

V.K. Bhalla*, E.S. Browne* and G.S. Sohal**

Department of Physiology and Endocrinology*
Department of Anatomy**
Medical College of Georgia
Augusta, GA 30912-3395

INTRODUCTION

To gain further knowledge of the mechanism of gonadotropin action in testes, a number of cell purification methods have been employed to isolate the morphologically distinct and biochemically responsive Leydig cells from the variety of cell types found in the testicular interstitium. Centrifugal fractionation of collagenase dispersed cells on density gradients of Metrizamide has been proven to be of considerable value (Conn et al., 1977; Dehejia et al., 1982). Typical Leydig cells were identified in a cell band (density 1.105 g/cm^3) which contained a large number of receptors for LH/hCG and responded to the hormone in vitro by producing cAMP and testosterone. A lighter cell fraction (density 1.085 g/cm^3) showed cells with a distinct morphology but little hCG binding or steroidogenic response. Similar observations were also reported in mouse testes (Schumacher et al., 1978). Payne et al. (1980a,b), on the other hand, suggested the existence of two distinct populations of hCG responsive testicular interstitial cells. The cells in population I (density 1.085–1.117 g/cm^3) produced less testosterone in response to hCG than cells in population II (density 1.128–1.148 g/cm^3). The binding affinity for hCG in both populations was, however, the same. All of these observations characterized typical Leydig cells as cells that exhibit high affinity binding of LH/hCG and produce testosterone in response to the hormone.

Although centrifugal fractionation on Metrizamide density gradients has been widely used to purify rat testicular interstitial cells, many other techniques for density gradient cell separation exist, all of which work on the basis of differences in buoyant densities and sedimentation rates (sizes) of different cell types. One of these is sedimentation at unit gravity, which was first developed for bone marrow cells (Peterson and Evans, 1967; for review, see Pretlow and Pretlow, 1982) but not applied to Leydig cell separation. Unit gravity sedimentation utilizing bovine serum albumin gradients was utilized to separate testicular germ cells in rodents (Lam et al., 1970; Lee and Dixon, 1972; Meistrich, 1972; and Romrell et al., 1976) and the ram (Loir and Lanneau, 1974). Recently enriched populations of pituitary somatotrophs, mammotrophs, thyrotrophs, and gonadtrophs (Hymer et al., 1973, 1974; Snyder et al., 1976; Leuschen et al., 1978; and Denef et al., 1978) have also been obtained with a two to sevenfold enrichment. This technique offers several advantages: (1) the cells can be separated under sterile conditions in sufficient quantities to allow extensive investigation, (2) the cells are not subject to the harsh force (> 3000 g) employed in the centrifugation techniques, (3) the cell separation can be performed at any desired temperature and, (4) necessary equipment is relatively inexpensive and can be readily made.

Along with the unit gravity technique, there are a wide variety of gradient materials available including BSA, sucrose, Ficoll, Metrizamide, and Percoll for use in unit gravity technique. In our first method for testicular interstitial cell separation we utilize Metrizamide and Percoll gradients at unit gravity. In our second method, the density gradient centrifugation technique commonly used by others is employed with certain modifications.

The two methods yield cell fractions with distinct morphological and biochemical characteristics unlike those reported in previous investigations employing centrifugation technique. We find one cell type (light cell) which is highly vacuolated and binds hCG, but does not produce testosterone in response to hCG. Another cell type (dark/heavier cell) appears similar in morphology to typical Leydig cells and produces testosterone in response to hCG; however, this cell type does not appear to bind hCG with high affinity.[1] We describe here the experimental data which led us to these conclusions.

METHODS AND RESULTS

Purification of Rat Testicular Interstitial Cells by Unit Gravity Sedimentation Procedure

To prepare rat testicular interstitial cells for the unit gravity sedimentation procedure, forty decapsulated testes were incubated with collagenase (to separate the interstitial cells from the seminiferous tubules) and deoxyribonuclease (to minimize aggregation due to cell damage) as described in Fig. 1. The seminiferous tubules were filtered out and the interstitial cells were removed from the incubation medium containing the enzymes by centrifugation. The cells were resuspended in fresh medium and counted.

---

[1]Relatively little is known about the manner in which the hormone recognizes and interacts with the membrane receptor to promote the biological response. In the classical concept, the receptor was originally considered to be a bifunctional molecule: it receives the information, and then transmits it to a secondary unit to promote biological response (Rodbell et al., 1969). These two characteristics of the receptor, which originated from the work of Langley (1906), were accepted and recognized until recent times. With advent of the radioligand binding technique (for review see ref. Hechter, 1978; Triggle, 1978; Pollet and Levey, 1980; Ariens et al., 1979) the classical definition of the term "receptor" changed. The binding assays are beginning to be accepted as a method which accurately measures functional receptors. Whether or not such molecules which bind the hormone with high affinity and specificity are truly the molecules promoting the biological response (at least in some systems) is unclear.

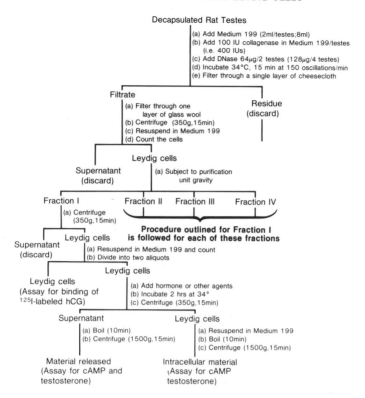

INCUBATIONS WITH DISPERSED LEYDIG CELLS

Decapsulated Rat Testes

(a) Add Medium 199 (2ml/testes;8ml)
(b) Add 100 IU collagenase in Medium 199/testes
   (i.e. 400 IUs)
(c) Add DNase 64µg/2 testes (128µg/4 testes)
(d) Incubate 34°C, 15 min at 150 oscillations/min
(e) Filter through a single layer of cheesecloth

Filtrate

(a) Filter through one
   layer of glass wool
(b) Centrifuge (350g,15min)
(c) Resuspend in Medium 199
(d) Count the cells

Residue
(discard)

Leydig cells

(a) Subject to purification
   unit gravity

Supernatant
(discard)

Fraction I

(a) Centrifuge
   (350g,15min)

Fraction II   Fraction III   Fraction IV

Leydig cells

**Procedure outlined for Fraction I
is followed for each of these fractions**

Supernatant
(discard)

(a) Resuspend in Medium 199 and count
(b) Divide into two aliquots

Leydig cells

Leydig cells
(Assay for binding of
^{125}I-labeled hCG)

(a) Add hormone or other agents
(b) Incubate 2 hrs at 34°
(c) Centrifuge (350g,15min)

Supernatant

(a) Boil (10min)
(b) Centrifuge (1500g,15min)

Leydig cells

(a) Resuspend in Medium 199
(b) Boil (10min)
(c) Centrifuge (1500g,15min)

Material released
(Assay for cAMP and
testosterone)

Intracellular material
(Assay for cAMP
testosterone)

Fig. 1. Schematic diagram of the preparation of different samples for the measurement of cAMP and testosterone released into the medium and intracellular cAMP and testosterone levels: After collagenase treatment, the dispersed Leydig cells were processed as shown and applied to unit gravity sedimentation procedure to obtain four pooled fractions. The cells in each fraction were resuspended in fresh buffer to a final concentration of 20×10^6 cells/ml. The cells were further diluted such that each tube in the binding assay contained 2×10^6 cells/250 µl of incubation volume. Ten million cells (0.5 ml) in each tube were stimulated with hCG (10 or 20 ng) in a total volume of 2 ml (1 or 2 ng/200 µl/2 x 10^6 cells). For hCG stimulation, the incubations were carried out at $34°C$ for 2 hours after which time the cells were centrifuged. The supernatant was boiled to recover the sample for the measurement of cAMP and testosterone released into the medium in the presence or absence of hormone. The cells free of medium were resuspended in 2 ml and boiled to break the cell membranes and obtain samples for measurement of intracellular cAMP and testosterone. cAMP and testosterone levels were measured by their respective radioimmunoassays.

This unpurified cell suspension was placed in a unit gravity sedimentation chamber, and a 0–10% gradient of Metrizamide was introduced into the chamber. The chamber was maintained at $4°C$ and the cells were allowed to settle for approximately one hour. The gradient was then siphoned into a fraction collector. The elution profiles are shown in Fig. 2. Each point on the graph represents a cell fraction of 15 ml. The twenty-one fractions were combined as shown to obtain four pooled fractions, henceforth referred to as fractions I, II, III, and IV.

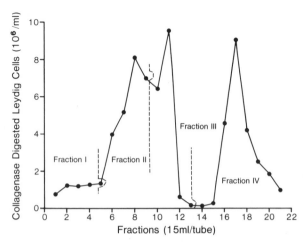

Fig. 2. Elution profile of cells after sedimentation at unit gravity by
Metrizamide gradient. A Metrizamide gradient (0-10% w/v) was used
and twenty-one fractions were collected. The cells (1587 x 10^6)
were applied to the column with 45% cell recovery in the collected
fractions.

Each of the four pooled fractions was tested for its responsiveness
to an hCG challenge *in* *vitro* to produce cAMP and testosterone in the
medium. Binding assays were also undertaken to measure hCG receptor
sites in each fraction to compare the binding properties of the cells with
their ability to produce cAMP and testosterone.

The Binding Characteristics of Cell Fractions Purified by Unit Gravity
Sedimentation Procedure Using Metrizamide Gradient

The binding of ^{125}I-labeled hCG to purified cell fractions is shown
in Fig. 3. Collagenase dispersed interstitial cells (2 x 10^6 cells/250
μl) were incubated with 2.5 ng ^{125}I-labeled hCG in the absence or
presence of increasing concentrations of unlabeled hCG (0.1-100 ng).
Nonspecific binding was determined in the presence of 2000 ng unlabeled
hCG. The specific binding of ^{125}I-labeled hCG to cells in fraction I was
maximal, but decreased sharply in the subsequent fractions. The specific
binding to cells in fraction IV was less than 60% of cells in fraction I
and was not inhibited by unlabeled hCG in a dose dependent manner. The
Scatchard analysis of the data revealed that cells in fraction I bound hCG
with high affinity. The equilibrium dissociation constant value (K_d)
was found to be 3.4 x 10^{-10} M with a total number of binding sites found
to be 1.32 x 10^{-5} nmoles/2 x 10^6 cells. The values for the affinity
constants were quite comparable to those reported by Payne et al., (1980)
who found K_a values of 6.4 x 10^9 M^{-1} and 6.7 x 10^9 M^{-1} for two populations of Leydig cells. The binding parameters for hCG binding to
cells in fraction IV, however, could not be calculated. The data did not

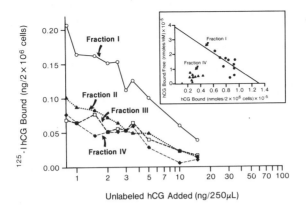

Fig. 3. The average specific binding of ^{125}I-labeled hCG to fractions collected from two different purifications using Metrizamide gradient. Cells in fractions I, II, III, and IV (2 x 10^6 cells/tube) from Metrizamide gradient were incubated with ^{125}I-labeled hCG (52522 CPM/2.5 ng) as described (Bhalla et al., 1987a). The specific binding of ^{125}I-labeled hCG to fractions I, II, III, and IV was 4322, 2135, 1439, 1624 CPM, respectively. The Scatchard plot of the binding of ^{125}I-labeled hCG to cells in fraction I is shown in the inset. An unweighted linear regression of the data ($r = 0.7$; $p < 0.05$) gives an abscissa intercept (B_{max}) value of 1.32 x 10^{-5} nmol/2 x 10^6 cells. The The experimental data of ^{125}I-labeled hCG to cells in fraction IV did not obey saturation kinetics. Each point is the mean of four determinations, two from each experiment (SE $\pm$ 5%).

obey saturation kinetics and the linear regression of a Scatchard plot of these data yielded a positive slope rather than a negative slope (Fig. 3, inset). It was thus inferred that the binding sites observed in this fraction were not of high affinity.

Production of cAMP and Testosterone in Cell Fractions Purified by Unit Gravity Sedimentation Procedure

The ability of the cells in fractions I and IV to produce cAMP and testosterone when stimulated with submaximal concentrations of hCG was tested in the following manner: Cells (500 µl; 10 x 10^6 cells) from each fraction were incubated with and without hCG as shown in Fig. 1. After centrifugation, the supernatant was heated in a boiling waterbath, recentrifuged, and kept frozen below 0°C until assayed for cAMP and testosterone by their respective radioimmunoassays. The pelleted cells were resuspended in the total incubation volume (1.25 ml) and heated in a boiling waterbath to break the cell membranes. After centrifugation, the supernatant was stored frozen at 0°C until assays for cAMP and testosterone could be performed. The intracellular and released cAMP and testosterone are shown in Figs. 4 and 5, respectively. The stimulation of

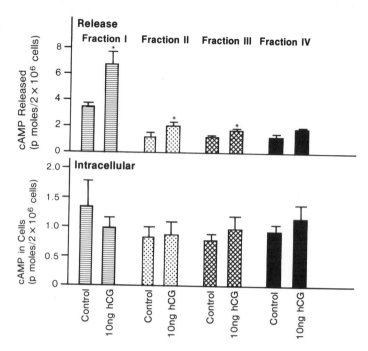

Fig. 4. Effect of hCG upon cAMP production in different cell fractions obtained by Metrizamide gradient: Ten million cells were incubated in the presence or absence of hCG in a total volume of 2 ml for 2 h at 34°C. After incubation, the cell suspension was processed as described (Bhalla et al., 1987a). The cyclic AMP released from cells (upper panel), and remaining in the cells (lower panel) was measured. Each point is the mean of four determinations, two from each experiment. Bars indicate Mean ± S.E. *Significant (p < 0.01) by Duncan analysis.

cAMP by hCG at low hormone concentrations (1.5×10^{-10}M) was maximal in cells in fraction I. The nonsignificant increase seen in fraction IV was attributed to insufficient concentrations of hCG (also see Bhalla et al., 1987b). The patterns of testosterone production are shown in Fig 5. The cells in fraction I were nonresponsive to hCG. This, incidently, was the fraction which contained high affinity binding sites for hCG (see Fig. 3 for comparison). The cells in fraction IV produced testosterone maximally (762% in comparison to the control levels). The release of testosterone into the medium was followed by a significant increase in the intracellular levels of testosterone, thus reflecting a de novo synthesis of the steroid in response to the hormone challenge.

It must be pointed out here that the concentration of hormone depicted in Figs. 4 and 5 as 10 ng hCG represents a final concentration of 1 ng/200 µl/2 $\times 10^6$ cells (1.5×10^{-10}M). This concentration of hormone, although sufficient to produce maximal stimulation of steroidogenesis, was much less effective in stimulating cAMP production. This is why a detectable increase in cAMP by cells in fraction IV was not seen in Fig. 4. An entire dose response curve for cAMP stimulation by cells in fraction IV is illustrated elsewhere (Bhalla et al., 1987b).

The data that we obtained by 0-10% Metrizamide gradient using unit gravity sedimentation deviated considerably from the reported

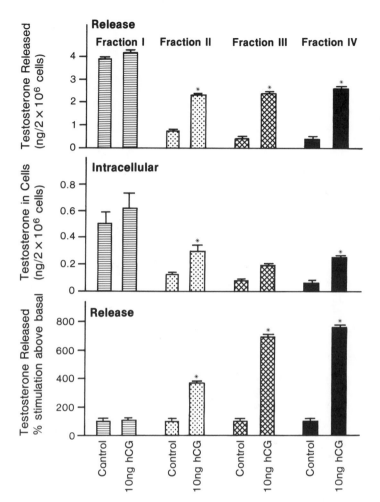

Fig. 5. Effect of hCG upon testosterone production in cell fractions obtained by Metrizamide gradient. The details of the experiment are discussed in the legend to Fig. 4. The testosterone levels released (upper panel) and testosterone remaining in the cells (middle panel) were measured by radioimmunoassay. Percent stimulation is also shown (lower panel). Each point is the mean of four determinations, two from each experiment. Bars indicate Mean ± S.E. *Significant (p < 0.01) by Duncan analysis.

observations. Two cell fractions with distinct biochemical features prompted us to verify these observations by using a gradient other than Metrizamide. We found that Percoll gradient (0–25% v/v or 0–50% v/v) was also as effective in separating the cells as Metrizamide gradient (Bhalla et al., 1987a). We found Percoll gradient to be more economical and effective. It is known to exert low osmotic pressure, which helps maintain better cell integrity and viability (Browning et al., 1981).

Electron Microscopy of Cells in Fractions I and IV

The fine structure of Leydig cells and localization of steroidogenic enzymes within their cytoplasm has been described (Christensen, 1965,

1975). The most prominent features of these cells include smooth endoplasmic reticulum, numerous mitochondria, lipid droplets, and nucleus with peripherally located heterochromatin. Cell morphologies in purified Leydig cell preparations designated as populations I and II were recently studied by Payne et al. (1982). The characteristics of typical Leydig cells in both populations were quite similar and no consistent differences in the morphology between the two populations were identified. There were some cells in each preparation, however, which were not identified. These cells were highly vacuolated and will be discussed later in this section. At about the same time, studies by Dehejia et al. (1982a, 1982b) provided evidence against the concept of Leydig cell heterogeneity. By using a continuous gradient of 0–14% Metrizamide, these investigators obtained five bands of cells. Two main binding peaks (peaks III and IV) were in the lower portion of the gradient with respective densities of 1.085 and 1.105 g/cm^3. These two peaks with the most binding/10^6 cells gave the largest response to acute hCG stimulation. The lack of hormonal responsiveness in other bands was correlated with the essential absence of hCG binding. Electron microscopical examination of different cell fractions revealed the presence of typical Leydig cells in peaks III and IV. The cells in the upper layer (peak II) were markedly heterogeneous with many being vacuolated, whereas the cells in the lower peaks (peaks III and IV) were identified as typical Leydig cells with morphological characteristics as described above. Thus, the vacuolated cells with a density of 1.078 (peak II) bound much less hCG than other active fractions and were much less responsive to hCG in producing cAMP and testosterone (vide supra).

The cell fractions we obtained by unit gravity sedimentation, quite surprisingly, displayed biochemical characteristics which were unique. It was of interest, therefore, to study cell morphology and compare with other preparations. The cells most frequently observed in fraction I were highly vacuolated and were quite similar in appearance to those described by Dehejia et al. (1982a, 1982b) as light cells with little binding. They contained electron dense nuclei and low density cytoplasm (Fig. 6). A typical cell in this fraction had an ovoid nucleus with one or more deep infoldings. The heterochromatin completely surrounded the peripheral portion of the nucleus. The nucleolus had a low density central mass surrounded by darker chromatin. These cells existed in various shapes, differing from each other by the presence of vacuoles of varying sizes and contours (Fig. 6B and C). The vacuolization was most extensive in the apical portion of the cell and appeared to travel towards the basal end. Moderate amounts of rough and smooth endoplasmic reticulum, ribosomes, lipid droplets, lysosomes, mitochondria, and Golgi complex were observed (Fig. 6D). Although the identity of this group of cells is not clearly evident, they were classified as indeterminate connective tissue cells by Kerr et al. (1985). On the basis of some similarities to typical Leydig cells, these authors suspected them to be the precursors of Leydig cells, although their true identity was not determined.

The cells in fraction IV (Fig. 7) displayed features generally attributed to Leydig cells (Fig. 7B and C). They contained a large ovoid or round nucleus containing a thin rim of heterochromatin at the periphery along with some patches of heterochromatin situated throughout the nucleoplasm. The nucleus was located eccentrically and contained a large dense nucleolus. The cytoplasmic features included numerous mitochondria with tubular cristae and extensively developed smooth endoplasmic reticulum, features typically found in steroid secreting cells (Fig. 7D). Well-developed Golgi complex, ribosomes, and lysosomes were apparent. In some cells, numerous lipid droplets or vacuoles were observed (Fig. 7B and C), whereas they were absent in others (Fig. 7A).

496

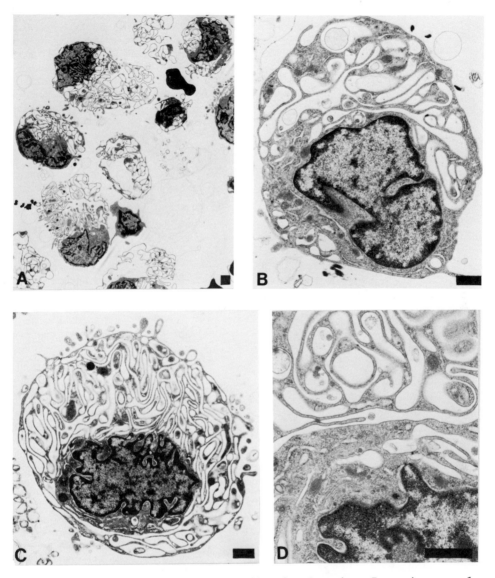

Fig. 6. Electron micrographs of cells in fraction I. A, a low magnification overview of cells. Majority of cells in this fraction were vacuolated as shown. In addition, the cell fraction contained broken cell membranes, germ cells, spermatids, and some blood cells (data not shown). B and C, light cells with varying numbers of vacuoles in the cytoplasm. D, a high magnification picture of a portion of a light cell. Bars = 1 μm.

In addition to vacuolated cells in fraction I and typical Leydig cells in fraction IV, some cells in fraction IV could not be positively identified on the basis of their morphological features (Fig. 8). These cells contained some of the features previously described for vacuolated cells in fraction I and Leydig cells in fraction IV. The most prominent feature of the cytoplasm was the presence of large numbers of coated vesicles and pits, primarily concentrated around the periphery and possibly indicative of exocytotic activity. Whether or not these

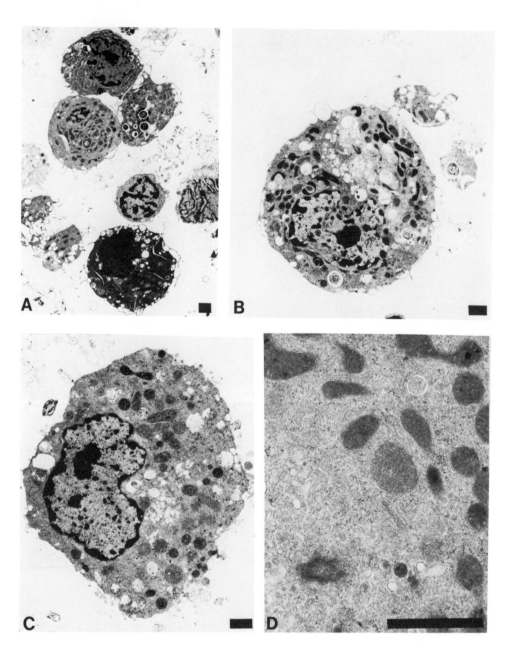

Fig. 7. Electron micrographs of cells in fraction IV. A, a low
magnification overview. In addition, the cell fraction contained
sperm tails, acrosomes, germ cells, and some blood cells (data not
shown). B and C, typical Leydig cells. D, a high magnification
picture of a Leydig cell cytoplasm showing large amounts of smooth
endoplasmic reticulum and numerous mitochondria. Bars = 1 μm.

represent typical Leydig cells in the process of being converted to
vacuolated cells is not quite clear at this time, although our biochemical
data seem to suggest this possibility (vide infra).

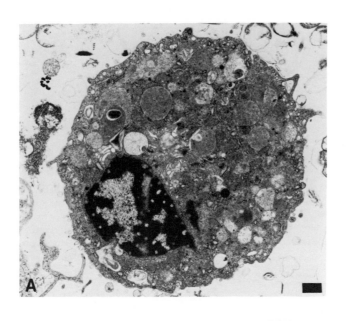

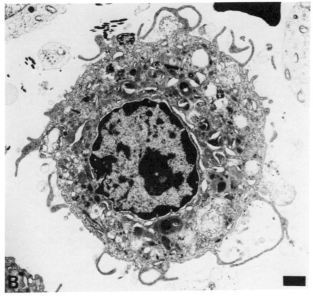

Fig. 8. Electron micrographs of additional cells in fraction IV which could not be positively identified. Bars = 1 µm.

Isolation of Testicular Interstitial Cells by Percoll or Metrizamide Density Gradient Centrifugation Technique:

Encouraged by the notable biochemical and morphological features of purified cell fractions in our study with unit gravity sedimentation, we attempted alternate methods to validate these findings. Monitoring the binding profiles of cells to localize typical Leydig cell bands was not our objective. Rather, our primary objective was to adapt experimental conditions necessary to isolate bands of cells with the biochemical functions previously established.

Several continuous and discontinuous gradients of Percoll and Metrizamide were tried (data not shown), and finally a procedure was developed using a discontinuous Percoll gradient made up of 5 ml of 25% Percoll (v/v) in Medium 199, and 20 ml of 20% Percoll gradient (v/v) in Medium 199. The collagenase dispersed interstitial cells were layered in 5 ml buffer and the tubes were centrifuged at 3800 x g for 30 min at 4° C. Fractions 1-10 (1 ml/fraction) were pooled to obtain Percoll fraction I, fractions 11-19 Percoll fraction II, fractions 20-26 Percoll fraction III, and fractions 27-30 Percoll fraction IV. These fractions were washed to remove Percoll. Fraction I, containing ^{125}I-labeled hCG binding sites, and fraction IV, containing cells that produce testosterone in response to hCG, were again layered on separate discontinuous Percoll gradients (see legend for Fig. 9). The majority of cells (~80%) from both initial fractions, I and IV, were recovered from the second gradient procedure. Each fraction was divided into 2 aliquots. Aliquot I (approximately 80-100 x 10^6 cells) was used for the binding assay to estimate the apparent equilibrium dissociation constant (K_d) and the apparent number of binding sites (B_{max}) for hCG. Aliquot II (all remaining cells) was used for investigation of cAMP and testosterone production and release.

Binding of ^{125}I-labeled hCG by Purified Interstitial Cells in Percoll Fractions I and IV

The binding of ^{125}I-labeled hCG to Percoll fractions I and IV in four different batches of collagenase dispersed interstitial cells is shown in Fig. 10. The conditions for the binding assay were identical to those to those previously described (vide supra). Nonspecific binding to cells in Percoll fraction I was approximately 15%, whereas Percoll fraction IV nonspecific binding was greater than 80%. ^{125}I-labeled hCG bound specifically to fraction I yielded some experimental variability with an average of fourteen experiments found to be 0.54 $\pm$ SD 0.14 ng/2 x 10^6 cells with a range of 0.36 to 0.78 ng/2 x 10^6 cells/250 µl. Specific binding in fraction IV was in the range of 0.07 to 0.19 ng/2 x 10^6 cells/250 µl.

The Scatchard analysis of the binding data to cells in fraction I invariably obeyed the saturation kinetics (Bhalla et al., 1987b). The apparent equilibrium dissociation constant was found to be 3.0 x 10^{-10} M with apparent number of binding sites (B_{max}) estimated to be 5.69 x 10^{-5} nmol/4 x 10^6 cells, slightly more than twofold greater than previously observed for cells purified by unit gravity sedimentation. The specific binding of ^{125}I-labeled hCG to cells in Percoll fraction IV did not obey saturation kinetics and the Scatchard plot of the binding data yielded a positive slope.

Production of Testosterone by Interstitial Cell Fractions Purified by Two Successive Discontinuous Percoll Gradients

To directly compare the binding characteristics of hCG with biological response in these four batches of interstitial cells, testosterone levels were measured (Fig. 11). hCG (4 ng/2 x 10^6 cells/250 µl) consistently failed to stimulate testosterone production in Percoll fraction I although this was the fraction which contained 80% of the receptor sites (refer to Fig. 10 for corresponding binding data). In contrast, the cells in Percoll fraction IV consistently produced high levels of testosterone in all four experiments. The mean basal testosterone value in controls for cells in Percoll fraction IV from eight experiments was found to be 0.23 $\pm$ 0.07 ng/2 x 10^6 cells/250 µl. hCG, at a concentration of 2.4 x 10^{-10} M, produced a twelvefold increase in biological response (2.7 $\pm$ 0.49 ng/2 x 10^6 cells/250 µl). The release of testosterone into the medium was also accompanied by an increase in

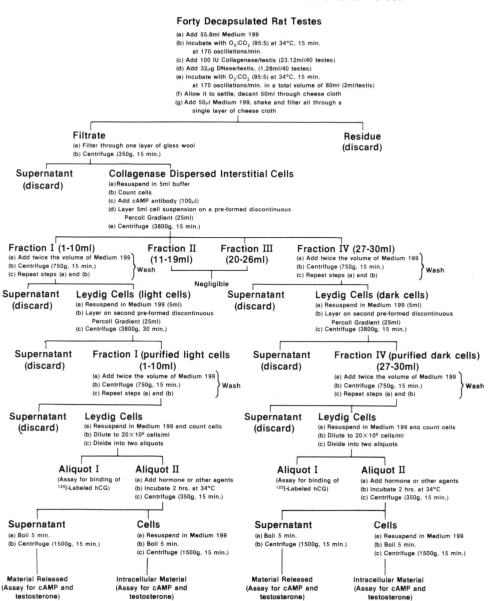

Separation of Collagenase Dispersed Leydig Cells by Centrifugation in Two Consecutive Discontinuous Gradients of Percoll

Forty Decapsulated Rat Testes
(a) Add 55.6ml Medium 199
(b) Incubate with O_2:CO_2 (95:5) at 34°C, 15 min.
    at 170 oscillations/min.
(c) Add 100 IU Collagenase/testis (23.12ml/40 testes)
(d) Add 32µg DNase/testis, (1.28ml/40 testes)
(e) Incubate with O_2:CO_2 (95:5) at 34°C, 15 min.
    at 170 oscillations/min. in a total volume of 80ml (2ml/testis)
(f) Allow it to settle, decant 50ml through cheese cloth
(g) Add 50µl Medium 199, shake and filter all through a
    single layer of cheese cloth

Filtrate
(a) Filter through one layer of glass wool
(b) Centrifuge (350g, 15 min.)

Residue
(discard)

Supernatant
(discard)

Collagenase Dispersed Interstitial Cells
(a) Resuspend in 5ml buffer
(b) Count cells
(c) Add cAMP antibody (100µl)
(d) Layer 5ml cell suspension on a pre-formed discontinuous
    Percoll Gradient (25ml)
(e) Centrifuge (3800g, 15 min.)

Fraction I (1-10ml)
(a) Add twice the volume of Medium 199
(b) Centrifuge (750g, 15 min.)
(c) Repeat steps (a) and (b)
} Wash

Fraction II (11-19ml)

Fraction III (20-26ml)

Negligible

Fraction IV (27-30ml)
(a) Add twice the volume of Medium 199
(b) Centrifuge (750g, 15 min.)
(c) Repeat steps (a) and (b)
} Wash

Supernatant
(discard)

Leydig Cells (light cells)
(a) Resuspend in Medium 199 (5ml)
(b) Layer on second pre-formed discontinuous
    Percoll Gradient (25ml)
(c) Centrifuge (3800g, 30 min.)

Supernatant
(discard)

Leydig Cells (dark cells)
(a) Resuspend in Medium 199 (5ml)
(b) Layer on second pre-formed discontinuous
    Percoll Gradient (25ml)
(c) Centrifuge (3800g, 15 min.)

Supernatant
(discard)

Fraction I (purified light cells) (1-10ml)
(a) Add twice the volume of Medium 199
(b) Centrifuge (750g, 15 min.)
(c) Repeat steps (a) and (b)
} Wash

Supernatant
(discard)

Fraction IV (purified dark cells) (27-30ml)
(a) Add twice the volume of Medium 199
(b) Centrifuge (750g, 15 min.)
(c) Repeat steps (a) and (b)
} Wash

Supernatant
(discard)

Leydig Cells
(a) Resuspend in Medium 199 and count cells
(b) Dilute to 20×10^6 cells/ml
(c) Divide into two aliquots

Supernatant
(discard)

Leydig Cells
(a) Resuspend in Medium 199 and count cells
(b) Dilute to 20×10^6 cells/ml
(c) Divide into two aliquots

Aliquot I
(Assay for binding of ^{125}I-Labeled hCG)

Aliquot II
(a) Add hormone or other agents
(b) Incubate 2 hrs. at 34°C
(c) Centrifuge (350g, 15 min.)

Aliquot I
(Assay for binding of ^{125}I-Labeled hCG)

Aliquot II
(a) Add hormone or other agents
(b) Incubate 2 hrs. at 34°C
(c) Centrifuge (350g, 15 min.)

Supernatant
(a) Boil 5 min.
(b) Centrifuge (1500g, 15 min.)

Cells
(a) Resuspend in Medium 199
(b) Boil 5 min.
(c) Centrifuge (1500g, 15 min.)

Supernatant
(a) Boil 5 min.
(b) Centrifuge (1500g, 15 min.)

Cells
(a) Resuspend in Medium 199
(b) Boil 5 min.
(c) Centrifuge (1500g, 15 min.)

Material Released
(Assay for cAMP and testosterone)

Intracellular Material
(Assay for cAMP and testosterone)

Material Released
(Assay for cAMP and testosterone)

Intracellular Material
(Assay for cAMP and testosterone)

Fig. 9. Schematic diagram of the preparation of purified cells by two step discontinuous Percoll Gradient: After collagenase digestion of rat testes (40 testes) at 34°C for 15 minutes in Medium 199 containing BSA and HEPES, the suspension was processed as described (Bhalla et al., 1987b). The collagenase dispersed interstitial cells were layered on two consecutive discontinuous Percoll gradients (1500×10^6 cells/gradient). When layered on the first gradient, the cells separated into three visibly distinguishable bands (fractions I, III, and IV). Fraction I cells and fraction IV cells were collected from each gradient and pooled, washed twice, and layered again on separate discontinuous Percoll gradients. On the second gradient, the cells banded at the same locations as in the initial gradient centrifugation. The cells from each fraction were collected, washed twice, counted, and diluted to 20×10^6 cells/ml for the studies indicated. The assays for hCG binding and hormone stimulated biological response were simultaneously conducted in each experiment to rule out artifacts arising due to cell variability in batch to batch preparations.

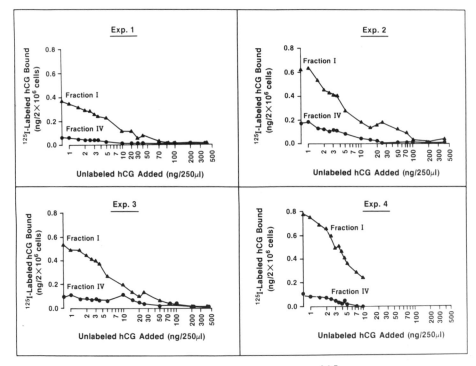

Fig. 10. Competitive inhibition binding data for ^{125}I-labeled hCG bound to purified fractions I and IV from four different experiments: Collagenase dispersed interstitial cells were purified on two consecutive discontinuous gradients of Percoll (Fig. 9) and 2×10^6 cells were incubated with ^{125}I-labeled hCG (2.5 ng; 65000 cpm) under the conditions defined (see text). Each experiment represents a different batch of cells from which purified fractions I and IV were obtained as shown in Fig. 9. The biological response of the cells to hCG for these experiments is shown in Fig. 11. Each point is the mean of duplicate determinations. S.E. $\pm$ 5%.

intracellular levels of testosterone (0.20 ± 0.06 ng in controls versus 0.60 ± 0.08 ng in hCG-treated cells) indicative of de novo synthesis of testosterone by these cells in the presence of hCG.

Effect of Bromo cAMP Upon Testosterone Production by Cells in Fractions I and IV Purified by Two Successive Percoll Gradients

The effect of bromo cAMP on testosterone production was determined in both cell fractions, I and IV. Cells in Percoll fraction I did not respond either to cAMP (0.14 ± 0.03 ng/2 x 10^6 cells/250 µl in control versus 0.14 ± 0.03 and 0.13 ± 0.03 ng/2 x 10^6 cells/250 µl in the presence of 1 mM bromo cAMP or dibutyryl cAMP, respectively) or hCG by producing testosterone. The cells in Percoll fraction IV, however, produced testosterone vigorously in the presence of bromo cAMP (1 mM and 2 mM, Fig. 12). Once again, 81% of the hCG binding sites were localized in fraction I, whereas only residual binding was observed in fraction IV (data not shown; see Bhalla et al., 1987b).

Autoradiographic Studies

Autoradiographic studies at the electron microscopic level demonstrated the accumulation of grains on cells in fraction I, more heavily than on cells in fraction IV (Fig. 13, A and B). The localization of grains on typical Leydig cells was minimal (Fig. 13C). The grains apparent on cells in fraction I represented specific binding since 80% of the binding was shown to be specific by in vitro studies (see Fig. 10). The grains on cells in fraction IV possibly represented nonspecific binding. Approximately 60–80% of the binding was found to be nonspecific as indicated by the data in Fig. 10. The accumulation of label on these micrographs resulted from an incubation of cells in each fraction with physiological concentrations of hCG (10 ng/2 x 10^6 cells/250 µl; 2.7 x 10^5 CPM). This concentration was more than sufficient to induce maximal steroidogenesis (4 ng/2 x 10^6 cells/250 µl; Fig. 11).

The experimental data illustrated hera are difficult to compare with the published reports for two main reasons: (1) in those studies a clear separation between cell types exhibiting the binding parameter and biological response was not achieved and (2) the experimental conditions for labeling the cells prior to autoradiographic studies either were not stated in the publications or supersaturating concentrations of ^{125}I-labeled hCG were utilized. Thus, the contribution of nonspecific binding is difficult to assess in those studies. A very light accumulation of grains at the Leydig cell membrane and also associated with microvilli has been published by others (Dehejia et al., 1982a,b).

Further Separation of Binding Sites from Steroidogenic Cell Fraction

The effectiveness of Percoll gradient in separating cells which bind hCG from cells which produce testosterone was evident from the preceding studies. A single purification step on collagenase dispersed interstitial cells resulted in a fraction (Percoll fraction IV) which contained residual binding sites for hCG (generally less than 30% of the binding activity observed in Percoll fraction I; Bhalla et al., 1987b). Further purification of Percoll fraction IV at unit gravity (Table I) or with a second discontinuous gradient centrifugation, as in Fig. 10, always resulted in the separation of a low density cell type with more specific binding activity than a greater density cell type which produced testosterone upon stimulation with hCG. The total (specific) hCG binding activity of cell fractions further purified from Percoll fraction IV on unit gravity sedimentation is shown in Table I. The sum of the total binding activities of each subfraction is approximately equal to the total activity of the starting material (27 ng for subfraction I + 11 ng for subfraction II + 30 ng for subfraction III + 21 ng for subfraction IV = 89 ng $\simeq$ 93 ng in the starting material). This indicates that the residual binding observed in fraction IV is due to the presence of hCG binding

TABLE I

^{125}I-labeled hCG Binding to Fractions I and IV Purified by Percoll Gradient Centrifugation and Subfractions obtained from Unit Gravity: ^{125}I-labeled hCG (2.5 ng/2 x 10^6 cells/250 µl) containing 67,397 CPM was incubated with and without hCG (2000 ng) with various fractions as indicated for 1 hr at 37 °C. Nonspecific binding was corrected and the binding data expressed as specific binding of ^{125}I-labeled hCG bound ng/2 x 10^6 cells/250 µl. In a separate experiment, the hCG stimulated testosterone released into the medium was measured as described (Bhalla et al., 1987b).

Samples	^{125}I-labeled hCG Binding ng/2 x 10^6 cells/250 µl	Cell Number Per Fraction (million cells)	Total hCG Binding Activity (specific) Per Fraction (ng)	Testosterone Released	
				Basal (ng/2 x 10^6 cells/250 µl)	hCG Stimulated(a) (ng/2 x 10^6 cells/250 µl)
Percoll Fraction I	0.49 ± 0.00	(320)**	(155)**	0.54 ± 0.12	1.15 ± 0.08
Percoll Fraction IV	0.15 ± 0.02	621 ± 20	93	0.20 ± 0.06	---*
Unit Gravity Subfraction I from Percoll Fraction IV	0.38 ± 0.01	72 ± 2	27	0.14 ± 0.00	0.20 ± 0.03
Unit Gravity Subfraction II from Percoll Fraction IV	0.20 ± 0.00	57 ± 2	11	---*	---*
Unit Gravity Subfraction III from Percoll Fraction IV	0.11 ± 0.00	274 ± 6	30	0.76 ± 0.10	0.95 ± 0.06
Unit Gravity Subfraction IV from Percoll Fraction IV	0.10 ± 0.00	218 ± 6	21	0.13 ± 0.01	0.62 ± 0.13

(a) hCG stimulation at 4 ng/2 x 10^6 cells/250 µl; *Not observed; **From a similar experiment

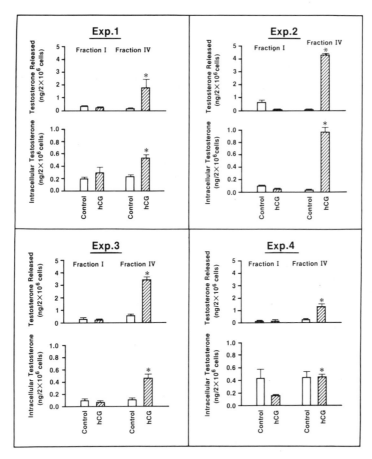

Fig. 11. hCG—stimulated testosterone released into the medium and intracellular testosterone levels in interstitial cells purified by
Percoll gradient. Collagenase dispersed interstitial cells were
purified by two consecutive discontinuous gradients of Percoll
and 10×10^6 cells from fractions I and IV were incubated with or
without hCG (20 ng) in a total volume of 1.25 ml. Testosterone
levels were measured in duplicate in these four representative
experiments, for which the ^{125}I-labeled hCG binding data are
previously shown (Fig. 10). (*Highly significant (1%) by Duncan
analysis; population variance).

sites in light cells. These cells do not synthesize testosterone upon hCG
stimulation (0.20 ng hCG stimulated testosterone vs. 0.14 ng basal in
subfraction I). These less dense cells may either be present as
contaminants or may possibly be generated during the time of further cell
purification steps (approximately 3 h). Their origin is under continued
investigation.

Densities of Cells in Fractions I and IV

The densities of cells purified by Percoll and Metrizamide
centrifugation were measured for comparison with the sedimentation
characteristics of preparations reported by other investigators (Table
II). Percoll fraction I was found to contain cells of density 1.029–
1.034 g/cm^3. Cells with similar biochemical and morphological

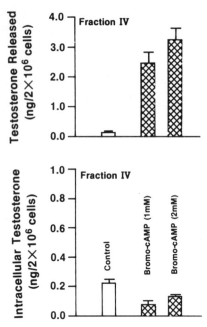

Fig. 12. Effect of bromo cAMP in stimulating testosterone production in cells purified on two consecutive discontinuous gradients of Percoll. Purified interstitial cells were incubated in the absence or presence of bromo cAMP (1 mM, 2 mM) under conditions described in the text for hCG stimulation. Testosterone released and remaining in the cells was measured by radioimmunoassay. The data are expressed as ng/2 x 10^6 cells/250 μl.

characteristics were also found when Metrizamide gradient was used (density 1.053–1.065 g/cm^3). Some of these vacuolated cells were observed in electron micorgraphs of cells from "Population I" of Payne et al. (1982), and also in a "lighter band" (density 1.085 g/cm^3) purified by Dehejia et al. (1982). Dehejia et al. (1982), in addition, described a band with density 1.078 g/cm^3 containing mostly cells similar in morphology to our Percoll fraction I (although their band apparently lacked receptors for hCG). Based upon these correlations on cell densities and our biochemical and morphological studies, we will henceforth refer to cells in fraction I (of either gradient centrifugation or unit gravity) as the light cells.

Percoll fraction IV contained cells which produced testosterone in response to hCG but apparently did not bind hCG with high affinity. These cells were of some density greater than 1.037 g/cm^3; a similar preparation obtained with Metrizamide gradient indicated a density of 1.090–1.110 g/cm^3. These cells were present in electron micrographs of all the steroidogenic preparations of Dehejia et al. (1982) and Payne et al. (1983), and have morphology which may be regarded as that of "typical" steroidogenic Leydig cells. We, henceforth, refer to these cells present in Percoll fraction IV from either gradient centrifugation or unit gravity as dark/heavier cells.

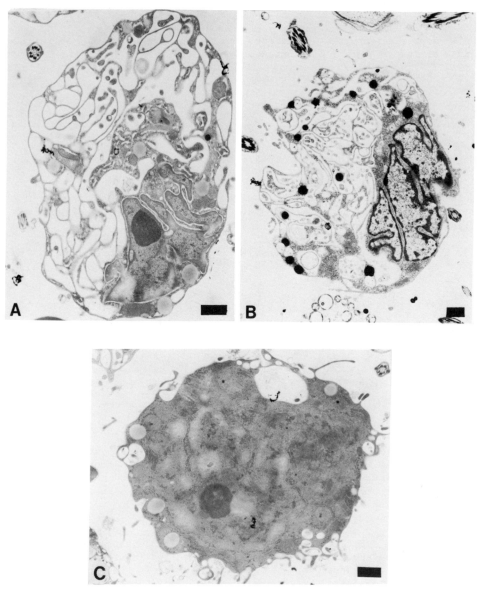

Fig. 13. Electron microscopical autoradiographic localization of [125]I-
labeled hCG. A and B, light cells from fraction I. C, a Leydig
cell from fraction IV. Bars = 1 μm.

DISCUSSION

    Androgens are secreted primarily by the Leydig cells in response to
luteinizing hormone (LH/hCG). Their crucial role in the maintenance of
spermatogenesis is well established, yet the mechanism by which
gonadotropin stimulates this synthesis is not fully understood. A well
regarded mechanism involves the occupancy of high affinity binding sites
on the physiologic receptors by the tropic hormone. This occupancy
promotes cAMP and testosterone production in the cell via adenylate
cyclase activation (For review, see Ariens et al., 1979; Hechter, 1978;

Table II

Cell Densities Reported by Various Investigators Based on the Refractive Index of Fractions from Gradient Centrifugation.

Method and Reference	Light Cells[a]	Heavier Cells[a]
14-32% continuous Metrizamide gradient, Dehejia et al., (1982a,b)	"Lighter Band" 1.085 g/cm^3	"Dense Band" 1.105 g/cm^3
0-40% continuous Metrizamide gradient, Payne et al., (1980a,b)	"Population I" 1.085-1.117 g/cm^3	"Population II" 1.128-1.145 g/cm^3
11, 20, 27% discontinuous Metrizamide gradient, O'Shaughnessy et al., (1981)	"Population I" 1.055-1.086 g/cm^3	"Population II" 1.093-1.116 g/cm^3
20, 25% discontinuous Percoll gradient, Bhalla et al., (1987b)	"Light Cells"[d] 1.029-1.034 g/cm^3	"Dark/Heavier Cells"[c&d] 1.037-1.046 g/cm^3
11, 23, 27% discontinuous Metrizamide gradient, Bhalla et al., (1987b)	"Light Cells" 1.053-1.065 g/cm^3	"Dark/Heavier Cells" 1.090-1.110 g/cm^3

(a) Names given to cell fractions differ in each report as shown.
(b) Method was reproduced by us (see Bhalla et al., 1987b).
(c) Cell fraction was located at the bottom of the centrifuge tube; density is probably greater than measured and shown here.
(d) Cell densities for light and heavier cells reported herein were also confirmed by the use of Pharmacia density marker beads for calibration of Percoll gradients.

Triggle, 1978; Pollet and Levey, 1980). This occupancy model was the simplest representation of events occurring at the level of the membrane since, among other reasons, the binding of ^{125}I-labeled hormone to its receptor was described by a single dissociation constant. Operational rules were established to define whether or not the specific binding units were indeed the receptors. The criteria were (1) homogeneity and structural specificity, (2) saturability, (3) cell specificity, (4) high affinity, (5) reversibility, and (6) parallelism between hormone binding and early chemical events (For review see Ariens, et al., 1979; Hechter, 1978; and Triggle, 1978).

In the course of our study, some of these stipulations required for the validity of the equilibrium binding model were not met (Chen et al., 1979; Bhalla et al., 1979). The binding sites were heterogeneous, the reaction between the hormone and receptor (high affinity sites) was irreversible, and the values for K_d and B_{max} were highly dependent upon the receptor site concentrations and the volume in which the reaction was carried out. These concerns were not limited to our study. A complete disagreement with criterion no. 6 was reported by several investigators (Catt and Dufau, 1973; Mendelson et al., 1975; Clark and Mennan, 1976; Mackie et al., 1972; Seelig and Sayers, 1973; and Boekaert et al., 1973;

also see reviews). A further complexity in the LH/hCG testicular receptor system was observed when pharmacological doses of hCG or LH were administered in vivo. Hormone given in these excessive doses failed to occupy the binding sites and negligible occupancy was seen in the first few hours during the period of active androgen secretion (Powell et al., 1981; Bhalla et al., 1981).

In 1975, Boeynaems and Dumont described how a quantitative analysis of the measurements of ligand binding data (hormone saturation curves) can be exploited to predict a theoretical model of the molecular organization of receptors. Based on their analyses, our experimental data supported the dissociation binding model. The basic characteristic of this model was a predictable increase in the equilibrium dissociation constant experimentally observed with increasing concentrations of total receptor sites (Bhalla et al., 1979, 1982). In the case of the occupancy model, this value does not change, reflecting a tight coupling between the regulatory and catalytic subunits of the receptor (Boeynaems and Dumont, 1975). Based upon the cumulative evidence, we proposed the dissociation binding model in which the binding subunit (R) is loosely coupled to the catalytic subunit (C) within the membrane, the catalytic subunit being adenylate cyclase (Bhalla et al., 1979, 1982). The concept of shedding of the regulatory subunit away from the membrane and into the surrounding fluid was proposed to accomodate two salient observations: (1) no binding was observed during the period of receptor activation in vivo (Powell et al., 1981; Bhalla et al., 1981) and (2) the binding sites were solubilized by buffer and were shown to have high affinity for the hormone in cell free preparations (Bhalla et al., 1976).

Despite this evidence, however, the existence of measurable high affinity binding sites in particulate form was not in dispute and indeed was demonstrated by several investigators. The question remained as to how or even whether such binding sites are coupled to the biological response and if so, how they might be related. If the dissociation model was the sole mechanism governing the biological response, cell separation of testicular interstitial cells might yield an hCG responsive cell type manifesting the biological response (testosterone production) without binding and another hCG responsive cell type containing the binding sites for hCG. A search for those cell types was thus undertaken.

The data presented here clearly demonstrate a cell type (light cell) which binds hCG selectively with high affinity and specificity. This cell type accounts for the observed binding phenomenon. Although the hormone does not stimulate steroidogenesis in this cell, the cell does react positively to Δ^5-3β-hydroxysteroid dehydrogenase. The activation of steroidogenesis in heavier/dark cells (typical Leydig cells) by the hormone without receptor occupancy is consistent with the theory of receptor dissociation whereby hormone promoted steroidogenesis is initiated. The association between these two cell types is intriguing. Whether or not cells are interconvertable remains to be established, but the preliminary evidence presented in Table I does support the cell conversion hypothesis to some extent. In view of the data presented, a working hypothesis by which LH/hCG interacts with the two cell types is described in Fig. 14. Stimulation of dark cells by the hormone possibly leads to cell activation when steroidogenesis is initiated. Once stimulated, the typical Leydig cell, being a secretory cell, begins to synthesize and release testosterone, altering cell shapes as exemplified by the intermediary cell shown in Fig. 8 and light cells (Fig. 6). At this stage, they acquire or synthesize the binding sites for LH/hCG which appears to be a step subsequent to hormone stimulated steroidogenesis. Thus, the dissociation model can be suitably applied to typical Leydig cells and the occupancy model to light cells (Fig. 14).

A. DISSOCIATION MODEL:

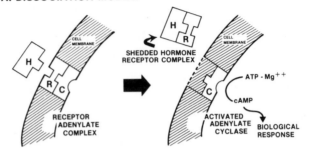

B. OCCUPANCY MODEL:

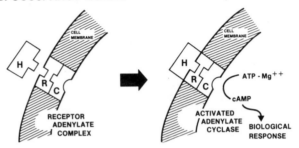

H=HORMONE; R=REGULATORY SUBUNIT; C=CATALYTIC SUBUNIT.

Fig. 14. A working hypothesis to illustrate the coexistence of two phenomena in unpurified cells experimentally resolved by the separation of light and heavier cells.

Insofar as the nature of receptors and their affinities for the ligand are concerned, the possibilities for the existence of low affinity hormone binding sites on typical Leydig cells cannot be excluded. It is unlikely, however, because of the extremely high sensitivity of these cells to minute concentrations of the hormone. The hypothesis proposed here merely provides the direction for continued study to obtain further evidence required to modify and elaborate on the mechanism of hormone activation of target cells.

SUMMARY

The mechanism by which luteinizing hormone (LH) promotes the production of testosterone in Leydig cells by binding to its high affinity sites was reinvestigated. Collagenase dispersed interstitial cells when purified by the application of a variety of techniques such as unit gravity sedimentation, gradient centrifugation, and a combination of the two procedures, were separated into two LH/hCG responsive cell fractions. The two types of interstitial cells displayed distinct biochemical and morphological characteristics. One cell type (the light cell) bound ^{125}I-labeled human chorionic gonadotropin (^{125}I-labeled hCG) with high affinity ($K_a \cong 3.33 \times 10^9$ M^{-1}) but testosterone was not produced by this cell type as a result of hCG target cell receptor interaction. On the other hand, hCG stimulated the production of testosterone in another cell type (the dark/heavier cell). Steroidogenesis was maximally stimulated (700-800 percent over basal) by concentrations of hCG in the range of 3×10^{-10}M, but high affinity binding sites for ^{125}I-labeled hCG were not detectable. The residual binding that occurred did not obey saturation kinetics and was predominantly nonspecific. The stimulation of steroidogenesis by hCG in dark/heavier cells was dose and time dependent. Addition of dibutyryl or bromo cAMP (1 mM) to the cell suspension resulted in production of testosterone demonstrating the involvement of an hCG sensitive adenylate cyclase system in the transfer signaling process. These observations suggest the lack of a direct association between the occupancy of high affinity binding sites by hCG and testosterone production in rat Leydig cells. The stimulation of a biological response by a pathway independent of hCG occupancy of high affinity binding sites on Leydig cell is discussed and morphology of light and dark/heavier cells is presented. Autoradiographic evidence substantiates the conclusions.

ACKNOWLEDGEMENTS

The authors wish to thank Dr. G. Bialy of the Center for Population Research and Contraceptive Branch of NICHHD for a generous gift of hCG. This work was supported by Grant HD 19097 from the National Institute of Child Health and Human Development, National Institutes of Health, Bethesda, MD. The expert assistance of Valerie E. Smith for typing this manuscript is gratefully acknowledged. Inadvertent omission of pertinent references is regretted.

REFERENCES

Ariens EJ, Beld AJ, Rodrigues de Miranda JF, Simonis AM, 1979. The Pharmacon—Receptor Effector Concept: A Basis for Understanding the Transmission of Information in Biological Systems. In: O'Brien RD (ed.), The Receptors, A Comprehensive Treatise. New York: Plenum Publishers, pp. 33—91

Bhalla VK, Flasch MV, Browne ES, Sohal GS, Sharawy MM, 1987b. Interstitial Cell Heterogeneity in Rat Testes: II. Purification of Cells by Percoll and Metrizamide Gradient Centrifugation with Preferential Localization of Gonadotropin Binding Sites in Light Cell Fraction and Hormone Induced Steroidogenesis in Heavier Cell Fraction. J Biol Chem 262:5322—5332

Bhalla VK, Haskell J, Grier H, Mahesh VB, 1976. Gonadotropin Binding Factors: Extraction of High Affinity Gonadotropin Binding Sites from Rat Testis and Partial Characterization of Their Interaction with Human Follitropin, Lutropin and Choriogonadotropin. J Biol Chem 251:4947—4957

Bhalla VK, Powell MM, Rojas FJ, Rajan VP, 1981. Gonadotropin Binding Model: Comparison of the Binding Characteristics of cAMP to Protein Kinase Holoenzyme and Gonadotropin to Particulate Receptors in Rat Testis. In: Mahesh VB, Muldoon TG, Saxena BB, Sadler WA (eds.), Functional Correlates of Hormone Receptors in Reproduction. New York: Elsevier North Holland Inc. pp. 419—436

Bhalla VK, Rajan VP, Burgett AC, Sohal GS, 1987a. Interstitial Cell Heterogeneity in Rat Testes: I. Purification of Collagenase Dispersed Leydig Cells by Unit Gravity Sedimentation and Demonstration of Binding Sites for Gonadotropin in Light Cells Versus Enhanced Steroidogenesis in Heavier Cells. J Biol Chem 262:5313—5321

Bhalla VK, Trowbridge CG, Chen CJH, Lindeman JG, Rojas FJ, 1979. Gonadal Receptors: II. Effect of Time and Reaction Volume Upon the Binding of Human Chorionic Gonadotropin and Human Luteinizing Hormone to Particulate Receptors. Biochim Biophys Acta 584:436—453

Bockaert J, Roy C, Rajerison R, Jard S, 1973. Specific Binding of [³H] Lysine—Vasopressin to Pig Kidney Plasma Membranes. J Biol Chem 248:5922—5931

Boeynaems JM, Dumont JE, 1975. Mini Review Quantitative Analysis of the Binding of Ligands to Their Receptors. Journal of Cyclic Nucleotide Res 1:123-142

Browning JY, D'Agata R, Grotjan HE, 1981. Isolation of Purified Rat Leydig Cells Using Continuous Percoll Gradients. Endocrinology 109:667-669

Catt KJ, Dufau ML, 1973. Spare Gonadotropin Receptors in Rat Testis. Nature New Biology 244:219-221

Chen CJH, Lindeman JG, Trowbridge CG, Bhalla VK, 1979. Gonadal Receptors: I. Evidence for Irreversibility in the Binding of Human Chorionic Gonadotropin and Human Luteinizing Hormone. Biochim Biophy Acta 584:407-435

Christensen AK, 1965. The Fine Structure of Testicular Interstitial Cells in Guinea Pigs. J Cell Biol 26:911-935

Clark MR, Menon KMJ, 1976. Regulation of Ovarian Steroidogenesis: The Disparity Between ^{125}I-labeled Choriogonadotropin Binding, cAMP Formation and Progesterone Synthesis in the Rat Ovary. Biochim Biophy Acta 444:23-32

Conn PM, Tsuruhara T, Dufau M, Catt KJ, 1977. Isolation of Highly Purified Leydig Cells by Density Gradient Centrifugation. Endocrinology 101:639-642

Cooke BA, Magee-Brown R, Golding M, Dix CJ, 1981. The Heterogeneity of Leydig Cells from Mouse and Rat Testes - Evidence for a Leydig Cell Cycle. Intern J Androl 4:355-366

Dehejia A, Nozu K, Catt KJ, Dufau ML, 1982. Luteinizing Hormone Receptors and Gonadotropin Activation of Purified Rat Leydig Cells. J Biol Chem 257:13781-13786

Dehejia A, Nozu K, Catt KJ, Dufau ML, 1982. Purification of Rat Leydig Cells: Functional and Morphological Evaluation. Ann NY Acad Sci 383:204-211

Denef C, Hautekeete E, Dewals R, 1978. Monolayer Cultures of Gonadotropns Separated by Velocity Sedimentation: Heterogeneity in Response to LHRH. Endocrinology 103:736-747

Hechter OM, 1978. The Receptor Concept: Prejudice, Prediction, and Paradox. In: Klachko DM, Forte LR, Franz JM (eds.), Hormone Receptors. New York: Plenum Press, pp. 1-44

Hymer WC, Evans WH, Kraicer J, Mastro A, Davis J, Griswold E, 1973. Enrichment of Cell Types from the Rat Adenohypophysis by Sedimentation at Unit Gravity. Endocrinology 92:275-287

Hymer WC, Snyder J, Wilfinger W, Swanson N, Davis J, 1974. Separation of Pituitary Mammotrophs from the Female Rat by Velocity Sedimentation at Unit Gravity. Endocrinology 95:107-122

Janszen FHA, Cooke BA, Van Driel MJA, Van Der Molen HJ, 1976. Purification and Characterization of Leydig Cells from Rat Testes. J Endocri 70:345-359

Kerr JB, Robertson DM, DeKretser DM, 1985. Morphological and Functional Characterization of Interstitial Cells from Mouse Testes Fractionated on Percoll Density Gradients. Endocrinology 116:1030-1043

Lam DM, Furrer R, Bruce WR, 1970. The Separation, Physical Characterization and Differentiation Kinetics of Spermatogonial Cells of the Mouse. Proc Natl Acad Sc 65:192-199

Langley JN, 1906. Croonian Lecture - On Nerve Endings and on Special Excitable Substances in Cells. Proc R Soc London, Series B 78:170-194

Lee IP, Dixon RL, 1972. Antineoplastic Drug Effects on Spermatogenesis Studied by Velocity Sedimentation Cell Separation. Toxicol Appl Pharm 23:20-41

Leuschen M, Tobin R, and Moriarty M, 1978. Enriched Populations of Rat Pituitary Thyrolrophs in Monolayer Culture Endocrinology 102:509-518

Mackie C, Richardson MC, Schulster D, 1972. Kinetics and Dose-Response Characteristics of cAMP Production by Isolated Rat Adrenal Cells Stimulated with ACTH. FEBS Letters 23:345-348

Meistrich ML, 1972. Separation of Mouse Spermatogenic Cells by Velocity Sedimentation. J Cell Physiol 80:299-312

Meistrich ML, Longtin J, Brock WA, Grimes SR, Mace ML, 1981. Purification of Rat Spermatogenic Cells and Preliminary Biochemical Analysis of These Cells. Biol Reproduction 25:1065-1077

Mendelson C, Dufau M, Catt K, 1975. Gonadotropin Binding and Stimulation of cAMP and Testosterone Production in Isolated Leydig Cells. J Biol Chem 250:8818-8823

O'Shaughnessy PJ, Wong K-L, Payne AH, 1981. Differential Steroidogenic Enzyme Activities in Different Populations of Rat Leydig Cells. Endocrinology 109:1061-1066

Payne AH, Downing JR, Wong K-L, 1980. Luteinizing Hormone Receptors and Testosterone Synthesis in Two Distinct Populations of Leydig Cells. Endocrinology 106:1424-1429

Payne AH, O'Shaughnessy PJ, Chase DJ, Dixon GEK, Christensen AK, 1982. LH Receptors and Steroidogenesis in Distinct Populations of Leydig Cells. Ann NY Acad Sci 383:174-203

Payne AH, Wong K-L, Vega MM, 1980. Differential Effects of Single and Repeated Administrations of Gonadotropins on Luteinizing Hormone Receptors and Testosterone Synthesis in Two Populations of Leydig Cells. J Biol Chem 255:7118-7122

Peterson EA, Evans WH, 1967. Separation of Bone Marrow Cells by Sedimentation at Unit Gravity. Nature 214:824-825

Pollet RJ, Levey GS, 1980. Principles of Membrane Receptor Physiology and Their Application to Clinical Medicine. Ann of Internal Med 92:663-680

Powell MM, Rajan VP, Cohen JD, and Bhalla VK, 1981. Gonadal Receptors: Upregulation in Response to Elevated Serum Gonadotropin Levels. Biol Reprod 25:708-718

Pretlow II TG, Pretlow TP (eds.), 1982. Cell Separation, Methods and Selected Applications. New York: Academic Press, Vol. I

Rodbell M, 1971. Hormones, Receptors, and Adenyl Cyclase Activity. In: Condliffe P, Rodbell M (eds.), Colloquium on the role of Adenyl Cyclase and cAMP in Biological Systems. Fogarty International Center, Gov't Printing Office, Washington, D.C. pp. 88-95

Romrell LJ, Bellve AR, Fawcett DW, 1976. Separation of Mouse Spermatogenic Cells by Sedimentation Velocity. Devel Biol 19:119-131

Schumacher M, Schafer G, Lichtenberg V, Hilz H, 1979. Maximal Steroidogenic Capacity of Mouse Leydig Cells. FEBS Letters 107:398-402

Seelig S, Sayers G, 1973. Isolated Adrenal Cortex Cells: ACTH Agonists, Partial Agonists, Antagonists; cAMP and Corticosterone Production. Arch Biochem Biophys 154:230-239

Snyder G, Hymer WC, Snyder J, 1977. Functional Heterogeneity in Somatotrophs Isolated from the Rat Anterior Pituitary. Endocrinology 101:788-799

Snyder J, Wilfinger W, Hymer WC, 1976. Maintenance of Separated Rat Pituitary Mammotrophs in Cell Culture. Endocrinology 98:25-32

Triggle DJ, 1978. The Receptor Theory. In: Smythies JR, Bradley RJ (eds.), Receptors in Pharmacology. New York: Marcel Dekker Inc., pp. 1-66

DIFFERENTIATION AND POLARIZED FUNCTION OF SERTOLI CELLS IN VITRO

M. Dym, M.A. Hadley, D. Djakiew, and S.W. Byers

Department of Anatomy and Cell Biology
School of Medicine - School of Dentistry
Georgetown University
Washington, DC 20007

INTRODUCTION

The seminiferous epithelium is composed of two populations of cell types, supportive epithelial elements, the highly polarized Sertoli cells and differentiating elements, the germ cells. The Sertoli cell is a tall, narrow, columnar cell that extends from the basal lamina of the seminiferous epithelium to the tubule lumen. It possesses numerous lateral branches which surround the developing germ cells and it appears to provide support for their differentiation. The basal aspects of the Sertoli cells are linked by specialized tight junctional complexes and these junctions subdivide the seminiferous epithelium into two compartments, basal and apical (Dym and Fawcett, 1970). The tight junctions form the morphological basis of the blood-testis barrier. The basal domain of the cell, in contact with extracellular matrix molecules, possesses receptors for circulating plasma constituents (e.g., FSH receptors - Orth and Christensen, 1977). The apical surface of the cell, in contact with the tubule lumen, is undoubtedly involved in sperm release (Fawcett, 1975). An examination of the ultrastructure of the Sertoli cell reveals a highly polar organization of the organelles (Fawcett, 1975). The Sertoli cell epithelium is specialized to perform a wide variety of vectorial functions, for example, it is likely that a number of Sertoli cell proteins are secreted into the tubule lumen whereas others are released from the basal surface of the cells into the interstitium.

In the mid-seventies several investigators succeeded in growing Sertoli cells in culture on plastic dishes (Dorrington et al., 1975; Steinberger et al., 1975; Welsh and Weibe, 1975). Although many functional parameters of the cultured cells appeared to be similar to the _in vivo_ Sertoli cells, their squamous appearance in conventional culture bore little resemblance to the highly polarized Sertoli cells observed _in vivo_. On plastic the cells form a monolayer at the bottom of the vessel and the culture medium, placed on top of the cells, bathes the apicolateral surface of the cells, but is prevented access to the basal plasma membranes which are firmly attached to the substrate. In direct contrast, _in vivo_ Sertoli cells gain access to circulating plasma constituents exclusively from their basal plasma membranes.

In an effort to obtain highly differentiated Sertoli cells, analogous to the _in vivo_ Sertoli cells, extracellular matrix molecules were used as substrate for the cultures (Suarez-Quian et al., 1984; Hadley et al., 1985). We also developed a dual compartment (bicameral) culture chamber

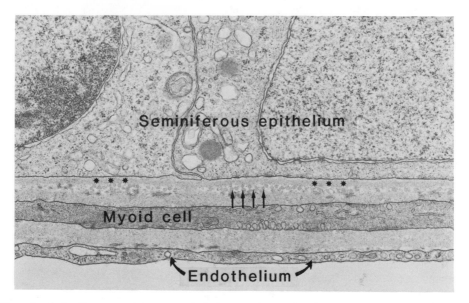

Fig. 1. A low power electron micrograph of a portion of the lamina propria
of rat seminiferous epithelium. The epithelium rests on a basement
membrane (*) deep to which collagen fibrils (arrows) are seen in
transverse section. The layer of myoid cells has an amorphous
ground substance on either side. The lymphatic endothelial layer
is the outermost (lower most) cell layer of the lamina propria.
The extracellular matrix zones between the seminiferous epithelium
and myoid cells and between the myoid and endothelial cells are
referred to as the inner and outer noncellular layers,
respectively. X2800. (From Dym and Fawcett, 1970).

so that Sertoli cells in vitro are able to establish two environments, one
at their basal surface and the second at the apical surface of the cells,
similar to the basal and apical compartments present in the in vivo
seminiferous epithelium (Byers et al., 1986). Dual compartment chambers
have been in use for a number of years by investigators examining transepithelial transport of ions in Madin-Darby canine kidney (MDCK) cells
(Misfeldt et al., 1976; Simmons et al., 1984). Other aspects of epithelial
cell polarity, such as intracellular sorting, have also been studied in
similarly designed chambers (Caplan and Anderson, 1986). Recently,
Janecki and Steinberger (1987) examined polarity of secretion of androgen
binding protein and transferrin by Sertoli cells in dual compartment
chambers. In these chambers one can separately manipulate and have
independent access to the media at the basal aspects of the cells as well
as to the media at the apical surfaces. Thus, for the first time in
Sertoli cell culture, one can examine a variety of polarized functions
such as transport of molecules across the epithelium and vectorial
secretion.

ULTRASTRUCTURAL LOCALIZATION OF EXTRACELLULAR MATRIX MOLECULES IN THE
LAMINA PROPRIA OF SEMINIFEROUS TUBULES

It has become clear that the structure and function of epithelial
cells are influenced by the substrate upon which they are growing
(Kleinman et al., 1981; Hay, 1984). In the testis the seminiferous
epithelium rests on a basement membrane which itself is surrounded by a
layer of peritubular myoid cells (Fig. 1). In order to determine the
molecular composition of the lamina propria of the seminiferous tubules we

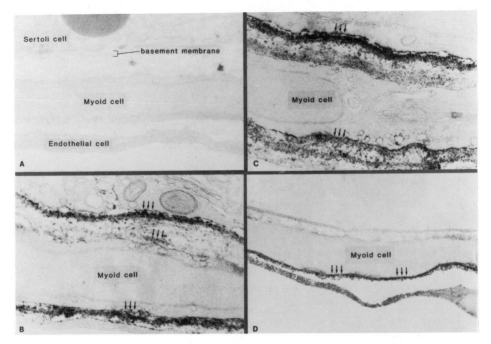

Fig. 2. Electron micrographs of the ultrastructural immunocytochemical localization of extracellular matrix molecules within the lamina propria of the rat testis. There was no background staining in normal rabbit serum or preimmune controls (A). Both laminin (B) and type IV collagen (C) antibodies stained within the inner and outer non-cellular layers. Fibronectin (D) staining was confined to the outer non-cellular layers. No intracellular staining was observed. X5000. (From Hadley and Dym, 1987).

undertook an ultrastructural immunocytochemical examination (Hadley and Dym, 1987). Laminin, type IV collagen, fibronectin, and heparan sulphate proteoglycan were localized in the testis (Fig. 2) using specific antibodies and the peroxidase-antiperoxidase method of Sternberger (1979). Fig. 3 is a montage summarizing the results of our immunocytochemical studies. Laminin, type IV collagen, and heparan sulphate proteoglycan were present in the basement membrane immediately deep to the seminiferous epithelium, whereas fibronectin was only localized at the external surface of the myoid cells. Heparan sulphate proteoglycan and type IV collagen were distributed adjacent to both sides of the myoid cell peritubular layer. Fibronectin and laminin were adjacent only to the interstitial side of the myoid cells.

The immunocytochemical localization of type IV collagen, laminin, and heparan sulphate proteoglycan immediately adjacent to the seminiferous epithelium is in accord with the reports that Sertoli cells secrete these three compounds in vitro (Skinner et al., 1985; Borland et al., 1986). There are conflicting reports concerning the production of fibronectin in the testis. Borland et al. (1986) report that cultured Sertoli cells produce this extracellular matrix molecule while Tung et al. (1984) and Skinner et al. (1985) indicate that contaminating fibroblasts or myoid cells are likely responsible for the fibronectin production. Our results demonstrating the lack of fibronectin localization immediately adjacent to the seminiferous tubules lends support to a myoid cell or fibroblast for this molecule.

517

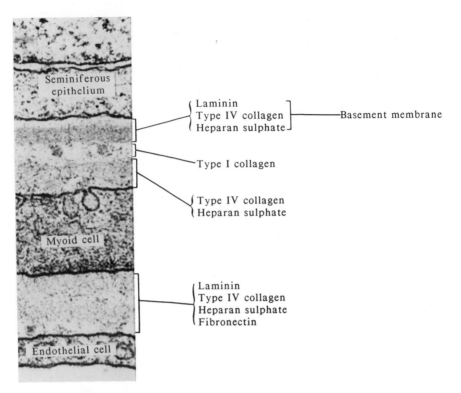

Fig. 3. A photographic montage of the testicular lamina propria summarizing the distribution of extracellular matrix molecules, as determined by electron microscopic immunocytochemistry. X7800. (From Hadley and Dym, 1987).

EFFECTS OF EXTRACELLULAR MATRIX ON SERTOLI CELL DIFFERENTIATION

On the basis of the immunocytochemical distribution of the extracellular matrix molecules in the testis, we decided to culture Sertoli cells on various ECM components. When cultured on plastic Sertoli cells were squamous and generally appears undifferentiated. However, on a laminin and type IV collagen substrate the Sertoli cells developed a cuboidal shape and resembled more closely the in vivo cells (Suarez-Quian et al., 1984). When we used a reconstituted basement membrane (RBM) as a substrate (Hadley et al., 1985) a dramatic cellular differentiation was obvious. The Sertoli cells became tall columnar and exhibited many other aspects of polarity and cellular differentiation that are present in vivo (Fig. 4). The nuclei of the cells became oval and occupied a basal position, basally located tight junctions were present, and the apical surface of the cells developed numerous projections. The nuclear envelope exhibited infoldings similar to those in the in vivo Sertoli cells. Secretion of total protein,, androgen binding protein, and transferrin were all markedly greater when Sertoli cells were grown on this RBM than when they were grown on plastic (Fig. 5). The conclusion from this part of our work is that Sertoli cells cultured on top of extracellular matrix molecules assume a phenotype and a morphology more characteristic of the in vivo differentiated Sertoli cells.

Normally a very thin (< 50 μm) film of RBM is used as substrate to maintain the differentiated functions of the Sertoli cells in culture. Germ cells that are trapped among the Sertoli cells are maintained in the

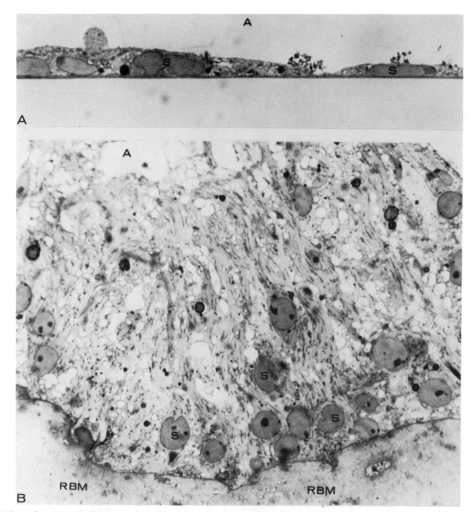

Fig. 4. A. A light micrograph of Sertoli cells cultured on plastic. Note
their squamous appearance. X1400.
B. A light micrograph of Sertoli cells cultured on reconstituted
basement membrane. Note the tall columnar appearance of the cells
and the basally located oval nuclei. The location of the apical
surface is shown as clear space (A).   Sertoli cell nuclei (S).
X1200. (From Hadley et al., 1985).

culture for at least five weeks, but they do not differentiate. However,
when Sertoli cells are plated within a thick (1.5 mm) layer of RBM they
rapidly reorganize to form seminiferous tubule-like cords (Fig. 6).   The
Sertoli cells of these in vitro cords demonstrate a highly differentiated
and polarized morphology similar to that which they acquire in monolayer
culture on RBM.   Basally located tight junctional complexes functionally
compartmentalize the cords into basal and apical compartments.   Cord
formation appears to be a laminin-dependent process since anti-laminin
antibodies, but not anti-type IV collagen, anti-fibronectin, anti-heparan
sulphate proteoglycan antibodies, nor normal rabbit serum, inhibit cord
formation.   Within the compartments of the in vitro cords spermatogonia
continue to differentiate to form late pachytene spermatocytes (Fig. 7).
Thus, it appears that the establishment of a specialized apical milieu by
Sertoli cells of the cultures cords is essential for germ cell
differentiation.

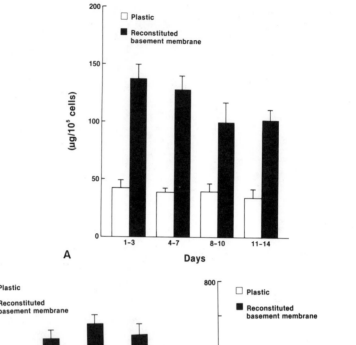

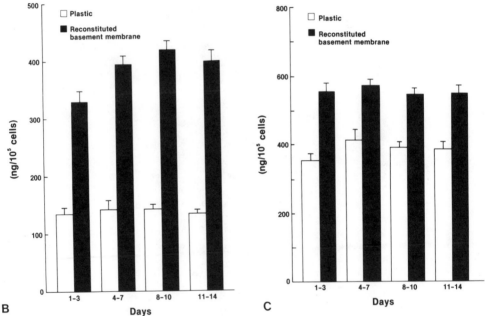

Fig. 5. Sertoli cells were cultured on either plastic or reconstituted basement membrane (RBM) matrix for 2 weeks. Culture media were collected at 3, 7, 10, and 14 days and aliquots were used for secretory protein assays. Data are presented as mean $\pm$ SD for three separate experiments.

A. Total protein (μg/10^5 cells) was measured by the Bio-Rad protein assay. Means are significantly different at $p < 0.01$.

B. Androgen-binding protein (ng/10^5 cells) was measured using radioimmunoassay. Means are significantly different at $p < 0.01$.

C. Transferrin (ng/10^5 cells) was measured using radioimmuno-assay. Means are significantly different at $p < 0.01$.

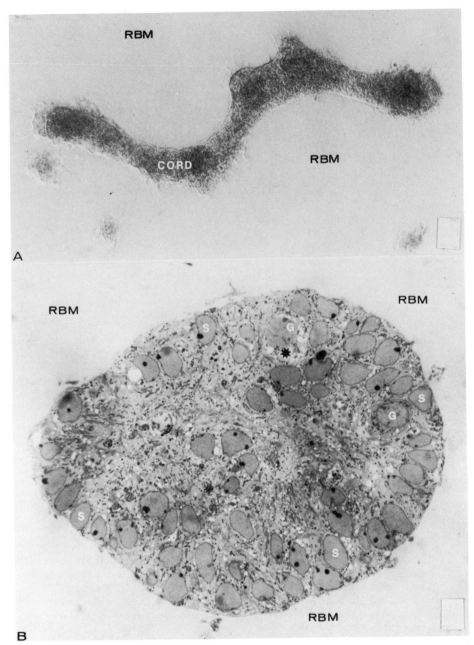

Fig. 6. A. A phase-contrast light micrograph of an _in vitro_ Sertoli cell
cord which has formed from dissociated Sertoli cells cultured for
24 h within RBM. X250.
B. A light micrograph showing a cross section through an _in
vitro_ Sertoli cell cord cultured for 14 days. Note the peripheral
location of most Sertoli cell nuclei (S).   Two germ cells (G) can
be seen within the cord, one of which is a pachytene spermatocyte
(*). RBM surrounds the cord. X1000.

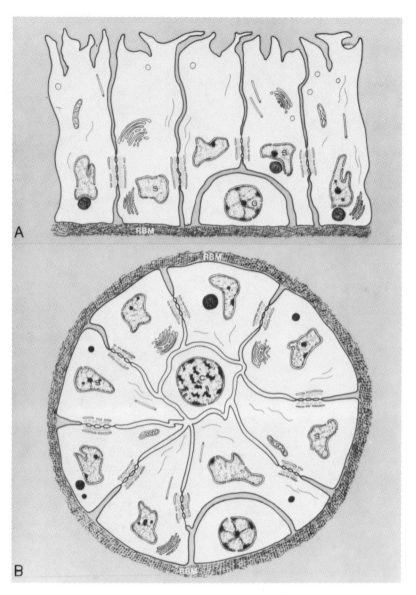

Fig. 7. A. This diagram of Sertoli cells on top of a <u>thin layer</u> of RBM
demonstrates how the culture media (grey) has equal access to both
the apical and basal regions of the cells.   Germ cells (G)   are
maintained in the basal region beneath the Sertoli-Sertoli tight
junctions, and have direct access to the culture medium which
flows through the RBM.   No compartmentalization of the Sertoli
cells exists.
B. This diagram of an <u>in vitro</u> Sertoli cell cord grown within a
<u>thick layer</u>   of RBM demonstrates that two environments are
established in these cultures.   The Sertoli-Sertoli   tight
junctions create a central (or apical) compartment (clear intercellular spaces) distinct from the peripheral (or   basal)
compartment beneath the tight junctions.   The culture medium
(grey) has free access to the basal compartment by filtering
through the RBM.   It does not have access to the apical
compartment.   Spermatocytes   (C)   that have developed   from
spermatogonia are found exclusively in the apical compartment.

DEVELOPMENT AND CHARACTERIZATION OF A BICAMERAL (DUAL COMPARTMENT) CELL CULTURE CHAMBER

Normal Sertoli cell function in vivo is dependent upon the creation of a unique luminal milieu in the seminiferous epithelium in which spermatocytes and spermatids can differentiate. The tight Sertoli cell junctional complexes separate this special apical environment from the basal compartment and circulating plasma constituents. Thus, substances from the blood have to pass through the Sertoli cell cytoplasm in order to reach the meiotic and spermiogenic germ cells. None of the culture systems previously developed (Dorrington et al., 1975; Steinberger et al., 1975; Welsh and Weibe, 1975) for Sertoli cell in vitro growth permits the establishment of the two seminiferous epithelial compartments that are present in vivo.

The in vitro Sertoli cell cord culture system described in the previous section provides compartmentalization and permits germ cells to differentiate to a more advanced stage than previously possible in vitro. However, this system is not easily controlled and manipulated. For example, it is impossible to sample or alter the environment in the central compartment of the cord without destroying it. To overcome this difficulty we designed a dual compartment (bicameral) culture vessel in which one could culture highly polarized and differentiated monolayers of Sertoli cells. Initially, we developed bicameral chambers by sawing the bottoms off small plastic centrifuge tubes (Fisher 4.978145) and attaching a 1.2 μm Millipore filter to one end of the tube using chloroform (Byers et al., 1986). A tripod of melted dental wax, 1 mm high, was added at the bottom of the chamber and the entire unit was inserted into a multiwell dish. The wax tripod served to raise the filter chamber off from the bottom of the multiwell dish. The culture chamber containing the filter was referred to as the inner chamber while the multiwell dish was called the outer chamber. The Millipore filters were impregnated with reconstituted basement membrane (Byers et al., 1985) prior to plating the cells. The principal advantage of this novel cell culture system is that for the first time in culture we have independent access to the media bathing the basal surface of the Sertoli cells as well as the media at the apical surface of the cells. Furthermore, we can separately manipulate and sample these two media which are separated by the side wall of the inner chamber. Fig. 8A is a schematic drawing of this culture chamber. More recently, we have collaborated with the Millipore Corporation to produce similar chambers (Millicell) and several of the experiments that will be described below utilized these chambers.

Prior to using this in vitro Sertoli cell epithelium for experiments it was essential to confirm that indeed we have a confluent layer of cells on top of the filter. Several approaches were utilized to characterize the system including morphology, permeability studies, and electrical resistance.

Lanthanum nitrate was used to test the patency of the narrow inter-cellular clefts between adjacent Sertoli cells. Prior to tight junction development basally added lanthanum nitrate permeated between the cells to reach their apical surfaces. After junction development, which usually occurred approximately five days after plating, the tracer compound was blocked by the tight junctions (Fig. 10) and prevented from penetrating deep into the epithelium (Byers et al., 1986; Djakiew et al., 1986). Thus, it appears that the "blood testis barrier" that was described in vivo (Dym and Fawcett, 1970) can be mimicked in Sertoli cell culture.

Tritiated inulin which is considered a more physiological tracer than lanthanum was added to the basal compartment of the filter chamber

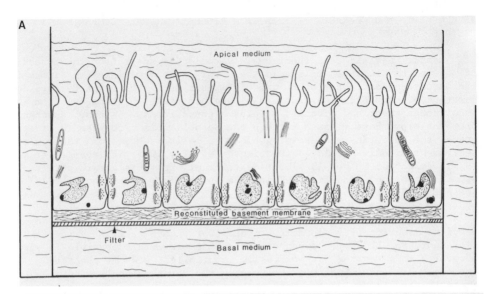

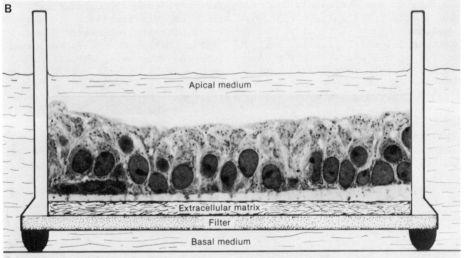

Fig. 8. A. A schematic diagram of a bicameral culture chamber used to grow Sertoli cells at confluence on a Millipore filter coated with reconstituted basement membrane matrix.

B. A photographic montage of an epithelial sheet of Sertoli cells in a bicameral culture chamber. Using this chamber it is possible to expose the cells to one culture environment at their basal surface (Basal medium) and to another at their apical surface (Apical medium). The apical and basal media may be sampled separately to measure Sertoli cell secretory products which are released in a vectorial manner.

containing Sertoli cells. The passage of ^{3}H-inulin across the cells was about 1% per hour. The filter alone or the filter coated with RBM (no cells) permitted close to 3% to pass each hour (Fig. 11) (Djakiew et al., 1986). These results are in accord with other published work showing confluency of MDCK cells in dual compartment chambers (Caplan and Anderson, 1986).

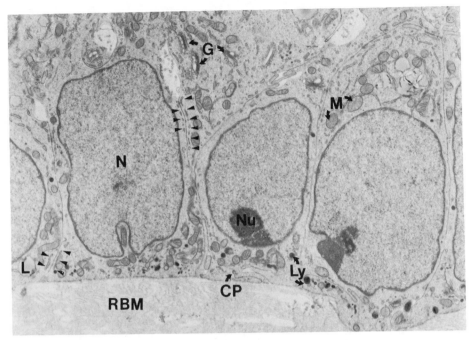

Fig. 9. An electron micrograph of several Sertoli cells cultured for 14 days in a bicameral chamber. Nucleus (N); nucleoli (Nu); mitochondria (M); Golgi apparatus (G); developing tight junction (arrowheads); lipid droplet (L); lysosome (Ly); reconstituted basement membrane (RBM); coated pit at the base (CP). X15000.

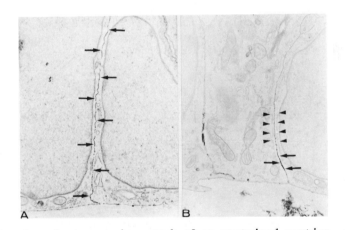

Fig. 10. A. An electron micrograph of an unstained section of Sertoli cells cultured for 3 days after seeding from 10 day old rats. Lanthanum nitrate has penetrated between the Sertoli cells (arrows) and there is no evidence of the formation of Sertoli cell tight junctions. X12000.
B. An electron micrograph of Sertoli cells cultured for 5 days after seeding from 10 day old rats. The newly formed Sertoli cell tight junctions (arrowheads) form a transepithelial permeability barrier to the passage of lanthanum nitrate (arrows). Cisternae of endoplasmic reticulum course parallel to apposing basolateral cell membranes of the nascent tight junction. X12000.

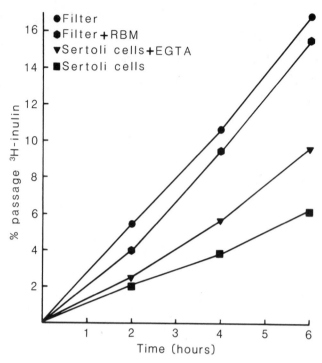

Fig. 11. The percentage of ^{3}H-inulin passing from the basal to the apical chambers across filters alone (circles), filters impregnated with RBM (hexagons), epithelial sheets of Sertoli cells (squares), and similarly prepared epithelial sheets preincubated with 2 mM EGTA (triangles). Medium containing 1 X 10^5 CPM of ^{3}H-inulin was placed in the basal chamber for a 6 h period. Passage of ^{3}H-inulin from the basal chamber to the apical chamber was monitored by collection of media from both chambers and the radioactivity in each solution was determined in a Beckman scintillation counter.

Electrical resistance studies indicated that immediately after seeding of the cells the resistance was only marginally above that of filters impregnated with RBM (no cells). The resistance of the epithelial sheet of Sertoli cells progressively increased reaching a peak of 490 milliohms/cm^2 after 1.5 days. The resistance then declined and remained at a low level until tight junctions developed at about 5 days. There was then a rapid increase in resistance that plateaued at close to 700 milliohms/cm^2. The resistance was maintained at this level (Djakiew et al., 1986). The peak of resistance noted at 1.5 days cannot be explained, however, it has been speculated that this could be due to some type of "cellular interactions" occurring at this time (Fig. 12).

Sertoli cells cultured for one or two weeks in the bicameral chambers on RBM coated filters remained highly polarized and differentiated and generally appeared similar to the in vivo Sertoli cell and to Sertoli cells cultured on RBM (Fig. 8B) (Hadley et al., 1985). A complete epithelial sheet of cells covered the Millipore filter and the cells exhibited many of the ultrastructural characteristics that are found in vivo. The nuclei occupied a basal position in the tall columnar Sertoli cell and basally located tight junctional complexes joined adjacent cells. The supranuclear cytoplasm contained an abundance of the typical organelles normally present in Sertoli cells (Fig. 9).

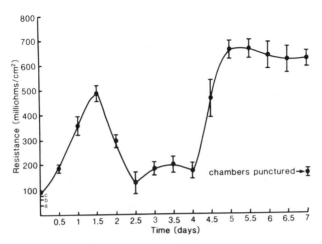

Fig. 12. The change in electrical resistance of Sertoli cells seeded from 10 day old rats and cultured for 7 days. Note the marked increase in electrical resistance after 5 days in culture. On the seventh day of culture, the filters were punctured with a needle resulting in a precipitous reduction in electrical resistance. The electrical resistance of media (a), a filter (b), and a filter impregnated with RBM (c) are also shown. The transient rise of resistance between 1 and 2 days remains unexplained.

[59]FE TRANSPORT ACROSS CONFLUENT EPITHELIAL SHEETS OF SERTOLI CELLS

The iron transport properties of transferrin have been well documented in the literature (Aisen and Litowsky, 1980; Frazier et al., 1982). In the testis the Sertoli cells produce a unique form of transferrin in that the carbohydrate moieties associated with the peptide chains differ from liver transferrin (Skinner and Griswold, 1980). It has been proposed that circulating iron is moved across the blood testis barrier by transfer from serum transferrin to testicular transferrin within the Sertoli cell cytoplasm (Huggenvik et al., 1984). Subsequently, the testicular transferrin bound to Fe is secreted apicolaterally to the differentiating germ cells and to the tubule lumen. To test this hypothesis we have carried out Fe transport studies across confluent sheets of Sertoli cells in bicameral cell culture chambers.

[59]Fe was incorporated into human serum apotransferrin according to the method described by van Renswoude et al. (1982). Human serum ^{59}Fe-transferrin (520 pM) in 750 μl serum-free defined medium was placed in the basal compartment of the dual environment culture chamber. The presence of ^{59}Fe in an immunoprecipitate of rat Sertoli cell transferrin secreted into the apical media of the chambers was monitored over 6 h at 33° C. Cross reactivity of the rat transferrin antibody with human transferrin was less than 0.01%.

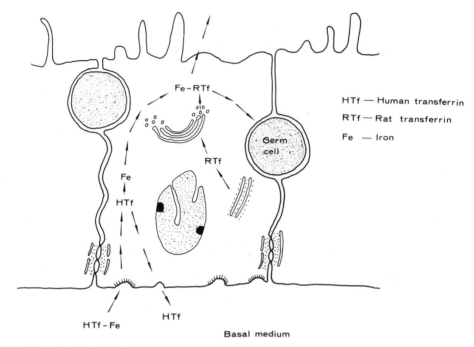

Apical medium

HTf — Human transferrin

RTf — Rat transferrin

Fe — Iron

HTf-Fe

HTf

Basal medium

Fig. 13. A schematic diagram of the pathway of the transcellular transport of ^{59}Fe from human serum ^{59}Fe-transferrin (Fe-HTf) in the basal medium to rat testicular ^{59}Fe-transferrin (Fe-RTf) in the apical medium of the bicameral culture chambers.

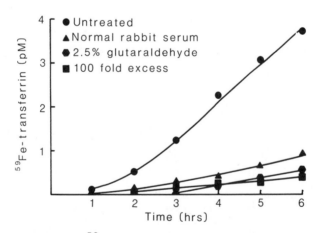

Fig. 14. The transport of ^{59}Fe from human serum transferrin (basal media) to immunoprecipitated rat testicular transferrin (apical media). (From Djakiew et al., 1986).

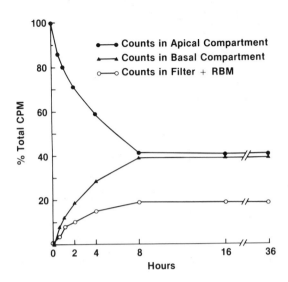

Fig. 15. ^{125}I-transferrin was added to the apical compartment of bicameral culture chambers without cells. The rate at which 3.0 X 10[5] cpm of the labeled protein equilibrated across the support (Millipore filter coated with RBM matrix) was measured by plotting the counts found in the apical and basal media and those trapped in the support, with time. The amount of label in each compartment was determined at the indicated h (abscissa). The ordinate indicates the percentage of the total counts in each of the compartments and in the RBM-coated filter. Note that as the counts diminish in the apical compartment they increase in the basal compartment, equilibrating between the two compartments by 8 h. The amount of counts trapped in the RBM-coated filter is represented by the open circles. The equilibration curves for transferrin and androgen-binding protein are identical.

Fig. 13 shows schematically the transfer of ^{59}Fe from human serum (^{59}Fe)-transferrin, added to the basal compartment, to testicular transferrin synthesized in the cytoplasm of Sertoli cells and subsequently secreted into the apical media. The experimental results are shown in Fig. 14 and indicate that there is a linear increase of ^{59}Fe in the apical compartment of the chamber over 6 h. Controls included the addition of a 100-fold excess of rat serum di-ferric transferrin in the basal media, prior fixation of the cells with glutaraldehyde, and substitution of normal rabbit serum for the primary antibody. When purified populations of pachytene spermatocytes or round spermatids were added to the apical compartment of the chambers (5 X 10[5] germ cells/chamber), the immunoprecipitable ^{59}Fe-Tf in the apical media was reduced by 50% and 20%, respectively, suggesting that some of the ^{59}Fe-Tf binds to the germ cells. These results demonstrate a unique mechanism for transporting Fe across Sertoli cells utilizing two forms of transferrin. In this context it is

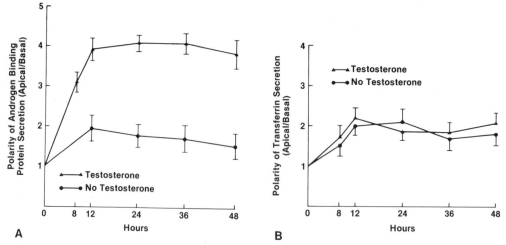

Fig. 16. The ratio of apically/basally secreted androgen-binding protein
(A) and transferrin (B) by Sertoli cells cultured in bicameral
chambers in the presence of serum-free defined medium is plotted
as a function of time. Results are shown for cells stimulated
and not stimulated with testosterone.
A. With testosterone, 4-fold more ABP is secreted apically than
basally. Without testosterone, only 1- to 2-fold more ABP goes
apically. All values are significantly different (p < 0.01).
B. Testosterone did not significantly influence the ratio of
transferrin secretion, which remained at 2-fold more apically
(p < 0.01).

interesting to note that in cerebral cortical neurons there is a special
form of transferrin that may also carry Fe across the blood-brain barrier
(Conner and Fine, 1986).

POLARITY OF SERTOLI CELL SECRETION

Sertoli cells are the principal secretory cells in the seminiferous
epithelium. They have been shown to secrete androgen binding protein
(Fritz et al., 1976), inhibin (Steinberger and Steinberger, 1976),
plasminogen activator (Lacroix et al., 1977), transferrin (Skinner and
Griswold, 1980), ceruloplasmin (Skinner and Griswold, 1983), clusterin
(Blaschuk et al., 1983), testibumin (Cheng and Bardin, 1986), and a large
number of other proteins that remain mostly unidentified (Wright et al.,
1981; Cheng et al., 1986). Gunsalus et al. (1980) have shown that androgen
binding protein (ABP) is secreted bidirectionally from the Sertoli cells.
Twenty percent of the protein is secreted basally into the systemic
circulation and 80% goes apically into the tubule lumen. It has also been
demonstrated that certain drugs (Spitz et al., 1985), hormones (Gunsalus
et al., 1980), or germ cell depletion (Gunsalus et al., 1985) may alter
this polarity of secretion.

In an effort to examine polarity of Sertoli cell secretion in vitro
we used the bicameral chambers for our cultures (Hadley et al., 1987).
For this system to be useful it is not only necessary that the cells form
confluent epithelial sheets, as demonstrated above, but that substances
secreted basally by the cells are free to pass through the reconstituted
basement membrane (RBM) coated filter into the basal compartment. Chambers
with RBM-coated filters but without cells were used to investigate the
equilibration of radiolabeled proteins between the apical and basal medium

compartments (Fig. 15). Experiments in which either 3.0 X 10^5 or 5.0 X 10^5 cpm of ^{125}I labeled ABP or Tf were added to the apical compartment medium showed that these proteins equilibrate across the RBM-coated filter support at the same rate (approximately 8 h, regardless of protein concentration). When 3.0 X 10^5 cpm were added approximately 18% of the counts bound the filter at equilibrium between the apical and basal compartments. When 5.0 X 10^5 cpm were added only 13% of the counts bound the filter at equilibrium, indicating that protein binding to the filter is a saturable event.

Over the course of 48 h in culture ABP and Tf are secreted in a vectorial manner, primarily from the apical surface of the Sertoli cells. The polarity of secretion (determined by the ratio of apical/basal secretion) plotted over time reveals that ABP secretion remains about 4:1 while Tf remains at 2:1. When testosterone is removed from the medium the ABP secretion ratio drops to 2:1 while the Tf ratio remains the same, indicating that ABP, but not Tf, secretion is androgen-dependent (Fig. 16). The apical/basal ratio of total secretory protein was 2.2:1, however, when testosterone was removed from the medium this ratio fell to 1.7:1. This suggests that, in addition to ABP, there are other Sertoli cell proteins whose secretory vector depends on testosterone.

CONCLUSIONS

Our work plus the work of several other laboratories demonstrate clearly that the extracellular matrix profoundly influences the expression of Sertoli cell phenotype in culture. Sertoli cells grown on a reconstituted basement membrane (RBM) resembled their in vivo counterparts very closely, both in structure and function. Highly differentiated Sertoli cells grown in bicameral chambers on filters coated with RBM can be used to investigate polarized cell functions such as transepithelial transport and bidirectional secretion. High cell density, the appropriate extracellular matrix, and basal diffusion of the medium all appear necessary for the maintenance of highly differentiated Sertoli cells in vivo. Finally, the bicameral culture chamber may provide a unique opportunity to investigate cell-cell interactions in the testis and germ cell differentiation in vitro.

REFERENCES

Aisen P, Listowsky I, 1980. Iron transport and storage protein. Ann Rev Biochem 49:357-393

Blaschuk O, Burdzy K, Fritz IB, 1983. Purification and characterization of a cell-aggregating factor (clusterin), the major glycoprotein in ram rete testis fluid. J Biol Chem 258:7714-7720

Borland K, Muffly KE, Hall PF, 1986. Production of components of extra-cellular matrix by cultured rat Sertoli cells. Biol Reprod 35:997-1008

Byers SW, Hadley MA, Dym M, 1985. Growth of polarized monolayers of epithelial cells from the male reproductive tract on permeable extra-cellular (ECM) matrix supports. Third Int Cong Androl, Boston

Byers SW, Hadley MA, Djakiew D, Dym M, 1986. Growth and characterization of polarized monolayers of epididymal epithelial cells and Sertoli cells in dual environment culture chambers. J Androl 7:59-69

Caplan MJ, Anderson HC, 1986. Intracellular sorting and polarized cell surface delivery of (Na^+, K^+)ATPase, and endogenous component of MDCK cell basolateral plasma membranes. Cell 46:623-631

Cheng YC, Mather JP, Byer AL, Bardin WC, 1986. Identification of hormonally responsive proteins in primary Sertoli cell culture medium by anion-exchange high performance liquid chromatography. Endocrinology 118:480-488

Cheng YC, Bardin WC, 1986. Rat testicular testibumin is a protein responsive to follicle stimulating hormone and testosterone that shares immunodeterminants with albumin. Biochemistry 25:5276-5288

Conner JR, Fine RE, 1986. The distribution of transferrin immunoreactivity in the rat central nervous system. Brain Res 368:319-328

Djakiew D, Hadley MA, Byers SW, Dym M, 1986. Transferrin-mediated transcellular transport of ^{59}Fe across confluent epithelial sheets of Sertoli cells grown in bicameral cell culture chambers. J Androl 7:355-366

Dorrington JH, Roller NF, Fritz IB, 1975. Effects of follicle-stimulating hormone on cultures of Sertoli cell preparations. Mol Cell Endocrinol 3:57-70

Dym M, Fawcett DW, 1970. The blood-testis barrier in the rat and the physiological compartmentation of the seminiferous epithelium. Biol Reprod 3:308-326

Fawcett DW, 1975. Ultrastructure and function of the Sertoli cell. In: Handbook of Physiology, Section 7, Vol 5. Washington, Am Physiol Soc

Frazier JL, Caskey JH, Yoffe M, Seligman PA, 1982. Studies of the transferrin receptor on both human reticulocytes and nucleated human cells in culture. Comparisons of factors regulating receptor density. J Clin Invest 69:853-865

Fritz IB, Griswold MD, Louis BG, Dorrington JH, 1976. Similarity of responses of cultured Sertoli cells to cholera toxin and FSH. Mol Cell Endocrinol 5:289-294

Gunsalus GL, Musto NA, Bardin CW, 1980. Bidirectional release of a Sertoli cell product, androgen binding protein, into the blood and seminiferous tubule. In: Steinberger A and Steinberger E (eds.), Testicular Development, Structure and Function. New York, Raven Press, pp. 291-298

Gunsalus GL, Musto NA, Bardin CW, Kunz HW, Gill TJ, 1985. Rats homozygous for the grc complex have defective transport of androgen-binding protein to the epididymis, but normal secretion into the blood. Biol Reprod 33:1057-1063

Hadley MA, Byers SW, Suarez-Quian CA, Kleinman HK, Dym M, 1985. Extracellular matrix regulates Sertoli cell differentiation, testicular cord formation, and germ cell development in vitro. J Cell Biol 101:1511-1522

Hadley MA, Dym D, 1987. Electron microscopic immunocytochemical localization of the extracellular matrix in the lamina propria of the rat testis. Biol Reprod

Hadley MA, Djakiew D, Byers SW, Dym M, 1987. Polarized secretion of androgen binding protein and transferrin by Sertoli cells grown in a bicameral culture system. Endocrinology 120:000-000

Hay ED, 1984. Cell-matrix interaction in the embryo: cell shape, cell surface, cell skeletons, and their role in differentiation. In: Trelstad RL (ed.), The Role of the Extracellular Matrix in Development. New York: Alan R. Liss, Inc., pp. 1-31

Huggenvik J, Sylvester SR, Griswold MD, 1984. Control of transferrin RNA synthesis in Sertoli cells. Ann NY Acad Sci 438:1-7

Janecki A, Steinberger A, 1987. Bipolar secretion of androgen-binding protein and transferrin by Sertoli cells cultured in a two compartment culture chamber. Endocrinology 120:291-298

Kleinman HK, Klebe RJ, Martin GR, 1981. Role of collagenous matrices in the adhesion and growth of cells. J Cell Biol 88:473-485

Lacroix M, Smith FE, Fritz IB, 1977. Secretion of plasminogen activator by Sertoli cell enriched cultures. Mol Cell Endocrinol 9:227-236

Misfeldt DS, Hamamoto ST, Pitelka DR, 1976. Transepithelial transport in cell culture. PNAS 73:1212-1216

Orth J, Christensen AK, 1977. Localization of ^{125}I-labeled FSH in the testes of hypophysectomized rats by autoradiography at the light and electron microscopic levels. Endocrinology 101:262-278

Simmons NL, Brown CDA, Rugg EL, 1984. The action of epinephrine on Madin-Darby canine kidney cells. Fed Proc 43:2225-2229

Skinner MK, Griswold MD, 1980. Sertoli cells synthesize and secrete transferrin-like protein. J Biol Chem 255:9523-9525

Skinner MK, Griswold MD, 1983. Sertoli cells synthesize and secrete a ceruloplasmin-like protein. Biol Reprod 28:1225-1229

Skinner MK, Tung PS, Fritz IB, 1985. Cooperativity between Sertoli cells and testicular peritubular cells in the production and disposition of extracellular matrix components. J Cell Biol 100:1941-1947

Spitz IM, Gunsalus GL, Mather JP, Thau P, Bardin WC, 1985. The effects of the indazole carboxylic acid derivative, tolnidamine, on testicular function: 1. Early changes in androgen binding protein secretion in the rat. J Androl 6:171-178

Steinberger A, Steinberger E, 1976. Secretion of an FSH-inhibiting factor by cultured Sertoli cells. Endocrinology 99:918-921

Steinberger A, Heindel JJ, Lindsay JN, Elkington JS, Sandborn BM, Steinberger E, 1975. Isolation and culture of FSH responsive Sertoli cells. Endocrinol Res Comm 2:261-272

Sternberger LA, 1979. Immunocytochemistry. NY: John Wiley and Sons.

Suarez-Quian CA, Hadley MA, Dym M, 1984. Effect of substrate on the shape of Sertoli cells "in vitro". Ann NY Acad Sci 438:417-434

Tung PS, Skinner MK, Fritz IB, 1984. Fibronectin synthesis is a marker for peritubular cell contaminants in Sertoli cell-enriched cultures. Biol Reprod 30:199-211

Van Renswoude J, Bridges KR, Harford JB, Klausner RD, 1982. Receptor-mediated endocytosis of transferrin and the uptake of Fe in K 562 cells: identification of a nonlysosomal acidic compartment. PNAS 79:6186-6190

Wright WW, Musto NA, Mather JP, Bardin CW, 1981. Sertoli cells secrete both testis specific and serum proteins. PNAS (USA) 78:7565-7569

Welsh MJ, Weibe JP, 1975. Rat Sertoli cells: a rapid method for obtaining viable cells. Endocrinology 96:618-624

SPERMATOGENESIS IN VITRO: SEARCHING FOR IN VIVO CORRELATES

Abraham L. Kierszenbaum, Munir Abdullah,
Hiroshi Ueda and Laura L. Tres

Departments of Anatomy and Pediatrics
University of North Carolina
Chapel Hill, NC 27514

INTRODUCTION

The development of methodology for the study of spermatogenesis in vitro has been a major task that remained elusive (Wolff and Haffen, 1965; Steinberger, 1975). Significant advances have been reported in the culture of meiotic cells of lilaceous plants (Hotta et al., 1966), and spermatogenic cells of Xenopus (Risley, 1983) and Drosophila (Liebrich, 1981). Attempts to maintain isolated spermatogenic cells viable for long periods of time (more than 48 h) have been relatively unsuccessful. In fact, the seminiferous epithelium contains spermatogenic cells that maintain spatial and functional relationships with Sertoli cells, and this cell-cell interaction is likely to play an important role in spermatogenesis in vivo and in vitro.

Among the various factors to be taken into consideration when developing methodology for the study of spermatogenesis in vitro are: (i) the minimal disruption of the structural relationship that Sertoli and spermatogenic cells maintain in vivo, (ii) the maintenance of Sertoli cell differentiated function, (iii) the use of serum-free culture media supplemented with factors known to facilitate long term viability and differentiation of spermatogenic cells, and (iv) the frequent or continuous replenishment of culture media and simultaneous removal of waste metabolic products from the immediate cell environment as it occurs in vivo.

The role of Sertoli cells in spermatogenesis has been increasingly emphasized throughout the years in many publications. However, although advances have been achieved in the hormonal regulation of Sertoli cell function, little is known regarding the precise mechanism by which Sertoli cells contribute to the spermatogenic process.

In this paper, we review work from our laboratories focused on the identification of bioregulatory molecules of Sertoli cell origin that have direct functional and structural roles in spermatogenesis. We have considered the possibility that a search for in vivo correlates of spermatogenesis in vitro is likely to provide understanding of the bioregulation of this complex and dynamic process. This search has resulted in the development of new methods (Kierszenbaum and Tres, 1987), and in the identification of proteins produced by Sertoli cells that behave as prospermatogenic molecules.

A brief description of the organization of the seminiferous tubule and dynamics of rat spermatogenesis is pertinent to this review. More detailed information can be found in original papers cited throughout this chapter.

STRUCTURAL ORGANIZATION OF THE SEMINIFEROUS TUBULE

Spermatogenesis results from the interaction of somatic Sertoli cells with spermatogenic cells. Spermatogenic cells are involved in mitotic (spermatogonia), meiotic (spermatocytes), and spermiogenic (spermatids) events. This somatic-spermatogenic cell interaction occurs in the seminiferous epithelium surrounded by a wall consisting of myoid-like cells with contractile properties. Both the seminiferous epithelium and the cellular wall contribute to the formation of the seminiferous tubule with a central luminal space (Fig. 1).

In the seminiferous epithelium, spermatogenic cells (spermatogonia, spermatocytes, and spermatids) display two important characteristics:

(i) spermatogenic cells are linked by intercellular bridges in a clone-like organization. Spermatogenic cell clones result from the progeny of single spermatogenic cells developing in an interconnected fashion. This structural clonal organization allows spermatogenic cells to display a degree of functional synchrony demonstrated by the DNA synthetic activities of spermatogonia and preleptotene spermatocytes both in vivo (Clermont, 1972) and in vitro (Tres and Kierszenbaum, 1983; Kierszenbaum and Tres, 1987).

(ii) spermatogenic cells are arranged in characteristic cellular compartments in the seminiferous epithelium (Kierszenbaum and Tres, 1981):

The mitotic cell compartment (Fig. 1) is closely associated with the peritubular cell wall and is represented by several classes of spermatogonia (types A, intermediate, and B, Clermont, 1972) resulting from successive mitotic cell divisions.

The meiotic cell compartment contains primary and secondary spermatocytes undergoing complex genetic and chromosomal events. These include the pairing of homologous autosomal and sex chromosomes, the exchange of genetic information between paired homologs, the disjunction of homologs following crossing-over, and final reduction of the genome size to a haploid state (Tres and Kierszenbaum, 1981).

The spermiogenic compartment consists of spermatids forming structures required for fertilization of the female gamete (acrosome) and for cell motility (tail or flagellum), and replacing somatic and testis-specific histones for more permanent basic protamines as chromatin condenses and changes its somatic nucleosomal, beaded structure into smooth chromatin fibers that associate side by side with each other (Kierszenbaum and Tres, 1975). These chromatin changes facilitate the structural stabilization of the spermatid genome in the epididymis.

The presence of basal occluding junctions between adjacent Sertoli cells allows a functional compartmentalization of the seminiferous epithelium into a basal compartment, containing spermatogonia and preleptotene spermatocytes being displaced to the meiotic cell compartment, and an adluminal compartment with spermatocytes and spermatids (Dym and Fawcett, 1970). The basal compartment is located below Sertoli occluding junctions whereas the adluminal compartment is located above the junctions.

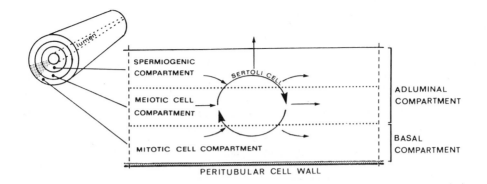

Fig. 1. Schematic representation of a segment of a seminiferous tubule with a central lumen and a peritubular cell wall. The seminiferous epithelium is represented by three cellular compartments: the <u>mitotic, meiotic, and spermiogenic cell compartments</u> containing spermatogonia, spermatocytes, and spermatids, respectively. The functional <u>basal compartment</u> comprises the spermatogonial mitotic cell compartment whereas the functional <u>adluminal compartment</u> comprises both the meiotic and spermiogenic cell compartments. A nondividing Sertoli cell provides a structural and functional link among the three spermatogenic cell compartments. A functionally cycling Sertoli cell is represented by clockwise arrows. Arrows projecting from and toward the Sertoli cell imply possible feedback mechanisms between Sertoli and associated spermatogenic cells (from Kierszenbaum and Tres, 1981).

Sertoli cells maintain a close relationship with proliferating and differentiating spermatogenic cells that is both spatial and functional. Because the association of spermatogenic cells with Sertoli cells varies along the length of the seminiferous tubule giving rise to distinct <u>cellular associations</u>, each representing a <u>stage of the spermatogenic cycle</u> (Clermont, 1972), it is possible to assume that the nondividing population of Sertoli cells responds in a cyclic manner to spermatogenic stage-related needs and to bioregulatory signals originating in the hypothalamic-pituitary axis (follicle-stimulating hormone, FSH; luteinizing hormone, LH) and in Leydig cells (testosterone) located outside the seminiferous tubule.

We have proposed that each cellular association, representing a stage of the spermatogenic cycle, can be regarded as a <u>functional domain</u> of Sertoli cells traversing alternate responsive and non-responsive phases of a functional cycle (Kierszenbaum and Tres, 1981). We also suggested the possibility that Sertoli cell <u>functional subdomains</u> can exist within a particular cellular association (Kierszenbaum and Tres, 1981). The concept of Sertoli cell subdomains takes into consideration the possible functional relationship that Sertoli cells maintain with individual spermatogonial, preleptotene, and spermatid clones in the intact testis (Dym and Fawcett, 1971) and in coculture (Tres and Kierszenbaum, 1983; Kierszenbaum and Tres, 1987).

Experimental data reviewed in this chapter provide further support to the existence of a Sertoli cell functional cycle that is dependent on the morphogenetic and functional activities of spermatogenic cells.

SPERMATOGENESIS IN THE RAT

The cellular stages that define the spermatogenic cycle in the rat were described about 35 years ago (Leblond and Clermont, 1952) and have been used ever since. Fourteen cellular associations can be observed in succession along the length of the rat seminiferous tubule and representing a <u>cycle</u> of the spermatogenic process. Each cellular association represents a stage of the spermatogenic cycle and its cellular composition and duration have been described (Clermont, 1972). Each spermatid developmental step, together with the various classes of spermatogonia and meiotic prophase spermatocytes (the predominant meiotic cell type found in all cellular associations), allows identification of each spermatogenic stage designated by Roman numeral (I to XIV). Nineteen spermatid developmental steps were defined by the shape and condensation of the nucleus and the morphogenetic sequence and topography of the acrosome (Perey et al., 1961).

Figure 2 is a modification of the original drawing reported by Perey et al. (1961). An imaginary axis between Stages VIII and IX (a stage where spermiation occurs) divides the rat spermatogenic cycle into two distinct groups of cellular associations: (i) a group consisting of Stages I to VIII in which at least <u>two</u> classes of spermatogonial cell types are observed (A, intermediate, or B), only <u>one</u> class of meiotic prophase spermatocytes, mainly pachytene spermatocytes, and <u>two</u> classes of developing spermatids (comprising <u>Golgi</u> and <u>cap</u> (steps 1-8), and <u>maturation</u> (steps 15-19) phases of spermiogenesis), and (ii) a group consisting of Stages IX to XIV in which only <u>one</u> class of spermatogonia type A is observed, <u>two</u> classes of spermatocytes (primary spermatocyte leptotene to diplotene stages; secondary spermatocytes in Stage XIV), and only <u>one</u> class of developing spermatids (<u>acrosomal</u> phase of spermiogenesis, steps 9-14).

RAT SPERMATOGENESIS

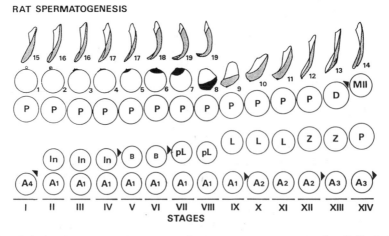

Fig. 2. Schematic representation of rat spermatogenesis (modified from Perey et al., 1961). Abbreviations: A, In, B: spermatogonial types A, intermediate, and B. pL, L, Z, P, and D: preleptotene, leptotene, zygotene, pachytene, and diplotene meiotic prophase I spermatocytes. M II: meiosis II (secondary spermatocytes). 1-19: steps of spermiogenesis. Black and gray areas indicate the acrosomal area. Arrowheads indicate the occurrence of cell divisions.

TEMPORAL APPEARANCE OF SPERMATOGENESIS IN THE RAT

The initiation and establishment of the spermatogenic process is dependent on continued and coordinated events in somatic and spermatogenic cells. Figure 3 illustrates the relationship between age (in days after birth) and the most advanced spermatogenic cell types found in cross sections of the developing seminiferous tubule. At birth (Day 0), the rat testis consists of seminiferous cords that contain only prospermatogonia and un-differentiated Sertoli cells (Clermont and Perey, 1957; Steinberger and Steinberger, 1970). Spermatogenesis starts by postnatal Day 4 and proceeds through Days 45-50 when the first spermatozoa appear in the seminiferous tubular lumen. By Day 65-70, all stages of the rat spermatogenic cycle are observed (Clermont and Perey, 1957; Knorr et al., 1970).

Coinciding with the onset of spermatogonial proliferation, Sertoli cells increase their population, and acquire structural and functional characte-ristics that are typical of the seminiferous epithelium: occluding junctions appearing between adjacent Sertoli cells contribute to the formation of a physiological barrier (Setchell, 1967) that separates the seminiferous epithelium into basal and adluminal compartments (Dym and Fawcett, 1970; Vitale et al., 1973).

Luminal spaces are visible in seminiferous tubules by Day 15 (Shabanowitz et al., 1986). By Day 20, the expansion of Sertoli cell population decreases (Nagy, 1972; Orth, 1982), as nonproliferating Sertoli cells establish close structural and functional relationships with developing spermatogenic cells that are characteristic of the adult testis.

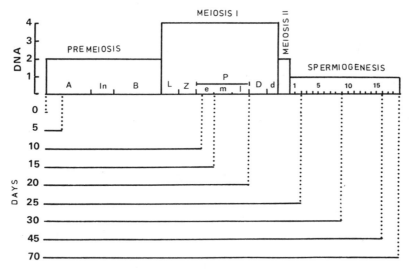

Fig. 3. Timing of spermatogenesis in the rat. The upper portion of the figure indicates the various stages of spermatogenesis. DNA content is indicated, being 1 = haploid DNA. Abbreviations: A, In, B: spermatogonia types A, intermediate, and B, respectively. L, Z, P (e, m, l), D, d: leptotene, zygotene, pachytene (early, middle, late), diplotene, diakinesis, respectively. Numbers in spermiogenesis box indicate steps of spermatid development. Dotted lines indicate most advanced spermatogenic cell type (from Kierszenbaum et al., 1986a; Shabanowitz et al., 1986).

SYSTEMS FOR THE STUDY OF SPERMATOGENESIS IN VITRO

The development of techniques for the isolation and culture of Sertoli cells has stimulated the search for factors that can maintain spermatogenic cells in long-term viable conditions when cocultured with Sertoli cells.

Figure 4 shows two major approaches for the study of spermatogenesis in vitro: (1) Sertoli-spermatogenic cell cocultures, and (2) isolated seminiferous tubular segments. In addition, enriched populations of spermatogenic cells can be fractionated and cells reaggregated with primary cultures of Sertoli cells.

Sertoli-spermatogenic cell cocultures maintained in Eagle's Minimum Essential Medium (lacking phenol red) supplemented with substances listed in Fig. 5, can sustain the proliferation and differentiation of spermatogenic cells for prolonged periods of time (up to 15 days) (Tres and Kierszenbaum, 1983; Tres et al., 1986). Serum-free, hormone/growth factor-supplemented medium (designated TKM, Tres et al., 1986) is changed twice daily. Discrete populations of spermatogenic cells are able to synchronously synthesize DNA (spermatogonia and preleptotene spermatocytes), and differentiate into meiosis II when cocultured with Sertoli cells (Tres and Kierszenbaum, 1983).

Isolated seminiferous tubules can be incubated in TKM for 24-48 h and provide useful experimental data. This approach has been very useful for determining the differentiated functions of Sertoli and peritubular cells in culture, and for monitoring the contribution of cell culture conditions to the protein secretory patterns that are cell specific (Kierszenbaum et al., 1986a).

A significant improvement in the manipulation of Sertoli and peritubular cell cultures, and cocultures of Sertoli cells with either peritubular cells or spermatogenic cells has been the development of an <u>automated cell perifusion system</u> (Kierszenbaum and Tres, 1987). Using this system, cells plated on a microporous substrate (HATF surfactant-free filters, Millipore) can be continuously perifused along the basal cell surfaces of the cultured cells using a tangential flow mode. The culture medium reaches the apical cell fluid compartment through small lateral openings, and cellular metabolites and secretory products reaching non-physiological concentrations can be gradually removed. Tangential flow prevents the effect of hydrostatic pressure on cultured cells and the build up of substances that are retained in the microporous substrate when a more conventional flow-through mode is used (Fig. 6 and Kierszenbaum and Tres, 1987).

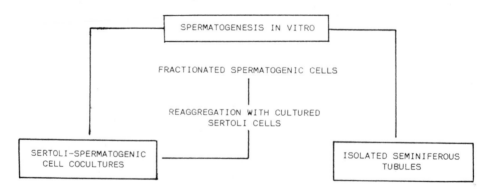

Fig. 4

COMPOSITION OF TKM (Tres et al., 1986; Tres and Kierszenbaum, 1986)

Eagle's Minimum Essential Medium (EMEM) lacking phenol red, supplemented with:

Insulin (zinc-free)	5 μg/ml
Transferrin	5 μg/ml
Epidermal growth factor	10 ng/ml
Growth hormone	133 μIU/ml
Follicle-stimulating hormone	500 ng/ml
Retinol	5 μM
Testosterone	0.1 μM
Dihydrotestosterone	0.1 μM
Glutamine	4 mM
Sodium pyruvate	1 mM
Non essential amino acids	0.1 mM
Penicillin	100 U/ml
Streptomycin	100 ug/ml

Fig. 5

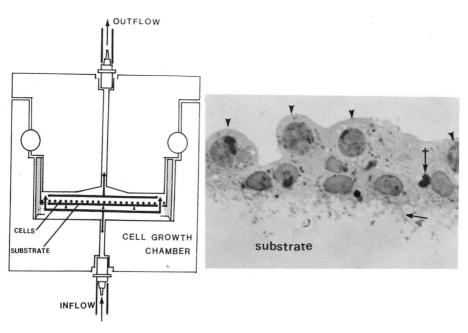

Fig. 6. Schematic diagram of the cell growth chamber containing cocultured Sertoli and spermatogenic cells on a microporous substrate. Arrows indicate the flow direction of culture medium. The photograph illustrates a coculture. Arrowheads indicate spermatogenic cells growing on apical surfaces of Sertoli cells with lipid droplets (crossed arrow) and basal cytoplasmic processes (arrow).

CELL CROSSCONTAMINATION AND FUNCTIONAL DIFFERENTIATION OF CULTURED CELLS

Two major difficulties are associated with the use of primary epithelial cell cultures: (1) the presence of contaminanting non-epithelial cells, and (2) the degree of differentiation that cells can maintain upon culturing.

These two aspects are of considerable importance for the identification of cell specific responses to hormones and growth factors.

A few secretory proteins have been used as markers for the identification of Sertoli and peritubular cells in culture. These include androgen-binding protein (ABP, Kierszenbaum et al., 1980) for Sertoli cells, and fibronectin for peritubular cells (Tung et al., 1984). However, neither ABP nor fibronectin provide data on the number of crosscontaminating cells. In fact, a reduced number of Sertoli cells in culture (Kierszenbaum et al., 1980) and in vivo (Feldman et al., 1981) display ABP immunoreactivity. The lack of ABP in the remaining cells may create uncertainty about their nature, although other morphological features, such as the presence of large lipid droplets in Sertoli cells (not observed in peritubular cells), facilitates their recognition (Fig. 6 and Kierszenbaum and Tres, 1981).

We have generated a monoclonal antibody (designated IFP-SC) against Triton-insoluble intermediate filament protein (IFP) extracted from cultured rat Sertoli cells (Kierszenbaum et al., 1986b). IFP-SC recognizes a form of intermediate filament protein in Sertoli cells (both in culture and in the intact seminiferous tubule) that is not present in peritubular cells. An advantage of IFP-SC is that the absolute number of Sertoli cells can be determined by monitoring by indirect immunofluorescence a structural component that is prevalent in Sertoli cells but not in peritubular cells.

IFP-SC generates a vimentin-like filamentous pattern that is not seen in peritubular cells (Kierszenbaum et al., 1986b). In addition, vimentin antibodies react with IFP in both Sertoli and peritubular cells in culture and in vivo using immunocytochemistry and Western blots. Clearly, monoclonal antibody IFP-SC differs from available polyclonal and monoclonal vimentin antibodies in its very specific property of generating a vimentin-like pattern in Sertoli cells. IFP-SC is routinely used in our laboratory to determine the purity of Sertoli and peritubular cell cultures.

A second relevant issue is the differentiation status of Sertoli cells in culture. We have reported that cultured Sertoli cells released from the physiological constraints that operate in vivo, modify their phenotypic expression and show functional characteristics that are newly expressed (Kierszenbaum et al., 1985) or enhanced (Lee et al., 1986) in vitro.

Sertoli cell function is regulated by FSH and androgens. FSH action on Sertoli cells is mediated, in part, by the activation of cyclic AMP-dependent protein kinases (Means et al., 1976; Fakunding and Means, 1977). Activation of cAMP-dependent protein kinases results in the phosphorylation of specific substrates which in turn modulate subsequent physiological responses. FSH (Spruill et al., 1983a) and calcium-calmodulin (Spruill et al., 1983b) stimulate the phosphorylation of various intracellular Sertoli cell proteins, including vimentin. FSH-dependent phosphorylation of vimentin is preceded by an increase in protein-bound cAMP levels (Spruill et al., 1984). Protein-bound cAMP represents the binding of cAMP to regulatory subunits of cAMP-dependent protein kinases (Spruill et al., 1984). Isoproterenol, a β-adrenergic analog, enhances cAMP levels in Sertoli cells.

The presence of β-adrenergic receptors in cultured Sertoli cells has been reported (Heindel et al., 1981) but the physiological significance of β-adrenergic agonist action on components of the seminiferous tubule is not well known. We asked the question: to what extent are isoproterenol responses induced in cultured Sertoli cells reproduced by this cell in the intact seminiferous tubule? (Kierszenbaum et al., 1985).

We have reported that FSH or isoproterenol acting on cultured Sertoli cells stimulate an increase in cAMP levels that activate cAMP-dependent protein kinases, and result in the phosphorylation of the IFP vimentin. However, in isolated seminiferous tubules from the same rats, FSH, but not isoproterenol, was able to phosphorylate vimentin (Fig. 7).

The radioligand (-)-[^{125}I]iodopindolol, a potent β-adrenergic antagonist with high affinity for β-adrenergic receptors and a low level of non specific binding (Toews and Perkins, 1984), was used to detect receptors in cultured Sertoli cells and in isolated seminiferous tubules. We reported that this radioligand has low binding affinity to homogenates of seminiferous tubules isolated from sexually-immature and mature rats. A time-course study demonstrated that isolated Sertoli cells at time 0 have almost undetectable binding affinity for radiolabeled iodopindolol. However, radioligand binding to Sertoli cells was detected at 4 h after plating, reaching maximum values after 12 h. Cycloheximide blocked the development of [^{125}I]iodopindolol binding to cultured Sertoli cells (Kierszenbaum et al., 1985).

Results of this study show that Sertoli cells in culture acquire both β-adrenergic receptors and responsiveness to β-adrenergic agonists.

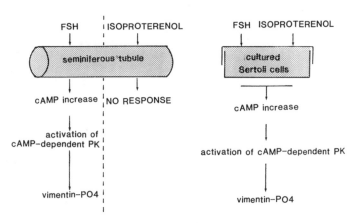

Fig. 7. Effect of FSH and isoproterenol on vimentin phosphorylation in seminiferous tubules and cultured Sertoli cells.

Another interesting observation is the contrasting levels of transferrin gene activity and expression in Sertoli cells in culture and in the intact seminiferous tubule. We have reported differences in the relative amount of newly-synthesized transferrin observed in fluids collected from seminiferous intertubular (TIF) and intratubular (SNF) compartments (Shabanowitz and and Kierszenbaum, 1986) following intratesticular [^{35}S]methionine labeling.

Figure 8 shows that, when [^{35}S]methionine was injected into testes of sexually-mature rats and allowed to circulate for 6 h, both the TIF and SNF fluids demonstrate radiolabeled transferrin as detected by two-dimensional polyacrylamide gel electrophoresis (PAGE), autoradiography, and transfer blots immunoreacted with anti-rat transferrin serum.

However, when testes were removed, injected with [^{35}S]methionine, and allowed to incubate in vitro for the same period of time, very little radio-labeled transferrin was observed in fluids from TIF and SNF prepared for auto-radiography, but abundant transferrin was detected in transfer blots immuno-reacted with rat anti-transferrin serum. These results suggested that: (i) a small amount of transferrin was produced in the testis, but (ii) bulk transferrin was of extratesticular origin, presumably from the liver.

A complementary approach was used to determine the levels of transferrin mRNA in total RNA from cultured rat Sertoli cells, whole testes, and isolated seminiferous tubules (Lee et al., 1986). This approach was based on results illustrated above, and on the two-dimensional PAGE and autoradiographic study of [^{35}S]methionine-labeled proteins from cultured Sertoli cells and isolated seminiferous tubules prepared from 25-day-old rats (Fig. 9).

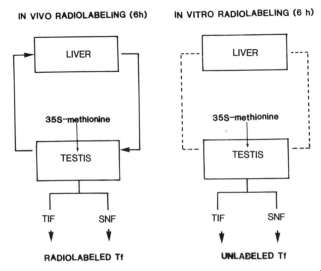

Fig. 8. Summary diagram illustrating differences in [^{35}S]methionine labeled and unlabeled transferrin recovered from testicular seminiferous intertubular (TIF) and intratubular (SNF) fluids following in vivo and in vitro radiolabeling of the testis.

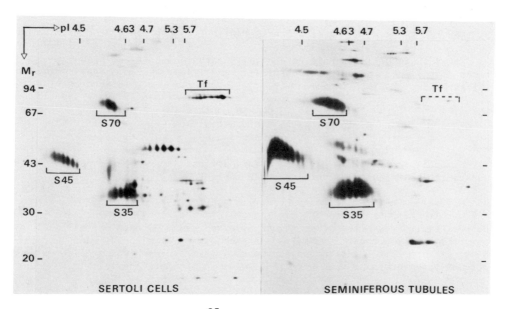

Fig. 9. Autoradiogram of [^{35}S]methionine-labeled proteins accumulated in the medium of cultured Sertoli cells and isolated seminiferous tubules from 25-day-old rats. S70, S45, and S35 designate a group of Sertoli cell secretory proteins. Tf indicates the position of transferrin (From Lee et al., 1986).

Radiolabeled transferrin is more conspicuous in cultured Sertoli cells than in incubated seminiferous tubules of the same rats. Furthermore, a striking contrast is the labeling intensity of three Sertoli cell secretory proteins (designated S70, S45, and S35) in Sertoli cell and seminiferous tubular samples. These results suggested that upon culturing, Sertoli cells enhanced transferrin expression and decreased the synthesis and secretion of proteins S70, S45, and S35.

Our next approach was the determination of transferrin mRNA levels. We used transfer blot hybridization of RNA to detect transferrin mRNA in total RNA extracted from Sertoli cells and seminiferous tubules by means of [^{32}P]-labeled rat transferrin cDNA (Lee et al., 1987). Using this method, we have observed that cultured Sertoli cells display a time-dependent activation of transferrin gene that is not observed in incubated seminiferous tubules (Lee et al., 1986), In addition, little transferrin mRNA was detected in whole testes and isolated seminiferous tubules of sexually immature and mature rats, when compared to abundant transferrin mRNA in cultured Sertoli cells and liver (Lee et al., 1986).

Contrasting levels of tranferrin mRNA have also been reported in cultured rat Sertoli cells and whole testes using a similar RNA transfer blot hybridization approach (Huggenvik et al., 1987). However, transferrin mRNA levels reported in poly(A)+ RNA extracted from whole testis (Huggenvik et al., 1987) are greater that the low amount reported by Lee et al., (1986) using total RNA from whole testes and isolated seminiferous tubules. The reason for this discrepancy is not clear at the present time.

PROTEIN SYNTHESIS AND SECRETION DURING TESTICULAR DEVELOPMENT

Rat Sertoli cells in culture and in isolated seminiferous tubules secrete three abundant proteins (DePhilip and Kierszenbaum, 1982) designated (Kierszenbaum et al., 1986a):

S70 (Mr/pI 72,000-70,000/4.4-4.6)
S45 (Mr/pI 45,000/4.0-4.5)
S35 (Mr/pI 35,000/4.4-4.65)

Similar proteins with different designations have been reported in the medium of rat Sertoli cell cultures (Wright et al., 1981; Kissinger et al., 1982).

Our experimental approach has been (i) the establishment of _in vivo_ correlates for secretory proteins observed in cultured Sertoli cells, and (ii) the study of their temporal appearance during the onset and establishment of the rat spermatogenic process. This approach was considered useful for selecting proteins with potential roles in spermatogenesis.

The presence of [^{35}S]methionine-labeled S70, S45, and S35 in the incubation medium of isolated seminiferous tubules deprived of their peritubular cell wall by enzymatic treatment indicates that these secretory proteins are not secreted by peritubular cells (Kierszenbaum et al., 1986a). Supporting this finding, recent immunocytochemical studies in our laboratory have shown that Sertoli cells, but not peritubular cells, have S70, S45, and S35 immunoreactive sites both _in vivo_ and in culture (Abdullah et al., submitted for publication).

An interesting aspect has been the demonstration that proteins S70, S45, and S35 display:

(i) a temporal appearance during the development of the rat seminiferous tubule, and
(ii) a correlation with the onset of spermatogenesis.

Figure 10 illustrates autoradiograms of [^{35}S]methionine-labeled proteins accumulated in the medium of incubated testes and seminiferous cords of 17.5-day-old rat fetuses and 0, 5, and 10 Day postnatal rats. S45 and S35 are begining to be visualized by Day 0 and are progressively enhanced from Day 5 on. A protein designated T35, of still undetermined cell origin, is synthesized by fetal testes, reaches a maximum of [^{35}S]methionine-labeling intensity at Day 0 postnatal, decreases at Days 5 and 10, and disappears thereafter (Fig. 10).

Autoradiograms displaying secretory proteins accumulated in the medium of incubated postnatal testes and seminiferous tubules isolated from pubertal rats are shown in Fig. 11. The rectangle in the autoradiogram labeled Day 0 indicates the position of Sertoli cell secretory proteins S70, S45, and S35 as well as protein designated T20 (undetermined cell origin).

A correlation with spermatogenic cell events occurring during this period of testicular development indicates the increasing radiolabeling intensity of S70, S45, and S35. High radiolabeling intensity is observed at Day 45 when step 16 of spermiogenesis appears in most seminiferous tubules. In addition, the gradual appearance of T20 is observed from about Day 27 on, thus suggesting that this protein is related to spermiogenesis.

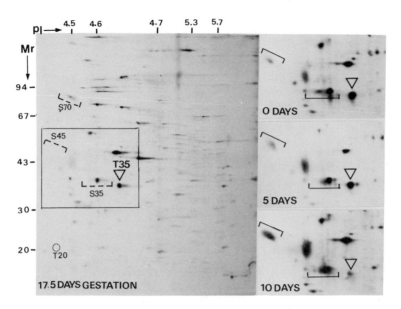

Fig. 10. Time course synthesis and secretion of proteins S45, S35, and T35 during fetal (17.5 days gestation) and early postnatal (Days 0, 5, and 10) stages of the development of the seminiferous tubule. Brackets with dashed lines indicate the expected position of S70, S45, and S35. Open arrowheads identify the location of T35. An open circle identifies the expected position of T20 (see Fig. 11). The rectangular box indicates the area of the autoradiogram depicted for 0, 5, and 10 Days (from Shabanowitz et al., 1986).

CYCLIC BEHAVIOR OF RAT SERTOLI CELL SECRETORY PROTEINS DURING SPERMATOGENESIS

Seminiferous tubular segments corresponding to four distinct rat spermatogenic stages were selected because of their different transillumination patterns (DePhilip et al., 1982). A pool of segments corresponding to the same stage was labeled with [^{35}S]methionine, and the radiolabeled proteins accumulated in the incubation medium were resolved by two-dimensional PAGE and autoradiography (Shabanowitz et al., 1986).

In spermatogenic Stage IV, neither S70, S45, nor S35 were detected in the autoradiograms. However, T20 was visible (Fig. 12). In spermatogenic Stage VII-VIII, S70, S45, and S35 were observed, but the intensity of T20 was very low. In spermatogenic Stages X and XIV, both S45 and S35 were detected but S70 was not. Differences in the labeling intensity of T20 can be detected in these two spermatogenic stages. It is interesting to point out that S45 and S35 always appear together. As will be discussed later, S45 and S35 are

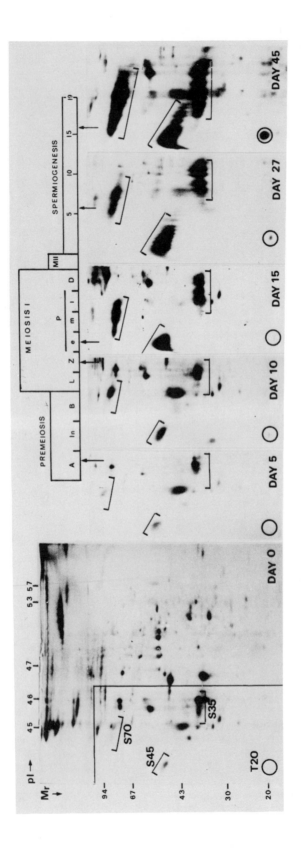

Fig. 11. Time course synthesis and secretion of S70, S45, and S35 (brackets) during postnatal and pubertal stages of seminiferous tubular development. For the use of abbreviations in the top diagram, see legend of Fig. 3. Arrows indicate the most advanced spermatogenic cell type observed at the indicated days in each of the autoradiograms (from Shabanowitz et al., 1986).

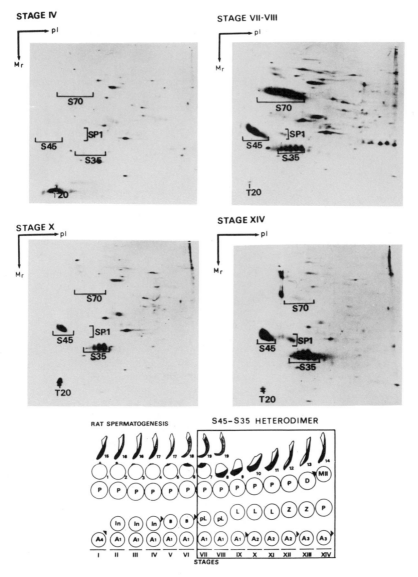

Fig. 12. Cyclic behavior of Sertoli cell secretory proteins S70, S45, and S35 during rat spermatogenesis. The diagram at the bottom indicates the spermatogenic stages in which the secretion of S45-S35 heterodimeric protein is predominant. Variations in SP1, a common secretory protein of Sertoli and peritubular cells is also indicated (from Shabanowitz et al., 1986).

components of a protein designated S45-S35 heterodimeric protein (Abdullah et al., submitted for publication). A similar finding was reported by Kissinger et al. (1982) and Sylvester et al. (1984).

An interesting relationship between S45-S35 heterodimeric protein and spermiogenesis emerges from an analysis of the spermatogenic cell composition of stages of spermatogenesis in which this protein appears. S45-S35 heterodimeric protein is present in spermatogenic stages where spermatids start to elongate and assemble various components of the flagellum or tail. These events start about Stage VII-VIII, continue until Stage XIV, and comprises the differentiation of spermatids steps 7 to 14. In fact, studies to be reported elsewhere (Abdullah et al., submitted for publication) support a role for S45-S35 heterodimeric protein in both the formation of outer dense fibers, a characteristic component of spermatozoal tail, and the acrosome.

The relationship that S70 maintains with spermatogenic cellular events during the fourteen stages of rat spermatogenesis is presently under investigation in our laboratory. Preliminary studies indicate a role for S70 in spermiogenesis.

PURIFICATION OF SERTOLI CELL SECRETORY PROTEINS

In the preceding sections of this chapter, we have shown that Sertoli cell secretory proteins S70, S45, and S35 are:

(i) · Sertoli cell-specific (DePhilip and Kierszenbaum, 1982; Kierszenbaum et al., 1986a),
(ii) temporally expressed during the development of the seminiferous tubule (Shabanowitz et al., 1986), and
(iii) spermatogenic stage-dependent (Shabanowitz et al., 1986).

We have have considered the possibility that S70 and S45-S35 heterodimeric protein are members of a family of prospermatogenic proteins (PSP) with functional and structural roles in spermatogenesis (Fig. 13).

An example of a functional PSP is Sc/Sm, a somatomedin-like peptide that accumulates in the medium of cultured Sertoli cells and has binding affinity for a selective group of pachytene spermatocytes as shown by immunofluorescence (Tres et al., 1986). Sc/Sm has been partially purified from medium conditioned by cultured rat Sertoli cells (Smith et al., 1987) but its possible effects on spermatogenic cells have not been determined. Sc/Sm has mitogenic effect on BALB/c 3T3 fibroblasts and binds to a carrier or binding protein produced by Sertoli cells (Smith et al., 1987). Sertoli cells have receptor sites for somatomedin-C (Borland et al., 1984) and, presumably, for its Sertoli cell equivalent Sc/Sm.

Sc/Sm peptide differs from the seminiferous growth factor reported by Feig et al. (1983). Seminiferous growth factor of Sertoli cell origin has Mr/pI 15,700/4.8-5.8. Native Sc/Sm (growth factor plus binding protein) has Mr/pI 25,000/7.5) and free Sc/Sm, Mr/pI >6,200/9.8 (Smith et al., 1987).

Protein purification of Sertoli cell secretory proteins S70, S45, and S35 was carried out under the assumption that these proteins have an important role in spermatogenesis, mainly in spermiogenesis.

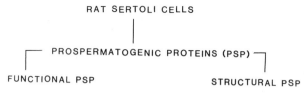

Fig. 13

FETUIN, AN ALBUMIN-LIKE CONTAMINANT

During our protein purification studies, we detected an abundant acidic glycoprotein with Mr/pI 68,000/4.2-3.5 (designated 68 kD) that accumulated in serum-free medium of cocultured Sertoli and spermatogenic cells (Abdullah et al., 1986). This 68kD protein disappeared gradually 5-6 days after plating.

Protein purification using anion exchange chromatography yielded a protein consisting of a major component (Mr 68,000) and two minor, low Mr components (Mr 62,000 and 56,000). A polyclonal antibody raised in rabbit against the Mr 68,000 component crossreacted with the two minor proteins, thus suggesting that they were antigenically related to the 68 kD protein.

The 68 kD protein was identified as fetuin, the major component of fetal bovine serum, based on similar Mr/pI, immunoreactivity, and trypsin inhibitory activity (Abdullah et al., 1986). The presence of 68 kD/fetuin in serum-free medium conditioned by Sertoli-spermatogenic cell cocultures was attributed to the initial use of fetal bovine serum-supplemented medium for achieving rapid attachment to the plastic substrate.

Figure 14 shows the time course accumulation of 68kD/fetuin in the medium of Sertoli-spermatogenic cell cocultures plated in TKM and for only 12 h in EMEM supplemented with 10% fetal bovine serum. Figure 15 demonstrates that Sertoli-spermatogenic cell cocultures plated for 6-12 h in EMEM supplemented with 10% fetal bovine serum, accumulated 68 kD/fetuin in serum-free TKM during a 5-6 day period. This protein was absent from media recovered from cells plated briefly in EMEM supplemented with adult bovine, chicken, or horse serum (Abdullah et al., 1986).

[^{14}C]amino acid-labeled secretory proteins accumulated in the medium of cocultured rat Sertoli and spermatogenic cells were resolved by two-dimensional PAGE. The radiolabeled sample contained unlabeled 68 kD/fetuin and was visualized by Coomassie Blue staining. Figure 16 shows that the 68 kD/fetuin tracer does not correlate with any of the radiolabeled secretory proteins, including adjacent S70, S45, and S35. Similar results were obtained when 68 kD/fetuin tracer and radiolabeled proteins were transferred to nitrocellulose film, immunoreacted with anti-68 kD/fetuin rabbit antibody and then exposed to X-ray film (not shown).

Indirect immunofluorescent studies using anti-68 kD/fetuin serum have shown immunoreactive sites in both Sertoli and spermatogenic cells (Fig.

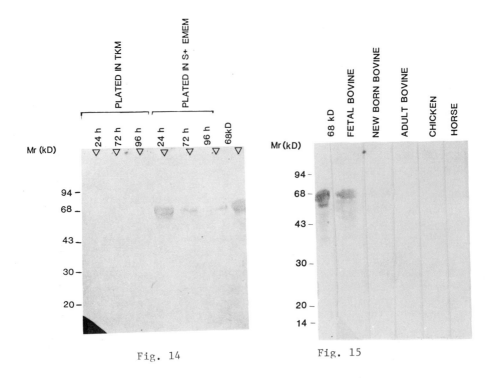

Fig. 14   Fig. 15

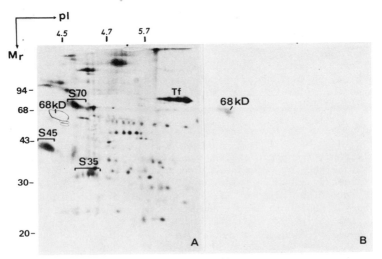

Fig. 16. A. Autoradiogram showing [14C]amino acid mixture-labeled secretory proteins S70, S45, S35 (brackets), and transferrin (Tf). The position of 68 kD is indicated in A and B . B is the corresponding Coomassie Blue-stained gel of the autoradiogram illustrated in A (from Abdullah et al., 1986).

17). In spermatogenic cells, immunofluorescent granules were associated with cell surfaces of most meiotic prophase spermatocytes (Fig. 17,C). A spot-like immunofluorescent product was seen in pachytene spermatocytes (Fig. 17,A and B). Cocultured Sertoli cells displayed a diffuse granular cytoplasmic pattern (Fig. 17, D). These immunofluorescent findings were detected in Sertoli-spermatogenic cells plated in fetal bovine serum-supplemented medium. Cocultures plated and maintained in TKM were not immunoreactive for 68 kD/ fetuin protein (Abdullah et al., 1986).

We have concluded that when seminiferous epithelial cell samples are plated for a brief period of time (6-12 h) in culture medium supplemented with fetal bovine serum, 68 kD/fetuin is taken up by Sertoli and spermato-genic cells, and then released into serum-free culture medium (Abdullah et al., 1986).

Fetuin contains growth promoting activity (Salomon et al., 1982; Feldman et al., 1984). The accepted view is that fetuin is a carrier protein for growth factors delivered to cultured cells. In fact, crude fetuin prepara-tions are associated with platelet-derived growth factor (Libby et al., 1985), and embryonin (Salomon et al., 1984; Feldman et al., 1984), two proteins with growth-promoting activity. Figure 18 provides an interpreta-tion of the pathway of fetuin in Sertoli cells and pachytene spermatocytes.

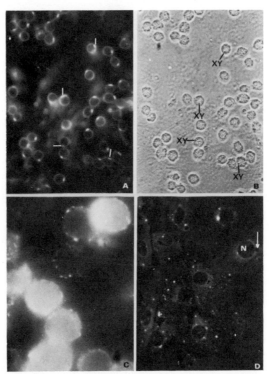

Fig. 17. Indirect immunofluorescent localization of 68 kD/fetuin in pachytene spermatocytes (A). B is the corresponding phase-contrast microscopy. The location of the XY pair is different from the spot-like, fetuin immunofluorescent site in the same cells. C shows cell surface location of fetuin in sperma-tocytes. D illustrates the diffuse cytoplasmic distribution of fetuin in Sertoli cells. N, Nucleus; arrow: lipid droplet in a Sertoli cell (from Abdullah et al., 1986).

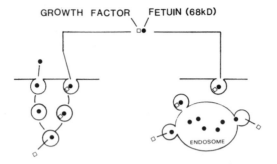

GROWTH FACTOR   FETUIN (68kD)

SERTOLI CELL   PACHYTENE SPERMATOCYTE

Fig. 18. Possible pathway of the 68 kD/fetuin complex in Sertoli cell and pachytene spermatocyte. The uptake of the complex by a Sertoli cell is followed by the release of growth factor from the complex and exocytosis of fetuin. In pachytene spermatocytes, fetuin may be accumulated in an endosome.

COPURIFICATION OF SERTOLI CELL SECRETORY PROTEIN S70, S45, AND S35.

We have used a high performance liquid chromatography (HPLC) procedure for purification of Sertoli cell secretory proteins S70, S45, and S35 accumulated in the medium of cultured Sertoli cells (manuscript submitted for publication). Sequential gel permeation, anion exchange, and hydroxylapatite columns have allowed the copurification of S70 and S45-S35 heterodimeric protein using unlabeled and [^{35}S]methionine-labeled Sertoli cell culture media.

Unreduced protein samples from hydroxylapatite column show that S70 and S45-S35 heterodimeric protein comigrate by gel electrophoresis in the Mr 72,000-70,000 range. Denatured and reduced protein samples resolved by gel electrophoresis showed that S70 retained its original electrophoretic migration position (Mr 70,000) whereas the S45 and S35 components of the heterodimeric protein migrated to Mr 45,000 and 35,000 positions, respectively (Fig. 19).

IMMUNOCYTOCHEMICAL LOCALIZATION OF S70, S45, AND S35 IN CULTURED SERTOLI CELLS AND SPERMATOZOA

Antibodies were raised in rabbits against proteins S70, S45, and S35 and a mixture of S45-S35. In Western blot experiments, anti-S70 serum crossreacted strongly with S70, S35, and less strongly with S45. Anti-S45 and anti-S35 sera reacted with their putative antigens and S70. Antisera to S45-S35 heterodimeric protein crossreacted strongly with S45 and S35 and less strongly with S70 (Abdullah et al., submitted for publication). We have concluded that S70, S45, and S35 share antigenic determinants with each other.

Immunocytochemical studies were carried out using cultured Sertoli cells. The immunohistochemical localization of S70, S45, and S35 in intact testes,

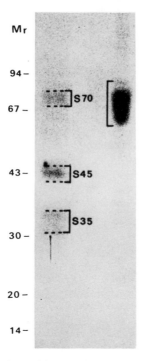

Fig. 19. Sertoli cell secretory proteins S70, S45,
and S35 purified by HPLC. Left lane:
Reduced protein sample. Right lane: Unreduced protein sample . Samples from
hydroxylapatite column (15 ug per lane).
Coomassie Blue staining.

spermatids, and spermatozoa (collected from the terminal portion of the epididymal duct) will be reported elsewhere (Abdullah et al., submitted for publication). Briefly, S70 immunoreactivity predominates in the acrosomal component
of differentiating spermatids, whereas S45 and S35 immunoreactivity is observed
in the tail or flagellum of maturing spermatids and epididymal spermatozoa.

Tail immunoreactivity is strong in the principal piece and correlates with
the location of outer dense fibers (ODF) (manuscript in preparation). ODF
are associated with microtubule doublets and first appear during Steps 8-14
of spermiogenesis (Irons and Clermont, 1982). It is interesting that spermiogenic Steps 8-14 are observed in spermatogenic Stages VIII-XIV, and that the
secretion of S45-S35 heterodimeric protein predominates in these particular
stageś (see Fig. 12). Crossreactivity of ODF with anti-S45 and S-35 has
been confirmed by immunogold labeling at the electron microscopic level
(manuscript in preparation).

In cultured Sertoli cells, immunoreactivity appears as a diffuse distribution of granules of variable diameter. This immunoreactive pattern can
be detected with either anti-S70, anti-S45, or anti-S35 sera. Figure 20,A
shows cultured rat Sertoli cells immunoreacted with antisera to a mixture of
S45 and S35. When examined at high resolution level using an immunogold procedure, immunoreactivity predominates in dense bodies, although immunoreactive small vesicles can also be detected (Fig. 20,B).

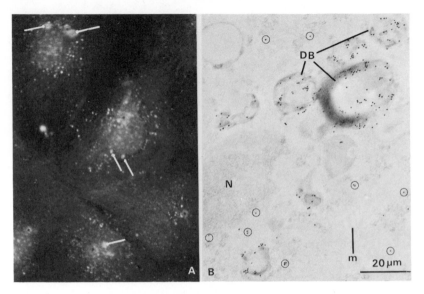

Fig. 20. A. Immunofluorescent localization of S45-S35 heterodimeric protein in cultured Sertoli cells. White lines indicate large immunofluorescent granules corresponding to dense bodies. B. Immunogold electron microscopic visualization of S45-S35 heterodimeric protein in Sertoli cells showing predominance in dense bodies (DB). Circles indicate small secretory vesicle. N, nucleus; m, mitochondrion.

We have used automated cell perifusion to determine whether the continuous removal of culture medium had an effect on the amount of immuno-reactive proteins in Sertoli cell cytoplasm. Sertoli cells were plated on surfactant-free microporous filters and cultures were maintained either under static conditions (TKM replaced every 24 h) or placed inside cell growth chambers (Fig. 6), and perifused with TKM during 24 h at a flow rate of 2 ml/h. After 24 h of culture or perifusion, samples were fixed and processed for immunocytochemistry using a plastic embedding medium (Kierszenbaum et al., 1986b).

Under static conditions, Sertoli cells display abundant S70 immuno-reactive granules of variable diameter (Fig. 21). However, under perifusion conditions, S70 immunoreactivity appears localized around the nuclear region of Sertoli cells (Fig. 21), and no immunoreactive dense bodies are detected.

CONCLUDING REMARKS

We have provided evidence in support of in vivo correlates for Sertoli cell secretory proteins S70 and S45-S35 heterodimeric protein previously detected in the medium of cultured Sertoli cells. Preliminary studies show that antibodies against these Sertoli cell secretory proteins crossreact with components of the acrosome and spermatozoal tail. Further studies should determine whether spermatids are capable of synthesizing proteins similar to S70 and S45-S35 heterodimeric protein or depend on this Sertoli cell "prospermatogenic proteins" for spermiogenesis.

Methods for the study of spermatogenesis _in vitro_ under static and automated cell perifusion conditions have been developed. Considerable potential exists for determining the effect of hormones and growth factors on the mitotic expansion of spermatogonial and preleptotene cell clones _in vitro_ under conditions that mimic the _in vivo_ situation. Probes for monitoring proliferation and differentiation of spermatogenic cells _in vitro_ and _in vivo_ are available. These include cDNAs for proteins that are specific for Sertoli cells and spermatogenic cells, including S70, S45-S35 heterodimeric protein (manuscript in preparation) as well as testis-specific or enriched histone variants (Kin, Y-J, Hwang, I, Tres, LL, Kierszenbaum, AL, Chae, C-B, submitted for publication).

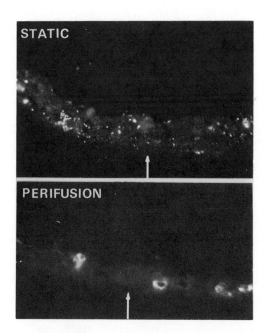

Fig. 21. Localization of S70 in cultured Sertoli cells maintained under static and perifusion conditions. The arrows indicate the basal surfaces of the cultured cells.

ACKNOWLEDGEMENTS

We want to thank Ms Liu Ping for her expert assistance in the immuno-fluorescent studies. We also want to express our appreciation to members of our laboratories who, through the years, have provided a productive and stimulating environment. Their contributions are cited in this chapter. This work was supported by USPHS grants HD11884 (ALK) and HD18315 (LLT) and, in part, by P-30 HD18968 (Cell Culture Core Facility, The Laboratories for Reproductive Biology).

REFERENCES

Abdullah, M, Crowell, JA, Tres, LL, Kierszenbaum, AL, 1986. Fetuin: A serum component associated with rat Sertoli and spermatogenic cells in coculture. J Cell Physiol 127:463-72

Borland, K, Mita, M, Oppenheimer, CL, Blinderman, LA, Massague, J, Hall, PH, Czech, MP, 1984. The actions of insulin-like growth factors I and II on cultured Sertoli cells. Endocrinology 114:240-46

Clermont, Y, 1972. Kinetics of spermatogenesis in mammals: seminiferous epithelium cycle and spermatogonial renewal. Physiol Rev 52:198-236

Clermont, Y, Perey, R, 1957. Quantitative study of the cell population of the seminiferous tubules in immature rats. Am J Anat 100:241-67

DePhilip, RM, Kierszenbaum, AL, 1982. Hormonal regulation of protein synthesis, secretion, and phosphorylation in cultured Sertoli cells. Proc Natl Acad Sci USA 79:6551-55

DePhilip, RM, Tres, LL, Kierszenbaum, AL, 1982. Stage-specific protein synthesis during rat spermatogenesis. Exp Cell Res 142:489-94

Dym, M, Fawcett, DW, 1970. The blood-testis barrier in the rat and the physiological compartmentalization of the seminiferous epithelium. Biol Reprod 3:308-26

Fakunding, JL, Means, AR, 1977. Characterization and follicle stimulating hormone activation of Sertoli cell cyclic AMP-dependent protein kinases. Endocrinology 101:1358-68

Feldman, SR, Gonias, SL, Ney, KA, Pratt, CW, Pizzo, SV, 1984. Identification of "embryonin" as bovine α_2-macroglobulin. J Biol Chem 259:4458-62

Feldman, M, Lea, OA, Petrusz, P, Tres, LL, Kierszenbaum, AL, French, FS, 1981. Androgen binding proteins: purification from rat epididymis, characterization, and immunocytochemical localization. J Biol Chem 256:5170-5175

Feig, L, Klagsbrun, GT, Bellvé, AR, 1983. Mitogenic polypeptide from the mammalian seminiferous epithelium: biochemical characterization and partial purification. J Cell Biol 97:1435-43

Heindel, JJ, Steinberger, A, Strada, SJ, 1981. Identification and characterization of a β-adrenergic receptor in rat Sertoli cells. Mol Cell Endocrinol 22:349-58

Hotta, Y, Ito, M, Stern, H, 1966. Synthesis of DNA during meiosis. Proc Natl Acad Sci USA 56:1184-91

Huggenvik, JI, Idzerda, RL, Haywood, L, Lee, DC, McKnight, GS, Griswold, MD, 1987. Transferrin messenger ribonucleic acid: molecular cloning and hormonal regulation in rat Sertoli cells. Endocrinology 120:332-40

Kierszenbaum, AL, Crowell, JA, Shabanowitz, RB, DePhilip, RM, Tres, LL, 1986a. Protein secretory patterns of rat Sertoli and peritubular cells are influenced by culture conditions. Biol Reprod 35:239-51

Kierszenbaum, AL, Crowell, JA, Shabanowitz, RB, Smith, EP, Spruill, WA, Tres, LL, 1986b. A monoclonal antibody recognizes a form of intermediate filament protein in rat Sertoli cells that is not present in seminiferous peritubular cells. Biol Reprod 35:227-38

Kierszenbaum, AL, Feldman, M, Lea, O, Spruill, WA, Tres, LL, Petrusz, P, French, FS, 1980. Localization of androgen-binding protein in proliferating Sertoli cells in culture. Proc Natl Acad Sci USA 77:5322-26

Kierszenbaum, AL, Spruill, WA, White, MG, Tres, LL. Perkins, JP, 1985. Rat Sertoli cells acquire a β-adrenergic response during primary culture. Proc Natl Acad Sci USA 82:2049-53

Kierszenbaum, AL, Tres, LL, 1975. Structural and transcriptional features of the mouse spermatid genome. J Cell Biol 65:258-70

Kierszenbaum, AL, Tres, LL, 1981. The structural and functional cycle of Sertoli cells in culture. In: Jagiello, G, Vogel, HJ (eds.), Bioregulators of Reproduction. New York: Academic Press, pp. 207-28

Kierszenbaum, AL, Tres, LL, 1987. An automated perifusion system for the study of rat spermatogenesis in vitro. Ann NY Acad Sci (in press)

Kissinger, C, Skinner, MK, Griswold, MD, 1982. Analysis of Sertoli cell secreted proteins by two-dimensional gel electrophoresis. Biol Reprod 27:233-40

Knorr, DW, Vanha-Perttula, T, Lipsett, MB, 1970. Structure and function of rat testis through pubescence. Endocrinology 86:1298-304

Leblond, CP, Clermont, Y, 1952. Definition of the stages of the cycle of the seminiferous epithelium in the rat. Ann NY Acad Sci 55:548-73

Lee, NT, Chae, CB, Kierszenbaum, AL, 1986. Contrasting levels of transferrin gene activity in cultured rat Sertoli cells and intact seminiferous tubules. Proc Natl Acad Sci USA 83:8177-81

Liebrich, W, 1981. In vitro spermatogenesis in Drosophila. I. Development of isolated spermatocyte cysts from wild-type D. hydei. Cell Tissue Res 220:251-62

Libby, P, Raines, EW, Cullinane, PM, Ross, R, 1985. Analysis of the mitogenic effect of fetuin preparations on arterial smooth muscle cells: The role of contaminant platelet-derived growth factor. J Cell Physiol 125:357-66

Means, AR, Fakunding, JL, Tindall, DJ, 1976. Follicle stimulating hormone regulation of protein kinase activity and protein synthesis in testis. Biol Reprod 14:54-63

Nagy, F, 1972. Cell division kinetics and DNA synthesis in the immature Sertoli cells of the rat testis. J Reprod Fertil 28:389-95

Orth, JM, 1982. Proliferation of Sertoli cells in fetal and postnatal rats: A quantitative autoradiographic study. Anat Rec 203:485-92

Perey, B, Clermont, Y, Leblond, CP, 1961. The wave of the seminiferous epithelium in the rat. Am J Anat 108:47-77

Risley, MS, 1983. Spermatogenic cell differentiation in vitro. Gamete Res 4:331-46

Salomon, DS, Bano, M, Smith, KB, Kidwell, WR, 1982. Isolation and characterization of a growth factor (embryonin) from bovine fetuin which resembles α2-macroglobulin. J Biol Chem 257: 14093-101

Setchell, BP, 1967. The blood-testicular fluid barrier in sheep. J Physiol 189:63-65

Shabanowitz, RB, Kierszenbaum, AL, 1986. Newly synthesized proteins in seminiferous intertubular and intratubular compartments of the rat testis. Biol Reprod 35:179-90

Shabanowitz, RB, DePhilip, RM, Crowell, JA, Tres, LL, Kierszenbaum, AL, 1986. Temporal appearance and cyclic behavior of Sertoli cell-specific secretory proteins during the development of the rat seminiferous tubule. Biol Reprod 35:745-60

Smith, EP, Svoboda, ME, Van Wyk, JJ, Kierszenbaum, AL, Tres, LL, 1987. Partial characterization of a somatomedin-like peptide from the medium of cultured rat Sertoli cells. Endocrinology 120:186-93

Spruill, WA, Steiner, AL, Tres, LL, Kierszenbaum, AL, 1983a. Follicle-stimulating hormone-dependent phosphorylation of vimentin in cultures of rat Sertoli cells. Proc Natl Acad Sci USA 80:993-97

Spruill, WA, Zysk, JR, Tres, LL, Kierszenbaum, AL, 1983b. Calcium/calmodulin-dependent phosphoylation of vimentin in rat Sertoli cells. Proc Natl Acad Sci USA 80:760-64

Steinberger, E, Steinerger, A, 1970. Replication patterns of Sertoli cells in maturing rats testis in vivo and in organ culture. Biol Reprod 18: 329-45

Sylvester, SR, Skinner, MK, Griswold, MD, 1984. A sulfated glycoprotein synthesized by Sertoli cells and by epididymal cells is a component of the sperm membrane. Biol Reprod 31:1087-1101

Tres, LL, Kierszenbaum, AL, 1983. Viability of rat spermatogenic cells in vitro is facilitated by their coculture with Sertoli cells in serum-free hormone-supplemented medium. Proc Natl Acad Sci USA 80:3377-81

Tres, LL, Kierszenbaum, AL, 1986. Effect of substrate and medium composition on rat Sertoli-spermatogenic cell cocultures maintained in an automated cell perifusion system. IV Decennial Tissue Culture Association Review Conference (Hershey, PA), p. 24 (abstract)

Tres, LL, Smith, EP, Van Wyk, JJ, Kierszenbaum, AL, 1986. Immunoreactive sites and accumulation of somatomedin-C in rat Sertoli-spermatogenic cell co-cultures. Exp Cell Res 162:33-50

Toews, ML, Perkins, JP, 1984. Agonist-induced changes in β-adrenergic receptors on intact cells. J Biol Chem 259:2227-35

Tung, PS, Skinner, MK, Fritz, IB, 1984. Fibronectin synthesis is a marker for peritubular cell contaminants in Sertoli cell-enriched cultures. Biol Reprod 30:199-211

Vitale, R, Fawcett, DW, Dym, M, 1973. The normal development of the blood testis barrier and the effects of clomiphene and estrogen treatment. Anat Rec 176:333-44

Wolff, E, Haffen, K, 1965. Germ cells and gonads. In: Willmer, EN (ed.), Cells and Tissues in Culture, Vol 2. New York: Academic Press, pp. 697-743

Wright, WW, Musto, NA, Mather, JP, Bardin, CW, 1981. Sertoli cells secrete both testis-specific and serum proteins. Proc Natl Acad Sci USA 78:7565-69

HORMONAL REGULATION OF SERTOLI CELL FUNCTION

B.M. Sanborn, L.A. Caston, S.W. Buzek, and K.K. Ussuf

Department of Biochemistry and Molecular
Biology and Department of Obstetrics
Gynecology and Reproductive Sciences
University of Texas Medical School at Houston
Houston, TX 77030

INTRODUCTION

The differentiation which occurs in the process of spermatogenesis, i.e., the series of events whereby a primitive germ cell gives rise to a spermatozoon, appears to be controlled in part by hormones and in part by a programmed series of events (Steinberger and Steinberger, 1972; Clermont and Harvey, 1967; Chemes et al., 1979; Chowdhury and Tcholakian, 1979). Key steps in spermatocyte division and spermiation are influenced by FSH, qualitatively in the immature rat and quantitatively in the adult. The completion of meiosis is critically dependent in vivo upon testosterone (or LH). FSH and testosterone (T) are required for both the initiation of spermatogenesis in immature rats and its reinitiation after prolonged regression such as occurs in adults after hypophysectomy. In contrast, qualitative maintenance of spermatogenesis in the adult can be achieved with testosterone alone. Other conditions, such as Vitamin A or insulin deficiency, which also cause arrest of spermatogenesis, may be the result of direct influences on cells in the seminiferous epithelium or indirect influences via changes in T or FSH (Murray et al., 1981; Unni et al., 1983).

The evidence accumulated to date has clearly established the Sertoli cell, a nondividing somatic cell in the seminiferous epithelium, as of major importance in controlling a number of aspects of testicular function. The tight Sertoli-Sertoli junctions contribute to the formation of the specialized environment of the intraluminal compartment, and specialized complexes are characteristic of areas of physical contact with germ cells (Russell and Peterson, 1985). Sertoli cell secretory activity is potentially bidirectional, but in vivo and in vitro studies indicate that, for androgen binding protein (ABP) and transferrin, at least, there is differential secretion into the basal and adluminal compartments (Mather et al., 1983; Cheng et al., 1986a; Janecki and Steinberger, 1986, 1987; Djakiew et al., 1986). A number of biochemical properties of Sertoli cells have been described and the influences of hormones on the expression of these parameters studied. It is the intention of this review to summarize the major conclusions that can be drawn from our own work and that of others concerning the hormonal regulation of Sertoli cell function and its relationship to the control of spermatogenesis in the rat, with primary emphasis on the effects of T and FSH.

Sertoli Cell Functions and Their Hormonal Regulation Encompass a Broad Spectrum of Responses

At the outset, it is important to realize that a broad spectrum of biochemical activities have been ascribed, directly or indirectly, to Sertoli cells. These are summarized in Table 1. They include general metabolic properties, specific enzymatic and metabolic activities, and secretory products.

Hormones influence a wide variety of Sertoli cell responses, as indicated also in Table 1. These include effects on such diverse parameters as cell division and DNA synthesis, general RNA and protein synthesis, and the activity of a number of specific intracellular proteins and secretory products.

A Variety of Experimental Approaches Have Been Used to Characterize Sertoli Cell Functions and Their Hormonal Regulation

Interpretation of data concerning the hormonal regulation of Sertoli cell function must be made in the context of the biological (i.e., in vivo or in vitro) and maturational state of the cell at the time of the measurement. With these points in mind, four general approaches have been used to assess Sertoli cell function. Specific examples of these approaches can be found in the references in Table 1.

(1) Parameters ascribed exclusively to Sertoli cells have been measured in total testis. This approach, while simple and least manipulative, has both positive and negative aspects. Because the Sertoli cell is in its natural environment, local testicular influences are present, and the activity measured represents the actual in vivo activity. On the other hand, because of the complex, changing cellular composition of the testis, expression of the data per mg protein or DNA is ambiguous. Thus estimation of maturational or hormonal patterns per Sertoli cell can only be extrapolated from the data, using information regarding the number of such cells per testis at a given age or in a given hormonal state. Variations on this approach use germ cell-depleted testis or seminiferous tubules, but here again, depending on the parameter being measured, some of the same considerations pertain, since a variety of cell types are still present.

(2) Sertoli cells have been isolated from animals of different ages, or hormonal status, purified by a brief period in culture, and parameters measured. Differential enzymatic digestion removes contamination with interstitial and peritubular cells (Steinberger et al., 1975; Tung et al., 1984) and yields a crude Sertoli cell preparation. Cell culture for 2-3 days, coupled with a brief hypotonic treatment, removes residual germ cells (Galdieri et al., 1981; Wagle et al., 1986). In our studies, we have chosen to use cells from animals between the ages of 15 and 40 days of age. These cells possess the morphological characteristics of Sertoli cells and are thus relatively easy to identify. For 15-35 day cells, we have achieved >95% purity by these methods (Sanborn et al., 1986). Residual germ cell contamination becomes an increasing problem after about 35 days of age.

The biochemical functions measured using the culture system can be attributed directly to the Sertoli cell and expressed per Sertoli cell protein or DNA, which is a distinct advantage. A disadvantage of this approach is that the Sertoli cells have been removed from their natural

TABLE 1   BIOCHEMICAL PARAMETERS MEASURED IN SERTOLI CELLS AND THE EFFECT OF HORMONAL STIMULATION

Parameter	FSH	T	A	Insulin	Other	References
General metabolism/metabolites						
DNA synthesis/content	+			+	IGF-I, II	1-6
RNA synthesis/content	+	+		+		4,7-9
Protein synthesis	+	+			IGF-II	5,7,8,10-13
Polyamine synthesis	+	+			*	14,15
Lipid synthesis/metabolism	+			+		16-18
Glycoprotein synthesis	+		+		Ad	11,19-21
Steroid metabolism	+				Ad	22-29
O2 consumption						30
Glucose, pyruvate metabolism	+			+		30-32
Glucose transport	+			+	IGF-I	33,34
Amino acid metabolism	+					32,35
Ca2+ redistribution	+					7,36
Protein phosphorylation	+					12,37
Nucleotide pools						9,38
Porphyrins						39
Enzymes						
γ-Glutamyl transpeptidase	+	±				8,40
Na/K ATPase						41,42
Succinate dehydrogenase						41,43
5'-Nucleosidase						41
Transglutaminase						44
Adenylyl cyclase/cAMP	+				Ad,αMSH,ACTH	7,26,45-47
Lactate dehydrogenase					Ad	8,43
Myoinositol-1-P synthase						48
Myoinositol-1-phosphatase						48
Aldose reductase						49
Glucose-6-phosphatase						41

(continued)

Table 1. (Continued)

Parameter	FSH	T	A	Insulin	Other	References
Acid phosphatase						41,50,51
Glucose-6-P dehydrogenase					Ad	43
Phosphodiesterases	+					7,36,52,53
Glycogen phosphorylase	+					54
Plasminogen activator	+					41,55
Branched amino acid aminotransferase						35
Ornithine decarboxylase	±					15,56
RNA polymerase II	+	+				57-59
Heme oxygenase						39
Biliverdin reductase						39
NADPH-cytochrome c reductase						39
Aminolevulinic acid synthetase						39
Uro-I synthetase						39
cAMP-dependent protein kinase	+					7,36,60
Phospholipid-dependent protein kinase						61
Receptors/binding proteins						
FSH receptor	+					7,62
Androgen receptor	+	+				59,63
Retinol binding protein	+					64
Vitamin D$_3$ receptor						65
β-Adrenergic receptor						26,46
Adenosine receptor						66
Opiate receptor						67
IGF receptors						5
cAMP binding proteins						60
GTP binding site						103
Other Proteins						
Calmodulin						36
Cell-specific chromatin antigens						68
Spectrin-like protein						69
Vimentin	+					12,37

Parameter	FSH	T	A	Insulin	Other	References
Proteoglycans	+					70
Nuclear basic proteins	+					4
High mobility group proteins						71
Actin, microfilaments						7
Tubulin, microtubules						7
Protein kinase inhibitor	+				Ad	7,72
Mitogenic peptide						73
Secreted proteins						
Androgen binding protein (ABP)	+	+	+	+	*,Ad	7,8,59,74-76
Transferrin	+	+	+	+	MSA,EGF	8,75-79
Inhibin	+	±				80-82
Sulfated glycoproteins 1&2	+	+				76,83
Clustrin	+					84,85
Plasminogen activator						41,55
Mullerian inhibitory substance						76,86
Testibumin	+	+				87
Lactalbumin-like substance						88
Ceruloplasmin						89
Growth-stimulating factor	+	+	+	+	*	90
Somatomedin C (IGF-I)						91
LHRH-like substance						92
H-Y antigen						93
CMB-2	+					94
Cyclic protein-2						95
Leydig cell stimulatory factors						6,96
Other proteins						see Table 2
Secreted metabolites/other substances						
Collagen IV						97
Laminin						97
Myoinositol						48

(continued)

Table 1. (Continued)

Parameter	FSH	T	A	Insulin	Other EGF,IGF-II	References
Lactate, pyruvate	+			+	*	5,8,98-101
Polyamines						14,15
Lipids,fatty acids						16,18
Proteoglycans						70,102
4-Methyl-2-oxovalerate						39

*-Combinations of hormones; A-vitamin A, Ad-adrenergic agents. References are representative and not exhaustive. (1) Orth, 1982; (2) Orth, 1986; (3) Griswold et al., 1977; (4) Welch et al., 1979; (5) Borland et al., 1984; (6) Saez et al., 1986; (7) Means et al., 1978, 1980; (8) Sanborn et al., 1986; (9) Griswold and Merryweather, 1982; (10) Galdieri et al., 1981; (11) Wilson and Griswold, 1979; (12) DePhilip and Kierszenbaum, 1982; (13) Wright et al., 1981; (14) Tsai and Lin, 1985; (15) Swift and Dias, 1987; (16) Beckman and Coniglio, 1979, 1980; (17) Fisher and Coniglio, 1983; (18) Jutte et al., 1985; (19) Galdieri and Monaco, 1983; (20) Galdieri and Nistico, 1986; (21) Griswold et al., 1986; (22) Wiebe, et al., 1982; (23) Tcholakian and Steinberger, 1979; (24) Tcholakian et al., 1983; (25) Wiebe et al., 1980; (26) Verhoeven et al., 1979; (27) Welsh and Wiebe, 1978; (28) Armstrong and Dorrington, 1977; (29) Tcholakian et al., 1984; (30) Floridi et al., 1983; (31) Robinson and Fritz, 1981; (32) Grootegoed et al., 1986; (33) Hall and Mita, 1984; (34) Mita et al., 1985; (35) Grootegoed et al., 1985; (36) Means et al., 1979; (37) Ireland et al., 1986; (38) Lamb et al., 1982; (39) Maines, 1984; (40) Lu and Steinberger, 1977; (41) Marzowski et al., 1985; (42) Muffly et al., 1985; (43) Cameron and Snydle, 1985; (44) Dias, 1985; (45) Steinberger et al., 1978a,b; (46) Heindel et al., 1981; (47) Bardin et al., 1984; (48) Robinson and Fritz, 1979; (49) Ludvigson et al., 1982; (50) Vanha-Perttula et al., 1986; (51) Chemes, 1986; (52) Verhoeven et al., 1981; (53) Conti et al., 1982; (54) Slaughter and Means, 1983; (55) Lacroix and Fritz, 1982; (56) Madhubala et al., 1986; (57) Sanborn et al., 1980; (58) Lamb et al., 1981; (59) Sanborn et al., 1984; (60) Spruill et al., 1984; (61) Galdieri et al., 1986; (62) Steinberger et al., 1978b; (63) Sanborn et al., 1985; (64) Huggenvik and Griswold, 1981; (65) Merke et al., 1985; (66) Monaco and Conti, 1986; (67) Fabbri et al., 1985; (68) Schmidt et al., 1983; (69) Borland et al., 1985; (70) Elkington and Fritz, 1980; (71) Bucci et al., 1986; (72) Tash et al., 1981; (73) Bellve et al., 1984; (74) Fritz, 1982; (75) Perez-Infante et al., 1986; (76) Griswold, 1987; (77) Skinner and Griswold, 1980; (78) Skinner and Griswold, 1982, 1983a; (79) Huggenvik et al., 1987; (80) Steinberger et al., 1983; (81) deJong and Robertson, 1985; (82) Ultee-VanGessel et al., 1986; Verhoeven and Franchimont, 1983; (83) Sylvester et al., 1984; (84) Blaschuk and Fritz, 1984; (85) Cheng, 1986; (86) Josso, 1986; (87) Bardin et al., 1986; (88) Skinner and Fritz, 1986; (89) Skinner and Griswold, 1983b; (90) Holmes et al., 1986; (91) Smith et al., 1987; (92) Sharpe et al., 1981; (93) Brunner et al., 1984; (94) Lee WM et al., 1986; (95) Wright and Luzarraga, 1986; (96) Verhoeven and Cailleau, 1986; (97) Skinner et al., 1985; (98) Oonk et al., 1985; (99) Jutte et al., 1983; (100) Mita et al., 1982; (101) Mallea et al., 1986; (102) Skinner and Fritz, 1985c; (103) Fletcher and Reichert, 1986.

566

environment. Normally influenced by signals provided by blood-borne substances, germ, peritubular and Leydig cells as well as the extracellular matrix, the cells may exhibit altered activity in the artificial environment of the culture dish. On a small scale, attempts at reproducing the microenvironment have had some success (Janecki and Steinberger, 1986, 1987; Byers et al., 1986). To date, however, for larger scale biochemical measurements these approaches have not proved practical or economically feasible. It is also possible that the cyclical properties of Sertoli cell function (Parvinen, 1982), as well as differences due to maturational age or hormonal status, would be lost in the culture state. Later in this manuscript we will present some data which suggest that qualitative differences are maintained, at least for up to 3 days in culture. Other morphological (Spruill et al., 1981) and immunological (Kierszenbaum et al., 1980) data also suggest uniqueness amongst Sertoli cells in a given culture. Consequently, while far from ideal, this approach has provided some useful information on Sertoli cell function.

(3) The ability of Sertoli cells in culture to mature or be influenced by hormones with respect to a given biochemical parameter has been measured. These studies have shown qualitative patterns similar to in vivo maturational patterns, but in general, the magnitude of the responses have been lower. The same advantages/disadvantanges as outlined in (2) above pertain. In addition, it must be kept in mind that the state of the cells in the dish is influenced to some extent by the hormonal status of the animal from which they are derived. This may magnify or blunt the in vitro response.

(4) Some parameters have been measured in situ in testicular sections using histochemical techniques. These are by nature semiquantitative. With the recent development of molecular probes for Sertoli cell products, in situ hybridization techniques could yield valuable information on Sertoli cell function in the testicular environment.

Sertoli Cell Functional Parameters Undergo Marked Differential Maturation

Since hormonal responsiveness is dependent, in part, on maturational state, it is important to understand Sertoli cell functions in the context of that maturation. Definition of the biochemical maturation of Sertoli cell function, particularly over the period of time when the first wave of spermatogenesis is taking place and the first meiotic divisions are being completed, i.e., before ~ 45 days of age, has yielded new insights with potential relevance to the control of spermatogenesis. The early waves of spermatogenesis are different from the adult waves both quantitatively (Alger et al., 1985) and with respect to hormonal dependence (Steinberger and Steinberger, 1972; Clermont and Harvey, 1967; Chemes et al., 1979). From the marked morphological differentiation of the rat Sertoli cell between day 1 and day 40 of postnatal life (Ramos and Dym, 1979; Hatier and Grignon, 1980), it is perhaps predictable that biochemical maturation should also occur over this period. The differentiation of nuclear morphology occurs by day 35 (Ramos and Dym, 1979) and the tight junctions constituting the blood-testis barrier form by days 16-19 (Vitale et al., 1973). Cytoskeletal differentiation occurs by day 14, with vimentin-type replacing cytokeratin-type intermediate filaments (Paranko et al., 1986). By 33-35 days of age, the cells are morphologically similar to those from adult animals (Chemes et al., 1979).

Using approach (2), i.e, in vivo maturation coupled with short term culture, we noted that Sertoli cell total protein/DNA, secreted protein/DNA and RNA/DNA ratios increased between 15 and 25 days of age (Sanborn et al., 1986), consistent with the increase in cytoplasmic/nuclear volume ratio which takes place over this period. Qualitatively, there were few changes in ^{35}S-Met-labeled soluble cellular or secreted proteins, as judged by 2D-PAGE, although the complexity of the isoelectric forms increased with age. However, two bands were consistently absent from 2D-PAGE maps of 15 day Sertoli cell secretory proteins but present in comparable maps from 25 and 35 day cells. We have termed these protein bands Band 1 (66 kd, pI 6.0-6.8) and Band 2 (56 kd, pI 5.3-6.0). This evidence suggested that Sertoli cell biochemical maturation was not a coordinated process but rather was differentially regulated.

Additional information came from the examination of specific biochemical parameters. The membrane-bound enzyme γ-glutamyl transpeptidase (GGT) increased approximately 6-fold, both per mg protein and per mg DNA in cells between 15 and 25 days of age (Sanborn et al., 1986). Sertoli cell androgen receptor, both that measured in the cytosol and nuclear fractions, increased per mg DNA between 15 and 25 days of age so that total receptor increased 2.5-fold over that interval (Buzek et al., 1987). Cytosol concentrations did not increase per mg protein, however, suggesting that the increase in cytosol receptor per cell simply paralleled the increase in cytoplasmic/nuclear ratio. A third pattern was exhibited by lactate dehydrogenase, which did not change in concentration per mg DNA and decreased slightly per mg soluble protein. These data again illustrate the fact that Sertoli cell biochemical maturation is markedly different, depending on the parameter measured.

The maturation of specific secretory activities was also regulated differentially. Lactate secretion, expressed per mg secreted protein or per mg DNA, increased markedly between 15 and 25 days of age. ABP and transferrin secretion increased between 15 and 25 days, and decreased somewhat, expressed per mg secreted protein, between 25 and 35 days. Examination of the secretion of these substances over the 3 day culture period also revealed different responses to the culture environment (Sanborn et al., 1986). Lactate secretion was essentially constant (a 2/1 ratio comparing concentrations in day 0-2 and day 2-3 media), while ABP secretion was lower in the second interval (5.8/1 ratio) and transferrin secretion greater (1/1 ratio).

Where other data are available, they are generally consistent with the above findings. Kissinger et al. (1982) noted that 2D-PAGE patterns of Sertoli cell secretory proteins were qualitatively similar from 20 and 60 day cells, although some differences between 10 and 20 day cell patterns were noted. Shabanowitz et al. (1986) noted maturational increases in the complexity of specific glycoprotein bands. Testicular ABP content increased markedly between 18 and 60 days of age (Gunsalus et al., 1978). The testicular concentration of ABP per mg of protein increased between 10 and 21 days of age and then decreased until 40 days (Tindall et al., 1985). Sertoli cells placed in culture for 4 days showed a pattern of increasing ABP secretion per cell when collected from 7-31 day animals (Rich et al., 1983). When left in culture, ABP secretion from Sertoli cells from 7-13 day animals continued to increase until a total age of 20 days and then declined.

Lactate secretion continued to increase per mg total protein in cells from animals between 21 and 42 days of age (Jutte et al., 1983). Skinner and Griswold (1982) reported that transferrin secretion per Sertoli cell increased progressively in 10, 20, and 60 day cells. In contrast, Perez-Infante et al. (1986) found that transferrin secretion per cell was relatively constant in cells from 7-16 day rats, with a marked increase between 19 and 31 days.

GGT concentrations per mg protein in the testis increased 3 to 6-fold from day 1 to day 20, then leveled off, suggestive of an enzyme residing in a nondividing receptor population such as the Sertoli cell (Hodgen and Sherins, 1973; Lu and Steinberger, 1977). Two reports have questioned the exclusive localization of this enzyme in Sertoli cells (Dierickx and Verhoeven, 1980; Niemi and Setchell, 1986). Nonetheless, the data of Lu and Steinberger (1977), Lipshultz et al. (1982) and ourselves (Sanborn et al., 1986) have demonstrated consistently high levels of enzyme in the rat Sertoli cell. Lu and Steinberger reported evidence of maturation of GGT per mg protein in the culture dish between day 30 at plating and 7 subsequent days (equivalent to 37 days). We have also noted an increase in GGT in 20 day cells in culture in the presence of insulin (10 µg/ml), epidermal growth factor (2.5 µg/ml), and transferrin (5 µg/ml) for 5-15 days (37 ± 10^a, 74 ± 6^a, and 130 ± 30^b units per mg DNA after 5, 10, and 15 days in culture (25, 30 and 35 total days),respectively (superscripts different at P< 0.05, n=3)).

We have measured androgen receptor concentration per testis by nuclear exchange assay and noted a gradual increase in concentration between 15 and 45 days of age, while the concentration per mg DNA dropped between 15 and 25 days of age and then increased (Buzek and Sanborn, 1986). The data obtained with Sertoli cells in culture are consistent with the testicular data. However, since Leydig and peritubular cells also contain androgen receptors (Gulizia et al., 1983; Nakhla et al., 1984; Isomaa et al., 1985; Buzek and Sanborn, 1986), the complete explanation of the testicular data requires maturational assessment of receptor concentrations in these other cell types as well.

Other maturational studies illustrate the broad diversity in Sertoli cell maturation. Solari and Fritz (1978) noted morphological maturation of Sertoli cells from 10 day rats in vitro. Borland et al. (1986) reported differences in the way that 15 and 25 day cells interact with a collagen matrix. FSH receptor concentration in cultured Sertoli cells increased per mg DNA and declined slightly per mg protein between 13 and 60 days (Steinberger et al., 1979); the decline in Sertoli cell clusters was more dramatic (Salhanick and Wiebe, 1980). Protein kinase inhibitor and phosphodiesterase forms and activity changed as a function of age in Sertoli cell-enriched testes (Means et al., 1980).Tsai and Lin (1985) have reported maturational changes in Sertoli cell polyamine concentrations, with greater spermine concentrations in 46 than in 13 day old Sertoli cells. Vanha-Perttula et al. (1986) reported a decline in Sertoli cell acid phosphatase between 10 and 30 days of age, and Galdieri et al. (1986) found a decline in phospholipid-dependent protein kinase and altered intracellular distribution between 8 and 20 days.

Steroidogenic activity in Sertoli cells was age-dependent, with the most pronounced activity between 10 and 17 days of age and conversion of testosterone and progesterone to C_{19} and C_{21} steroids (Welsh and Wiebe, 1978). 5α-Dihydroprogesterone production peaked around 17 days of age and declined thereafter (Weibe, 1982). The ability to metabolize T to estradiol or estradiol-like metabolites was age-dependent, with significant activity at 5 days and virtually no such activity by 30 days of age (Armstrong and Dorrington, 1977). Inhibin secretion was independent of age of the Sertoli cell between 18 and 90 days (Steinberger et al., 1983).

Lamb et al. (1982) noted an increase in RNA content per Sertoli cell between 18 and 35 days. There was also an increase in size of the nucleotide pools, with disproportionate increases in UTP and CTP per mg DNA.

It should also be noted that hormonal responsiveness of the Sertoli cell is altered by its maturational status. The decrease in FSH responsiveness with respect to events related to cAMP elevation beginning at 19 d of age and resulting in refractoriness by 36 days of age is well documented in the rat (Steinberger et al., 1978a,b; Means et al., 1980) and could be prevented by hypophysectomy (Heindel et al., 1982). The refractoriness has been attributed to changes in receptor-cyclase coupling, phosphodiesterase activity, and protein kinase inhibitor concentration (Means et al., 1980; Conti et al., 1982). However, effects of FSH on Ca^{2+} distribution and cell shape are independent of age (Means et al., 1978). Less is known about the maturation of responsiveness to T, although 18 day Sertoli cells responded to T with an increase in RNA polymerase II activity, as did 25 day cells (Sanborn et al., 1983), despite a lower concentration of androgen receptor in the former.

The Effects of Hormones on Sertoli Cell Function are Both Quantitative and Qualitative in Nature

Studies of hormonal influences on Sertoli cell function have employed the same general approaches as described above. FSH has been reported to affect cell shape (Hutson, 1978; Means et al., 1980; Spruill et al., 1981), while T and vitamin A affect nuclear morphology (Chemes et al., 1979; Unni et al., 1983). Junctional complex formation may be facilitated by T (Means et al., 1980; Hagenas et al., 1978) but does not require FSH or T for maintenance. Hormonal effects on Sertoli cell biochemical parameters are summarized in Table 1. They encompass such diverse effects as changes in phosphorylation or glycosylation of specific proteins and changes in activity and/or concentration of proteins, only some of which have been characterized.

We have used primarily the in vivo hormonal manipulation model of hypophysectomy at an age prior to full maturation (20 days) followed by regression with or without hormonal supplementation and, in some cases, reintroduction of hormones at a later date. In the case of ABP secretion, where we already knew the in vivo responses to such manipulation (Fritz, 1982; Sanborn et al., 1983), the secretion of ABP by the Sertoli cells in culture exhibited the same qualitative changes (discussed in more detail below). Consequently, we concluded that the approach was useful for studying hormonal influences on these cells.

Total, soluble, and secretory protein/DNA ratios were decreased significantly in cells from hypophysectomized rats, as was ^{35}S-Met incorporation into insoluble cellular protein and secreted protein (Sanborn et al., 1986) (Figure 1). Treatment with TP or TP+FSH, from the time of hypophysectomy, partially prevented the decline in these parameters. 2D-PAGE patterns of ^{35}S-Met-labeled soluble intracellular and secreted proteins (loading equal numbers of counts on each gel in order to detect qualitative differences) showed remarkably few qualitative changes following hormonal deprivation and/or hormonal supplementation (Sanborn et al., 1986; Figure 2A,B). Band 1, noted previously to be maturation-dependent, was missing in maps from cells isolated from hormone-deprived and hormone-supplemented animals. It may be regulated by factors other than gonadotropins. In contrast, although Band 2 was absent in maps from cells from hypophysectomized rats, the most acidic protein in Band 2 was observed following treatment with TP or FSH+TP. The bands tentatively identified as the subunits of a major sulfated glycoprotein SGP-2 by analogy with the electrophoretic mobility characteristics described by Kissinger et al. (1982) did not show major

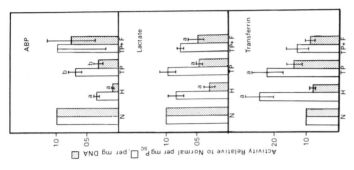

Figure 1B Effect of hypophysectomy (H) and daily maintenance with TP or TP plus FSH (TP + F) on ABP, lactate, and transferrin secretion. Experimental details are described in Figure 1A. Data are expressed as ratios, relative to normal values (N), of values per mg P (open bars) and per mg DNA (stipled bars). a) Different from N at P < 0.05; b) different from N or H at P < 0.05. (Sanborn et al., 1986).

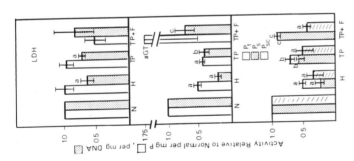

Figure 1A Effect of hypophysectomy (H) and daily treatment with testosterone propionate (TP) or TP plus FSH as described in Figure 2 on lactate dehydrogenase (LDH) and GGT (γGT) activity in Sertoli cells in culture. Data were expressed relative to intact control values (N) in each experiment and pooled as the mean ± SE. In the LDH and γGT panels, the open bars represent data expressed per mg soluble (P_s) or total (P_t) protein respectively, and the stippled bars represent data expressed per mg DNA. In the lower panel, the open bars represent secreted P_t to DNA ratios, the stippled bars represent P_s to DNA ratios, and the hatched bars represent secreted protein (P_{sc}) to DNA ratios. a) Different from N; b) different from N or H; c) different from N, H, or TP (P < 0.05). (Sanborn et al., 1986).

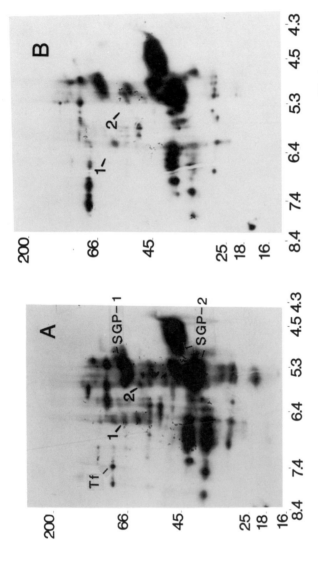

Figure 2A-B Two-dimensional gel autoradiographs of newly synthesized, [^{35}S]methionine-labeled proteins secreted by Sertoli cells from normal 35-day-old rats (A) and rats hypophysectomized on day 20 and killed 15 days later (B) (Sanborn et al., 1986). Approximately 200,000 cpm were loaded onto each gel. The autoradiographs are representative of data from four experiments, with duplicate or triplicate individual determinations per experiment. Only changes that were consistent within and between experiments are discussed in the text. The positions where Band 1 (mol wt, 66,000; pI 6.0-6.8) and Band 2 (mol wt, 56,000; pI 5.3-6.0) appear or should appear are indicated. Transferrin (Tf) was identified by Western blot. The tentative identification of SGP-1 and SGP-2 subunits as suggested from the data of Kissinger et al., (1982) are also indicated.

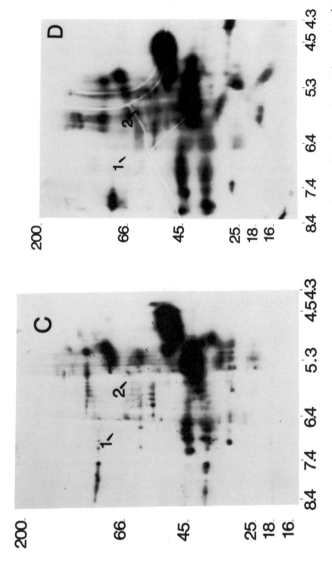

Figure 2C-D Autoradiographs of secretory proteins as in Figure 2A from hypophysectomized rats treated with TP(0.5 mg/day) (C) or TP + FSH (35 μg twice daily) (D) (Sanborn et al., 1986).

changes in intensity or isoelectric heterogeneity in cells from rats of different hormonal status. Only a small decrease in intensity in the putative SGP-1 was noted. These data led us to conclude that, aside from the overall quantitative changes, the qualitative influences of FSH and T on Sertoli cell function represent changes in relatively minor abundance proteins. In the case of T, this situation contrasts markedly with the androgenic control of major secretory proteins in organs such as the prostate (Parker et al., 1978) and is more similar to the influence of androgens on the kidney, which involves induction of specific enzymes, none of which predominate as major proteins (Catterall et al., 1986).

We noted that the concentrations of a number of specific Sertoli cell proteins were affected by hormones. Sertoli cell androgen receptor concentrations were reduced by hypophysectomy and partially maintained by TP , FSH, and FSH+TP (Sanborn et al., 1984). Sertoli cell GGT concentrations were likewise reduced by hypophysectomy and partially maintained by TP, FSH, or FSH+TP (Sanborn et al., 1986; Caston and Sanborn, submitted) (Figure 1A). In addition, we have recently found that testicular GGT is increased following 15 days of hormonal ablation by administration of 35 µg FSH twice daily (5.8 ± 0.7 and 18.7 ± 1.2 units/testis without and with FSH supplementation, respectively; n=3, P<0.01). In contrast, lactate dehydrogenase concentration per mg protein was not affected by hypophysectomy and decreased only slightly per mg DNA. Treatment with TP or TP+FSH did not prevent the decrease. Thus, as with maturational effects, hormones affect specific parameters.

Secretory proteins are perhaps the best studied of Sertoli cell functions. Even 2D-PAGE of ^{35}S-Met-labeled proteins has been reported to be not sensitive enough to detect ABP, yet there is considerable evidence to indicate that ABP secretion is regulated both in vivo and in vitro by FSH and T (Fritz, 1982; Sanborn et al., 1983; Griswold, 1987). In fact, it was the confirmation using cultured Sertoli cells from hormonally manipulated animals that hypophysectomy decreased ABP secretion but maintenance with FSH or TP partially prevented this decrease (Figure 1B) that convinced us that in vivo hormonal influences persisted for long enough in culture to make this a useful experimental model. In the same cells, lactate secretion per DNA was decreased and not affected by FSH + TP treatment. Transferrin secretion per mg DNA was not affected by hypophysectomy. Consequently, since the secreted protein/DNA ratio decreased with hypophysectomy, transferrin secretion per mg secreted protein increased following hypophysectomy but was restored to control levels by treatment with FSH+TP (Sanborn et al., 1986).

These findings differ somewhat from those obtained using in vitro hormone stimulation, where lactate secretion per mg protein was variably affected by FSH and insulin (Robinson and Fritz, 1981; Jutte et al., 1983; Mita et al., 1982). Skinner and Griswold (1982) reported that transferrin secretion per cell was increased by FSH, T, retinol, and insulin in a complex fashion, while Perez-Infante et al. (1986) found its secretion was primarily increased by vitamin A. Huggenvik et al. (1987) have recently reported that transferrin mRNA in cultured cells was increased by retinol and insulin to a greater extent than by FSH. NT Lee et al. (1986) found little evidence for expression of the Sertoli cell gene in vivo. ABP mRNA has been reported to be increased by FSH or FSH plus T (Hall et al., 1986; Reventos et al., 1986). Differences in in vivo and in vitro requirements and/or maturational stage at the time of stimulation may account for the apparent disparities in results between laboratories (Sanborn et al., 1986; NT Lee et al., 1986).

A number of effects of T on Sertoli cells have been described (Table 1). These may be both direct and indirect. Unlike FSH, which appears to bind almost exclusively to Sertoli cells (Steinberger et al., 1978b) T binds to androgen receptors in Sertoli, Leydig, and peritubular cells (Sanborn et al., 1983; Gulizia et al., 1983; Nakhla et al., 1984; Isomaa et al., 1985; Buzek and Sanborn, 1986). Skinner and Fritz (1985a,b) have found that peritubular cells secrete at least one substance, the concentration of which is stimulated in the dish by androgen treatment. This substance, termed P-Mod-S, stimulates ABP and transferrin secretion. We failed to detect significant peritubular cell contamination in our cultures, using the presence of ^{35}S-Met-labeled fibronectin in 2D-PAGE maps (Tung et al., 1984) and nuclear morphology (Bucci et al., 1986) as criteria. Consequently, we are reasonably certain that T exerts some its effects directly on the Sertoli cell. However, documentation of culture composition has not always been provided in the literature, and the alternative explanation for the effects of T must be kept in mind.

A number of proteins have been reported to be secretory products of Sertoli cells (Table 1). Properties of those characterized and the status of knowledge concerning their hormonal regulation are summarized in Table 2. By far the vast majority of these proteins are not yet ascribed definitively to Sertoli cells, given the ambiguities discussed above concerning variability in culture purity between laboratories. In addition, in some cases it has not yet been established that the proteins are the result of new protein synthesis. Relatively few of these proteins have been identified as analogous to or equivalent to known proteins. The major glycoproteins SGP-1 and SGP-2 are visible in most 2-D-PAGE maps reported by all investigators, but other differences may relate to age, hormonal state of the cells at isolation and while in culture, and the purity of the cultures themselves. Hypophysectomy has been reported by Shabanowitz and Kierszenbaum (1986) to decrease SGP-1 secretion by tubule cultures. Examination of our gels (Figure 2) suggests some decline in SGP-1 secretion by Sertoli cells isolated from hypophysectomized rats. However, this was not of the order of magnitude as the decline in Band 1 and Band 2, which are clearly distinct from SGP-1 and SGP-2.

Cheng et al. (1986b) have used a combination of HPLC separation followed by 1D-PAGE to characterize the hormonal control of Sertoli cell secretory activity. They used cells from 20 day animals and stimulated them for 6 days in vitro in the presence of insulin, transferrin, and epidermal growth factor. By loading equal amounts of medium, they were able to delineate quantitative effects of FSH and T, alone and in combination, on more than 28 proteins. These are not necessarily newly synthesized proteins, however, and the purity of the cultures was not detailed. Nonetheless, the added fractionation afforded by this approach has enabled this group to identify testibumin, a protein the concentration of which was increased by FSH and T (Bardin et al., 1986), clustrin, increased by FSH and T (Cheng, 1986), and CMB-2, a protein increased by FSH (Lee, WM et al., 1986). The analysis also revealed other proteins regulated differentially by FSH and by T (Table 2). In preliminary work, we have performed similar separations on medium from 15 and 35 day cells, with and without stimulation with hormones in vitro. We have some evidence for maturational patterns between 15 and 35 days using this approach, with less apparent hormonal dependence at 35 than at 15 days. This suggests either that, once initiated, the secretion of some proteins no longer requires the continued presence of hormones, or that the influence of hormones in vivo persists long enough in the culture situation to be reflected in protein secretion. Both may be true. We know from our work on the effects of hypophysectomy, that hormonal influences on Sertoli cell function persist for at least 3 days in culture.

TABLE 2   PHYSICAL PROPERTIES AND HORMONAL REGULATION OF SECRETORY PROTEINS ASCRIBED TO RAT SERTOLI CELLS

Protein	SU/MW,kd	pI	Tentative Identity	FSH	T	Others	References
Ceruloplasmin	130	5.5-6.0	Prot-1				1,2
Prot-2	120	5.5					2
SCm2	110	4.6-5.2	+				3,4
SCm1	80	4.6-5.5					3,4
Transferrin	75	7.2-7.4	Prot-3	+		I,A	2,5,6
SGP-1	70/140	4.2-4.6	S70, S4, Prot-4	+		HPX↓	2-4, 6-8
Band 1	66	6.0-8.0				HPX↓	9
Band 2	56	5.3-6.0				HPX↓	9
Testibumin	68		CMB-1	+	+		10,11
ABP	45 >85	4.7-5.5		+	+	I,A	6,12
	41	4.7-5.5					
SGP-2	46 >70	4.2-4.6	S45, S8, Prot-5 (DAG)				3,4,6,7,13
	33 >70	4.4-4.6	S35, S7, Prot-6				
Clustrin	34-45/70-90		CMB-2†	+	+		6,11,14
Prot-5b	39	6.5-7.1		+			2
Plasminogen activator	38-40/75			+			15
Somatomedin C	25	7.5		+			16
CMB-2	22			+			17
Cyclic protein-2	20	4.9-6.2					7,18
αLactalbumin- like protein	20						19
Seminiferous growth factor	14-20				+		20
Leydig cell stimulatory factors	10-30				+		21,22
Sertoli cell growth factor	8			+			23
CMB-14,18,20,27	30-150						11
CMB-7,8,22,23,24,28	<20-35				+		11
CMB-3,4,10,13,16,19	<20-120			+			11
CMB-9,12,21,25,26	<20-90			+	+		11

SU/NMW - subunit/native molecular weight in kilodaltons; I-insulin; A-Vitamin A.
(1) Skinner and Griswold, 1983b; (2) Kissinger et al., 1982; (3) DePhilip and Kierszenbaum, 1982; (4) Kierszenbaum et al., 1986; (5) Skinner and Griswold, 1982; (6) Griswold, 1987; (7) Wright and Luzarraga, 1986; (8) Shabanowitz and Kierszenbaum, 1986; (9) Sanborn et al., 1986; (10) Bardin et al., 1986; (11) Cheng et al., 1986b; (12) Mather et al., 1983; (13) Griswold et al., 1986; (14) Blaschuk and Fritz, 1984; (15) Marzowski et al., 1985; Lacroix and Fritz, 1982; (16) Smith et al., 1986; (17) Lee, WM et al., 1986; (18) Wright et al., 1981; (19) Skinner and Fritz, 1986; (20) Bellve et al., 1984; (21) Verhoeven and Cailleau, 1986; (22) Sharpe et al., 1981; (23) Holmes et al., 1986.

The Mechanisms Underlying Hormonal Responses in Sertoli Cells Are Only Partially Understood

In general, the mechanisms by which FSH affects the Sertoli cell are incompletely defined but are thought to involve stimulation of adenylyl cyclase, activation of cAMP-dependent protein kinase, and phosphorylation of important proteins (Means et al., 1980). To date, however, phosphorylation has yet to be linked with any specific response. FSH also affects Ca^{2+} redistribution and thus indirectly may affect the cytoskeleton and ultimately cellular processes (Means et al., 1980), but the details of these processes are also not well delineated. Finally, the recent acquisition of molecular probes for ABP allowed the determination that FSH or FSH plus T increased the mRNA species coding for this protein (Joseph et al., 1985; Hall et al., 1986; Reventos et al., 1986). Transferrin mRNA is apparently increased by treatment of cells with FSH, T, insulin, or retinol (Huggenvik et al., 1987) but may not be transcribed in vivo (Lee, NT et al., 1986). The links between cAMP stimulation, protein phosphorylation, Ca^{2+} dynamics and the stimulation of specific RNA synthesis remain to be established. In fact, additional mechanisms may pertain.

The mechanisms by which T influences Sertoli cell function are also not well understood. T binds to Sertoli cell androgen receptors, increases tightly-bound nuclear forms of the receptor bound to chromatin acceptor sites, stimulates RNA polymerase II activity, and increases the concentration of specific proteins (Sanborn et al., 1980, 1983, 1986). In other steroid sensitive systems, specific DNA binding domains have been found in the 5' region of hormonally regulated genes (Groner et al., 1984). In addition, a number of soluble factors, some of which are developmentally or regionally expressed, appear to regulate gene transcription (Kadonaga et al., 1986). Sertoli cell androgen receptors have yet to be obtained in significant quantity in highly purified form and Sertoli cell androgen-responsive genes have yet to be isolated, but this is clearly a direction in which mechanistic studies will develop.

Even less is understood about the mechanisms by which Vitamin A and insulin influence the Sertoli cell. Transferrin mRNA is increased by both of these hormones in Sertoli cells in culture (Huggenvik et al., 1987). Retinol binding protein has been described in Sertoli cells and may mediate the action of this hormone or serve to transport it to the lumen of the seminiferous tubule. The action of insulin in any tissue is not well understood. In the Sertoli cell, it apparently increases glucose transport (Mita et al., 1985) and influences nucleotide pool size (Griswold and Merryweather, 1982), but the mechanisms remain to be delineated.

A Number of Other Factors Influence Sertoli Cell Function

Besides the direct effect of hormones on the Sertoli cells, their function is regulated by a complex interplay of signals from the local environment. It is important to keep in mind in this discussion that most of the effects to be described have been studied in vitro and that the in vivo relevance in many cases has yet to be established. Leydig cells secrete T, which probably influences the Sertoli cell directly and indirectly as well via effects on peritubular cells. Leydig cells also secrete POMC (preopiomelanocortin) derivatives which may stimulate or inhibit ABP secretion (Bardin et al., 1984; Fabbri et al., 1985; Gerendai et al., 1986). The Sertoli cell in turn secretes a GnRH-like substance and perhaps other factors which have both positive and negative effects on Leydig cell steroidogenesis and factors which mediate FSH-stimulated increases in LH receptor and steroidogenic activity in Leydig cells (Sharpe et al., 1981; Verhoeven and Cailleau, 1986; Saez et al., 1986).

Peritubular cells influence ABP and transferrin secretion by Sertoli cells either directly (Hutson and Stucco, 1981) or indirectly via secretory products (Skinner and Fritz, 1985a,b;1986). They also secrete unique components of the extracellular matrix comprising the basal lamina of the seminiferous tubule (Skinner et al., 1985). Extracellular matrix components influence Sertoli cell secretory properties in the culture dish (Mather et al., 1984; Hadley et al., 1985) and may do so in vivo as well. The Sertoli cell also contributes unique components to the basal lamina (Skinner et al., 1985). There is one report of metabolic cooperation at the level of RNA (Hutson, 1983).

Finally, germ cells influence Sertoli cell function. They alter the expression of Sertoli cell membrane glycoprotein (Galdieri and Monaco, 1986), influence ABP secretion (Galdieri et al., 1984; LeMagueresse et al., 1986) and apparently stimulate Sertoli cell adenylyl cyclase (Welsh et al., 1985). The complex environment of the cycling seminiferous epithelium (Parvinen, 1982) may be important in providing influences for regulating Sertoli cell function and vice versa.

The Relationship of Specific Sertoli Cell Functions to the Spermatogenic Process is Incompletely Understood at Present

The complex interactions outlined above suggest that intracellular communication between the components of the testis is important in maintaining the proper milieu for the spermatogenic process. This may be the reason why the only in vitro models in which completion of meiosis have been reported have been tubule culture (Parvinen et al., 1983) and aggregates of germ cells in contact with Sertoli cells in a specialized hormonal milieu (Tres and Kierszenbaum, 1983). The cyclical pattern of seminiferous tubule synthetic activity and associated Sertoli cell activity may be critical to the process (Parvinen, 1982).

The influence of specific Sertoli cell products on the spermatogenic process is an area of active interest at present. Clearly, spermatocytes utilize lactate or pyruvate as an energy source in vitro (Jutte et al., 1981, 1982). The secretion of lactate by Sertoli cells may be a key function, but it is clearly not sufficient to allow meiosis to proceed. Spermatocytes bind ABP specifically (Steinberger et al., 1984), suggesting some possible role in androgen transport, but this is yet to be demonstrated. Transferrin binds to spermatocytes and round spermatids (Holmes et al., 1983; Sylvester and Griswold, 1984; Steinberger et al., 1984) and a role in iron transport to these cells has been suggested (Morales and Clermont, 1986; Djakiew et al., 1986; Griswold, 1987). Somatomedin-C, which has been reported to be secreted by Sertoli cells, also binds to pachytene spermatocytes (Tres et al., 1986). The Sertoli cell apparently secretes substances which increase RNA and DNA synthesis in germ cells (Saez et al., 1986). It has been suggested that plasminogen activator is associated with the migration of Sertoli cell cytoplasmic extensions and in the restructuring of the seminiferous tubule during the spermatogenic process (Hettle et al., 1986). The dimeric acidic glycoproteins are associated with epididymal and ejaculated spermatozoa (Sylvester et al., 1984), but their functional role is not known at present. GGT may be involved in salvaging the components of oxidized glutathione from the tubular lumen (Hanigan and Pitot, 1985).

SUMMARY

   The Sertoli cell is clearly influenced, directly and indirectly, by
hormones. Among these are FSH, T, insulin and Vitamin A, but others may
also be involved. Mechanisms are still not well understood. The
biochemical effects of these hormones can be divided into quantitative
and qualitative influences, with the former predominating. Specific
cellular and secretory proteins and metabolites are affected, in many
cases by more than one hormone. Often these same functions are influenced
by other factors in the environment of the testis as well. Hormonal
responsiveness of the Sertoli cell is determined in part by the
maturational state of the cell. Some secreted products bind to specific
cell types in the testis and epididymis and may influence the process of
spermatogenesis. However, detailed mechanisms are not known at the
present time. Understanding Sertoli cell function at the biochemical
level and its control by hormones is clearly of key importance in
understanding the control of the spermatogenic process.

ACKNOWLEDGEMENTS   The authors express appreciation to Dr. M. Griswold
for providing his manuscript prior to publication and to K. Silva for
preparation of this manuscript. Supported in part by HD017795.

REFERENCES

Alger NE, Nequin LG, Russell LD, 1985. Spermatogenesis is inefficient
     during pubertal development in the normal rat. J. Androl 6:36-P
Armstrong DT, Dorrington JH, 1977. Estrogen biosynthesis in the
     ovaries and testes. In: Thomas JA, Singhal RL (Eds), Regulatory
     Mechanisms Affecting Gonadal Hormone Action. Baltimore: University
     Park Press, Vol 3, pp. 217-58
Bardin CW, Shaha C, Grima J, Cheng CY, 1986. The use of testibumin, a
     new marker for Sertoli cells, to study testicular physiology. Endoc
     Soc A239, 90
Bardin CW, Shaha C, Mather JP, Salomon Y, Margioris AN, Liotta AS,
     Gerendai I, Chen CL, Krieger DT, 1984. Identification and possible
     function of proopiomelanocortin-derived peptides in the testis. Ann
     NY Acad Sci 438:346-64
Beckman JK, Coniglio JC, 1979. A comparative study of the lipid
     composition of isolated rat Sertoli and germinal cells. Lipids
     14:262-7
Beckman JK, Coniglio JC, 1980. The metabolism of polyunsaturated
     fatty acids in rat Sertoli and germinal cells. Lipids 15:389-94
Bellve AR, Feigg LR, 1984. Cell proliferation in the mammalian testis.
     Biology of the seminiferous growth factor (SGF). Rec Prog Horm Res
     40:531-67
Blaschuk O, Fritz IB, 1984. Isoelectric forms of clusterin isolated
     from ram rete testis fluid and from secretions of primary cultures of
     ram and rat Sertoli-cell enriched cultures. Can J Biochem Cell Biol
     62:456-61
Borland K, Mita M, Oppenheimer CL, Blinderman LA, Massague J, Hall PF,
     Czech MP, 1984. The actions of insulin-like growth factors I and II
     on cultured Sertoli cells. Endocrinology 114:240-6
Borland K, Osawa S, Kew D, Coleman DB, Hall PF, 1985. Identification
     of a spectrin-like protein in Sertoli cells. Biol Reprod 32:1143-56
Borland K, Ehrlich HP, Muffly K, Dills WL, Hall PF, 1986. Interaction
     of rat Sertoli cells with a collagen lattice in vitro. In Vitro
     22:661-9
Brunner M, Moreira-Filho CA, Wachtel G, Wachtel S, 1984. On the
     secretion of H-Y antigen. Cell 37:615-9

Bucci LR, Brock WA, Johnson TS, Meistrich ML, 1986. Isolation and biochemical studies of enriched populations of spermatogonia and early primary spermatocytes from rat testes. Biol. Reprod. 34:195-206

Buzek SW, Caston LA, Sanborn BM, 1987. Evidence for age-dependent changes in Sertoli cell androgen receptor concentration. J. Androl, in press

Buzek SW, Sanborn BM, 1986. Nuclear androgen receptor (AR) increases in rat testis during sexual maturation. Endocrine Soc A854, 244

Byers SW, Hadley MA, Djakiew D, Dym M, 1986. Growth and characterization of polarized monolayers of epididymal epithelial cells and Sertoli cells in dual environment culture chambers. J Androl 7:59-68

Cameron DF, Snydle E, 1985. Selected enzyme histochemistry of Sertoli cells 2. Adult Sertoli cells in co-culture with peritubular fibroblasts. Andrologia 17:185-93

Catterall JF, Kontula KK, Watson CS, Seppanen PJ, Funkenstein B, Melanitou E, Hickok NJ, Bardin CW, Janne OA, 1986. Regulation of gene expression by androgens in murine kidney. Rec Prog Hor Res 42:71-109

Chemes H, 1986. The phagocytic function of Sertoli cells: a morphological, biochemical, and endocrinological study of lysosomes and acid phosphatase localization in the rat testis. Endocrinology 119:1673-81

Chemes HE, Dym M, Raj HGM, 1979. Hormonal regulation of Sertoli cell differentiation. Biol Reprod 21:251-62

Cheng CY, 1986. A FSH and testosterone (T) responsive glycoprotein isolated from rat primary Sertoli cell-enriched cultures (SCCM) shares immunodeterminants with ram clusterin. Endoc Soc A238, 90

Cheng CY, Gunsalus GL, Morris ID, Turner TT, Bardin CW, 1986a. The heterogeneity of rat androgen binding protein (rABP) in the vascular compartment differs from that in the testicular tubular lumen. Further evidence for bidirectional secretion of rABP. J Androl 7:175-9

Cheng CY, Mather JP, Byer AL, Bardin CW, 1986b. Identification of hormonally responsive proteins in primary Sertoli cell culture medium by anion-exchange high performance liquid chromatography. Endocrinology 118:480-8

Chowdhury AK, Tcholakian RK, 1979. Effects of various doses of testosterone propionate on intratesticular and plasma testosterone levels and maintenance of spermatogenesis in adult hypophysectomized rats. Steroids 34:151-62

Clermont Y, Harvey SC, 1967. Effects of hormones on spermatogenesis of the rat. Ciba Foundation Colloquia on Endocrinology 16:173-85

Conti M, Toscano MV, Petrelli L, Geremia R, Stefanini M, 1982. Regulation by follicle-stimulating hormone and dibutyryl adenosine 3',5'-monophosphate of a phosphodiesterase isoenzyme of the Sertoli cell. Endocrinology 110:1189-96

DeJong FH, Robertson DM, 1985. Inhibin: 1985 update on action and purification. Mol Cell Endocrinol 42:95-103

DePhilip RM, Kierszenbaum AL, 1982. Hormonal synthesis, secretion, and phosphorylation in cultured rat Sertoli cells. Proc Natl Acad Sci USA 79:6551-5

Dias JA, 1985. Transglutaminase activity in testicular homogenates and serum-free Sertoli cell cultures. Biol Reprod 33:835-43

Dierickx P, Verhoeven G, 1980. Effect of different methods of germinal cell destruction on rat testis. J Reprod Fert 59:5-9

Djakiew D, Hadley MA, Byers SW, Dym M, 1986. Transferrin-mediated transcellular transport of ^{59}Fe across confluent epithelial sheets of Sertoli cells grown in bicameral cell culture chambers. J Androl 7:355-66

Elkington JSH, Fritz IB, 1980. Regulation of sulfoprotein synthesis by rat Sertoli cells in culture. Endocrinology 107:970-6

Fabbri A, Tsai-Morris C, Luna S, Fraioli F, Dufau M, 1985. Opiate receptors are present in the rat testis. Identification and localization in Sertoli cells. Endocrinology 117:2544-7

Fisher DM, Coniglio JG, 1983. Composition of, and [^{14}C]acetate incorporation into, lipids of rat Sertoli cells in culture. Biochim Biophys Acta 751:27-32

Fletcher PW, Reichert LE, 1986. Guanine triphosphate binding site regulation by follicle-stimulating hormone and guanine diphosphate in membranes from immature rat Sertoli cells. Endocrinology 119:2221-6

Floridi A, Marcante ML, D'Atri S, Feriozzi R, Menichini R, Citro G, Cioli V, De Martino C, 1983. Energy metabolism of normal and lonidamine-treated Sertoli cells of rats. Exper Molec Path 38:137-47

Fritz IB, 1982. Comparison of granulosa and Sertoli cells at various stages of maturation: similarities and differences. Adv Exp Med Biol 147:357-84

Galdieri, M, Caporele C, Adamo S, 1986. Calcium-phospholipid-dependent protein kinase activity of cultured rat Sertoli cells and its modifications by Vitamin A. Mol Cell Endocrinol 48:213-20

Galdieri M, Monaco L, 1983. Retinol increases synthesis and secretion of Sertoli cell mannose-containing glycoproteins. Cell Biol Int Rep 7:219-26

Galdieri M, Monaco L, 1986. Sertoli cell membrane polypeptide composition is modulated by germ cells. Cell Differ 18:51-5

Galdieri M, Monaco L, Stefanini M, 1984. Secretion of androgen binding protein by Sertoli cells is influenced by contact with germ cells. J Androl 5:409-15

Galdieri M, Nistico L, 1986. Vitamin A modifies the glycopeptide composition of cultured Sertoli cells. J Androl 7:303-9

Galdieri M, Ziparo E, Palombi F, Russo MA, Stefanini M, 1981. Pure Sertoli cell cultures: A new model for the study of somatic-germ cell interactions. J Androl 2:249-54

Gerendai I, Shaha C, Gunsalus GL, Bardin CW, 1986. The effects of opiate antagonists suggest that testicular opiates regulate Sertoli cell and Leydig cell function in the neonatal rat. Endocrinology 118:2039-44

Griswold MD, 1987. Protein secretions of Sertoli cells. Int Rev Cytol, in press

Griswold MD, Merryweather J, 1982. Insulin stimulates the incorporation of ^{32}P into ribonucleic acid in cultured Sertoli cells. Endocrinology 111:661-7

Griswold MD, Roberts K, Bishop P, 1986. Purification and characterization of a sulfated glycoprotein secreted by Sertoli cells. Biochemistry 25:7265-70

Griswold MD, Solari A, Tung PS, Fritz IB, 1977. Simulation by FSH of DNA synthesis and of mitosis in cultured Sertoli cells prepared from testes of immature rats. Mol Cell Endocrinol 7:151-65

Grootegoed JA, Jansen R, van der Molen HJ, 1985. Intercellular pathway of leucine catabolism in rat spermatogenic epithelium. Biochem J 226:889-92

Grootegoed JA, Oonk RB, Jansen R, van der Molen HJ, 1986. Metabolism of radiolabelled energy-yielding substrates by rat Sertoli cells. J Reprod Fert 77:109-18

Groner B, Kennedy N, Skroch P, Hynes NE, Ponta H, 1984. DNA sequences involved in the regulation of gene expression by glucocorticoid hormones. Biochim Biophys Acta 78:1-6

Gulizia S, D'Agata R, Sanborn BM, Steinberger E, 1983. Evidence for the presence of androgen receptors in purified rat Leydig cells. J Androl 4:248-52

Gunsalus GL, Musto NA, Bardin CW, 1978. Factors affecting blood levels of androgen binding protein in the rat. Int J Androl 2:424-33

Hadley MA, Byers SW, Suarez-Quian CA, Kleinman HK, Dym M, 1985. Extra-cellular matrix regulates Sertoli cell differentiation, testicular cord formation, and germ cell development. J Cell Biol 101:1511-22

Hagenas L, Ploen L, and Ekwall H, 1978. Blood-testis barrier: Evidence for intact inter-Sertoli cell junctions after hypophysectomy in the adult rat. J Endocr 76:87-91

Hall PF, Mita M, 1984. Influence of follicle-stimulating hormone on glucose transport by cultured Sertoli cells. Biol Reprod 31:863-9

Hall SH, French FS, Conti M, Joseph DR, 1986. Characterization and regulation of androgen binding protein mRNA in rat testis. Endoc Soc A855, 244

Hanigan H, Pitot HC, 1985. Gamma-glutamyl transpeptidase--its role in hepatocarcinogenesis. Carcinogenesis 6:165-72

Hatier R, Grignon G, 1980. Ultrastructural study of Sertoli cells in rat seminiferous tubules during intrauterine life and the postnatal period. Anat Embryol 160:11-27

Heindel JJ, Berkowitz A, Steinberger A, Strada SJ, 1982. Modification of Sertoli cell responsiveness to FSH by cryptorchidism and hypophysectomy in immature and adult rats. J Androl 3:337-43

Heindel JJ, Steinberger A, Strada SJ, 1981. Identification and characterization of a β_1-adrenergic receptor in the rat Sertoli cell. Mol Cell Endocrinol 22:349-58

Hettle JA, Waller EK, Fritz IB, 1986. Hormonal stimulation alters the type of plasminogen activator produced by Sertoli cells. Biol Reprod 34:895-904

Hodgen GD, Sherins RJ, 1973. Enzymes as markers of testicular growth and development in the rat. Endocrinology 93:985-9

Holmes SD, Bucci LR, Lipshultz LI, Smith RG, 1983. Transferrin binds specifically to pachytene spermatocytes. Endocrinology 113:1916-8.

Holmes SD, Spotts G, Smith RG, 1986. Rat Sertoli cells secrete a growth factor that blocks epidermal growth factor (EGF) binding to its receptor. J Biol Chem 261:4076-80

Huggenvik J, Griswold MD, 1981. Retinol binding protein in rat testicular cells. J Reprod Fert 61:403-8

Huggenvik JI, Idzerda RL, Haywood L, Lee DC, McKnight GS, Griswold MD, 1987. Transferrin messenger ribonucleic acid: Molecular cloning and hormonal regulation in rat Sertoli cells. Endocrinology 120:332-40

Hutson JC, 1978. The effects of various hormones on the surface morphology of testicular cells in culture. Am J Anat 151:55-69

Hutson JC, 1983. Metabolic cooperation between Sertoli cells and peritubular cells in culture. Endocrinology 112:1375-81

Hutson JC, Stucco DM, 1981. Peritubular cell influence on the efficiency of androgen-binding protein secretion by Sertoli cells in culture. Endocrinology 108:1362-8

Ireland ME, Rosenblum BB, Welsh MJ, 1986. Two-dimensional gel analysis of Sertoli cell protein phosphorylation: Effect of short term exposure to follicle-stimulating hormone. Endocrinology 118:526-32

Isomaa V, Parvinen M, Janne OA, Bardin CW, 1985. Nuclear androgen receptors in different stages of the seminiferous epithelial cycle and the interstitial tissue of rat testis. Endocrinology 116:132-7

Janecki A, Steinberger A, 1986. Polarized Sertoli cell functions in a new two-compartment culture system. J Androl 7:69-71

Janecki A, Steinberger A, 1987. Bipolar secretion of androgen-binding protein and transferrin by Sertoli cells cultured in a two-compartment culture chamber. Endocrinology 120:291-8

Joseph DR, Hall SH, French FS, 1985. Identification of complementary DNA clones that encode rat androgen binding protein. J Androl 6:392-5

Josso N, 1986. AntiMullerian hormone: New perspectives for a sexist molecule. Endorcine 7:421-33

Jutte NH, Eikvar L, Levy FO, Hansson V, 1985. Metabolism of palmitate in cultured rat Sertoli cells. J Reprod Fert 73:497-503

Jutte NHPM, Grootegoed JA, Rommerts FFG, van der Molen HJ, 1981. Exogenous lactate is essential for metabolic activities in isolated rat spermatocytes and spermatids. J Reprod Fert 65:399-405

Jutte NHPM, Jansen R, Grootegoed JA, Rommerts FFG, Clausen OPF, van der Molen HJ, 1982. Regulation of survival of rat pachytene spermatocytes by lactate supply from Sertoli cells. J Reprod Fert 65:431-8

Jutte NHPM, Jansen R, Grootegoed JA, Rommerts FFG, van der Molen HJ, 1983. FSH stimulation of the production of pyruvate and lactate by rat Sertoli cells may be involved in hormonal regulation of spermatogenesis. J Reprod Fert 68:219-26

Kadonaga JT, Jones KA, Tjian R, 1986. Promoter-specific activation of RNA polymerase II transcription by Spl. Trends Biochem Sci 11:20-3

Kierszenbaum AL, Crowell JA, Shabanowitz RB, DePhilip RM, Tres LL, 1986. Protein secretory patterns in rat Sertoli and peritubular cells are influenced by culture conditions. Biol Reprod 35:239-51

Kierszenbaum AL, Feldman M, Lea O, Spruill WA, Tres LL, Petrusz P, French FS, 1980. Localization of androgen-binding protein in proliferating Sertoli cells in culture. Proc Natl Acad Sci USA 77:5322-6

Kierszenbaum AL, Spruill WA, White MG, Tres LL, Perkins JP, 1985. Rat Sertoli cells acquire a β-adrenergic response during primary culture. Proc Natl Acad Sci 82:2049-53

Kissinger D, Skinner MK, Griswold MD, 1982. Analysis of Sertoli cell-secreted proteins by two-dimensional electrophoresis. Biol Reprod 27:233-40

Lacroix M, Fritz IB, 1982. The control of the synthesis and secretion of plasminogen activator by rat Sertoli cells in culture. Mol Cell Endocrinol 26:247-58

Lamb DJ, Kessler MJ, Shewach DS, Steinberger A, Sanborn BM, 1982. Characterization of Sertoli cell RNA synthetic activities in vitro at selected times during sexual maturation. Biol Reprod 27:374-82

Lamb DJ, Tsai YH, Steinberger A, Sanborn BM, 1981. Sertoli cell RNA polymerase: Characterization and stimulation by FSH and testosterone in vitro. Endocrinology 1089:1020-6

Lee NT, Chae CB, Kierszenbaum AL, 1986. Contrasting levels of trans-ferrin gene activity in cultured rat Sertoli cells and intact seminiferous tubules. Proc Natl Acad Sci USA 83:8177-81

Lee WM, Cheng CY, Bardin CW, Gunsalus GL, Musto NA, 1986. Measurement of a follicle-stimulating hormone-responsive protein of Sertoli cell origin using an enzyme-linked immunoblot assay. Endocrinology 119:1914-21

LeMagueresse B, LeGac F, Loir M, Jegou B, 1986. Stimulation of rat Sertoli cell secretory activity in vitro by germ cells and residual bodies. J Reprod Fert 77:489-98

Lipshultz LI, Murthy L, Tindall DJ, 1982. Characterization of human Sertoli cells in vitro. J Clin Endocrinol Metab 82:228-37

Lu C, Steinberger A, 1977. Gamma-glutamyl transpeptidase activity in the developing rat testis. Enzyme location in isolated cell types. Biol Reprod 17:84-8

Ludvigson MA, Waites GMH, Hamilton DW, 1982. Immunocytochemical evidence for the specific localization of aldose reductase in Sertoli cells. Biol Reprod 26:311-7

Madhubala R, Steinberger A, Tsai YH, 1986. Inhibiton of ornithine decarboxylase activity by follicle stimulating hormone in primary culture of rat Sertoli cells. J Cell Biol 103:485A

Maines MD, 1984. Characterization of heme oxygenase activity in Leydig cells and Sertoli cells of the rat testis. Differential distribution of activity and response to cadmuim. Biochem Pharmacol 33:1493-502

Mallea LE, Machado AJ, Narvaroli F, Rommerts FFG, 1986. Epidermal growth factor stimulates lactate production and inhibits aromatization in cultured Sertoli cells from immature rats. Int J Androl 9:201-8

Marzowski J, Sylvester SR, Gilmont RR, Griswold MD, 1985. Isolation and characterization of Sertoli cell plasma membranes and associated plasminogen activator activity. Biol Reprod 32:1237-45

Mather JP, Gunsalus GL, Musto NA, Cheng CY, Parvinen M, Wright W, Perez-Infante V, Margioris A, Liotta A, Becker R, Krieger DT, Bardin CW, 1983. The hormonal and cellular control of Sertoli cell secretion. J Steroid Biochem 19:41-51

Mather JP, Wolpe SD, Gunsalus GL, Bardin CW, Phillips DM, 1984. Effect of purified and cell-produced extracellular matrix components on Sertoli cell function. Ann NY Acad Sci 438:572-5

Means AR, Dedman JR, Fakunding JL, Tindall DJ, 1978. Mechanism of action of FSH in the male rat. In: Birnbaumer L, O'Malley BW (Eds), Receptors and Hormone Action, Vol. III. New York: Academic Press, pp. 363-92

Means AR, Dedman JR, Tash JS, Tindall DJ, van Sickle M, Welsh MJ, 1980. Regulation of the testis Sertoli cell by follicle stimulating hormone. Ann Rev Physiol 42:59-70

Means AR, Dedman JR, Welsh MJ, Marcum MK, Brinkley BR, 1979. FSH and calcium as modulators of Sertoli cell differentiation and function. In: Hamilton TH, Clark JH, Sadler WA, (Eds.) Ontogeny of Receptors and Reproductive Hormone Action. New York: Raven Press, pp. 207-24

Merke J, Hugel U, Ritz E, 1985. Nuclear testicular 1,25-dihydroxy-vitamin D_3 receptors in Sertoli cells and seminiferous tubules of adult rodents. Biochem Biophys Res Commun 127:303-9

Mita M, Borland K, Price JM, Hall PF, 1985. The influence of insulin and insulin-like growth factor-I on hexose transport by Sertoli cells. Endocrinology 116:987-92

Mita M, Price JM, Hall PF, 1982. Stimulation by follicle-stimulating hormone of synthesis of lactate by Sertoli cells from rat testis. Endocrinology 110:1535-40

Monaco L, Conti M, 1986. Localization of adenosine receptors in rat testicular cells. Biol Reprod 35:258-66

Morales C, Clermont Y, 1986. Receptor-mediated endocytosis of trans-ferrin by Sertoli cells of the rat. Biol Reprod 35:393-405

Muffly KE, Turner TT, Brown M, Hall PF, 1985. Content of K+ and Na+ in seminiferous tubule and rete testis fluids from Sertoli cell-enriched testes. Biol Reprod 33:1245-51

Murray FT, Orth J, Gunsalus G, Musto NA, Weisz J, Li J, Jefferson LS, Bardin CW, 1981. The pituitary-testicular axis in the streptozotocin diabetic male rat: evidence for gonadotroph, Sertoli cell and Leydig cell dysfunction. Int J Androl 4:265-80

Nakhla AM, Mather JP, Janne OA, Bardin CW, 1984. Estrogen and androgen receptors in Sertoli, Leydig, myoid, and epithelial cells: effects of time in culture and cell density. Endocrinology 115:121-8

Niemi M, Setchell BP, 1986. Gamma glutamyl transpeptidase in the vasculature of the rat testis. Biol Reprod 35:385-91

Oonk RB, Grootegoed JA, van der Molen HJ, 1985. Comparison of the effects of insulin and follitropin on glucose metabolism by Sertoli cells from immature rats. Mol Cell Endocrinol 42:39-48

Orth JM, 1982. Proliferation of Sertoli cells in fetal and postnatal rats: a quantative autoradiographic study. Anat Rec 203:485-92

Orth JM, 1986. FSH-induced Sertoli cell proliferation in the developing rat is modified by ß-endorphin produced in the testis. Endocrinology 119:1876-8

Paranko J, Kallajoki M, Pelliniemi LJ, Lehto VP, Virtanen I, 1986. Transient coexpression of cytokeratin and vimentin in differentiating rat Sertoli cells. Devel Biol 117:35-44

Parker MG, Scrace GT, Mainwaring WIP, 1978. Testosterone regulates the synthesis of major proteins in rat ventral prostate. Biochem J 170:115-21

Parvinen M, 1982. Regulation of the seminiferous epithelium. Endocrine Rev 3:404-17

Parvinen M, Wright WW, Phillips DM, Mather JP, Musto NA, Bardin CW, 1983. Spermatogenesis in vitro: completion of meiosis and early spermiogenesis. Endocrinology 112:1150-2

Perez-Infante V, Bardin CW, Gunsalus GL, Musto NA, Rich KA, Mather JP, 1986. Differential regulation of testicular transferrin and androgen-binding protein secretion in primary cultures of rat Sertoli cells. Endocrinology 118:383-92

Ramos AS, Dym M, 1979. Ultrastructural differentiation of rat Sertoli cells. Biol Reprod 21:909-22

Reventos J, Hammond GL, Musto NA, Gunsalus GL, Bardin CW, 1986. Testosterone and FSH induction of rat androgen binding protein in Sertoli cells as studied by complementary DNA. Endoc Soc A618, 185

Rich KA, Bardin CW, Gunsalus GL, Mather JP, 1983. Age-dependent pattern of androgen-binding protein secretion from rat Sertoli cells in primary culture. Endocrinology 113:2284-93

Robinson R, Fritz IB, 1979. Myoinositol biosynthesis by Sertoli cells, and levels of myoinositol biosynthetic enzymes in testis and epididymis. Can J Biochem 57:962-7

Robinson R, Fritz IB, 1981. Metabolism of glucose by Sertoli cells in culture. Biol Reprod 24:1032-41

Russell LD, Peterson RN, 1985. Sertoli cell junctions: morphological and functional correlates. Int Rev Cytol 94:177-211

Saez JM, Tabone E, Perrard-Sapori MH, Rivarola MA, 1986. Paracrine role of Sertoli cells. Med Biol 63:225-35

Salhanick AI, Wiebe JP, 1980. FSH receptors in isolated Sertoli cells: changes in concentration of binding sites at the onset of sexual maturation. Life Sciences 26:2281-8

Sanborn BM, Lamb DJ, Tsai YH, Steinberger A, 1980. Androgen action in the Sertoli cell. In: Mahesh VB, Muldoon TG, Saxena BB, Sadler WA, (Eds.), Functional Correlates of Receptors in Reproduction. New York: Elsevier/North Holland, pp. 205-20

Sanborn BM, Wagle JR, Steinberger A, 1984. Control of androgen cytosol receptor concentrations in Sertoli cells: effect of androgens. Endocrinology 114:2388-93

Sanborn BM, Wagle JR, Steinberger A, Greer-Emmert D, 1986. Maturational and hormonal influences on Sertoli cell function. Endocrinology 118:1700-9

Sanborn BM, Wagle JR, Steinberger A, Lamb DJ, 1983. The Sertoli cell as an androgen target. In: D'Agata R, Lipsett MB, Polosa P, van der Molen HJ, (Eds.), Recent Advances in Male Reproduction: Molecular Basis and Clinical Implications. New York: Raven Press, pp. 69-78

Schmidt WN, Briggs RC, Hnilica LS, 1983. Tissue and cell-specific antigens in chromatin from cultured rat Sertoli cells. Arch Biochem Biophys 220:584-93

Shabanowitz RB, DePhillip RM, Crowell JA, Tres LL, Kierszenbaum AL, 1986. Temporal appearance and cyclic behavior of Sertoli cell-specific secretory proteins during the development of the rat seminiferous tubule. Biol Reprod 35:745-60

Shabanowitz RB, Kierszenbaum AL, 1986. Newly synthesized proteins in seminiferous intertubular and intratubular compartments of the rat testis. Biol Reprod 35:179-90

Sharpe RM, Fraser HM, Cooper I, Rommerts FFG, 1981. Sertoli-Leydig cell communication via an LHRH-like factor. Nature 290:785-87

Skinner MK, Fritz IB, 1985a. Androgen stimulation of Sertoli cell function is enhanced by peritubular cells. Mol Cell Endocrinol 40:115-22

Skinner MK, Fritz IB, 1985b. Testicular peritubular cells secrete a protein under androgen control that modulates Sertoli cell functions. Proc Natl Acad Sci USA 82:114-18

Skinner MK, Fritz IB, 1985c. Structural characterization of proteoglycans produced by testicular peritubular cells and Sertoli cells. J Biol Chem 260:11874-83

Skinner MK, Fritz IB, 1986. Identification of a non-mitogenic paracrine factor involved in mesenchymal-epithelial cell interactions between testicular peritubular cells and Sertoli cells. Mol Cell Endocrinol 44:85-97

Skinner MK, Griswold MD, 1980. Sertoli cells synthesize and secrete transferrin-like protein. J Biol Chem 255:9523-25

Skinner MK, Griswold MD, 1982. Secretion of testicular transferrin by cultured Sertoli cells is regulated by hormones and retinoids. Biol Reprod 27:211-21

Skinner MK, Griswold MD, 1983a. Multiplication stimulating activity (MSA) can substitute for insulin to stimulate the secretion of testicular transferrin by cultured Sertoli cells. Cell Biol Int Rep 7:441-6

Skinner MK, Griswold MD, 1983b. Sertoli cells synthesize and secrete a ceruloplasmin-like protein. Biol Reprod 28:1225-9

Skinner MK, Tung PS, Fritz IB, 1985. Cooperativity between Sertoli cells and testicular peritubular cells in the production and deposition of extracellular matrix components. J Cell Biol 100:1941-7

Slaughter GR, Means AR, 1983. Follicle-stimulating hormone activation of glycogen phosphorylase in the Sertoli cell-enriched rat testis. Endocrinology 113:1476-85

Smith EP, Svoboda ME, Van Wyk JJ, Kierszenbaum AL, Tres, LL, 1987. Partial characterization of a somatomedin-like peptide from the medium of cultured rat Sertoli cells. Endocrinology 120:186-93

Solari AJ, Fritz IB, 1978. The ultrastructure of immature Sertoli cells. Maturation-like changes during culture and maintenance of mitotic potentiality. Biol Reprod 18:329-45

Spruill WA, Steiner AL, Tres LL, Kierszenbaum AL, 1984. Characterization of cyclic AMP-binding proteins in rat Sertoli cells using a photoaffinity ligand. Mol Cell Biochem 60: 147-57

Spruill WA, White MG, Steiner AL, Tres LL, Kierszenbaum AL, 1981. Temporal sequence of cell shape changes in cultured rat Sertoli cells after experimental elevation of intracellular cAMP. Exp Cell Res 131:131-48

Steinberger A, Dighe RR, Dias R, 1984. Testicular peptides and their endocrine and paracrine functions. Arch Biol Med Exp 17:267-71.

Steinberger A, Heindel JJ, Lindsey JN, Elkington JSH, Sanborn BM, Steinberger E, 1975. Isolation and culture of FSH responsive Sertoli cells. Endoc Res Commun 2:261-72

Steinberger A, Hintz M, Heindel JJ, 1978a. Changes in cyclic AMP responses to FSH in isolated rat Sertoli cells during sexual maturation. Biol Reprod 19:566-72

Steinberger A, Seethalakshmi L, Kessler M, Steinberger E, 1983. Sertoli cell factor. In: McCann SM, Dhindsa DS (Eds.), Role of Peptides and Proteins in Control of Reproduction. New York: Elsevier Science Pub Co, pp. 215-36

Steinberger A, Sanborn BM, Steinberger E, 1978b. FSH and the Sertoli cell. In: McKerns KW. (Eds.), The Gonadotropins: Structure and Function, Raven Press: New York, pp. 517-51

Steinberger E, Steinberger A, 1972. Testis: Basic and clinical aspects. In: Balin H, Glasser S (Eds.), Reproductive Biology. Amsterdam: Excerpta Medica, pp. 144-67

586

Steinberger A, Walther J, Heindel JJ, Sanborn BM, Tsai YH, Steinberger E, 1979. Hormone interactions in the Sertoli cells. In Vitro 15:23-31

Swift TA, Dias JA, 1987. Stimulation of polyamine biosynthesis by follicle-stimulating hormone in serum-free cultures of rat Sertoli cells. Endocrinology 120:394-400

Sylvester SR, Griswold MD, 1984. Localization of transferrin and transferrin receptors in rat testes. Biol Reprod 31:195-203

Sylvester SR, Skinner MK, Griswold MD, 1984. A sulfated glycoprotein synthesized by Sertoli cells and by epididymal cells is a component of the sperm membrane. Biol Reprod 31:1087-101

Tash JS, Welsh MJ, Means AR, 1981. Regulation of protein kinase inhibitor by follicle-stimulating hormone in Sertoli cells in vitro. Endocrinology 108:427-34

Tcholakian RK, Berkowitz AS, Newaz SN, 1984. Metabolism of (^{3}H) 5α-androstane-3α,17ß-diol by cultures of isolated rat Sertoli cells and the effect of LH and FSH. Steroids 43:457-67

Tcholakian RK, Steinberger A, 1979. In vitro metabolism of testosterone by cultured Sertoli cells and the effect of FSH. Steroids 33:495-526

Tcholakian RK, Steinberger A, St. Pyrek J, 1983. Metabolism of ^{3}H-estradiol-17β by cultures of isolated rat Sertoli cells and the effect of FSH: presence of 16 α-hydroxylase. Steroids 42:55-62

Tindall DJ, Rowley DR, Murthy L, Lipshultz LI, Chang CH, 1985. Structure and biochemistry of the Sertoli cell. Int Rev Cytol 94:127-49

Tres LL, Kierszenbaum AL, 1983. Viability of rat spermatogenic cells in vitro facilitated by their coculture with Sertoli cells in serum-free hormone-supplemented medium. Proc Nat Acad Sci USA 80:3377-81

Tres LL, Smith EP, Van Wyk JP, Kierszenbaum AL, 1986. Immunoreactive sites and accumulation of somatomedin-C in rat Sertoli spermatogenic cell co-cultures. Exp Cell Res 162:33-50

Tsai YH, Lin SN, 1985. Differential release of polyamines by cultured rat Sertoli cells. J Androl 6:348-52

Tung PS, Fritz IB, 1984. Extracellular matrix promotes rat Sertoli cell histotypic expression in vitro. Biol Reprod 30:213-29

Tung PS, Skinner MK, Fritz IB, 1984. Fibronectin synthesis is a marker for peritubular cell contaminants in Sertoli cell-enriched cultures. Biol Reprod 30:199-211

Unni E, Rao MRS, Ganguly J, 1983. Histological and ultrastructural studies on the effect of vitamin A depletion and subsequent repletion with vitamin A on germ cells & Sertoli cells in rat testis. Ind J Exp Biol 21:180-92

Ultee-vanGessel AM, Leemborg FG, deJong FH, van der Molen HJ, 1986. In vitro secretion of inhibin-like activity by Sertoli cells from normal and prenatally irradiated immature rats. J Endoc 109:411-8

Vanha-Perttula T, Mather JP, Bardin CW, Moss SB, Bellve AR, 1986. Acid phosphatases in germinal and somatic cells of the testes. Biol Reprod 35:1-9

Verhoeven G, Cailleau J, 1986. Specificity and partial purification of a factor in spent media from Sertoli cell enriched cultures that stimulates steroidogenesis in Leydig cells. J Steroid Biochem 25:393-402

Verhoeven G, Cailleau J, de Moor P, 1981. Hormonal control of phosphodiesterase activity in cultured rat Sertoli cells. Mol Cell Endocrinol 24:41-51

Verhoeven G, Dierickx P, deMoor P, 1979. Stimulation effect of neuro-transmitters on the aromatization of testosterone by Sertoli cell-enriched cultures. Molec Cell Endocrinol 13:241-53

Verhoeven G, Franchimont P, 1983. Regulation of inhibin secretion by Sertoli cell-enriched cultures. Acta Endocrinol 102:136-43

Vitale R, Fawcett DW, Dym M, 1983. The normal development of the blood-testis barrier and the effect of clomiphene and estrogen treatment. Anat Rec 176:331-44

Wagle JR, Heindel JJ, Sanborn BM, Steinberger A, 1986. Effect of hypotonic treatment on Sertoli cell purity and cell function in culture. In Vitro 22:325-31

Welch JP, Mitchell RB, Davis JC, 1979. Cytophotometric measurements of DNA, RNA and basic protein in FSH treated rat Sertoli cells in culture. Biol Reprod 21:69-74

Welsh MJ, Ireland ME, Treisman GJ, 1985. Stimulation of rat Sertoli cell adenylate cyclase by germ cells in vitro. Biol Reprod 33:1050-56

Welsh MJ, Wiebe JP, 1978. Sertoli cell capacity to metabolize C_{19} steroids: variation with age and the effect of follicle-stimulating hormone. Endocrinology 103:838-94

Wiebe JP, 1982. Identification of a unique Sertoli cell steroid as 3α-hydroxy-4-pregnen-20-one (3α-dihydroprogesterone: 3α-DHP). Steroids 39:259-78

Wiebe JP, Tilbe KS, Buckingham KD, 1980. An analysis of the metabolites of progesterone produced by isolated Sertoli cells at the onset of gametogenesis. Steroids 35: 561-77

Wilson RM, Griswold MD, 1979. Secreted proteins from rat Sertoli cells. Exp Cell Res 123:127-35

Wright WW, Luzarraga ML, 1986. Isolation of cyclic protein-2 from rat seminiferous tubules fluid and Sertoli cell culture medium. Biol Reprod 35:761-72

Wright WW, Musto NA, Mather JP, Bardin CW, 1981. Sertoli cells secrete both testis-specific and serum proteins. Proc Natl Acad Sci USA 78:7565-9

PART V

SHORT PAPERS

CHRONIC EFFECTS OF INSULIN AND INSULIN-LIKE GROWTH FACTOR-1

ON STEROIDOGENESIS BY RAT LEYDIG CELLS

William H. Moger[a] and Paul R. Murphy[b]

[a]Departments of Physiology and Biophysics
  and of Obstetrics and Gynecology
  Dalhousie University
  Halifax, Nova Scotia
[b]Department of Physiology
  University of Manitoba
  Winnipeg, Manitoba

INTRODUCTION

There are no apparent acute effects of insulin on Leydig cell glucose transport or androgen production (Moger and Murphy, 1986). However, chronic exposure of primary cultures of rat or mouse testicular cells to insulin has been reported to increase LH-stimulated androgen production (Adashi et al., 1982; Murphy and Moger, 1982). Insulin-like growth factor-1 (IGF-1) receptors have been reported to be present on Leydig cells (Handelsman et al., 1985; Lin et al., 1986a) and recent reports have suggested an intratesticular site of IGF-1 production (D'Ercole et al., 1984; Handelsman et al., 1985). As with insulin, IGF-1 did not acutely alter androgen production by Leydig cells (Handelsman et al., 1985). The present study has compared the chronic effects of insulin and IGF-1 on androgen production by purified rat Leydig cells.

MATERIALS AND METHODS

NIH-LH-B9 (0.7 x NIH-LH-S1) was the gift of Dr. Leo Reichert Jr. and the NIAMDD Pituitary Distribution Program. Crystalline bovine insulin was obtained from Sigma Chemical Corp. (St. Louis, MO). IGF-1, prepared by recombinant DNA technology, was purchased from Amersham (Oakville, Ontario).

Purified Leydig cells were prepared from 275-300 g Sprague-Dawley rats on two-step Percoll gradients and established in culture, 1 x 10^5 cells per well, as recently described (Anakwe and Moger, 1986). Insulin and/or IGF-1 were included in the modified medium 199 (Gibco Burlington, Ontario) as indicated in the Results section. The cells were maintained in an atmosphere of 1% O_2, 5% CO_2, and 94% N_2. After 24 h the medium was replaced with fresh medium containing 0.1 mM 3-isobutyl-1-methyl-xanthine (MIX) and LH, dibutyryladenosine 3',5'-monophosphate (bt$_2$cAMP), or pregnenolone as indicated in the Results section. The plates were returned to the incubator for 3 h at which time the media were removed and stored at -40°C until assayed for androgens. Cell protein and medium androgen content was determined as described previously (Murphy and Moger, 1982). DNA was measured as described by West et al. (1985). The results presented

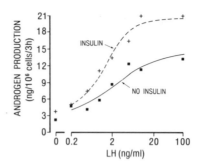

Fig. 1. Effect of 24 h insulin (μM) exposure on the response of Leydig cells to various concentrations of LH.

are from representative experiments. Concentrations for half-maximum effect (EC_{50}) were calculated by the program ALLFIT (DeLean et al., 1978).

RESULTS

The LH concentration-response relationship was altered in cells exposed to insulin (1 μM) for 24 h and then stimulated with LH for 3 h (Fig. 1). ALLFIT analysis demonstrated that insulin significantly increased androgen secretion in response to maximum LH stimulation (14.7 $\pm$ 2.9 vs. 20.6 $\pm$ 2.8 ng/10^6 cells/3h for control and insulin treated cultures, respectively. Basal androgen production and the EC_{50} for LH stimulation were not significantly affected by insulin treatment.

Both insulin and IGF-1 increased the steroidogenic response to maximum (100 ng/ml) LH stimulation in a concentration dependent fashion (Fig. 2). Although the slope of the insulin concentration-response curve was less steep than that of IGF-1, the EC_{50} for the two (IGF-1 = 2.5 nM; insulin = 4 nM) were similar. With maximum concentrations (> 10 nM IGF-1; > 100 nM insulin) LH-stimulated androgen production was enhanced to a similar extent. As shown in Fig. 3, there was no additive or synergistic effect when maximum concentrations of insulin (100 nM) and IGF-1 (10 nM) were combined.

Basal androgen secretion was not affected by maximal stimulatory concentrations of either IGF-1 or insulin. The augmentation of steroido-genesis by IGF-1 and insulin was not, however, restricted to an effect on the response to LH stimulation. Androgen secretion stimulated by bt_2cAMP and the conversion of the steroid precursor pregnenolone to androgens were also enhanced by both IGF-1 and insulin (Table 1).

The effects of insulin and IGF-1 on androgen production were not due to effects on cell numbers as measured by DNA content (Fig. 4). Insulin and IGF-1 caused small increases in the amount of cellular protein in some experiments but these increases, when present, were usually not statisti-cally significant (Fig. 4).

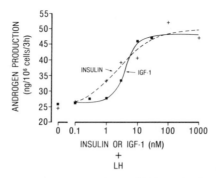

Fig. 2. Leydig cell response to maximum (100 ng/ml) LH stimulation after 24 h exposure to various concentrations of insulin or IGF-1.

Table 1. Effect of Insulin (100 nM) or IGF-1 (10 nM) on Androgen Production Induced with LH, Dibutyryladenosine 3',5'-monophosphate (bt$_2$cAMP), or Pregnenolone

	Treatment		
Stimulus	None	IGF-1	Insulin
None	3.9 ± 0.2	3.4 ± 0.7	3.6 ± 1.0
LH (100 ng/ml)	12.9 ± 0.6	20.1 ± 1.2	22.6 ± 1.1
Bt$_2$cAMP (2.5 nM)	16.4 ± 1.4	24.4 ± 0.8	28.4 ± 2.0
Pregnenolone (300 nM)	31.7 ± 0.6	44.3 ± 2.0	38.5 ± 4.1

DISCUSSION

Chronic exposure of Leydig cells to insulin increases the capacity of Leydig cells to produce androgens in response to LH stimulation without a pronounced effect on the cells sensitivity to LH (Adashi et al., 1982; Murphy and Moger, 1982). These observations have now been extended to demonstrate a similar effect of chronic IGF-1 treatment on steroidogenesis. Lin et al. (1986a) have recently reported very similar findings with IGF-1. As both insulin and IGF-1 are effective in nM concentrations it can be postulated that the effects of each may be physiologically relevant and that each may be acting via its own receptor. This is in keeping with the evidence from receptor binding studies that indicate high-affinity low capacity binding of both insulin and IGF-1 by Leydig cells (Handelsman et al., 1985; Lin et al., 1986a,b). Despite the suggestion that insulin and IGF-1 affect Leydig cells via separate receptors, the effects of the two agents are clearly similar. Both insulin and IGF-1 increase LH- and bt$_2$cAMP-stimulated and pregnenolone supported androgen production but in a non-additive manner. Over the time period of the experiments neither agent affected cell survival and both had only minimal effects on the amount of cellular protein. Although the site(s) in the steroidogenic process affected by insulin and IGF-1 have

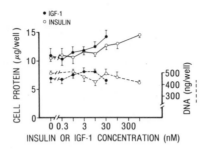

Fig. 3. Effect of separate and combined treatment with insulin (100 nM) and IGF-1 (10 nM) on LH (100 ng/ml) stimulated androgen production.

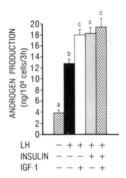

Fig. 4. Effect of 24 h treatment with various concentrations of insulin or IGF-1 on cell protein and DNA.

not been identified, it is possible that they are necessary to maintain the differentiated state of the Leydig cell. This suggestion would be in keeping with the effect of IGF-1 on differentiation of other mesenchymal cells (Froesch et al., 1985).

ACKNOWLEDGEMENT

    This work was supported by a grant from the MRC of Canada. The technical assistance of Mrs. Alice Giles is gratefully acknowledged. The testosterone antiserum was provided by Dr. D.T. Armstrong, University of Western Ontario.

REFERENCES

Adashi EY, Fabics C, Hsueh AJW, 1982. Insulin augmentation of testosterone production in a primary culture of rat testicular cells. Biol Reprod 26:270-280
Anakwe OO, Moger WH, 1986. Catecholamine stimulation of androgen production by rat Leydig cells. Interactions with luteinizing hormone and luteinizing hormone-releasing hormone. Biol Reprod In Press

DeLean A, Munson PJ, Rodbard D, 1978. Simulatneous analysis of families of sigmoidal curves; application to bioassay, radio-ligand assay and physiological dose-response curves. Am J Physiol 235:E97-E102

D'Ercole AJ, Stiles AD, Underwood LE, 1984. Tissue concentrations of somatomedian C: further evidence for multiple sites of synthesis and paracrine or autocrine mechanisms of action. Proc Natl Acad Sci USA 81:935-939

Froesch ER, Schmid C, Schwander J, Zapf J, 1985. Actions of insulin-like growth factors. Ann Rev Physiol 47:443-467

Handelsman DJ, Spaliviero JA, Scott CD, Baxter RC, 1985. Identification of insulin-like growth factor-1 and its receptors in the rat testis. Acta Endocr 109:543-549

Lin T, Haskell J, Vinson N, Terracio L, 1986a. Direct stimulatory effects of insulin-like growth factor-I on Leydig cell steroidogenesis in primary culture. Biochem Biophys Res Comm 137:950-956

Lin T, Haskell J, Vinson N, Terracio L, 1986b. Characterization of insulin and insulin-like growth factor I receptors of purified Leydig cells and their role in steroidogenesis in primary culture: a comparative study. Endocrinology 119:1641-1647

Moger WH, Murphy PR, 1986. Fuel requirements for Leydig cell steroidogenesis. Prog Soc Obst Gyn Canada p. 126

Murphy PR, Moger WH, 1982. Short-term primary culture of mouse interstitial cells: effects of culture conditions on androgen production. Bio Reprod 27:38-47

West DC, Sattar A, Kumar S, 1985. A simplified in situ solubilization procedure for the determination of DNA and cell number in cultured mammalian cells. Anal Biochem 147:289-295

ISOLATION AND CHARACTERIZATION OF SEMINAL FLUID PROTEINS

THAT BIND HEPARIN

D.J. Miller, N.L. First, and R.L. Ax

Endocrinology-Reproductive Physiology Program
University of Wisconsin
Madison, WI

INTRODUCTION

The molecular mechanism by which sperm are capacitated is poorly understood. Heparin has been shown to facilitate the acrosome reaction in bull cauda epididymal (Handrow et al., 1982) and ejaculated (Miller and Hunter, 1986; Parrish et al., 1985) sperm as well as rabbit sperm (Lenz et al., 1983). Heparin is structurally similar to heparan sulfate (Gallagher and Walker, 1985) and heparan-like material is found in follicular fluid (Ax et al., 1985) and oviductal fluid (Lee et al., 1986). Heparin binds to bull, rabbit and monkey sperm in a saturable, reversible, temperature-, pH- and Ca^{2+}-dependent fashion (Handrow et al., 1984). A heparin-binding component was isolated by affinity chromatography of hypotonic $MgCl_2$ extracted ejaculated bull sperm. The primary protein isolated was a small 15-18 kilodalton (kD), acidic (pI from 4.2-5.0) protein (Lavin et al., 1986). Rabbit ejaculated, but not epididymal, sperm contain a 40 kD component that reacts with anti-fibronectin affinity-purified anti-sera (Koehler et al., 1980). Plasmin and chymotryspin digestion of fibronectin demonstrates 2 heparin-binding domains that are 29 and 31 kD (Woods et al., 1986). These fragments may be present in sperm or seminal plasma and may regulate capacitation. Since seminal plasma alters heparin binding of epididymal sperm (Lee et al., 1985) and inhibits follicular fluid proteoglycan-induced acrosome reactions (Lenz et al., 1982) there may be components in seminal plasma that bind directly to heparin and alter heparin-binding capacitation and the acrosome reaction in that fashion. The objective of this study was to isolate heparin-binding proteins in seminal plasma and characterize them by electrophoresis.

MATERIALS AND METHODS

Seminal Plasma Collection

Semen was collected from 6 bulls used for artificial insemination, immediately cooled to 5 C in a cold room, pooled and immediately centrifuged. Seminal fluid from the semen was harvested by centrifuging twice for 10 min at 500 xg and finally 10 min at 2,000 xg. The final supernatant was stored at -70 C for the subsequent chromatographic step.

Heparin-Sepharose Chromatography

Heparin (Calbiochem, LaJolla, CA, a sodium salt from porcine intestinal mucosa) was coupled to cyanogen bromide–activated Sepharose 4B (Pharmacia, Piscataway, NJ) by a modification of the procedure recommended for proteins by the manufacturer. Three g of Sepharose were reacted with 300 mg of heparin for 2 h and then filtered to remove the unbound. Active groups remaining were blocked with TRIS and the gel was equilibrated in 40 mM Tris, pH 7.35, 2 mM $CaCl_2$, 0.05% sodium azide, 1 mM phenylmethylsulfonylfluoride, 1 μM pepstatin A, (TC buffer) and poured into a 1.5 x 10 cm column.

Seminal plasma was ultra–filtered against running buffer on an Amicon Stirred Cell using a YM–5 membrane (5 kD cutoff). Approximately 5 mg of seminal plasma protein were applied to the column. Unbound protein was eluted with 50 ml of TC buffer at 0.33 ml/min collecting 3 ml fractions. Bound material was eluted with a 200 ml linear 0 to 2 M NaCl gradient at 0.33 ml/min. Eluates were monitored for protein by absorbance at 280 nm and for conductivity to determine NaCl concentrations.

Fractions were pooled that were contained in the area of the peaks from the apex to half the distance to the base. Protein was quantitated by the bicinchoninic acid protein microassay, as described by Pierce (Rockford, IL).

One-Dimensional Polyacrylamide Gel Electrophoresis (SDS-PAGE)

The isolates were characterized by electrophoresis in polyacrylamide slab gels with 5 mM dithiothreitol (DTT) and 14.4% sodium dodecyl sulfate (SDS). A range of 5 to 15 μg of protein was loaded. The stacking gel was 4% polyacrylamide and the separating gel was 15% polyacrylamide. The gels were silver-stained (Goldman et al., 1981). The mol wt standards were (Sigma, St. Louis, MO): bovine serum albumin (66 kD), ovalbumin (45 kD), glyceraldehyde 3-phosphate dehydrogenase (36 kD), carbonic anhydrase (29 kD), trypsin (24 kD), β-lactaglobulin (18.4 kD) and lysozyme (14.3 kD).

RESULTS

Heparin Affinity Chromatography

Seminal plasma collected from supernatants of semen contained several polypeptides that bound to heparin–Sepharose. Approximately 70% of the total protein was bound to the column under those conditions. Fig. 1 shows an elution profile from the column. The first peak that remained bound to the column after washing with 5 bed volumes of running buffer was eluted with 0.12 M NaCl (peak 1). A second higher affinity peak, peak 2, was eluted with 0.46 M NaCl. A relatively long trailing edge was observed on peak 2 which was saved for later characterization. Baseline separation was not achieved by altering chromatographic conditions. If the column was allowed to equilibrate in 2 M NaCl overnight, a third high affinity peak was observed. However, the amount of protein in that peak was very small.

SDS-PAGE

The peaks eluted from the heparin affinity column were characterized by SDS-PAGE under reducing conditions (using DTT). Fig. 2 shows each of the peaks after electrophoresis and silver staining. In lane 1 is seminal plasma. Bands were found that migrated similar to proteins with molecular weights of 65, 44, 29.5, 19.7, 13.2 kD along with two especially prominent bands which were resolved in less intensely stained gels as 15.5 and 16.5 kD. In lane 2 is the middle half of peak 1. The predominant bands found there were at 15.5 and 16.5 kD, corresponding to the two major bands in seminal plasma. A smaller band of 13.2 kD was also present. In lane 3 is

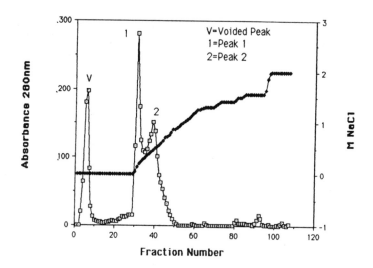

Fig. 1. Elution profile of seminal plasma from the heparin affinity column. The peaks, as numbered, show that many of the polypeptides in seminal plasma bind heparin and can be fractionated by affinity chromatography.

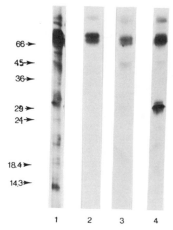

Fig. 2. One dimensional SDS-PAGE, under reducing conditions, of seminal plasma and peaks from the affinity column. Samples are as follows: in lane 1 is seminal plasma, lane 2-peak 1, lane 3-peak 2, lane 4-peak 2, trailing edge. Molecular weight markers are identified with arrows.

protein from peak 2. The same 3 bands as found in peak 1 were also found in peak 2. An additional band in peak 2 was found corresponding to a molecular weight of 19.7 kD. The trailing edge of peak 2 was applied to lane 4. Another band of 29.5 kD was observed. The 19.7 kD band was less evident in the trailing edge.

Polypeptides that migrated similarly to the major bands in seminal plasma, the 15.5 and 16.5 kD bands, were present in all peaks. The less prominent band corresponding to 13.2 kD was also present in all peaks and the trailing edge. Peak 2 had an additional 19.7 kD polypeptide and its trailing edge contained a 29.5 kD protein. Small higher affinity peaks, occasionally observed, had gel profiles similar to the trailing edge of peak 2.

DISCUSSION

Since the 15.5 and 16.5 kD polypeptides were in every peak obtained from the affinity column, they may be a part of a multimeric complex that was reduced by DTT during electrophoresis. Electrophoresis under non-reducing conditions will be used to investigate this possibility. Alternatively, affinity chromatography may be insufficient to separate those proteins due to their extremely high concentrations. Higher resolution affinity chromatography using high-performance liquid chromatography will be used to attempt to separate them.

Two families of these proteins are remarkedly similar to those found on ejaculated bull sperm (Lavin et al., 1986). Membranes removed from washed sperm membranes removed by hypotonic $MgCl_2$ buffer contained heparin-binding proteins with a mol wt range of 15-18 kD. Another heparin-binding sperm protein was found that had a mol wt of approximately 20 kD and a pI of 7.5-8.0. These proteins in seminal plasma may bind to sperm at ejaculation or may possibly be released from sperm in large quantities into seminal plasma. The former is most likely since many of the proteins on sperm are contributed by accessory fluids (Russell et al., 1984). The affect of heparin appears to be on capacitation (Parrish et al., 1985; Miller and Hunter, 1986). The binding of heparin to sperm may be necessary for capacitation since protamine, a heparin-binding cationic polypeptide, abolished heparin's ability to capacitate sperm (JJ Parrish, personal communication).

Seminal plasma inhibits follicular fluid proteoglycan-induced capacitation (Lenz et al., 1982). The simplest explanation is that seminal plasma does so by binding and inactivating heparan on the proteoglycan, since many of the seminal plasma polypeptides were found to bind heparin (Fig. 1). These polypeptides may be the active component(s) that inhibit capacitation. Seminal plasma also inhibits the head to head agglutination that heparin normally induces in the bull sperm (DJ Miller, unpublished observation).

In summary, several polypeptides found at high concentrations in bull seminal plasma bind to heparin-Sepharose. These polypeptides were characterized by SDS-PAGE and some were similar to polypeptides found on washed, ejaculated sperm membranes. Since heparin capacitates bull sperm, these proteins may play a role in regulating capacitation.

SUMMARY

Heparin facilities the acrosome reaction in bull and rabbit sperm. It binds to sperm saturably, and the binding affinity and the susceptibility to heparin are related to fertility. Isolation of heparin-binding proteins from washed ejaculated sperm reveals proteins mostly in the 14-18 kilodalton (kD) range. Seminal plasma inhibits heparin-induced capacitation, possibly by binding and inactivating heparin. These experiments focused on isolating and characterizing heparin-binding proteins in seminal plasma. Semen samples from 6 bulls were pooled, centrifuged twice for 10 min at 500 xg and finally at 2000 xg for 10 min. The supernatant was ultrafiltered on an Amicon YM-5 (5 kD cutoff) membrane against 40 mM Tris with 2 mM Ca^{2+} plus protease, inhibitors. Five mg of protein were applied to a 1.5 x 10 cm column of heparin-Sepharose. Bound proteins were eluted with 200 ml of a 0 to 2 M NaCl gradient at 0.33 ml/min collecting 3 ml fractions. Two major peaks eluted at 0.12 and 0.46

M NaCl. Those were subjected to SDS-PAGE under reducing conditions. The 0.12 M peak was composed of small (mol wt 15.5 and 16.5 kD) polypeptides which were similar to those found on sperm. The 0.46 M peak included those polypeptides plus a 19.5 kD polypeptide and its trailing edge had a 29.5 kD polypeptide. These polypeptides may regulate capacitation by viture of their ability to bind heparin.

REFERENCES

Ax RL, Bellin ME, Grimek HJ, 1985. Properties and regulation of synthesis of glycosaminoglycans by the ovary. In: Toft DO, Ryan RJ (eds), Proc of the 5th Ovarian Workshop. Published by Ovarian Workshop, Champaign, IL

Gallagher JT, Walker A, 1985. Molecular distinctions between heparan sulfate and heparin. Biochem J 230:665-675

Goldman D, Sedman SA, Ebert MH, 1981. Ultrasensitive stain for proteins in polyacrylamide gels shows regional variation in cerebrospinal fluid protein. Science 211:1437-1438

Handrow RR, Lenz RW, Ax RL, 1982. Structural comparisons among glycosaminoglycans to promote an acrosome reaction in bovine spermatozoa. Bioch Bioph Res Comm 107:1326-1332

Handrow RR, Boehm SK, Lenz RW, Robinson JA, Ax RL, 1984. Specific binding of the glycosaminoglycan ^{3}H-heparin to bull, monkey and rabbit spermatozoa in vitro. J Androl 5:51-63

Koehler JK, Nudelman ED, Hakomori S, 1980. A collagen-binding protein on the surface of ejaculated rabbit spermatozoa. J Biol Chem 86:529-536

Lavin CA, Robinson JA, Ax RL, 1986. Characterization of heparin binding domains from monkey and bull spermatozoa. 42nd Ann Mtg, Amer Fert Soc, Abstr P-094

Lee CN, Handrow RR, Lenz RW, Ax RL, 1985. Interactions of seminal plasma and glycosaminoglycans on acrosome reactions in bovine spermatozoa in vitro. Gamete Res 12:345-355

Lee CN, Clayton MK, Bushmeyer SM, First NL, Ax RL, 1986. Glycosaminoglycans in ewe reproductive tracts and their influence on acrosome reactions in bovine spermatozoa in vitro. J Animal Sci 63:861-867

Lenz RW, Ax RL, Grimek HJ, First NL, 1982. Proteoglycan from bovine follicular fluid enhances an acrosome reaction in bovine spermatozoa. Bioch Bioph Res Comm 106:1092-1098

Lenz RW, Bellin ME, Ax RL, 1983. Rabbit spermatozoa undergo an acrosome reaction in the presence of glycosaminoglycans. Gamete Res 8:11-19

Miller DJ, Hunter AG, 1986. Effect of osmolarity and glycosaminoglycans on motility, capacitation, acrosome reaction and in vitro fertilization of bovine ejaculated sperm. J Dairy Sci 69:2915-2924

Parrish JJ, Susko-Parrish JL, First NL, 1985. Effect of heparin and chondroitin sulfate on the acrosome reaction and fertility of bovine sperm in vitro. Theriogenology 24:537-549

Russell LD, Peterson RN, Hunt W, Strack LE, 1984. Posttesticular surface modifications and contributions of reproductive tract fluids to the surface polypeptide composition of boar spermatozoa. Biol Reprod 30:959-978

Woods A, Couchman JR, Johanson S, Hook M, 1986. Adhesion and cytoskeleton organisation of fibroblasts in response to fibronectin fragments. The EMBO Journal 5:665-670

INDUCTION OF 3β-HYDROXYSTEROID DEHYDROGENASE ACTIVITY BY INSULIN-LIKE

GROWTH FACTOR-I IN PRIMARY CULTURE OF PURIFIED LEYDIG CELLS

T. Lin, N. Vinson, J. Haskett, and E.P. Murono

Medical and Research Service
WJB Dorn Veterans' Hospital
Departments of Medicine, Pharmacology
and Physiology
University of South Carolina School of Medicine
Columbia, SC 29201

INTRODUCTION

Recent evidence suggests that insulin-like growth factor-I (IGF-I)/ somatomedin C (Sm C) may be involved in modulating gonadal steroidogenesis. High affinity IGF-I binding sites were found in a membrane fraction of whole testes and crude interstitial cells (Takano et al., 1976; Handelsman et al., 1985). Patients with isolated growth hormone deficiency have delayed puberty and low IGF-I levels (Zapf et al., 1981). Adequate therapy with growth hormone during childhood permits the occurrence of normal spontaneous adolescence (Laron et al., 1983). Recently we have reported that IGF-I has direct stimulatory effects on Leydig cell steroidogenesis (Lin et al., 1986a). hCG-stimulated testosterone was markedly enhanced in the presence of IGF-I. Furthermore, IGF-I also potentiated 8-bromo cyclic AMP-induced testosterone formation and hCG-stimulated cyclic AMP formation. This suggests that IGF-I has effects both at the hCG receptor sites and at the steps beyond cyclic AMP formation. In the present study, we have evaluated the effect of IGF-I on 3β-hydroxysteroid dehydrogenase activity (3β-HSD). 3β-HSD is one of the key enzymes associated with the smooth endoplasmic reticulum (SER) which converts Δ^5-3β-hydroxysteroids to Δ^4-3-ketosteroids. 3β-HSD activity is localized exclusively in the Leydig cells and its activity appears to be regulated by LH and hCG (Hafiez et al., 1974).

MATERIALS AND METHODS

Materials

Dulbecco's modified Eagle's medium (DMEM) and Ham's F12 nutrient mixture were from Gibco (Grand Island, NY). Pure biosynthetic IGF-I was from Amgen Biological (Thousand Oaks, CA). $(7-^3H)$-pregnenolone (19.3 Ci/mmol) and $(4-^{14}C)$-progesterone (57.2 mCi/mmol) were from New England Nuclear (Boston, MA). Each of the radioactive steroids was checked for purity by column or thin layer chromatography before use. Non-radioactive progesterone and pregnenolone were from Steraloid, Inc. and were recrystalized prior to use. Chromatography sheets (ITLC-SA), Gelman) were from Fisher Scientific Co.

Isolation and Culture of Purified Leydig Cells

Adult Sprague-Dawley rats (60-90-days-old) were obtained from Zivic-Miller Laboratories (Allison Park, PA). Purified Leydig cells were prepared as previously reported (Lin et al., 1986a). All procedures were carried out under sterile conditions. Purified Leydig cells (10^5 cells/ml) were plated in a 1:1 mixture of DMEM-F12 nutrient mixture with the addition of 1.2 g/liter sodium bicarbonate, 15 mM Hepes buffer, 100 U/ml penicillin, 100 µg/ml streptomycin with 0.1% fetal calf serum under 95% air-5% CO_2.

Assay of 3β-hydroxysteroid Dehydrogenase Activity

Cells were cultured with or without IGF-I for 24 h. Medium was removed and fresh medium was added with or without IGF-I and hCG for additional 24 h. Medium was then removed and saved for testosterone assay. Fresh medium with ^{3}H-pregnenolone (1 µM, 0.5 µCi) was added with 10 µM spironolactone to prevent further metabolism of pregnenolone. The incubation was carried out for 20 min. Reactions were terminated by adding 0.1 ml 1 N NaOH. ^{14}C-progesterone (25 µg, 4,000 cpm) and 25 µg each of unlabeled pregnenolone and progesterone were then added to estimate recovery and to serve as carriers. The media were removed and steroids extracted with 5 vol diethylether. The solvent was dried under vacuum and the residue chromatographed using chloroform (99.5 ml)/methanol (0.5 ml) as the solvent system to quantitate progesterone (Murono, 1983).

RESULTS

We found that 3β-HSD activity was linear up to 50 min of incubation (Fig. 1). Therefore, 20 min incubation time was chosen for all subsequent studies. We first evaluated the effect of hCG on 3β-HSD activity during short term culture. 3β-HSD of freshly isolated Leydig cells was 73.3 pmol progesterone formed/10^5 cells/20 min and the activity remained unchanged after 24 h in culture (80.9 pmol prog/10^5 cells/20 min). hCG (10 ng/ml) was then added and 3β-HSD activities were measured after varying time periods. 3β-HSD activity increased significantly at 3 h and remained elevated at 24 h (basal control, 140.1; and hCG treated group, 238 pmol prog/10^5 cells/20 min). Increased 3β-HSD activity correlated well with increased testosterone formation (Fig. 2). In the presence of IGF-I (50 ng/ml), 3β-HSD activity increased 50% above the control. Concomitant addition of IGF-I and hCG further increased 3β-HSD activity more than two fold (Fig. 3). IGF-I had no effect on cellular DNA contents and had no effect on ^{3}H-thymidine incorporation into Leydig cells (data not shown). Therefore, increased 3β-HSD activity was not due to increased cell number.

DISCUSSION

The insulin-like growth factors or somatomedins are a family of low molecular weight circulating growth factors that share considerable structural and functional similarities with proinsulin (Rechler and Nissley, 1985). IGF-I has a variety of effects including acute insulin-like metabolic actions and long-term growth promoting effects. The liver is the major site of synthesis of IGF-I and its circulating level is highly GH dependent. Recent report suggests that IGF-I can be synthesized by extrahepatic tissues and mRNA for IGF-I has been localized in testes (D'Ercole et al., 1984; Zangger et al., 1986). It was also reported that IGF-I was released by Sertoli cells and peritubular cells in culture and the secretion of IGF-I was dependent on GH and FSH (Tres et al., 1986). Therefore, IGF-I may be produced locally and have paracrine effects.

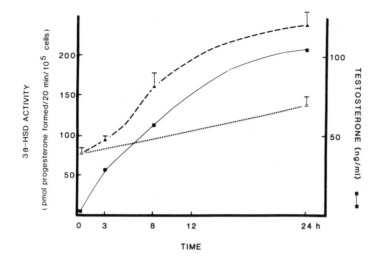

Fig. 1. Time course of 3β–HSD activity. Purified Leydig cells (10^5 cells/ ml) were cultured for 24 h. Medium was removed and fresh medium with ^{3}H–pregnenolone was added with 10 μM spironolactone. Reactions were terminated after varying time periods and 3β –HSD activity was measured as described in Materials and Methods. Results are the mean of duplicate incubations.

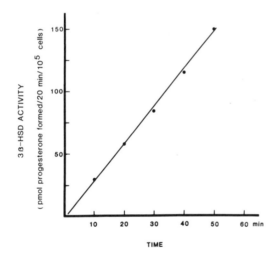

Fig. 2. The effects of hCG on Leydig cell 3β–HSD activity and testosterone formation. Purified Leydig cells (10^5 cells/ml) were cultured for 24 h. Medium was removed and fresh medium was added with (▲) or without (△) hCG (10 ng/ml), 3β–HSD activity was then estimated after varying time periods. Results are the mean ± SE of triplicate incubations.

We have reported previously that IGF-I in concentrations as low as 1 ng/ml significantly enhanced hCG–stimulated testosterone formation (Lin et al., 1986a). IGF-I also enhanced 8-bromo cyclic AMP–induced steroido- genesis and hCG–stimulated cyclic AMP formation. We have further identified high affinity IGF-I binding sites in purified Leydig cells by competitive binding and affinity cross linking. The binding of ^{125}I-IGF-I

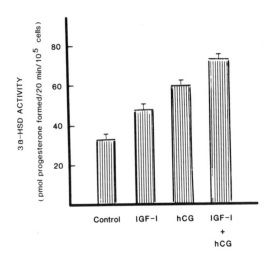

Fig. 3. The effects of IGF-I on Leydig cell 3β -HSD activity. Purified Leydig cells (10^5 cells/ml) were cultured with or without IGF-I (50 ng/ml) for 24 h. Fresh medium was then added with or without IGF-I and/or hCG (10 ng/ml). Cultures were continued for additional 24 h and 3β-HSD activity was measured. Results are the mean ± SE of triplicate incubations.

to purified Leydig cells was linear with a binding affinity of 0.56×10^9 M^{-1} and a capacity of 167 fmol/mg protein. Type I IGF receptor of purified Leydig cells has a mol wt of over 250,000 under non-reduced conditions and 130,000 under reduced conditions consistent with α-subunits of type I IGF receptor (Lin et al., 1986b). In the present study, we found that the stimulatory effects of IGF-I on Leydig cell steroidogenesis is associated with enhanced 3β-HSD activity, a postreceptor event. The effect of IGF-I on Leydig cell steroidogenesis is specific and cannot be mimicked by other growth factors. Transferrin in a concentration of 5 μg/ml has no effect on either basal or hCG-stimulated testosterone formation. EGF actually inhibited hCG-stimulated steroidogenesis (Lin et al., 1986a).

Even though 3β -HSD is not the rate limiting step in Leydig cell steroidogenesis, it is localized exclusively in the Leydig cells and its activity correlated well with increased testosterone formation. Using crude interstitial cells from adult hypophysectomized rats, it was reported that hCG initially stimulated 3β -HSD activity, then followed by a time-dependent inhibition of the enzyme activity (Ruiz de Galarreta et al., 1983). We did not observe this biphasic effect of hCG in our primary culture of Leydig cells from normal adult rats. 3β -HSD activity was increased in a time-dependent manner after the addition of hCG and correlated well with increased testosterone formation. Its activity was further enhanced in the presence of IGF-I. In conclusion, we have provided evidence that IGF-I can modulate Leydig cell steroidogenesis and increased testosterone formation is associated with induction of 3β-HSD activity.

ACKNOWLEDGEMENT

This study was supported by VA Medical Research Funds (T.L. and E.P.M.) and Biomedical Research Small Grant (E.P.M. and J.H.).

REFERENCES

D'Ercole AJ, Stiles AD, Underwood LE, 1984. Tissue concentrations of somatomedin C: further evidence for multiple sites of synthesis and paracrine or autocrine mechanism of action. Proc Natl Acad Sci 81:935-939

Hafiez AA, Philpott JE, Bartke A, 1971. The role of prolactin in the regulation of testicular function: the effect of prolactin and luteinizing hormone on 3β-hydroxysteroid dehydrogenase activity in the testes of mice and rats. J Endocrinol 50:619-623

Handelsman DJ, Spaliviero JA, Scott CD, Baxter RC, 1985. Identification of insulin-like growth factor-I and its receptors in the rat testis. Acta Endocrinol (Copenh) 109:543-549

Laron Z, Minouni F, Pertzelan A, 1983. Effects of human growth hormone therapy on penile and testicular size in boys with isolated growth hormone deficiency: first year of treatment. Isr J Med Sci 19:338-344

Lin T, Haskell J, Vinson N, Terracio L, 1986a. Direct stimulatory effects of insulin-like growth factor-I on Leydig cell steroidogenesis in primary culture. Biochem Biophys Res Commun 137:950-956

Lin T, Haskell J, Vinson N, Terracio L, 1986b. Characterization of insulin and insulin-like growth factor-I receptors of purified Leydig cells and their role in steroidogenesis: a comparative study. Endocrinology 119:1641-1647

Murono EP, 1983. Ethanol directly inhibits Δ^5-3β-hydrosteroid dehydro-genase-isomerase activity of rat testis interstitial cells. Steroids 42:457-468

Rechler MM, Nissley SP, 1985. The nature of regulation of the receptors for insulin-like growth factors. Annu Rev Physiol 47:425-442

Ruiz de Galarreta CM, Fanjul LF, Meidan R, Hsueh AJW, 1983. Regulation of 3β -hydroxysteroid dehydrogenase activity by human chorionic gonado-tropin, androgens and anti-androgens in cultured testicular cells. J Biol Chem 258:10988-10996

Takano K, Hall K, Frylund K, Sievertsson H, 1976. Binding of somatomedins and insulin to plasma membrane prepared from rat and monkey tissue. Horm Metab Res 8:16-24

Tres LL, Smith E, Van Wyk JJ, Kierszenbaum AL, 1986. Immunoreactive sites and accumulation of somatomedin C in rat Sertoli-spermatogenic cell co-culture. Exp Cell Res 162:33-50

Zapf J, Walter H, Froesch ER, 1981. Radioimmunological determination of insulin-like growth factor I and II in normal subjects and in patients with growth disorders and extrahepatic tumor hypoglycemia. J Clin Invest 68:1321-1330

Zangger I, Eberwine J, Barchas JD, Janson M, Rosenfeld RG, Hoffman AR, Hintz RL, 1986. Tissue distribution and abundance of IGF-I mRNA. Clin Res 34:68A

THE EFFECTS OF LOW O_2 AND ANTIOXIDANTS ON STEROIDOGENIC CAPACITY

IN CULTURED RAT LEYDIG CELLS

T.O. Abney and R.B. Meyers

Department of Physiology and Endocrinology
Medical College of Georgia
Augusta, GA 30912

The development of methods for Leydig cell culture in a chemically defined medium has proven of considerable value to numerous investigators in recent years. It is now possible to study hormone action and control of Leydig cell steroidogenesis using such a system. Most investigators report that mature rat Leydig cells in culture remain viable for 7-10 days, however, function (LH/hCG binding to membrane receptors and testosterone production) is lost after 3-4 days (Mather et al., 1981; Browning et al., 1983).

Steroidogenic cells in culture are known to be susceptible to O_2-mediated damage. For example, the activities of mitochondrial 11β-hydroxylase (Hornsby, 1980), as well as microsomal 17α-hydroxylase and 17-20 lyase (Quinn and Payne, 1984) are reduced by O_2 generated free radicals. The present study was an attempt to optimize the culture conditions by examining the effects of reduced O_2 tension and antioxidants on the steroidogenic capacity of cultured Leydig cells from the mature rat.

Adult (55-65-day-old) male Sprague-Dawley rats were utilized. Purified Leydig cells were isolated by collagenase dispersion – Percoll gradient centrifugation and cell number, determined by hemocytometer, was adjusted prior to culture. The cells were plated in 16 mm culture wells at a concentration of 2.5×10^5 cells/ml as described by Lin et al. (1986). Cultures were maintained at 37° in a humidified atmosphere under standard conditions of 5% CO_2 and 95% air, which is equivalent to 19% O_2, or with 5% CO_2 : 5% O_2 : 90% N_2. Ascorbate (0.1 mM) or dimethylsulfoxide (DMSO, 100 mM) were used as antioxidants. Cell function and viability were assessed by (1) ^{125}I-hCG binding, (2) gonadotropin-stimulated steroidogenesis and (3) cell attachment to the culture wells. Steroids were measured by radioimmunoassay.

As shown in Table 1, the number of cells, expressed as a percent of the cells initially adhered at 3 hr, were relatively constant through day 4. In contrast, the membrane receptors (^{125}I-hCG binding) decreased dramatically during the first 24 hr and then fell slowly to less than 1.0% by day 4. Thus our culture system supports cell viability, however, as reported by others (Mather et al., 1981), specific cell functions, in this case gonadotropin binding, appears to be lost.

Table 1. Leydig Cell Viability and hCG-binding Capacity

	Day of Culture	hCG-binding % of Control	Cell number % of Control
Control	0	100	100
	1	31.7 + 4.2	91.5 + 2.5
	2	25.7 + 7.8	92 + 13.5
	3	11.3 + 2.0	82 + 17.5
	4	<1	75 + 15

Controls represent values obtained 3 hr after plating the cells at which time > 85% of the cells had adhered.

As shown in Fig 1, hCG (10 mIU)-stimulated testosterone production in the presence of 19% O_2 (95% air) fell markedly during the first four days and was not detectable by RIA after day 5. In the presence of 100 mM DMSO, this decrease in testosterone production was less during each day of culture through day 4. Low O_2 (5%) and DMSO offered an even greater protective effect and significantly prolonged the steroidogenic capacity of the cultures. No such effects were observed with ascorbate. In essence, testosterone production, expressed as ng/24 hr/well, in the presence of 5% O_2 and DMSO was equivalent on day 3 (16.4 ng) to DMSO alone on day 2 (15.3 ng) which was twice the control level on day 2 (7.8 ng). As a result of these experiments, we have standardized the system to include 100 mM DMSO in all media and cultures are conducted in an atmosphere of 5% O_2: 5% CO_2: 90% N_2.

One of the most interesting observations made in these studies was that progesterone production, expressed as ng/24 hr/ml, was responsive to hCG through day 10 of culture as illustrated in Fig 2. On day 2, hCG-stimulated steroid production was predominantly testosterone. However, while testosterone production decreased as a function of culture period, progesterone production rose markedly. hCG-stimulated progesterone production was maintained throughout the culture period, while basal progesterone levels were undetectable after day 1. These results suggest that Leydig cells in our culture system not only remain viable, but also remain responsive to gonadotropin hormone and retain steroidogenic capacity.

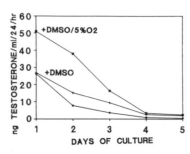

Fig. 1. Effects of DMSO and reduced oxygen tension on hCG-stimulated testosterone production. Leydig cells were cultured under conditions of 95% air, 5% CO_2 ± DMSO or under conditions of low O_2 (5% O_2, 90% N_2, 5% CO_2) with DMSO.

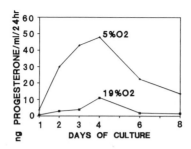

Fig. 2. hCG stimulated progesterone production in the presence of standard (19%) or reduced (5%) O_2 levels.

This suggests that the gonadotropin receptors are retained at least in sufficient numbers to render the cells sensitive to hCG action.

Another interesting observation was the effect of reduced O_2 levels on hCG stimulated progesterone production. As shown in Fig. 2, cells cultured under conditions of reduced O_2 (5%) produced 8 to 10 fold more progesterone than cells cultured under conditions of 19% O_2. Unlike the effect of reduced O_2 on testosterone production, progesterone production was enhanced on days 1 through 8 of culture.

Cells cultured under conditions of 19% O_2 displayed an immediate decrease in total steroid (testosterone plus progesterone) production as shown in Fig 3. However cells cultured under reduced O_2 tension (5% O_2) exhibited a stabilization of total steroid production during the first four days of culture. Although steroid levels fell after day 4 they remained elevated above those of cultures incubated under conditions of 19% O_2.

These data show that reduction of O_2 tension and the use of the antioxidant DMSO result in enhanced steroidogenic capacity in the cultured rat Leydig cell. This suggests that the cytochrome P-450 steroidogenic enzymes of the cultured rat Leydig cell are susceptible to free radical damage. The increased progesterone production in the presence of reduced O_2 indicates for the first time that O mediated damage also occurs at the level of the mitochondrial cholesterol side chain cleavage enzyme (a cytochrome P-450 complex).

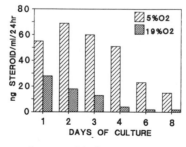

Fig. 3. hCG-stimulated total steroid (testosterone and progesterone) production in the presence of standard (19%) or reduced (5%) O_2 levels.

REFERENCES

Browning JY, Heindel JJ, Grotjan HE, 1983. Primary culture of purified Leydig cells isolated from adult rat testes. Endocrinology 112:543-549

Hornsby PJ, 1980. Regulation of cytochrome P-450-supported 11β -hydroxylation of deoxycortisol by steroids, oxygen and antioxidants in adrenocrotical cell cultures. J Biol Chem 255:4020-4027

Lin T, Haskell J, Vinson N, Terracio L, 1986. Direct stimulatory effects of insulin-like growth factor-I on Leydig cell steroidogenesis in primary culture. Biochem Biophys Res Commun 137:950-955

Mather JP, Saez JM, Haour F, 1981. Primary cultures of Leydig cells from rat, mouse and pig: advantages of porcine cells for the study of gonadotropin regulation of Leydig cell function. Steroids 38:35-44

Quinn PG, Payne AH, 1984. Oxygen-mediated damage of microsomal cytochrome P-450 enzymes in cultured Leydig cells. J Biol Chem 259:4130-4135

DIRECT EFFECTS OF HORMONES OF THE HYPOTHALAMIC-PITUITARY-THYROID

AXIS ON TESTICULAR STEROIDOGENESIS IN HAMSTERS

A.G. Amador, A. Bartke, and R.W. Steger

Department of Physiology
School of Medicine
Southern Illinois University
Carbondale, IL 62901-6512

INTRODUCTION

Hormones of the hypothalamic-pituitary-thyroid axis have been shown to influence testicular function in rodents directly or indirectly. Panke et al. (1978) reported that experimental induction of hypo- or hyper-thyroidism in the rat led to reductions in plasma testosterone and in testicular LH receptors without altering peripheral concentrations of LH, FSH or PRL. There is a well known relationship between hypo- and hyper-thyroidism, and fertility disorders in both men and women (Ingbar, 1985). Vriend and Wasserman (1986) showed that the thiourea-induced increase in TSH can partially reverse the short-photoperiod related testicular regression in Syrian hamsters. Furthermore, in genetically hypothyroid mice in which thyroid hormones are absent and plasma TSH levels are extremely high, Leydig cell responsiveness to hCG in vivo is signifi-cantly increased by TSH (Amador et al., 1986). Moreover, specific receptors for both TSH and thyroid hormones have been detected in rodent testes (Oppenheimer et al., 1974; Davies et al., 1978). Therefore, the present study was designed to study the effects of TRH, TSH, calcitonin (CTH) and thyroid hormones in vitro on testicular steroidogenesis in Syrian and Siberian hamsters.

METHODS

Syrian Hamsters

Adult (> 3 months old) Syrian (Golden) hamsters (Mesocricetus auratus) (Lak:LVG[SYR]; Charles River Lakeview) used in this study were maintained in a room with controlled temperature (20 ± 2C) and light (16 h light:8 h dark).

To study in vitro steroidogenesis, testes were decapsulated and divided into 12 fragments of similar weight, and incubated in 2 ml Krebs-Ringer bicarbonate containing glucose (1 mg/ml) for 4 h at 32 ± 2C under oxygen:carbon dioxide (95:5), in the presence of 0, 3.125 or 12.5 mIU hCG/ml. At each of these doses of hCG, the incubation media contained also, depending on the experiment: 0, 2, 20 or 200 mIU bTSH/ml; 0, 0.36, 3.6 or 36 ng TRH/ml; 0, 0.2, 2 or 20 ng thyroxine/ml; 0, 0.03, 0.3, or 3 ng triiodothyronine/ml; 0, 3.4, 34, or 340 ng hCTH/ml. All hormones were obtained from Calbiochem (San Diego, CA). The supplier states that the bTSH preparation does not "cross react" with LH. Incubation media testosterone levels were measured by radioimmunoassay (Wolfe et al., 1981).

Siberian Hamsters

Adult (> 3 months old) Siberian (Djungarian) hamsters (Phodopus sungorus) from our animal colony were maintained since birth under controlled light (16 h L:8 h D) and temperature (20 ± 2C) conditions.

The study of in vitro steroidogenesis was undertaken in conditions similar to those described for Syrian hamsters. Testes were divided into 6 fragments and incubated in the presence of 0 or 12.5 mIU hCG/ml. One fragment from each of these groups received, in addition, 0, 2 or 20 mIU bTSH/ml.

Statistics

Data were analyzed by two-way analysis of variance. Data were tested for normality of distribution by the Kolmogorov-Smirnov test and for homogeneity of variance by Barlett's test. Log and square root transformations were utilized as needed (Sokal and Rohlf, 1981). Since the present in vitro system as low intra-assay variability, and a high inter-assay variation, only qualitative comparisons can be made among assays.

RESULTS

In the incubation of Syrian hamster testes hCG produced the expected increase in the accumulation of testosterone in the media. The presence of bTSH at any dose elevated testosterone to levels higher or similar to those measured in the presence of hCG alone. The combined addition of bTSH and hCG did not cause any further increase in testosterone levels (Table 1a). Incubation with TRH, thyroxine, triiodothyronine (not shown) or hCTH did not increase media testosterone levels, and did not alter the steroidogenic response to hCG (Table 1b, c, d).

In Siberian hamsters incubation with bTSH elevated media testosterone to levels similar to those measured in the presence of hCG. Incubation with both hCG and bTSH did not increase testosterone above levels obtained in the presence of either hormone alone (Table 1e).

DISCUSSION

Of all the hormones of the hypothalamic-pituitary-thyroid axis used in the present study, only TSH showed an effect on in vitro steroidogenesis in Syrian hamster testes. However, this effect was maximal at all doses used. It was equal or superior to that obtained with the highest dose of hCG used, and a combined treatment with both TSH and hCG did not cause a further increase in the production of testosterone. In Siberian hamsters, TSH had the same effects on in vitro testosterone synthesis as those observed in Syrian hamsters. Addition of TSH, at the doses employed, increased testosterone levels in the same extent as 12.5 mIU hCG/ml, and the combination of both hormones caused no further increase in testosterone accumulation. The present data agrees with our previous findings in genetically hypothyroid mice in which, in the absence of triiodothyronine and thyroxine, and in the presence of extremely high levels of TSH, in vivo stimulation of testosterone synthesis by hCG is enhanced compared to normal mice. Also, in these mice testicular LH receptor synthesis, after hCG treatment, appears to be enhanced (Amador et al., 1986). In Syrian hamsters with short photoperiod-induced testicular

Table 1. Media testosterone levels from Syrian and Siberian hamsters testes incubated with hCG, hormones of the hypothalamic-pituitary-thyroid axis or both. Six Syrian or seven Siberian hamsters were used per incubation. Values are expressed as pg/mg (mean ± SEM). Values with the same letter are statistically similar (Student-Newman-Keuls multiple range test).

	hCG (mIU/ml)		
	0	3.125	12.5

Syrian Hamsters

A) bTSH (mIU/ml)

	0	3.125	12.5
0	123.5 ± 17.5 (a)	206.5 ± 27.4 (b)	689.5 ± 111.8 (c)
2	1085.0 ± 120.2 (d)	1066.2 ± 114.0 (cd)	1100.7 ± 161.5 (d)
20	717.0 ± 99.4 (cd)	707.3 ± 39.1 (cd)	1071.2 ± 172.5 (cd)
200	767.0 ± 157.7 (cd)	963.3 ± 79.7 (cd)	1020.8 ± 193.0 (cd)

B) TRH (ng/ml)

	0	3.125	12.5
0	187.7 ± 23.2 (ab)	249.2 ± 28.3 (bc)	291.3 ± 25.0 (c)
0.36	190.3 ± 19.9 (ab)	241.8 ± 25.9 (bc)	311.7 ± 36.8 (c)
3.6	194.2 ± 22.1 (ab)	269.8 ± 19.6 (c)	333.3 ± 11.4 (c)
36	170.7 ± 18.5 (a)	284.2 ± 42.4 (c)	326.7 ± 26.4 (c)

C) Thyroxine (ng/ml)

	0	3.125	12.5
0	311.2 ± 17.4 (a)	594.8 ± 46.5 (c)	554.6 ± 80.1 (c)
0.2	310.7 ± 24.4 (a)	477.8 ± 59.6 (bc)	588.0 ± 53.3 (c)
2	286.8 ± 23.4 (a)	484.3 ± 43.3 (bc)	551.5 ± 79.4 (c)
20	355.5 ± 39.9 (ab)	484.3 ± 35.4 (bc)	595.0 ± 69.4 (c)

D) hCTH (μg/ml)

	0	3.125	12.5
0	184.5 ± 42.2 (abc)	266.5 ± 29.3 (bcd)	444.2 ± 67.5 (de)
3.4	142.8 ± 26.5 (a)	289.7 ± 25.3 (cde)	458.0 ± 40.0 (de)
34	158.0 ± 35.4 (a)	248.0 ± 26.8 (bcd)	510.3 ± 45.1 (e)
340	157.2 ± 28.6 (ab)	284.2 ± 35.7 (cde)	435.8 ± 64.8 (de)

Siberian Hamsters

E) bTSH (mIU/ml)

	0	3.125	12.5
0	13.7 ± 2.0 (a)	----	71.2 ± 7.5 (b)
2	66.3 ± 6.4 (b)	----	57.8 ± 5.8 (b)
20	70.1 ± 5.7 (b)	----	77.9 ± 6.4 (b)

atrophy and reduction in plasma TSH levels, the thiourea-induced elevation in plasma TSH has been shown to partially reverse the testicular atrophy (Vriend and Wasserman, 1986). Therefore, since testicular TSH receptors have been measured in rodents (Davies et al., 1978), the present results argue strongly for a direct stimulatory role of TSH on Leydig cell function.

ACKNOWLEDGEMENTS

The present study was supported by NIH grants HD20001 and HD20033 (AB). We thank Ms. Sherie Hodges and Ms. Rita Meyers for their technical assistance, and Ms. Karen Graham for her help in preparing the manuscript.

REFERENCES

Amador AG, Parkening TA, Beamer WG, Bartke A, Collins TJ, 1986. Testicular receptors and circulating hormone levels in three mouse models for inherited diseases (Tfm/y, lit/lit and hyt/hyt). Endocr Exp 20, in press

Davies TF, Smith BR, Hall R, 1978. Binding of thyroid stimulators to guinea pig testis and thyroid. Endocrinology 103:6–10

Inbar SH, 1985. The thyroid gland. In: Wilson JD, Foster DW (eds.), Williams Textbook of Endocrinology, Philadelphia: WB Saunders Co., pp. 682–815

Oppenheimer JH, Schwarz HL, Surks MI, 1974. Tissue differences in the concentration of triiodothyronine nuclear binding sites in the rat: liver, kidney, pituitary, heart, brain, spleen and testis. Endocrinology 95:897–903

Panke VG, Brandt A, Leidenberger F, 1978. Gonadal function of male and female rats in experimentally induced hypo- and hyperthyroidism. Acta Endocrinol Suppl 215:99–100

Sokal RR, Rohlf FJ, 1981. Biometry, 2nd Edn. San Francisco: WH Freeman and Co.

Virend J, Wasserman RA, 1986. Effects of afternoon injections of melatonin in hypothyroid male Syrian hamsters. Neuroendocrinology 42:498–503

Wolfe HG, Bartke A, Amador A, Van Sickle M, Dalterio S, Brown D, 1981. Testicular function in strains of mice selected for differences in gonadotropin-induced ovulation rate. J Endocrinol 90:367–373

HOW DOES THE GONAD AFFECT LHRH SECRETION? EFFECTS OF GONADECTOMY ON LHRH
RELEASE FROM MEDIAN EMINENCE NERVE TERMINALS INCUBATED IN VITRO AND ON
LHRH CONCENTRATION IN HYPOPHYSEAL PORTAL BLOOD

M.M. Valenca, M. Ching, C. Masotto, and A. Negro-Vilar

Reproductive Neuroendocrinology Section, LRDT
National Institute of Environmental Health Sciences
National Institutes of Health
Research Triangle Park, NC 27709

INTRODUCTION

It is well established that the central nervous system exerts an
important influence on many aspects of reproductive function, particularly by controlling the secretion of pituitary gonadotropins. This
is accomplished by way of specific brain messengers which, in turn,
regulate pituitary endocrine function. The primary hypothalamic hormonal
messenger regulating luteinizing hormone (LH) secretion is the decapeptide LHRH, which is synthetized as a large molecular weight prohormone in
the neuronal cell bodies. After axonal transport, the mature decapeptide
LHRH is stored in nerve terminals in the median eminence (ME) from which
it is secreted into the hypophyseal portal circulation. The secretion of
LHRH seems to be under the influence of multifactorial regulatory mechanisms in which numerous extracellular messengers may participate,
including the catecholamines (Negro-Vilar et al., 1979), peptides
(Valenca et al., 1985; Kalra and Kalra, 1984) and sex steroid hormones
(Rudenstein et al., 1979; Kalra and Kalra, 1984). A review of some of
these regulatory mechanisms is included in this volume (see chapter by
Negro-Vilar, Valenca and Culler).

A great deal of information has been published on the influence
exerted by sex steroids and other gonadal factors on the hypothalamic-
hypophyseal axis. Gonadectomy leads to a rise in plasma LH levels caused
by the removal of the well-known inhibitory influence of gonadal hormones
on the hypothalamic-pituitary axis. Gonadal hormones may modulate LH
secretion directly by affecting the gonadotrophs, or indirectly, by
inducing changes in the synthesis, transport, processing, storage,
degradation and/or neurosecretion of LHRH.

Our studies were designed to evaluate the changes in the secretion of
LHRH and LH which might occur after orchidectomy (ORDX), using a combination of in vivo and in vitro approaches, in order to obtain a better
estimate of the relative contribution of central (hypothalamic) vs.
peripheral (pituitary) sites mediating the rise in gonadotropins.

METHODOLOGY

In order to quantitate changes in the LHRH content of select brain regions after ORDX, groups of adult Sprague-Dawley rats were sacrificed at different times following ORDX. Brain regions (preoptico-anterior-hypothalamus, mediobasal hypothalamus without ME, and ME) were dissected, homogenized in 0.1M HCl, and heated for 10 minutes in a boiling water bath. The samples were then stored at -70C until assayed for LHRH immunoreactivity by RIA. Additionally, hypothalamic median eminence fragments were incubated in vitro in a Krebs-Ringer bicarbonate glucose buffer, and the release of LHRH was evaluated by RIA as previously described by Negro-Vilar et al. (1979). Hypophyseal portal blood was collected as described elsewhere (Ching, 1982). Briefly, under Althesin (Glaxo, Research Triangle Park, N.C.) anesthesia, the hypothalamo-pituitary region was exposed using a parapharyngeal surgical approach. Samples of portal blood were collected at 10-minute intervals for up to 90-120 minutes, the peptide extracted using C-18 cartridges, and the LHRH concentrations determined by RIA. Trunk blood samples were obtained in certain experiments by decapitation. LH and FSH were measured using RIA kits provided by NIADDK. In order to evaluate pituitary responsiveness after ORDX, hemipituitaries obtained from intact and orchidectomized adult rats were incubated for 90 minutes in the presence or absence of LHRH (5 nM).

RESULTS AND DISCUSSION

The characteristic post-orchidectomy increase in plasma LH and FSH was accompanied by a substantial decrease in the LHRH content of the preoptico-mediobasal hypothalamic region (not shown) and even more clearly at the level of the ME (see Fig. 1A). These changes in content have been interpreted by many laboratories as an indication of an increased secretion of LHRH into the hypophyseal portal circulation which, in turn, would account for the enhanced secretion of the gonadotropins. However, evaluation of LHRH secretion from ME nerve terminals incubated in vitro showed a decrease in basal LHRH secretion after ORDX, such that by 3 days post-castration LHRH release was significantly lower than that of the intact controls. The lowest levels of LHRH secretion in vitro were observed on day 10 post-ORDX (Fig. 1B). To corroborate this unexpected finding, other experiments were designed to obtain an in vivo measure of the changes in LHRH secretion after ORDX. With this purpose, hypophyseal portal blood was collected from intact and 10-day ORDX rats. As illustrated in Fig. 1C, significantly lower levels of LHRH were observed in portal plasma from castrated rats. This would suggest that ORDX diminishes the rate of LHRH secretion in vivo, a result which agrees with that described in our in vitro studies. Subcutaneous implantation of dihydrotestosterone (DHT) or testosterone (T) pellets at the time of ORDX reversed this effect (data not shown). Additional studies indicate that while both intact and ORDX animals present a pulsatile pattern of LHRH secretion, ORDX reduces the mean level of LHRH in hypophyseal portal plasma without significantly changing the nadir-to-peak amplitude or frequency of the LHRH pulses, at least during the interval post-ORDX examined in the study (Ching, Valenca and Negro-Vilar, unpublished results).

Recently, Dluzen and Ramirez (1985) investigated the in vivo LHRH output from the mediobasal hypothalamus with the aid of a push-pull cannula. In agreement with our results, they observed a lower overall mean release of LHRH with attenuated pulse amplitude in castrated rats as compared with intact animals. In contrast, Eskay et al. (1977) described LHRH levels of 27-33 pg/ml in hypophyseal portal plasma of intact male rats, and by eight weeks after ORDX those levels increased to 48-60 pg/ml. The authors hypothesized, however, that LHRH release was impaired by the pentobarbital anesthetic that they had used. Indeed, in the present studies we found that under althesin anesthesia the LHRH levels in

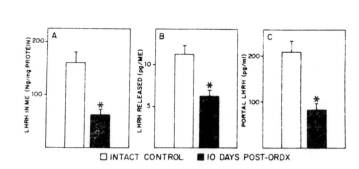

Fig. 1. Effects of orchidectomy (ORDX) on median
eminence (ME) LHRH content and secretion.
A: LHRH content in the ME; B: LHRH release from ME nerve terminals incubated
in vitro for 30 minutes; C: Concentration
of LHRH in the hypophyseal portal plasma
of intact and orchidectomized rats.

the portal plasma of intact rats were at least 5-fold higher than those
reported by Eskay et al. (1977).

We also evaluated changes in pituitary responsiveness after ORDX in
vitro by incubating hemipituitaries obtained from 10-day ORDX or intact
rats. Basal LH secretion was 2-fold higher from pituitaries obtained
from ORDX animals, and the response to 5 nM LHRH was enhanced 5-fold over
that of pituitaries from intact rats (data not shown). It is well
established that gonadectomy increases the number of pituitary LHRH
receptors (Weiland et al., 1986). In fact, it seems that the amount of
LH released correlates well with the number of LHRH receptors present in
that gland (Clayton et al., 1982), an observation in agreement with the
enhanced responsiveness to LHRH observed in our study.

Sex steroids may affect the synthesis, transport, processing, and,
consequently, the secretion of LHRH by at least two possible mechanisms:
(a) through the classical genomic activation and (b) by a non-genomic
mediated effect, which is still not completely understood. Several
groups have found that certain neuronal populations involved in LHRH
regulation, such as β-endorphin and dynorphin (Morrel et al., 1984),
and catecholamine (Sar, 1984) neurons, are capable of accumulating sex
steroids. Thus, the influence of the gonadal steroids on LHRH neurons
could be mediated through peptidergic and/or catecholaminergic neurons
rather than by a direct action of the sex hormones on the LHRH cells
(Morrel et al., 1984). Furthermore, the rise in the gonadotropins
following gonadectomy may also participate in the suppressive effect of
ORDX on LHRH release through an enhancement of the negative short loop
feedback of pituitary gonadotropins on LHRH neurons. In this regard, an
inverse relationship between LHRH and LH levels has been described in
intrasellar-collected pituitary blood from humans (Carmel et al., 1979).
A schematic representation of how gonadal and pituitary factors might
affect LHRH secretion is shown in Fig. 2.

Rudenstein et al. (1979) demonstrated that in vitro LHRH release
from hypothalami of rats 7 days post-ORDX was less than from hypothalami
obtained from intact animals. In addition, T, DHT, and estradiol all
prevented the decrease in LHRH release observed in ORDX animals. These

619

observations led the authors to postulate that the sex steroids would maintain, rather than inhibit, LHRH secretion in vivo. Similar observations and conclusions were reached by Kalra et al. (1987). Carmel et al. (1976) called the failure of estrogen injections to suppress hypophyseal portal LHRH concentration in monkeys a "surprising paradoxical effect," since similar doses of estradiol, in amounts mimicking early follicular phase levels, decrease LH levels and suppress "circhoral" LH variations.

Even though LHRH concentrations in hypophyseal portal plasma decrease at 10 days post-ORDX, the rise in plasma LH levels following ORDX seems to be mediated by LHRH since immunoneutralization of LHRH or the use of an LHRH antagonist can reverse this effect (Clayton et al., 1982; Culler and Negro-Vilar, 1986). These observations suggest that the post-ORDX elevation in plasma LH levels may be due to increased pituitary sensitivity to LHRH. The exaggerated responsiveness of the pituitary to LHRH after ORDX may be explained by several mechanisms: (a) the absence of the direct negative influence of gonadal factors over the gonadotrophs, (b) changes in the pattern of LHRH secretion that modify the responsiveness of the pituitary, or (c) a combination of both. With regard to FSH, it appears that a putative FSH-releasing factor(s) may have some role on the regulation of FSH secretion (Culler and Negro-Vilar, 1986), although further studies are necessary to clarify this issue.

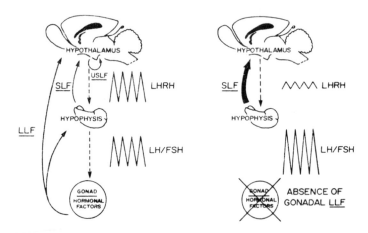

Fig. 2. This diagram shows the hypothetical feedback control mechanisms modulating the hypothalamic-hypophyseal-testicular axis. The neurohormone LHRH is released in a pulsatile pattern from ME nerve terminals into the hypophyseal portal blood. At the pituitary level LHRH activates specific receptors, and as a result, the gonadotropins (LH and FSH) are also secreted in an episodic fashion. The actions of the gonadotropins on the testes promote the production of sex steroids and other gonadal factors which, in turn, influence brain LHRH secretion directly and/or indirectly (left panel). Please note that in the absence of the gonadal long loop feedback (LLF) after ORDX (right panel), high amplitude gonadotropin pulses are present in the plasma in contrast to the diminished output of LHRH in the specialized portal circulation. USLF = ultra-short loop feedback of LHRH on LHRH neurons; SLF = short loop feedback of pituitary gonadotropins on LHRH neurons.

The data presented here indicate that the elevated circulating LH levels, detected several days after ORDX, are not the result of a sustained increased secretion of LHRH into the portal circulation but, rather, may primarily reflect an enhanced sensitivity of the gonadotrophs to the action of this decapeptide. In addition, our data also suggest that, contrary to current thinking, removal of gonadal steroids and/or other gonadal factors may result in a decreased secretion of LHRH from ME nerve terminals into the hypophyseal portal blood. Additional studies to determine differences in the synthesis and processing of the LHRH prohormone molecule between intact and gonadectomized animals may also help to clarify this issue.

REFERENCES

Carmel PW, Antunes JL, Ferin M, 1979. Collection of blood from the pituitary stalk and portal veins in monkeys, and from the pituitary sinusoidal system of monkey and man. J Neurosurg 50:75-80

Carmel PW, Araki S, Ferin M, 1976. Pituitary stalk portal blood collection in rhesus monkeys: evidence for pulsatile release of gonadotropin-releasing hormone (GnRH). Endocrinology 99:243-48

Ching M, 1982. Correlative surges of LHRH, LH and FSH in pituitary stalk plasma and systemic plasma of rat during proestrus: effect of anesthetics. Neuroendocrinology 34:279-84

Clayton RN, Popkin RM, Fraser HM, 1982. Hypothalamic regulation of pituitary gonadotropin-releasing hormone receptors: effects of gonadotropin-releasing hormone immunoneutralization. Endocrinology 110:1116-23

Culler MD, Negro-Vilar A, 1986. Evidence that pulsatile follicle-stimulating hormone secretion is independent of endogenous luteinizing hormone-releasing hormone. Endocrinology 118:609-12

Dluzen DE, Ramirez VD, 1985. In-vivo activity of the LH-releasing hormone pulse generator in castrated and intact male rats. J Endocr 107:331-40

Eskay RL, Mical RS, Porter JC, 1977. Relationship between luteinizing hormone releasing hormone concentration in hypophyseal portal blood and luteinizing hormone release in intact, castrated, and electrochemically stimulated rats. Endocrinology 100:263-70

Kalra PS, Crowley WR, Kalra SP, 1987. Differential in vitro stimulation by Naloxone and K+ of luteinizing hormone-releasing hormone and catecholamine release from the hypothalami of intact and castrated rats. Endocrinology 120:178-85.

Kalra SP, Kalra PS, 1984. Opioid-adrenergic-steroid connection in regulation of luteinizing hormone secretion in the rat. Neuroendocrinology 38:418-26

Morrell JI, Schwanzel-Fukuda M, Fahrbach SE, Pfaff DW, 1984. Axonal projections and peptide content of steroid hormone concentrating neurons. Peptides 5:227-39

Negro-Vilar A, Ojeda SR, McCann SM, 1979. Catecholaminergic modulation of luteinizing hormone-releasing hormone release by median eminence terminals in vitro. Endocrinology 104:1749-57

Rudenstein RS, Bigdeli H, McDonald MH, Snyder PJ, 1979. Administration of gonadal steroids to the castrated male rat prevents a decrease in the release of gonadotropin-releasing hormone from the incubated hypothalamus. J Clin Invest 63:262-67

Sar M, 1984. Estradiol is concentrated in tyrosine hydroxylase-containing neurons of the hypothalamus. Science 223:938-40

Valenca MM, Johnston CA, Negro-Vilar A, 1985. Gonadal inhibitory effects of LHRH analogs: evidence for a hypothalamic site of action mediated by opiate peptides. J Andrology (supp) 6:41

Weiland NG, Barraclough CA, Catt KJ, 1986. Effects of long- and short-term gonadectomy on the hypothalamo-hypophysial (LH-releasing hormone-LH) system in oestrogen-treated male and female rats. J Endocr 110:367-73

ORCHIDECTOMY INDUCES TEMPORAL AND REGIONAL CHANGES IN THE

SYNTHESIS AND PROCESSING OF THE LHRH PROHORMONE IN THE RAT BRAIN

M.C. Culler, W.C. Wetsel, M.M. Valenca, C.A. Johnston,
C. Masotto, M. Sar*, and A. Negro-Vilar

Reproductive Neuroendocrinology Section, LRDT
National Institute of Environmental
   Health Sciences, NIH
Research Triangle Park, NC 27709
*Department of Anatomy
University of North Carolina
Chapel Hill, NC 27514

INTRODUCTION

Recently, the sequence of the cDNA which encodes the LHRH-prohormone was elucidated from human placenta and human and rat hypothalamus and the corresponding amino acid sequence deduced (Seeburg et al., 1984 and Adelman et al., 1986). In addition to LHRH, the prohormone contains a 56 amino acid sequence, designated gonadotropin-releasing hormone associated peptide (GAP), which is attached to the C-terminus of the LHRH decapeptide. Although not yet confirmed, the human GAP sequence has been reported to possess both gonadotropin-releasing and prolactin inhibiting activity (Nikolics et al., 1985). Additionally, a 13-amino acid fragment of the human GAP sequence (proLHRH 14-26) has been reported to stimulate gonadotropin release (Millar et al., 1986). Regardless of whether the non-LHRH portion of the LHRH prohormone contains biological activity, this sequence can serve as a valuable marker for studies of LHRH prohormone synthesis, processing and degradation. In order to initiate these types of studies, we have generated specific antisera (MC-1, 2 and 3) against a fragment of the human GAP sequence (proLHRH 38-66) and developed a radioimmunoassay procedure for the quantitation of GAP (Culler and Negro-Vilar, 1986). The antisera are specific for midportion sequences of the GAP molecule and do not cross-react with any other known brain peptide.

GAP-like immunoreactivity (GAP-LI) can be detected in extracts of brain regions, known to contain LHRH nerve perikarya, fibers and terminals, taken from intact male rats. The quantity of GAP-LI is approximately equimolar with the LHRH-like immunoreactivity (LHRH-LI) in these same extracts (Culler and Negro-Vilar, 1986). Using immunocytochemical techniques, the anti-GAP sera are able to stain neuronal structures, especially perikarya, under conditions of tissue preparation which have often made immunostaining of LHRH difficult or undetectable (Sar et al., 1987). GAP-LI and LHRH-LI are distributed similarly in both male and female rat brains at all neuronal levels (perikarya, fibers and terminals). Using a double staining procedure, both GAP-LI and LHRH-LI can be demonstrated to be present within the same neuron (Sar et al., 1987).

These observations agree with the immunocytochemical findings of the Genentech group using a different anti-GAP serum (Phillips et al., 1985).

We have also examined brains from the mutant hypogonadal (hpg) mouse for the presence of GAP-LI. The hpg mouse is characterized by a congenital absence of LHRH. In extracts of various hypothalamic regions from hetero-zygous control mice, both LHRH-LI and GAP-LI are readily detectable using RIA. Immunocytochemical localization of GAP-LI in the control mice again revealed a common distribution with LHRH-LI. Neither LHRH-LI nor GAP-LI, however, are detectable in brains from homozygous hpg mutant mice using either technique, suggesting that the LHRH prohormone is also absent in these animals (Culler MD, Sar M and Negro-Vilar A; unpublished results). Recently, this finding has been substantiated by the discovery that a large segment of the gene sequence encoding the GAP peptide is absent in the hpg mouse (Mason et al., 1986a,b). Interestingly, the gene sequence encoding the LHRH decapeptide is intact and transcriptionally active in the hpg mouse, suggesting that the GAP sequence may be required for normal translation, synthesis and/or processing.

Collectively, the results of these studies support the concept that GAP represents a portion of the LHRH prohormone. Use of the GAP sequence as a marker for studies of LHRH prohormone processing should, therefore, be appropriate and should provide useful information concerning the regulation of the LHRH system.

EFFECTS OF ORCHIDECTOMY ON THE LHRH-PROHORMONE SYSTEM

Gonadal factors, both steroid and peptide, play a major role in the regulation of LHRH although the exact mechanisms remain largely unknown. Using parallel quantitative of the free, intact LHRH decapeptide and the GAP sequence recognized by the anti-GAP serum as an index, we have initiated studies to determine the influence of the gonadal factors on LHRH-prohormone processing in select brain regions.

Adult, male Sprague-Dawley rats (approx. 300-350 g), orchidectomized for 0, 1, 2, 3, 5, 7, 14, and 40 days, were decapitated and the median eminence (ME), hypothalamus (HYP), and preoptic-anterior hypothalamic areas (POA) rapidly dissected from the brain. The tissue fragments were immediately homogenized in 0.1 N acetic acid and heated to destroy enzy-matic activity. After centrifugation, the supernatants were lyophilized and resuspended in assay buffer. Both GAP-LI and LHRH-LI were quantitated by RIA using anti-GAP serum MC-2 and Rice's anti-LHRH serum #5 as previ-ously described (Culler and Negro-Vilar, 1986). This anti-LHRH serum has been demonstrated to recognize only the intact, free LHRH decapeptide (Hedger et al., 1985). Trunk blood, collected at the time of sacrifice, was assayed for LH.

Extracts of brain regions from intact, male rats were found to have similar molar quantities of GAP-LI and LHRH-LI as previously observed (Culler and Negro-Vilar, 1986). One day after orchidectomy, the level of LHRH-LI in the POA rose slightly while the level of GAP-LI began to decrease. This opposite shift resulted in a drop in the GAP/LHRH molar ratio (Fig. 1). After 2 days, the ratio gradually increased and, with time, returned to the intact ratio value. The actual levels of GAP-LI and LHRH-LI in the POA, fluctuated slightly over the remaining times examined after orchidectomy but remained somewhat lower (the level of LHRH having fallen slightly on day 2) than the intact levels through 40 days.

In the HYP, the level of both LHRH-LI and GAP-LI began to decrease 2 days after orchidectomy although the decrease in GAP-LI was greater than that of LHRH-LI. After the initial decrease, the peptides remained at

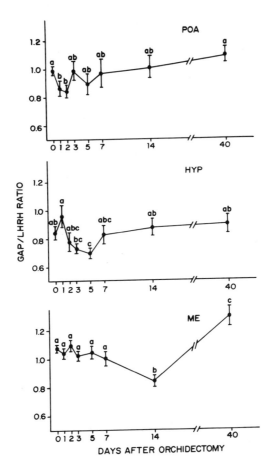

Fig. 1. Changes in the molar GAP/LHRH ratio, as a function of time following orchidectomy, in extracts of male rat preoptic-anterior hypothalamic areas (POA), hypothalamus (HYP), and median eminence (ME). Means $\pm$ SE with the same letter superscript are not significantly different ($P \leq 0.05$).

about the same level through day 5 and then continued to decline together through day 14. The GAP/LHRH ratio was decreased on days 2 through 5, but then gradually returned to intact ratio values (Fig. 1).

Similar to the HYP, a decline in both GAP-LI and LHRH-I was evident in the ME beginning 2 days after orchidectomy, however, the rate of decrease was similar for both peptides and continued through day 14. The GAP/LHRH ratio remained relatively unchanged until day 14 at which time the ratio was decreased (Fig. 1). By 40 days after orchidectomy, the GAP/LHRH ratio was significantly higher than in the intact controls.

Plasma LH levels were significantly elevated 1 day after orchidectomy. Between days 1 and 5, plasma LH continued to rise but at a much slower rate than during the first 24 hrs. After day 5, plasma LH rose more rapidly reaching an increase of over 50-fold by day 40. These changing rates of increase may be partially reflective of the observed changes in the brain GAP/LHRH ratios.

The early changes in the POA and HYP GAP/LHRH ratio, following orchidectomy, may indicate changes in the synthesis, transport and/or processing of the prohormone. The later time at which the decrease in the GAP/LHRH ratio occurred in the HYP as compared with the POA, is in keeping with the HYP containing predominately LHRH nerve fibers coming from the perikarya located in the POA. THe gradual return of both POA and HYP GAP/LHRH molar ratios to intact values may reflect the intervention of homeostatic mechanisms to restore the imbalance caused by orchidectomy.

The lack of early ratio shifts in the ME may also be a reflection of the anatomy of the LHRH system in this region since the ME contains predominantly nerve cell fibers and terminals. Altered rates of release and/ or enzymatic degradation at the level of the terminals could easily alter or mask any ratio shifts reaching this level from the more rostral fiber tracts. The later ratio shifts in the ME may also reflect alterations in release and/or degradation and may be part of a homeostatic mechanism. The progressive decline in both GAP-LI and LHRH-LI in the HYP and ME after orchidectomy, suggests either that synthesis of the prohormone gradually decreased or that the rate(s) of release, transport and/or degradation gradually increased. Further studies utilizing microdissection techniques and replacement of specific gonadal factors are now underway to further elucidate the nature of these changes.

Because the GAP antiserum recognizes only a portion of the GAP sequence, the assay may detect not only the 56 amino acid GAP molecule but also smaller, processed forms which also contain the appropriate antigenic the different molecular weight forms of LHRH and GAP present in the brain of intact and orchidectomized rats. Four-week orchidectomized and intact adult Sprague-Dawley rats were killed and the ME and combined POA-HYP tissue blocks (POH) dissected. Pools of tissues from 20 rats were defatted and then homogenized under acidic conditions in the presence of enzyme inhibitors. Peptides were extracted from the homogenates using Sep-Pak C18 cartridges (Waters Corp.) and lyophilized. The lyophilized extracts were reconstituted in a buffer containing 5 M guanidine hydrochloride and 6% 2-mercaptoethanol and the peptides separated according to molecular weight by high performance liquid chromatography (HPLC). Under these conditions, peptides larger than 25000 MW eluted in the void volume while peptides smaller than 500 MW eluted in salt volume. Fractions were collected, desalted, and assayed for GAP-LI and LHRH-LI as previously described (Culler and Negro-Vilar, 1986).

In both the ME and POH of the intact rat, only one peak of LHRH-LI was observed which had a MW of approximately 1200 and which eluted in the same position as synthetic LHRH. Approximately 2-fold more LHRH-LI was found in the ME than in the POH. When the fractions from the POH were examined for GAP-LI, peaks were observed at MWs of approximately 14500, 6500 and less than 1000. The major peak at 6500 MW, which corresponded to the elution position of synthetic, human GAP 1-56, had a shoulder indicating immunoreactive peptides between 6500 and 1000 MW. This 6500 MW peak was more prominent in the ME than in the POH.

As in the intact rats, extracts of both POH and ME from orchidectomized rats contained a single peak of LHRH-LI, however, the amount of LHRH-LI in both regions was at least 3-fold less than in the intact rats. The POH from orchidectomized rats contained the 6500 MW peak which, as in intact rats, had a shoulder indicating that smaller MW peptides were present, however, the 14000 and 1000 MW peaks were absent. Extracts of POH from orchidectomized rats contained approximately 2-fold less GAP-LI in the 6500 MW region than extracts from intact rats. In the ME of orchidectomized rats, 2 peaks of GAP-LI were observed with MWs of 14500 and 6500 compared with the ME from intact rats which contained only the

6500 MW material. This 6500 MW peak was 2-fold lower in the ME from orchidectomized rats than from intact rats.

These data indicate that multiple molecular forms of GAP-LI exist in the POH of the intact animal. The 14500 MW peptide peak probably represents the intact LHRH prohormone. The lack of LHRH-LI in the region of this peak is most likely due to our choice of anti-LHRH serum (Rice's anti-LHRH #5) which is specific for the free LHRH decapeptide (Hedger et al., 1985). This 14500 MW prohormone seems to be cleaved into at least two peptides which correspond to GAP and the LHRH decapeptide. Since both of these peptide species were observed in POH extracts, some processing of the prohormone may occur in the region of the perikarya and/or nerve fibers. The higher levels of these peptides in the ME may represent final processing of the prohormone and/or storage in the nerve terminals. The significance of the GAP-immunoreactive shoulder on the 6500 MW peak and the peak of GAP-LI with MW less than 1000 is unknown at this time.

The lower levels of both GAP-LI and LHRH-LI in the orchidectomized rat agrees with the results of our temporal study. The presence of the suspected 14500 MW prohormone in the ME of orchidectomized rats, but not of intact rats, suggests that processing of the prohormone may be decreased after orchidectomy. Additional studies are now underway to further resolve these findings.

The observed alterations that occur in 1) brain levels of GAP-LI and LHRH-LI, 2) the molar ratio of GAP/LHRH and 3) the amounts and profile of the LHRH and GAP immunoreactive species in the brain following orchidectomy, demonstrate that the testes exert substantial influence over the regulation and processing of the LHRH prohormone. The full significance of these findings, in relation to the changes that occur in pituitary function, will require additional investigation. Understanding how these processes are altered by various physiologic and pathologic conditions will greatly enhance our understanding of the role of LHRH and its prohormone in regulating reproductive function.

REFERENCES

Adelman JP, Mason AJ, Hayflick JS, Seeburg PH, 1986. Isolation of the gene and hypothalamic cDNA for the common precursor of gonadotropin-releasing hormone and prolactin release-inhibiting factor in human and rat. Proc Natl Acad Sci USA 83:179-183

Culler MD, Negro-Vilar A, 1986. Development of specific antisera and a radioimmunoassay procedure for the gonadotropin-releasing hormone associated peptide (GAP) of the LHRH prohormone. Brain Res Bull 17:219-223

Hedger MP, Robertson DM, Browne CA, deKretser DM, 1985. The isolation and measurement of luteinizing hormone-releasing hormone (LHRH) from the rat testis. Mol Cell Endocrinol 42:24-29

Mason AJ, Hayflick JS, Zoeller RT, Young WS, Phillips HS, Nikolics K, Seeburg PH, 1986a. A deletion truncating the gonadotropin releasing hormone gene is responsible for hypogonadism in the hpg mouse. Science 234:1366-1371

Mason AJ, Pitts SL, Nikolics K, Szonyi E, Wilcox JN, Seeburg PH, Stewart TA, 1986b. The hypogonadal mouse: reproductive functions restored by gene therapy. Science 234:1372-1378

Millar RP, Wormald PJ, Milton RC de L, 1986. Stimulation of gonadotropin release by a non-GnRH peptide sequence of the GnRH precursor. Science 232:68-70

Nikolics K, Mason AJ, Szonyi E, Ramachandran J, Seeburg PH, 1985. A prolactin-inhibiting factor within the precursor for human gonadotropin-releasing hormone. Nature 316:511-517

Phillips HS, Nikolics K, Branton D, Seeburg PH, 1985. Immunocytochemical localization in rat brain of a prolactin release-inhibiting sequence of gonadotropin-releasing hormone prohormone. Nature 316:542-545

Sar M, Culler MD, McGimsey WC, Negro-Vilar A, 1987. Immunocytochemical localization of the gonadotropin-releasing hormone associated peptide of the LHRH-precursor. Neuroendocrinology 45:172-175

Seeburg PH, Adelman JP, 1984. Characterization of cDNA for precursor of human luteinizing hormone releasing hormone. Nature 311:666-668

PULSATILE HORMONE SECRETION LATE IN THE LUTEAL PHASE OF NORMAL

AND INFERTILE WOMEN DURING DIURNAL HOURS

C.L. Huges, R. Fleming, J.R.T. Coutts, and M.C. Macnaughton

Division of Reproductive Endocrinology and Infertility
Department of Obstetrics and Gynecology
Duke University Medical Center
Durham, NC 27706
     and
Department of Obstetrics and Gynaecology
Glasgow University
Glasgow, Scotland

INTRODUCTION

Human luteal function can be assessed by endometrial biopsy (Noyes et al., 1950) or by progesterone (P) levels. Protocols for estimation for luteal function by P include a single determination (Israel et al., 1972), summation of three values (Abraham et al., 1974), daily measures alone (Abraham et al., 1974) or with ultrasonic monitoring of periovulatory changes in follicle morphology (Coutts et al., 1981a, 1982). Recent evidence suggests that secretion of P in rhesus monkeys (Healy et al., 1984) and normal women (Filicori et al., 1983) is episodic with approximately 4 pulses of P per 24 h. These studies suggest that proper characterization of a luteal phase as normal or abnormal may require assessment of frequency, amplitude, duration or diurnal dependency of such pulses. The practice of evaluating ovulation and the luteal phase in infertility patients with random luteal phase P levels could be erroneous if significant phasic changes are not considered in diagnostic testing. Whereas patients with ovulatory dysfunction due to the "luteinized unruptured follicle syndrome" (LUF) (Marik and Hulka, 1978; Koninckx et al., 1978) commonly have associated luteal phase inadequacy described as "poor progesterone surge" (PPS) (Coutts et al., 1981a, 1982), we compared women with normal menstrual cycles or LUF/PPS cycles to determine whether the secretory patterns of P, estradiol (E), luteinizing hormone (LH) and follicle stimulating hormone (FSH) differed.

PATIENTS AND METHODS

Five infertile patients with LUF/PPS had been studied for two or more menstrual cycles by frequent ovarian ultrasonic scans and daily blood sampling for measurement of hormones. Six patients with infertility due to causes other than ovulatory or luteal dysfunction (tubal or male factors) and one normal volunteer had been previously assessed for 1 or more cycles. Menstrual cycle day was assigned by menses, levels of LH, E, and P and serial ultrasonic scans of the ovaries (Adam, 1981; Coutts et al., 1982). Subjects arrived on the ward at 0800 h. An arm vein was catheterized by 0830 h and flushed with 10 ml of heparinized saline.

Subjects were then permitted normal activities excluding coffee and smoking, but all had mid-morning and mid-afternoon snacks and a meal at 1200 h. From 0900 h – 1700 h, 5 ml blood samples were drawn and 5 ml flushes of heparinized saline were injected every 20 min. Plasma samples were collected and stored at -20°C. P, E, LH, and FSH levels were determined by RIA (Coutts et al., 1981b). Intra-assay and inter-assay coefficients of variation for each hormone were as follows: P (4.1 to 8.6% and 4.6%), E (3.5 to 9.0% and 6.5%), LH (4.1% and 6.4%) and FSH (6.1% and 8.7%). Data were compared by analysis of variance or Mann-Whitney U test as appropriate. Receiver operating characteristics (ROC) analysis (Metz, 1982) was used to compare decision thresholds for plasma P levels. We have applied the criteria of Soules et al. (1984) for detection of hormonal pulses, requiring an increment/decrement between nadir and peak of > 5 S.D. in the assay variability. Pulse amplitude is the difference between the nadir and the peak value in the pulse. Pulse duration is 20 min X number of values in the pulse. Whereas amplitude and duration may be underestimated for pulses which appear to extend beyond the sam-pling interval, we have taken these instances to be only the maximal quantitated values and have made comparisons with non-parametric statistical methods.

RESULTS AND DISCUSSION

Relative to the mid-cycle LH surge, the menstrual cycle day for normal subjects was from +8 to +11 with X $\pm$ S.D. = 9.4 $\pm$ 1.1 days. Menstrual cycle day for LUF/PPS subjects was from +9 to +11 with X $\pm$ S.D. = 9.4 $\pm$ 0.9 days. While mean morning (AM) P levels in normal cycles were 17.8 to 20.7 ng/ml, afternoon (PM) levels were lower (p < 0.01) and were from 12.5 to 16.8 ng/ml (Fig. 1). In LUF/PPS cycles, mean AM levels of P were 10.9 to 12.6 ng/ml which differ (p < 0.05) from PM levels of 8.4 to 11.7 ng/ml. AM and PM levels of P in normal cycles were greater (p < 0.01) than the corresponding levels in LUF/PPS cycles. ROC analysis of 120 AM and 180 PM values (Table 1) yield estimates of accuracy, true positive fraction (TPF) and true negative fraction (TNF) and show that a P level in the AM is a "better" test than in the PM. Since the AM ROC curve (Fig. 2) shows a more extreme displacement above and to the left of the diagonal and the populations of FPF values show a significant AM versus PM differ-ence (p < 0.02), the AM test is confirmed to be superior to the PM test. The optimal PM threshold appears to be 9 ng/ml with an accuracy of 0.73.

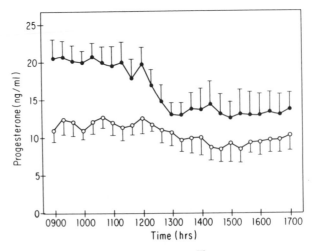

Fig. 1. Levels of plasma progesterone ($\overline{X} \pm$ SE) in women with normal menstrual cycles (n = 7; $\bullet$) or LUF/PPS cycles (n = 5; 0).

Table 1. Decision Performance Indices for Plasma Progesterone Levels in Normal and LUF/PPS Cycles.

THRESHOLD (ng/ml) (< implies disease state)	TPF	FPF	TNF	FNF	Accuracy
Morning (0900h – 1200h)					
6	0	0	1	1	0.58
7	0.04	0	1	0.96	0.60
8	0.10	0.01	0.99	0.90	0.62
9	0.24	0.04	0.96	0.76	0.66
10	0.36	0.04	0.96	0.64	0.71
11	0.42	0.07	0.93	0.58	0.72
12	0.46	0.09	0.91	0.54	0.72
13	0.60	0.13	0.87	0.40	0.76
14	0.68	0.16	0.84	0.32	0.77
15	0.78	0.17	0.83	0.22	0.81
16	0.90	0.20	0.80	0.10	0.84
17	0.96	0.24	0.76	0.04	0.84
18	0.98	0.30	0.70	0.02	0.82
19	1	0.41	0.59	0	0.76
Afternoon (1220h – 1700h)					
4	0	0	1	1	0.58
5	0.08	0.05	0.95	0.92	0.58
6	0.11	0.11	0.89	0.89	0.56
7	0.17	0.14	0.86	0.83	0.57
8	0.35	0.14	0.86	0.65	0.65
9	0.57	0.15	0.85	0.43	0.73
10	0.61	0.26	0.74	0.39	0.69
11	0.69	0.30	0.70	0.31	0.70
12	0.73	0.41	0.59	0.27	0.65
13	0.79	0.52	0.48	0.21	0.61
14	0.83	0.59	0.41	0.17	0.59
15	0.91	0.66	0.34	0.09	0.58
16	0.93	0.74	0.26	0.07	0.60
17	0.96	0.79	0.21	0.04	0.58
18	0.97	0.81	0.19	0.03	0.52
19	1	0.86	0.14	0	0.50

For the superior AM test an optimal threshold appears to be 16 ng/ml with an accuracy of 0.84, TPF of 0.90 and TNF of 0.80. It must be noted that selection of a threshold is arbitrary but decision performance indices allow judgements about the accuracy of detection of the disease or disease-free state.

Pulse analysis of plasma levels of P, E, LH and FSH are shown in Table 2. Number, duration, amplitude, and diurnal pattern of E and FSH pulses did not differ in normal and LUF/PPS cycles. While the diurnal pattern and number of P or LH pulses in the two groups did not differ, AM pulses of P in normal cycles were of greater amplitude than either AM pulses in LUF/PPS cycles (p < 0.02) or PM pulses in normal cycles (p < 0.02). While AM pulses of P in normal cycles were of greater duration than PM pulses within the same group (p < 0.02), differences in pulse duration were not detected between groups or within the LUF/PPS group (e.g., AM versus PM pulses). In the interval of 0900 h – 1700 h, LH pulses in LUF/PPS cycles were of greater amplitude than pulses in normal cycles (p < 0.02). No diurnal differences in amplitude of LH pulses were detected within either group. No differences in duration of LH pulses were detected within or between groups. Relative to normal cycles, the overall pattern suggested by pulse analysis of LUF/PPS cycles is increased amplitude of LH pulses throughout the day, but decreased amplitude of P pulses, particularly during AM hours. Consequently, the phasic diurnal pattern of P secretion which is prominent in normal cycles is attenuated in LUF/PPS cycles.

Dating of a luteal phase endometrial biopsy has an accuracy of 80% (within 2 days) among observers and in comparison to basal temperature and menstrual dates (Noyes and Haman, 1953). Alternatively, luteal function

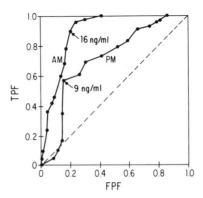

Fig. 2. Morning (0900 h – 1200 h) and afternoon (1220 h – 1700 h) ROC
curves for plasma progesterone levels in normal and LUF/PPS
cycles.

Table 2. Summary of Pulse Analysis of Plasma P, LH, E and FSH in Normal
and LUF/PPS Cycles Late in the Luteal Phase. (Duration in
min. Amplitude in ng/ml for P, IU/1 for LH and FSH and pg/ml
for E).

		AM (0900h-1200h)	FM (1220h-1700h)	Overall (0900h-1700h)
Progesterone				
Normal	pulses/patient	1.14 ± 0.38	1.43 ± 1.27	2.57 ± 1.51
	duration of pulses	205 ± 84.7	58.0 ± 62.1	123.3 ± 103.2
	amplitude of pulses	10.6 ± 2.56	4.46 ± 3.56	7.19 ± 4.39
LUF/PPS	pulses/patient	1.00 ± 0.71	1.20 ± 0.45	2.20 ± 1.10
	duration of pulses	96.0 ± 100.4	36.7 ± 32.0	63.6 ± 74.2
	amplitude of pulses	4.92 ± 3.41	3.87 ± 1.72	4.35 ± 2.54
LH				
Normal	pulses/patient	1.57 ± 0.53	2.29 ± 0.76	3.86 ± 1.07
	duration of pulses	48.8 ± 56.6	65.5 ± 58.0	55.6 ± 56.6
	amplitude of pulses	2.30 ± 1.72	1.96 ± 2.57	2.10 ± 2.23
LUF/PPS	pulses/patient	1.60 ± 0.89	2.80 ± 1.30	4.40 ± 1.67
	duration of pulses	77.5 ± 49.5	57.1 ± 33.2	64.6 ± 40.0
	amplitude of pulses	5.15 ± 2.20	4.99 ± 2.53	5.05 ± 2.36
Estradiol				
Normal	pulses/patient	1.00 ± 1.15	0.57 ± 0.79	1.57 ± 1.81
	duration of pulses	25.7 ± 9.76	28.0 ± 11.0	25.5 ± 9.34
	amplitude of pulses	39.3 ± 4.50	48.8 ± 19.3	42.7 ± 12.1
LUF/PPS	pulses/patient	1.60 ± 0.89	1.00 ± 1.00	2.60 ± 1.52
	duration of pulses	100.0 ± 102.0	50.0 ± 44.1	69.2 ± 72.4
	amplitude of pulses	48.8 ± 19.8	44.0 ± 16.0	46.9 ± 17.9
FSH				
Normal	pulses/patient	0.71 ± 0.95	0.71 ± 0.95	1.43 ± 1.62
	duration of pulses	40.0 ± 24.5	54.3 ± 57.4	50.0 ± 49.2
	amplitude of pulses	0.66 ± 0.18	0.84 ± 0.40	0.75 ± 0.31
LUF/PPS	pulses/patient	0.40 ± 0.89	0.60 ± 0.89	1.00 ± 1.73
	duration of pulses	30.0 ± 14.1	26.7 ± 11.6	28.0 ± 11.0
	amplitude of pulses	0.50 ± 0.0	0.60 ± 0.20	0.56 ± 0.15

can be meticulously evaluated by daily blood samples for hormone levels
with serial ultrasonic scans for monitoring of follicle morphology (Adam,
1981; Coutts et al., 1981b, 1982). While the utility of a single P level
in detection of luteal deficiency has been challenged (Abraham et al.,
1974; Jones et al., 1974), other data suggest a single P level is useful

in assessment of normal ovulatory function (Hull et al., 1982). Our data confirm the utility of P levels for assessment of luteal function with an accuracy that can equal or exceed that of an endometrial biopsy.

Backstrom et al. (1982) demonstrated fluctuations in the concentrations of E, FSH, LH, prolactin and P in the human menstrual cycle. Other studies in the human (Filicori et al., 1983; Soules et al., 1984) and the monkey (Healy et al., 1984) have confirmed these results. Pulsatile secretion of LH in the luteal phase shows reduced frequency as the luteal phase progresses. The relationship of luteal phase LH pulses to P pulses is not clear since 70% of P pulses in the monkey were independent of LH or prolactin pulses (Healy et al., 1984) while correlation of LH and P pulses was reported in the late luteal phase in 2 women (Filicori et al., 1983). Our results suggest that normal luteal function is not a prerequisite to manifestation of the P oscillator, since pulsatile P secretion did occur in our patients with LUF/PPS. We observed an inverse relationship between luteal phase P and LH secretion, since diminished P secretion was associated with enhanced LH secretion in LUF/PPS cycles. Soules et al. (1984) suggested that the LH secretory pattern is mediated by feedback effects of P and our results support but do not prove this contention. An ovarian luteal phase cause for the abnormal hormonal profiles in LUF/PPS cycles would seem to be further supported by the ultrasonic criterion of "luteinized unruptured follicle", however failure of ovulation per se and/or consequent abnormal luteal function may be more appropriately attributed to abnormal antecedent follicular development, whether due to intrinsic ovarian or extra-ovarian causes.

ACKNOWLEDGEMENT

We gratefully acknowledge the expert technical assistance of Mrs. S. Finnie and Mrs. E. McNally.

REFERENCES

Adam AH, 1981. Infertile patients - ovarian ultrasound in relation to their management. In: Christie AD (ed.), Ultrasound and Infertility. Bromley: Chartwell-Bratt, pp. 47-57

Abraham GE, Maroulis GB, Marshall JR, 1974. Evaluation of ovulation and corpus luteum function using measurements of plasma progesterone. Obstet Gynecol 44:522-525

Backstrom CT, McNeilly AS, Leask RM, Baird DT, 1982. Pulsatile secretion of LH, FSH, prolactin, oestradiol and progesterone during the human menstrual cycle. Clin Endocrinol 17:29-42

Coutts JRT, Adam AH, Fleming R, 1981a. Ovarian ultrasound and endocrine profiles in women with unexplained infertility. In: Christie AD (ed.), Ultrasound and Infertility. Bromley: Chartwell-Bratt, pp. 89-105

Coutts JRT, Gaukroger JM, Kader AS, Macnaughton MC, 1981b. Steroidogenesis by the human Graafian follicle. In: Coutts JRT (ed.), Functional Morphology of the Human Ovary. Lancaster: MTP Press, pp. 53-72

Coutts JRT, Adam AH, Fleming R, 1982. The deficient luteal phase may represent an anovulatory cycle. Clin Endocrinol 17:389-394

Filicori M, Butler J, Crowley W, 1983. Pulsatile progesterone secretion from the human corpus luteum. Program and Abstracts 65th Annual Meeting of the Endocrine Society, San Antonio, Abstract #459, p. 195

Healy DL, Schenker RS, Lynch A, Williams RF, Hodgen GD, 1984. Pulsatile progesterone secretion: its relevance to clinical evaluation of corpus luteum function. Fertil Steril 41:114-121

Hull MGR, Savage PE, Bromham DR, Ismail AAA, Morris AF, 1982. The value of a single serum progesterone measurement in the midluteal phase as a criterion of a potentially fertile cycle ("ovulation") derived from treated and untreated conception cycles. Fertil Steril 37:355-360

Israel R, Mishell DR, Stone SC, Thorneycroft IH, Moyer DL, 1972. Single luteal phase serum progesterone assay as an indicator of ovulation. Am J Obstet Gynecol 44:26-34

Jones GS, Aksel S, Wentz AC, 1974. Serum progesterone values in the luteal phase defects. Obstet Gynecol 44:26-34

Koninckx PR, Heyns W, Corveleyn PA, Brosens IA, 1978. Delayed onset of luteinization as a cause of infertility. Fertil Steril 29:266-269

Marik J, Hulka J, 1978. Luteinized unruptured follicle syndrome: a subtle cause for infertility. Fertil Steril 29:270-274

Metz CE, 1978. Basic principles of ROC analysis. Sem Nuc Med 8:283-298

Noyes RW, Haman JO, 1953. Accuracy of endometrial dating. Fertil Steril 4:504-517

Noyes RW, Hertig AT, Rock J, 1950. Dating the endometrial biopsy. Fertil Steril 1:3-25

Soules MR, Steiner RA, Clifton DK, Cohen NL, Aksel S, Bremner WJ, 1984. Progesterone modulation of pulsatile luteinizing hormone secretion in normal women. J Clin Endocrinol Metab 58:378-383

LH AND ESTRADIOL SECRETION AFTER HCG AND GNRH TREATMENTS IN GILTS

J.E. Tilton and A.J. Ziecik*

Department of Animal and Range Sciences
North Dakota State University
Fargo, ND 58105
*Visiting Professor
University of Agricultural and Technology
Olsztyn, Poland

INTRODUCTION

Two different mechanisms have been proposed to regulate luteinizing hormone (LH) secretion patterns: a long-loop feedback of estradiol-17β (E) from the ovaries (Brinkley, 1981) and(or) a short-loop or internal feedback of pituitary hormones themselves (Motta et al., 1969; Docke and Glaser, 1971). In some species the increase in circulating estrogen to a critical level in late follicular phase elicits a surge release of LH by a positive feedback mechanism (Miyake et al., 1977). Results presented by Concannon et al. (1979) suggested that the preovulatory surge of LH in the female dog may be a response to withdrawal of the negative feedback effect of estrogens. Also, this withdrawal might be facilitated by a preovulatory increase in progesterone. Previous work in our laboratory (Ziecik et al., 1987) demonstrated that 1000 and 1500 IU of hCG blocked the post-synchronization preovulatory LH surge in 33% and 50%, respectively, of gilts previously treated with altrenogest and PMSG. The objective of the following study was to investigate the effect of higher doses of hCG in combination with bolus injections of GnRH on the secretion of E and LH.

MATERIALS AND METHODS

Fourteen cyclic gilts were individually fed 15 mg altrenogest hd^{-1} day^{-1} for 14 consecutive days. On the 12th day of altrenogest all gilts were fitted with indwelling cephalic vein catheters. On day 14 of altrenogest, all gilts received 750 IU PMSG. Seventy-two h after PMSG injections, they were assigned to treatment: I-2000 IU hCG, II-2000 IU hCG plus 500 μg GnRH, and III-3000 IU hCG plus 500 μg GnRH. All injections were given intramusclarly with GnRH given 48 h after hCG. Plasma samples were collected every 6 h beginning at the time of PMSG injection. Sampling frequency was increased to bi-hourly collections two h prior to hCG and continued at that frequency until 24 h after the GnRH injection. Estradiol and LH were analyzed by radioimmunoassay.

Plasma hormone data were analyzed as a split plot design with repeated measurements over time using the General Linear Models (GLM) procedure of the statistical analysis computer program (SAS, 1982).

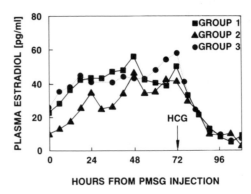

HOURS FROM PMSG INJECTION

Fig. 1. Concentrations of estradiol-17β after PMSG injection. Arrow shows time of hCG injection.

RESULTS

The secretion patterns of E in blood plasma for all groups are shown in Fig. 1. PMSG caused a similar E rise in all groups with peak concentrations in Group 3 slightly higher than Groups 1 or 2. In all groups, plasma E concentrations rapidly dropped (P < .01) after hCG injection reaching basal levels of less than 10 pg/ml 18–24 h later.

In Group 1, the preovulatory LH surge occurred approximately 10 h before hCG injections in 2 of 5 gilts (Fig. 2) with a surge duration of 18 h. In Group 2 (Fig. 3) all gilts exhibited an LH surge immediately following the GnRH treatment. Amplitude of these surges was similar to the spontaneous surge observed in Group 1 but the duration was shorter (8.0 ± 1.8 h). Two of five gilts in Group 3 had an LH surge 7 and 12 h (duration 18 and 24 h, respectively) before the GnRH bolus injection (Fig. 4). The remaining 3 gilts had an LH surge (duration –8.3 ± 1.1 h) immediately after the GnRH injection. The average duration of preovulatory LH surges occurring in Groups 1 and 3 (N = 4) was 18.1 ± .7 h, approximately 10 h larger (P < .01) than LH surges induced by GnRH (8.1 ± 1.1 h). LH peak amplitudes were not different (P > .01) between spontaneous (3.5 ±

DISCUSSION

Previously we observed a failure of some hCG treated gilts to exhibit a preovulatory LH surge (Ziecik et al., 1987). The present data support those results. Some gilts will exhibit a spontaneous preovulatory LH surge despite application of high doses of hCG (2000 and 3000 IU), however, the majority of gilts will not. We propose two possible hypotheses to explain these results; 1) a short-loop regulation or 2) a lack of sufficient time interval from the estrogen trigger to enable endogenous LH to be released. The decrease of E immediately after hCG injection may reflect the inhibition of steroidogenesis in preovulatory follicles. It appears that PMSG caused sufficient estrogen stimulation of the hypothalamus-pituitary unit although the possibility also exists that thresholds were not reached in some animals. However, the same PMSG stimulation in altrenogest-treated gilts receiving 0 and 500 IU hCG resulted in preovulatory LH surges in all animals (N = 12) with a duration of 24 and 27 h, respectively (Ziecik et al., 1987).

Superovulation induced by exogenous gonadotropins in the rhesus monkey (Schenken and Hodgen, 1983; Collins et al., 1984) and women (Collins et al., 1984) is reportedly followed by a failure of the spontaneous preovulatory LH surge. Their results were believed due to

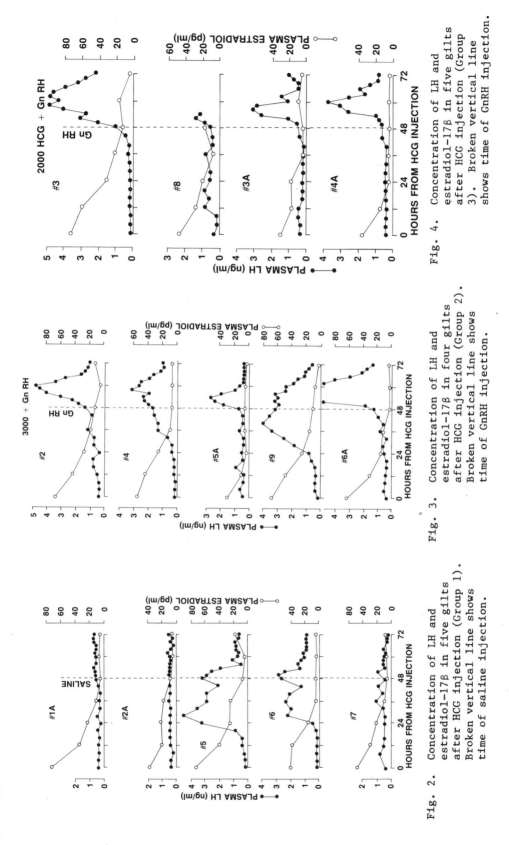

Fig. 2. Concentration of LH and estradiol-17β in five gilts after HCG injection (Group 1). Broken vertical line shows time of saline injection.

Fig. 3. Concentration of LH and estradiol-17β in four gilts after HCG injection (Group 2). Broken vertical line shows time of GnRH injection.

Fig. 4. Concentration of LH and estradiol-17β in five gilts after HCG injection (Group 3). Broken vertical line shows time of GnRH injection.

supraphysiological levels of estradiol (unlikely in our experiment) or by the hypothesis of Hodgen (1982) that some ovarian factor(s) stimulated by exogenous gonadotropin inhibit the positive feedback of estradiol. Kuhl and Taubert (1975) reported that administration of LH can increase the activity of hypothalamic L-cystine arylamidase, a rate limiting enzyme which has been demonstrated to inactivate GnRH bioactivity. Further localization of LH in the subcellular fraction of the hypothalamus (Emanuele et al., 1985) provided a framework for speculation on the role of brain-based LH. It is proposed that hypothalamic LH may serve as a mediator of the short-loop feedback mechanism therein regulating release of LH from the anterior pituitary (Motta et al., 1969) by interneuronal communication between LH- and GnRH-containing cells (Emanuele et al., 1985). Since hCG crosses the blood-brain barrier (Bagshawe et al., 1968) we suggest the failure of preovulatory LH surges observed in hCG treated females in our experiments was caused by an inhibitory effect of exogenous gonadotropin on the hypothalamo-pituitary unit. The observations are further supported by an ongoing study in ovariectomized gilts in which hCG inhibits positive feedback of LH release after estradiol administration (Ziecik and Britt - unpublished data). Since GnRH (500 µg) did release LH from the pituitary our results support suggestions that exogenous gonado-tropins effect endogenous LH release at the level of the hypothalamus rather than the pituitary.

In summary, the results of this experiment support previous findings concerning the effect of hCG on endogenous LH release. The area of modulation as indicated by the response to GnRH subsequent to hCG appears to be at the hypothalamic level and likely not at the pituitary.

ACKNOWLEDGEMENTS

Published with approval of the Director of the North Dakota Argic. Exp. Sta. as Article No. 1519. Supported by Hatch Project ND01779.

The authors wish to thank RM Weigl for assay of the samples and J Berg for the preparation of this manuscript.

REFERENCES

Bagshawe KD, Orr AH, Rushworth AG, 1968. Relationship between concentrations of human chorionic gonadotropin in plasma and cerebrospinal fluid. Nature 217:950

Brinkley HJ, 1981. Endocrine signaling and female reproduction. Biol Reprod 24:22

Collins RL, Williams RF, Hodgen GD, 1984. Endocrine consequences of prolonged ovarian hyperstimulation: hyperprolactinemia, follicular atresia and premature luteinization. Fertil Steril 42:436

Concannon PR, Cowan R, Hansel W, 1979. LH release in ovariectomized dogs in response to estrogen withdrawal and its facilitation by progesterone. Biol Reprod 20:523

Docke F, Glaser D, 1971. Internal feedback of luteinizing hormone in cyclic female rats. J Endocrinol 51:403

Emanuele N, Lipov S, Anderson J, Anderson E, Kirsteins L, Nishimura N, Lawrence AM, 1985. Subcellular localization of hypothalamic luteinizing hormone. Neuroendocrinology 40:339

Hodgen GD, 1982. The dominant ovarian follicle. Fertil Steril 38:281

Kuhl H, Taubert HD, 1975. Short-loop feedback mechanism of luteinizing hormone: LH stimulates hypothalamic L-cystine arylamidase to inactivate LH-RH in the rat hypothalamus. Acta Endocrinol 78:649

Miyake A, Aono T, Tamizawa O, Kinugasa T, Kurachi K, 1977. Influence of human chorionic gonadotropin on the response of luteinizing hormone to luteinizing hormone releasing hormone in gonadectomized women. J Endocrinol 74:499

Motta M, Frashini F, Martini L, 1969. Short feedback mechanism in the control of anterior pituitary function. In: WF Ganong and L Martini (ed.), Frontiers in Neuroendocrinology, Oxford University Press, London, pp. 211-253

SAS, 1982. SAS User's Guide: Statistics. Statistical Analysis System Institute, Inc., Cary, NC

Schenken RS, Hodgen GD, 1983. Follicle stimulating hormone induced ovarian hyperstimulation in monkeys: blockade of the luteinizing hormone surge. J Clin Endocrinol Metab 57:50

Ziecik A, Tilton JE, Espana F, Weigl RM, 1987. Effect of human chorionic gonadotropin on preovulatory luteinizing hormone surge and ovarian hormone secretion in gilts. J Anim Sci (In Press)

SEVERAL GnRH STIMULATION FREQUENCIES DIFFERENTIALLY RELEASE FSH

AND LH FROM ISOLATED, PERFUSED RAT ANTERIOR PITUITARY CELLS

R.T. Savoy-Moore and K.H. Swartz

Departments of OB/GYN and Physiology
Wayne State University
Detroit, MI 48201

INTRODUCTION

Gonadotropin-releasing hormone (GnRH) releases both follicle-stimulating hormone (FSH) and luteinizing hormone (LH) from the anterior pituitary, but not always at the same time. Both hormones surge on the day of proestrus in the rat, but only the FSH surge persists through estrous morning (Butcher et al., 1975; Smith et al., 1975). One way that a single releasing hormone could differentially affect two gonadotropins is by variation of the GnRH pulse pattern. Both frequency and amplitude have been altered to release more FSH than LH (Wildt et al., 1981; Wise et al., 1978).

We have found that isolated, perfused rat anterior pituitary cells respond to hourly GnRH pulses by releasing more FSH than LH after GnRH stimulation ceased (Savoy-Moore and Swartz, 1986; unpublished observations). In the present study, we examined whether different GnRH frequencies could elicit different patterns of FSH and LH.

MATERIALS AND METHODS

Isolated anterior pituitary cells were obtained as previously described (Savoy-Moore et al., 1980) from female Sprague-Dawley rats (Charles River) decapitated at random estrous cycle stages. Cells were resuspended in culture media [Dulbecco's modified Eagle's media (DMEM; GIBCO) plus 10% fetal bovine serum (FBS, FLOW), penicillin and strepto-mycin (100 U/ml, GIBCO)] and mixed with Cytodex-1^R (Pharmacia) microcarrier beads (2.5 mg/million cells), then incubated in 60 mm petri dishes at 37°C, under 95% air and 5% CO_2. Culture media was renewed at 48 h. At 120 h, cells were rinsed and packed in plastic columns (5 million cells per column) and perfused at 37°C under air at 0.17 ml/min. The perfusion media is composed of DMEM with 25 mM HEPES (Sigma), 3.55 mM bicarbonate, 2% charcoal-stripped FBS, 5 μg/ml insulin (Sigma), 5 μg/ml transferrin, 30 pM triiodothyronine (T_3), and 0.5 ng/ml parathyroid hormone (US Biochemicals). After a 3-3.5 h equilibration period, perfusate effluent was collected every 15 min. A 1 or 2 h period of basal release collection preceeded GnRH stimulation, which began at time 0. After GnRH stimulation, perfusate was collected for at least 4 h.

Experimental Design: Experiment 1. GnRH was pulsed (5 nM, 10 min) every 60, 90, or 120 min for 13 h. At h 16, an additional GnRH pulse was administered; samples were collected through h 21. One column (COL) did not receive GnRH. Experiment 2. The 60, 90 and 120 min GnRH stimulation frequencies were repeated; one COL was unstimulated. Perfusate was collected through h 20. An additional COL was perfused with media plus 0.5% charcoal-treated porcine serum and pulsed hourly. Experiment 3. A series of 9 and 12 pulses administered every 60 min were compared to 5 and 6 pulses given every 120 min. One COL of each frequency was also pulsed at h 18. One COL received a GnRH pulse only at h 18.

Radioimmunoassays: LH was measured in 4–100 μl of perfusate using an ovine:ovine assay (Niswender et al., 1968) with RP-2 as standard. Within assay coefficient of variation (CV) averaged 8.7%, while between assay CV was 18.9%. FSH was assayed in 400 μl of perfusate using the NIADDK kit with RP-2 standard. Within assay CV averaged 8.7%, between assay CV 13.8%.

Statistical Analyses: An FSH to LH ratio was calculated for each sample from each column. Repeated measures ANOVA was used to compare FSH, LH and their ratio before and after the GnRH pulse series (or comparable times in unstimulated columns). In addition, the FSH and the LH released were totaled for each GnRH pulse and the ratio calculated for each period of GnRH-induced release. Profile analysis was used to compare the effects of frequency on GnRH-induced release over time (Morrison, 1967). A significance level of $p < .05$ was chosen.

RESULTS

When GnRH was pulsed every 60, 90 or 120 min, concomitant FSH and LH surges were elicited (Fig. 1, left panel). After h 5, FSH was elevated above baseline between GnRH pulses and following their cessation. When FSH during h −1 to 0 was compared to h 14–18 for GnRH-stimulated columns, FSH rose 4.5-fold; LH was unchanged [0.77 ± 0.08, n = 15; vs. 1.01 ± 0.13 (n = 36) ng/ml; mean $\pm$ SE]. The FSH to LH ratio also increased by 5.3-fold. In the unstimulated column (COL) 4, FSH rose while LH fell, elevating the FSH to LH ratio 3.3-fold. At h 16, GnRH released FSH which appeared comparable to earlier surges, but LH release was depressed.

Similar results were obtained in experiment 2, (Fig. 1, right panel). In response to all three frequencies, LH decreased from 2.78 ± 0.30 (n = 36) during h −2 to 0 to 1.30 ± 0.14 ng/ml (n = 108) during h 13–20, while FSH rose 2-fold, increasing the FSH to LH ratio 5.9-fold. FSH declined from h 13 thru 20, but had not returned to pre-GnRH levels even 7 h after GnRH. In the unstimulated column (COL 5), neither FSH nor LH was altered during comparable times, but the ratio increased 1.5-fold.

In experiment 3, GnRH was pulsed every 60 or 120 min for 10 or 13 h (Fig. 2). Comparing h −2 to 0 with h 13–18 in COLs 1–4, FSH rose 1.7-fold while LH declined 2.6-fold. The FSH to LH ratio rose 4-fold. In COL 5, both FSH and LH fell during those periods, but the greater drop in LH increased the FSH to LH ratio 2.3-fold. Stimulation of COL 5 at h 18 elicited an LH surge apparently comparable to the first surge in COL 1; FSH appeared to be lower than any initial surge. In COLs 2 and 4, GnRH at h 18 seemed to evoke the same magnitude FSH release as earlier, but the LH response was distinctly lower.

The effect of these frequencies on GnRH-induced release was evaluated by totalling the gonadotropin release from the beginning of one GnRH-induced surge to the beginning of the next. A ratio of FSH/LH was then calculated for GnRH-induced release, and plotted over time (Fig. 3). For all frequencies, the lines approximate horizontal thru h 7; after that, all rise. Profile analysis of the common times (1, 7 and 13 h) indicated

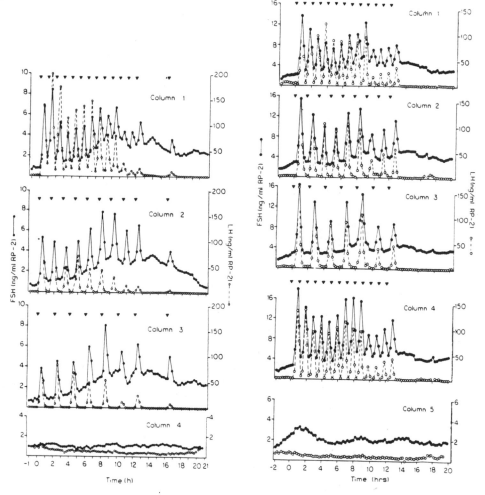

Fig. 1. Left panel, exp. 1; right panel, exp. 2. Arrowheads indicate GnRH pulses (5 nM, 10 min) administered every 60 (COLs 1 and COL 4, right panel), every 90 (COLs 2) or every 120 min (COLs 3). COL 4 (1eft panel) and COL 5 (right panel) did not receive GnRH.

that mean values were different over time, but that effect was not due to frequency (p = .17).

DISCUSSION

Differential: FSH and LH release occurred in response to GnRH regardless of frequency. FSH rose even though columns pulsed every 120 min received half as much GnRH as columns treated every 60 min. A single pulse of GnRH to COL 5 (previously unstimulated) during experiment 3 confirmed that these cells could release a normal LH pulse even after 23 h of perfusion. Therefore, the lowered LH pulses in columns 2 and 4 may reflect depletion of the readily releasable LH pool.

On proestrus, GnRH release is elevated between 1400-2000 h (Sarkar et al., 1976). Since GnRH periodicity is 48 min (Levine and Ramirez, 1982), 7 to 8 GnRH pulses normally occur on proestrus afternoon. Our data indicate that this periodicity is sufficient to elevate FSH subsequently without increased LH. The 3-fold increase in FSH in these experiments

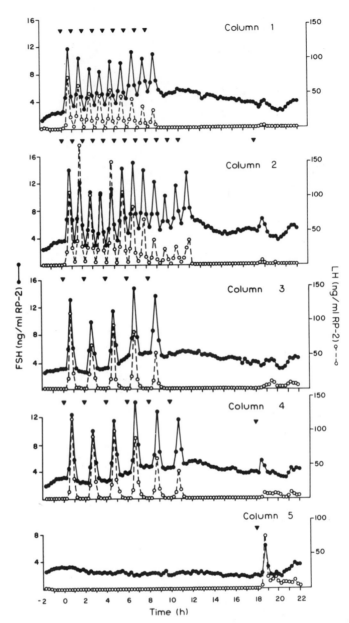

Fig. 2. Arrowheads indicate GnRH pulses (5 nM, 10 min) for exp. 3. Only 60 (COL 1 & 2) and 120 (COL 3 & 4) min frequencies were used.

also compares well with the 3 to 4-fold rise during the secondary FSH surge.

These experiments suggest that the GnRH frequency may not be very important to elevated FSH release without LH, as long as cells have been stimulated for several hours. McIntosh and McIntosh (1986) also concluded that frequency was not important, but they found no differential release, no matter what GnRH frequency was used to stimulate freshly dissociated ovine anterior pituitary cells. This effect might be species specific, or the absence of normal hormonal feedback for the 6 days these cells were in culture and perfusion may be crucial to the observation of differential FSH and LH release in the rat anterior pituitary cell perfusion system.

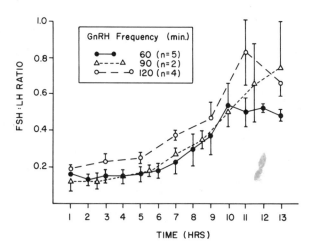

Fig. 3. FSH and LH release summated per GnRH stimulation, then a ratio was calculated. Mean (+ SE) values for the three experiments are plotted at the start of GnRH stimulation.

REFERENCES

Butcher RL, Collins WE, Fugo NW, 1975. Altered secretion of gonadotropins and steroids resulting from delayed ovulation in the rat. Endocrinology 96:576-586

Levine JE, Ramirez VD, 1982. Luteinizing hormone-releasing hormone release during the rat estrous cycle and after ovariectomy, as estimated with push-pull cannulae. Endocrinology 111:1439-1448

McIntosh J, McIntosh RP, 1986. Varying the patterns and concentrations of GnRH stimulation does not alter the ratio of LH and FSH released from perifused sheep pituitary cells. J Endocrinol 109:155-161

Morrison DF, 1967. Multivariate Statistical Methods. New York:McGraw-Hill, pp. 141-148

Sarkar DK, Chiappa SA, Fink G, Sherwood NM, 1976. Gonadotropin-releasing hormone surge in pro-oestrous rats. Nature 264:461-463

Savoy-Moore RT, Landefeld T, Marshall JC, 1980. Hormonal measurement in rat anterior pituitary cell cultures: loss of immunoreactive LH counteracted by fetal calf serum and bacitracin. Mol Cell Endocrinol 18:11-20

Savoy-Moore RT, Swartz KH, 1986. Isolated, perfused rat anterior pituitary cells can differentially release FSH following GnRH stimulation. Abstract 128, Society for the Study of Reproduction 19th Annual Meeting

Smith M, Freeman M, Neill J, 1975. The control of progesterone secretion during the estrous cycle and early pseudopregnancy in the rat: prolactin, gonadotropin and steroid levels associated with rescue of the corpus luteum of pseudopregnancy. Endocrinology 96:219-226

Wildt L, Hausler A, Marshall G, Hutchison JS, Plant TM, Belchetz PE, Knobil E, 1981. Frequency and amplitude of GnRH stimulation and gonadotropin secretion in the Rhesus monkey. Endocrinology 109:376-385

Wise PM, Rance N, Barr GD, Barraclough CA, 1979. Further evidence that luteinizing hormone-releasing hormone also is follicle-stimulating hormone-releasing hormone. Endocrinology 104:940-947

OPIOID MODULATION OF FOLLICLE STIMULATING HORMONE (FSH)

AND PROLACTIN (PRL) SECRETION IN THE POSTPARTUM SOW

C.R. Barb, R.R. Kraeling, G.B. Rampacek, and L.S. Leshin

Animal Physiology Research Unit
R B Russell Research Center
USDA-ARS
  and
University of Georgia
Athens, GA 30613

INTRODUCTION

Lactation in the sow is characterized by an anestrous period that usually lasts up to 6 wk (Cole et al., 1975). Serum concentrations of luteinizing hormone (LH) are suppressed during lactation (Stevenson et al., 1981; Stevenson and Britt, 1980) while prolactin (PRL) concentrations are elevated (Bevers et al., 1978; Stevenson et al., 1981; Dusza and Krzymowska, 1981). Serum follicle stimulating hormone (FSH) concentrations increase gradually beyond the third week of lactation which appears to be related to a reduction in suckling frequency of the litter (Stevenson et al., 1981). Weaning generally results in an increase in serum LH and FSH (Cox and Britt, 1982; Shaw and Foxcroft, 1985) concentrations and a decrease in serum PRL concentrations (Bevers et al., 1978; van Landeghem and van de Wiel, 1978).

In the postpartum rat, suckling elicited an increase in serum concentrations of β-endorphin (Riskind et al., 1984). The suckling induced suppression of serum LH concentrations and elevation of serum PRL concentrations was blocked by naloxone (NAL), an opioid antagonist, in rats (Sirinathsinghji and Martini, 1984). In the postpartum cow, NAL elevated serum LH concentrations but failed to alter PRL concentrations (Whisnant et al., 1986). In the postpartum sow, NAL increased serum LH concentrations (Barb et al., 1986; Mattioli et al., 1986; Armstrong et al., 1986) and decreased serum PRL concentrations during lactation (Mattioli et al., 1986), but the LH response to NAL was absent following weaning (Barb et al., 1986). The above studies did not examine the role of endogenous opioid peptides (EOP) in modulating FSH secretion during lactation and after weaning or in modulating PRL secretion after weaning in the sow. Therefore, the objective of the following experiment was to determine if EOP are involved in modulating FSH and PRL secretion in lactating and weaned sows.

MATERIALS AND METHODS

Eleven multiparous crossbred sows, 210.0 $\pm$ 9.3 kg body weight at 22.4 $\pm$ 0.8 days postpartum and nursing 10.2 $\pm$ 0.7 pigs, were used. Sows and their litters were penned in individual farrowing crates. A cannula was placed into the jugular vein (Kraeling et al., 1982) of all sows the day prior to treatment. Sows received either 1 (n = 3), 2 (n = 4), or 4 (n = 4) mg/kg body weight of NAL (Sigma Chemical Co., St. Louis, MO, U.S.A.[4]) in saline i.v. Blood was collected at 15 min intervals for 2 h before and 4 h after NAL treatment. All sows were then given 100 μg GnRH in saline i.v. and blood collection continued for an additional h. Pigs were weaned after blood sampling. At 40 h postweaning, sows were treated and blood collected as during suckling. Blood samples were allowed to clot for 24 h at 4°C, and serum was harvested after centrifugation and stored at $-20 \pm$ °C. Serum FSH was measured by a double antibody RIA validated in our laboratory as described by Guthrie and Bolt, 1983. Serum PRL was quantified in all samples by a RIA described previously (Kraeling et al., 1983). Serum progesterone was quantitated by RIA (Kraeling et al., 1981) only on the first sample collected during the suckling and weaned periods. Serum estrogen for each sow was quantitated (Lutz et al., 1984) on a pooled sample representing samples collected during the pretreatment bleeding period. Serum FSH and PRL concentrations were subjected to a split plot in time analysis of variance procedure of the Statistical Analysis System (SAS, 1982). Sampling time was divided into six periods. Period one represented the mean of samples collected for 2 h before treatment with NAL. The remainder of the sampling time was divided into five 1-h periods. Differences between period means were determined by the Bonferroni test (Gill, 1978).

RESULTS

Serum concentrations of progesterone were similar for suckled (1.7 $\pm$ 0.3 ng/ml) and weaned (1.5 $\pm$ 0.2 ng/ml) sows. Serum estrogen was undetectable (< 3 pg/ml) for all suckled sows and weaned sows except for two weaned sows that had 6 pg/ml of serum. Serum FSH and PRL concentrations following NAL were similar for all doses, therefore data was pooled across doses. During suckling, serum FSH concentrations averaged 25.7 $\pm$ 1.1 ng/ml during the pretreatment period and increased (P < 0.05) to an average of 32.2 $\pm$ 2.4 ng/ml during the first h after NAL treatment and remained elevated (33.2 $\pm$ 2.1 ng/ml) above pretreatment concentrations an additional hour (Fig. 1). Pretreatment serum FSH concentrations increased (P < 0.05) from 25.7 $\pm$ 1.1 ng/ml during suckling to 33.2 $\pm$ 2.6 ng/ml after weaning. After weaning, NAL treatment failed to alter serum FSH concentrations (33.2 $\pm$ 2.6 ng/ml before NAL and 33.1 $\pm$ 3.8 ng/ml for the first h after NAL). Gonadotropin-releasing hormone elevated (P < 0.05) serum FSH concentrations above pretreatment concentrations for suckled (25.7 $\pm$ 1.1 vs 70.1 $\pm$ 5.2 ng/ml) but not weaned sows (33.2 $\pm$ 2.6 ng/ml vs 39.7 $\pm$ 5.1 ng/ml). During suckling, serum PRL concentrations averaged 39.7 $\pm$ 1.7 ng/ml during the pretreatment period and decreased (P < 0.05) to an average of 25.7 $\pm$ 3.3 ng/ml, during the second h after NAL and remained depressed for an additional 2 h (Fig. 2). Pretreatment serum PRL concentrations decreased (P < 0.05) from 39.7 $\pm$ 1.7 ng/ml during suckling to 6.4 $\pm$ 0.2 ng/ml after weaning. After weaning, serum PRL concentrations were similar for the pretreatment period (6.4 $\pm$ 0.2 ng/ml) and for the first 60 min after NAL (6.2 $\pm$ 0.2 ng/ml).

DISCUSSION

We have previously reported that the EOP modulate the suckling induced suppression of LH secretion in the postpartum sow (Barb et al., 1986). The results of the present study indicate the EOP modulate the suckling induced increase in PRL and decrease in FSH secretion in the postpartum sow. These data are in agreement with the work in the rat (Sirinathsinghji and Martini, 1984) and sow (Mattioli et al., 1986) in which systemic administration of NAL blocked the suckling induced increase in PRL secretion, but do not agree with work in the postpartum rat

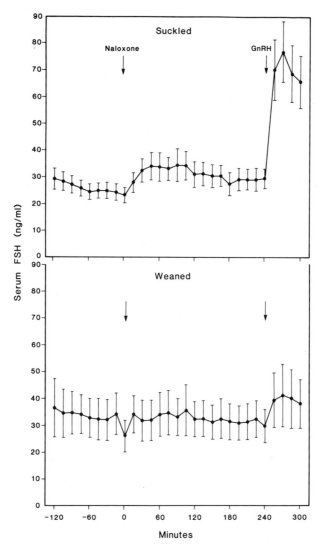

Fig. 1. Serum concentrations of FSH across doses before and after naloxone and GnRH. Each point represents the mean ± SEM.

(Sirinathsinghji and Martini, 1984) in which NAL treatment failed to alter FSH secretion. In the present study, the FSH and PRL response to NAL was absent after weaning suggesting that: 1) either weaning decreases EOP inhibition of FSH secretion or weaning induced a decrease in responsiveness of pituitary FSH to endogenous GnRH released by NAL, and 2) weaning decreased EOP stimulation of PRL secretion.

Dopamine secreted from the tuberoinfundibular neurons of the hypothalamus is now well established as a PRL inhibiting factor (Neill, 1980). Sirinathsinghji and Martini (1984) reported that bromocriptine, a dopamine receptor agonist, suppressing effect on PRL secretion in the lactating rat was similar to that of NAL. Similarly, in the lacting sow, the suppressing effect of bromocriptine on PRL secretion (Kraeling et al., 1982) is similar to that of NAL observed in the present experiment. Therefore, the above data suggests that the EOP stimulate PRL secretion by suppressing the tuberoinfundibular dopaminergic neurons. However, the rapid increase

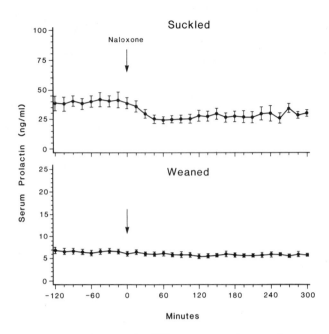

Fig. 2. Serum concentrations of PRL across doses before and after
naloxone. Each point represents the mean ± SEM.

in PRL secretion during suckling could also be due to an enhanced release
of PRL releasing factors, i.e., thyrotropin releasing hormone (de Greef
and Visser, 1981).

Kalra and Kalra (1984) presented a model for the EOP modulation of
GnRH secretion which is based on a postulated axo-axonic link between the
EOP and catecholaminergic neurons in the hypothalamus. Wilkes and Yen
(1980) and Rasmussen et al. (1983) reported that NAL increased GnRH
release from rat and human hypothalamus _in vitro_. Perfusion of rat hypothalamic fragments _in vitro_ with NAL resulted in a concurrent release of
GnRH, epinephrine, norepinephrine and dopamine (Leadem et al., 1985),
therefore, suggesting a hypothalamic site of action for the EOP.

It is proposed that suckling may trigger an increase in hypothalamic
concentrations of EOP. The EOP may, therefore, play a fundamental role in
the modulation of LH, FSH and PRL secretion in the lactating sow.

ACKNOWLEDGEMENT

This research was supported by USDA funds and State and Hatch funds
allocated to the Georgia Agricultural Experiment Station. Mention of
trade name does not constitute a guarantee or warranty by the USDA.

REFERENCES

Armstrong JD, Britt JH, Kraeling RR, 1978. Endogenous opioid peptides
    mediate the effects of suckling on LH during lactation in sows. J
    Anim Sci 63:Suppl 1 (abstract)
Barb CR, Kraeling RR, Rampacek GB, Whisnant CS, 1986. Opioid inhibition
    of luteinizing hormone secretion in the postpartum lactating sow.
    Biol Reprod 35:368-71

Bevers MM, Willemse AH, Kruip Th.AM, Van de Wiel DFM, 1981. Prolactin levels and the LH response to synthetic LH-RH in the lactating sow. Anim Reprod Sci 4:155-63

Cole DJA, Varley MA, Hughes PE, 1975. Studies in sow reproduction 2. The effect of lactation length on the subsequent reproductive performance of the sow. Anim Prod 20:401-06

Cox NM, Britt JH, 1982a. Relationships between endogenous gonadotropin-releasing hormone, gonadotropins, and folliclar development after weaning in sows. Biol Reprod 27:70-78

de Greef WJ, Visser TJ, 1981. Evidence for the involvement of hypothalamic dopamine and thyrophin-releasing hormone in suckling induced release of prolactin. J Endocrinol 91:213-23

Dusza L, Krzymowska H, 1981. Plasma prolactin levels in sows during pregnancy, parturition and early lactation. J Reprod Fert 61:131-34

Gill JL, 1978. Design and analysis of experiments in the animal and medical sciences. Vol. 1. Ames:Iowa State University Press, pp. 176

Guthrie HD, Bolt DJ, 1983. Changes in plasma estrogen, luteinizing hormone, follicle stimulating hormone and 13, 14-dihydro-15-keto prostaglandin $F_{2\alpha}$ during blockade of luteolysis in pigs after human chorionic gonadotropin treatment. J Anim Sci 57:993

Kalra SP, Kalra PS, 1984. Opioid-adrenergic-steroid connection in regulation of luteinizing hormone secretion in the rat. Neuroendocrinology 38:418-26

Kraeling RR, Rampacek GB, Cox NM, Kiser TE, 1982. Prolactin and luteinizing hormone secretion after bromocryptine (CB-154) treatment in lactating sows and ovariectomized gilts. J Anim Sci 54:1212-20

Kraeling RR, Rampacek GB, Kiser TE, 1981. Corpus luteum function after indomethacin treatment during the estrous cycle and following hysterectomy in the gilt. Biol Reprod 25:511-18

Leadem CA, Crowley WR, Simpkins JW and Kalra SP, 1985. Effects of naloxone on catecholamine and LHRH release from the perifused hypothalamus of the steroid-primed rat. Neuroendocrinol 40:497-500

Lutz JB, Rampacek GB, Kraeling RR, Pinkert CA, 1984. Serum luteinizing hormone and estrogen profiles before puberty in the gilt. J Anim Sci 58:686-91

Mattioli M, Conte F, Galeati G, Seren E, 1986. Effect of naloxone on plasma concentrations of prolactin and LH in lactating sows. J Reprod Fert 76:167-73

Neill JD, 1980. Neuroendocrine regulation of prolactin secretion. In: Ganong WF, Martini L (eds.), Frontiers in Neuroendocrinology, Vol. 6. New York: Raven Press, pp. 129-55

Rasmussen DD, Liu JH, Wolf PL, Yen SSC, 1983. Endogenous opioid regulation of GnRH release from human fetal hypothalamus in vitro. J Clin Endocrinol Metab 57:881-84

Riskind PN, Millard WJ, Martin JB, 1984. Opiate modulation of the anterior pituitary hormone response during suckling in the rat. Endocrinology 114:1232-37

SAS, 1982. SAS User's Guide. Statistical Analysis System Institute, Cary, NC

Shaw JH, Foxcroft GR, 1985. Relationships between LH, FSH and prolactin and reproductive activity in the weaned sow. J Reprod Fert 75:17-28

Sirinathsinghji DJS, Martini L, 1984. Effects of bromocriptine and naloxone on plasma level of prolactin, LH and FSH during suckling in the female rat: Response to gonadotropin releasing hormone. J Endocrinol 100:175-82

Steel RGD, Torrie JH, 1960. Principles and Procedures of Statistics. New York, Toronto, and London: McGraw-Hill

Stevenson JS, Cox NM, Britt JH, 1981. Role of the ovary in controlling luteinizing hormone, follicle stimulating hormone, and prolactin secretion during and after lactation in pigs. Biol Reprod 24:341-53

Stevenson JS, Davis DL, 1984. Influence of reduced litter size and daily litter separation on fertility of sows at 2 to 5 weeks postpartum. J Anim Sci 59:284-93

van Landeghem AAJ, van de Wiel DFM, 1978. Radioimmunoassay for porcine prolactin: plasma levels during lactation, suckling and weaning and after TRH administration. Acta Endocr Copenh 88:653-67

Wilkes MM, Yen SSS, 1980. Augmentation by naloxone of efflux of LRF from superfused medial basal hypothalamus. Life Sci 28:2355-59

ON THE SITE OF ACTION OF THE ESTRADIOL-INDUCED RELEASE

OF LUTEINIZING HORMONE (LH) IN PIGS

J.S. Kesner, R.R. Kraeling, G.B. Rampacek, C.R. Barb,
M.S. Estienne, R.D. Kineman, and C.E. Estienne

Animal Physiology Research Unit
R B Russell Research Center
USDA-ARS
  and
University of Georgia
Athens, GA 30613

INTRODUCTION

The site of action by which estradiol triggers the preovulatory LH surge varies widely among species. In sheep (Clarke and Cummins, 1984), cattle (Kesner et al., 1981) and rats (Fink, 1979), estradiol is thought to sensitize the pituitary to gonadotropin releasing hormone (GnRH) stimulation and then to accelerate the secretion rate of GnRH. In primates, estradiol stimulates the pituitary directly to occasion the LH surge (Knobil, 1980). Herein we report the effect of the pharmacological block of central nervous system (CNS) inputs on the estradiol-induced LH surge in gilts, and describe the dynamic changes in pituitary responsiveness to GnRH after estradiol exposure.

MATERIALS AND METHODS

Mature crossbred gilts (90 to 145 kg) were ovariectomized (OVX) 2 to 17 weeks before use. Jugular catheters were used for sampling blood and infusing GnRH. AIMAX [methallibure; ICI 33828; N-methyl-N'-(1-methyl-2-propenyl)-1,2-hydrazinedicarbothioamide; Ayerst Laboratories, NY, NY [1]] was fed daily at a dose of 125 mg in 1.8 kg ration for 7 or 8 (Expt 1) days to block CNS stimulation of gonadotropin release (Malven, 1971). β-estradiol-3-benzoate (10 µg E_2B/kg BW) in corn oil was injected im on the fifth day of AIMAX feeding.

First, each of three gilts were fed a ration with or without AIMAX for 7 days. From gilts fed AIMAX, blood was sampled every 15 min for 4 h before and 1, 2 and 3 days after starting AIMAX. On day 7, all six gilts were given 15 µg GnRH iv and blood was sampled every 15 min for 2 to 4 h before and 2 h after GnRH. Second, each of four gilts received: 1) AIMAX and E_2B, 2) AIMAX and oil, or 3) untreated feed and E_2B. Blood was sampled every 4 h for 96 h after E_2B or oil, and every 15 min for 4 h before AIMAX (day −5), before E_2B (day 0) and 6 days after stopping AIMAX (day 9). Third, each of four gilts were given: 1) AIMAX, E B and GnRH, 2) AIMAX, oil and GnRH, 3) AIMAX, E_2B and citrate, or 4) untreated feed, E_2B and citrate. One gilt in group 2 was eliminated from analysis due to illness. Five min pulses of 1 µg GnRH or 3.5% sodium citrate vehicle, infused iv

every 45 min, were begun 7 h after the first feeding of AIMAX and were
stopped 96 h after E_2B. Blood was sampled every 4 h for 96 h after E_2B or
oil; when applicable, these samples were collected 15 to 30 min after a
GnRH pulse. Samples were also collected every 15 min for 4 h before
starting treatments (day –5) before E_2B or oil (day 0), and every 15 min
for 1.5 h at 12, 24, 48, 72, 96 and 100 h after E_2B or oil. The catheter
dead space was refilled with GnRH infusate after collecting a sample that
preceded a GnRH pulse.

Serum concentrations of LH, follicle stimulating hormone (FSH), prolactin, growth hormone, estradiol–17β, cortisol, 3,5,3'–triiodothyronine
and thyroxine were measured using specific RIAs validated in our
laboratory. An LH pulse was defined as an increment exceeding two standard
deviations of the intraassay variation that occurred within 15 min of the
previous nadir. Pulse amplitude equalled the difference between the pulse
peak and the preceding nadir. Data were analyzed by split–plot in time
analysis of variance (ANOVA) using the general linear model procedure of
the Statistical Analysis System (SAS, 1985). Specific, preselected
comparisons between least square means were elevated using the least
significant difference procedure (SAS, 1985).

RESULTS

AIMAX reduced serum LH concentrations from 1.25 ± 0.13 ng/ml (mean
± SE) to 0.29 ± 0.08 ng/ml within 24 h (P < 0.001), and to undetectable
levels (< 0.18 ng/ml) after 7 days of treatment. LH response to a GnRH
stimulus, however, was not diminished (P > 0.1) in these gilts (6.95 ±
0.63 ng/ml) compared to untreated controls (4.07 ± 0.04 ng/ml; Fig. 1).

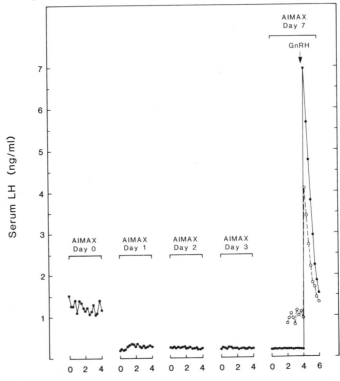

Fig. 1. Mean serum LH concentrations in OVX gilts fed AIMAX daily. GnRH
(15 μg) was injected iv to treated (●——●) and untreated (○- -○)
gilts.

AIMAX reduced serum FSH concentrations 27% (P < 0.001), but did not alter serum concentrations of prolactin, growth hormone, thyroid hormones, or cortisol (P < 0.1). Behavior was not modified except for a slightly reduced appetite.

Serum estradiol concentrations peaked at 78 ± 9 pg/ml by 12 h after E_2B, and then declined to 10 to 25 pg/ml at 48 to 96 h after E_2B. The LH surge occurred 48 to 80 h after E_2B and was completely blocked by AIMAX in three of four gilts (data not shown). Normal pulsatile LH profiles in serum were present 6 days after AIMAX was withdrawn. In a third experiment, pulsatile GnRH restored mean serum concentrations of LH as well as frequency and amplitude of LH pulses in AIMAX-treated gilts. LH pulses in similarly treated gilts were muted or blocked (P < 0.001) within 12 h after E_2B, but returned to control levels by 24 h and did not vary thereafter (Fig. 2 and 3). Prior to the surge, LH levels were suppressed for 36 to 52 h after E_2B, when AIMAX and GnRH were not given, while AIMAX alone blocked the LH surge in two of four gilts given E_2B (Fig. 3).

DISCUSSION

Our results indicate that AIMAX blocked the release of LH, and to a lesser extent FSH, at a site other than the pituitary. That the efficacy of GnRH to release gonadotropins was not reduced by AIMAX, that small doses of pulsatile GnRH chronically restored normal pulsatile LH release in gilts treated with AIMAX, and that AIMAX did not alter behavior or secretions of other pituitary hormones, supports the view that AIMAX acted with relative specificity to inhibit GnRH secretion in pigs.

The peak serum estradiol concentrations for approximately one-third of the gilts given E_2B in this study fell within the normal range for periovulatory pigs. The LH responses of this group of gilts did not differ from those to their treatment-mates that experienced higher steroid

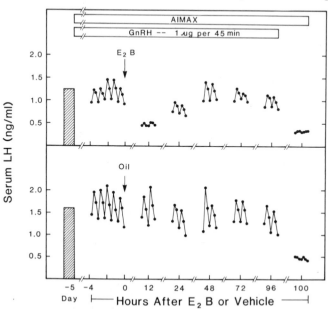

Fig. 2. LH pulses in OVX gilts given AIMAX and GnRH were defined before and after E_2B (top; n = 4) or oil (n = 3) and after stopping GnRH. Blood was sampled at 15 min intervals. Bars depict means of samples collected before starting treatments. SE derived from the ANOVA error mean square were 0.08 for pulse amplitude.

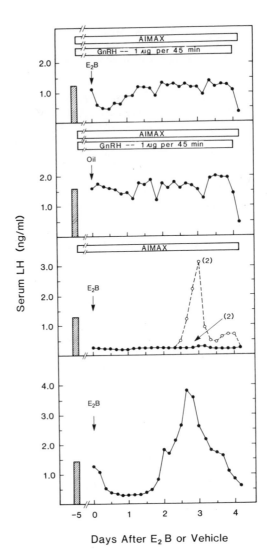

Fig. 3. Mean serum LH concentrations in OVX gilts given (top to bottom); AIMAX, GnRH and E₂B; AIMAX and GnRH; AIMAX and E₂B; or E₂B. Bars depict means of samples collected before starting treatments (SE = 0.10 to 0.34). SE derived from the ANOVA error mean square for the two groups given GnRH was 0.13.

levels. Variation whereby LH surges were blocked by AIMAX in some gilts, but not in others is not accounted for by the amount of AIMAX consumed (appetite), serum estradiol concentrations achieved, body weight, or days from OVX to experimentation. The AIMAX dose may have been marginal for complete blockade in all animals.

AIMAX blocked the LH surge in pigs, albeit not in all pigs, indicating that expression of the LH surge in this species required CNS input as suggested by others (Docke and Busch, 1974; Parvizi et al., 1976; Elsaesser and Foxcroft, 1978; Elsaesser and Parvizi, 1979). That the central signal is merely permissive, serving only to support steroidal stimulation of the pituitary directly as in monkeys (Knobil, 1980) is unlikely since estradiol did not augment LH release in the presence of GnRH.

Estradiol alone inhibited LH release for 36 to 52 h before emergence of the LH surge, although gonadotrope responsiveness to GnRH was inhibited for only 20 h, in accordance with another report (Cox and Britt, 1982), and was not stimulated during the period that precedes or coincides with the anticipated LH surge; a unique finding amongst spontaneously ovulating mammals. We propose that estradiol inhibited LH release by first reducing pituitary responsiveness to GnRH, and then by inhibiting GnRH secretion. Emergence of the LH surge may reflect the resumption of GnRH secretion. Whether the prolonged secretory hiatus of LH and/or GnRH is a necessary antecedant to the surge, or whether the surge of LH release reflects a surge of GnRH stimulation is unknown.

ACKNOWLEDGEMENTS AND FOOTNOTES

[1]AIMAX is no longer produced by Ayerst or its affiliates.

[2]Supported by USDA funds and State and Hatch funds allocated to the Georgia Agricultural Research Station. Mention of a trade name, proprietary product, or specific equipment does not constitute a guarantee or warranty by the USDA or University of Georgia and does not imply its approval to the exclusion of other products which may be suitable.

[3]Authors are indebted to M. Joyce Minor for her secretarial skills.

REFERENCES

Clarke IJ, Cummins JT, 1984. Direct pituitary effects of estrogen and progesterone on gonadotropin secretion in the ovariectomized ewe. Neuroendocrinology 39:267-274

Cox NM, Britt JH, 1982. Effect of estradiol on hypothalamic GnRH and pituitary and serum LH and FSH in ovariectomized pigs. J Anim Sci 55:901-908

Docke F, Busch W, 1974. Evidence for anterior hypothalamic control of cyclic gonadotropin secretion in female pigs. Endokrinologie 63:415-421

Elsaesser F, Foxcroft GR, 1978. Maturational changes in the characteristics of oestrogen-induced surges of luteinizing hormone in immature domestic gilts. J Endocrinol 78:455-456

Elsaesser F, Parvizi N, 1979. Estrogen feedback in the pig: sexual differentiation and the effect of prenatal testosterone treatment. Biol Reprod 20:1187-1193

Fink G, 1979. Feedback actions of target hormones on hypothalamus and pituitary with special reference to gonadal steroids. Ann Rev Physiol 41:571-585

Kesner JS, Convey EM, Anderson CR, 1981. Evidence that estradiol induces the preovulatory LH surge in cattle by increasing pituitary sensitivity to LHRH and then increasing LHRH release. Endocrinology 108:1386-1391

Knobil E, 1980. The neuroendocrine control of the menstrual cycle. Rec Prog Horm Res 36:53-88

Malven PV, 1971. Hypothalamic sites of action for methallibure (ICI 33828) inhibition of gonadotropin secretion. J Anim Sci 32:912-918

Parvizi N, Elsaesser F, Smidt D, Ellendorf F, 1976. Plasma luteinizing hormone and progesterone in the adult female pig during the estrous cycle, late pregnancy and lactation, after ovariectomy and pentobarbitone treatment. J Endocrinol 69:193-203

SAS/STAT Guide, 1985. SAS Institute Inc., Cary, NC, pp. 183-260

INDUCTION OF FOLLICULAR CYSTS IN RAT OVARIES BY PROLONGED

ADMINISTRATION OF HUMAN CHORIONIC GONADOTROPIN

Katryna Bogovich
University of South Carolina
School of Medicine
Columbia, SC 29208

INTRODUCTION

Polycystic ovarian disease (PCOD) appears to arise from an inappropriate peripubertal response within the hypothalamic-pituitary-ovarian axis. For review see Yen (1980), Coney (1984) and Futterweit (1984). The association of elevated serum concentrations of luteinizing hormone (LH), androgens and estrogens with this syndrome has led to the hypothesis that overstimulation of the ovary by LH leads to the development of ovarian follicular cysts. Pharmacologic amounts of androgens and estrogens appear capable of inducing the formation of ovarian cysts. However, such quantities of these steroids also disrupt hypothalamic-pituitary interactions (Barraclough, 1966; Shulster et al., 1984) and, therefore, the potential for these steroids to affect the development of cysts by inducing inappropriate secretion of gonadotropins cannot be ruled out. Furthermore, until now, the time- and dose-related effects of prolonged, chronic exposure to LH-like activity on ovarian follicular cyst formation had not been determined. This work was undertaken to determine: 1) if chronic administration of LH-like activity is capable of inducing the formation of cystic follicles in intact animals; and 2) if such treatment is effective in inducing cystic follicles in ovaries at different stages of development/differentiation.

MATERIALS AND METHODS

Immature rats were obtained from Harlan Sprague-Dawley (Indianapolis, IN) at 22 days of age. Silestrus progesterone implants (the generous gift of Dr. JoAnne Richards, Baylor College of Medicine, Houston, TX) were administered sc on day 24 of life and remained in situ during subsequent treatments. Beginning on day 27, these animals received 0 (controls), 0.1, 0.5 or 1.5 IU human chorionic gonadotropin (hCG) sc twice daily for 9 days. First-time pregnant rats were obtained from Harlan Sprague-Dawley with the data of insemination designated as day 1. These animals received 0 (controls), 1 or 3 IU hCG twice daily for 9 days beginning on day 13 of pregnancy. Immature rats hypophysectomized at 21 days of age were obtained from Charles River (Boston, MA). These animals received 0, 0.1 or 0.5 IU hCG twice daily for 9 days beginning on day 27 of life.

For all treatment groups, 5-7 animals were decapitated, blood was collected, and the ovaries were excised between 0800 and 0900 on each day of treatment and on the morning after the final injection of hCG. At least one ovary from each group was fixed in formalin and embedded in paraffin.

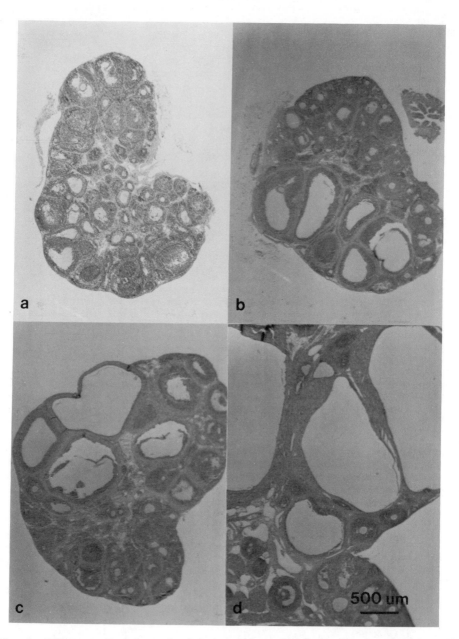

Fig. 1. Effects of prolonged administration of increasing doses of LH-like
activity on the ovaries of immature rats with sc progesterone
implants: (a) 0 IU hCG (control), day 36; (b) 0.1 IU hCG x 18, day
36; (c) 0.5 IU hCG x 18, day 36; (d) 1.5 IU hCG x 14, day 34.

Sections were cut, stained with hematoxylin-eosin and photomicrographs
were obtained. Although the largest follicles and cysts from the
remaining ovaries were incubated with or without 8-bromo cAMP and steroid
substrates, the limited nature of this presentation dictates that these
results and the serum steroid and gonadotropin profiles be presented
elsewhere.

RESULTS

Fig. 1 illustrates the effects of 7-9 days of treatment with increasing quantities of hCG on the ovaries of immature rats with sc progesterone implants. Figure 1(a) shows a cross-section from the ovary of a control animal (0 IU hCG) on day 36: ten days after the start of the injection regimen. These ovaries possess small antral, preantral and primordial follicles with many of the small antral follicles showing signs of atresia. In addition, the ovaries appear virtually identical to those of 36 day old animals without progesterone implants (not shown). Fig. 1(b) illustrates the effects of 0.1 IU hCG twice daily for 9 days. These ovaries possess large antral follicles with many granulosa cells and with stimulated theca. Although the stromal-interstitial tissue also appears stimulated, the cross-sectional area of these ovaries appears to be the same as that of ovaries from control animals. In contrast, Fig. 1(c) demonstrates that ovaries from immature rats receiving 0.5 IU hCG for 9 days possess, in addition to pre- and small antral follicles, many follicles approximately 1 mm in diameter which might be described as "precystic". Only a thin layer of granulosa cells remain in these follicles (suggestive of atresia) while the theca and the ovarian stromal-interstitial tissue appear stimulated. In addition, the cross-sectional area of these ovaries is approximately 50% larger than that of ovaries from control animals. Fig. 1(d) shows that ovaries from immature rats receiving 1.5 IU hCG twice daily possess fully cystic follicles by day 34 (after only 14 injections of hCG). The cysts are approximately 2 mm in diameter and possess few, if any, granulosa cells. Both theca and ovarian stromal-interstitial tissue appear stimulated although the stromal-interstitial tissue also has a "collagenized" appearance. In addition, the cross-sectional area of these ovaries is almost 5 times larger than that of ovaries from control rats. Still, these ovaries do possess some pre- and small antral follicles - with many displaying signs of atresia.

Fig. 2 illustrates the effects of 9 days of treatment with increasing doses of hCG on the ovaries of pregnant rats. Fig. 2(a) shows a portion of an ovary from a control animal (0 IU hCG) on day 22 of pregnancy. These ovaries possess large corpora lutea (approximately 2 mm in diameter), small antral follicles showing signs of atresia, and preovulatory follicles (approximately 0.8 mm in diameter). Fig. 2(b) illustrates the effects of 1 IU hCG twice daily for 9 days on follicular development in these animals. The ovaries still possess large corpora lutea but also display follicles with larger diameters than those of preovulatory follicles in control ovaries. These follicles (approximately 1.3 mm in diameter) appear to possess decreasing numbers of granulosa cells while their theca appear stimulated. In addition, the ovarian stromal-interstitial tissue seems stimulated and has a "collagenized" appearance. Fig. 2(c) shows the effects of 3 IU hCG on ovaries from pregnant rats. Corpora lutea appear healthy even though follicular cysts, ranging in size from 1.5 to 3 mm in diameter, are present. As shown in this panel, the smaller follicular structures might still be considered "precystic" since small numbers of granulosa cells are still present. However, the larger cyst-like structures possess virtually no granulosa cells.

Fig. 3 illustrates the effect of 9 days of treatment with increasing doses of hCG on the ovaries of hypophysectomized immature rats. Fig. 3(a) shows that ovaries from such animals are quite small and, as expected, contain only primordial and preantral follicles. Figs. 3(b) and 3(c) show the effects of 0.1 and 0.5 IU hCG twice daily for 9 days, respectively. Neither dose induces cystic follicles in these animals although the cross-sectional areas of the hCG-treated ovaries are 2 and 3 times, respectively, that of control hypophysectomized animals. The ovarian stromal-interstitial tissue seems stimulated for both doses of hCG, while

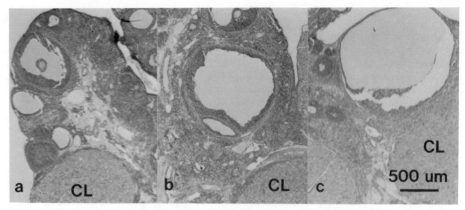

Fig. 2. Effects of prolonged administration of increasing doses of LH-like
activity on the ovaries of pregnant rats (day 22): (a) 0 IU hCG
(control); (b) 1 IU hCG x 18; (c) 3 IU hCG x 18. CL: corpus
luteum.

only thecal tissue from animals receiving 0.5 IU hCG appears stimulated.
In addition, the ovaries from hypophysectomized rats receiving 0.5 IU hCG
seem to possess fewer follicles than ovaries from similar animals
receiving 0 or 0.1 IU hCG.

DISCUSSION

These results clearly demonstrate, for the first time, that prolonged
stimulation by chronically maintained serum concentrations of LH-like
activity is capable of inducing ovarian follicular cysts in a dose-
dependent and time-related manner in intact immature and pregnant rats.
In contrast, such stimulation is not sufficient for the induction of
ovarian cysts in immature hypophysectomized rats. This observation agrees
with that of Leathem (1958) and suggests that LH is not the only pituitary
hormone required for the development of ovarian cysts. Until now, no
animal model has been able to differentiate between the role of peptide
hormones and the effects of steroids in the initiation of the development
of ovarian cysts. Pharmacologic amounts of estrogens or androgens induce
cystic ovaries in immature and young adult rats but may do so by
disrupting normal hypothalamic-pituitary interactions (Barraclough, 1966;
Shulster et al., 1984). Hypothyroid rats respond to 20 days of ovulatory
doses of hCG with luteinized, cyst-bearing ovaries (Leathem, 1958).
However, the effects of such treatment on intact animals were not
determined. In our hands, intact immature rats without progesterone
implants, which received only 1.5 IU hCG twice daily for 7 days, possessed
luteinized ovaries containing follicular cysts by the eighth day of treatment (data not shown). This suggests that chronically elevated serum
progesterone in the hypothyroid rat, rather than the transient rise in
serum prolactin cited by Lee et al. (1986), may be responsible for the
acyclic state of this animal model. The acyclic models used in the
present study allow the effects of exogenous hormones on ovarian function
to be determined (Bogovich et al., 1981; Richards and Bogovich, 1982) and,
therefore, provide a means of investigating 1) how peptide hormones elicit
the initial changes in follicular function which lead to the development
of ovarian cysts; and 2) how ovarian responses to such stimulation might
perpetuate PCOD. Although these models cannot determine the primary
etiological factor(s) in the development of PCOD, induction of follicular
cysts in the highly differentiated ovaries of pregnant rats does support
indirectly the concept of a central lesion providing the impetus for the
onset of PCOD.

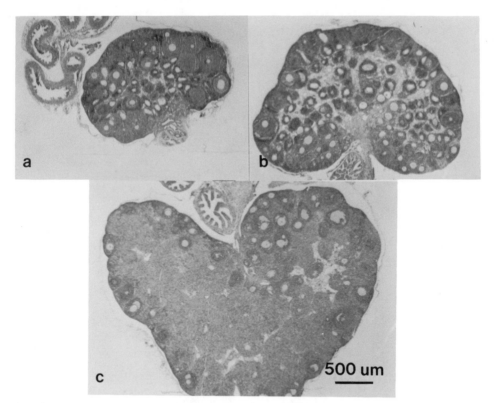

Fig. 3. Effects of prolonged administration of increasing doses of LH-like
activity on the ovaries of hypophysectomized immature rats by day
36 of life: (a) 0 IU hCG (control); (b) 0.1 IU hCG x 18; (c) 0.5
IU hCG x 18.

REFERENCES

Barraclough CS, 1966. Modification in the CNS regulation of reproduction
    after exposure of prepubertal rats to steroid hormones. Recent Prog
    Horm Res 22:503-515
Bogovich K, Richards JS, Reichert LE, 1981. Obligatory role of luteinizing
    hormone (LH) in the initiation of preovulatory follicular growth in
    the pregnant rat: specific effects of human chorionic gonadotropin
    and follicle-stimulating hormone on LH receptors and steroidogenesis
    in theca, granulosa, and luteal cells. Endocrinology 109:860-867
Coney P-J, 1984. Polycystic ovarian disease: current concepts of patho-
    physiology and therapy. Fertil Steril 42:667-682
Futterweit W, 1984. Clinical Perspectives in Obstetrics and Gynecology.
    New York: Springer-Verlag
Leathem JH, 1958. Hormonal influences on the gonadotropin sensitive
    hypothyroid rat ovary. Anat Rec 131:487-500
Lee M-T, Bruot BC, Adams WC, 1986. Hormonal changes in the early
    development of ovarian cysts in the rat. Biol Reprod 35:542-548
Richards JS, Bogovich K, 1982. Effects of human chorionic gonadotropin
    and progesterone on follicular development in the immature rat.
    Endocrinology 111:1429-1438
Shulster A, Farookhi R, Bawer JR, 1984. Polycystic ovary condition in
    estradiol treated rats: spontaneous changes in characteristic
    endocrine features. Biol Reprod 31:587-593
Yen SSC, 1980. The polycystic ovary syndrome. Clin Endocrinol 12:177-208

PARACRINE CONTROL OF 3β-HYDROXYSTEROID DEHYDROGENASE, 17,20-LYASE

AND AROMATASE ENZYME SYSTEMS IN PORCINE THECAL CELLS

Sharon A Tonetta

University of Southern California Medical School
Livingston Biological Research Laboratory
Los Angeles, CA 90033

INTRODUCTION

Thecal cells are important intraovarian sources of androgens, providing substrate necessary for the synthesis of estradiol by granulosa cells (Armstrong and Dorrington, 1977). However, in addition to androgens, progesterone and estradiol are secreted by thecal cells of sows (Stoklosowa et al., 1982), cows (McNatty et al., 1984), monkeys (Vernon et al., 1983) and humans (McNatty et al., 1979). Although thecal steroidogenesis is regulated partly by circulating levels of luteinizing hormone (Baird and McNeilly, 1981; Merz et al., 1981; Carlson et al., 1984), the intrafollicular control of thecal steroidogenesis is unclear. As local steroidal and nonsteroidal factors can modify granulosa cell steroidogenesis in vitro (Hsueh et al., 1984; Tonetta and diZerega, 1986), the effects of physiological concentrations of steroids and gonadotropin on cultured porcine thecal cells were determined.

The formation of steroids involves key systems in the steroidogenic pathway. The formation of Δ^4-3 ketosteroids is mediated by the enzyme system 3β-hydroxysteroid dehydrogenase (3β-HSD; Samuels et al., 1951) while conversion of progestins to androgens is mediated by the 17-hydroxylase/17,20-lyase enzyme system (Tsang et al., 1979). Further conversion of androgens to estrogens is mediated by the aromatase enzyme system (Dorrington et al., 1975; Moon et al., 1975). As thecal cell steroidogenesis is stimulated by LH (McNatty et al., 1980; Stoklosowa et al., 1982), human chorionic gonadotropin (hCG) was used as a biological probe of enzyme activity. Porcine thecal cells exist in a milieu rich in progesterone (P), estradiol-17B ($E_{17\beta}$) and dihydrotestosterone (DHT); therefore, the effects of these steroids on 3β-HSD, 17,20-lyase and aromatase activities were evaluated. Evidence is presented that regulation of enzyme activities in thecal cells may involve ultra short-loop feedback by steroids.

METHODS

Porcine ovaries were obtained from a local abattoir. Thecal tissue was excised from large follicles (> 8 mm), placed in sterile, serum-free Medium 199 (M199) and processed as previously described (Tonetta et al., 1986). Thecal cells were plated (5 x 10^5 cells/ml/well) in M199 + epidermal growth factor (2.5 ng/ml) + insulin (10 ug/ml) + transferrin (5 ug/ml) and incubated at 37°C in 5% CO_2, 95% air.

Twenty-four h after plating, cells were treated as follows: 1) control (M199); 2) hCG (5 mIU); 3) P (300 ng) or $E_{17\beta}$ (400 ng) or DHT (100 ng); 4) hCG + P or $E_{17\beta}$ or DHT. Media was removed at 3,6,12,24, and 48 h after treatment and cells were used to measure enzyme activity. Cultures were repeated 3 times with 3-5 replicates per treatment.

Measurement of 3β-HSD activity was performed as described by Battin and diZerega (1985). Briefly, cells were scraped off the plates, soni-cated, and centrifuged (10,000 x g, 4°C) for 30 min. The resulting super-natant was further centrifuged at 100,000 x g for 60 min (4°C) to obtain the microsomal pellet. M199 + 5 uM NAD^{++} 10^{-6} M pregnenolone was added to the microsomes and incubated at 37°C for 1 h. The reaction was stopped by heating at 60°C for 30 min. 3β-HSD activity was measured by conversion of pregnenolone to progesterone.

Lyase activity was determined by a modification of the method of Quinn and Payne (1985). After media was removed, 1 ml of M199 containing 0.5 mM NADPH was added to each well. ^{3}H-progesterone (1 uCI/0.1 ml) was then added and the plates incubated for 1 h at 37°C in 95% air, 5% CO_2. The reaction was stopped by the addition of 0.1 N NaOH. All tubes were extracted 2x with ether, dried under nitrogen and reconstituted in methanol. Samples and standards were spotted onto thin layer silica gel plates and chromatographed for 1 h in 7:1 chloroform: ether. Spots were identified under ultraviolet light, cut out and counted in 9.5 ml of scintillation fluid.

Aromatase activity was determined by a radiometric assay (Gore-Langton and Dorrington, 1981). Since the conversion of ^{3}H-T to $E_{17\beta}$ results in the release of ^{3}H-H_2O, the percent conversion of ^{3}H-T to ^{3}H-H_2O reflects aromatase activity (expressed as % control).

Statistical analysis was determined by one way analysis of variance. Dunnet's test was used to compare treatments with controls. Bonferroni-T was used to compare treatments.

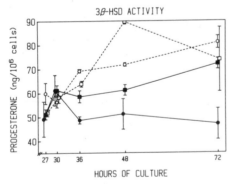

Fig. 1. Effects of steroids on 3β-HSD activity in porcine thecal cells. Cells were treated at 24 h with M 199 (■—■), E17β (400 ng; ●—●), P (300 ng; o—o) or DHT (100 ng; □—□). 3β-HSD activity was measured at 27, 30, 36, 48 and 72 h of culture (mean ± SE).

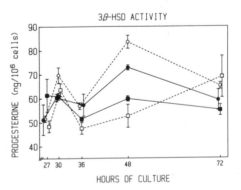

Fig. 2. Effects of steroids plus hCG on 3β-HSD activity in porcine thecal cells. Cells were treated with hCG (5 mIU, ■—■), hCG + E17β (●—●), hCG + P (o—o) or hCG + DHT (□—□) at 24 h of culture. 3β-HSD activity was measured at 27, 30, 36, 48 and 72 h of culture (mean ± SE).

RESULTS

3β-HSD activity was unchanged from 27 to 72 h. hCG had little effect on 3β-HSD activity at all times (Fig. 2). Progesterone alone increased (p < 0.05) 3β-HSD activity after 36 h (Fig. 1) while hCG + progesterone enhanced activity after 48 h (Fig. 2). Estradiol alone inhibited (p < 0.05) activity at 36 and 72 h (Fig. 1) while DHT enhanced activity at 48 h.

Lyase activity was unaffected by administration of hCG from 27 to 72 h of culture (Fig. 3). Progesterone alone significantly (p < 0.05) inhibited lyase activity at 27 h while progesterone + hCG inhibited lyase activity at 27 and 30 h (Fig. 3, 4). In addition, estradiol with or without hCG significantly (p < 0.05) inhibited lyase activity at 30 and 36 h (Fig. 3, 4). DHT had little effect on lyase enzyme activity (Fig. 3, 4).

Thecal aromatase activity in controls was unchanged from 27 to 72 h. hCG significantly increased (p < 0.05) aromatase activity at 27 and 30 h of culture but had little effect at other times (Fig. 6). Estradiol enhanced (p < 0.05) aromatase activity at 27 h (p < 0.05; Fig. 5) while hCG + estradiol measured aromatase activity at 27 and 30 h (Fig. 6). HCG + progesterone or DHT stimulated aromatase activity at 30 h (Fig. 6).

DISCUSSION

The formation of Δ^4-3 ketosteroids, a key step in steroidogenesis, is mediated by the enzyme system 3β-HSD (Samuels et al., 1951). In granulosa cells, FSH enhances 3β-HSD activity (Jones and Hsueh, 1982). Although hCG had little effect on thecal 3β-HSD activity, progesterone secretion was enhanced after hCG administration. In addition, progesterone either alone

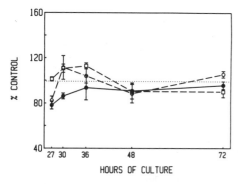

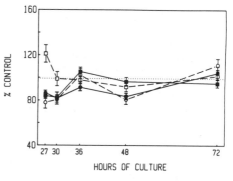

Fig. 3. Effects of steroids on lyase activity in porcine thecal cells. Cells were treated at 24 h with E17β (400 ng; ●—●), P (300 ng; ○—○) or DHT (100 ng; □—□). Lyase activity was measured at 27, 30, 36, 48 and 72 h of culture (mean ± SE).

Fig. 4. Effects of steroids plus hCG on lyase activity in porcine thecal cells. Cells were treated with hCG (5 mIU, ■—■), hCG + E17β (●—●), hCG + P (○—○) or hCG + DHT (□—□) at 24 h of culture. Lyase activity was measured at 27, 30, 36, 48 and 72 h of culture (mean ± SE).

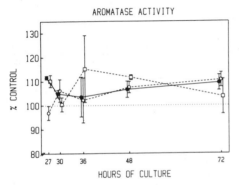

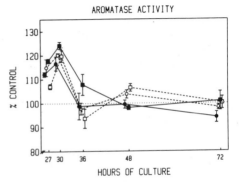

Fig. 5. Effects of steroids on aromatase activity in porcine
thecal cells. Cells were
treated at 24 h with E17β
(400 ng; ●—●), P (300 ng; o—o),
or DHT (100 ng; □—□), aromatase
activity was measured at 27,
30, 36, 48 and 72 h of culture
culture (mean ± SE).

Fig. 6. Effect of steroids plus
hCG on aromatase activity
in porcine thecal cells.
Cells were treated with
hCG (5 mIU, ■—■), hCG +
E17β(●—●), hCG + P (O—O),
hCG + DHT (□—□).
Aromatase activity was
measured at 27, 30, 36,
48 and 72 h of culture
(mean ± SE).

or in conjunction with hCG enhanced thecal 3β-HSD activity indicating a
positive feedback. However, in a previous study, large doses of
progesterone inhibited 3β-HSD activity in theca cells from large porcine
follicles (Tonetta et al., in review). Additionally, estradiol inhibited
thecal 3β-HSD activity (Tonetta et al., 1986). In granulosa cells from
sows, hamsters and rats, progesterone secretion is enhanced by DHT and
inhibited by estradiol (Haney and Schomberg, 1978; Schomberg et al., 1976;
Lucky et al., 1977).

Thecal cells are the major producers of follicular androgens (Makris
and Ryan, 1975; Tsang et al., 1979). Previous studies of thecal steroidogenesis using explants or ovarian homogenates focused on the lyase enzyme
system. Androgen production is stimulated by LH in rat ovaries (Erickson
and Ryan, 1976), rat and rabbit follicles (Lieberman et al., 1975; Mills,
1975) and thecal preparations from hamsters and rabbits (Makris and Ryan,
1975; Lieberman et al., 1975). Lyase activity in thecal cultures from
large porcine follicles was not affected by hCG administration. However,
lyase activity in these cells was inhibited by progesterone and estradiol
while DHT had little affect. In rat ovarian homogenates (Johnson and
Griswold, 1984), androstenedione, testosterone and dihydrotestosterone
have little effect on lyase activity while progesterone was inhibitory.

The conversion of androgens to estrogens is mediated by the aromatase
enzyme system (Dorrington and Armstrong, 1975; Moon et al., 1975). In
granulosa cells, the aromatase enzyme system is primarily induced by FSH
while thecal production of estradiol is enhanced by LH/hCG treatment
(Stoklosowa et al., 1982; Tsang et al., 1985). In porcine thecal cells
from large follicles, estradiol with or without hCG enhanced aromatase
activity. However, large doses of estradiol as well as progesterone
inhibited thecal aromatase activity (Tonetta et al., 1986). In granulosa
cells, progesterone can inhibit FSH-induced estrogen production (Schreiber
et al., 1980, 1981; Fortune and Vincent, 1983) while estradiol alone or
with FSH potentiates aromatase activity (Tseng, 1984).

Although thecal steroidogenesis is enhanced by gonadotropin stimulation, the fine control of enzyme activity appears to be regulated by the steroidal environment. During steroidogenesis, 3β-HSD is enhanced by progesterone thereby insuring the continuant of steroidogenesis. Similarly, estradiol production is enhanced by estradiol. However, large quantities of progesterone and estradiol in the environ inhibits 3β-HSD, lyase as well as aromatase activities. Therefore, thecal steroidogenesis is regulated by paracrine and autocrine mechanisms.

REFERENCES

Armstrong DT, Dorrington JH, 1977. Estrogen biosynthesis in the ovaries and testis. In: Thomas JA, Singhal RL (eds.), Regulatory Mechanisms Affecting Gonadal Hormonal Action. Baltimore: University Park Press, 3:217

Baird DT, McNeilly AS, 1981. Gonadotropic control of follicular development and function during the estrous cycle of the ewe. J Reprod Fertil 30:119-133

Battin DA, diZerega GS, 1985. Effect of human menopausal gonadotropin and follicle regulatory protein(s) on 3β-hydroxysteroid dehydrogenase in human granulosa cells. J Clin Endocrinol Metab 60:1116-19

Carlson RS, Findlay JK, Clarke IJ, Burger HG, 1984. Estradiol, ketosterone, and androstenedione in ovine follicular fluid during growth and atresia of ovarian follicles. Biol Reprod 24:105-13

Dorrington JH, Moon YS, Armstrong DT, 1975. Estradiol-17β biosynthesis in cultured granulosa cells from hypophysectomized immature rats: stimulation by follicle-stimulating hormones. Endocrinology 97:1328-31

Erickson GF, Ryan KJ, 1976. Stimulation of ketosterone production in isolated rabbit thecal tissue by LH/FSH, dibutyrl cyclic AMP, PGFα, and PGE2. Endocrinology 99:452-58

Fortune JE, Vincent SE, 1983. Progesterone inhibits the induction of aromatase activity in rat granulosa cells in vitro. Biol Reprod 28:1078-89

Gore-Langton RE, Dorrington JH, 1981. FSH induction of aromatase in cultured rat granulosa cells measured by a radiometric assay. Mol Cell Endocrinol 22:135-51

Haney AF, Schomberg DW, 1978. Steroidal modulation of progesterone secretion by granulosa cells from large porcine follicles: a role for androgens and estrogens in controlling steroidogenesis. Biol Reprod 19:252-48

Hsueh AJW, Adashi EY, Jones PBC, Welsh Jr TH, 1984. Hormonal regulation of the differentiation of cultured ovarian granulosa cells. Endocrine Rev 5:76-127

Johnson DC, Griswold T, 1985. Ovarian C17-20-lyase: changes in intact and hypophysectomized immature rats treated with pregnant mare's serum gonadotropin. J Steroid Biochem 20:733-39

Jones PBC, Hsueh AJW, 1982. Regulation of ovarian 3β-hydroxysteroid dehydrogenase activity by gonadotropin-releasing hormone and follicle-stimulating hormone in cultured rat granulosa cells. Endocrinology 110:1663-71

Lieberman ME, Barnea A, Bauminger S, Tsafriri A, Collins WP, Lindner HR, 1975. LH effect on the pattern of steroidogenesis in cultured Graafian follicles of the rabbit: dependence on macromolecular synthesis. Endocrinology 96:1533-42

Lucky AW, Schreiber JR, Hillier SG, Schulman JD, Ross GT, 1977. Progesterone production by cultured preantral rat granulosa cells: stimulation by androgens. Endocrinology 100:128-33

Makris A, Ryan KJ, 1975. Progesterone, androstenedione, testosterone, estrone and estradiol synthesis in hamster ovarian follicle cells. Endocrinology 96:694-701

McNatty KP, Makris A, Osathanondh R, Ryan KJ, 1980. Effects of luteinizing hormone on steroidogenesis by thecal tissue from human ovarian follicles in vitro. Steroids 36:53-63

McNatty KP, Heath DA, Lun S, Fannin JM, McDiarmid JM, Henderson KM, 1984. Steroidogenesis by bovine theca internal in an in vitro perfusion system. Biol Reprod 30:159-70

McNatty KP, Smith DM, Makris A, Osathanondh R, Ryan KJ, 1979. The micro-environment of the human antral follicle: interrelationships among the steroid. Fertil Steril 32:433-438

Merz EA, Hauser ER, England BG, 1981. Ovarian function in the cycling cow: relationship between gonadotropin binding to theca and granulosa and steroidogenesis in individual follicles. J Anim Sci 52:1457-68

Mills TM, 1975. Effect of luteinizing hormone and cyclic adenosine 3',5'-monophosphate on steroidogenesis in the ovarian follicle of the rabbit. Endocrinology 96:440-54

Moon YS, Dorrington JH, Armstrong DT, 1975. Stimulating action of follicle stimulating hormone on estradiol-17β secretion by hypophysectomized rat ovaries in organ culture. Endocrinology 97:244-47

Quinn PG, Payne AH, 1985. Oxygen-mediated damage of microsomal cytochrome P-450 enzymes in cultured Leydig cells. J Biol Chem 259:4130-35

Samuels LM, Helreich M, Lasater M, Reich H, 1951. An enzyme in endocrine tissues which oxidizes Δ^5-3-hydroxysteroid to a β-unsaturated ketones. Science 113:490-91

Schomberg DW, Stouffer RL, Tyrey L, 1976. Modulation of progestin secretion in ovarian cells by 17β-hydroxy-5α-androstan-3-one (dihydrostenedione): a direct demonstration in monolayer culture. Biochem Biophys Res Commun 68:77-8

Schreiber JR, Nakamura K, Erickson GF, 1980. Progestins inhibit FSH-stimulated steroidogenesis in cultured rat granulosa cells. Mol Cell Endocrinol 19:165-73

Schreiber JR, Nakamura K, Erickson GF, 1981. Progestins inhibit FSH-stimulated granulosa cell estrogen production at a postcyclic AMP site. Mol Cell Endocrinol 21:161-70

Stoklosowa S, Gregoraszcsuk E, Channing CP, 1982. Estrogen and progesterone secretion by isolated cultured porcine thecal and granulosa cells. Biol Reprod 26:943-52

Tonetta SA, De Vinna RS, diZerega GS, 1986. Modulation of porcine thecal cell aromatase activity by human chorionic gonadotropin, progesterone, estradiol-17β and dihydrotestosterone. Biol Reprod 35:785-91

Tonetta SA, diZerega GS, 1986. Paracrine regulation of follicular maturation in primates. Clinics in Endocrinology and Metabolism 15:135-36

Tsang BK, Ainsworth L, Downey BR, Marcus GJ, 1985. Differential production of steroids by dispersed granulosa and theca interna cells from developing preovulatory follicles of pigs. J Reprod Fertil 74:459-71

Tsang BK, Moon YS, Simpson CW, Armstrong DT, 1979. Androgen biosynthesis in human ovarian follicles: cellular source, gonadotropic control, and adenosine 3',5'-monophosphate mediation. J Clin Endocrinol Metab 48:153-58

Tseng L, 1984. Effect of estradiol and progesterone on human endometrial aromatase activity in primary cell culture. Endocrinology 115:833-35

Vernon MW, Dierschke DJ, Sholl SA, Wolf RC, 1983. Ovarian aromatase activity in granulosa and theca cells of rhesus monkeys. Biol Reprod 28:342-49

STIMULATION OF INTRACELLULAR FREE Ca^{2+} BY LUTEINIZING

HORMONE IN ISOLATED BOVINE LUTEAL CELLS

John S. Davis

James A. Haley Veterans Hospital and
Departments of Internal Medicine and
Pharmacology and Theraptutics
University of South Florida College of Medicine
Tampa, FL 33612

INTRODUCTION

The mechanisms by which gonadotropins regulate the development and function of the corpus luteum are not completely understoon. Cyclic AMP is recognized as the primary 'second messenger' of luteinizing hormone (LH) action in the corpus luteum. We have recently reported that gonadotropins also stimulate the accumulation of 'second messengers' derived from phospholipase C (PLC) catalysed hydrolysis of phosphatidylinositol 4,5-bisphosphate (Davis et al., 1986a,c, 1987). Hydrolysis of this phospholipid liberates diacylglycerol and inositol 1,4,5-trisphosphate (IP_3) which serve to activate protein kinase C (Nishizuka, 1986) and mobilize intracellular Ca^{2+} (Berridge, 1984), respectively. This newly described 'second messenger' system may be responsible for regulating cellular growth, metabolism and secretion in hormone, growth factor and neurotransmitter responsive tissues. The present studies were performed to examine the effect of LH on intracellular Ca^{2+} concentrations in isolated bovine luteal cells. LH-induced increases in the levels of IP_3 and cAMP were correlated with increases in intracellular Ca^{2+}.

METHODS

Corpora lutea were obtained from cows during early pregnancy, sliced, and dispersed with collagenase as previously described (Davis et al., 1981). Intracellular free Ca^{2+} concentrations ($[Ca^{2+}]i$) were measured by the method of Tsien et al. (1982) with minor modifications (Davis et al., 1986b,c). After preincubation for 15 min at $37\,^{\circ}C$, Quin 2/AM was added to luteal cell suspensions (10^7 cells/ml) from a 20 mM stock solution in DMSO. The final concentrations of Quin 2/AM and DMSO were 0.1 mM and 0.5%, respectively. After 15 min, 3 volumes of media were added and the incubation was continued for another 30 min. Control cells were treated in parallel with similar DMSO concentrations. The unhydrolysed ester was removed by two washes followed by suspending the cells in media without BSA. An aliquot of the Quin 2-loaded cells was suspended in 1.5 ml of medium ($3-5 \times 10^6$ cells.ml) in a quartz cuvette (1 x 1 cm). Cellular fluorescence signals derived from the Quin $2-Ca^{2+}$ complex and reduced pyridine nucleotides were measured at an excitation wave length of 339 nm (slit 5 nm) and emission wave lengths of 492 nm and 440 nm, respectively (slit 10 nm). All measurements were made using a Perkin-Elmer MPF-44A spectrofluorometer equipped with a magnetic stirrer and a thermostated cell holder. Maximum (Fmax) and minimum (Fmin) fluorescence were measured by rapidly saturating intracellular Quin 2 with Ca^{2+} (Fmax) by permeabilizing the cells with 0.2 % Triton X-100, and by removing the Ca^{2+} with 5 mM EGTA in Tris buffer, pH 8.5, to determine the basal fluorescence (Fmin) where virtually no Ca^{2+} was bound to Quin 2.

Levels of cAMP and IP_3 were also measured. Luteal cell preparations $(1-1.5 \times 10^7$ cells/ml) were preincubated for 3 hr in medium containing myo[$2-^3H$]inositol (50 μCi/ml) under an atmosphere of 95 % O_2, 5 % CO_2 at 37 C. After this prelabeling period, the cells were washed and incubations were performed in triplicate with 5×10^5 cells/0.5 ml. After a 15 min preincubation with Ca^{2+}-free medium containing 1 mM $CaCl_2$ or 1 mM EGTA, control medium or LH was added and the incubations were continued for up to 5 min. Incubations were terminated by the addition of 10 % TCA. The acid-soluble fraction was extracted with ether to remove the TCA and aliquots were taken for determinations of cAMP and IP_3. cAMP was measured by radioimmunoassay after acetylation and [^3H]IP_3 was purified on columns of BioRad AG 1-X8 ion exchange resin (Davis et al., 1986a,b).

RESULTS AND DISCUSSION

Quin 2/AM was readily taken into bovine luteal cells and de-esterified trapping the Ca^{2+}-sensitive dye, Quin 2, in the soluble portion of the cell (Fig. 1). LH increased fluorescence in Quin 2-loaded cells but had no effect in DMSO-treated cells. Emission spectra (400-600 nm) in Quin 2-loaded luteal cells revealed a single peak of fluorescence at 492 nm in both untreated and LH-treated cells, however, the intensity of the fluorescence was greater in the presence of LH. Changes in the fluorescence of reduced pyridine nucleotides (at 440 nm) were very small (< 8 %) and much slower than the Quin 2-Ca^{2+} changes (at 492 nm). No corrections were required for the intrinsic fluorescence of the cells, media, hormones or detergents. The concentration of intracellular free calcium, [Ca^{2+}]i associated with Quin 2 fluorescence in bovine luteal cells is shown in Fig 2. Treatment with LH approximately doubled [Ca^{2+}]i. The response to LH was rapid with maximal increases occurring within the first 15-30 sec of treatment and was sustained for at least 10 min. The response to LH was concentration dependent; 1 μg/ml of LH provoked maximal increases and 10 ng/ml of LH stimulated small but significant increases in [Ca^{2+}]i (Davis et al., 1987).

To determine whether the response to LH was due to an increase in Ca^{2+} influx or an increase in intracellular Ca^{2+} mobilization, experiments were also performed in Ca^{2+}-free medium (∿ μM free Ca^{2+}). LH stimulated similar increases in [Ca^{2+}]i under these conditions (Fig. 3) indicating that extracellular Ca^{2+} is not the primary source of the Quin 2-Ca^{2+} signal. Luteal cells were also incubated with Ca^{2+}-free medium containing 2 mM EGTA (∿ nM free Ca^{2+}). These conditions were severe enough to begin depleting intracellular Ca^{2+} ; note the reductions in basal [Ca^{2+}]i (Fig. 3). LH also increased [Ca^{2+}]i under these conditions of modest Ca^{2+} depletion although to a lesser extent and for a shorter duration. The data again demonstrate that LH increases intracellular Ca^{2+} mobilization and but also suggest that the sustained increases in [Ca^{2+}]i observed in response to LH involve the stimulation of a Ca^{2+}-dependent process and/or an increase in Ca^{2+} influx.

Other stimuli increased [Ca^{2+}]i in Quin 2-loaded bovine luteal cells. Fig. 4 compares the effect of the Ca^{2+} ionophore, ionomycin, to the effect of LH. High concentrations of ionomycin (1-10 μM) provoked near maximal increases in cellular fluorescence whereas lower concentrations (10 nM) provoked increases similar to those produced by LH. 8-Bromo-cAMP (5 mM) and forskolin (5 μM) also increased [Ca^{2+}]i, but the effects were small

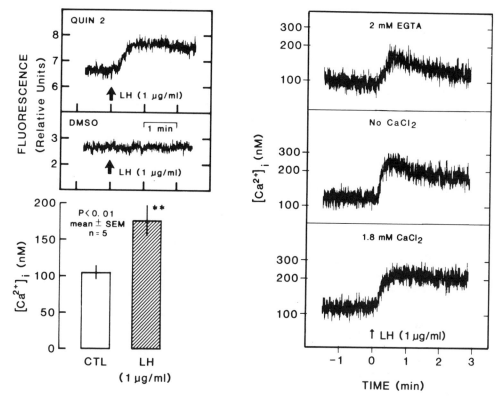

Fig. 1. Fluorescence changes in Quin 2- and DMSO-treated bovine luteal cells (upper left). Luteal cells were pretreated with Quin 2 or DMSO as described in METHODS prior to the addition of LH (arrow).

Fig. 2. Effect of LH on intracellular Ca^{2+} concentrations in bovine luteal cells (lower left). Luteal cells were loaded with Quin 2 and fluorescence signals were quantified before (CTL) and after addition of LH (1 μg/ml). Results are expressed as means $\pm$ SEM from five separate cell preparations.

Fig. 3. Effect of extracellular Ca^{2+} on increases in intracellular Ca^{2+} concentrations in LH-treated bovine luteal cells (right). Luteal cells were loaded with Quin 2 and were resuspended in Ca^{2+}-free media. Fluorescence signals were monitored after equilibrating luteal cell suspensions for 5 min with 1.8 mM $CaCl_2$, no additional $CaCl_2$, or 2 mM EGTA. LH (1 μg/ml) was added at time zero (arrow).

(5-20 nM increases in $[Ca^{2+}]i$) with maximal increases occurring after 1-2 min of treatment. Other experiments were performed to determine whether the Quin 2-Ca^{2+} responses observed in ionomycin- and cAMP-treated luteal cells were associated with increases in IP_3. In 5 min incubations no increases in IP_3 were observed in response to 8-Bromo-cAMP (0.1-5 mM), forskolin (0.1-10 μM), or ionomycin (1 nM - 10 μM). Ionomycin (10 μM), however, provoked small increases in IP_3 during 30 min incubations. The data suggest that the LH-induced increases in $[Ca^{2+}]i$ may be partially attributed to inceases in cAMP. The results of these studies also indicate that increases in IP_3 accumulation are not the result of increased levels of cAMP and $[Ca^{2+}]i$.

Since both IP_3 and cAMP potentially increase $[Ca^{2+}]i$, experiments were performed to determine whether the LH-induced increases in these 'second messengers' were correlated with the changes observed in $[Ca^{2+}]i$.

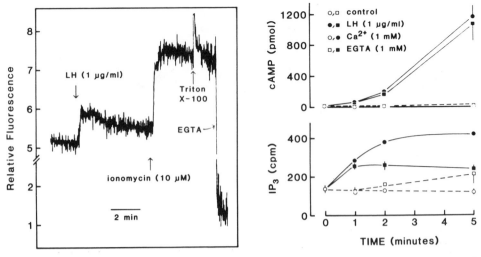

Fig. 4. Effects of LH and ionomycin on cellular fluorescence in Quin 2 loaded bovine luteal cells (left). Luteal cells were sequentially treated with LH (1 $\mu g/ml$) and the Ca^{2+} ionophore, ionomycin (10 μM). Maximum and minimum fluorescence signals were determined by the addition of Triton X-100 and EGTA, respectively.

Fig. 5. Effects of extracellular Ca^{2+} on LH-stimulated accumulation of cAMP and IP_3 in bovine luteal cells. Luteal cells were suspended in Ca^{2+}-free media and were preincubated for 15 min in the presence of either 1 mM $CaCl_2$ (o,●) or 1 mM EGTA (□,■) prior to incubation with (●,■) or without (o,□) LH (1 $\mu g/ml$). Results are means $\pm$ SEM of triplicate incubations in a representative experiment.

When luteal cells were incubated in Ca^{2+}-free medium containing 1 mM $CaCl_2$ or 1 mM EGTA the effect of LH on cAMP accumulation was not altered (Fig. 5). The initial (up to 1 min) increase in IP_3 in response to LH was also similar in either medium. Thereafter, the effect of LH on IP_3 accumulation was reduced in the presence of EGTA, indicating that continued IP_3 formation is dependent on a source of extracellular Ca^{2+}. These results taken together with data demonstrating a requirement for extracellular Ca^{2+} for sustained elevations of $[Ca^{2+}]i$ in LH-treated luteal cells (Fig. 3) suggest that the effect of LH on Ca^{2+} mobilization is largely independent of cAMP.

These studies further implicate Ca^{2+} in the action of LH and support the hypothesis that gonadotropins provoke their cellular responses by stimulating at least two 'second messenger' systems; the adenylate cyclase-cAMP system and the phospholipase C-IP_3 and diacylglycerol system. The similar Ca^{2+}-dependencies observed in these experiments and the similar dose-response and temporal relationships (Davis et al., 1987) between LH-induced increases in IP_3 accumulation and increases in $[Ca^{2+}]i$ suggest that IP_3 is a 'second messenger' for Ca^{2+} mobilization in bovine luteal cells. Although the exact role of increased $[Ca^{2+}]i$ remains to be determined, Ca^{2+} participates in ovarian steroid synthesis (Hansel and Dowd, 1986; Higuchi et al., 1976; Sawyer et al., 1979; Shemesh et al., 1984; Veldhuis et al., 1984) and secretion (Sawyer et al., 1979), ornithine decarboxylase induction (Veldhuis and Hammond, 1981), proteoglycan production (Lenz et al., 1982), and Ca^{2+}-dependent protein phosphorylation (Clark et al., 1985; Davis and Clark, 1983). Additionally, increases in intracellular Ca^{2+} may also serve to modulate gonadotropin-stimulated cAMP accumulation (Bedwani and Wong, 1975; Dorflinger et al., 1984).

Supported by the Veterans Administration and by NIH grant HD-22248.

REFERENCES

Bedwani JR, Wong PYD, 1975. Ionic dependence of luteinizing hormone-induced steroidogenesis in the rabbit ovary. J Physiol 250:669-679

Berridge MJ, 1984. Inositol trisphosphate and diacylglycerol as second messengers. Biochem J 220:345-360

Clark MR, Kawai Y, Davis JS, LeMaire WJ, 1985. Ovarian protein kinases. In: Toft DO, Ryan RJ (eds), Proceedings of the Fifth Ovarian Workshop. Illinois: Ovarian Workshops, pp. 383-401

Davis JS, Clark MR, 1983. Activation of protein kinase in the bovine corpus luteum by phospholipid and Ca^{2+}. Biochem J 214:569-574

Davis JS, Farese RV, Marsh JM, 1981. Luteinizing hormone stimulates phospholipid labeling and progesterone synthesis in isolated bovine luteal cells. Endocrinology 109:469-475

Davis JS, Weakland LL, Farese RV, West LA, 1987. Luteinizing hormone increases inositol trisphosphate and cytosolic free Ca^{2+} in isolated bovine luteal cells. J Biol Chem, in press

Davis JS, Weakland LL, West LA, Farese RV, 1986a. Luteinizing hormone stimulates the formation of inositol trisphosphate and cyclic AMP in rat granulosa cells: evidence for phospholipase C generated second messengers in the action of luteinizing hormone. Biochem J 238:597-604

Davis JS, West LA, Farese RV, 1986b. Gonadotropin-releasing hormone (GnRH) rapidly stimulates the formation of inositol phosphates and diacylglycerol in rat granulosa cells: further evidence for the involvement of Ca^{2+} and protein kinase C in the action of GnRH. Endocrinology 118:2561-1571

Davis JS, West LA, Weakland LL, Farese RV, 1986c. Human chorionic gonadotropin activates the inositol 1,4,5-trisphosphate-Ca^{2+} intracellular signalling system in bovine luteal cells. FEBS Lett 208:287-291

Dorflinger LJ, Albert PJ, Williams AT, Behrman HR, 1984. Calcium is an inhibitor of luteinizing hormone-sensitive adenylate cyclase in the luteal cell. Endocrinology 114:1208-1215

Hansel W, Dowd JP, 1986. New concepts of the control of corpus luteum function. J Reprod Fert 78:755-768

Higuchi T, Kaneko A, Abel JH Jr, Niswender GD, 1976. Relationship between membrane potential and progesterone release in ovine corpora lutea. Endocrinology 99:1023-1032

Lenz RW, Ax RL, First NL, 1982. Proteoglycan production by bovine granulosa cells in vitro is regulated by calmodulin and calcium. Endocrinology 110:1052-1054

Nishizuka Y, 1986. Studies and perspectives of protein kinase C. Science 233:305-312

Sawyer HR, Abel JH Jr, McMlellan MC, Schmitz M, Niswender GD, 1979. Secretory granules and progesterone secretion by ovine corpora lutea in vitro. Endocrinology 104:476-486

Shemesh M, Hansel W, Strauss JF, III, 1984. Calcium-dependent, cyclic nucleotide-independent steroidogenesis in the bovine placenta. Proc Natl Acad Sci USA 81:6403-6407

Tsien RY, Pozzan T, Rink TJ, 1982. Calcium homeostasis in intact lymphocytes: cytoplasmic free calcium monitored with a new intracellularly trapped fluorescent indicator. J Cell Biol 94:325-344

Veldhuis JD, Hammond JM, 1981. Role of calcium in the modulation of ornithine decarboxylase activity in isolated pig granulosa cells in vitro. Biochem J 196:795-801

Veldhuis JD, Klase PA, Demers LM, Chafouleas JG, 1984. Mechanisms subserving calcium's modulation of luteinizing hormone action in isolated swine granulosa cells. Endocrinology 114:441-449

THE EFFECT OF PROSTAGLANDIN $F_{2\alpha}$ ON MITOCHONDRIAL ELECTION DENSE

INCLUSIONS AND SECRETORY GRANULES OF THE BOVINE LARGE LUTEAL

CELL DURING LATE PREGNANCY

M.J. Fields, W. Dubois, B.A. Ball, M. Drost, and P.A. Fields*

Animal Science Department
University of Florida
College of Medicine
Gainesville, FL 32610

*Department of Anatomy
University of South Alabama
Mobile, AL 36688

INTRODUCTION

The bovine corpus luteum has been shown to contain two populations of electron dense granules which are morphologically distinct (Fields et al., 1985). One population consists of small (100-300 nm diameter) membrane bounded granules that are readily observed to undergo exocytosis (Singh, 1975; Heath et al., 1983; Fields et al., 1985). The product of these small luteal granules has not been identified in the pregnant cow. However, in the nonpregnant cow luteal secretory granules have been shown to contain oxytocin associated neurophysin (Fields and Fields, 1986).

The second population of granules is a larger (500-1800 nm diameter) membrane-bounded granule that forms in the mitochondria (Sorenson and Singh, 1973; Singh, 1975; Fields et al., 1985; Weber et al., 1987). These granules are not observed to undergo exocytosis. The product of these mitochondrial granules has not been identified in either the pregnant or nonpregnant cow.

This study was designed to determine whether the luteal secretagogue $PGF_{2\alpha}$, which has been shown to induce exocytosis of small luteal granules in the nonpregnant cow (Heath et al., 1983) would increase the number or induce the exocytosis of these large granules.

MATERIALS AND METHODS

Four Angus cows were randomly assigned to receive intravenously either 5 ml saline (n = 2 control) or 25 mg $PGF_{2\alpha}$ (n = 2 treated) within 1-2 weeks of parturition. The $PGF_{2\alpha}$ (Lutalyse) was obtained from Upjohn Co., Kalamazoo, MI. The ovary bearing the corpus luteum was removed through a flank incision 60 min post-injection and processed for electron microscopy.

677

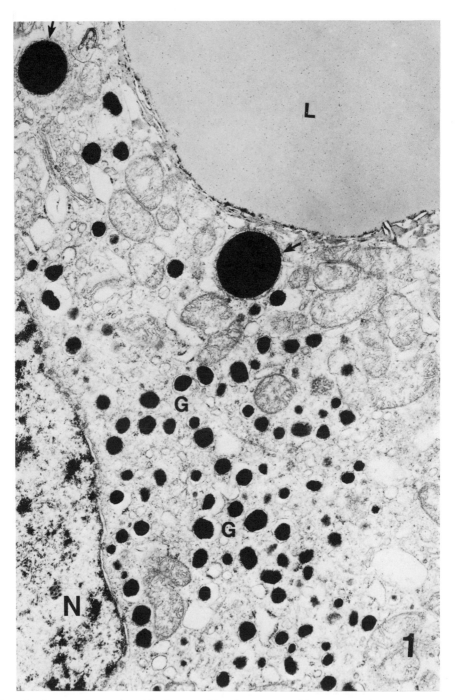

Fig. 1. Large luteal cell from a 60 min post-saline treated late pregnant cow. Secretory granules (G) are clustered in a region near the Nucleus (N) of the cell. The large granules (arrows) and lipid droplets (L) are also characteristic of large luteal cells from late pregnant cows. X 27,300.

RESULTS AND DISCUSSION

Two populations of electron dense granules were observed in large luteal cells of the late pregnant cow. In saline treated cows small secretory granules, 100–300 nm in diameter in any one large luteal cell, were observed in a single cluster. These clusters, although distributed throughout the cytoplasm from cell to cell, more often were seen in a paranuclear organization (Fig. 1). In cows treated with $PGF_{2\alpha}$ in which the corpus luteum was removed 60 min post-treatment, the luteal cell was primarily found to lack secretory granules (Fig. 2). However, when granules were observed they were found in the region of the cell's plasma membrane. Exocytosis of these luteal granules were readily observed. These observations are in agreement with Heath et al. (1983) that in the nonpregnant cow, $PGF_{2\alpha}$ depleted the luteal cell of secretory granules. The hormonal contents of secretory granules in the pregnant cow has not yet been determined. Although luteal relaxin has been characterized in the cow (Fields et al., 1980), efforts to localize relaxin in these granules has not been successful.

Electron dense inclusions in the mitochondria were similar in large luteal cells of both the saline (Fig. 1) and $PGF_{2\alpha}$ treated cows (Fig. 2 and 3). These large membrane bound granules, 500–1800 nm in diameter, are similar to those previously reported in luteal cells of pregnant cows (Sorensen and Singh, 1973; Singh, 1975; Fields et al., 1985; Weber et al., 1987). Although morphometric analysis was not attempted, there was no apparent change in response to PGF in the number of mitochondrial

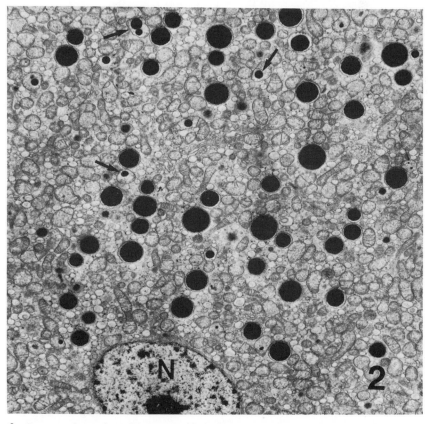

Fig. 2. Large luteal cell from a 60 min post-$PGF_{2\alpha}$ treated late pregnant cow in which the small secretory granules have been depleted. The only electron dense bodies remaining in the cell appear to be the early developing large mitochondrial granules (arrows). X 7,500.

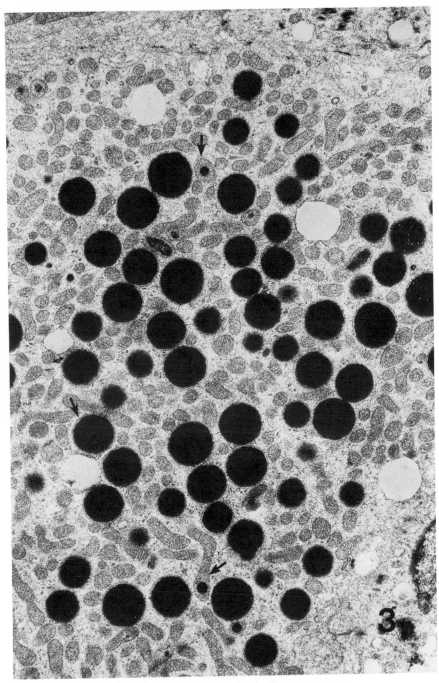

Fig. 3. Large luteal cell from a 60 min post—PGF$_{2\alpha}$ treated late pregnant
cow. Large granules which apparently originate in the mitochondria expand in size to several fold larger than the
mitochondria which contain cristae. X 13,500.

granules or in their localization throughout the cell. No exocytosis of
mitochondrial granules was observed. The electron density of these
granules indicates the material is of a protein nature that when fully

expanded obliterates the mitochondrial cristae. These inclusions in the mitochondria have not been observed in corpora lutea from early to mid-pregnant cows (Fields et al., 1985), nor have they been observed in small luteal cells. Although the role of these inclusions are unknown they appear to be associated with a protracted involutive state of the large luteal cell in the later third of pregnancy.

ACKNOWLEDGEMENTS

This study was supported by grants from the National Science Foundation (PCM 8110062 and 8409304) and the National Institutes of Health (RO1 HD 15773 and 18075). A special thanks to Jan C. Lauer for help in preparation of this manuscript. Florida Agriculture Experiment Station Journal Series No. 7885.

REFERENCES

Fields MJ, Fields PA, 1986. Luteal neurophysin in the nonpregnant cow and ewe: immunocytochemical localization in membrane-bounded secretory granules of the large luteal cell. Endocrinology 118:1723-1725

Fields MJ, Dubois W, Fields PA, 1985. Dynamic features of luteal secretory granules: ultrastructural changes during the course of pregnancy in the cow. Endocrinology 117:1675-1682

Fields MJ, Fields PA, Castro-Hernandez A, Larkin LH, 1980. Evidence for relaxin in corpora lutea of late pregnant cows. Endocrinology 107:869-876

Heath E, Weinstein P, Merritt B, Shanks R, Hixon J, 1983. Effects of prostaglandins on the bovine corpus luteum: granules, lipid inclusions and progesterone secretion. Biol Reprod 29:977-985

Singh UB, 1975. Structural changes in the granulosa lutein cells of pregnant cows between 60 and 245 days. Acta Anat 93:447-457

Sorenson VW, Singh UB, 1973. On mitochondrial inclusions in granulosa lutein cells of pregnant cows. Experientia 29:592-593

Weber DM, Fields PA, Romrell LJ, Tumwasorn S, Ball BA, Drost M, Fields MJ, 1987. Functional differences between small and large luteal cells of the late-pregnant vs nonpregnant cow. Biol Reprod (In Press)

IN VITRO PRODUCTION OF ANGIOTROPIC FACTOR BY BOVINE CORPUS LUTEUM:

PARTIAL CHARACTERIZATION OF ACTIVITIES THAT ARE CHEMOTACTIC

AND MITOGENIC FOR ENDOMETRIAL CELLS

D.A. Redmer, J.D. Kirsch, and A.T. Grazul*

Department of Animal and Range Sciences
North Dakota State University
Fargo, ND 58105
*Animal Physiology Department
University of Agriculture and Technology
Olsztyn, Poland

INTRODUCTION

In conjunction with the tremendous physiological changes that occur in the corpus luteum (CL) during the estrous cycle are equally dynamic intrinsic vascular changes (Bassett, 1943). Since the CL receives one of the greatest rates of blood flow of any organ (Bruce and Moore, 1976; Ford et al., 1982), factors that regulate luteal angiogenic processes may be important determinants of subsequent luteal function.

Proliferation of capillary vessels consists of at least 3 processes: fragmentation of existing capillary basement membrane, migration of endothelial cells from existing vessels, and proliferation of endothelial cells (Shepro and D'Amore, 1984). Several reports have shown that luteal tissue contains an endothelial cell mitogen (Jakob et al., 1977; Gospodarowicz and Thakral, 1978; Heder et al., 1979). In addition, Redmer et al. (1985a,b) showed that cultured primate CL and follicles secrete a factor that stimulates endothelial cell migration in vitro. Our recent findings (unpublished observations) have shown that bovine luteal tissues secrete factor(s) that stimulate endothelial cell mitogenesis and migration. We report herein further investigation of this factor(s) secreted in vitro by bovine luteal tissues.

MATERIALS AND METHODS

Minced samples of bovine CL (n = 10) were incubated separately for 6 h with or without luteinizing hormone (LH; 1 µg/ml USDA LH b5) in serum-free media (200 mg tissue/3 ml media per tube). For control, tubes were incubated without tissue but with serum-free media (+ LH). Control media will be referred to as unconditioned media (UCM). Conditioned media from luteal tissues incubated with and without LH were pooled separately. Reported herein are preliminary studies on dose response and partial characterization of endothelial cell mitogenic and migration-stimulating activities from these two pools of luteal conditioned media (LCM). A report on angiotropic activities of individual LCM has been submitted for publication and a preliminary communication presented (Grazul et al., 1986).

To evaluate mitogenic activity of LCM, endothelial cells were obtained from aortas of mature cows and identified by using procedures described by Redmer et al. (1985a). Confluent monolayer cultures of endothelial cells were trypsinized and aliquoted to 24-well plates (20 x 10^3 cells/ml per well). After a 24 h preincubation, media were changed to include 20% LCM or UCM. Samples were assayed in triplicate wells and number of cells/well was determined after 72 h incubation by using a Coulter counter. To test the ability of LCM to stimulate migration of endothelial cells, chemotaxis assays were conducted as described by Redmer et al. (1985a). Bottoms of wells of 48-well microchemotaxis chambers were filled with UCM or UCM containing 30% LCM. Endothelial cells were suspended in UCM and placed into tops of wells (70 x 10^3 cells/well). Number of cells migrating across a polycarbonate membrane (8 µm pores) was determined after 5 h incubation. To determine if activity stimulated chemotaxis or chemokinesis, a 2 x 2 factorial design was used as previously described (Redmer et al., 1985a). Each sample was assayed in triplicate wells.

Aliquots (2 ml) of each pool of LCM were fractionated by using an ultrafiltration cell (Amicon) to obtain filtrates with mol wt of <10,000, <100,000, and <300,000. Retentates were resuspended in 2 ml UCM to obtain fractions >100,000 and >300,000 mol wt. Fractionated media then were utilized in endothelial cell migration and mitogenesis bioassays as described above. In addition, mitogenic and migration-stimulating activities were tested for heat stability by incubating LCM or UCM at 80°C for 30 min.

RESULTS

With increasing concentrations of LCM (up to 40%), a dose-dependent increase in endothelial cell mitogenesis was observed (Fig. 1). Likewise, increasing concentrations of LCM (up to 10%) resulted in increased numbers of endothelial cells migrating (Fig. 2).

As depicted in Fig. 3, a chemokinetic effect which increased endothelial cell migration from 2.1 x 10^3 cells/well (U/U) to 4.9 x 10^3 cells/well (C/C), was observed in the absence of a concentration difference

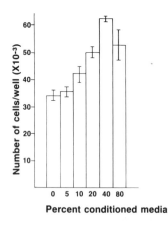

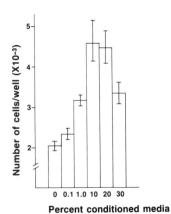

Fig. 1. Endothelial cell mitogenesis in response to increasing doses of a pool of luteal conditioned media (+LH).

Fig. 2. Endothelial cell migration in response to increasing doses of a pool of luteal conditioned media (+LH).

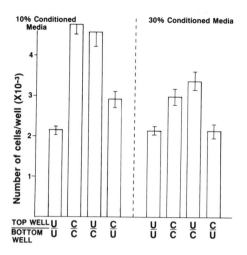

Fig. 3. Demonstration of chemotactic and chemokinetic effects of luteal
conditioned (C) vs unconditioned (U) media on endothelial cell
migration. Effects were determined with either 10% or 30%
conditioned media (+ LH) in tops and/or bottoms of wells.

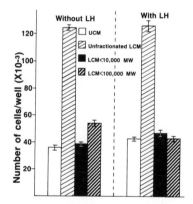

Fig. 4. Effect of pooled luteal conditioned media (LCM; + LH) fractionated
by ultrafiltration through 10,000 or 100,000 mol wt cutoff
membranes on endothelial cell mitogenesis. Unconditioned media
(UCM) used as control.

across the membrane when 10% LCM was used. Endothelial cell motility was
increased to a comparable degree when LCM were present only in tops of
wells (U/U vs C/U). Similar chemokinetic and chemotactic responses were
observed with 30% LCM.

The pools of LCM (+ LH) exhibited mitogenic activity and stimulated a
200% increase in numbers of endothelial cells at 72 h compared with UCM
(Fig. 4). When pools of LCM were fractionated, activity was lost in

fractions <10,000 and <100,000 mol wt. When one of the pools of LCM (–LH) was fractionated at 100,000 and 300,000 mol wt, the mitogenic activity remained in the retentates (Fig. 5). Pools of LCM (+ LH) caused a 75% increase in numbers of endothelial cells migrating compared with UCM (Fig. 6). When subjected to ultrafiltration, activity was lost in fractions <10,000 and <100,000 mol wt in the LCM pool without LH, but was present in both filtrates in the pool containing LH (Fig. 6). Heat-treatment of unfractionated LCM pools (+ LH) inactivated mitogenic and migration-stimulating factor(s) (Table 1).

DISCUSSION

Data presented herein demonstrate *in vitro* secretion of angiotropic activity from bovine luteal tissues that is stimulatory to both endothelial cell migration and mitogenesis. The observation that bovine

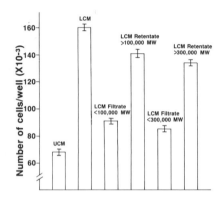

Fig. 5. Effect of a pool of luteal conditioned media (LCM; –LH) fractionated by ultrafiltration through 100,000 or 300,000 mol wt cutoff
membranes on endothelial cell mitogenesis. Unconditioned media
(UCM) used as control.

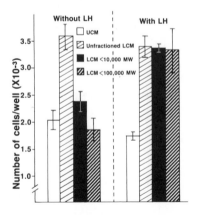

Fig. 6. Effect of pooled luteal conditioned media (LCM; + LH) fractionated
by ultrafiltration through 10,000 or 100,000 mol wt cutoff
membranes on endothelial cell migration. Unconditioned media
(UCM) used as control.

Table 1. Effect of Heat Treatment on Mitogenic and Migration–Stimulating
Activities of Unconditioned and Luteal Conditioned Media

Treatment[a]	Migration[b]	Mitogenesis[b]
UCM	2.0 ± 0.2	36.6 ± 0.8
UCM heat treated	1.8 ± 0.1	39.2 ± 2.7
LCM	3.5 ± 0.2	124.3 ± 2.3
LCM heat treated	1.6 ± 0.1	44.1 ± 0.2
LCM + LH	3.1 ± 0.2	125.7 ± 4.3
LCM + LH heat treated	1.6 ± 0.1	43.6 ± 2.8

[a]UCM = unconditioned media, LCM = luteal conditioned media.
[b]Mean number of cells/well ($x\ 10^{-3}$) $\pm$ SE.

luteal tissues are able to induce a complete neovascularization response
on the chick chorioallantoic membrane (Koos and LeMaire, 1983; Grazul et
al., 1986) suggests that mitogenic and migration–stimulating activities
observed in the present study represent true angiogenic factor(s).

Angiotropic factor from bovine CL appears to be of large molecular
weight (greater than 300,000) and heat labile. Angiotropic factors similar
in molecular weight and heat liable have been found in other reproductive
tissues, including ovarian follicles (Koos, 1986), and bovine (Redmer and
Reynolds, unpublished) and human (Burgos, 1983) placentas. Data on
molecular weight should be interpreted with caution, however, since angiotropic activities from other tissues may be molecules of 200–800 mol wt
normally bound to a carrier of 100,000 mol wt (Weiss et al., 1979; Burgos,
1983). In the present study, addition of LH to luteal tissue cultures
resulted in secretion of a factor <10,000 mol wt which stimulated endothelial cell migration but not endothelial cell mitogenesis. It is
unlikely that this factor is steroidal since preliminary investigations in
our laboratory and previous work by Redmer et al. (1985a) have shown that
the non–steroidal fraction of LCM contains the majority of the angiotropic
activity. Likewise, data from the present study shows that the small
molecular weight fraction is heat labile, and therefore unlikely to be a
steroid.

Our findings also show that endothelial cell migration–stimulating
activity is chemotactic and not just chemokinetic. Although endothelial
cell motility increased in the presence of angiotropic factor(s), net cell
movement occurred in response to a concentration gradient. These
observations agree with those of Redmer et al. (1985a) in which similar
responses were observed from primate LCM.

In summary, the results from these studies provide additional
evidence that secretion of luteoangiotropic factor(s) may be important in
luteal vascular development and subsequent luteal functions and that
several factors may be involved in regulating the various components of
luteal angiogenesis. Formation and secretion of these factors, in turn,
can be modulated by extraovarian hormones.

ACKNOWLEDGEMENTS

The author wishes to thank Dr. LP Reynolds for pertinent suggestions
and comments and J Berg for the preparation of this manuscript.

Published with approval of the Director of the North Dakota Agricultural Experiment Station as Article No. 1517. Supported by Hatch Project ND01780.

REFERENCES

Bassett DL, 1943. The changes in the vascular pattern of the ovary of the albino rat during the estrous cycle. Am J Anat 73:251-291

Bruce NW, Moor RM, 1976. Capillary blood flow to ovarian follicles, stroma and corpora lutea of anaesthetized sheep. J Reprod Fertil 46(2):299-304

Burgos H, 1983. Angiogenic and growth factors in human amino-chorion and placenta. Eur J Clin Invest 13(4):289-296

Ford SP, Reynolds LP, Magness RR, 1982. Blood flow to the uterine and ovarian vascular beds of gilts during the estrous cycle or early pregnancy. Biol Reprod 27:878-885

Gospodarowicz D, Thakral KK, 1978. Production of a corpus luteum angiogenic factor responsible for proliferation of capillaries and neovascularization of the corpus luteum. Proc Natl Acad Sci 75(2):847-851

Grazul A, Kirsch JD, Redmer DA, 1986. Angiogenic activity at several stages of luteal development in the cow. J Anim Sci 63(Suppl 1):355 (Abstract)

Heder G, Jakob W, Halle W, Mauersberger B, Kambach G, Jentzsch DK, Oehme P, 1979. Influence of porcine corpus luteum extract on DNA synthesis and proliferation of cultivated fibroblasts and endothelial cells. Exp Path 17:493-497

Jakob W, Jentzsch KD, Mauersberger B, Oehme P, 1977. Demonstration of angiogenesis-activity in the corpus luteum of cattle. Exp Pathol 13:231-236

Koos RD, LeMaire WJ, 1983. Factors that may regulate the growth and regression of blood vessels in the ovary. Semin Reprod Endocrinol 1:295

Koos RD, 1986. Stimulation of endothelial cell proliferation by rat granulosa cell-conditioned medium. Endocrinology 119:481-489

Redmer DA, Rone JD, Goodman AL, 1985a. Evidence for a non-steroidal angiotropic factor from the primate corpus luteum: Stimulation of endothelial cell migration in vitro. Proc Soc Exp Biol Med 179:136-140

Redmer DA, Rone JD, Goodman AL, 1985b. Detection of angiotropic activity from primate dominant follicles. 67th Annu Mtg Endocr Soc, Baltimore

Shepro D, D'Amore PA, 1984. Physiology and biochemistry of the vascular wall endothelium. In: Renkin EM, Michel CC (eds) Handbook of Physiology, Section 2, Vol IV, Part 1. Waverly Press, Baltimore, p 103

Weiss JB, Brown RA, Kumar S, Phillips P, 1979. An angiogenic factor isolated from tumours: A potent low-molecular-weight compound. Brit J Cancer 40:493-496

ACTION OF PROGESTERONE AND RU38486 ON PROTEIN

SYNTHESIS IN RAT PLACENTA

T.F. Ogle, T.M. Nosek, and T.M. Mills

Department of Physiology and Endocrinology
Medical College of Georgia
Augusta, GA 30912-3395

INTRODUCTION

Progesterone action in maintenance of pregnancy includes inhibition
of myometrial contractility and local suppression of the maternal immune
system.   In addition to these actions, this study indicates that
progesterone (P) also directly maintains placental function.   We propose
that this action is mediated by the placental nuclear progesterone
receptor (Rpn) characterized previously (Ogle, 1986a,b).   This study
examines the effects of ovariectomy (OVX) and the antiprogestin, RU38486
(RU), on the Rpn and on the regulation of protein synthesis.   Capacity for
protein synthesis is used in this report as a physiological indicator of
placental function.

MATERIALS AND METHODS

Pregnant Long–Evans rats were used.   Placenta extirpation and
preparation have been described (Ogle, 1983).   Concentration of P was
determined in serum and placental tissues by RIA (Mills and Osteen, 1977).
Receptor sites were determined by saturation assay using six
concentrations of ^{3}H–P (Ogle, 1986a,b).   Protein synthesis was measured
in minced placental tissues by incubation in 2uCi ^{3}H–leucine and 1.75 ml
Eagle's basal medium with Earle's salts, L–glutamine (2 nM), leucine (108
mg/ml), and 25 mM HEPES buffer at 37°C for 1.5 h in 93% O_2 and 7% CO_2,
pH 7.2–7.4.   Protein was precipitated by homogenization of tissue in
perchloric acid at 0°C.   The pellet was washed,   solubilized, and then
emulsified in Scintiverse II.   The amount of ^{3}H–leucine incorporated was
normalized for protein content.   The procedure has been validated in this
laboratory according to the precautions noted by Rannels et al. (1982).

Uterine pressure was determined by inserting an inflatible bulb 2 cm
into the left horn of the uterus of rats anesthetized with chloral
hydrate.   The bulb was inflated to a volume of 50 ul with saline and
connected to a transducer which measured uterine contractions as pressure
changes in the bulb.   Animals were given 5 min to stabilize, after which
uterine pressure was recorded for 15 min.   At the end of this period
isoproterenol (40 ug, i.p.), a potent myometrial relaxant, was
administered to determine baseline values of uterine pressure.   Uterine
contractility was quantified by integrating the area under the pressure
curve of each record.   Pressure records were stored and analysed with a
Branden digital oscilloscope.

All values are reported as means ± S.E.M. Statistical analysis was by ANOVA and the Student-Newman-Keul's multirange test.

RESULTS AND DISCUSSION

Rats were OVX on day 11 of pregnancy to examine the role of the ovary in supporting placental function. Fig. 1 shows that the placental concentration of P fell 87% by 2 h post-OVX. Rpn declined 90% by 4 h of OVX. The ability of the placenta to synthesize new proteins exhibited a more gradual decline, falling 45% by 8 h and more than 90% by 16 h. These findings reinforce the notion that the placenta requires intense and continuous stimulation by ovarian steroids for normal function. The next experiment examined the action of P in preventing these effects. Fig. 2 shows that when serum P levels were maintained in OVX rats (OVX-P), Rpn levels were increased more than 5-fold (P < .05) and protein synthetic capacity was increased 10-fold (P < .05). Therefore, chronic treatment with P acts to maintain the Rpn in rat placenta and maintains capacity of this tissue for protein synthesis. We next examined the possibility that P action on placental function may have been mediated by changes in uterine contractility. Fig. 3 presents typical responses of the uterus to these treatments. The non-pregnant pattern of uterine contraction (Panel A), characterized by maintenance high tonic tension and interrupted by brief episodes of relaxation, were elevated over that of pregnant rats (Panel B). OVX did not alter the pattern (Panel C) from the pregnant control group or the OVX-P group (Panel D). The total pressure generated by the uterus (pressure-time integral) exhibited the same pattern: high tonic pressure in non-pregnant rats (5,152 cm H_2O-t) and relaxation in pregnant (2,544 cm H_2O-t) and OVX rats (OVX: 2,945 cm H_2O-t; OVX+P: 3,323 cm H_2O-t). Therefore, it is unlikely that changes in placental function after OVX and OVX-P are secondary to changes in uterine contractility. We propose, instead, that the Rpn system mediates these changes. To test this hypothesis more directly we administered RU to intact pregnant rats (Fig. 4). Since the antiprogestin competes with P for the receptor binding site (unpublished observations; Horwitz, 1985), we predicted that if changes in placental protein synthesis are mediated by the Rpn, RU would have the same effect as OVX by effectively withdrawing P from placental tissues. Fig. 4 shows that protein synthetic capacity was drastically curtailed by RU (P < .05). Rpn declined 47-95% within 6-24h of RU treatment. Thus, RU mimics the effect of OVX even though serum P was

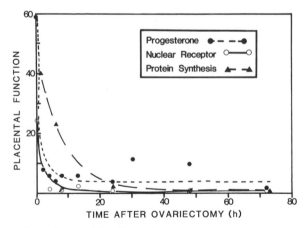

Fig. 1. Changes in placental function after OVX on D11 of pregnancy. Concentration of P in placenta (ng/placenta x 4). Concentration of Rpn_3 (pmol bound/placenta x 10^2). Rate of protein synthesis (nmol ^{3}H-leucine incorp./mg prot-h).

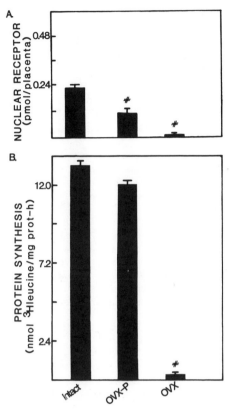

Fig. 2. The effects of P on the Rpn (A) and the capacity of the placenta
for protein synthesis (B). Rats were OVX on D11, the OVX-P group
received a 50 mg P pellet, i.p., immediately after surgery. Intact
pregnant rats served as controls. Animals were killed 96 h later.
n = 3 experiments. *Different from Intact and OVX, P < .05.

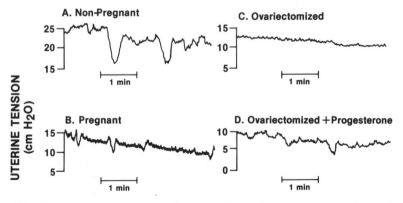

Fig. 3. Uterine pressure in non-pregnant and pregnant rats: A. Non-
pregnant; B. Pregnant, D15; C. OVX on D11, recorded on D15; D. OVX
implanted with 50 mg P pellet on D11, recorded on D15.

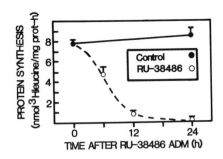

Fig. 4. Effects of RU 38486 on protein synthesis in rat placenta. Rats received a 20 mg RU pellet, i.p., on D11. Capacity for protein synthesis was decreased at all time periods after RU treatment, P < .05. n = 3 experiments.

unaffected. These findings show that P directly up-regulates Rpn and the capacity of the placenta for protein synthesis. Any manipulation that withdraws P from the placenta, whether by OVX or RU, produces the identical sequelae of events: loss of Rpn and capacity for synthesis of new proteins. Thus P appears to maintain pregnancy, at least in part, by supporting placental synthesis of new proteins, a function mediated by Rpn.

ACKNOWLEGEMENTS

This work was supported by a grant from the Medical College of Georgia Research Institute. The excellent technical assistance of Ms. Veronica V Woodward is gratefully acknowledged.

REFERENCES

Horwitz KB, 1985. The antiprogestin RU 38486: receptor mediated progestin versus antiprogestin actions screened in estrogen-insensitive T47D$_{CO}$ human breast cancer cells. Endocrinology 116:2236-2245

Mills TM, Osteen KJ, 1977. 17β-Estradiol receptor and progesterone and 20α-hydroxy-4-pregnene 3-one content of the developing corpus luteum of the rabbit. Endocrinology 101:1744-1750

Ogle, TF, 1983. Action of glycerol and sodium molybdate in stabilization of the progesterone receptor from rat trophoblast. J Biol Chem 258:4982-4988

Ogle, TF, 1986a. Characterization of progesterone binding to nuclear receptors in rat placenta. J Steroid Biochem 24:945-951

Ogle, TF, 1986b. Evidence for nuclear processing of progesterone receptors in rat placenta. J Steroid Biochem 25:183-190

Rannels DE, Wartell SA, Watkins CA, 1982. The measurement of protein synthesis in biological systems. Life Sciences 30:1679-1690

THE INTRAOVARIAN EFFECT OF PROGESTERONE ON FOLLICLE DEVELOPMENT

IN THE RABBIT OVARY

T. Mills and V. Stopper

Department of Physiology and Endocrinology
Medical College of Georgia
Augusta, GA 30912

INTRODUCTION

In most mammalian species examined, the growth of follicles seems to be blunted during periods of high circulating progesterone (pregnancy, luteal phase, pseudopregnancy). Reports of follicular profiles in rat (Greenwald, 1966; Bogovich et al. 1981), monkey (DiZerega and Hodgen, 1981) and human being (Baird et al. 1984) show that there is an inverse relationship between the luteal production of progesterone and the sizes of follicles present in the ovaries. In some species such as the rat, the inhibitory effect of the progestin may be exerted at the level of the pituitary-hypothalamus with a blockage of the secretion of FSH (Banks and Freeman, 1980; DePaolo and Barraclough, 1979). In other species such as the monkey and the human being, the effect of the progestin is likely exerted at the level of the ovary itself (DiZerega and Hodgen, 1981; Baird et al, 1984).

We have recently demonstrated that progesterone partially inhibits follicular growth in the rabbit ovary (Setty and Mills, 1987). When progesterone is elevated during pregnancy or after the implantation of pellets of pure progesterone, follicles of external diameter greater than 1.7 mm are completely absent from the ovary, although such follilces are common in nonpregnant and untreated rabbits. Other experiments showed that the progesterone effect was exerted directly on the ovary and was not indirect via the suppression of gonadotropin secretion; blood levels of neither LH nor FSH were affected by the high circulating levels of the progestin in the pregnant or pellet treated animals (Setty and Mills, 1987).

The present study was undertaken to examine further the action of progesterone on the growth and development of follicles in the ovary of the rabbit by exploring the following two questions: a) does progesterone act to inhibit follicular development by an intraovarian route? b) if the progesterone effect is exerted directly on the ovary, is it caused by simple diffusion from the corpus luteum or is the progesterone transported via blood, lymph, or some other system?

METHODS

Collection of Follicles

Each rabbit (New Zealand White) received a single subcutaneous injection of 50 IU PMSG 3 days before injection of 25 IU hCG to induce ovulation. Rabbits were killed by cervical dislocation or lethal injection, follicles of 1 mm or greater (external diameter) were collected from the ovaries by microdissection under 7X magnification and the external diameter measured with an ocular micrometer.

Measurement of Progesterone in Blood and Lymph

Ovarian vein blood and ovarian lymph were collected by direct cannulation of the vessels. The collection of lymph was facilitated by the infusion of 1% Evans Blue in saline directly into the ovary; this dye is taken up in the lymph, colors the external vessels blue and thereby makes the cannulation possible. Samples of peripheral blood, ovarian vein blood and ovarian lymph were quantified for content of progesterone and 20 dihydroprogesterone as previously described (Mills and Osteen, 1977).

Surgical Procedures

In these studies, cautery was used to destroy recently ovulated follicles and thereby prevent the development of corpora lutea; 24 h after injection of an ovulatory dose of hCG (12 h after ovulation), does were laporatomized and all ovulated follicles on one ovary were destroyed by cautery. In some studies, all ovulated follicles in the opposite ovary were allowed to remain and develop into CL; in other studies, only one ovulated follicle on the opposite ovary was allowed to remain and further develop. On day 10 - 12 after the surgery, ovaries were collected for the measurement of the sizes and numbers of follicles present in ovaries with no CL, a single CL or several CL.

RESULTS AND DISCUSSION

Follicle Development in the Presence and Absence of Corpora Lutea

In order to determine whether progesterone can affect the development of ovarian follicles via an intraovarian route, ovaries were created with a full complement of CL, with a single CL, or with no CL; the growth of follicles was then measured in the ipsilateral (with CL) and contralateral (without CL) ovaries. The results presented in Fig. 1 (upper panel) show that when there are no CL in the contralateral ovary but several CL in the ipsilateral ovary, follicular growth in both ovaries is halted before the follicles reach large size. The fact that blood levels of progesterone in these animals remains above 2-3 ng/ml suggests that several CL in one ovary can secrete sufficient progesterone to interrupt the follicular growth in both the ipsilateral and contralateral ovaries. In further investigations, ovaries were prepared in which the ipsilateral ovary had one CL while the contralateral ovary had no CL. In this case, the single CL clearly interrupted the follicular growth in the ipsilateral ovary but was without effect in the contralateral ovary (Fig. 1, lower panel). Thus, it would seem that luteal progesterone can exert a local, intraovarian suppressive action on follicular growth. However, under most circumstances in normal, pregnant animals, several CL are present and follicle growth in both ovaries is likely attenuated by both the high circulating levels of progesterone and the local, intraovarian action of the progestin.

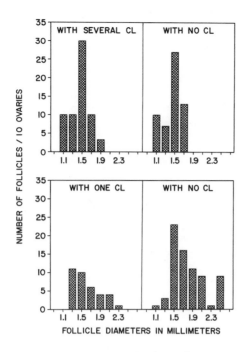

Fig. 1. The intraovarian effect of luteal progesterone on follicle growth in the rabbit ovary.

Progesterone in Ovarian Blood and Lymph

The experiments presented in the previous sections prove that progesterone suppresses the development of large ovarian follicles and that at least part of this action of the progestin is exerted via an intraovarian route. That the intraovarian route involves diffusion of the progestin through the ovary seems to be ruled out by the fact that there is no apparent gradient of effect along the length of the ovary. If diffusion was involved, a zone would be expected around the corpus luteum in which follicles were smaller than follicles at greater distances away from the luteal source of the steroid. Such zones were not seen in the ovaries.

Two other systems for progesterone transport are known to exist in the ovary: blood and lymph. Even though the structure of the blood system has been partially described for the rabbit ovary (Burr and Davis, 1951) and other species, the nature of ovarian lymph remains largely unstudied in any mammal with the exception of the sheep in which lymph has been shown to play a role in the progesterone secretory process (Lindner et a. 1964). These same authors also reported that the corpus luteum is likely the major source of ovarian lymph during the luteal phase and during ovine pregnancy (Morris and Sass, 1964). In the present study, both blood and lymph were sampled as possible carriers of ovarian progesterone. Table 1 shows that ovarian lymph and ovarian vein blood and blood within the ovaries contain high but variable concentrations of progesterone. Based on this finding, it seems possible that both blood and lymph could play a role in the intraovarian movement of progesterone in the rabbit. Current studies in the laboratory have been undertaken to investigate the patterns of blood and lymph flow within the ovary and to determine which system is likely the more important in the distribution of the luteal steroids within the ovary.

Table 1. Progesterone in Blood and Lymph (ng/ml)

Ovarian Vein	Intraluteal	Perifollicular	Intraovarian	Lymph
126 ± 58	224 ± 100	107 ± 32	98 ± 30	17 ± 12

The concentration of progesterone (mean + SEM) in blood collected directly from the Ovarian Vein, from corpora lutea (Intraluteal), from the blood spaces surrounding large follicles (Perifollicular), from regions of the ovary away from corpora lutea and follicles (Intraovarian) and Lymph collected from the ovarian lymphatic drainage. Peripheral blood levels of progesterone in these animals ranged from 4 to 7 ng/ml.

CONCLUSION

The results of these studies show that progesterone inhibits the development of follicles by acting within the ovary (intraovarian action) in addition to the action through the general circulation. The mechanisms by which the progestin exerts this action and the route of intraovarian progesterone delivery are not understood. However, in an ovary which contains a corpus luteum, both ovarian vein blood and ovarian lymph contain high levels of the progestin and both of these systems may be involved in the intraovarian distribution of the steroid.

ACKNOWLEDGEMENTS

These studies were supported in part by a grant from the Medical College of Georgia Research Institute. The authors wish to acknowledge the excellent technical assistance of Maurine Sorrell.

REFERENCES

Baird DT, Backstrom T, McNeilly AS, Smith SK, Wathem CG, 1984. Effect of enucleation of the corpus luteum at different stages of the luteal phase of the human menstrual cycle on subsequent follicular growth. J Reprod Fertil 70:615-624

Banks JA, Freeman ME, 1980. Inhibition of daily LH release mechanism by progesterone acting at the hypothalamus. Biol Reprod 22:217-222

Bogovich K, Richards J, Reichert L, 1981. Obligatory role of luteinizing hormone (LH) in the initiation of preovulatory follicular growth in the pregnant rat: specific effects of human chorionic gonadotropin and follicle stimulating hormone on LH receptors and steroidogenesis in theca, granulosa and luteal cells. Endocrinology 109:860-867

Burr JH, Davies JI, 1951. The vascular system of the rabbit ovary and its relationship to ovulation. Anat Rec 111:273-297

DePaolo LV, Barraclough CA, 1979. Dose dependent effects of progesterone on the facilitation and inhibition of spontaneous gonadotropin surges in estrogen treated ovariectomized rats. Biol Reprod 21:1015-1023

DiZerega GS, Hodgen GD, 1981. Folliculogenesis in the primate ovarian cycle. Endocine Rev 2:27-49

Greenwald GS, 1966. Ovarian follicular development and pituitary FSH and LH content in the pregnant rat. Endocrinology 79: 572-578

Lindner HR, Sass MB, Morris B, 1964. Steroids in the lymph and blood of conscious ewes. J Endocrinol 30:361-376

Mills TM, Osteen KG, 1977. 17β -estradiol receptor and progesterone and 20α -hydroxy-4-pregnen-3-one content of the developing corpus luteum of the rabbit. Endocrinology 101:1744-1750

Morris B, Sass MB, 1966. The formation of lymph in the ovary. Proc Roy Soc B 164:577-591

Setty SL, Mills TM, 1987. The effects of progesterone on follicular growth in the rabbit ovary. Biol Reprod (in press)

THE RELATIONSHIPS OF OOCYTE QUALITY AND FOLLICULAR FLUID PROLACTIN

AND PROGESTERONE IN SUPEROVULATED BEEF HEIFERS WITH AND

WITHOUT NORGESTOMET IMPLANTS

T. Wise, U. Suss*, and R.R. Maurer

USDA Meat Animal Research Center
Clay Center, NE

*Institut fur Nutztierwissen Schaften
Gruppe Tierzucht ETH-Zentrum
CH-8092, Zurich, Switzerland

INTRODUCTION

Limited application of in vitro fertilization and embryo transplant in the animal industry has been primarily due to biological rather than technical aspects. The successful collection of mature, fertilizable oocytes is a prime requirement for in vitro fertilization but the variability of oocytes that develop into viable embryos has restricted the usefulness of this tool. Increased concentrations of progesterone and prolactin in human follicular fluid have been reported to be associated with oocyte maturity and successful in vitro fertilization (Fishel et al., 1983; Laufer et al., 1984). The objectives of this study were to (1) relate follicular fluid prolactin and progesterone concentrations to oocyte quality in superovulated beef cattle and (2) document the changes in prolactin and progesterone in conjunction with oocyte quality throughout the ovulatory period in both superovulated animals and in superovulated animals with the ovulatory LH spike suppressed with progestin implants.

MATERIALS AND METHODS

Crossbred heifers (n = 103) were estrous synchronized with prostaglandin $F_{2\alpha}$ (PGF2α; Lutalyse, Upjohn Co.) and superovulated with FSH (FSH-P, Burns Biotech) by the methods of Wise et al. (1986). Animals (7-9/time) were ovariectomized every 12 h (12-108 h post $PGF_{2\alpha}$) and follicles harvested. To study the effects of progesterone on oocyte quality and follicular maturation, a group of superovulated animals (n = 28) was implanted with silastic Norgestomet implants (Synchromate-B, CEVA) 12 h before the initial $PGF_{2\alpha}$ injection. Seven progestin-implanted animals were ovariectomized at 72, 84, 96 and 108 h after the $PGF_{2\alpha}$ injection. Large follicles ($\geq$ 4 mm diameter) were measured for diameter, follicular fluid harvested and then flushed for the collection of oocytes. In follicles $\leq$ 4 mm diameter, follicular fluid was pooled and the syringe rinsed to collect oocytes. Oocytes were evaluated (1) under the dissecting microscope and then (2) stained with Orcein for examination under phase microscopy. Luteinizing hormone (LH) concentrations in peripheral serum, progesterone and prolactin content in follicular fluid were quantitated with radioimmunoassay. Data were analyzed by analysis of variance (split plot model) in which time, heifer (time), follicular size and size x time interaction were included in the model. For all three dependent variables (prolactin, progesterone and oocyte quality) the time x size interaction was significantly different (P < .01; Fig. 1 and 2).

RESULTS

Sixty-five percent of the animals exhibited on LH-surge, thus hormonal changes and oocyte characteristics were divided into two animal groups (those with or without the LH-surge). The LH-surge occurred at 42 h after the initial $PGF_{2\alpha}$ injection which initiated luteal regression. None of the animals treated with the Norgestomet implant exhibited an LH-surge. First ovulations were detected at 60 h after the initial $PGF_{2\alpha}$ injection. Follicular fluid prolactin concentrations in small ($\leq$ 4 mm diameter), medium sized (5-8 mm) and large follicles ($\geq$ 9 mm) are depicted in Fig. 1. Follicular fluid prolactin concentrations increased up to the time of the LH-surge then steeply declined (P < .05). The decline in follicular fluid prolactin concentrations do not seem to be directly induced by the LH-surge. In animals in which no LH-surge was detected or suppressed with Norgestomet implants, (Fig. 1b and 1c) follicular prolactin concentrations were similar to those exhibiting on LH-surge (Fig. 1a). After the LH-surge (Fig. 2a) follicular progesterone concentrations increased (P < .01) in large and medium sized follicles as the cellular steroidogenic capability changed to primarily luteal in function. In animals in which no LH-surge was detected or suppressed with Norgestomet implants (Fig. 2b and 2c), follicular progesterone concentrations remained low in all follicular sizes.

Oocyte recovery rate was 76.8% from 2470 follicles in 1379 oocytes were classified as viable or degenerate. Oocyte quality increased from 60 to 70% good oocytes up to the LH-surge then tended to remain around 60% for the rest of the period analyzed (60-108 h). Oocyte quality was significantly better (80% good; p < .05) in the larger follicular size group ($\geq$ 9 mm diameter). In animals which no LH-surge was detected, oocyte quality increased up to 60 h after the initial $PGF_{2\alpha}$ injection and then declined to 30-40% good oocytes. In Norgestomet implanted animals, 36% of the oocytes rated good at 72 h post $PGF_{2\alpha}$ and 19% at 108 h. Thus, without the LH-surge, oocyte quality quickly decreased 72-84 h after the initial $PGF_{2\alpha}$ injection. In animals that exhibited an LH-surge, a peak of follicular prolactin at 48 h was detected in follicles producing high quality oocytes and this peak was not seen in follicles producing low quality oocytes or in follicles from animals that did not exhibit an LH-surge. Thus, prolactin concentrations were increased (P < .01) in follicles producing high quality oocytes (partially before and accompanying the LH-surge). After the LH-surge prolactin decreased in all follicles and, as an index of oocyte quality, is questionable. No follicular progesterone relationship was detected with oocyte quality. Follicular progesterone and prolactin concentrations were inversely related.

DISCUSSION

Unlike LH, circulating prolactin concentrations do not seem to increase in conjunction with estrus in the cow (Dieleman et al., 1986). Only animals that exhibited an LH-surge had an additional increase in prolactin noted at 48 h which is probably physiologically tied to the LH-surge that accompanies ovulation. Oocytes collected from animals that exhibited an LH-surge remained consistently high in quality throughout the time periods analyzed, but oocytes collected from animals that had no LH-surge sharply declined in quality by 84 h after the $PGF_{2\alpha}$. Prolactin

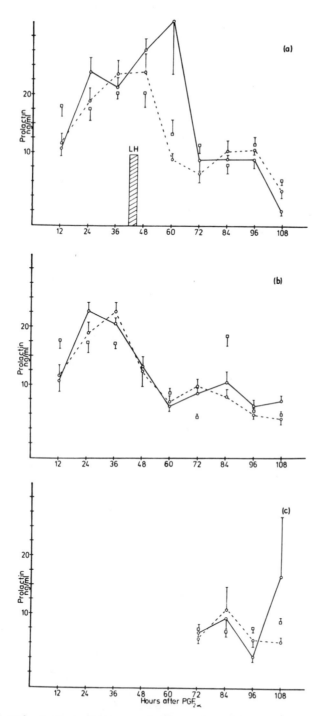

Fig. 1. Pre- and post-ovulatory changes of follicular fluid prolactin
concentrations (mean $\pm$ SEM) in small (□ ; < 4 mm), medium
(O---O; 5-8 mm) and large follicles ($\geq$ 9 mm; O- - -O) in animals
exhibiting an LH-surge (panel a; n = 56), no LH-surge (panel b;
n = 48) and animals with Norgestomet implants (panel c; n = 28).

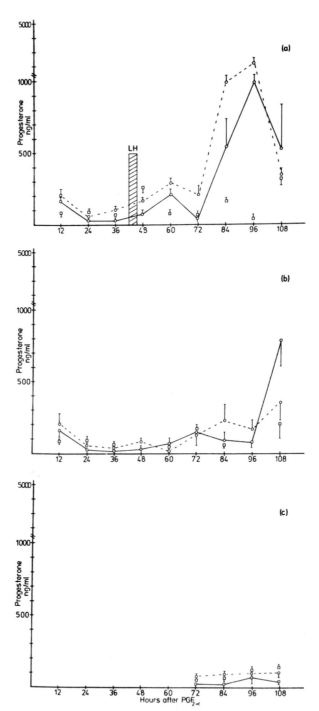

Fig. 2. Pre- and post-ovulatory changes of follicular fluid progesterone concentrations (mean ± SEM) in small (□ ; ≤ 4 mm), medium (O---O; 5-8 mm) and large follicles (≥ 9 mm; O- - -O) in animals exhibiting a LH-surge (panel c; n = 56), no LH-surge (panel b; n = 40), and animals with Norgestomet implants (panel c; n = 28).

concentrations in follicular fluid have been shown to reflect oocyte maturity and may indicate that follicular prolactin plays an active role in preovulatory oocyte development (Laufer et al., 1984). Prolactin receptors are heavily concentrated in the ovum and surrounding granulose cells during the final stages of follicular maturation (Nolin, 1980; Dunaif et al., 1982). The role prolactin may play in oocyte maturation could be either inhibitory or stimulatory (Baker and Hunter, 1978; Nolin, 1980). High follicular progesterone concentrations have been related to successful in vitro fertilizations (Fishel et al., 1983) but no distinct relationships were noted in this study between oocyte quality and follicular fluid progesterone concentrations. McNatty et al. (1974) suggested that prolactin may be involved in the modulation of follicular progesterone secretion by the follicle thus the negative correlation noted between follicular prolactin and progesterone at ovulation. As bovine follicle size increases, follicular fluid prolactin concentrations increase and progesterone is also inversely related to prolactin concentrations (Henderson et al., 1982). The aspect of prolactin regulation of progesterone synthesis is complex since the literature indicates both inhibitory and luteotropic effects upon granulosa cells (McNatty et al., 1974; Veldhuis and Hammond, 1980). The effect of prolactin on oocyte maturation may be indirect through modifying follicular steroid synthesis and essential neurochemical events involved in LHRH and LH release. The characterization of the endocrine and biochemical events associated with follicle and oocyte maturation may eventually elucidate the correct stimulatory relationships to maximize oocyte quality and numbers.

REFERENCES

Baker TG, Hunter RHF, 1978. Interrelationships between the oocyte and somatic cells within the Graafian follicle of mammals. Ann Biol An Biochem Biophys 18:419-426

Dieleman SJ, Bevers MM, VanTol HTM, Willemse AH, 1986. Peripheral plasma concentrations of oestradiol, progesterone, cortisol, LH and prolactin during the oestrous cycle in the cow with emphasis on the perioestrous period. Anim Reprod Sci 10:275-292

Dunaif AE, Zimmerman EA, Friesen HG, Frantz AG, 1982. Intracellular localization of prolactin receptor and prolactin in the rat ovary by immunocytochemistry. Endocrinology 11:1465-1471

Fishel SB, Edwards RG, Walters DE, 1983. Follicular steroids as a prognosticator of successful fertilization of human oocytes in vitro. J Endocrinol 99:335-344

Henderson KM, McNeilly AS, Swanston IA, 1982. Gonadotropin and steroid concentrations in the bovine follicular fluid and their relationship to follicular size. J Reprod Fertil 65:467-473

Laufer N, Botero-ruiz W, DeCherney AH, Haseltine F, Polan ML, Behrman HR, 1984. Gonadotropin and prolactin levels in follicular fluid of human ova successfully fertilized in vitro. J Clin Endocrinol Metab 58:430-434

McNatty KP, Sawers RS, McNeilly AS, 1974. A possible role for prolactin in control of steroid secretion by the human Graafian follicle. Nature 250:653-655

Nolin JM, 1980. Incorporation of endogenous prolactin by granulosa cells and dictyate oocytes in the postpartum rat: effects of estrogen. Biol Reprod 22:417-422

Veldhuis JD, Hammond JD, 1980. Estrogens regulate divergent effects of prolactin in the ovary. Nature 284:262-264

Wise T, Vernon MW, Maurer RR, 1986. Oxytocin, prostaglandins E and F, estradiol, progesterone, sodium and potassium in preovulatory follicles either developed normally or stimulated by follicle stimulating hormone. Theriogenology 26: In Press

COORDINATED RECEPTOR REGULATION IN OVARIAN GRANULOSA CELLS

T.T. Chen, K. Gao, M.A. Shelton, and D.D. Hatmaker

Department of Zoology
University of Tennessee
Knoxville, TN 37996-0810

INTRODUCTION

The concept of receptor down regulation by its own hormone through internalization of the hormone-receptor complex is well established (for reviews, see Blecher and Bar, 1981; Kaplan, 1981). However, the cellular basis for heterologous receptor regulation is less well defined. The process is important in endocrine cells, particularly in the ovarian cell, because granulosa and luteal cells are under the control of at least a dozen different hormones and growth factors (Hseuh et al., 1984). To further complicate matters, any one of these agents may positively or negatively regulate receptors for other hormones. For example, physiologic levels of LH increase prolactin (Prl) receptor number, while pharmacologic levels of LH induce a transient decrease of Prl receptors (Chen et al., 1979). Likewise, physiologic levels of Prl result in an increase of LH receptors in these cell types (Richards, 1979), but the effect of high levels of Prl on LH receptor has not been well studied.

High circulating levels of Prl, i.e., during lactation and in hyperprolactinemia, are known to be associated with disturbances of the female reproductive cycle and ovulation, which lead to infertililty (Barbieri and Ryan, 1983). High Prl has also been shown to adversely affect the steroidogenic response of the granulosa cell to gonadotropins (McNeilly et al., 1982). In this study, we used three approaches to establish the direct effect of high Prl on granulosa cell function and to test our hypothesis that high Prl induces down regulation of both Prl and LH receptor through a coordinated internalization process. We postulate that this process may be responsible, at least in part, for the high Prl-induced unresponsiveness of the cell to gonadotropic stimulation.

MATERIALS AND METHODS

Granulosa cell culture

To prepare granulosa cell cultures for use in these studies, ovaries from PMSG-primed immature rats were excised and follicles were pierced with a needle to release the granulosa cells into PC-1 medium (Ventrex). Granulosa cells were plated at 1×10^6 cells per dish in 1.5 ml of PC-1 (serum-free) medium. The cells were cultured in the presence of 200 ng of rFSH and culture medium was changed every 24 to 48 h.

Two studies were initially undertaken (experiments 1 and 2) to assess the direct effect of high Prl on ovarian granulosa cell function: In experiment 1, cultured rat granulosa cells were incubated with graded doses of ovine Prl (oPrl) and its effects on steroidogenesis were monitored. In experiment 2, granulosa cells were co-cultured with a Prl-secreting tumor cell line (rat GH$_3$ cells) to mimic an __in vitro__ hyper-prolactinemic condition and granulosa cell function was evaluated. Finally, the intracellular pathway of Prl and LH receptors in granulosa cells was traced using colloidal gold labeled LH/hCG and Prl.

Experiment 1. oPrl (31 iu/mg) ranging from 10 ng to 10 ug was added daily to granulosa cell cultures. Media were collected every 24 h and steroid production was determined by radioimmunoassay.

Experiment 2. Forty-eight h after plating of granulosa cells, a Prl-secreting tumor cell line, GH$_3$ cells (ATCC, Rockville, MD) was added to the granulosa cell cultures (375,000 GH$_3$ cells/dish). Cultures consisting of either GH$_3$ or granulosa cells only were cultured in separate dishes as controls. Media was collected every 48 h over six days and hormones in the medium were assayed.

Experiment 3. Localization of LH and Prl receptors. In order to determine if high levels of Prl induce down regulation and internalization of both Prl and LH receptors, the hormones were labeled with two different sizes of colloidal gold (Au$_5$ and Au$_{16}$, respectively) and used as tracers for Prl and LH receptor.

Granulosa cells were cultured for three days before use. On day 3, Au$_5$-Prl and Au$_{16}$-hCG were added to granulosa cell cultures either separately or together and pre-incubation was carried out at 4°C for 1.5 h to allow binding without internalization. Following the preincubation period, the temperature was raised to 37°C and incubation was continued for up to 24 h. At the end of each incubation, cells were fixed with glutaraldehyde and processed for electron microscopy.

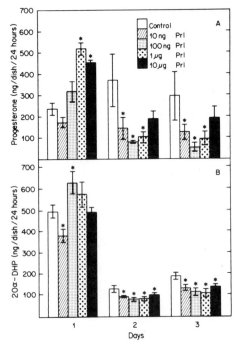

Fig. 1. Relationship of steroid production by cultured granulosa cells to graded doses of oPrl.

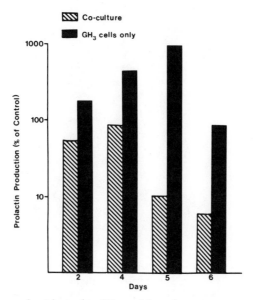

Fig. 2. Prolactin production by GH3 cells alone or co-cultured with rat granulosa cells. Prolactin in media was determined by RIA. Results were expressed as percent of control. (Control equals to Prl present in granulosa cell culture medium).

RESULTS

Experiment 1. Prolactin at low doses initially stimulated progesterone production by cultured granulosa cells (Fig. 1A, Day 1). However, continued exposure of the cells to Prl (Days 2 and 3) caused an inhibition of progesterone production. This inhibition appears to be the result of reduced synthesis of progesterone, as 20α-dihydroprogesterone (20α-DHP) levels were also depressed (Fig. 1B).

Experiment 2. Although GH3 cells produced less Prl in co-culture (Fig. 2), this level of Prl was sufficient to inhibit granulosa cell progesterone production (Fig. 3A, Day 2). This effect became more pronounced as the co-culture was continued (Days 4 and 6). Again, the reduction of progesterone can be attributed to reduced synthesis of progesterone because 20α-DHP was correspondingly decreased (Fig. 3B).

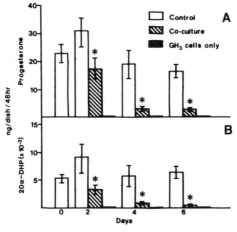

Fig. 3. Steroid production by granulosa cells cultured alone or with GH3 cells. The steroid production by GH3 cells alone was negligible.

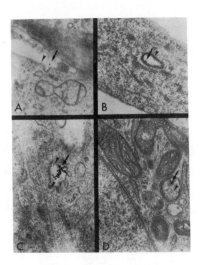

Fig. 4. Localization of Au_{16}-hCG and Au_5-Prl in cultured granulosa cells.
Granulosa cells pre-incubated with Au_{16}-hCG (arrows) and Au_5-Prl
(arrowheads) for 1.5 h at $4^{\circ}C$ followed by $37^{\circ}C$ incubation for A)
10 min; B) 1 h; C) 3 h; and D) 24 h.

Experiment 3. When gold-labeled hCG and Prl was incubated with
cultured granulosa cells at $4^{\circ}C$, the labels were initially on the cell
surface (Fig. 3, panel A). However, when internalization was allowed to
proceed at $37^{\circ}C$, both hCG and Prl tracers were found to become associated
with identical endosomes (Fig. 4, panels B, C and D).

DISCUSSION

These results show that high Prl causes an impaired steroidogenic
response of the granulosa cell. This defect may be due, in part, to high
Prl-induced internalization of LH receptor in the granulosa cell.

Our findings are consistent with a recent report by Fortune and
Vincent (1986) who showed that Prl inhibits FSH-stimulated progesterone
secretion by rat granulosa cells cultured in serum-free medium. In
addition, our tracer study provided a possible mechanism for the
inhibitory effect of Prl on granulosa cell function.

Although co-culture of GH_3 cells with granulosa cells is an over-
simplified model of the hyperprolactinemic condition, we consider this to
provide a reasonable approach to study the disease at the cellular level.
The drastic reduction of both progesterone and 20α-DHP synthesis after
prolonged co-culture of granulosa cells with GH_3 cells strongly suggests
that high Prl inhibits steroid synthesis by this cell type. The possibility that this inhibition was due to the more presence of GH_3 cells or
its product(s) is unlikely for two reasons: 1) We have co-cultured
granulosa cells with Hela cells and found that progesterone production by
the granulosa cell was not drastically affected (data not shown); 2) The
only protein product present in the GH_3 cell culture medium was a broad
single band with M_r 22,000 on SDS gel (data not shown), indicating that
Prl and growth hormone were the major proteins secreted.

It is interesting that after four days of co-culture, the production
of Prl by GH_3 cells was diminished (in contrast to GH_3 cells cultured
alone). This suggests that the presence of granulosa cells or their
secretory product(s) may inhibit GH_3 cell function.

Coordinated internalization of receptors has been shown in a number of systems (Carpentier et al., 1982; Willingham et al., 1983). This concept has evolved and been recognized as an important process in cell physiology. The phenomenon is not unexpected in the granulosa cell considering that the cell is under multiple hormonal controls, the result of which is either enhancement or suppression of steroid secretion. The fact that both Prl and LH/hCG tracers were found in identical endosomes suggests that high concentrations of one hormone will result in loss of other receptors from the cell surface through a coordinated endocytic process. Whether or not these endosomes contain unbound LH receptors is currently under investigation.

In conclusion, our results show that high Prl induces functional impairment of granulosa cells. This may be attributable to the loss of Prl and LH receptors by internalization. These findings suggest a mechanism through which high Prl renders the cell unresponsive to gonadotropic stimulation.

ACKNOWLEDGEMENTS

The authors wish to thank Linda Turner for manuscript preparation and Dr. Gordon D. Niswender of Colorado State University and the National Hormone and Pituitary Program for providing antisera used in this study. This work was supported by a grant from the American Cancer Society (BC-543) and NIH grant HD-14505.

REFERENCES

Barbieri RL, Ryan KJ. 1983. Bromocryptine:endocrine pharmacology and therapeutic applications. Fertil Steril 39:727-41

Blecher M, Bar RS, 1981. Introduction to receptors. In: Blecher M and Bar RS (eds.), Receptors and Human Disease. Baltimore: Williams and Wilkins, pp. 2-23

Carpentier J, Gordon P, Anderson RGW, Goldstein JL, Brown MS, Cohen S, Orci L, 1982. Localization of ^{125}-I epidermal growth factor and ferritin-low density lipoprotein in coated pits. J Cell Biol 95:73-77

Chen TT, Harwood JP, Katt KJ, 1981. In: Schwartz NB and Hunzicker-Dunn M (eds.), Dynamics of Ovarian Function. New York: Raven Press, pp. 167-73

Fortune JE, Vincent SE, 1986. Prolactin modulates steroidogenesis by rat granulosa cells: 1. effects on progesterone. Biol Reprod 35:84-91

Hseuh AJW, Adashi EY, Jones PBC, Welsh TH, 1984. Hormonal regulation of the differentiation of cultured ovarian granulosa cells. Endocrine Review 5:76-127

Kaplan J, 1981. Polypeptide-binding membrane receptors: analysis and classification. Science 212:14-20

McNeilly AS, Glasier A, Jonassen J, Howie PW, 1982. Evidence for direct inhibition of ovarian function by prolactin. J Reprod Fertil 65:559-69

Richards JS, 1978. Hormonal control of ovarian follicular development. Rec Prog Horm Res 35:343-68

Willingham MC, Haigler HT, Fitzgerald DJP, Gallo MG, Rutherford AV, Pastan IH, 1983. The morphologic pathway of binding and internalization of epidermal growth factor in cultured cells. Exp Cell Res 146:163-75

OVARIAN AND DECIDUAL RELAXINS IN HUMAN PREGNANCY

G. Bryant-Greenwood, S. Ali, M. Mandel, and F. Greenwood

University of Hawaii, Honolulu, Hawaii

INTRODUCTION

The corpus luteum of pregnancy is one source of relaxin in the human (Weiss et al., 1976). Concentrations of relaxin-like immunoactivities in the systemic circulation have been measured by a porcine relaxin RIA (Quagliarello et al., 1979) and more recently by a RIA for synthetic human relaxin (Eddie et al., 1986). The levels of human relaxin in plasma are low in comparison to those in the sow. Peak values attained at between 12-20 weeks are approximately 1 ng/ml, declining thereafter and remaining at between 0.2-0.6 ng/ml for the rest of gestation.

The physiological roles of luteal relaxin in early pregnancy appear to be two-fold, to synergize with progesterone to cause myometrial quiescence (Beck et al.., 1982), and to cause a loosening of the connective tissue in the pelvic joint (Abramson et al., 1934). Evidence for a myometrial action is limited by the difficulty in obtaining human myometrium in early pregnancy and in testing with human rather than porcine relaxin. An action on the pelvic joint is likewise difficult to study, but from work carried out before the dangers of x-rays in pregnancy were appreciated, it was shown that a marked but small increase in the width of the pelvic gap occurs in early pregnancy with little further change prior to parturition (Abramson et al., 1934) – an interesting correlation to the subsequent measurements of plasma relaxin in pregnancy. Human luteal relaxin in early pregnancy may thus exhibit systemically two of its biological activities demonstrated in other species (Bryant-Greenwood, 1982).

We have suggested that the simultaneous expression of these activities by a paracrine, decidual relaxin is significant in the synchronization of the mechanical events of parturition: uterine relaxation, fetal membrane stretch and rupture, cervical dilatation. On this interpretation, relaxin, like progesterone produced by the corpus luteum in early pregnancy, is replaced by its intrauterine counterpart as pregnancy progresses. However, progesterone produced by the fetal-placental unit is also a systemic hormone, whereas relaxin produced by the decidua and placenta fails to enter the general circulation. We have suggested a paracrine role for decidual/placental relaxin (Koay et al., 1985a), intercalated with other paracrine systems in the intrauterine compartments. Since myometrial activity and cervical dilatation are not controlled by a single hormone, we have sought other paracrine systems in the fetal membranes (Koay et al., 1986). Fig. 1 summarizes the data obtained in our laboratory over the last five years on the existence and expression of intrauterine relaxins.

IDENTIFICATION OF RELAXIN PRODUCING CELLS

Different methodologies have been used to show the presence of relaxin in human decidua and placenta (Bigazzi et al., 1980; Fields and Larkin, 1981; Yamamoto et al., 1981; Schmidt et al., 1984). The general concensus is that the amounts remaining at the end of various isolation procedures are very low and there is no data on recovery. By immunocyto-chemical localization however, there are considerable areas of chorionic cytotrophoblast and decidua which stain positively with antiserum to porcine relaxin (Koay et al., 1985b). In frozen tissue sections we have been able to detect differences between tissues collected after normal spontaneous delivery and Cesarean section with no labor (unpublished observations). These results support our isolation studies (Yamamoto et al., 1981), suggesting that the intrauterine sources are depleted after spontaneous labor and delivery. The evidence for the production of relaxin rather than its presence, by isolation or immunolocalization currently rests upon the positive localization of the prohormone and/or relaxin C-peptide with an antiserum to a synthetic 14 amino-acid peptide derived from the amino-acid sequence of human prorelaxin (unpublished observations).

A human placenta cDNA library (Clonetech) has been screened for relaxin and prolactin gene sequences with 5'-end labelled oligonucleotides (18 mer, relaxin and 15 mer, prolactin). The relaxin probe had 9 thymines (50%) and 6 adenines (33%) and was not specific: the 4 clones from 16,000 plaques screened were not confirmed using a 48 mer oligonucleotide and the library is being re-screened. In contrast, of the 4 clones positive for prolactin, 3 gave a positive signal with the larger probe. The λgt 11 recombinant cDNA clones positive for relaxin and prolactin gene sequences will be probed for antigen expression: histochemical localization by _in situ_ hybridizations to specific cell types of relaxin and prolactin messages is being undertaken.

Table 1. Decidual Relaxin: Paracrine Role in Late Pregnancy

MYOMETRIAL CERVIX	DECIDUA	CHORION	AMNION	AMNIOTIC FLUID
		Plasminogen Activator Collagenolytic Activity		Inhibited and latent enzymes
[R] ←	RELAXIN SYNTHESIS →	[R] →	[R] →	RELAXIN
			PGE PRODUCTION	

[R] = relaxin receptor.

——→ path of steroid relaxin or enzymes.

- - -→ biological consequence of relaxin; relaxin receptor.

IDENTIFICATION OF RELAXIN RECEPTORS

Table 1 shows a decidual relaxin acting on both the cervix and myometrium. Porcine relaxin, applied topically to the cervix of preparturient women, has been shown to cause cervical dilatation and hasten labor (MacLennan et al., 1980; Evans et al., 1983). On the other hand, the human myometrium in late pregnancy appears to be refactory to porcine relaxin in vitro (MacLennan et al., 1986); the experiment has yet to be performed with human relaxin. The action on the cervix is to alter the collagenous structure of this tissue in the same way that we have proposed it to act upon the collagen of the amnion and chorion (Koay et al., 1986). Conventional studies with particulate cell preparations have demonstrated a relaxin receptor distinct from the insulin receptor (Koay et al., 1986). we have applied the method used by Chegini and Rao (1985) to localize more precisely the receptors for relaxin in the amnion and chorion. In preliminary experiments we have established that the binding of ^{125}I-labelled monoacylated porcine relaxin is time and temperature dependent and is fully displaceable with excess cold relaxin. Maximum binding occurs after a 30 min incubation at $24^{\circ}C$ and appears associated with the chorion and chorionic cytotrophoblast. Further work with this method on tissues from different stages of pregnancy will give information on the ontogeny of the relaxin receptor in these tissues. This, together with concomitant localization of relaxin within the cells of synthesis, should provide more conclusive evidence for its paracrine nature.

THE BIOLOGICAL ACTIONS OF INTRAUTERINE RELAXIN

We have proposed that relaxin is one of the hormonal determinants of connective tissue remodelling in the human fetal membranes. This is based upon the differing effects of relaxin and other hormones in vitro, on dispersed amnion and chorion cells with respect to the secretion of plasminogen activator (PA) and collagenolytic enzyme activities (Koay et al., 1986). Pericellular collagenolysis is controlled by a number of hormones acting to release a number of enzymes and their precursors (Reynolds, 1985). The central event in the process of collagen degradation is the production of active collagenase from latent collagenase. We have obtained evidence that this system is in operation within the human fetal membranes and thus may be essential for the continuation and termination of human pregnancy (Bryant-Greenwood and Greenwood, 1987).

We have two correlates in vivo to justify this proposal. The levels of inhibited and latent forms of PA and collagenolytic enzyme activities in amniotic fluid increase as parturition approaches (Koay et al., 1986). Components of pericellular collagenolysis are indeed present within the fetal membranes as shown by immunocytochemical localization of the specific components (unpublished observations). Their spatial arrangement suggests that the system is one of interaction between the elements of the fetal membranes. Final interpretation on control, however, awaits the completion of this work.

Prostaglandin production, particularly by the amnion, increases during labor and delivery (Olson et al., 1983). If relaxin exerts its effect on the collagen of the membranes over a period of weeks, it is likely that decidual relaxin is secreted prior to prostaglandin production; indeed we have detected immunoreactive relaxin in human amniotic fluids obtained from 16-42 weeks gestation (Bryant-Greenwood and Greenwood, 1987). Thus an effect on amniotic PGE production was sought in vitro (Lopez Bernal et al., 1987). We have demonstrated in this study that the effect of relaxin depends upon the physiological status of the amnion used. Thus a 40% decrease in PGE production obtained from patients after term Cesarean section without labor, on exposure to relaxin in vitro, contrasted with a dose-dependent and significant increase in PGE production by amnion collected after spontaneous labor and delivery. Preliminary data further suggests that this effect of relaxin may depend upon the availability of free arachidonic acid for PG synthesis. Further

711

work is in progress to clarify the mechanism by which relaxin alters prostaglandin production in these tissues.

CONCLUSION

The coexistence of endocrine and paracrine systems using the same hormonal messages should not be surprising. The human fetal membranes, subject to selected signals from the maternal and fetal compartments, provide a model to study paracrine:paracrine and endocrine:paracrine interactions which may be highly significant in the fine tuning of human parturition.

ACKNOWLEDGEMENTS

In acknowledging the stimulation, collaboration, discussions and joint publications with Professor A.C. Turnbull, Dr. M. Rees, Dr. A. Lopez Bernal and Dr. S. Cederhold-Williams, we must also state that the views expressed here are not necessarily in accord with those of our distinguished colleagues.

REFERENCES

Abramson D, Sumner M, Roberts M, Wilson PD, 1934. Relaxation of the pelvic joints in pregnancy. Surg Gynecol Obstet 58:595-613
Beck P, Adler P, Szlachter N, Goldsmith LT, Steinetz BG, Weiss G, 1982. Synergistic effect of human relaxin and progesterone on human myometrial concentrations. Int J Gynaec Obstet 20:141-144
Bigazzi M, Nardi E, Bruni P, Petrucci F, 1980. Relaxin in human decidua. J Clin Endocrinol Metab 51:939-941
Bryant-Greenwood GD, 1982. Relaxin as a new hormone. Endocr Rev 3:62-90
Bryant-Greenwood GD, Greenwood FC, 1987. Postulated roles for luteal or decidual relaxins at parturition in the pregnant sow and woman. In: Mitchell BF (ed.), The Human Fetal Membranes. Perinatology Press, In Press
Chegini N, Rao ChV, 1985. Epidermal growth factor binding to human amnion, chorion, decidua and placenta from mid and term pregnancy: quantitative light microscope autoradiographic studies. J Clin Endocrinol Metab 61:529-535
Eddie LW, Lester A, Bennet G, Bell RJ, Grier M, Johnston PD, Niall HD, 1986. Radioimmunoassay of relaxin in pregnancy with an analogue of human relaxin. The Lancet 1:1344-1346
Evans MI, Dougan MB, Moawad AH, Evans WJ, Bryant-Greenwood GD, Greenwood FC, 1983. Ripening of the human cervix with porcine ovarian relaxin. Amer J Obstet Gynecol 147:420-414
Fields PA, Larkin LH, 1981. Purification and immunohistochemical localization of relaxin in human term placenta. J Clin Endocrinol Metab 52:79-85
Koay ESC, Greenwood FC, Bryant-Greenwood GD, 1985a. Relaxin: a local hormone in human parturition. In: Jaffe R and DelAqua S (eds.), The Endocrine Physiology of Pregnancy and the Peripartal Period. Raven Press, New York, pp 247-253
Koay ESC, Bagnell CA, Bryant-Greenwood GD, Lord SB, Cruz AC, Larkin LH, 1985b. Immunocytochemical localization of relaxin in human decidua and placenta. J Clin Endocrinol Metab 60:859-863
Koay ESC, Bryant-Greenwood GD, Yamamoto SY, Greenwood FC, 1986. The human fetal membranes: a target tissue for relaxin. J Clin Endocrinol Metab 62:513-521
Lopez Bernal A, Bryant-Greenwood GD, Hansell DJ, Hicks BR, Greenwood FC, Turnbull AC, 1987. Effect of relaxin on prostaglandin E production by human amnion: changes in relation to the onset of labor. Brit J Obstet Gynecol, In Press

MacLennan AH, Green RC, Bryant-Greenwood GD, Greenwood FC, Seamark RF, 1980. Ripening of the human cervix and induction of labor with purified porcine relaxin. The Lancet 1:220-223

MacLennan AH, Grant P, Ness D, Down A, 1986. Effect of porcine relaxin and progesterone on rat, pig and human myometrial activity in vitro. J Reprod Med 31:43-49

Olson DM, Opavsky MA, Challis JRG, 1983. Prostaglandin output in relation to parturition by cells dispersed from human intrauterine tissues. J Clin Endocrinol Metab 57:694-699

Quagliarello J, Szlachter N, Steinetz BG, 1979. Serial relaxin concentrations in human pregnancy. Am J Obstet Gynecol 135:43-44

Reynolds JJ, 1985. The molecular and cellular interactions involved in connective tissue destruction. Brit J Dermatol 112:715-723

Schmidt CL, Sarosi P, Steinetz BG, O'Byrne EM, Tyson JE, Horvath V, Sas M, Weiss G, 1984. Relaxin in human decidua and term placenta. Europ J Obstet Gynecol Reprod Biol 17:171-182

Weiss G, O'Byrne EM, Steinetz BG, 1976. Relaxin: a product of the human corpus luteum of pregnancy. Science 194:948-949

Yamamoto S, Kwok SCM, Greenwood FC, Bryant-Greenwood GD, 1981. Relaxin purification from human placental basal plates. J Clin Endocrinol Metab 52:601-604

THE EFFECT OF PROSTAGLANDIN $F_{2\alpha}$ ON RELAXIN RELEASE DIFFERS

IN THE CYCLIC AND PREGNANT SOW

C.A. Bagnell, J.P. McMurtry*, and G.S. Lewis* [**]

Pacific Biomedical Research Center
University of Hawaii
Honolulu, Hawaii

*Nonruminant Nutrition and **Reproduction Labs
Animal Science Institute
USDA, Agricultural Research Service
Beltsville, MD

INTRODUCTION

Relaxin is known classically as a hormone of pregnancy with high levels produced by the corpus luteum (CL) of the sow. However, recently relaxin has been localized in luteal tissue of the pseudopregnant pig (Fields and Fields, 1985) and in the nonpregnant pig throughout the cycle (Ali et al., 1986). Luteal relaxin immunoactivity was highest from days 7 to 15 of the cycle with a decline by day 18 coincident in time with luteolysis, defined functionally as a decline in serum progesterone and occurring between days 14 and 18 of the sow estrous cycle (Cavazos et al., 1969). Whether prostaglandin $F_{2\alpha}$ ($PGF_{2\alpha}$), postulated as the major signal for luteolysis in this species (Moeljono et al., 1976), is also responsible for relaxin release from the CL of the cycle is not known and was an objective in this study using exogenous prostaglandin.

Administration of $PGF_{2\alpha}$ (10 mg, i.m.) to pigs during late pregnancy causes luteolysis, as demonstrated by a rapid decline in serum progesterone, the release of relaxin and parturition (Sherwood et al., 1976). However, Nara et al. (1982) showed that a nonluteolytic dose of $PGF_{2\alpha}$ (i.e. no change in serum progesterone) was effective in releasing relaxin in late pregnant pigs. This suggests that relaxin release in the pregnant sow may be linked to a fall in progesterone accompanying luteolysis.

The objective of these studies was to compare the effects of $PGF_{2\alpha}$ in the pregnant and cyclic sow and determine whether this stimulus for relaxin secretion in late pregnancy is also effective during the cycle. The day 15 cyclic sow was chosen since it has been shown that luteal relaxin immunoactivity is elevated at this time and exogenous $PGF_{2\alpha}$ will induce luteolysis after day 12 of the cycle (Diehl and Day, 1974).

METHODS

Animals

Multiparous Duroc sows housed at the U.S. Department of Agriculture, Beltsville, MD were used in these studies. Animals exhibiting at least two estrous cycles averaging 21 days were checked for estrus in the presence of a boar. The day of onset of estrus was designated as day 0 of the cycle and twelve animals were used on cycle day 15. Six animals were mated on the first day of estrus (day 0 of pregnancy) and maintained until day 108 of pregnancy. Anesthetized sows were given a single bolus injection of $PGF_{2\alpha}$ (10 or 100 µg; Sigma) or saline via the right ovarian artery. In other experiments PGF (100 µg) was given via the jugular vein. Multiple blood samples were taken from the right ovarian vein both before and during a 90 min period after $PGF_{2\alpha}$/saline treatment. Right (treated) and left (untreated) ovaries were collected and processed for immunohistochemistry to determine relaxin content. Progesterone and relaxin were measured in blood samples by radioimmunoassay (RIA).

Radioimmunoassay

Relaxin levels were determined with a homologous porcine relaxin RIA (Afele et al., 1979). Concentrations of progesterone were measured by RIA in plasma samples according to methods described by Guthrie (1977).

Immunohistochemistry

Bouin's-fixed, paraffin-embedded tissues were immunostained for relaxin antigenic sites using the avidin-biotin immunoperoxidase method (Hsu et al., 1981). The antiserum used to CM-a' porcine relaxin, raised in a rabbit by the method of Vaitukaitis et al. (1971) has been characterized (Bryant-Greenwood and Greenwood, 1979). The optimum dilution of this antiserum necessary to achieve adequate staining with minimal background has been shown to be 1:10,000 and 1:500 for corpora lutea from pregnant and day 15 cyclic sows, respectively (Ali et al., 1986). This reflects the markedly different concentrations of relaxin present in luteal tissue in these two reproductive states.

RESULTS

Day 108 Pregnant Sow

Ovarian venous progesterone levels in day 108 pregnant sows were unchanged after administration of 100 ug $PGF_{2\alpha}$ via the ovarian artery and when compared with saline-treated controls (data not shown). However, the $PGF_{2\alpha}$ treatment induced an immediate release of relaxin from the ovary and serum relaxin remained elevated above pretreatment levels throughout the sampling period (pre-$PGF_{2\alpha}$ = 25.6 +/- 8.0 ng/ml, post $PGF_{2\alpha}$ = 202.8 +/- 47.2 ng/ml). Relaxin immunostaining indicated no loss of relaxin from the corpora lutea of sows 90 min after treatment with $PGF_{2\alpha}$.

Day 15 Cyclic Sow

The effects of $PGF_{2\alpha}$/saline treatment on ovarian venous progesterone of selected animals is shown in Fig. 1. Within 10 min of injection of 10 µg $PGF_{2\alpha}$ into the right ovarian artery there was a rise in progesterone at 15-25 min followed by a decline at 30 min (Fig. 1b). Progesterone levels remained lower than pretreatment values throughout the remainder of the 90 min sampling period. This is in contrast to an animal given saline in which there was no evidence for a decline in progesterone (Fig. 1a). When $PGF_{2\alpha}$ was administered systemically via the jugular vein there was a 40 min delay before a rise in progesterone was observed (Fig. 1c). However, there was no evidence for a decline in progesterone associated with luteolysis in the 90 min sampling period.

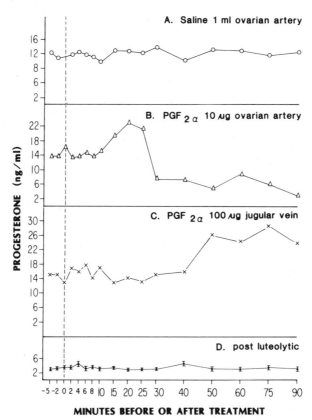

Fig. 1. Ovarian venous progesterone after administration of $PGF_{2\alpha}$ /saline in day 15 cyclic sows. A. Saline via ovarian artery; B. $PGF_{2\alpha}$ (10 µg) via ovarian artery; C. $PGF_{2\alpha}$ (100 µg) via jugular vein; D. Animals with low pretreatment progesterone levels indicating luteolysis had already occurred, n = 8 mean +/- SEM.

In these studies a total of 12 animals were used (3-4 animals per treatment group) however, 8/12 animals tested had pretreatment progesterone levels between 2-4 ng/ml indicating that the corpora lutea had already undergone luteolysis. There was no evidence for a further decline in progesterone after any treatment in these animals (Fig. 1d).

Relaxin was not detectable in ovarian venous plasma throughout the sampling period in any of the day 15 cyclic sows studied. In addition, the immunohistochemistry showed that luteal relaxin after the $PGF_{2\alpha}$ treatment was no different from that of the untreated ovary or the ovaries of saline-treated sows.

DISCUSSION

Luteal regression occurs in the cyclic sow between day 13-18 with ovarian venous progesterone declining to basal levels by day 17 (Gomes et al., 1965). This is reflected in the present studies in which 8 of 12 animals tested had already undergone luteolysis as evidenced by low serum progesterone values at day 15 of the cycle. In sows prior to CL regression, with elevated serum progesterone prior to treatment, the administration of $PGF_{2\alpha}$ via the ovarian artery was effective in inducing a decline in serum progesterone. This supports other studies which indicate that $PGF_{2\alpha}$ will induce functional luteolysis when administered after day 12 of the cycle (Diehl and Day, 1974; Douglas and Ginther, 1975). However, in the day 15 cyclic animal, exogenous $PGF_{2\alpha}$ treatment failed to generate a rise in ovarian venous relaxin when administered either via the ovarian artery or jugular vein and on a time scale similar to that used to demonstrate a brisk response in the pregnant sow. Nara et al. (1982) showed that jugular vein infusion of $PGF_{2\alpha}$ in the day 108 pregnant sow caused a rise in peripheral relaxin which peaked at 90 ng/ml within 10 min postinfusion. The present work correlates well with those studies by showing that direct injection of $PGF_{2\alpha}$ into the ovarian artery of day 108 pregnant sows caused an abrupt rise in ovarian venous relaxin which was 2-fold higher than Nara et al. (1982) described in peripheral plasma.

The difference in tissue content of relaxin in the CL of pregnancy and the cycle is considerable. Anderson et al. (1973) reported that levels of relaxin bioactivity in extracts of ovaries from cyclic sows was more than 100-fold lower than in late pregnancy. Although immunoactive relaxin is detectable in the CL between days 11-15 of the sow cycle, a 20-fold greater concentration of antibody was needed to localize relaxin at this time when compared with the pregnant animal (Ali et al., 1986). Based on these observations, if endogenous $PGF_{2\alpha}$ was indeed involved in the release of relaxin from the CL of the cycle, one would expect serum relaxin levels to be up to 100-fold lower after an exogenous $PGF_{2\alpha}$ challenge when compared with similar treatment of the pregnant animal. Using the peripheral relaxin levels reported by Nara et al. (1982) within 90 min after $PGF_{2\alpha}$ treatment (30-90 ng relaxin/ml), one would expect picogram concentrations of relaxin/ml to be released into the systemic circulation from the CL of the cycle. However, relaxin was not detected in blood draining the ovary of cyclic animals using a RIA with a sensitivity of 156 pg/ml. Therefore, the lower tissue content of relaxin in the cycle when compared with pregnancy does not fully explain our inability to demonstrate a release of relaxin into circulation after a $PGF_{2\alpha}$ challenge.

Attempts to show depletion of immunoreactive relaxin in the pregnant or cyclic sow CL after prostaglandin administration were not successful irrespective of whether relaxin was detected by RIA in the ovarian vein or not.

The present studies indicate that in the cyclic animal luteolysis, either occurring naturally or induced with exogenous prostaglandin, does not result in a detectable release of relaxin into the blood. These results in the cyclic cow are in contrast to the rise in plasma relaxin associated with luteolysis occurring naturally or induced with exogenous $PGF_{2\alpha}$ in the late pregnant sow (Sherwood et al., 1976 and 1979). Thus, there appears to be a different control of relaxin secretion from the CL of the cycle and the CL of pregnancy.

REFERENCES

Afele S, Bryant-Greenwood GD, Chamley WA, Dax EM, 1979. Plasma relaxin immunoactivity in the pig at parturition and during nuzzling and suckling. J Reprod Fert 56:451-7

Ali SM, McMurtry JP, Bagnell CA, Bryant-Greenwood GD, 1986. Immunocytochemical localization of relaxin in corpora lutea of sows throughout the estrous cycle. Biol Reprod 34:139-43

Anderson LL, Ford JJ, Melampy RM, Cox DF, 1973. Relaxin in porcine corpora lutea during pregnancy and after hysterectomy. Am J Physiol 225(5):1215-9

Bryant-Greenwood GD, Greenwood FC, 1979. Specificity of radioimmunoassay for relaxin. J Endocrinol 81:239-47

Cavazos LF, Anderson LL, Belt WD, Hendricks DM, Kraeling RR, Melampy RM, 1969. Fine structure and progesterone in the corpus luteum of the pig during the estrous cycle. Biol Reprod 1:83–106

Diehl JR, Day BN, 1974. Effect of prostaglandin $F_{2\alpha}$ on luteal function in swine. J Anim Sci 39:392–6

Douglas RH, Ginther OJ, 1975. Effect of prostaglandin $F_{2\alpha}$ on estrous cycle or corpus luteum in mares and gilts. J Anim Sci 40:518–21

Fields PA, Fields MJ, 1985. Ultrastructural localization of relaxin in the corpus luteum of the nonpregnant, pseudopregnant, and pregnant pig. Biol Reprod 32:1169–79

Gomes WR, Herschler RC, Erb RE, 1965. Progesterone levels in ovarian venous effluent of the non-pregnant sow. J Anim Sci 24:722–7

Guthrie HD, 1977. Induction of ovulation and fertility in prepubertal gilts. J Anim Sci 45:1360–7

Hsu SM, Raine L, Fanger H, 1981. The use of avidin-biotin-peroxidase complex (ABC) in immunoperoxidase techniques: a comparison between ABC and unlabeled antibody (PAP) procedures. J Histochem Cytochem 29:557–80

Moeljono MPE, Bazer FW, Thatcher WW, 1976. A study of prostaglandin $F_{2\alpha}$ as the luteolysin in swine: I. Effect of prostaglandin $F_{2\alpha}$ in hysterectomized gilts. Prostaglandins 1:737–43

Nara BS, Ball GD, Rutherford JE, Sherwood OD, First NL, 1982. Release of relaxin by a nonluteolytic dose of prostaglandin F2 in pregnant swine. Biol Reprod 27:1190–5

Sherwood OD, Chang CC, BeVier GW, Diehl JR, Dziuk PJ, 1976. Relaxin concentrations in pig plasma following the administration of prostaglandin $F_{2\alpha}$ during late pregnancy. Endocrinology 98:875–9

Sherwood OD, Nara BS, Crnekovic VE, First NL, 1979. Relaxin concentrations in pig plasma after the administration of indomethacin and prostaglandin $F_{2\alpha}$ during late pregnancy. Endocrinology 104:1716–21

Vaitukaitis JL, Robbins JB, Nieschlag E, Ross GT, 1971. A method for producing specific antisera with small doses of immunogen. J Clin Endocrinol Metab 33:988–91

LEVELS OF FOLLICLE REGULATORY PROTEIN IN REGULAR MENSTRUATING

AND AMENORRHEIC PATIENTS

G.S. diZerega, S.A. Tonetta, K. Fujimori, and G. Westhof

University of Southern California Medical School
Livingston Biological Research Laboratory
Los Angeles, CA 90033

INTRODUCTION

Many investigators have hypothesized that a timely induction or activation of the granulosal aromatase system and the concomitant changes in the intrafollicular sex steroid profiles are major determinants of the dominant follicle during its selection early in the ovarian cycle. Adashi and Hsueh (1982) suggested that intrafollicular estrogen can enhance the actions of FSH through aromatase stimulation. Once a chosen follicle is producing a significant amount of estrogen, it then would have the capacity to produce more estrogen than neighboring follicles and the selection of the dominant follicle would be assured. In contrast, follicles destined to undergo atresia may not be able to produce enough estrogen to counteract the adverse effects of FSH deprivation. Moor et al. (1978), Carson et al. (1981), and Tsonis et al. (1985) concluded that reduced aromatase activity in atretic follicles was due to a loss of existing aromatase activity rather than a failure to acquire this activity initially, and suggested that a decrease in aromatase activity was an early event in the atretic degeneration of antral follicles. It is the lack of granulosal aromatase and not androgen substrate per se which limits estrogen production in human atretic follicles. Thus, atresia appears to be, at least in part, an active event, mediated by the intra-ovarian suppression of FSH responsive aromatase activity.

Using inhibition of granulosa cell aromatase as the reference bioassay, we identified a protein referred to as follicular regulatory protein (FRP) which was purified to homogeneity from human (Ono et al., 1985) and porcine (Holmberg et al., 1986) follicular fluid as proteins having similar molecular weights (15,000) and isoelectric points (pH 4.5). When FRP was tested in an in vitro inhibin assay (Channing, Wehrenberg and Campeau, personal communications) or injected into male rats (Tsutsumi et al., 1987), no reduction in FSH levels was detectable. FRP is secreted by human granulosa, but not theca cells (diZerega et al., 1983), and is undetectable in serum or urine from women after oophorectomy or during the menopause (Lew et al., 1987). In response to gonadotropins administered in vitro or in vivo, or prolongation of culture in vitro, granulosal secretion of FRP activity declines (diZerega et al., 1983; Kling et al., 1984). Similar observations were reported by Saito and Hiroi using porcine follicular fluid to inhibit PMSG-stimulation of ovarian weight in the mouse (1986), Cahill et al. (1985) using ovine follicular fluid to inhibit PMSG-responsive folliculogenesis in the ewe, McNatty et al. (1985)

using bovine follicular fluid to inhibit FSH-induced aromatase activity in granulosa cells from the large bovine follicle and Chari et al. (1985) who used the Orange A bound fraction of porcine follicular fluid to inhibit granulosa cell aromatase activity in vitro and to inhibit the PMSG-induced increases in peripheral serum estradiol levels in the rat.

In spontaneously cycling, untreated women, there is a positive correlation between the bioactivities of FRP and inhibin and estradiol concentrations in follicular fluid (Marrs et al., 1985). The correlations between these protein activities and steroid concentrations in follicular fluid are lost after development of multiple follicles by the hyper-stimulation of follicular development used for in vitro fertilization-embryo transfer (IVF-ET). This suggests that the intrafollicular closed loop feedback relationships regulating the production of these hormones and paracrine modulation of folliculogenesis are opened by pharmacologic stimulation (diZerega et al., 1983b).

FRP SECRETION

FRP is secreted from small and medium-sized porcine follicle granu-losa cells, but not theca cells as determined by both biological and immunological assays (Tonetta et al., 1986). Further enlargement of the follicle to preovulatory size (8-10 mm in diameter) leads to a marked reduction in FRP secretion. These findings are also reminiscent of the diminished secretion of plasminogen activator (Shaw et al., 1985), inhibin (Marrs et al., 1984) and oocyte maturation inhibitor (OMI) (Weimer et al., 1984) activities in follicular fluid from preovulatory compared to more immature follicles. Together, these findings suggest that as the follicle luteinizes, changes in protein secretion accompany the well-known alterations in steroidogenesis.

High levels of intrafollicular FSH reduce the action of FRP in granulosa cells which may, in turn, allow for enhanced granulosal aroma-tization and estrogen production. Previously, Schreiber and diZerega (1986) reported that increasing exposure of rat granulosa cells to FSH reduced FRP-associated aromatase inhibition. In the presence of constant FSH stimulation, increasing concentrations of FRP reduced in a dose-dependent manner aromatase activity of granulosa cells in vitro. Taken together with the previous finding that FRP modulates other FSH-responsive activities in granulosa cells (i.e., LH/hCG receptor induction) (Montz et al., 1984), cAMP formation (Ujita et al., 1987) and 3β-HSD activity Chicz et al. (1986), these data indicate that recruitment and selection of the preovulatory follicle may occur as a result of early follicular exposure to FSH and a reduction in sensitivity to and/or production of FRP by that follicle.

The inhibitory effects of FRP on granulosal aromatase activity depend upon the response of the cell to FSH: large amounts of FSH can partially overcome FRP inhibition while relatively small amounts of FSH sensitize the granulosal aromatase system to FRP. Although androgens potentiate FSH-mediated granulosal functions, they also sensitize granulosa cell steroidogenic enzymes to inhibition by FRP. The demonstration that FRP acts primarily on granulosa cells of less mature antral follicles to inhibit aromatase supports the hypothesis that FRP may facilitate follicle selection and suggests a role for FRP in atresia. In response to increased exposure to FSH, the preovulatory follicle may decrease its own sensitivity to FRP in contrast to other follicles, thereby allowing for a selection advantage expressed, in part, as enhanced estrogen synthesis. The other follicles in the cohort would rapidly undergo atresia as a result of the FRP-mediated reduction in FSH-responsivity. Most of the effects of FRP on granulosal activities reflect an interplay between the systemic endocrine and local paracrine systems. That FRP functions, at least in part, by modulating follicular response to FSH is consistent with the hypothesis that paracrine effectors are the principle mediators of folliculogenesis in the presence of gonadotropins (diZerega and Hodgen, 1981; Channing et al., 1982; diZerega et al., 1983c, 1985).

MENSTRUAL CYCLE

FRP levels rise beginning in the mid follicular phase to reach their zenith by the midluteal phase; a pattern which correlates with the decrease in the number of healthy follicles during the periovulatory and midluteal phases of the menstrual cycle (Fig. 1, diZerega and Katt, unpublished results). The source of the increasing levels of FRP beginning in the mid follicular phase and continuing through the midluteal phase is of interest. Immunohistochemical evaluation of the corpus luteum demonstrated binding to the large luteal cells which are thought to be the successors of the granulosa cells from the preceding preovulatory follicle (Fujimori and diZerega, unpublished observations). Luteal cells may therefore be a principle source of FRP. The nadir of FRP levels occurred during the early follicular phase, a time which is characterized by the largest number of nonatretic follicles during the menstrual cycle. Non-atretic follicles are not likely to be a predominant source of FRP in the luteal phase since their number decreases during this time. McNatty (1982) and Chikazawa et al. (1986) found 1-3 nonatretic follicles per pair of normal ovaries extirpated from the luteal phase. The number of atretic follicles increases during the luteal phase, especially those 4-6 mm in diameter. This temporal correlation between rising levels of FRP and atretic follicles provides a new focus for further studies of FRP secretion. The small number of healthy follicles available for recruitment by the rise in serum FSH levels during the late luteal -early follicular phase may be determined by the number of follicles which survive the early-midluteal phase. McNatty et al. (1983) reported a marked increase in FSH responsive aromatase activity by granulosa cells aspirated from human follicles collected during the mid-late luteal phase. Thus, FRP produced in the periovulatory interval and subsequent luteal phase may play a role in the reduction of the potential follicular population to that number consistent with the ovulatory quota. Recruitment and selection of the preovulatory follicle may occur as a result of an FRP mediated reduction in the number of healthy follicles during the luteal phase followed by an FSH mediated stimulation of the remaining viable follicle(s) in the subsequent intercycle interval.

ANOVULATION

Inappropriate exposure to elevated FRP levels could limit the ability of follicles to respond to FSH resulting in suppressed follicular maturation and estradiol production. A group of anovulatory patients was recently found to have elevated levels of immunoreactive FRP in their peripheral serum (Lew et al., 1987). A differential ovarian response to clomiphene citrate therapy by these anovulatory patients was observed on days 11-12 of the menstrual cycle. There appeared to be abnormalities of follicular recruitment during the intercycle interval as demonstrated by reduced serum estradiol levels and concomitant elevations in serum FRP levels 3-5 days after the onset of the last menstrual period. From these limited data it is tempting to speculate that anovulation in these patients may result from an over production of FRP, presumably by the granulosa cells, and subsequent reduction in follicular responsivity to gonadotropin stimulation. That additional gonadotropin exposure was not successful in overcoming this putative FRP-induced block in follicular maturation is evidenced by the elevated levels of FRP in other anovulatory patients 11-12 and 22-23 days after the onset of the last menstrual period

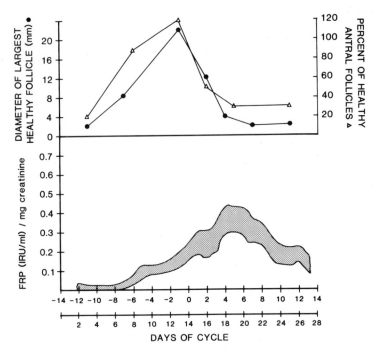

Fig. 1. Top Panel: Diameter of largest healthy follicle and percent of healthy antral follicles determined from human ovaries collected throughout the normal menstrual cycle. Adapted from McNatty (1981) and Chikazawa et al. (1986). Bottom Panel: Urinary levels of follicle regulatory protein (expressed in IRU: immunoreactive units) determined via ELISA in daily collections throughout the cycle from normally menstruating women. Shaded area represents 2 standard errors of the mean (Katt and diZerega, unpublished results).

even after receiving clomiphene citrate therapy on days 3-7 of the cycle. A separate group of anovulatory patients was identified by both relatively low serum estradiol and FRP levels determined on days 3-5 of the cycle (Lew et al., 1987). Follicles in these patients may be understimulated by gonadotropins. Anovulatory patients who did respond to clomiphene citrate therapy by increasing their peripheral estradiol levels on days 11-12 and presumably ovulated as evidenced by elevated serum progesterone levels on days 22-23 did not have elevated FRP levels during these intervals. Thus, follicular dysfunction expressed clinically as anovulatory cycles may result from abnormalities in ovarian protein as well as from gonadotropin secretion.

REFERENCES

Adashi EY, Hsueh ASW, 1982. Estrogens augment the stimulation of ovarian aromatase activity by follicle-stimulating hormone in cultured rat granulosa cells. J Biol Chem 257:6077-6083
Cahill LP, Driancourt MA, Chamley WA, Findlay JK, 1985. Role of intrafollicular regulators and FSH in growth and development of large antral follicles in sheep. J Reprod Fertil 75:1-9
Carson RS, Findlay JK, Clarke IJ, Burger HG, 1981. Estradiol, testosterone and androstenedione in ovine follicular fluid during growth and atresia of ovarian follicles. Biol Reprod 24:105-112

Channing CP, Anderson LD, Hoover DJ, Kolena J, Osteen KG, Pomerantz SH, Tanabe K, 1982. The role of nonsteroidal regulators in control of oocyte and follicular maturation. Recent Prog Horm Res 38:331-404

Chari S, Daume E, Strum G, Vaupel H, Schuler I, 1985. Regulators of steroid secretion and inhibin activity in human ovarian follicular fluid. Mol Cell Endocrinol 41:137-145

Chicz R, Nakamura RM, Goebelsmann U, Campeau JD, Tonetta SA, Frederick JJ, diZerega GS, 1986. Follicle regulatory protein noncompetitively inhibits microsomal 3B-ol dehydrogenase activity. J Steroid Biochem 23:662-668

Chikazawa K, Araki S, Tamada T, 1986. Morphological and endocrinological studies on follicular development during the human menstrual cycle. J Clin Endocrinol Metab 62:305-313

diZerega GS, Hodgen GD, 1981. Folliculogenesis in the primate ovarian cycle. Endocrine Rev 2:27-49

diZerega GS, Marrs RP, Campeau JD, Nakamura RM, Kling OR, 1983a. Human granulosa cell secretion of protein(s) which suppress follicular response to gonadotropins. J Clin Endocrinol Metab 56:147-155

diZerega GS, Campeau JD, Lobo RA, Nakamura RM, Ujita EL, Marrs RP, 1983b. Activity of a human follicular fluid protein(s) during normal and stimulated ovarian cycles. J Clin Endocrinol Metab 57:838-846

diZerega GS, Campeau JD, Ujita EL, Kling OR, Marrs RP, Lobo RA, Nakamura RM, 1983c. Possible role for a follicular fluid protein in the intraovarian regulation of folliculogenesis. Sem Reprod Endocrinol 1:309-322

Hillier SG, 1985. Sex steroid metabolism and follicular development in the ovary. Oxf Rev Reprod Biol 7:168-222

Holmberg EA, Campeau JD, diZerega GS, 1986. Purification of follicle regulatory protein by preparative gel electrophoresis. Prot Biol Fluid 34:911-916

Kling OR, Roche PC, Campeau JD, diZerega GS, 1984. Identification of protein(s) in porcine follicular fluid which suppress follicular response to gonadotropins. Biol Reprod 30:564-572

Lew MW, Katt EL, Rodgers KE, diZerega GS, 1987. Alteration of follicle regulatory protein levels in human reproductive disorders: Anovulation. Obstet Gynecol 67:000 (In Press)

Marrs RP, Lobo RA, Campeau JD, Ujita EL, Nakamura RM, diZerega GS, 1984. Correlation of human follicular fluid inhibin activity with spontaneous and stimulated follicular development. J Clin Endocrinol Metab 58:704-709

McNatty KP, 1982. Ovarian follicular development from the onset of luteal regression in humans and sheep. In: Rolland R, vanHall EV, Hillier SG, McNatty KP, Schoemaker J (eds.), The proceedings of the IVth Reinier De Graaf Symposium, "Follicular Maturation and Ovulation". Elsiever, Amsterdam, pp 1-18

McNatty KP, Hillier SG, van den Boogaard AMJ, Trimbus-Kemper TCM, Reichert LE, vanHall EV, 1983. Follicular development during the luteal phase of the human menstrual cycle. J Clin Endocrinol Metab 56:1022-1031

McNatty KP, Hudson N, Gibb M, Ball K, Henderson KM, Heath DA, Lun S, Kieboom LE, 1985. FSH influences follicle viability, oestradiol biosynthesis and ovulation rate in Romney ewes. J Reprod Fertil 75:121-131

Montz FJ, Ujita EL, Campeau JD, diZerega GS, 1984. Inhibition of LH/hCG binding to porcine granulosa cells by a follicular fluid protein(s). Am J Obstet Gynecol 148:436-441

Ono T, Campeau JD, Holmberg EA, Nakamura RM, Ujita EL, Devereaux DL, Tonetta SA, DeVinna R, Ugalde M, diZerega GS, 1986. Biochemical and physiological characterization of follicle regulatory protein: a paracrine regulator of folliculogenesis. Am J Obstet Gynecol 154:709-716

Saito H, Hiroi M, 1986. Correlation between the follicular gonadotropin inhibitor and the maturity of the ovum-corona-cumulus complex. Fertil Steril 46:66-72

Schreiber JR, diZerega GS, 1986. Porcine follicular fluid protein(s) inhibits rat ovary granulosa cell steroidogenesis. Am J Obstet Gynecol 155:1281-1288

Shaw KJ, Campeau JD, Roche PC, diZerega GS, 1985. Porcine granulosa cell production of plasminogen activator: Disparity between the effects of hCG and FSH. Clin Exp Endocrinol 8:26-34

Tonetta SA, Yanagihara DL, Bryant S, diZerega GS, 1986. Quantitation of FRP secretion by porcine granulosa cells. Biol Reprod 35(suppl 1) 1-3

Tsonis CG, Carson RG, Findlay JK, 1985. Relationships between aromatase activity, follicular fluid oestradiol-17B and testosterone concentrations, and diameter and atresia of individual ovine follicles. J Reprod Fertil 72:153-163

Tsutsumi I, Fujimori K, Ono T, Nakamura RM, diZerega GS, 1987. Inhibition of spermatogenesis in the rat with follicle regulatory protein. Bio Reprod 36:000 (In Press)

Tsutsumi I, Toppari J, Campeau JD, diZerega GS. Inhibition of fertility in the male rat by systemic treatment with follicle regulatory protein. Fertil Steril (in review)

Ujita EL, Campeau JD, diZerega GS, 1986. Inhibition of porcine granulosa cell adenylate cyclase activity by an ovarian protein. Expl Clin Endocrinol 87:000 (In Press)

Weimer S, Campeau JD, diZerega GS, 1984. Alteration of human follicular fluid plasminogen activator activity by ovarian hyperstimulation. IVF 4:263-266

Winer-Sorgen S, Brown J, Ono T, Gale J, Campeau JD, Marrs RP, diZerega GS, 1986. Correlation of oocyte maturation inhibitor activity in human follicular fluid: quantitative determination in unstimulated, clomiphene citrate and human menopausal gonadotropin stimulated ovarian cycles. IVF 4:218-223

OVARIAN FOLLICULAR FLUID PEPTIDE (OFFP): EFFECTS ON FOLLICULAR

MATURATION, OVULATION AND LUTEINIZATION IN MICE AND MARMOSETS

T.D. Nandedkar, A.L. Kadam, S.D. Kholkute, N.J. Joshi, and S.B. Moodbidri

Institute for Research in Reproduction (ICMR)
Jehangir Merwanji Street, Parel
Bombay 400 012, INDIA

ISOLATION OF A SHEEP FOLLICULAR FLUID PEPTIDE

Intraovarian concentrations of steroids and non-steroidal factors have recently been reported by Demoalin and Franchimont (1985) to play an important role in deciding the fate of the follicle. We have attempted to partially purify and characterize a peptide from sheep follicular fluid (OFFP). The pooled follicular fluid was chromatographed on Sephadex G-100 and later on Sephadex G-25 column (Kadam et al., 1984). The bioactive fraction (GF$_2$) inhibited maturation of large follicles and ovulation in cycling mice. It also decreased plasma progesterone (P) levels but did not affect FSH levels significantly in these mice (Nandedkar et al., 1986). It reduced 3β hydroxysteroid dehydrogenase (3β HSD) activity in mouse granulosa cells. Further, the purified Fraction 1 on high performance liquid chromatography also inhibited 3β HSD activity in mouse granulosa cells. The trypsin digestion and heat inactivation of the peptide indicated the heat labile nature of OFFP (Kadam et al., 1984).

Studies of the Effect GF$_2$ in Maromsets: Binding of ^{125}I-hFSH to marmoset granulosa cells was inhibited by GF$_2$. Heat inactivation of the peptide GF$_2$ could partially overcome its inhibitory effect on ^{125}I-hFSH binding (Fig. 1) confirming the heat labile nature of the peptide.

In view of the inhibitory effect of GF$_2$ on FSH binding to marmoset granulosa cells, a dose of 100 ug/day of GF$_2$ was injected for 4 days during late follicular phase into 5 normal cycling marmosets. Blood samples were collected every alternate day and plasma estradiol (E$_2$) and P levels were estimated by RIA. The E$_2$ levels failed to show any change in the treatment cycle compared to the control (pretreatment) cycle in all the marmosets. However, the area under the curve for P was significantly decreased in the treatment cycle in 4 out of 5 marmosets (Fig. 2, representative marmoset #101). The length of luteal phase in the treatment cycle was reduced by 2-4 days (mean 2.8 ± 0.14) in all the marmosets.

The results thus reveal that the low molecular weight, heat labile peptide from sheep follicular fluid inhibits FSH binding to granulosa cells thereby decreasing plasma progesterone levels causing luteal deficiency in marmosets. This peptide has some similarity with the FSH binding inhibitor described by Darga and Reichert (1978), follicular regulating protein described by Tonetta and diZerega (1986) and luteinizing inhibitor described by Ledwitz-Rigby and Rigby (1981).

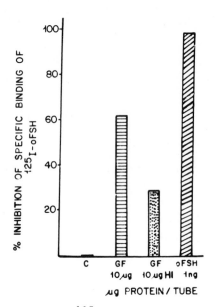

Fig. 1. Inhibition of binding of ^{125}I-hFSH to granulosa cells of marmoset ovarian follicles by GF_2 (partially purified fraction from sheep follicular fluid) and heat inactivated GF_2 (HIGF_2).

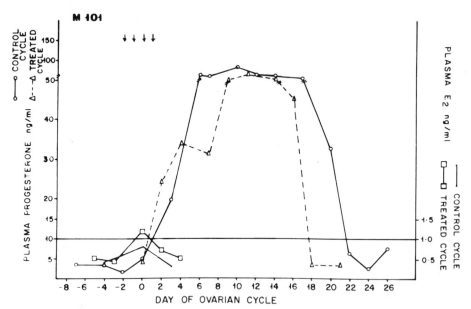

Fig. 2. Plasma E_2 levels in the control (pretreatment) cycle (●——●) and treatment cycle (□——□) and plasma P levels in control (○——○) and treatment (△---△) cycle.

REFERENCES

Darga NC, Reichert LE Jr, 1978. Some properties of the interaction of follicular stimulating hormone with bovine granulosa cells and its inhibition by follicular fluid. Bio Reprod 19:235-240

Demoalin A, Franchimont P, 1985. Ovarian hormones and cybermins. In: Runnebaum B, Rabe T, Kiesel L (eds.), Future Aspects in Contraception. Boston: MTP Press, pp 281-288

Kadam AL, Moodbidri SB, Nandedkar TD, 1984. Inhibition of 3β hydroxy-steroid dehydrogenase (3β HSD) activity in the granulosa cells of mouse ovarian follicles by follicular fluid peptide. Steroids 44:497-509

Ledwitz-Rigby F, Rigby BW, 1981. Ovarian inhibitors and stimulators of granulosa cells maturation and luteinization. In: Franchimont P, Channing CP (eds.), Intragonadal Regulation of Reproduction. New York: Academic Press, pp 97-131

Nandedkar TD, Kadam AL, Moodbidri SB, 1986. Control of follicular maturation in mouse by a non-steroidal regulator from sheep follicular fluid. Int J Fertil (In Press)

Tonetta SA, diZerega GS, 1986. Paracrine regulation of follicular maturation in primates. Clinics in Endocrinol and Metab 15:135-150

CHEMICAL CHARACTERISTICS OF FOLLICULAR GLYCOSAMINOGLYCANS

M.E. Bellin and R.L. Ax

Department of Dairy Science
University of Wisconsin-Madison
Madison, WI

INTRODUCTION

As follicles mature, they shift from follicle stimulating hormone (FSH) to luteinizing hormone (LH) dependence and atretic follicles generally exhibit reduced capacity to bind gonadotropins (Nimrod et al., 1977; Erickson et al., 1979; Amsterdam et al., 1981). In vitro assays revealed that binding of gonadotropins to rat granulosa could be inhibited by addition of bovine follicular fluid (FF). Fluid from small follicles was more effective than large follicles at inhibiting that binding (Amsterdam et al., 1979). The glycosaminoglycans (GAGs) chondroitin sulfate and heparin effectively inhibited hormonal stimulation of adenylyl cyclase in a dose-dependent manner in fragmented granulosa cell preparations (Salomon et al., 1978; Nimrod and Linder, 1980). Dermatan sulfate (DS), also referred to as chondroitin sulfate-B, and heparan sulfate (HS) are the predominant GAGs found in follicular fluid and secreted by granulosa cells in vitro (Ax et al., 1985). Heparin is structurally similar to HS except it possesses a unique oligosaccharide sequence which provides its characteristic anticoagulant activity (Lindahl et al., 1977). Concentrations of FF, DS and HS decreased with follicular enlargement and were significantly elevated in atretic follicles regard- less of follicle size in the bovine (Grimek et al., 1984; Bellin and Ax, 1984; Bushmeyer et al., 1985). Therefore, changes in concentrations of GAGs in FF contrast patterns for gonadotropin binding to granulosa. The purpose of this study was to examine the chemical characteristics as well as the concentrations of follicular fluid GAGs isolated from follicles at different stages of maturation.

MATERIAL AND METHODS

Follicular Fluid

Bovine ovaries were obtained within 30 min of slaughter, granulosa cells were removed by centrifugation and FF was frozen immediately (Champion Packers, Waterloo, WI). The GAGs were isolated from FF aspirates from 2633 small (5 mm), 1702 medium (6-10 mm) and 491 large (11-20 mm) follicles. Those follicles were previously characterized by FF estradiol-17β (E) concentrations (Bushmeyer et al., 1985) and the populations of follicles were normally distributed among the E concentrations for each of the three follicle sizes. Fluid samples from follicles distributed in the lowest 15% and highest 15% by E content was pooled and categorized as atretic and non-atretic respectively, for each follicle size.

Glycosaminoglycan Isolation

Follicular fluid GAGs were isolated as previously described (Bellin and Ax, 1984) with minor modifications. Aliquots of 0.5 ml FF were extracted using a ratio of FF:chloroform:methanol of 1:3:1 (v/v/v). After thoroughly mixing, samples were centrifuged (200 g) for 20 min at 4°C. The clear aqueous top layer was aspirated, and brought to a final concentration of 5% trichloroacetic acid.

Anion-Exchange Chromatography

Purification of DS and HS isolated from FF was achieved by ion exchange chromatography on a diethylaminoethyl (DEAE)-cellulose HPLC column. The eluting buffer was 50 mM sodium acetate containing 7 M urea and 0.5% (w/v) Triton X-100, pH 6.0 (Fosang et al., 1984). The equivalent of 0.5 ml of FF was applied to a Waters Associates (Milford, MA) protein pak DEAE-5PW Sephacel (7.5 mm x 7.5 cm) column. Linear gradients to 1.0 M NaCl in the same eluting buffer were run at flow rates of 1 ml/min. One min fractions were collected during a total run time of 60 min. Content of GAGs in 100 µl from each DEAE fraction was determined by gel filtration HPLC using a Waters Associates I-125 column (Bellin and Ax, 1984). The resulting peak area was compared to peak areas of known concentrations of bovine FF HS and DS that were previously purified, extensively dialyzed against deionized distilled water, lyophilized, and weighed.

Hexosamine and Disaccharide Analyses

Amino sugar constituents of N-acetylglucosamine (gluN) and N-acetylgalactosamine (galN) residues were quantitated as described by Grimek et al. (1984). DS isolated from the six groups of FF was dissolved in 0.2 M TRIS, pH 7.35 (100 µl) and divided into two 50 µl aliquots. One aliquot was treated with 0.01 unit of chondroitinase ABC from Protease vulgaris (Sigma). The other aliquot was combined with 0.01 unit of chondroitinase AC from Arthrobacter aurescens (Sigma). The disaccharide products were analyzed according to Grimek et al. (1984).

RESULTS

Adaption of an anion exchange chromatography procedure (Fosang et al., 1984) to HPLC allowed rapid simultaneous separation of GAGs without chemical destruction of either HS or DS. Heparan sulfate from atretic and non-atretic follicles eluted between .01 and .16 M NaCl with the apex of the peak at .071 M NaCl, and DS eluted between .323 and .442 M NaCl with the apex of the peak at .353 M NaCl, regardless of follicle size. In the sample obtained from atretic large follicles, GAG eluting at 0.004 M NaCl was also observed. That GAG was not chemically characterized. Concentrations of DS decreased as follicle size increased from 1.29 in small to .52 mg/ml in large follicles and corresponding HS concentrations ranged from .24 to 0.5 mg/ml (Table 1). Concentrations of DS and HS were greater in atretic compared to non-atretic follicles (Table 1). The proportions of DS were similar among the follicle sizes.

Dermatan sulfate contains the amino sugar galactosamine, sulfated at the 4' position, whereas HS contains predominantly a 2' and 6' sulfated glucosamine. Disaccharide analyses of DS treated with chondroitinase-ABC indicated that only disaccharides sulfated in the 4' position were present. The follicular fluid DS samples were not hydrolyzed by chondroitinase AC. That finding confirmed that DS, also referred to as chondroitin sulfate-B, was the major chondroitin sulfate isomer in FF, regardless of the E content of the fluid. The ratios of the amino sugars in the FF GAGs are summarized in Table 2. The ratio of glucosamine to galactosamine

Table 1. Concentrations of Follicular Fluid GAGs From Small (5 mm), Medium (6-10 mm), and large (11-20 mm) bovine follicles. Data are from triplicate estimates. E = estradiol 17-β (ng/ml).

Follicle Size	Endocrine Status E=(ng/ml)	Heparan Sulfate mg/ml (SEM)	Dermatan Sulfate mg/ml (SEM)	Dermatan Sulfate % of total GAG
Small	Atretic E = 6.8	.24 (.09)	1.29 (.08)	84.3
	Non-atretic E = 60.7	.10 (.02)	.79 (.01)	88.8
Medium	Atretic E = 1.5	.06 (.02)	1.16 (.09)	95.1
	Non-atretic E = 133	.07 (.02)	.80 (.06)	92.0
Large	Atretic E = 2.3	.08 (.01)	.89 (.31)	91.8
	Non-atretic E = 295	.05 (.01)	.52 (.04)	91.2

Table 2. The Ratios of the Amino Sugars From N-acetlygalactosamine (galN) and N-acetylglucosamine (gluN) From Isolated Heparan Sulfate and Dermatan Sulfate From Atretic and Non Atretic Follicles.

Follicle Size	Heparan Sulfate (gluN/galN) Atretic	Non-atretic	Avg. Mean (S.D.)	Dermatan Sulfate (galN/gluN) Atretic	Non-atretic	Avg. Mean (S.D.)
Small	2.39	1.4	1.90 (.50)	3.1	3.8	3.44 (.35)
Medium	5.10	1.78	3.44 (1.7)	10.0	5.3	7.65 (2.4)
Large	2.49	3.19	2.84 (.35)	19.7	12.1	15.9 (3.8)
Avg.	3.22 (1.25)	2.1 (.78)		10.93 (6.8)	7.06 (3.61)	

for HS was 3.22 ± 1.25 and 2.1 ± .78 for atretic and non-atretic follicles, respectively, and the ratio did not vary significantly among follicle sizes. The ratios of galactosamine to glucosamine were 10.93 ± 6.80 and 7.06 ± 3.61 for DS from atretic and non-atretic follicles. The ratio of galactosamine to glucosamine in DS increased as follicle size increased being 3.44 ± .35, 7.65 ± 2.35, and 15.9 ± 3.8 for small, medium and large follicles, respectively (p < .01), but did not differ between atretic and non-atretic follicles.

DISCUSSION

The changes in concentrations of HS and DS paralleled results from previous studies in which GAGs decreased with follicular enlargement and were elevated in atretic compared to non-atretic bovine follicles. In those studies GAG was selectively treated to remove either HS or DS

(Bellin and Ax, 1984; Grimek et al., 1984; Bushmeyer et al., 1985). The proportion of DS reported in this study agrees with the proportion of DS secreted by rat, porcine and bovine granulosa cells in vitro (Ax et al., 1985).

The concentrations of HS in FF were approximately 1/13 those reported by Bushmeyer et al. (1985) for the same samples (Table 1). That discrepancy can be accounted for because HS purified from bovine FF possessed a higher absorbance of 205 nm per unit mass compared to heparin which was the reference standard in the study of Bushmeyer et al. (1985). The slope for follicular HS was 13.5 fold greater than the slope for commercial heparin but the slopes of the lines for the commercial chondroitin sulfate and follicular DS were not different.

The chemical composition of the amino sugars and the disaccharide products provided additional information on the identities of the GAGs. The presence of only 4' sulfated disaccharide products confirmed that atretic and non-atretic follicles posses only one isomer of chondroitin sulfate, DS. The galactosamine to glucosamine ratio for bovine FF DS attached to a protein core (proteoglycan) was 2.06 and 7.64 for small and large follicles respectively (Grimek et al., 1984). In that study comparisons between atretic and non-atretic follicles were not performed.

In conclusion, with follicular enlargement, changes in the concentrations and chemical characteristics of the follicular fluid GAGs occurred. Concentrations of DS decreased as follicle size increased and the ratio of galactosamine to glucosamine in the DS increased as follicle size increased. The concentrations of the HS were 1/13 the values previously reported and the amino sugar composition of HS did not differ significantly among the various follicle sizes. Atretic compared to non-atretic follicles contained elevated concentrations of DS and HS. The amino sugar composition of the DS and HS did not differ between atretic and non-atretic follicles. Collectively these changes may play an important role in differentiation of granulosa cells, and future studies should center on metabolism of GAGs during follicular maturation.

SUMMARY

The two glycosaminoglycans found in FF, DS and HS were isolated from fluid of atretic and non-atretic bovine follicles and simultaneously separated by anion exchange high performance liquid chromatography. Heparan sulfate eluted with the apex of the peak at .071 M NaCl. Dermatan sulfate eluted with the apex of the peak at .353 M NaCl. The concentrations of DS and HS were greater in atretic follicles compared to non-atretic follicles and decreased as follicles size increased, as previously reported. The proportion of DS was approximately 90% of the total glycosaminoglycan concentration of all FF samples. The ratio of galactosamine to glucosamine in DS increased as follicle size increased, being $3.44 \pm .35$, 7.65 ± 2.4, and 15.9 ± 3.8 ($P < .05$) for small, medium and large follicles, respectively.

REFERENCES

Amsterdam A, Knecht M, Catt KJ, 1981. Hormonal regulation of cyto-differentiation and intercellular communication in cultured granulosa cells. Proc Natl Acad Sci USA 78:3000-3004

Amsterdam A, Riesel R, Mintz Y, Shemesh M, Salomon Y, 1979. Inhibition of gonadotropin sensitive adenylate cyclase by ovarian follicular fluid. Biochem Biophys Res Commun 87:505-512

Ax RL, Bellin ME, Grimek HJ, 1985. Properties and regulation of synthesis of glycosaminoglycans by the ovary. In: Toft DO, Ryan DJ (eds.), Proceedings of the Fifth Ovarian Workshop. Ovarian Workshops, Champaign, IL, pp. 451–480

Bellin ME, Ax RL, 1984. Chondroitin sulfate: an indicator of atresia in bovine follicles. Endocrinology 114:428–434

Bushmeyer SM, Bellin ME, Brantmeier SA, Boehm SK, Kubajak CL, Ax RL, 1985. Relationships between bovine follicular glycosaminoglycans and steroids. Endocrinology 117:870–885

Erickson GF, Wang C, Hseuh AJW, 1979. FSH induction of functional LH receptors in granulosa cells cultured in a chemically defined medium. Nature 279:336–338

Fosang AJ, Handley CJ, Santer V, Lowther DA, Thorburn GD, 1984. Pregnancy-related changes in the connective tissue of the ovine cervix. Biol Reprod 30:1223–1235

Grimek HJ, Bellin ME, Ax RL, 1984. Characteristics of proteoglycans isolated from small and large bovine follicles. Biol Reprod 30:397–409

Lindahl U, Hook M, Backstrom G, Jacobsson I, Riesenfeld J, Malmstrom A, Roden L, Feingold DS, 1977. Structure and biosynthesis of heparin-like polysaccharides. Fed Proc 36:19–23

Nimrod A, Linder HR, 1980. Heparin facilitates the induction of LH receptors by FSH in granulosa cells cultures in serum-enriched medium. FEBS Lett 110:155–157

Nimrod A, Tsafriri A, Linder HR, 1977. In vitro induction of binding sites for hCG in rat granulosa cells by FSH. Nature 267:632–633

Salomon Y, Amir Y, Azulai R, Amsterdam A, 1978. Modulation of adenylate cyclase activity by sulfated glycosaminoglycans. I. Inhibition by heparin of gonadotropin-stimulated ovarian adenylate cyclase. Biochim Biophys Acta 554:262–272

ON THE INTRINSIC OVARIAN CONTROL OF THE DEVELOPMENTAL

TRANSITION FROM PRIMORDIAL TO PRIMARY FOLLICLE

G.W. Mulheron, S.L. Quattropani, and J.M. Nolin*

Department of Anatomy
Medical College of Virginia
Richmond, VA 23298

*Department of Biology
University of Richmond
Richmond, VA 23173

INTRODUCTION

In studies designed to immunocytochemically map the distribution of endogenous pituitary FSH and LH in preovulatory follicles in the rat, we discovered that both gonadotropins were detectable in both primordial and primary follicles. These findings were explored and the results are reported here.

MATERIALS AND METHODS

Adult Sprague-Dawley rats purchased from Charles River, Inc., MA were maintained in a central animal facility in accordance with Public Health guidelines. They were housed 2 per cage, under controlled temperature ($22\,^{\circ}$ C $+/- 2^{\circ}$) and light (14/10) and provided food and water ad libitum. Daily vaginal smears were taken and only those animals that exhibited 2 consecutive 4-day-cycles were used. Ovaries were taken under pentobarbital anesthesia at 9:00 am on the morning of the next projected vaginal estrus, diestrus, diestrus II, and proestrus. Ovaries were also collected at 1:00 pm and 6:00 pm on the afternoon of proestrus. Pituitaries were collected 3 months later and used to characterize the antisera. Tissues were fixed in Bouin's fluid, embedded in paraffin, and cut to 5 µm serial sections.

Rabbit antisera to rat pituitary FSH (AFSH S-11) and LH (ALH S-9) were diluted in phosphate buffered saline (PBS, pH 7.0) containing .05 M EDTA. Highly purified rat FSH I-6 and rat LH I-6, reconstituted and diluted with PBS, were used to test for absorption and cross-absorption. Antisera and hormones were gifts from the National Pituitary Agency. The immunocytochemical protocol used was originally designed for hormone target cells by Nolin and Witorsch (1976).

Absorption tests were done on diestrous II ovaries and pituitary glands, using the lowest concentration of antiserum that was reactive within the incubation scheme. In each test, 8 µl of antiserum were applied, using an Eppendorf pipettor, directly to the section. Hormones were then added similarly, in 2 µl aliquots. Sections were incubated for 4 days. Cross absorption tests for AFSH were done by substituting rat LH

I-6 for rat FSH I-6, and for ALH by substituting rat FSH I-6 for rat LH I-6 at hormone concentrations sufficient to fully absorb their respective antiserum.

RESULTS

In the anterior pituitary, both gonadotrophins were restricted to the morphologically distinct, giant gonadotrophs from the oophorectomized rats (Fig. 1).

Follicles in the ovary were identified as follows: primordial follicles were comprised of an oocyte encircled by a single layer of squamous somatic cells, whereas follicles with an oocyte surrounded by one or more layers of cuboidal somatic cells were considered primary.

In primordial follicles, FSH was confined to the oocyte, but was present in both cytoplasm and germinal vesicle, GV (Fig. 2). In early primary follicles, FSH was still present in these compartments but also became detectable in granulosa cells (GC), both in cytoplasm and nuclei. In older primary follicles, with multilayered GC, FSH was lost from the GV (Fig. 3).

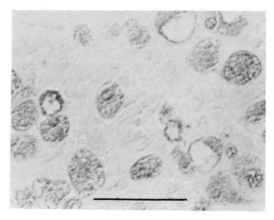

Fig. 1. ALH/AFSH staining of the giant gonadotrophs that are characteristic of pituitaries from long-term oophorectomized rats. Bar = 50 μm.

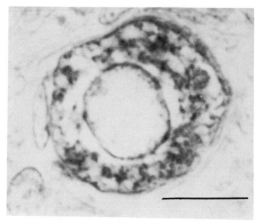

Fig. 2. FSH-positive oocyte cytoplasm and nuclei in a primordial follicle. (AFSH 1:100; absorption at 24 μg FSH/ml AFSH). Bar = 10 μm.

In contrast, in primordial follicles, LH was present in both cytoplasm and nuclei of both oocyte and GC (Fig. 4). In small primary follicles, LH remained detectable in oocyte cytoplasm and GV, but only occasionally appeared in GC (Fig. 5). In larger primary follicles LH had disappeared completely from GC and in the oocyte, like FSH it was present only in the cytoplasmic compartment (Fig. 6).

While the individual distribution patterns of both FSH and LH did not vary with the estrous cycle, most of the primordial follicles on any given section exhibited no immunoreactivity. By comparison, most primary follicles were immunoreactive. The only discernable estrous cycle dependent difference was the presence of a larger number of hormone-positive primordial follicles on the day of proestrus.

DISCUSSION

Clearly the gonadotrophins are essential for late-stage development of follicles and ovulation of ova. However, their contribution to the initiation of primordial follicle growth is a question that was essentially put to rest long ago when, during the 1930's and 1940's, all the available evidence led to the conclusion that FSH and LH play no part in early developmental processes. We think that the findings described here warrant a re-opening of that old question. Our arguments for that follow.

Our data meet the classical criteria of antiserum specificity in the ovary and are strengthened both by the similar data in the pituitary gland, as well as the limitation of immunopotency in the pituitary to the specific cell population known to be the source of the antigens. These specificity data suggest therefore the lack of shared epitopes between the two hormones. Any possibility that either or both antisera were reacting only with epitopes shared with functionally, very different molecules (Jagiello and Mesa-Tejada, 1979) remains, but allows equally for the interpretation that interaction with those otherwise shared epitopes was due, in fact, to their presence as part of LH/FSH rather than as epitopes of another molecule. In any event, the hormone-hormone and primordial-primary heterogeneities in distribution patterns strongly suggest physiological relevance.

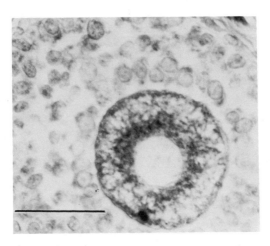

Fig. 3. FSH-positive oocyte cytoplasm and GC cytoplasm and nuclei in a mature primary follicle. Bar = 35 μm.

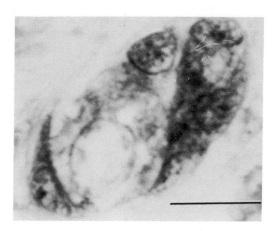

Fig. 4. LH-positive oocyte and GC cytoplasm and nuclei in a primordial and nuclei in a primordial follicle. (ALH 1:375; absorption at 200 μg LH/ml ALH). Bar = 10 μm.

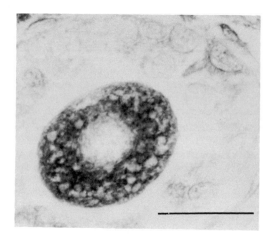

Fig. 5. LH-positive oocyte cytoplasm and GV, with the occasionally LH-positive GC typical of the early primary follicle. Bar = 20 μm.

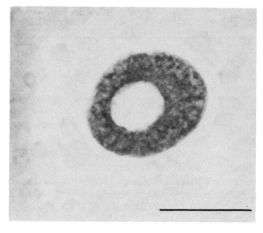

Fig. 6. LH-negative GC and LH-positive oocyte cytoplasm in the mature primary follicle. Bar = 30 μm.

Therefore, it seems that we can conclude with certain confidence that the present work allows the construction of the following hypothesis. During the estrous cycle, probably on the day of proestrus, certain primordial follicles become gonadotrophin targets, with LH being taken up and translocated to nuclei by both somatic and germ cells, and FSH being taken up only by the oocyte where its ultimate destination also appears to be nuclear. Not unprecedented, nuclear uptake of other polypeptide hormones has been shown previously (inter alia, Nolin and Witorsch, 1976). Oocyte uptake of FSH seems likely to be direct, i.e., from blood vessels with which it may be in intimate contact at this point of development (Zamboni, 1974). LH may also reach the oocyte via this route, but its presence in primordial somatic cells suggests another possibility, uptake via a GC transcellular route. Although not abundant, other evidence of gonadotrophin target status for developing follicles does exist (Petrusz, 1974; Oxberry and Greenwald, 1982; Tesoriero, 1986). During the growth transition from primordial to primary stages, both FSH and LH are lost from oocyte GV but remain sequestered in oocyte cytoplasm, while LH is replaced by FSH in GC. Only LH and FSH in oocyte cytoplasm are constant features of the transition.

The potential of oocytes to sequester FSH and LH during their early development could explain how follicles develop to preantral size in the absence of a continuing pituitary source of gonadotrophins. This is especially noteworthy because earlier experiments which "established" the pituitary independence of early follicular growth were based largely on findings in hypophysectomized animals that were old enough to have already been exposed to both gonadotrophins. FSH and LH are immunodetectable in rat primordial and primary follicles even as early as 8 days of age (Mulheron, et al., 1986), when circulating LH/FSH levels are known to have begun their first post-partum rise (Döhler and Wuttke, 1974).

There is a third constant with regard to the presence of gonadotrophins in the oocyte cytoplasm of developing follicles. Prolactin is found exclusively in that compartment during early growth (Nolin, 1982). This suggests either direct uptake from the blood by the primordial follicle and subsequent retention or since PRL is found only in cytoplasm, synthesis of PRL by the oocyte is perhaps, governed by a direct action of LH/FSH on oocyte GV during the primordium.

Central to our current hypothesis then is the presence of all three gonadotrophins, FSH, LH, and PRL in the oocyte during early follicular development. We propose that it is the oocyte that ultimately governs its own destiny and that it does so in concert with both pituitary and ovarian gonadotrophins. This concept of oocyte-directed follicle development is, of course, not new. It was proposed by Edgar Allen 65 years ago (Allen, 1922) and had found support in at least one other set of experiments since then. Nekola and Nalbandov (1971) showed that the luteinization of granulosa cells that occurs spontaneously in culture can be prevented by co-culture with oocytes.

ACKNOWLEDGEMENTS

This work was supported by NIH HD16505. Thanks to Michael Bonaduce for his technical assistance and thanks to Anthony Padua.

REFERENCES

Allen E, 1922. The oestrous cycle in the mouse. Am J Anat 30:297-371
Döhler KD, Wutke W, 1974. Serum LH, FSH, prolactin and progesterone from birth to puberty in female and male rats. Endocrinology 25:1003-1008

Jagiello G, Mesa-Tejada R, 1979. Cross-reactivity of human luteinizing hormone (β-subunit) and a serine protease demonstrated by immunoperoxidase staining in human oocytes. Endocrinology 104:302–306

Mulheron GW, Nolin JM, Quattropani SL, 1986. Immunodetectable FSH and LH in ovarian follicles of immature rats. J Cell Biol 103(5):480a

Nekola MV, Nalbandov AV, 1971. Morphological changes of rat follicular cells as influenced by oocytes. Biol Reprod 4:154–160

Nolin JM, Witorsch RJ, 1976. Detection of endogenous immunoreactive prolactin in rat mammary epithelial cells during lactation. Endocrinology 99:949–58

Nolin JM, 1982. Molecular homology between prolactin and ovarian peptides: evidence for physiological modification of the present molecule by the target. Peptides 3:823–831

Oxberry BA, Greenwald GS, 1982. An autoradiographic study of the binding of ^{125}I-labeled follicle-stimulating hormone, human chorionic gonadotrophin and prolactin to the hamster ovary throughout the estrous cycle. Biol Reprod 27:505–516

Petrusz P, 1974. Demonstration of gonadotropin-binding sites in the rat ovary by an immunoglobulin-enzyme bridge method. Europ J Obstet Gynec Reprod Biol 4/1 Supplement, S3–S9

Tesoriero JV, 1986. ^{3}H-uridine in small ovarian follicles, in vitro. J Cell Biol 103(5):478a

Zamboni L, 1974. Fine morphology of the follicle wall and follicle cell-oocyte association. Biol Reprod 10:125–149

STEREOCHEMICAL COMPLEMENTARITY BETWEEN ANTIESTROGENS AND DNA

L.B. Hendry, T.G. Muldoon, and V.B. Mahesh

Departments of Medicine, Physiology and Endocrinology
Medical College of Georgia
Augusta, GA 30912

Stereochemical Genetics, Inc.
P.O. Box 11649
Augusta, GA 30907

INTRODUCTION

Why do relatively few small molecules exist in nature in comparison to the almost infinite number of structures which are conceivable? The answer to this question appears to lie in the discovery of stereochemical complementarity of nucleic acids and a variety of natural products (Hendry et al., 1977). Small molecules which exist in nature exhibit complementarity. Small molecules which do not exhibit complementarity do not exist in nature.

One type of complementarity is manifest in the ability of various molecules to insert between base pairs in double stranded DNA and form stereospecific donor/acceptor linkages to the helix. Many phytohormones, disaccharides, thyroid hormones, prostaglandins, cyclic nucleotides and hormonal steroids display such complementarity. Molecular modeling studies have also demonstrated that within any given class of compounds, type and degree of complementarity correlates with type and degree of biological activity (Hendry et al., 1984b).

In the case of hormonal steroids, operational criteria were developed which have proven to be reasonable predictors of agonist activity (Hendry et al., 1986a,b; Uberoi et al., 1985). Herein, we describe operational criteria for antagonists and discuss the application of the criteria to antiestrogens.

METHODOLOGY

The term, stereochemical complementarity, has been employed to describe complexes of nucleic acids and a variety of small molecules in which there are stereospecific hydrogen bonds and optimum associations of nonpolar groups, including stacking of aromatic moieties. To exhibit complementarity, a molecule should ideally: 1) be accommodated completely in the space between base pairs; and 2) form stereospecific linkages between each functional group and hydrogen bonding positions bordering the cavities. These two criteria have been referred to as fit in the topographical domain and hydrogen bonding domain, respectively. The operational criteria for agonists require that the candidate molecule fit into DNA at the same site and in the manner of the parent compound. The operational criteria for antagonists are the following. The candidate molecule must contain: 1) structural components which fit into and recognize the same site in DNA as the parent hormone; and 2) functional groups which form donor/acceptor linkages that are different from the parent hormone. Such linkages can occur within the topographical domain or along the outside surface of the helix in the major or minor grooves.

The modeling techniques employed in this study included: 1) Corey-Pauling—Koltun (CPK) space filling models of the candidate ligand inserted into CPK models of DNA; 2) CPK models of the ligand inserted into silastic polymer cavities of partially unwound DNA constructed from x-ray space filling coordinates (see Hendry et al., 1984; 1986a; for description); and 3) Kendrew skeletal models of the ligand inserted into Kendrew models of DNA. The results reported herein were consistent among the three techniques.

In previous studies using the above modeling techniques, each of the mammalian steroid hormones (estrogens, progestins, androgens, mineralo-corticoids and glucocorticoids) was shown to fit in a "lock and key" fashion between base pairs at the site 5'-dTdG-3'·5'-dCdA-3' (Hendry et al., 1986b). When inserted into this site, each hormone formed at least two donor/acceptor hydrogen bonds linking phosphate oxygens on adjacent strands of double stranded DNA. The pattern of linkages was unique for each hormone. Hormone agonists fit our operational criteria for agonist activity, i.e. the structure formed complementary stereochemical complexes with DNA at the same site and in the manner of the parent hormone. Degree of fit correlated with degree of activity. In the case of estrogens, estradiol fit into the site and formed two hydrogen bonds (proton donors) to negatively charged phosphates on adjacent strands (Fig. 1 and 2). Thus, to be considered an antiestrogen, a candidate molecule should fit into 5'-dTdG-3'·5'-dCdA-3' and form different linkages to the DNA than does estradiol.

RESULTS

A series of antiestrogens (Sutherland and Jordan, 1981) (trans-hydroxytamoxifen, tamoxifen, nafoxidine, and LY117018) and one inactive geometrical isomer (cis-tamoxifen) was examined by the operational criteria for an antagonist (Fig. 2). Estradiol is included in the table for comparison; note that the two +'s in the hydrogen bonding domain represent the two stereospecific hydrogen bonds to DNA and the 5 +'s reflect the relative fit in the topographical domain.

All the active antiestrogens fit the operational criteria for an estrogen antagonist. Each molecule is capable of fitting between base pairs at 5'-dTdG-3'·5'-dCdA-3' and forming a different hydrogen bond than estradiol. The hydrogen bond is common to each of the antagonists and involves an amino group attached to a side chain which extends out of the helix. (The outside surface of the helix provides multiple potential sites for hydrogen bonding to heteroatoms on the base pairs). A description of the fit of each molecule follows.

Nafoxidine inserts into the helix at 5'-dTdG-3'·5'-dCdA-3' in a different manner than estradiol in that no hydrogen bonds are formed within the site. A pyrrolidinylethoxyphenyl moiety extends out of the helix and forms a hydrogen bond in the major groove. Similarly, trans-tamoxifen (Fig. 1) fits into the site and contains a dimethylaminoethoxyphenyl group which extends into the major groove and links to the outside surface of the DNA. Trans-hydroxytamoxifen fits into the site in the same manner as trans-tamoxifen but possesses an additional linkage within the site. This linkage is analogous to that of estradiol.

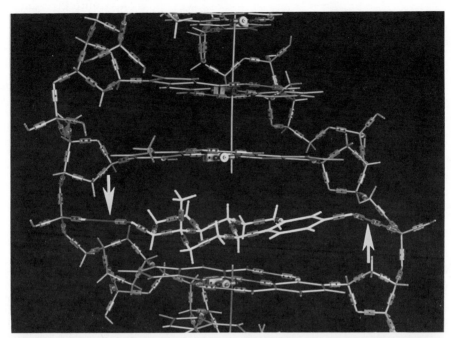

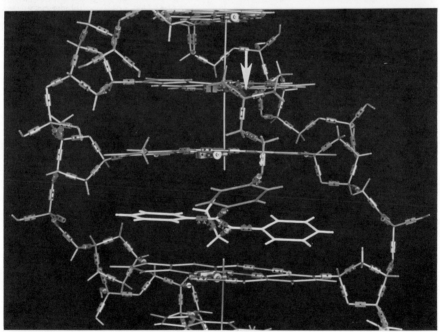

Fig. 1. Kendrew models depicting the fit of estradiol (top) and tamoxifen (bottom) between base pairs in the site 5'-dTdG-3'·5'-dCdA-3'. view of the DNA is from the major groove with the 5'-dTdG-3' strand on the left. Arrows indicate positions of hydrogen bonds (see text for discussion).

Compound	Fit In Hydrogen Bonding Domain	Fit In Topographical Domain	Extension Out Of Helix	Hydrogen Bonding To Helix Surface	Antagonist Activity
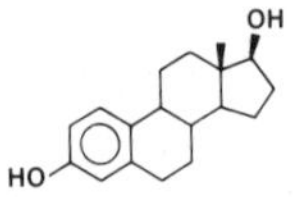 Estradiol	+ +	+ + + +	0	0	
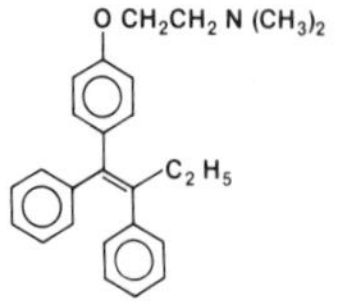 Cis-Tamoxifen	+ 0	+ + + − −	+	−	−
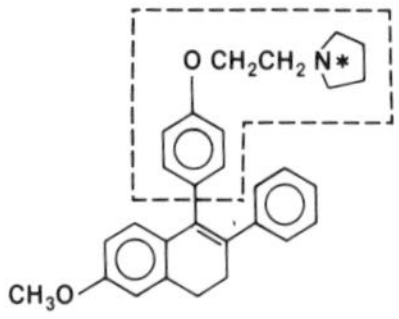 Nafoxidine	0 0	+ + + +	+ +	+[a]	+
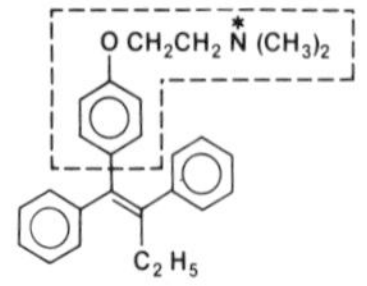 Trans-Tamoxifen	0 0	+ + + + + +	+ +	+[a]	+
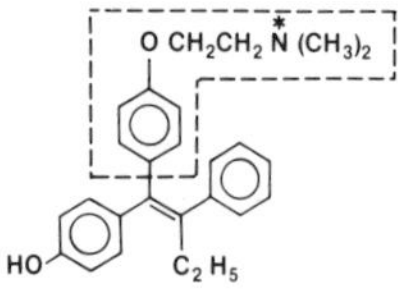 Trans-Hydroxytamoxifen	+ 0	+ + + + +	+ +	+[a]	+
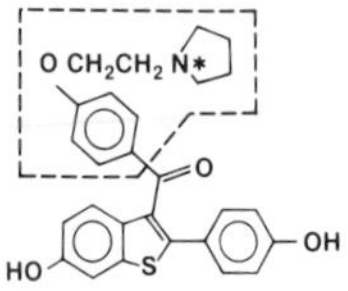 LY117018	+ +	+ + + +	+ +	+[b]	+

Fig. 2. Results of fit of compounds into DNA at 5'–dTdG–3'·5'–dCdA–3' according to operational criteria for an estrogen antagonist (see text for discussion). The portion of each antagonist which extends out of the helix is enclosed by dotted lines. Functional groups which form stereospecific hydrogen bonds to the surface of the DNA are denoted by asterisks: a – major groove; b – minor groove.

Note that both trans-tamoxifen and trans-hydroxytamoxifen appear to fit better into DNA in the topographical domain than estradiol. LY117018 fits into 5'-dTdG-3'·5'-dCdA-3' and forms two hydrogen bonds within the site analogous to estradiol. LY117018 also contains a pyrrolidinylethoxyphenyl group which extends out of the helix and forms a hydrogen bond to DNA. LY117018 differs from the other molecules in that the pyrrolidinylethoxyphenyl side chain extends into the minor groove.

Cis-tamoxifen is a relatively poor fit into the DNA. In addition to exhibiting poor complementarity in the topographical domain, cis-tamoxifen cannot form a stereospecific hydrogen bond to the outside surface of the DNA. Thus, the inactivity of this isomer as an antiestrogen is consistent with its poor complementarity in the estrogen site.

CONCLUSION

The stereochemical complementarity exhibited between natural products and nucleic acids has led us to postulate that a physicochemical blueprint for small molecules exists within the structure of DNA. For lack of a better term, we have named this phenomenon a "stereochemical logic" in gene structure. The logic has been defined as those unique features of DNA structure that ultimately dictate constraints on the structure, function, activity and metabolism of small molecules (Hendry et al., 1984). The findings presented herein suggest that the logic may be applied to our understanding of agonists and should prove to be useful in the design of novel antiestrogens. Whether these observations reflect a stage in the receptor mediated mode of action of estrogens and/or antiestrogens remains to be established.

REFERENCES

Hendry LB, Bransome ED Jr, Hutson MS, Campbell LK, 1984a. A newly discovered stereochemical logic in the structure of DNA suggests that the genetic code is inevitable. Perspect Biol Med 27:623-651

Hendry LB, Bransome ED Jr, Hutson MS, Campbell LK, 1984b. Method of predicting biological activity of compounds by nucleic acid models. US Patent 4,461,619

Hendry LB, Bransome ED Jr, Lehner AF, Muldoon TG, Hutson MS, Mahesh VB, 1986a. The stereochemical complementarity of DNA and reproductive steroid hormones correlates with biological activity. J Steroid Biochem 24:843-852

Hendry LB, Muldoon TG, Bransome ED Jr, Lehner AF, Hutson MS, Uberoi NK, Myers RB, Segaloff A, Mahesh VB, 1986b. Predicting hormonal activity from the stereochemistry of nucleic acids: a prospective study with novel estrogenic steroids. In: Puett D, Ahmad F, Black S, Lopez DM, Melner MH, Scott WA, Whelan WJ (eds.), Advances in Gene Technology: Molecular Biology of the Endocrine System, New York: Cambridge University Press, pp. 292-293

Hendry LB, Witham FH, Chapman OL, 1977. Gene regulation: the involvement of stereochemical recognition in DNA-small molecule interactions. Perspect Biol Med 21:120-130

Sutherland RL, Jordan VC, 1981. Non-steroidal Antioestrogens, New York: Academic Press

Uberoi NK, Hendry LB, Muldoon TG, Myers RB, Segaloff A, Bransome ED, Mahesh VB, 1985. Structure-activity relationships of some unique estrogens related to estradiol are predicted by fit into DNA. Steroids 45:325-340

INVITED CONTRIBUTORS

Everett Anderson, Ph.D.
     Professor
     Department of Anatomy & Cellular
     Biology & Associate Director of
     the Laboratory of Human Reproduction
     & Reproductive Biology
     Harvard Medical School
     45 Shattuck Street
     Boston, Massachusetts 02115

Lloyd L. Anderson, Ph.D.
     Professor
     Department of Animal Science
     Iowa State University
     11 Kildee Hall
     Ames, Iowa 50011

Andrzej Bartke, Ph.D.
     Professor and Chairman
     Department of Physiology
     Southern Illinois University
     School of Medicine
     Carbondale, Illinois 62901

Vinod K. Bhalla, Ph.D.
     Professor of Endocrinology
     Department of Physiology & Endocrinology
     Medical College of Georgia
     Augusta, Georgia 30912-3395

Dharam S. Dhindsa, D.V.M., Ph.D.
     Executive Secretary/Referral Officer
     Reproductive Biology Study Section
     Division of Research Grants
     National Institutes of Health
     Westwood Building, Room 307
     Bethesda, Maryland 20205

Martin Dym, Ph.D.
     Professor and Chairman
     Department of Anatomy
     Georgetown University
     School of Medicine
     3900 Reservoir Road, N.W.
     Washington, DC 20007

William H. Fletcher, Ph.D.
Professor
Department of Anatomy
Loma Linda University
School of Medicine &
J. L. Pettis Memorial Veterans Medical Center
Loma Linda, California 92357

Robert V. Gallo, Ph.D.
Professor of Endocrinology
Department of Physiology & Neurobiology
Box U-42, Room TLS 416
University of Connecticut
75 North Eagleville Road
Storrs, Connecticut 06268

Geula Gibori, Ph.D.
Professor
Department of Physiology & Biophysics
University of Illinois Medical Center
P.O. Box 6998
Chicago, Illinois 60680

Stuart Handwerger, M.D.
Professor of Pediatrics
Department of Pediatrics
Division of Endocrinology
Duke University Medical Center
Box 3080
Durham, North Carolina 27710

Anne N. Hirshfield, Ph.D.
Associate Professor of Anatomy
Department of Anatomy
University of Maryland
School of Medicine
655 West Baltimore Street
Baltimore, Maryland 21201

Tomas Hökfelt, Ph.D.
Professor
Department of Histology
Karolinska Institute
Stockholm, Sweden

Aaron J.-W. Hseuh, Ph.D.
Professor
Department of Reproductive
Medicine M-025
University of California, San Diego
La Jolla, California 92093

Satya P. Kalra, Ph.D.
Professor
Department of Obstetrics & Gynecology
The University of Florida
College of Medicine
Box J-294, JHMHC
Gainesville, Florida 32610

P. Landis Keyes, Ph.D.
    Professor of Physiology
    Department of Physiology
    University of Michigan
    School of Medicine
    744 Medical Science II Building
    Ann Arbor, Michigan 48109

Abraham L. Kierszenbaum, M.D., Ph.D.
    Professor of Anatomy & Associate Director
    The Laboratories for Reproductive Biology
    University of North Carolina at Chapel Hill
    Medical Research Building - 206H
    Chapel Hill, North Carolina 27514

Virendra B. Mahesh, Ph.D., D.Phil.
    Regents Professor and Chairman
    Department of Physiology & Endocrinology
    Medical College of Georgia
    Augusta, Georgia 30912-3395

Luciano Martini, M.D.
    Professor and Chairman
    Department of Endocrinology
    University of Milano
    VIA Tranquillo Cremona 29
    1-20145 Milano, Italy

Thomas G. Muldoon, Ph.D.
    Professor of Endocrinology
    Department of Physiology & Endocrinology
    Medical College of Georgia
    Augusta, Georgia 30912-3395

Andres F. Negro-Vilar, M.D., Ph.D.
    Head, Section Reproductive Neuroendocrinology
    National Institute of Environmental Health Sciences
    National Institutes of Health
    P. O. Box 12233
    Research Triangle Park, North Carolina 27709

Anita H. Payne, Ph.D.
    Professor of Biological Chemistry
    Department of Obstetrics & Gynecology
    University of Michigan
    School of Medicine
    Women's Hospital L-1224
    Ann Arbor, Michigan 48109

Kenneth J. Ryan, M.D.
    Professor and Chairman
    Department of Obstetrics & Gynecology
    Director of the Laboratory of Human Reproduction
    & Reproductive Biology
    Harvard Medical School
    45 Shattuck Street
    Boston, Massachusetts 02115

Barbara M. Sanborn, Ph.D.
    Professor
    Department of Biochemistry
    & Molecular Biology
    The University of Texas
    Health Science Center at Houston
    School of Medicine
    P. O. Box 20708
    Houston, Texas 77225

Frederick Stormshak, Ph.D.
    Professor
    Department of Animal Science
    Oregon State University
    Corvallis, Oregon 93731

Edward E. Wallach, M.D.
    Professor and Chairman
    Department of Obstetrics & Gynecology
    Johns Hopkins University
    School of Medicine
    600 North Wolfe Street
    Baltimore, Maryland 21205

Luteinizing process, 192, 193, 263,
   727–729
Luteolysis, 442, 449, 715, 718
   and prostaglandin F, 342–345
Luteostasin, 341
Luteotropin, 331
   embryonic, 335–340, 362–369,
      379–397
17-20 Lyase, 467–473, 477, 478,
      665–670
Lymph, ovarian, 695

Maturation
   follicular, 211–218
   sexual, 153
Mauthner neuron, 299, 300
Median eminence, 21, 34, 87, 88,
      617–625
   hypothalamic, 94, 96
   nerve fibers, 32
   nucleus, arcuate complex, 23–27
Medoxyprogesterone, 362–365
Membrane
   fetal, 709
   fluid, 328, 344
   of organelle, 328
Mesometrium, 391
Messenger
   intracellular, 85–108
   multiple, 35
   second, see AMP
Mestranol, 133
Met-enkephalin, 3, 25, 122
Methallibure, 653–656
1-Methyl-3-isobutly xanthine (MIX),
      281, 435
17α-Methyltestosterone, 472
Methyltrienolone (R1881), 472–476
Microspectrofluorometry, 21
Migration-stimulating factor, 686
Monkey, 346
Morphine, 4–8, 73–75, 110, 112
Morula, 179, 192
Motor synapse, giant of crayfish,
   299
Mullerian inhibiting substance (MIS),
      204–205
Myocyte, cardiac, 305
Myoinositol, 672
Myosin, 266, 435
   light chain kinase (MLCK), 435

Nafoxidine, 744, 746
Naloxone, 7–11, 13, 74, 76, 91,
      111–114, 119, 121, 122, 160,
      451, 647–650
Neovascularization, 687
Nerve fiber in median eminence,
      32–33
Neuraminidase, 277
Neuropeptide tyrosine, 25

Neuropeptide Y, 27, 66, 70–73, 88
Neurophysin, 28, 29, 677
Neurotensin, 25–28
Neurotoxin, 22
Nicotine, 190
Nordihydroguaiaretic acid (NGDA),
      93, 103
Norepinephrine, 66, 70, 72, 86, 87,
      90, 91, 156, 159, 165
Norgestomet implant, 697–701
Northern blot analysis, 302
Norikoff hepatome, 301, 304

Obesity, 237
Oleic acid, 328
Oocyte, 179, 182–184, 188–191, 203,
      226–229, 316, 697–701, 722,
      738–741
Opioid peptide, endogenous (EOP), 3,
      25–28, 66, 73–76, 109, 113,
      114, 119, 122, 123, 160,
      647, 649
Orcein dye, 697
Orchidectomy, 7, 111, 617, 619,
      623–627
Ornithine decarboxylase, 402, 403
Ovarian disease, polycystic, 659
Ovariectomy, 110, 243, 244, 442, 689
Ovulation, 7–17, 179–186, 219,
      240–241, 245–246, 703,
      727–729
   regulation of, 179–323
Oxygen
   -mediated damage, 469, 609
   tension, 469, 610, 611
Oxytocin, 338, 342, 345–347, 435,
      440, 446, 450, 451, 677

Parturition, 119, 421
PCO, see Polycystic ovarian syndrome
Penicillim, 260
Pentobarbital, 91
Pergonal, 288–289
Perfusion, computer-assisted, 99–103
Periovularory, interval, 179–201
Phenobarbital, 132, 138
Phentolamine, 87, 91
Phorbol 12,13–dibutyrate, 92, 94
Phosphatidic acid, 96
Phosphatidylcholine, 328, 344, 413
Phosphatidylethanolamine, 328, 344
Phosphatidylinositol, 96, 328, 343,
      344
Phosphatidylserine, 328
Phosphodiesterase, 189, 344, 569, 570
Phosphoinositide, 347
Phospholipase A, 93, 96, 344
   C, 88, 92–95, 100–103, 671
Phospholipid, 328
Photoperiod, 153–166

Rabbit (continued)
  vein, ovarian, 181
Rat
  androgen-sterilized, 240
  corpus luteum, 426–427
  cycling, 7–11, 141–144, 211–236,
    242–244, 307
  cyst, ovarian, follicular,
    659–663
  hypothyroid, 239, 259
  hypophysectomy, 659, 661, 663
  hysterectomy, 433
  immature, 259–274, 659–663, 703
  kidney cell, 302
  ovary, 426–428, 659–663
  placenta, 689–692
  pregnancy, 379–397, 431–437,
    659–662
    and luteopropin, 379–397
  pseudopregnancy, 433
  relaxin secretion, 431–437
  steroid, gonadal, 67–70
  stimulation
    electrochemical, 142, 144
    sham, 144
  suckling, 647–648
Receiver operating characteristic
    (ROC) analysis, 630
Receptor
  blocker, alpha-adrenergic, 91
  dissociation, 509
  internalization, 703
  occupancy, 299–323
  regulation, heterologous, 703–707
Relaxin, 218, 424–453, 679, 709–719
  in animal species, 421–423
  assay, 425, 430
  decidual, 709–713
  gel filtration, 431
  immunoactivity, 423
  immunohistochemistry, 716
  and insulin, 421–422
  localization, cellular, 424–430
  ovarian, 424–429, 709–713
  paracrine, 709
  in placenta, 428
  in pregnancy, 421–463
  radioimmunoassay, 716
  regulation, 431
  release, 715–719
  mRNA, 427
  secretion, 431
  in species, 421–423
  storage granules, 424–430
Renewal tissue, 211, 213, 229
Response
  immune, 399
  peripubertal, 659
Rhodamine phalloidin, 266
Ribonucleic acid (RNA), 382, 383, 427
Rous sarcoma virus, mutant, 302

RU *38486* 689–692

Scleroblast, 272
Sedimentation of bone marrow cells,
    489–493
Seminal fluid, *see* Fluid, seminal
Serotonin, 14, 86
Sertoli cell, 515–536, 547–551,
    554–556, 561–588
Signal
  transduction, 87–88, 92–95
  transmembrane, 85–108
Somatomedin, 399, 405, 410, 603
Somatostatin, 25, 34, 90
Sperm, 515 554, 597, 600
Spermatocyte, 554, 561
Spermatogenesis, 535–561
  protein, secretory, 547–551, 554
Sphingomyelin, 328
Spironolactone, 605
Stalk section, hypophyseal, 335
Stearic acid, 328
Steelman-Pohley assay, 277, 278
Steroid, 660, 704
  feedback, 65, 74–75, 153–175
  gonadal, 65, 71–77, 86, 91, 111,
    113, 153–155
  neuropeptide, 65–83
  ovarian, 110, 112, 117
  replacement, 68
  testicular, 71
  see separate steroids
  see Oocyte
Steroidogenesis, 263, 329, 332, 473,
    489–513, 591–595, 603, 606,
    609–616, 636, 665–668, 722
  antioxidants, 609–612
  gonadotropin-stimulated, 489–513
Streptomycin, 260
Stress, 28, 31, 272
Substance P, 22
Suckling, 647, 648
Superovulation, induced, 636
Surfactant, pulmonary, 399, 413
Syncytistrophoblast, 337

Tammar wallaby, 337
Tamoxifen, 745, 746
Tanycyte, 32
Testosterone, 13, 15, 74, 75,
    153–159, 166, 192, 239, 244,
    310, 433, 467–484, 489, 491,
    493–496, 500, 503–507, 531,
    561, 569, 603, 606, 610–615,
    618, 668
Testosterone propionate, 259, 571
3α,5α-Tetrahydroprogesterone,
    135–141
Tetrahymena pyriformis relaxin,
    422–423
Theca cell, 226–229, 329, 330,
    665–670, 721

6-Thioguanine, 300
Thromboxane, 95, 434
Thyroid hormone, 613, 655
Thyrotropin, 27
Thyrotropin-releasing hormone, 23,
    54, 613
Thyrotropin-stimulating hormone,
    4, 613-614
Throxine, 613, 614, 654
Tracer, fluorescent, 32
Transferrin, 520, 527, 528, 529,
    530, 541, 544, 545, 561,
    568, 571, 606
Transforming growth factor (TGF),
    204-205
Transmission, electrical
    intercellular, 299
3,5,3'-Triiodothyronine, 613, 614,
    654
Trophoblast, 337-338
Trypan blue, 386, 387
Tubule, seminiferous, 516-518,
    536-537, 540
  diagram, 537
Tyrosine hydroxylase, 23, 27, 28, 33

Ultrafiltration, 684, 686
Unit gravity sedimentation,
    489-493
Uterus motility, 421
  see Contractility

Vasopressin, 90
Verapamil, 343
Vesicle, germinal
  breakdown (GVBD), 188
Vimentin, 260
VIP see Polypeptide
Vitamin A, 561

Xenopus embryo, 303

Zona pellucida, 223, 225